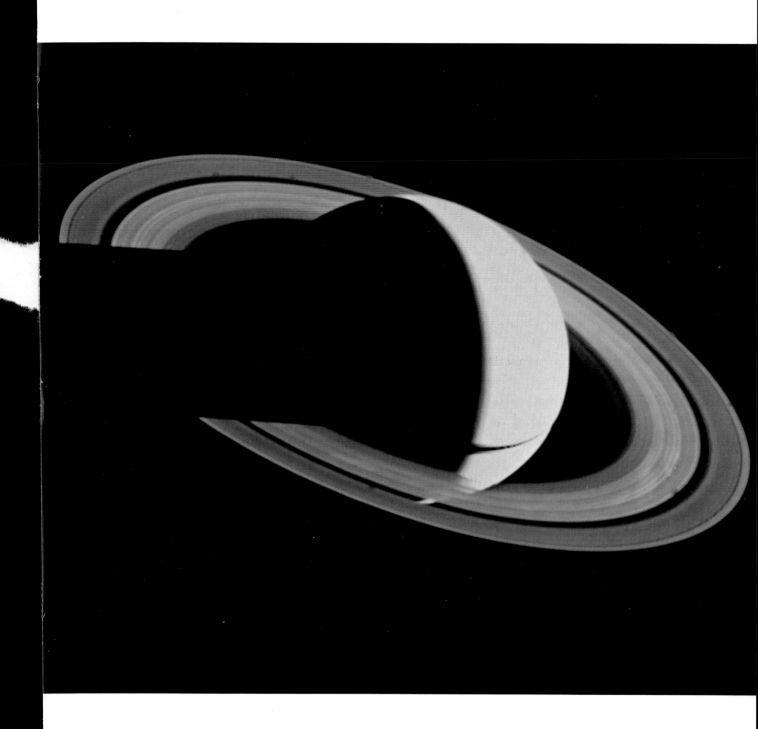

Exploration of the Universe

fourth edition

George O. Abell

University of california, los angeles

SAUNDERS GOLDEN SUNBURST SERIES
SAUNDERS COLLEGE PUBLISHING

Philadelphia   New York   Chicago
San Francisco   Montreal   Toronto
London   Sydney   Tokyo   Mexico City
Rio de Janeiro   Madrid

Address orders to:
383 Madison Avenue
New York, NY 10017

Address editorial correspondence to:
West Washington Square
Philadelphia, PA 19105

This book was set in Plantin and Serif Gothic by The Clarinda Company.
The editors were John Vondeling, Lloyd Black, Janis Moore, and Michael Fare.
The art and design director was Richard L. Moore.
The text design was done by Nancy E.J. Grossman.
The cover design was done by Richard L. Moore.
The artwork was drawn by Vantage Art, Inc.
The production manager was Tom O'Connor.
The printer was R. R. Donnelley and Sons.

Front cover: NGC 5128, a peculiar galaxy that is a strong radio emitter. *(Color composite by R. J. Dufour of Rice University, from three photographs taken with the 4-m telescope at the Tololo Inter-American Observatory. Copyright © 1980 by R. J. Dufour, Rice Unive*

Half-title page: Saturn, observed by Voyager 1 four days after passing the planet. *(NASA/JPL)*

Frontispiece: The Trifid nebula in Sagittarius. The blue region on the left is starlight reflected b interstellar dust, and the red region on the right is light emitted by ionized gas. *(Kitt Peak National Observatory)*

Back cover: The open cluster M16 and associated nebulosity in Serpens. *(Palomar Observatory/California Institute of Technology)*

EXPLORATION OF THE UNIVERSE (4/e)   ISBN 0-03-058502-3

234   039   9876543

CBS COLLEGE PUBLISHING
Saunders College Publishing
Holt, Rinehart and Winston
The Dryden Press

PREFACE

These have been exciting times for science in general and for astronomy in particular. We have learned about the continents of Venus by probing her dense atmosphere with radar. We think we have discovered the dying glow of the primeval fireball that began the expansion of the universe. We have identified gravitational lenses in space, and we think we may have found black holes. Especially thrilling have been the discoveries from our exploration of the solar system with our space program: dry rivers on Mars, belching volcanoes on Jupiter's satellite Io, and rings around Jupiter. We have seen that the rings of Saturn, so smooth-appearing when viewed with terrestrial telescopes, are actually made up of at least hundreds of thousands of ringlets. We have mapped more than a dozen new worlds, each comparable to our own moon — worlds in orbits about the giant planets of the solar system. At this writing, our Voyager 2 spacecraft is en route to an encounter with Uranus in 1986.

I am grieved that the U.S. program for the exploration of the solar system is being phased out, with that activity being left to scientists of other lands. I hope this setback is temporary. But I am pleased that interest in the new frontiers of astronomy has not waned. On the contrary, our classes in astronomy are more crowded than ever. Students are becoming more sophisticated and more fervent in their desire to understand as much as they can of what we have learned about the cosmos. People seem, somehow, to be aware of what astronomy can offer to the human perspective. What they are *not* aware of, though, is the *method* of science — of the exacting procedure and rigid rules of the scientific method. It is here that a gap in communications exists between scientists and nonscientists.

That communications gap becomes especially obvious when we note that many people, thirsty for knowledge about the new frontiers, have turned to all manner of unreliable sources for their information. We scientists, in my opinion, have an obligation to the public to increase our efforts in presenting an honest view of our science. The universe is a wondrous place, and it is, I think, the purpose of a liberal arts science course to open the window to those wonders. We should show not only how the universe *is*, but also how, by simple rational processes, we can probe its mysteries. The great value of science is its approach to understanding nature.

That is what *Exploration of the Universe* is about. Of course, it is a textbook in astronomy — specifically for a rather comprehensive course in astronomy for the liberal arts student. But if the reader finds the book to be *only* an astronomy text, rather than an introduction to the rational exploration of nature, I have failed to convey a large part of my message.

I realize that *Exploration* can be a rather large dose for a first course in some schools for the nonscience student, and especially so for one taking less than a full-year course; that is why there is also a brief version, *Realm of the Universe* (Saunders College Publishing, 1980). An even more "basic" book is on the drawing board. Still, I have tried, as in the three previous editions of *Exploration*, to be as comprehensive as I could, not just for the occasional science student who may find the book a useful reference, but for the lay nonspecialist — like my father and grandmother, both of whom were interested in the world about them, and to whom new questions always arose amid the answers to old ones. But if this book is used for a one-semester or one-quarter course, the instructor will probably want to make many sections optional. Certainly, these will include those sections in special type (set off by double horizontal lines). Probably also, in a short course, the instructor will want to make optional major parts or all of Chapters

6, 9, 13, 19, 20, 21, 23, 26, 30, and 35; of course, individual preferences will vary. However, I urge against too thorough a coverage of the earlier material at the expense of not reaching the grand messages of the final chapters.

This fourth edition of *Exploration of the Universe* is a complete revision. Much out-of-date material has been deleted, and much new material has been added. All chapters have been updated, and Chapters 1, 2, 11, 12, 17 through 19, 28, 29, and 31 through 39 either are new or have been substantially rewritten. Of course, new illustrations, especially those from our spacecraft, have been added. As with previous editions, I have avoided mathematics beyond the simplest algebra throughout the text. This applies to the exercises as well, although most will take some original thought, in that their solutions cannot be found by thumbing back through the chapters. (The more challenging exercises are marked with a star.) Also, as before, I have tried to stress that astronomy is a very human endeavor and have related it to those men and women who created our science. Finally, because of my concern over the widespread confusion between science and pseudoscience, I have addressed several popular fads.

Many people have reviewed parts of this book and have offered valuable suggestions. They include L. H. Aller, R. L. Bishop, A. D. Fiala, T. Gehrels, O. Gingerich, C. H. Jerred, M. Jura, W. M. Kaula, W. J. Luyten, D. M. McCarthy, D. Mihalas, J. S. Miller, D. Morrison, P. J. E. Peebles, D. M. Popper, W. M. Sinton, R. Ulrich, and J. Wasson. Special thanks are due Mr. Bradley Wood, our electronics genius at UCLA, who assisted greatly in the preparation of the material on new detectors in Chapter 11. Also, I continue to thank three old and dear friends whose help and encouragement in the preparation of the first edition of this book have continued to be felt strongly even in this edition. They are Dan Popper, Emeritus Professor of Astronomy at UCLA; Paul Routly, now at the U.S. Naval Observatory; and the late Paul Wylie, whom I replaced at the time of his retirement from UCLA and who encouraged me to write an astronomy text in the first place.

Obviously, none of the reviewers has seen the final version of the manuscript. Although I have tried to incorporate their suggestions, I have probably goofed here and there. There are surely typos and trivial mistakes, some of which have been faithfully reproduced through the three previous editions! Many good friends, known and unknown, have written me, calling my attention to such slips in the past. I greatly appreciate their thoughtfulness and will continue to be grateful to all of you — students, professors, and others — who let me know of mistakes in this book, so that they can be corrected in future printings.

I also want to express appreciation for the great assistance I have received from my editors at Saunders, especially Lloyd Black and Janis Moore. Finally, I thank my dear wife, Phyllis, who has joined me in the production of the book by reading, correcting, pasting, criticizing, and in general trying to make it more meaningful to you all. If it works, it is largely her doing; if not, it is because I didn't listen to her enough. Mainly, though, I hope you enjoy *Exploration* and gain from it new insight and a new perspective of the universe.

G. O. Abell

Encino, California
December 1981

CONTENTS

APPENDICES

John Muir (1828–1914), Scots-American naturalist, was largely responsible for the establishment of Yosemite and other national parks. He wrote, "When we try to pick out anything by itself, we find it hitched to everything else in the universe." *(Sierra Club)*

1

THE UNITY OF THE UNIVERSE

The ancients believed in a sort of unity between the heavens and the earth. To be sure, in classical Greece the earth was thought to be composed of base stuff—the four "elements," earth, water, air, and fire—and the heavens of crystalline material, but the heavens were the realms of the gods; the planets *were* gods in some early cultures. And the gods, presumably, controlled or influenced human affairs. Earthly events seemed chaotic and unpredictable, but the Ancients recognized a regularity in the motions in the heavens and quite naturally hoped that by understanding the motions of their planet gods, they would better understand the individual lots of men and women. They thus sought a unity between the earth and heavens through the primitive religion of astrology. Ironically, today, in our 20th-century enlightenment, a large fraction of all people still believe in that ancient religion.

But there *is* a real unity, and one far grander and more beautiful than our ancestors could possibly have imagined. That real unity is in the basic structure of matter everywhere in the universe, and in the laws of nature—the rules that govern how everything works.

1.1 THE UNITY OF STRUCTURE

We have learned that all matter is made of the same stuff—the matter of the earth and moon (by direct analyses of their rocks), of the other planets (through analysis by space flybys and landers) (Figure 1.1), and of the stars and even the remotest galaxies (from studying their spectra). This stuff is not the elements earth, water, air, and fire of antiquity, but approximately a hundred different kinds of atoms that make up the hundred or so naturally occurring elements and, in various combinations, the molecules of the billions of kinds of chemical compounds.

Each atom has a *nucleus* with a positive electrical charge, and surrounding it negatively charged *electrons*, carrying a combined charge equal, but opposite in sign, to that of the nucleus. These atoms are held together, and different atoms are bound in molecules, by the *electromagnetic* forces that act between electrically charged particles. The electromagnetic force is the second strongest of the four forces known in nature (Figure 1.2). In its binding of molecules together in solids, that force accounts for the rigidity of steel!

1

Figure 1.1 Surface of Mars photographed July 24, 1976 by the Viking 1 Lander. The structure in the foreground is part of the Lander itself. (*NASA/JPL*)

The sizes of atoms are exceedingly tiny, typically about one hundred-millionth of a centimeter in radius. Yet atoms are almost empty space, far more so than is the solar system. The earth's distance from the sun is about a hundred times the sun's diameter, but on the average the electron in a hydrogen atom is distant from the atomic nucleus by about 100,000 times the size of that nucleus. By volume, atoms are emptier than the inner part of the solar system by a thousand million times.

But even the tiny nucleus is not a single entity; it is composed of still smaller particles called *protons* and *neutrons*. They are similar in size, and each has a mass of almost 2000 times that of an electron. But while the neutron is, as its name implies, electrically neutral, the proton carries a positive charge equal in magnitude to the negative charge carried by the electron. The number of protons in the nucleus of an atom (equal to the number of external electrons) determines what kind of atom it is and what kind of element that kind of atom makes up. An atom of hydrogen, for example, has a single proton in its nucleus, usually no neutrons (that is, its nucleus is a proton), and one external electron. The helium atom has a nucleus with 2 protons and usually 2 neutrons. Lithium has 3 protons, oxygen has 8, and uranium has 92. Except for hydrogen, the number of neutrons in a nucleus is usually roughly comparable to the number of protons.

The positively charged protons in an atomic nucleus all repel each other (like electrical charges repel, and opposite charges attract). Then why doesn't the nucleus fly apart? Because its particles are bound together by the *strong nuclear force;* it is about 100 times as strong as the electromagnetic force of repulsion between two protons. The strong force is the strongest of the forces of nature.

Neutrons themselves are stable in an atomic nucleus, but outside the nucleus after about 11 minutes, on the average, a neutron spontaneously

Figure 1.2 While Nature binds matter together with the electromagnetic force, man has learned to harness it for myriad commercial applications. Here a powerful electromagnet is used to drop bales of steel cans into a recycling furnace scrap bucket. *(Bethlehem Steel)*

breaks up into a proton, an electron, and another particle called an *antineutrino* (more about these later). The force that is involved in the decay of the neutron is the *weak force;* this third of the known forces is about a thousand times weaker than the electromagnetic force.

Up to now we have been on pretty familiar ground, or at least familiar to the physicist. But are there still more fundamental particles that make up the protons, neutrons, and a host of other subatomic particles known to exist for brief times? Most physicists think there are. It is widely believed that a few kinds of even smaller particles, called *quarks,* in various combinations make up all of the other nuclear particles. The quark theory helps us understand many of the properties of the atomic nucleus, but quarks have never been isolated and are still only hypothetical.

Nevertheless, the deeply significant point is that everything, everywhere, is basically the same. So far as we know, from the time when the very first stable matter formed in the universe, it was made up of the same things: mainly protons, elec-

trons, and neutrons. All matter since then has been subjected to the same kinds of forces and has obeyed the same laws.

1.2 THE UNITY OF BEHAVIOR

The seeds of modern science began to sprout following the Reformation. The 17th-century German astronomer Johannes Kepler discovered for the first time certain simple mathematical rules that describe accurately the motions of the planets (for example, that they move in precise elliptical paths about the sun). His contemporary, the Italian physicist Galileo Galilei, discovered some other precise rules that describe the behavior of bodies on earth (for example, that all freely falling bodies near the earth's surface pick up speed—*accelerate*—at the same rate) (Figure 1.3). Later in the same century, Isaac Newton showed that Kepler's celestial rules and Galileo's terrestrial ones are united by the same underlying laws. Newton had the insight to recognize that the force that makes planets fall in ellipses about the sun and the force that makes apples fall with uniform acceleration near the earth's surface, are different manifestations of the same thing: gravitation.

Gravitation is the fourth and weakest force of nature, yet it was the first to be discovered. Gravitation is an incredibly weak force; the gravitational attraction between the proton and electron in a hydrogen atom is weaker than the electromagnetic attraction between their opposite (plus and minus)

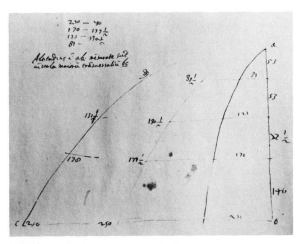

Figure 1.3 A reproduction of a page of Galileo's notes concerning his investigation of the paths of falling bodies.

charges by 10^{39} (one followed by 39 zeros) times! Yet Newton's discovery of gravitation (Figure 1.4) preceded by nearly two centuries a comparable understanding of any of the other forces. Why?

Nuclear forces act only over the very tiny dimensions of the atomic nucleus; the electromagnetic force acts only between electrically charged particles, and most of bulk matter is electrically neutral. Nature seems to have provided equal quantities of positive and negative charge in the universe, and the very electromagnetic force that acts between charges has assured that plus and minus charges are distributed more or less uniformly. Thus over large distances matter appears neutral, and only gravitational forces are important. Very locally, over the dimensions of atoms and molecules, the positive and negative charges of nearby protons and electrons *do not* completely cancel, so they pull on each other and bind atoms and molecules together and to adjacent molecules. But bulk matter is electrically neutral, so it exerts no net electrical attraction or repulsion on its surroundings.

On the other hand, we are all aware of the gravitational force, despite the great weakness of its attraction between ordinary masses, because we are very close to a body of astronomical mass—the earth. Gravitation depends only on the total amount (mass) of matter, not on charge, so if enough mass is collected together, as in the earth, its gravitational pull can be quite significant, as everyone knows who has tumbled on an ice rink.

1.3 THE UNIVERSE

The earth is a small planet in the solar system. Four of the system's nine planets are enormous in comparison; Jupiter has more than 300 times the amount of matter that the earth does (Figure 1.5). But the solar system is dominated by a far larger body, with 1000 times the mass of Jupiter: the sun. The sun's gravitation is great enough to keep all nine planets, a hundred thousand or more asteroids (or minor planets), thousands of millions of comets, and a far, far greater number yet of tiny chunks of ice and/or rock and metal—the meteoroids—all revolving about it.

The highest speed possible (as we shall see in Chapter 13) is the speed of light—3×10^{10} cm/s (300,000 km/s, or 186,000 miles/s). It takes light just over a second to reach earth from our nearest celestial neighbor, the moon; the moon is just over one *light second* away. The sun is 400 times the moon's distance, and it takes light 8 minutes to come from the sun; it is 8 *light minutes* away. The other planets range from light minutes to light hours away.

The sun is an enormous ball of tremendously hot gas, so hot that all of its chemical elements, including those like iron and tungsten that are solid on earth, are vaporized to the gaseous state (Figure 1.6). At its surface, the sun's temperature is about 6000 K, but at its center the solar temperature

Figure 1.4 Newton's home, Woolsthorpe Manor, where he claimed to have conceived his ideas on gravitation. The legend is that he was inspired by the fall of an apple from the tree in the foreground. (*Photograph by the author*)

ranges up to many millions of degrees.* Deep in the sun's interior, the thermonuclear conversion of hydrogen to helium gives rise to the sun's enormous outpouring of energy. The sun has a third of a million times the mass of the earth and more than a million times the earth's volume. Yet the sun is an ordinary star.

The thousands of stars we see around us in the sky, and the many millions revealed by telescopes, are among the few hundred thousand million stars that make up our *Galaxy*. Those stars are suns, more or less like our own sun, but at distances of *light years*. (A *light year*, abbreviated LY, is the distance light travels in one year: 9.46 million million kilometers, or about 6 million million miles.) The entire Galaxy is a wheel-shaped system about 100,000 LY across, with the sun far from its center, perhaps two-thirds to three-quarters of the way to an edge of the wheel. As we look into the sky in

Figure 1.5 Jupiter and its four largest satellites, assembled (not to the same scale) from photographs by Voyager 1 into this composite picture. *(NASA/JPL)*

*In astronomy, we almost always express temperature in Kelvins (K), which are Celsius degrees but measured from absolute zero: −273°C; see appendix 5.

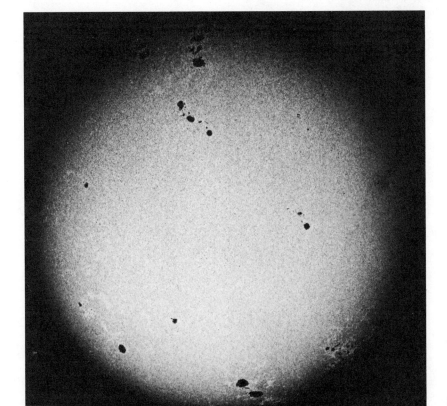

Figure 1.6 The sun. *(Mount Wilson and Las Campañas Observatories)*

directions that take our line of sight edge-on through the Galaxy, the very many remote stars in those directions produce a faint glow of light—the *Milky Way,* an irregular luminous band completely circling the sky (Figure 1.7). Our Galaxy is therefore often called the Milky Way Galaxy, or simply the Milky Way.

The Galaxy rotates, just as the planets revolve about the sun, but the sun, carrying our solar system with it, revolves about the center of the Milky Way Galaxy in its galactic orbit, at a speed of 200 to 300 km/s, in about 200 million years. Probably many of the other stars revolving about the nucleus of our Galaxy also have planets revolving about them, but those other stars are so remote that we have not yet been able to detect any planets associated with them.

Our Galaxy is not the end of the story. There are millions, probably thousands of millions, of other galaxies in the observable universe (Figure 1.8). These galaxies, like remote islands, each with its thousands of millions of suns, are separated from each other by distances many times their own diameters. Still, even galaxies tend to group in *clusters of galaxies,* and those clusters of galaxies, perhaps also with individual galaxies not members of clusters, are parts of still larger systems called *superclusters.*

I have been tossing off words describing objects of incredible mass and dimensions, and distances of unimaginable extent, as though they were apples and inches. There is no possibility that someone not already familiar with these concepts can grasp them on a first reading. That, of course, is one purpose of studying a whole course in astronomy. Still, it is worthwhile to try to give a feeling for the scale of the universe, no matter how hard it may be to grasp it thoroughly in one run-through.

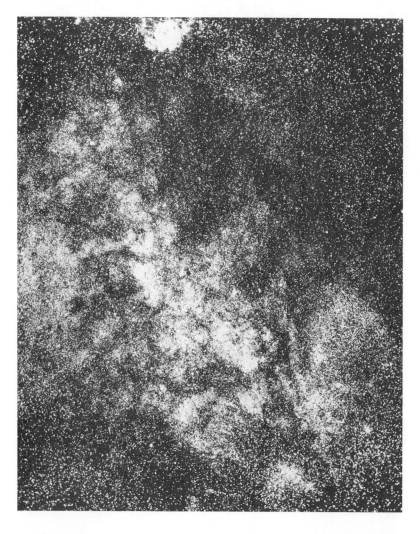

Figure 1.7 A portion of the Milky Way in Sagittarius. (*Yerkes Observatory*)

Figure 1.8 The neighboring spiral galaxy in Andromeda, M31. (*Mt. Wilson and Las Campanas Observatories*)

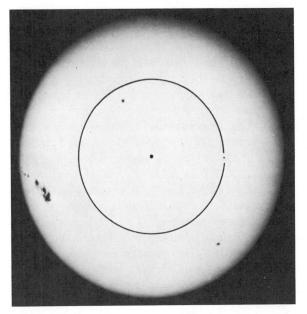

Figure 1.10 The moon's orbit, to scale, superimposed on a photograph of the sun. (*Griffith Observatory*)

1.4 THE SCALE OF THE UNIVERSE

The earth is a nearly spherical body about 13,000 km (8000 miles) in diameter. The moon, about one-fourth the earth's diameter, is 30 of those earth diameters away. We can actually present a scale drawing of the earth and moon on a page of this book (Figure 1.9).

The average distance of the earth from the sun is called an *astronomical unit* (AU). That radius of the earth's nearly circular (actually elliptical) orbit

is about 100 times the diameter of the sun, but the whole orbit of the moon would fit easily inside the sun itself (Figure 1.10). We cannot, therefore, show the earth, sun, and moon to scale on the same page. But if we show the sun as a small dot (Figure 1.11a), we can see the relative sizes of the orbits of the inner planets (Mercury, Venus, Earth, and Mars) to the correct scale, even though those planetary bodies themselves would not show up if drawn to the correct size. Even then, we must change the scale (Figure 1.11b) to show the orbits of the outer planets (Jupiter, Saturn, Uranus, Neptune, and Pluto) with the right sizes relative to the size of the orbit of Mars. Pluto's average distance from the sun is 40 times that of the earth; that is, Pluto is 40 AU from the sun.

The nearest star beyond the sun is, in contrast, 300,000 AU away. On the scale of Figure 1.11b, its distance would be more than a kilometer. That star is 4 LY away; most visible stars are hundreds or even thousands of light years away.

Suppose we make a rough scale drawing, showing the stars within 10 LY of the sun. In Fig-

Figure 1.9 The earth (*left*) and moon drawn to scale.

ure 1.12a, the circle represents a sphere of 10 LY radius centered on the sun. Roughly ten stars are included. Now we change scale. In Figure 1.12b, the sphere of 10 LY radius is the small circle in the center, and the larger circle represents a sphere of 100 LY radius—ten times as great in radius, but 1000 times as great in volume. In that sphere we would find approximately 10,000 stars. Similarly, the sphere of 100 LY radius is the small circle in Figure 1.12c, while the larger circle represents a sphere 1000 LY in radius within which there are ten million (10^7) stars. In our next change of scale, Figure 1.12d, the stars do thin out, but in some dimensions before others. We have begun to reach the boundaries of our Galaxy, but at first only in its thin dimension. The entire wheel-shaped Galaxy, in an edge-on cross section, is shown in Figure 1.12d, and the sphere of 1000 LY radius is the small circle on the left side, with the invisible sun deep in its center.

As are most (if not all) galaxies, our own Milky Way Galaxy is part of a cluster, actually a collection of about two dozen galaxies called the *Local Group*. The Local Group is about 3 million light years across and is shown, roughly to correct scale, in Figure 1.13. Far beyond its boundaries are other such groups, and rare in space, typically tens to hundreds of millions of light years apart, are the great clusters of hundreds or even thousands of member galaxies each. One such is the Hercules cluster (Figure 1.14).

The most remote clusters of galaxies yet identified are one to two thousand times as far away from us, on the scale of Figure 1.13, as the size of a page of this book, perhaps half a kilometer or so. Even farther off are the remote *quasars* (Chapter 37), and yet beyond the quasars is the remote glow from the past—a glow of radio radiation that has been traveling through space to reach us from all directions for at least 10 thousand million years. When that radiation began its journey through space it was light, not radio waves. It has been transformed by the expansion of the universe and is our best evidence today that the universe has evolved from a hot, dense state that existed far in the past.

(a) The Big Bang

The radio waves mentioned above are, in a very real sense, the dying glow of that explosive begin-

ning of our universe called the *big bang* (Chapter 39). During the first few minutes after the big bang, atomic nuclei formed, but mainly only those of hydrogen and helium. It was nearly another million years before the universe cooled (because of its expansion) enough to permit electrons to join nuclei and make atoms. At this stage the universe became transparent, so light could flow freely through it; it is just this radiation that we now observe as radio waves.

It was probably another thousand million years before matter collected together to form galaxies and stars. In the centers of stars atoms were reheated, and thermonuclear reactions fused hydrogen into helium, providing the energy by which the stars shine. Later, in the interiors of certain stars, atoms of helium fused into those of heavier elements—carbon, oxygen, nitrogen, and silicon, for example—those elements that can form rocky planets.

Subsequently, these stars ejected some of their matter, so enriched in heavier elements, into interstellar space, to eventually condense into new stars.

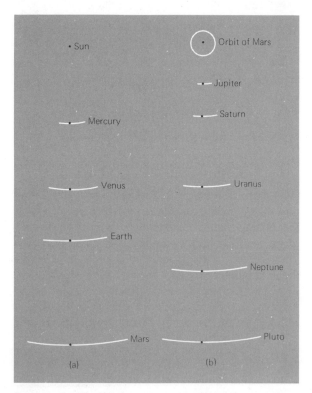

Figure 1.11 The distances of the planets. (a) The inner planets. The sun's size is shown to the same scale as the planets' orbits, but the planets themselves could not be seen on this scale. (b) The outer planets to the scale of the orbit of Mars.

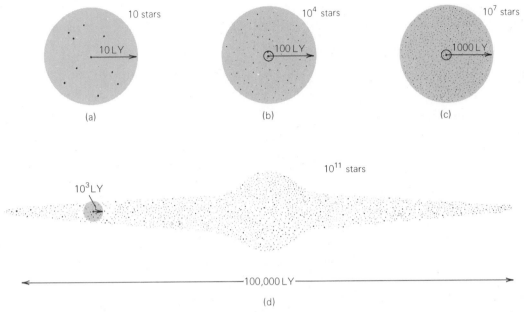

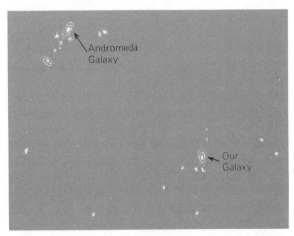

Figure 1.12 The distribution of stars around the sun within (a) 10 LY; (b) 100 LY; (c) 1000 LY; (d) the Galaxy.

One such star that formed about 4.6 thousand million years ago, with its system of planets, is our sun. Much of the material of the sun and, we think, virtually all of that of the earth, consists of atoms synthesized by nuclear reactions in earlier-generation stars. We discuss all these goings-on in later chapters.

1.5 THE NATURE OF SCIENCE

There are large gaps in our full understanding of the origin of the universe through the big bang, as well as of the life cycles and evolution of stars, just as there are gaps in our knowledge of the evolution of life on earth. But the general picture is fairly clear and becomes clearer as research continues to advance the scientific frontier.

That is not to say, however, that science ever provides, or even attempts to provide, the absolute truth or the ultimate answers to everything. For example, we can apply known physical laws—our best theory—to the conditions of the early universe, but that does not tell us where its matter and energy came from in the first place, or how the matter and radiation got into that hot dense state. Perhaps our ideas about the big bang are even completely wrong; we are, after all, extrapolating known theory to quite a limit in space and time.

Very many hypotheses at the scientific frontier (and the big bang is certainly at that frontier) will turn out to be wrong. It is far less likely that our ideas of electricity and magnetism, of gravitation (as formulated in the general theory of relativity), or of the behavior of planets in the solar system will be wrong, for these ideas have been extraordinarily well tested.

It is the testing that is the part and parcel of science. Science is not just a collection of knowledge, of figures and test tubes. It is an organized

Figure 1.13 Schematic diagram of the Local Group, approximately to scale.

Figure 1.14 A cluster of galaxies in Hercules. *(Mt. Wilson and Las Campañas Observatories)*

method for exploring nature. Science involves three steps: (1) An observation or an experimental result is noted (say, the falling of a body with uniform acceleration). (2) A hypothesis is advanced to describe that result in terms of a more general model (a theory of gravitation). (3) The model is then used to predict new observations or the results of new experiments (that the moon's orbit must be an ellipse, or that an unknown planet, Neptune, is causing irregularities in the motion of Uranus). Then we must check whether these predictions hold up. If not, the hypothesis must be discarded or modified. If so, more tests of additional predictions must be devised, and with ever broader applications, until the hypothesis has been subjected to the most ruthless scrutiny. Even then, it may later prove wrong or at least to have limited usefulness, but the longer it survives and the more documented it becomes, the better are its chances to become a part of established theory. As we shall see in Chapter 39, the big bang has survived three crucial tests and is certainly worthy of being taken very seriously, but it is by no means a *fact*—not yet. And the fabric of even the best of scientific theories may begin to tear when it is pushed to limits beyond the regime in which it is well tested, as did Newtonian gravitational and mechanical theory in the 20th century, when it was challenged with speeds near that of light or with gravitational fields of enormous strength. Even so, when applied with proper re-

spect for its limitations, Newtonian theory is of incalculable value.

So it is that third step in the *scientific method*, the test of the hypothesis, that is crucial. Without such tests and checks, a model is mere speculation and not part of science. For example, an assertion that a civilization of people exactly like us exists on a planet revolving about a star in a remote galaxy,

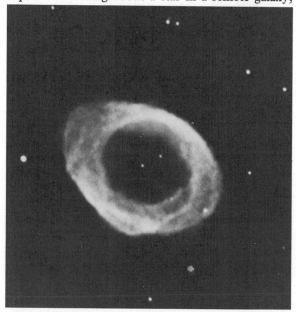

Figure 1.15 The Ring nebula in Lyra, an example of a shell of gas ejected from a star. *(Mt. Wilson and Las Campañas Observatories)*

or the contention that far beyond the limits of our observations are other universes, are not scientific hypotheses or theories—at least not today—for there is no way to test them. Perhaps the assertions are true, but in the absence of any possibility of verifying them, their truth or error is irrelevant and certainly has nothing to do with science.

To be sure, science has limitations. It must operate by strict rules, and very many questions cannot be asked, let alone answered, in science. Many other valid and highly valuable areas of human activity, art and music for example, have great importance even though they lie outside of science. Still, science has proved itself to be a remarkably successful way of studying nature. It has provided models by which we can understand a great deal, and which have enabled us to develop a highly advanced technology.

And if it has not yet uncovered the whole truth of the universe (and it never can), science has revealed a marvelous unity in the universe; as in a Greek tragedy, it has a unity of time, place, and action. Time and place (time and space), as we shall see, are inextricably interrelated, for even at this moment we observe the remote past of the universe as we look to its remote parts. And everywhere—then, there, here, and now—we find the same kinds of stuff: atoms, electrons, and so on. Unity of action is even more remarkable. The laws of nature, so far as we can tell, are truly universal.

Nor can they be suspended, even for a moment. It was when man stopped believing in magic that science began to advance. Magic is the antithesis of science. The fake psychics and fortunetellers notwithstanding, if we could really suspend nature's laws we would have utter chaos. Far from everything being possible, *nothing* would be possible!

As a final reminder of the great unity of the universe, recall that, according to the results of our current research, the very atoms that comprise our own bodies were formed in the centers of past-generation stars. We are, ourselves, quite literally, made of stardust!

EXERCISES

1. If a gas is *ionized*—that is, if the electrons are stripped from the nuclei of its atoms, so that each freed electron moves about in the gas in the same way an individual molecule would—the gas can be compressed to a far higher density than ordinary solids can. Explain why. (Assume that high temperatures or other effects prevent ions from reuniting with electrons, thereby becoming neutral atoms.)

2. The electromagnetic force causes a repulsion between the positively charged protons in an atom, but the strong nuclear force still binds them together in the atomic nucleus, along with a comparable number of neutrons. But the strong force is effective only between particles in virtual contact, while the hundred-times-weaker electromagnetic force acts over a considerable distance. Does this suggest to you why atomic nuclei with roughly 100 or more protons are not stable, and therefore disintegrate? Explain.

3. If we were to try to communicate, say by radio waves (which travel with the speed of light), with a hypothetical inhabitant of a planet revolving about a star 100 LY away, how long would we have to wait after transmitting a question before we could expect to receive an answer?

4. From the data given in this chapter, calculate how many circles the size of the earth's orbit would be required to reach across the length of a diameter of the Galaxy, if they were laid out barely touching each other, as in a chain. How many such circles would it take to reach a remote cluster of galaxies, say, 300 million LY distant?

5. Review how many changes of scale are required to prepare a series of scale drawings on standard pieces of paper (8½ by 11 inches), beginning with a diagram that shows the earth as a circle and ending with one that shows the correct relative distance for a remote cluster of galaxies. By what factor must the linear size of the largest distance on one diagram be reduced to show it as the smallest one on the next diagram in the series?

6. Give at least three examples of questions that are improper in the realm of science.

7. For each of the following phenomena, make up at least one hypothesis that is certainly wrong because it violates other experimental or observational evidence:
   (a) Objects, when dropped, fall to the ground.
   (b) The room lights go on when the wall switch is pushed up.
   (c) Water boiling in a pan eventually disappears if left over the fire long enough.
   (d) Airplanes can fly even though they are heavier than air.
   (e) Blue-eyed parents have blue-eyed children.
   (f) Drunk drivers are more likely to be involved in accidents than sober ones.

Claudius Ptolemaeus (Ptolemy) (2nd century A.D.) was one of the great astronomers of antiquity. He devised a system of cosmology that described the motions of the planets so satisfactorily that there was no substantial change until the time of Copernicus. (*Burndy Library, photograph by Owen Gingerich*)

2

EARLY ASTRONOMY: MYTH AND SCIENCE

2.1 EARLIEST ASTRONOMERS

Speculations on the nature of the universe must date from prehistoric times. It is difficult to state definitely when the earliest observations of a more or less quantitative sort were made or when astronomy as a science began. Certainly, in many of the ancient civilizations the regularity of the motions of celestial bodies was recognized, and attempts were made to keep track of and predict celestial phenomena. In particular, the invention of and keeping of a calendar requires at least some knowledge of astronomy—the basic units of the calendar being the day, the month (originally, the 29- and 30-day cycling of the moon's phases), and the year of seasons (the *tropical year*).

The Chinese had a working calendar and had determined the length of the year several centuries before Christ (B.C.). About 350 B.C., the astronomer Shih Shen prepared what was probably the earliest star catalogue, containing about 800 entries. The Chinese also kept rather accurate records of comets, meteors, and fallen meteorites from 700 B.C. Records were made of sunspots visible to the naked eye and of what the Chinese called "guest stars," stars that are normally too faint to be seen but suddenly flare up to become visible for a few weeks or months (such a star is now called a *nova*). The most significant of the Chinese observations of nova outbursts was of the great supernova of 1054 A.D. in the constellation of Taurus. Today's remnant of that cosmic explosion is the Crab nebula, a chaotic, expanding mass of gas (see Chapter 34).

The Babylonians, Assyrians, and pre-Christian Egyptian astronomers also knew the approximate length of the year from early times. By a few centuries B.C., the Egyptians had adopted a calendar based on a 365-day year. Of particular significance to them was the date when the bright star Sirius could first be seen in the dawn sky, rising just before the sun. This *heliacal* rising of Sirius coincided fairly well with the average time of the annual flooding of the Nile, which gave the astronomer-priests the ability to predict very roughly when this economically important event could be expected to occur.

There is evidence of ancient astronomical knowledge in other parts of the world as well. The Maya in Central America developed a sophisticated calendar and made astronomical observations in a period contemporary with early European civilizations (Chapter 8), and the Polynesians learned to navigate, evidently by means of celestial observations, over hundreds of kilometers of ocean separating their islands.

Particularly interesting are monuments left by Bronze Age people in northwestern Europe, especially in the British Isles. The best-preserved of the monuments is Stonehenge (Figure 2.1), about 13 km from Salisbury in southwest England. It is a complex array of stones, ditches, and holes arranged in concentric circles. Carbon dating and other studies show that Stonehenge was built during three periods ranging from about 2500 B.C. to about 1700 B.C. Some of the stones are aligned with the directions of the sun during its rising and setting at critical times of the year (such as the beginnings of summer and winter), and it is widely thought that at least one function of the monument was connected with the noting of these occasions. However, only some of the many hundreds of other monuments have alignments that can be interpreted as astronomical.

2.2   EARLY GREEK ASTRONOMY

The high point in ancient science was in the Greek culture from 600 B.C. to 400 A.D. The earliest Greeks were not scientists in the modern sense. They were often more interested in solving abstract geometrical problems, reasoning from given axioms, than in making original observations. Yet, in that Greek reservoir of ideas and inspiration, many observations were carried out, with the result that science in general and astronomy in particular were raised to a level unsurpassed until the sixteenth century.

(a)   Early Concepts of the Sky

The Celestial Sphere

If we gaze upward at the sky on a clear night, we cannot avoid the impression that the sky is a great hollow spherical shell with the earth at the center. The early Greeks regarded the sky as just such a *celestial sphere;* some apparently thought of it as an actual sphere of a crystalline material, with the stars embedded in it like tiny jewels. The sphere, they reasoned, must be of very great size, for if its surface were close to the earth, as one moved from place to place he would see an apparent angular displacement in the directions of the stars.

Of course, at any one time we see only a hemisphere overhead, but with the smallest effort of imagination we can envision the remaining hemisphere, that part of the sky that lies below the horizon. The fact that the sky appears as a full hemisphere and not merely part of one was further evidence that the celestial sphere must be extremely large—infinite as far as the eye can tell—compared to the earth. If we watch the sky for several hours, we see that the celestial sphere is gradually and continually changing its orientation. The effect is caused simply by the rotation of the earth, which carries us under successively different portions of the sphere. Following along with us must be our *horizon,* that line in the distance at which the ground seems to dip out of sight, providing a demarcation between earth and sky. (The horizon may, of course, be hidden from view by mountains, trees, buildings or, in large cities, smog.) As

Figure 2.1   Stonehenge. *(Courtesy E. C. Krupp, Griffith Observatory)*

our horizon tips down in the direction that the earth's rotation carries us, stars hitherto hidden beyond it appear to rise above it. In the opposite direction the horizon tips up, and stars hitherto visible appear to set behind it. Analogously, as we round a curve in a mountain road, new scenery comes into view while old scenery disappears behind us.

The direction around the sky toward which the earth's rotation carries us is *east;* the opposite direction is *west.* The Greeks, unaware of the earth's rotation, imagined that the celestial sphere rotated about an axis that passed through the earth. As it turned, it carried the stars up in the east, across the sky, and down in the west.

Celestial Poles

A careful observer will notice that some stars do not rise or set. As seen from the Northern Hemisphere, there is a point in the sky some distance above the northern horizon about which the whole celestial sphere appears to turn. As stars circle about that point, those close enough to it can pass beneath it without dipping below the northern horizon. A star exactly at the point would appear motionless in the sky. Today the star *Polaris* (the North Star) is within 1° of this pivot point of the heavens.

Figure 2.2 Time exposure showing trails left by stars as a consequence of the apparent rotation of the celestial sphere. *(Lick Observatory)*

The Greeks regarded that pivot point as one end of the axis about which the celestial sphere rotates. We know today that it is the earth that spins about an axis through its North and South Poles. An extension of the axis would appear to intersect the sky at points in line with the North and South Poles of the earth but, because of the virtually infinite size of the celestial sphere, immensely far away. As the earth rotates about its polar axis, the sky appears to turn in the opposite direction about those *north* and *south celestial poles.*

An observer at the North Pole of the earth would see the north celestial pole directly overhead (at his *zenith*). The stars would all appear to circle about the sky parallel to the horizon, none rising or setting. An observer at the earth's equator, on the other hand, would see the celestial poles at the north and south points on his horizon. As the sky apparently turned about these points, all the stars would appear to rise straight up in the east and set straight down in the west. For an observer at an arbitrary place in the Northern Hemisphere (for example, in Greece), the north celestial pole would appear at a point between the zenith and the north point on his horizon, its location depending on his relative distances from the equator and North Pole of the earth (see Chapter 7). The stars that were not always above the horizon would rise at an oblique angle in the east, arc across the sky in a slanting path, and set obliquely in the west.

Rising and Setting of the Sun

The sun is always present at some position on the celestial sphere. When the apparent rotation of the sphere carries the sun above the horizon, the brilliant sunlight scattered about by the molecules of the earth's atmosphere produces the blue sky that hides the stars that are also above the horizon. The early Greeks were aware that the stars were there during the day as well as at night.

Annual Motion of the Sun

The Greeks were also aware, as were the Chinese, Babylonians, and Egyptians before them, that the sun gradually changes its position on the celestial sphere, moving each day about 1° to the east among the stars. Of course, the daily westward rotation of the celestial sphere (or eastward rotation of the earth) carries the sun, like everything else in the heavens, to the west across the sky. Each day, however, the sun rises, on the average, about four

Figure 2.3 Time exposure showing star trails in the region of the north celestial pole. The bright trail near the center was made by Polaris (the North Star). *(Yerkes Observatory)*

minutes later with respect to the stars; the celestial sphere (or earth) must make just a bit more than one complete rotation to bring the sun up again. The sun, in other words, has an independent motion of its own in the sky, quite apart from the daily apparent rotation of the celestial sphere.

In the course of one year the sun completes a circuit of the celestial sphere. The early peoples mapped the sun's eastward journey among the stars. This apparent path of the sun is called the *ecliptic* (because eclipses can occur only when the moon is on or near it—see Chapter 9). The sun's motion on the ecliptic is in fact merely an illusion produced by another motion of the earth—its annual revolution about the sun. As we look at the sun from different places in our orbit, we see it projected against different stars in the background, or we would, at least, if we could see the stars in the daytime; in practice, we must deduce what stars lie behind and beyond the sun by observing the stars visible in the opposite direction at night. After a year, when we have completed one trip around the sun, it has apparently completed one circuit of the sky along the ecliptic. We have an analogous experience if we walk around a campfire at night; we see the flames appear successively in front of each of the people seated about the fire.

It was also noted by the ancients that the ecliptic does not lie in a plane perpendicular to the line between the celestial poles, but is inclined at an angle of about $23\frac{1}{2}°$ to that plane. This angle is called the *obliquity* of the ecliptic and was measured surprisingly accurately by several ancient observers. The obliquity of the ecliptic, as we shall see, is responsible for the seasons (Chapter 7) and also for the invariable tilt in the axes of terrestrial globe maps.

Fixed and Wandering Stars

The sun is not the only moving object among the stars. The moon and each of the five planets visible to the unaided eye—Mercury, Venus, Mars, Jupiter, and Saturn—change their positions in the sky from day to day. The moon, being the earth's nearest celestial neighbor, has the fastest apparent motion; it completes a trip around the sky in about 1 month. During a single day, of course, these objects rise and set, as do the sun and the stars. We are referring here to their independent motions among the stars, superimposed on the daily rotation of the celestial sphere. The Greeks distinguished between what they called the *fixed* stars, the real stars that appeared to maintain fixed patterns among themselves throughout many generations, and the *wandering stars* or *planets*. The Greek word *planet* means "wanderer." Today, we do not regard the sun and moon as planets, but the Greeks applied the term to all seven of the moving objects in the sky. Much of ancient astronomy was devoted to observing and predicting their motions. In fact, they give us the names for the seven days of our week; Sunday is the sun's day, Monday the moon's day, and Saturday is Saturn's day. We have only to look at the names of the other days of the week in the Romance languages to see that they are named for the remaining planets.

The Zodiac

The individual paths of the moon and planets in the sky all lie close to the ecliptic, although not exactly on it. The reason is that the paths of the planets about the sun, and of the moon about the earth, are all in nearly the same plane, as if they were marbles rolling about on the top of a table. The planets and moon are always found in the sky within a narrow belt 18° wide centered on the ecliptic, called the *zodiac*. The apparent motions of the planets in the sky result from a combination of their actual motions and the motion of the earth about the sun, and consequently they are somewhat complex.

Constellations

The backdrop for the motions of the "wanderers" in the sky is the canopy of stars themselves. Like the Chinese and the Egyptians, the Greeks had divided the sky into *constellations*, apparent configurations of stars. Modern astronomers still make use of these constellations to denote approximate locations in the sky, much as geographers use political areas to denote the locations of places on the earth. The boundaries between the modern constellations are imaginary lines in the sky running north-south and east-west, so that every point in the sky falls in one constellation or another.

Many of the 88 recognized constellations are of Greek origin and bear names that are Latin translations of those given them by the Greeks. Today, the lay person is often puzzled because the constellations seldom resemble the people or animals for which they were named. In all likelihood, the Greeks themselves did not name groupings of stars because they resembled actual people or objects, but rather named sections of the sky in *honor* of the characters in their mythology, and then fitted the configurations of stars to the animals and people as best they could.

(b) The First Greek Astronomers

The Ionian School

The earliest Greek scientists were the Ionians, who lived in Asia Minor. Pythagoras (who died ca. 497 B.C.) was originally an Ionian, but he later founded a school of his own in southern Italy. He pictured a series of concentric spheres, in which each of the seven moving objects—the planets, the sun, and the moon—was carried by a separate sphere from the one that carried the stars, so that the motions of the planets resulted from independent rotations of the different spheres about the earth. These motions gave rise to harmonious sounds, the *music of the spheres*, which only the most gifted ear could hear.

Pythagoras also believed that the earth, moon, and other heavenly bodies were spherical. It is doubtful that he had a sound reason for this belief, but it may have stemmed from the realization that the moon shines only by reflected sunlight, and that the moon's sphericity is indicated by the curved shape of the *terminator*, the demarcation line between its illuminated and dark portions. If he had

so reasoned that the moon is round, the sphericity of the earth might have seemed to follow by analogy.

Another member of the Pythagorean school was Philolaus, who lived in the following century. He may have been the first person to introduce the concept that the earth is in motion. Apparently he held that it is too base to occupy the center of the universe; he assigned a *central fire* to that position. About the fire revolved the earth and other planets. The earth and seven planets or "wandering stars" made eight moving objects, and the sphere of "fixed" stars made a ninth object. Philolaus, however, as a confirmed Pythagorean, believed 10 to be the most perfect number because it is the sum of 1, 2, 3, and 4 (and perhaps also because he had 10 fingers); therefore, he believed there must be another body. This, he proposed, was a *counter earth*, which revolved around the central fire exactly between it and the earth—thus hiding the fire from view from any place on earth. (The central fire itself seems not to have been counted in the numbering of celestial objects.) The period of revolution of earth was 1 day, and the earth rotated as it revolved about the fire so as to keep Greece always turned away from it. Philolaus regarded the celestial sphere as motionless and its apparent rotation as the result of the revolution and rotation of the earth. He proposed that the sun, the moon, and the planets moved in their respective spheres outside the orbit of the earth. It was an imaginative concept based entirely on fancy and cannot be regarded as a forerunner of the heliocentric theory.

Nevertheless, the concept of a moving earth had been introduced, although in a completely erroneous manner. It was a bold idea that may have had some influence on later Greek thought. Other Greek philosophers of the sixth to fourth centuries B.C. who are said to have believed in a moving earth are Hicetas, Heracleides, and Ecphantus. Centuries later Copernicus, in his *De Revolutionibus* (Chapter 3), quoted the Pythagoreans as authorities for his own doctrines.

In their invention of cosmological schemes, the Greeks did not always necessarily attempt to describe what they regarded as reality. Rather, they were often trying to find a scheme—a model—that would *describe the phenomena* and would predict events (eclipses, configurations of the planets, and so on). The epicycles of Ptolemy, developed later, may similarly be regarded as mathematical representations of the motions of planets in the sky.

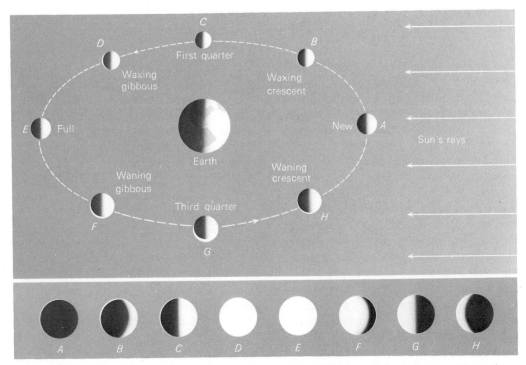

Figure 2.4 Phases of the moon. The moon's orbit is viewed obliquely. *(Below)*, the appearance of the moon from the earth.

(c) The Moon's Phases

Aristotle (384–322 B.C.), most famous of the Greek philosophers, wrote encyclopedic treatises on nearly every field. Aristotle's writings tell us that such phenomena as phases of the moon and eclipses were understood at least in the fourth century B.C. The basic concepts are so important to the development of astronomy that we shall consider them here rather than in the later chapters that deal more directly with these topics.

The moon's changing shape during the month results from the fact that it is not itself luminous but is illuminated by sunlight. Because of its sphericity, only half of the moon is illuminated, that is, having daylight, at one time—the half turned toward the sun. The apparent shape of the moon in the sky depends simply on how much of its daylight hemisphere is turned to our view.

Even in Aristotle's time it was known that the sun is more distant than the moon. This was surmised from the sun's slower apparent motion among the stars on the celestial sphere and also from the fact that the moon occasionally passes exactly between the earth and sun and temporarily hides the sun from view *(solar eclipse)*. Thus when the moon is in the same general direction from earth as the sun (position *A* in Figure 2.4), its day-

light side is turned away from the earth. Because its night side—the side turned toward us—is dark and invisible, we do not see the moon in that position. The phase of the moon is then *new*. (Perhaps it would seem more reasonable to call it "no moon" instead of "new moon," for we do not see any moon at all.) To appear silhouetted in front of the sun, producing a solar eclipse, the moon must be at the new phase and must also lie on the line joining the earth and sun (see Chapter 9). A solar eclipse does not occur at every new moon because the plane of the moon's orbit is inclined slightly (about 5°) to the plane of the ecliptic; hence the new moon usually lies above or below the earth-sun line.

A few days after new moon, the moon reaches position *B*, and from the earth we see a small part of its daylight hemisphere. The illuminated crescent increases in size on successive days as the moon moves farther and farther around the sky away from the direction of the sun. During these days the moon is in the *waxing crescent* phase. About a week after new moon, the moon is one quarter of the way around the sky from the sun (position *C*) and is at the *first quarter* phase. Here the line from the earth to the moon is at right angles to the line from the earth to the sun and half of the moon's daylight side is visible—it appears as a half moon. (The moon is seen as half full at this point because the sun is very much farther away than the moon, and the sun's rays that illuminate the moon and earth are essentially parallel.)

During the week after the first quarter phase we see more and more of the moon's illuminated hemisphere, and the moon is in the *waxing gibbous* phase (position *D*). Finally, about two weeks after new moon, the moon (at *E*) and the sun are opposite each other in the sky; the side of the moon turned toward the sun is also turned toward the earth; we have *full moon*. During the next two weeks the moon goes through the same phases again in reverse order—through *waning gibbous, third* (or *last*) *quarter*, and *waning crescent*. Occasionally the full moon passes through the earth's shadow, which of course extends outward in space in the direction opposite the sun. This is a *lunar eclipse.*

If you find difficulty in picturing the phases of the moon from this verbal account, try a simple experiment: Stand about six feet in front of a bright electric light outdoors at night and hold in your hand a small round object such as a tennis ball or an orange. If the object is then viewed from various sides, the portions of its illuminated hemisphere that are visible will represent the analogous phases of the moon.

(d) The Spherical Shape of the Earth

Another important topic discussed by Aristotle was the shape of the earth. He cited two convincing arguments for the earth's sphericity. First is the fact that during a lunar eclipse, as the moon enters or emerges from the earth's shadow, the shape of the shadow seen on the moon is always round (Figure 2.5). Only a spherical object always produces a round shadow. If the earth were a disk, for example, there would be some occasions when the sunlight would be striking the disk edge on, and the shadow on the moon would be a line.

As a second argument, Aristotle explained that northbound travelers observe the stars near the north celestial pole to be higher in the sky than is observed at home, and different stars pass through the zenith. Conversely, when one travels to more southern latitudes the stars near the north celestial pole are seen lower in the sky, and some stars that are never above the horizon at home are seen to rise and move across the southern sky. The only possible explanation is that the travelers' horizons had tipped to the north or south, respectively, which

indicates that they must have moved over a curved surface of the earth. As a third piece of evidence that the earth is round, Aristotle mentioned that elephants had been observed to the east in India and also to the west in Morocco; evidently, those two places must not be far apart! But he also advanced a theoretical argument that material falling to a center would take on a spherical shape—an idea consistent with the gravitational theory of Newton two millennia later.

(e) The Motion of the Earth

It is interesting that Aristotle pointed out that the apparent daily motion of the sky can be explained by a hypothesis of the rotation of either the celestial sphere or the earth. He rejected the latter explanation. He also considered the possibility that the earth revolves about the sun rather than the sun about the earth. He discarded this *heliocentric* hypothesis in the light of an argument that has been used many times since. Aristotle explained that if the earth moved about the sun we would be observing the stars from successively different places along our orbit, and their apparent directions in the sky would then change continually during the year.

Any apparent shift in the direction of an object as a result of motion of the observer is called *parallax*. An annual shifting in the apparent directions of the stars that results from the earth's orbital motion is called *stellar parallax*. For the nearer stars it is observable with modern telescopes (see Chapter 22), but it is impossible to measure with the naked eye because of the great distances of even the nearest stars. Indeed, Tycho Brahe in the sixteenth century was unable to detect stellar parallax and concluded that the earth is stationary.

2.3 LATER GREEK ASTRONOMY

The early Greeks, as we have seen, were aware of, and to some extent understood, the phenomena of the sky. Remarkable progress, however, was made in the centuries following Aristotle, especially by the school of astronomers centered in Alexandria, where Greek science attained its greatest heights.

(a) Aristarchus of Samos

Especially interesting is Aristarchus of Samos (ca. 310–230 B.C.), who is reported to have believed that the earth revolves about the sun. We know of this, however, only from the writings of others, for only one manuscript of Aristarchus survives: "On the Sizes and Distances of the Sun and Moon." But this document alone is remarkable and deserves some discussion.

Aristarchus opens his treatise with several postulates, the "givens" that are needed to proceed with a geometrical proof. The most essential of these are (1) that the moon receives its light from the sun, (2) that it appears half full when the angle in the sky between the moon and sun is 3° less than a right angle (that is, 87°), and (3) that the diameter of the earth's shadow at the moon's distance is twice the size of the moon. (He also implicitly assumed that the moon's orbit about the earth is a perfect circle.) From these assumptions, Aristarchus, using the rules of Euclidean geometry, derives that (1) the distance of the sun is more than 18 but less than 20 times the distance of the moon and (2) the ratio of the sun's diameter to that of the earth is more than 19 to 3 but less than 43 to 6.

In other words, Aristarchus found that the sun is about 19 times as far away as the moon is (the correct figure is about 400) and that the sun's diameter is about 7 times the earth's (the correct value is 109). A reading of his account suggests that to Aristarchus the entire exercise is no more than an interesting geometry problem. There is no mention at all of where the numbers used in his postulates came from. Perhaps he made crude estimates; perhaps they were someone else's estimates; we do not know. But there is no suggestion that Aristarchus himself actually made any careful observations or measurements. To him, it was, I repeat, an exercise in geometry.

On the other hand, the basic ideas were ingenious and beautiful in their simplicity. Moreover these ideas were applied later by other astronomers—especially by Hipparchus—to attempt an accurate determination of the size and distance of the moon. It is interesting, therefore, to see how the method can work. The following, I emphasize, is *not* the procedure or reasoning of Aristarchus (whose geometry is actually rather tedious) but is a description of how *we* would be able to derive these astronomical dimensions, given Aristarchus' assumptions.

The moon appears exactly *half full* (first and last quarters) when the terminator—the line dividing the light and dark halves—is a perfectly straight line as viewed from the earth. But the moon is spherical, and the terminator, being a line upon its surface, must be curved. Thus, the only way it can appear straight is for us to view it exactly edge on. That is, the plane of the terminator must contain the line of sight from the earth to the center of the moon. When that is true, the line from the moon to the earth must be at right angles to the line from the moon to the sun. In Figure 2.6 these right angles are *EMS* and *EM'S*. Now we see that because the sun is not infinitely far away, by assumption, the points *M'*, *E*, and *M* do not lie along a straight line. Hence the moon, moving at a uniform rate, should require a shorter time to go from *M'* to *M* than from *M* to *M'*. We could use the difference between these intervals from third quarter to first quarter moon and from first quarter to third quarter to determine the angle *M'EM*. For example, if the period from *M* to *M'* were, say, twice that from *M'* to *M*, the angle *M'EM* would be a third of a circle, or 120°, and the angle *SEM* would be 60°.

Figure 2.5 Partially eclipsed moon moving out of the earth's shadow. Note the curved shape of the shadow. (*Yerkes Observatory*)

We have no idea how Aristarchus arrived at the figure 87°. Even with our modern equipment of the late 20th century, we could not observe the instants of quarter moon with sufficient accuracy to determine the ES/EM ratio meaningfully, because of the sun's great distance.

However determined, the angle $M'EM$ can be constructed inside a circle representing the moon's orbit; the lines MS and $M'S$, drawn tangent to the circle at M and M', intersect at S, thus determining the position of the sun and hence its distance in terms of the size of the moon's orbit.

To find the relative sizes of the sun and moon, we use the information that the earth's shadow at the moon's distance is twice the size of the moon (the correct ratio is about 8 to 3). Now it is well known that the sun and moon appear to be the same *angular size* in the sky. By angular size we mean the angle subtended by the diameter of an object, that is, the angle of intersection between two lines drawn from a point on the earth (for example, the observer's eye) to opposite ends of a diameter of the object. The sun and the moon each has an angular size of about $1/2°$. If, as Aristarchus had determined, the sun is 19 times as distant as the moon, it must also be 19 times as big to appear the same size. Aristarchus grossly overestimated the angular sizes of the sun and moon to be about 2° each (perhaps the error was intentional to emphasize the geometry rather than reality). With such data we could find the relative sizes of the earth, the moon, and the sun by geometrical construction.

We illustrate the geometrical principles of the construction in Figure 2.7. First, at E, which represents the center of the earth, we draw two lines that intersect at an angle of $1/2°$. During a lunar eclipse the sun and moon are opposite in the sky; thus in direction s the $1/2°$ angle can be considered as representing the angular diameter of the sun, and in direction m the angular diameter of the moon. The sun, S, and moon, M, can now be drawn in, and at arbitrary distances from E as long as the distance ES is 19 times the distance EM. Now at M, the diameter of the earth's shadow, AA', can be constructed at twice the size of the moon. Because the rays of sunlight, in which the earth casts its shadow, travel in straight lines, the lines AB and $A'B'$, drawn tangent to the sun at B and B', must also be tangent to the earth. Thus, finally, the sphere of the earth can be drawn in to proper scale at E. We have now constructed a scale drawing of the earth, moon, and sun. We need only measure with a ruler to obtain their relative sizes.

Perhaps it was his finding that the sun was seven times the earth's diameter that led Aristarchus to the conclusion that the sun, not the earth, was at the center of the universe. At any rate, he is the first person of whom we have knowledge who professed a belief in the heliocentric hypothesis—that the earth goes about the sun. He also postulated that the stars must be extremely distant to account for the fact that their parallaxes could not be observed.

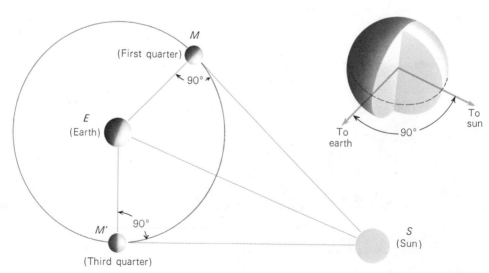

Figure 2.6 Aristarchus' method of determining the relative distances of the sun and moon.

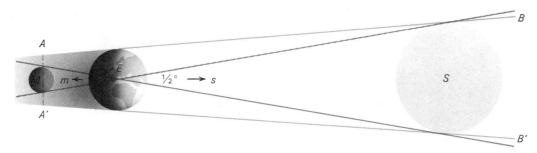

Figure 2.7 The principle by which Aristarchus could determine the relative sizes of the sun, moon, and earth.

(b) Measurement of the Earth by Eratosthenes

Aristarchus had derived the dimensions of the sun and moon, but only in terms of the size of the earth. The latter was not accurately known to him. The first fairly accurate determination of the earth's diameter was made by Eratosthenes (276–195 or 196 B.C.), an astronomer of the Alexandrian school.

To appreciate Eratosthenes' technique for measuring the earth, which is in principle the same as many modern methods, we must understand that the sun is so distant from the earth compared with its size, even by Aristarchus' value, that the sun's rays intercepted by all parts of the earth approach it along sensibly parallel lines. Imagine a light source near the earth, say at position A in Figure 2.8. Its rays strike different parts of the earth along diverging paths. From a light source at B, or at C, still farther away, the angle between rays that strike extreme parts of the earth is smaller. The more distant the source, the smaller the angle between the rays. For a source *infinitely* distant, the rays travel along parallel paths. The sun is not, of course, infinitely far away, but light rays striking the earth from a point on the sun diverge from each other by at most an angle of less than one third of a minute of arc ($^1/_3'$), far too small to be observed with the unaided eye. As a consequence, if people

all over the earth who could see the sun were to point at it, their fingers would all be pointing in the same direction—they would all be parallel to each other. The concept that rays of light from the sun, planets, and stars approach the earth along parallel lines is vital to the art of celestial navigation—the determination of position at sea.

Eratosthenes noticed that at Syene, Egypt, now modern Aswân, on the first day of summer, sunlight struck the bottom of a vertical well at noon, which indicated that Syene was on a direct line from the center of the earth to the sun. At the corresponding time and date in Alexandria, 5000 stadia north of Syene (the *stadium* was a Greek unit of length), he observed that the sun was not directly overhead but slightly south of the zenith, so that its rays made an angle with the vertical equal to $^1/_{50}$ of a circle (about 7°). Yet the sun's rays striking the two cities are parallel to each other. Therefore (see Figure 2.9), Alexandria must be one fiftieth of the earth's circumference north of Syene, and the earth's circumference must be 50 × 5000, or 250,000, stadia. The figure was later revised to 252,000, so that each degree on the earth's surface would have exactly 700 stadia.

It is not possible to evaluate precisely the accuracy of Eratosthenes' solution because there is doubt as to which of the various kinds of Greek stadia he used. If it was the common Olympic stad-

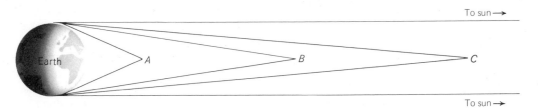

Figure 2.8 The more distant an object, the more nearly parallel are the rays of light coming from it.

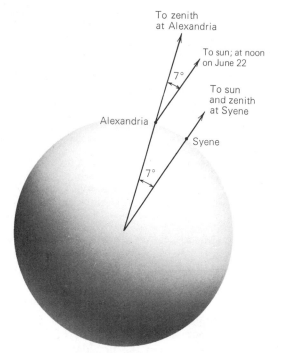

Figure 2.9 Eratosthenes' method of determining the size of the earth.

about 850 entries. He designated for each star its celestial coordinates, that is, quantities analogous to latitude and longitude that specify its position (direction) in the sky. He also divided the stars according to their apparent brightnesses into six categories, or *magnitudes*, and specified the magnitude of each star. In the course of his observations of the stars, and in comparing his data with older observations, he made one of his most remarkable discoveries: the position in the sky of the north celestial pole had altered over the previous century and a half. Hipparchus correctly deduced that the direction of the axis about which the celestial sphere appears to rotate continually changes. The real explanation for the phenomenon is that the direction of the earth's rotational axis changes slowly because of the gravitational influence of the moon and the sun, much as a top's axis describes a conical path as the earth's gravitation tries to tumble the top over. This variation in orientation of the earth's axis, called *precession*, requires about 26,000 years for one cycle (Chapter 6).

ium, his result was about 20 percent too large. According to another interpretation, he used a stadium equal to about $^1/_6$ km, in which case his figure was within 1 percent of the correct value of 40,000 km. The diameter of the earth is found from the circumference, of course, by dividing the latter by π.

(c) Hipparchus

The greatest astronomer of pre-Christian antiquity was Hipparchus, who was born in Nicaea in Bithynia. The dates of his life are not accurately known, but he carried out his work at Rhodes, and possibly also at Alexandria, in the period from 160 to 127 B.C. Many of the phenomena Hipparchus detected are quite subtle, and the measurements he made—all without optical aid—were remarkable for the time.

Hipparchus' Star Catalogue

Hipparchus erected an observatory on the island of Rhodes and built instruments with which he measured as accurately as possible the directions of objects in the sky. He compiled a star catalogue of

Other Measurements

Hipparchus, refining the technique first applied by Aristarchus, also obtained a good estimate of the moon's size and distance. He used the correct value of $^1/_2°$ for the angular diameters of the sun and moon and also the correct value $^8/_3$ for the ratio of the diameter of the earth's shadow to the diameter of the moon. He tried several values for the relative distances of the sun and the moon, including the value found by Aristarchus, but found that the exact distance assumed for the sun, provided it was large, did not have much effect on the figures he derived for the moon. He found the moon's distance to be 59 times the earth's radius; the correct number is 60.

He determined the length of the year to within 6 minutes and even analyzed his possible errors, estimating that he could not be farther off than about 15 minutes. He also carefully observed the motions of the sun, moon, and planets and found a method by which he could predict the position of the sun on any date of the year with an accuracy equal to the best observations and the position of the moon with somewhat less accuracy. His work made possible the reliable prediction of eclipses, and with the information he left, astronomers thereafter could predict a lunar eclipse to within an hour or so.

The Motions of the Sun and the Moon

Hipparchus' study of the motion of the sun deserves special mention. The earth's true orbit around the sun is not a circle but an ellipse; the earth's distance from the sun and its orbital speed both vary slightly. Now we can account for the apparent motion of the sun by imagining it to move around the earth in an elliptical path of exactly the same shape as the earth's orbit. This apparent path of the sun, as we have seen, is the ecliptic. Because we see the sun's apparent orbit edge on (from the inside), the ecliptic is a circle around the sky. Moreover, the sun's eastward rate of motion on the ecliptic varies, exactly as the earth's orbital speed varies. The variation in speed is slight but is observable.

Eudoxus of Cnidas (ca. 408–355 B.C.) had accounted for the sun's motion approximately by representing it with a series of rotating spheres pivoted one on the other. Later the mathematician Apollonius of Perga (latter half of the third century B.C.) suggested that the motions of all the heavenly bodies could be represented equally well by a combination of uniform circular motions. By uniform circular motion is meant a motion at a uniform speed about the circumference of a circle. Because the circle is the simplest geometrical figure, and because uniform motion seemed the most natural kind, Hipparchus, following the suggestion of Apollonius, attempted to find a combination of uniform circular motions that would account for the sun's apparently irregular behavior.

The plan he adopted was to represent the sun's orbit by an *eccentric*, a circle, but with the earth slightly off center (Figure 2.10). The scheme was highly successful because the true orbit of the earth is very close to a circle with the sun just off center. Now, one effect of the sun's variable speed on the ecliptic is to produce an inequality in the lengths of the seasons. Although the inequality had been known before, Hipparchus remeasured the small differences between the seasons' durations and from them deduced that the earth's distance from the center of the sun's orbit must be 1/24 of the sun's distance. He found further that the earth and sun were nearest each other in early December, which was correct at that time. (The date has changed over the thousands of years because of precession and, to a lesser extent, because of a slow motion of the long axis of the earth's elliptical orbit; the closest approach now occurs in early January.)

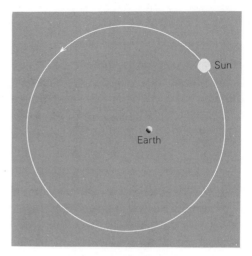

Figure 2.10   The eccentric.

Hipparchus pointed out that he could also have represented the sun's apparent motion by presuming it to move on the circumference of a portable circle called an *epicycle*, whose center, in turn, revolves about the earth in a circle called a *deferent* (Figure 2.11). He considered the eccentric a simpler and thus preferable system.

The moon's motion is more complicated, and Hipparchus was not quite so successful in finding a geometrical scheme to describe it. According to the model he adopted, the moon went in a circle about a point near the earth (an eccentric), but the center of the eccentric also revolved slowly about the earth. Hipparchus measured the nine-year period of this revolution, as well as a 19-year period during

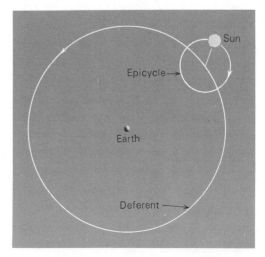

Figure 2.11   The deferent and epicycle.

which the intersections of the moon's orbit with the ecliptic slide completely around the ecliptic, and the 5° inclination between the moon's orbit and the ecliptic. The apparent motions of the planets are even more complicated than that of the moon. Hipparchus thus declined to fit the planets into a cosmological scheme but rather made careful observations of their positions for use by later investigators.

(d) Ptolemy

The last great Greek astronomer of antiquity was Claudius Ptolemy (or Ptolemaeus), who lived around 140 A.D. He compiled a series of 13 volumes on astronomy known as the *Almagest*. All of the *Almagest* does not deal with Ptolemy's own work, for it includes a compilation of the astronomical achievements of the past, principally of Hipparchus. In fact, it is our main source of information about Greek astronomy. The *Almagest* also contains the contributions of Ptolemy himself.

The Distance to the Moon

One of Ptolemy's accomplishments was a new measurement of the distance to the moon. The method he used, the principle of which is illustrated in Figure 2.12, makes use of the moon's parallax, discussed by Hipparchus in connection with solar eclipses. Suppose we could observe the moon directly overhead. We would have to be, then, at position *A* on the earth, on a line between the center of the earth *E*, and the center of the moon *M*. Suppose that at the same time someone else at position *B* were to observe the angle *ZBM* between the moon's direction and the point directly over his head, *Z*. The angle *MBE* would then be determined in the triangle *MBE* (it is 180° minus angle *ZBM*). The distance from *A* to *B* determines the

angle *BEM*. For example, if *A* is one twelfth of the way around the earth from *B*, the angle *BEM* is 30°. The side *BE* is of course the radius of the earth. We therefore know two angles and an included side of the triangle *MBE*. It is now possible to determine, either by trigonometry or geometrical construction, the distance *EM* between the centers of the earth and moon. This is an example of the principle of *surveying*. We shall discuss it further in Chapter 22.

In practice, we do not need another observer at *B*, for the rotation of the earth will carry us over there in a few hours anyway, and we can observe the angle *ZBM* then. We shall have to correct, however, for the motion of the moon in its orbit during the interval between our two observations; the moon's motion being known, the correction is a detail easily accomplished. Using the principle described, Ptolemy determined the moon's distance to be 59 times the radius of the earth or 29½ times the earth's diameter—very nearly the correct value.

Ptolemy's Scheme of Cosmology

Ptolemy's most important original contribution was a geometrical representation of the solar system that predicted the motions of the planets with considerable accuracy. Hipparchus, having determined by observation that earlier theories of the motions of the planets did not fit their actual behavior, and not having enough data on hand to solve the problem himself, instead amassed observational material for posterity to use. Ptolemy supplemented the material with observations of his own and with it produced a cosmological hypothesis that endured until the time of Copernicus.

The complicating factor in the analysis of the planetary motions is that their apparent wanderings in the sky result from the combination of their own

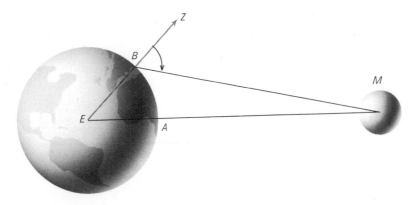

Figure 2.12 Ptolemy's method of finding the distance to the moon.

Armillary Sphere of Antonio Santucci delle Pomerance, made for the Grand Duke Ferdinando I Medici in 1593. *(Istituto e Museo di Storia della Scienza di Firenze)*

Telescopes donated by Galileo to the Grand Duke Ferdinando II and to his brother, the Prince Leopoldo. The longest has a wooden tube covered with paper, a focal length of 1.33 m, and an aperture of 26 mm. *(Istituto e Museo di Storia della Scienza di Firenze)*

Newton's birthplace at Woolsthorpe. The apple tree in the foreground grew from the stump of the one standing in Newton's time. *(Photograph by the author)*

First editions of some books of great historical interest. *(Crawford Library; Courtesy Astronomer Royal for Scotland, Royal Observatory Edinburgh)*

motions and the earth's orbital revolution. Notice, in Figure 2.13, the orbit of the earth and the orbit of a hypothetical planet farther from the sun than the earth. The earth travels around the sun in the same direction as the planet and in nearly the same plane, but has a higher orbital speed. Consequently, it periodically overtakes the planet, like a faster race car on the inside track. The apparent directions of the planet, seen from the earth, are shown at successive intervals of time along lines $AA'A''$, $BB'B''$, and so on. In the right side of the figure we see the resultant apparent path of the planet among the stars. From positions B to D, as the earth passes the planet, it appears to drift backward, to the *west* in the sky, even though it is actually moving to the *east*. Similarly, a slowly moving car appears to drift backward with respect to the distant scenery when we pass it in a faster-moving car. As the earth rounds its orbit toward position E, the planet again takes up its usual eastward motion in the sky. The temporary westward motion of a planet as the earth swings between it and the sun is called *retrograde* motion. (During and after its retrograde motion, the planet's apparent path in the sky does not trace exactly over itself because of the slight inclinations between the orbits of the earth and other planets. Thus, the retrograde path

is shown as an open loop in Figure 2.13.) Obviously, we need a different explanation for retrograde motion on the hypothesis that the planet is revolving about the earth.

Ptolemy solved the problem by having a planet P (Figure 2.14) revolve in an epicyclic orbit about C. The center of the epicycle C in turn revolved in the deferent about the earth. When the planet is at position x, it is moving in its epicyclic orbit in the same direction as the point C moves about the earth, and the planet appears to be moving eastward. When the planet is at y, however, its epicyclic motion is in the opposite direction to the motion of C. By choosing the right combination of speeds and distances, Ptolemy succeeded in having the planet moving westward at the right speed at y and for the correct interval of time. However, because the planets, as does the earth, travel about the sun in elliptical orbits, their actual behavior cannot be represented accurately by so simple a scheme of uniform circular motions. Consequently, Ptolemy made the deferent an eccentric, centered not on the earth, but slightly away from the earth at A. Furthermore, he had the center of the epicycle, C, move at a uniform angular rate, not around A, or E, but at point B, called the *equant*, on the opposite side of A from the earth.

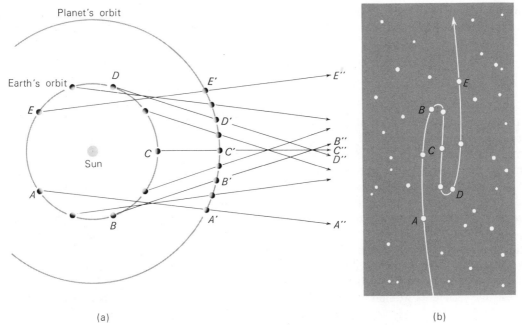

(a)   (b)

Figure 2.13 Retrograde motion of a superior planet. (a) Actual positions of the planet and earth. (b) The apparent path of the planet as seen from the earth, against the background of stars.

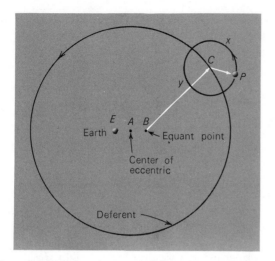

Figure 2.14 Ptolemy's system of deferent, epicycle, eccentric, and equant.

It is a tribute to the genius of Ptolemy as a mathematician that he was able to conceive such a complex system to account successfully for the observations. His hypothesis, with some modifications, was accepted as absolute authority throughout the Middle Ages, until it finally gave way to the heliocentric theory in the 17th century. In the *Almagest*, however, Ptolemy made no claim that his cosmological model described reality. He intended his scheme rather as a mathematical representation to predict the positions of the planets at any time. Modern astronomers do the same thing with algebraic formulas. Our modern mathematical methods were not available to Ptolemy; he had to use geometry.

2.4 ASTROLOGY

Modern research has shown that all matter in the universe is composed of atoms—and the same kinds of atoms. Thus our Viking space probes of Mars and our telescopic spectra of the light from the most remote quasars indicate that, whatever we do not yet understand about Mars and the quasars, at least they are made of the same stuff that makes up our own bodies.

Still, we cannot fault the ancients for assuming that the luminous orbs in the sky, the stars and planets, are made of "heavenly" substances and not of the "earthly" elements we find at home. In fact,

the realization that celestial worlds are actually worlds and not ethereal substance is relatively recent in the history of science. Small wonder, then, that the ancients regarded the planets (including the sun and moon), which alone moved about among the stars on the celestial sphere, as having special significance. Thus the planets came to be associated with the gods of ancient mythologies; in some cases, they were themselves thought of as gods. Even in the comparatively sophisticated Greece of antiquity, the planets had the names of gods and were credited with having the same powers and influences as the gods whose names they bore. From such ideas grew the religion of astrology.

Astrology began, we think, in the valley of the Euphrates and Tigris Rivers a millennium or so before Christ. The Mesopotamians and the Babylonians, believing that the planets and their motions influenced the fortunes of kings and nations, practiced what we call *mundane* astrology. When the Babylonian culture was absorbed by the Greeks, their astrology gradually influenced the entire western world and eventually spread to the Orient as well. By the third or second century B.C. the Greeks democratized astrology by developing the tradition that the planets influenced the life of every individual. In particular, they believed that the configuration of the planets at the moment of a person's birth affected his personality and fortune. This form of astrology, known as *natal* astrology, reached its acme with Ptolemy in the second century A.D. Ptolemy, as famous for his astrology as for his astronomy, compiled the *Tetrabiblos*, a treatise on astrology that remains the "bible" of the subject even today.

(a) The Horoscope

The key to natal astrology is the *horoscope*, a chart that shows the positions of the planets in the sky at the moment of an individual's birth. The charting of a horoscope, as of any map, requires the use of coordinates. The celestial coordinates used by astrology, in antiquity as well as today, are analogous to, and share a common origin with, those used by astronomers.

First, the planets (including the sun and the moon—classed as planets by the ancients) are located in the sky with respect to the fixed stars on the celestial sphere by specifying their positions in the zodiac—the belt centered on the ecliptic that contains the planets. For the purposes of astrology, the zodiac is divided into twelve sectors called *signs*, each 30° long. Second, the constantly turning celestial sphere, with its stars and the planets, must have its orientation specified with respect to the earth at the time and place of the subject's birth. For this purpose the sky is divided into twelve regions, called *houses*, that are fixed with respect to the horizon. Each day the turning sky carries the

planets and signs through all of the houses. An example of a horoscope (my own) is shown in Figure 2.15.

(b) Interpretation of the Horoscope

There are more or less standardized rules for the interpretation of the horoscope, many or most of which (at least in Western schools of astrology) are derived from the *Tetrabiblos* of Ptolemy. Each sign, each house, and each planet, the latter supposedly

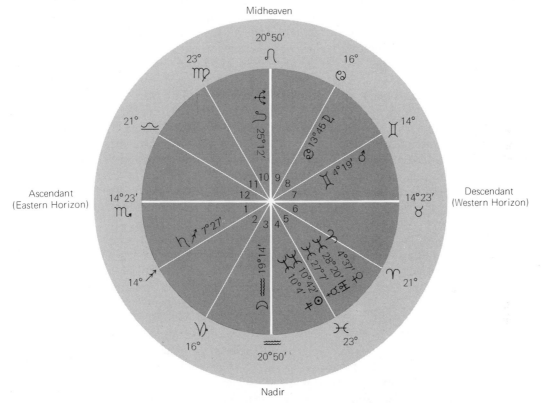

Figure 2.15 Natal horoscope of the author, who was born in Los Angeles, California, on March 1, 1927, at 10:50 PM, PST. The 12 pie-shaped sectors represent the 12 houses, and the outer circular zone represents the zodiac. The definition of houses used in preparing this horoscope is that of Placedus, in which, as the rotating celestial sphere carries the planets around the sky, each place in the zodiac spends equal time in each of the six houses above the horizon (*diurnal houses*) and also in the six houses below the horizon (*nocturnal houses*); however, the time required for an object to pass through a diurnal house is not the same as that required for it to pass through a nocturnal house except for objects on the celestial equator. The boundaries between the houses (*cusps*) intersect the ecliptic in the zodiacal signs indicated by their symbols in the outer circular zone. The number beside each sign symbol is the angular distance of the cusp from the beginning of that sign. The position of each planet is shown in the house it occupied at the instant of the author's birth. Beside the symbol for the planet is the symbol of the zodiacal sign it was also in at that time, and the angular distance of the planet from the beginning of that sign. The places where the horizon intersects the zodiac are shown, and also the highest point of the ecliptic in the sky (*midheaven*) and its lowest point below the horizon, or *nadir* (this astrological definition of the nadir is different from the astronomical one, in which the nadir is directly opposite the zenith).

acting as a center of force, is associated with particular matters.

The interpretation of a horoscope is a very complicated business, and whereas the rules may be standardized, how each rule is to be weighed and applied is a matter of judgment—and "art." It also means that it is very difficult to tie astrology down to specific predictions.

The interpretation of an individual's horoscope, charted for the time and place of his birth is *natal* astrology; his characteristics and fortunes, presumably, depend on his natal horoscope. Another branch of the subject is *horary* astrology, which purports to answer direct specific questions by casting a horoscope for the time and place at which the question was first posed. Horary astrology might be used, for example, to find whether the coming Monday would be a good time for a particular business deal.

Because of precession, the signs of the zodiac are slowly slipping westward with respect to the constellations. The traditional astrology, nevertheless, is based on the moving *signs*, not the constellations; it is called *tropical astrology*. It may have a logical basis, because the seasons themselves depend on the sun's position with respect to the equinoxes and solstices on the ecliptic. On the other hand, there is a school of astrology that is based on the positions of the planets in the *constellations* rather than in the signs; it is called *sidereal astrology*.

(c) Value of Astrology

Today, with our knowledge of the nature of the planets as physical bodies, composed as they are of rocks and fluids, it is hard to imagine that the directions of these planets in the sky at the moment of one's birth could have anything to do with his personality or future. The gravitational influence of the moon and sun on tides is unquestionable, but tides produced on a person by a book in his hand are millions of times as strong as those produced by all the planets combined. The sun's light and heat are obviously of great importance to us, but even minute variations in the sun's irradiation are millions of times as great as the combined light of the planets. Jupiter (and to a lesser extent, the other planets) has a strong magnetic field and emits radio waves, but their detection requires magnetometers carried on space probes and large radio telescopes.

The feeble radio signals from a small 1000-watt transmitter 100 miles away reach us with a strength millions of times as great as the radio waves from Jupiter, and can be picked up by a pocket transistor radio. Even the magnet in the loudspeaker of that radio produces around the listeners a magnetic field enormously stronger than does Jupiter. Moreover, the distances of the planets from the earth vary greatly, and any gravitational and radiation effects would vary as the inverse square of their distances—factors ignored by astrology.

Astrology would have to argue that there are unknown forces exerted by the planets that depend on their configurations with respect to each other and with respect to arbitrary coordinate systems invented by man—forces for which there is not a whit of solid evidence. Are astronauts on the moon similarly affected by the same kind of force exerted by the earth? Or is the earth, alone, subject to these unknown laws of nature?

In the most orthodox astrology, one's entire life (and death) is predetermined by his natal horoscope. If a man dies in an auto accident at the age of 63 because someone else ran a stoplight, are we supposed to assume that all of the complicated chain of events that led to the circumstances of his being in that accident were blueprinted by the planets at the instant of his birth, but that all would have been different if he had been born two hours later? Most of us would find this assumption so incredible that we would need the most overwhelming evidence of its validity before taking it seriously. In the tens of centuries of astrology, no such evidence has been presented.

One could argue, on the other hand, that astrology only works statistically; that other influences—heredity and environment, for example—are important too, and that astrological influences are only important as tendencies, everything else being equal. In that case the reality of astrological effects could only be tested statistically. From time to time astrologers have presented statistical "proofs" of astrology, but not one survives objective scientific scrutiny. A recent and exhaustive study[1] of the astrological literature by an Australian

---

[1]*Recent Advances in Natal Astrology*, by Geoffrey Dean *et al.*, published under the aegis of the Astrological Association, 1977. Distributed in North America by Para Research, Inc., Whistlestop Mall, Rockport, MA 01966.

astrologer failed to turn up a single piece of evidence that he regarded as verification of traditional astrology.

In retrospect, we can understand the belief in astrology on the part of ancient peoples who thought the heavenly bodies to be made of celestial material different from the elements that compose the earth, and to be placed in the sky by their gods for the benefit of mankind. In the light of modern knowledge, the astrological claims seem so farfetched as to be ludicrous. Because we would not expect the supposed influences, even in a statistical sense, we would want solid evidence and demonstrable predictions. Physical scientists and others who have investigated the subject with the hope of finding some grain of validity in it have found negative results. Virtually all scientists reject astrology as an unfounded superstition. Yet it continues to appeal to the popular fancy. The hope of predicting the future by magical or mystical means, and perhaps of transferring one's responsibilities and the

blame for one's failures and misfortunes to an omnipotent power, continues to be a strong attraction. Moreover, it may simply be "fun" to speculate about the unknown and unprovable no matter how little basis there may be for it. Many astrologers today acknowledge that astrology cannot be proven by statistics or by experiment, but assert that it must be "known" or "realized" as knowledge or truth. In this context, it is outside the realm of science, and no rational argument based on the rules of science is relevant. To many astrology is still a religion, and hence is outside the scope of our consideration here.

One fact remains: The practice of astrology in ancient times required the knowledge of the motions of the planets in order to construct horoscopes for past or future events. The quest to find a mechanism for charting the planets, joined with a natural curiosity about nature, stimulated centuries of observations and calculations, leading—as we shall see—to our modern technology.

EXERCISES

1. Where on earth are all stars above the horizon at one time or another?

2. Where on earth is only half the sky ever above the horizon?

3. Look up the names of the days of the week in French, Italian, and Spanish, and compare them with the names of the planets.

4. Why was Philolaus' hypothesis not scientific?

5. Show by a diagram how a solar eclipse *can* occur at new moon but does not usually occur.

6. About what time of day or night does the moon rise when it is full? When it is new?

7. Why can an eclipse of the moon never occur on the day following a solar eclipse?

8. As seen by a terrestrial observer, which (if any) of the following can never appear in the opposite direction in the sky from the sun? in the same direction? at an angle of 90° from the sun? (a) Mars; (b) a star; (c) the sun; (d) Earth; (e) Jupiter; (f) the moon; (g) Venus; (h) Mercury.

9. Suppose, in applying a method based on Aristarchus' procedure, we found that the interval from third quarter to first quarter moon was one week, and that the interval from first quarter to third quarter moon was three weeks. Then what distance would he have derived for the sun (in terms of the moon's distance from the earth)?

10. Suppose the sun were twice as distant as the moon. What fraction of the month would the moon spend between third quarter and first quarter?

11. The earth's diameter is about three and two-thirds times the diameter of the moon. What is the angular diameter of the earth as seen by an observer on the moon?

12. Suppose Eratosthenes had found that at Alexandria at noon on the first day of summer the line to the sun makes an angle of 30° with the vertical. What then would he have found for the earth's circumference?

13. Suppose Eratosthenes's results for the earth's circumference were quite accurate. If the diameter of

the earth is 12,740 km, evaluate the length of his stadium in kilometers.

14. Why would Eratosthenes's method not have worked if the earth were flat, like a pancake?

15. You are on a strange planet. You note that the stars do not rise or set, but that they circle around parallel to the horizon. Then you travel over the surface of the planet in one direction for 10,000 km, and at that new place you find that the stars rise straight up from the horizon in the east and set straight down in the west. What is the circumference of the planet?
Answer: 40,000 km

16. Is retrograde motion observed for an inferior planet? Explain.

17. One aspect of the planets that is regarded favorably by astrology is the *trine*—when two planets are 120° away from each other in the zodiac. Give two examples of trine aspects that can never occur. (There is a hint in Appendix 9.)

18. Many people try to use pseudostatistical arguments to justify their beliefs in a pseudoscience. Try the experiment of flipping a coin ten times and then recording the number of heads that turn up. Do this experiment 100 or more times (several people can flip coins at the same time and then pool results, thereby saving labor). Prepare a table showing how many times no head was obtained (ten tails in a row), how many times one head was obtained, how many times two heads, and so on. Make a graph showing the same data. What was the most frequent number of heads? What fraction of the time were less than three or more than seven heads obtained? If an event occurring only one percent of the time is enough to arouse your suspicions, how many heads would you have to obtain in a single experiment to question the honesty of the coin?

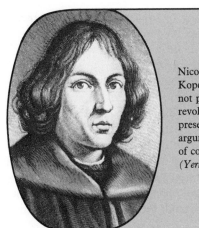

Nicolaus Copernicus (Mikolaj Kopernik) (1473–1543) did not prove that the earth revolves about the sun, but he presented compelling arguments that turned the tide of cosmological thought. *(Yerkes Observatory)*

THE HELIOCENTRIC HYPOTHESIS: THE COPERNICAN COSMOLOGICAL PRINCIPLE

In the 13 centuries following Ptolemy the most significant astronomical investigations were made by the Hindus and Arabs. The Hindus invented our system of numbers with place counting by tens. The Arabs brought the Hindu system of numbers to Europe and developed trigonometry. They also had access to some of the records of the Greek astronomers. Their greatest contribution was to provide continuity between ancient astronomy and the development of modern astronomy in the Renaissance.

Astronomy made no major advances in medieval Europe, where the prevailing philosophy was acceptance of the dogma of authority. Medieval cosmology combined the crystalline spheres of Pythagoras (as perpetuated by Aristotle) with the epicycles of Ptolemy. Astrology was widely practiced, however, and an interest in the motions of the planets was thus kept alive. Then came the Renaissance; in science the rebirth was clearly embodied in Nicholas Copernicus.

3.1 COPERNICUS

Nicholas Copernicus (in Polish, Mikolaj Kopernik, 1473–1543) was born in Torun on the Vistula in Poland. His training was in law and medicine, but Copernicus' main interest was astronomy and mathematics. By the time he had reached middle age, he was well known as an authority on astronomy.

Copernicus' great contribution to science was a critical reappraisal of the existing theories of cosmology and the development of a new model of the solar system. His unorthodox idea that the sun, not the earth, is the center of the solar system had become known by 1515, chiefly through an early manuscript circulated by him and his friends.

His ideas were set forth in detail in his *De Revolutionibus*, published in the year of his death (Figure 3.1). Supervision over the publication of the book fell into the hands of a Lutheran preacher Andrew Osiander, who was probably responsible for the augmented title of the work—*De Revolutionibus Orbium Celestium* (On the Revolutions of the Celestial Spheres). Osiander wrote a preface, which he neglected to sign, expressing the view that science presented only an abstract mathematical hypothesis and implying that the theory set forth in the book was only a convenient calculating scheme. The preface was almost certainly in contradiction to Copernicus' own feelings.

In *De Revolutionibus*, Copernicus sets forth certain postulates from which he derives his system of planetary motions. His postulates include the assumptions that the universe is spherical and that the motions of the heavenly bodies must be made up of combinations of uniform circular motions; thus Copernicus was not free of all traditional prejudices. Yet, he evidently found something orderly and pleasing in the heliocentric system, and his defense of it was elegant and persuasive. His ideas, although not widely accepted until more than a century after his death, never disappeared and were ultimately of immense influence.

(a) Planetary Motions According to Copernicus

A person moving uniformly is not necessarily aware of his motion. We have all experienced the phenomenon of seeing an adjacent train, car, or ship appear to change position, only to discover that it is we who are moving (or vice versa). Copernicus argued that the apparent annual motion of the sun about the earth could be equally well represented by a motion of the earth about the sun, and that the rotation of the celestial sphere could be accounted for by assuming that the earth rotates about a fixed axis while the celestial sphere is stationary. To the objection that if the earth rotated about an axis it would fly into pieces, Copernicus answered that if such motion would tear the earth apart, the even faster motion (because of its greater size) of the celestial sphere required by the alternative hypothesis would be even more devastating to it.

The important point that Copernicus made in *De Revolutionibus* is that the earth is but one of six (then known) planets that revolve about the sun. Given this, he was able to work out the correct general picture of the solar system. He placed the planets, starting nearest the sun, in the order Mercury, Venus, Earth, Mars, Jupiter, and Saturn. Further, he deduced that the nearer a planet is to the sun, the greater is its orbital speed. Thus the retrograde motions of the planets (Section 2.3d) were easily understood without the necessity for epicycles. Also, Copernicus worked out the correct approximate scale of the solar system. To understand how, it will be helpful to define a few terms that describe the positions of planets in their orbits. These are illustrated in Figure 3.2.

A *superior planet* is any planet whose orbit is larger than that of the earth, that is, a planet that is farther from the sun than the earth is (Mars, Jupiter, and Saturn). An *inferior planet* is a planet closer to the sun than the earth is (Venus and Mercury).

Every now and then, the earth passes between a superior planet and the sun. Then that planet appears in exactly the opposite direction in the sky from the sun—or at least as nearly opposite as is allowed by the slight differences of inclination among the planes of the orbits of the planets. At such time, the planet rises at sunset, is above the horizon all night long, and sets at sunrise. We look one way to see the sun, and in the opposite direction to see the planet. The planet is then said to be in *opposition*.

On other occasions, a superior planet is on the other side of the sun from the earth. It is then in the same direction from the earth as the sun is, and of course is not visible. At such time, the planet is said to be in *conjunction*.

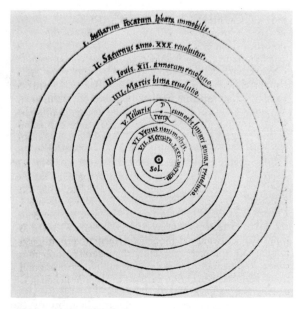

Figure 3.1 Plan of the solar system in the first edition of Copernicus' *De Revolutionibus*. (*Crawford Collection, Royal Observatory Edinburgh*)

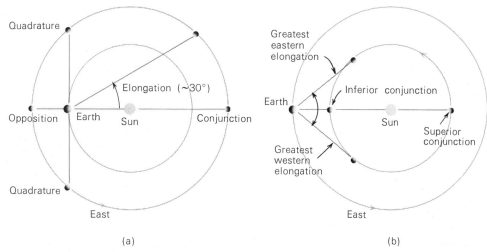

Figure 3.2 (a) Configurations of a superior planet; (b) configurations of an inferior planet.

In between these extremes (but not halfway between), a superior planet may appear 90° away from the sun in the sky, so that a line from the earth to the sun makes a right angle with the line from the earth to the planet. Then the planet is said to be at *quadrature*. At quadrature, a planet rises or sets at either noon or midnight.

The angle formed at the earth between the earth-planet direction and the earth-sun direction is called the planet's *elongation*. In other words, the elongation of a planet is its angular distance from the sun as seen from the earth. At conjunction, a planet has an elongation of 0°, at opposition 180°, and at quadrature 90°.

An inferior planet can never be at opposition, for its orbit lies entirely within that of the earth. The greatest angular distance from the sun, on either the east or west side, that the inferior planet can attain is called its *greatest eastern elongation* or *greatest western elongation*.

When an inferior planet passes between the earth and sun, it is in the same direction from earth as the sun and is said to be in *inferior conjunction*. When it passes on the far side of the sun from the earth, and is again in the same direction as the sun, it is said to be at *superior conjunction*.

Sidereal and Synodic Periods of a Planet

Copernicus recognized the distinction between the *sidereal period* of a planet—that is, its actual period of revolution about the sun with respect to the fixed stars—and its *synodic period*, its apparent period of revolution about the sky with respect to the sun. The synodic period is also the time required for it to return to the same configuration, such as the time from opposition to opposition or from conjunction to conjunction.

Consider two planets, A and B, A moving faster in a smaller orbit (Figure 3.3). At position (1), planet A passes between B and the sun S. Planet B is at opposition as seen from A, and A is in inferior conjunction as seen from B. When A has made one revolution about the sun and has returned to position (1), B has, in the meantime, moved on to position (2). In fact, A does not catch up with B until both planets reach position (3). Now planet A has gained one full lap on B. Planet A has revolved in its orbit through 360° *plus* the angle that B has described in traveling from position (1) to position (3) in its orbit. The time required for the faster-moving planet to gain a lap on the slower-moving one is the synodic period of one with respect to the other. If B is the earth and A an inferior planet, the synodic period of A is the time required for the inferior planet to gain a lap on the earth; if A is the earth and B a superior planet, the synodic period of B is the time for the earth to gain a lap on the superior planet.

What is observed directly from the earth is the synodic, not the sidereal period of a planet. By reasoning along the lines outlined in the last paragraph, however, we can deduce the sidereal periods of the planets from their synodic periods. Let a planet's sidereal period be P years and its synodic period S years. In S years, the earth, completing one revolution per year, must make S trips around

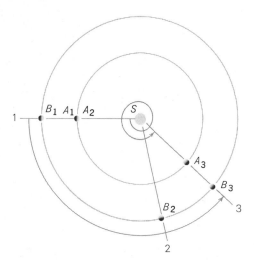

Figure 3.3 Relation between the sidereal and synodic periods of a planet.

the sun. (The quantity S, of course, can be less than 1, in which case the earth would complete less than one circuit.) The other planet, completing one revolution in P years, would make, in S years, S/P trips around the sun. Consider first an inferior planet. It has made one more trip around the sun during its synodic period than has the earth, so $S + 1 = S/P$, which, by rearrangement of terms, can be written

$$\frac{1}{P} = 1 + \frac{1}{S} \qquad \text{for an } inferior \ planet.$$

For a superior planet, it is the earth that gains the extra lap, and $S = S/P + 1$, which can be written

$$\frac{1}{P} = 1 - \frac{1}{S} \qquad \text{for a } superior \ planet.$$

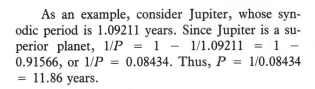

As an example, consider Jupiter, whose synodic period is 1.09211 years. Since Jupiter is a superior planet, $1/P = 1 - 1/1.09211 = 1 - 0.91566$, or $1/P = 0.08434$. Thus, $P = 1/0.08434 = 11.86$ years.

Relative Distances of the Planets

Copernicus was able to find the planets' distances from the sun relative to the earth's. For the sake of illustration, let us assume that the orbits of the planets are precisely circular, even though that assumption is an oversimplification. The problem is particularly simple for the inferior planets. When an inferior planet is at greatest elongation (Figure 3.4), the line of sight from the earth to the planet, EP, must be tangent to the orbit of the planet, and hence perpendicular to the line from the planet to the sun, PS. We have, therefore, a right triangle, EPS. The angle PES is observed (it is the greatest elongation), and the side ES is the earth's distance from the sun. The planet's distance from the sun can then be found, in terms of the earth's distance, by geometrical construction or by trigonometric calculation.

As a simple illustration of the procedure by which the distance of a superior planet can be found, suppose (Figure 3.5) the planet P is at opposition. We can now time the interval until the planet is next at quadrature; the planet is then at P' and the earth at E'. With a knowledge of the sidereal periods of the planet and the earth, we can calculate the fractions of their respective orbits that have been traversed by the two bodies. Thus the angles PSP' and ESE' can be determined, and subtraction gives the angle $P'SE'$ in the right triangle $P'SE'$. The side SE' is the earth's distance

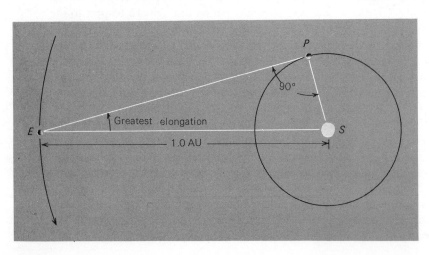

Figure 3.4 Determination of the distance of an inferior planet from the sun, relative to the earth's distance.

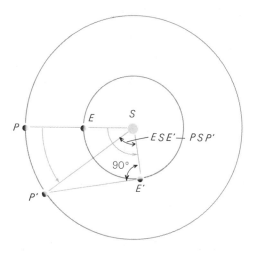

Figure 3.5 Determination of the distance of a superior planet from the sun, relative to the earth's distance.

from the sun, so enough data are available to solve the triangle and find the planet's distance from the sun, $P'S$ (again in terms of the earth's distance), by construction or calculation.

The values obtained by Copernicus for the distances of the various planets from the sun, in units of earth's distance, are summarized in Table 3.1. Also given are the values determined by modern measurement.

Table 3.1 **Distances of Planets from the Sun**

PLANET	COPERNICUS	MODERN
Mercury	0.38	0.387
Venus	0.72	0.723
Earth	1.00	1.00
Mars	1.52	1.52
Jupiter	5.22	5.20
Saturn	9.18	9.54

So far, we have discussed the Copernican theory as though Copernicus regarded the planets as having circular orbits centered on the sun. However, we recall that centuries earlier Ptolemy had introduced epicycles, eccentrics, and equants to account for those minor irregularities that arise because of deviations from uniform circular motion (actually because the true orbits of planets are ellipses). Copernicus rejected the equants of Ptolemy as unworthy of the perfection of heavenly bodies and instead introduced a system of eccentrics and small epicycles to take care of the irregularities.

Harvard astronomer Owen Gingerich, who has a special interest in the history of astronomy, has searched out in various libraries, public and private, more than 500 copies of early editions of *De*

Revolutionibus to inspect the handwritten marginal annotations of the original owners of the books. Many of these readers turned out to be famous astronomers themselves. Gingerich has noted that the early part of *De Revolutionibus*, in which Copernicus sets forth his general plan of the heliocentric hypothesis, usually have rather few annotations; evidently this part, with its radical new cosmology, was not very attractive to astronomers of the sixteenth century. Well read and marked-up portions of the books were the later ones, dealing with the rather dull details by which Copernicus was able to account for the motions of the planets without equants. It is an interesting comment on the insights of scholars of the generation following Copernicus. But of course not *all* astronomers took Copernicus' cosmological ideas lightly.

Philosophically, the main point of Copernicus' idea is that the earth is not something special, but merely one of the several planets in revolution about the sun. The idea that we are at a *typical*, rather than a *special* place in the universe, is sometimes referred to as the *Copernican cosmological principle*. It will be brought up again and again in our investigation of the universe as a whole.

Contrary to popular belief, Copernicus did not *prove* that the earth revolves about the sun. In fact, with some adjustments the old Ptolemaic system could have accounted as well for the motions of the planets in the sky. But the Ptolemaic cosmology was clumsy and lacked the beauty and coherence of its successor. Copernicus made the earth an astronomical body, which brought a kind of unity to the universe. It was, to borrow from Neil Armstrong, a "giant step for mankind." It is fitting that the quinquecentennial celebration of the birth of Copernicus was honored by scientists throughout the world in 1973, for the Copernican revolution marked the origin of modern science and of our contemporary understanding of the universe.

3.2 TYCHO BRAHE

Three years after the publication of *De Revolutionibus*, Tycho Brahe (1546–1601) was born of a family of Danish nobility. Tycho (as he is generally known) developed an early interest in astronomy and as a young man made significant astronomical observations.

In 1572 he observed a nova or "new star" (now believed to be a supernova—see Chapter 34) that rivaled the planet Venus in brilliance. Tycho observed the star for 16 months until it disappeared from naked-eye visibility. Now, we have seen (Section 2.3c) that the moon exhibits a diurnal parallax, or apparent displacement in direction, because of the rotation of the earth, which constantly shifts our position of observation. The effect is the same whether we regard it as being caused by the earth's rotation or a rotation of the celestial sphere carrying the moon about us. Tycho, despite the most careful observations, was unable to detect any parallax of his nova and accordingly concluded that it must be more distant than the moon. This conclusion was of the utmost importance, for it showed that changes can occur in the celestial sphere, generally regarded as perfect and unchanging, apart from the regular motions of the planets.

The reputation of the young Tycho Brahe as an astronomer gained him the patronage of Frederick II, and in 1576 Tycho was able to establish a fine astronomical observatory on the Danish island of Hveen. The chief building of the observatory was named *Uraniborg*. The facilities at Hveen included a library, laboratory, living quarters, workshops, a printing press, and even a jail. There, for 20 years, Tycho and his assistants carried out the most complete and accurate astronomical observations yet made.

Unfortunately, Tycho was both arrogant and extravagant, and after Frederick II died, the new king, Christian IV, lost patience with the astronomer and eventually discontinued his support. Thus, in 1597 Tycho was forced to leave Denmark. He took up residence near Prague, taking with him

Figure 3.6 Tycho Brahe's observatory, Uraniborg. *(Yerkes Observatory)*

some of his instruments and most of his records. There, as court astronomer for Emperor Rudolph II of Bohemia, Tycho Brahe spent the remaining years of his life analyzing the data accumulated over 20 years of observation. In 1600, the year before his death, he secured the assistance of a most able young mathematician, Johannes Kepler, who, like Tycho, was in exile from his native land.

(a) Tycho's Observations

Tycho, like others of his time and before him, believed that comets were luminous vapors in the earth's atmosphere. In 1577, however, a bright comet appeared for which he could observe no parallax. Tycho concluded that the comet was at least three times as distant as the moon and guessed that it probably revolved around the sun, in contradiction to earlier beliefs. Other comets were observed by him or his students in 1580, 1582, 1585, 1590, 1593, and 1596.

Tycho is most famous for his very accurate observations of the positions of the stars and planets. With instruments of his own design, he was able to make observations accurate to the limit of vision with the naked eye. The positions of the nine fundamental starts in his excellent star catalogue were accurate in most cases to within 1'. Only in one case was he off by as much as 2', and this was a star whose position was distorted by atmospheric refraction (see Chapter 10).

Tycho's observations included a continuous record of the positions of the sun, moon, and planets. His daily observations of the sun, extending over years and comprising thousands of individual sightings, led to solar tables that were good to within 1'. He reevaluated nearly every astronomical constant and determined the length of the year to within one second. His extensive and precise observations of planetary positions enabled him to note that there were variations in the positions of the planets from those given in published tables, and he even noted regularities in the variations.

(b) Tycho's Cosmology

Tycho rejected the Copernican heliocentric hypothesis on what seemed at the time to be very sound grounds. First, he found it difficult to reconcile a moving earth with certain Biblical statements, nor could he even imagine an object as heavy and "sluggish" as the earth to be in motion. The fact that he could not detect a parallax for even a single star, moreover, meant that the stars would have to be enormously distant if the earth revolved around the sun. The great void that would be required between the orbit of Saturn and the stars would alone have been enough to make him doubt the motion of the earth; even more convincing to Tycho was the fact that he believed that he could measure the angular sizes of stars. The brightest of them he thought to be 2' across. Now, the farther away an object is, the larger must be its true size in order that it have a given angular diameter. Tycho could not detect as much as 1' of parallax for any star, so it followed that the stars were so distant that, to have angular diameters of 2', their actual sizes would have to be twice the size of the entire orbit of the earth. If they were still farther away, their diameters would have to be proportionally greater. (Later telescopic observations showed that the stars, unlike the planets, appear as luminous points; their disklike appearance to the naked eye is illusory.)

Tycho did, however, suggest an original system of cosmology, although it was not worked out in full detail. He envisioned the earth in the center, with the sun revolving about the earth each year, and with the other planets revolving about the sun in the order Mercury, Venus, Mars, Jupiter, and Saturn (Figure 3.7).

3.3 KEPLER

Johannes Kepler (1571–1630) was born in Weil-der-Stadt, Württemberg (southwestern Germany). He attended college at Tübingen and studied for a theological career. There he learned the principles of the Copernican system. He became an early convert to the heliocentric hypothesis and defended it in arguments with his fellow scholars.

In 1594, because of his facility as a mathematician, he was offered a position teaching mathematics and astronomy at the high school at Graz. As part of his duties at Graz, he prepared almanacs that gave astronomical and astrological data. Eventually, however, the power of the Catholic church in Graz grew to the point where Kepler, a Protestant, was forced to quit his post. Accordingly, he went to Prague to serve as an assistant to Tycho Brahe.

Tycho set Kepler to work trying to find a satisfactory theory of planetary motion—one that was compatible with the long series of observations

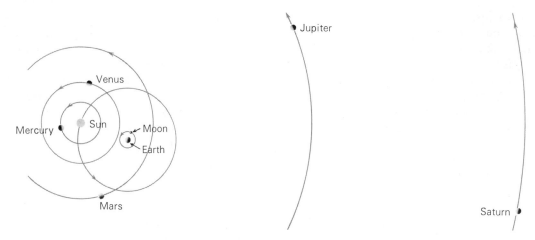

Figure 3.7 Tycho's model for the solar system.

made at Hveen. Brahe, however, was reluctant to supply Kepler with enough data to enable him to make substantial progress; perhaps Brahe was afraid of being "scooped" by the young mathematician. After Tycho's death, though, Kepler succeeded him as mathematician to the emperor Rudolph and obtained possession of the majority of Tycho's records. Their study occupied most of Kepler's time for more than 20 years.

(a) The Investigation of Mars

Kepler's most detailed study was of Mars, for which the observational data were the most extensive. He published the first results of his work in 1609 in *The New Astronomy*, or *Commentaries on the Motions of Mars*. He had spent several years trying to fit various combinations of circular motion, including eccentrics and equants, to the observed motion of Mars, but without success. At one point he found a hypothesis that agreed with observations to within 8′ (about one quarter the diameter of the full moon), but he believed that Tycho's observations could not have been in error by even this small amount, and so, with characteristic integrity, he discarded the hypothesis. Finally, Kepler tried to represent the orbit of Mars with an oval, and soon discovered that the orbit could be fitted very well by a curve known as an *ellipse*.

Properties of the Ellipse

Next to the circle, the ellipse is the simplest kind of closed curve. It belongs to a family of curves known as *conic sections* (Figure 3.8). A conic section is simply the curve of intersection between a hollow cone (whose base is presumed to extend downward indefinitely) and a plane that cuts through it. If the plane is perpendicular to the axis of the cone (or parallel to its base), the intersection is a circle. If the plane is inclined at an arbitrary angle, but still cuts completely through the surface of the cone, the resulting curve is an ellipse. If the plane is parallel to a line in the surface of the cone, it never quite cuts all the way through the cone, and the curve of intersection is open at one end. Such a curve is called a *parabola*. If the plane is inclined at an even smaller angle to the axis of the cone, an open curve results that is called a *hyperbola*. The ellipse, then, ranges from a circle at one extreme to a parabola at the other. The parabola separates the family of ellipses from the family of hyperbolas.

An interesting and important property of an ellipse is that from *any point* on the curve the sum of the distances to two points inside the ellipse, called the *foci* of the ellipse, is the same. This property suggests a simple way to draw an ellipse. The ends of a length of string are tied to two tacks pushed through a sheet of paper into a drawing board, so that the string is slack. If a pencil is then pushed against the string, so that the string is held taut, and then slid against the string around the tacks (Figure 3.9), the curve that results is an ellipse; at any point where the pencil may be, the sum of the distances from the pencil to the two tacks is a constant length—the length of the string. The tacks, of course, are at the two foci of the ellipse.

The maximum diameter of the ellipse is called its *major axis*. Half the distance, that is, the distance from the center of the ellipse to one end, is the *semimajor axis*. The *size* of an ellipse depends on the length of the major axis. The *shape* of an

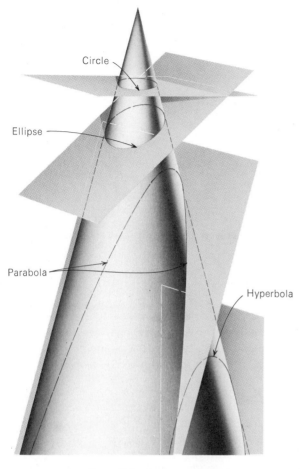

Figure 3.8 Conic sections.

distance between the tacks divided by the length of the string. If the foci (or tacks) coincide, the ellipse is a circle; a circle is, then, an ellipse of eccentricity zero. Ellipses of various shapes are obtained by varying the spacing of the tacks (as long as they are not farther apart than the length of the string). If one tack is removed to an infinite distance, and if enough string is available, "our end" of the resulting, infinitely long ellipse is a parabola. A parabola has an eccentricity of one. An ellipse is completely specified by its major axis and its eccentricity. Figure 3.10 shows several ellipses.

Kepler found that Mars has an orbit that is an ellipse and that the sun is at one focus (the other focus is empty). The eccentricity of the orbit of Mars is only about 0.1; the orbit, drawn to scale, would be practically indistinguishable from a circle. It is a tribute to Tycho's observations and to Kepler's perseverance that he was able to determine that the orbit was an ellipse at all.

ellipse depends on how close together the two foci are compared to the major axis. The ratio of the distance between the foci to the major axis is called the *eccentricity* of the ellipse. If an ellipse is drawn as described above, the length of the major axis is the length of the string, and the eccentricity is the

The Varying Speed of Mars

Before he saw that the oribt of Mars could be represented accurately by an ellipse, Kepler had already investigated the manner in which the planet's orbital speed varied. After some calculation, he found that Mars speeds up as it comes closer to the sun and slows down as it pulls away from the sun. Kepler expressed this relation by imagining that the sun and Mars are connected by a straight, elastic line. As Mars travels in its elliptical orbit around the sun, in equal intervals of time the areas swept out in space by this imaginary line are always equal (Figure 3.11). This relation is commonly called the *law of areas*.

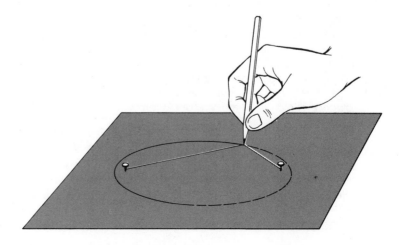

Figure 3.9 Drawing an ellipse.

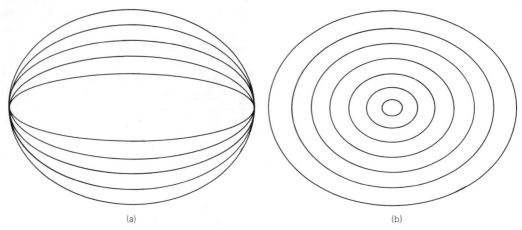

(a)   (b)

Figure 3.10 (a) Ellipses of the same major axis but various eccentricities; (b) ellipses of the same eccentricity but various major axes.

HOW KEPLER DETERMINED THE ORBIT OF MARS

Kepler determined the distance between Mars and the sun at various positions of the planet in its orbit by the process of *triangulation.* In Figure 3.12, S represents the sun and M represents Mars at some point in its path around the sun. Suppose we observe Mars when the earth is at E_1. The angle SE_1M at the earth between Mars and the sun is observable. Since the sidereal period of Mars is 687 days, after 687 days Mars will return to point M. The earth, meanwhile, will have completed nearly two full revolutions around the sun and will be at E_2. Angle SE_2M can now be observed. In exactly 2 years, or $730\frac{1}{2}$ days, the earth will have returned to E_1. The earth is short by $730\frac{1}{2} - 687 = 43\frac{1}{2}$ days of completing two revolutions about the sun. Thus the angle E_1SE_2 is known—it is the angle through which the earth moves in $43\frac{1}{2}$ days. Lines SE_1 and SE_2 are each the earth's distance from the sun. Thus two sides and an included angle of the triangle E_1SE_2 are known, and the triangle can be solved for the side E_1E_2, in terms of the distance from the earth to the sun, and for the angles SE_1E_2 and SE_2E_1.

Subtraction of angles SE_1E_2 and SE_2E_1 from SE_1M and SE_2M, respectively, gives the angles E_2E_1M and E_1E_2M, both in the triangle E_1ME_2. In that latter triangle since two angles and an included side are now known, sides E_1M and E_2M and the third angle can be found. Finally, the distance of Mars from the sun (but in terms of the earth's distance) can be found from either triangle SE_1M or SE_2M.

Kepler found the distance of Mars from the sun at five points along its orbit by choosing from Tycho's records the elongations of Mars on each of five pairs of dates separated from each other by intervals of 687 days.

Kepler's First Two Laws of Planetary Motion Summarized

We may summarize the most important contributions in *The New Astronomy,* or *Commentaries on the Motions of Mars,* by stating what are now known as Kepler's first two *laws of planetary motion:*

> *KEPLER'S FIRST LAW: Each planet moves about the sun in an orbit that is an ellipse, with the sun at one focus of the ellipse.*
>
> *KEPLER'S SECOND LAW (THE LAW OF AREAS): The straight line joining a planet and the sun sweeps out equal areas in the orbital plane in equal intervals of time.*

At the time of publication of the *New Astronomy* (1609), Kepler appears to have demonstrated the validity of these two laws only for the case of Mars. However, he expressed the opinion that they held also for the other planets.

(b) The Harmony of the Worlds

Kepler believed in an underlying harmony in nature, and he constantly searched for numerological relations in the celestial realm. It was a great personal triumph, therefore, that he found a simple algebraic relation between the lengths of the semimajor axes of the planets' orbits and their sidereal periods. Because planetary orbits are elliptical, the distance between a given planet and the sun varies. Now, the major axis of a planet's orbit is the sum of its maximum and minimum distances from the sun. Therefore, half of this sum, the semimajor axis, can be thought of as the *average* distance of a

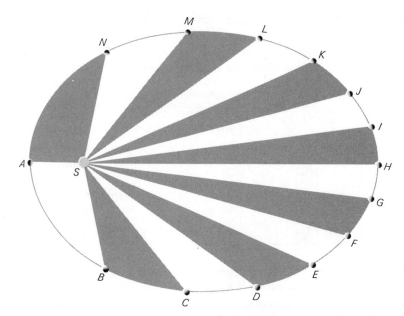

Figure 3.11 Law of equal areas. A planet moves most rapidly on its elliptical orbit when it is at position A, nearest the focus of the ellipse, S, where the sun is. The planet's orbital speed varies in such a way that in equal intervals of time it moves distances AB, BC, CD, and so on, so that regions swept out by the line connecting it and the sun (shaded and clear zones) are always the same in area.

planet from the sun. In a circular orbit, the semimajor axis is simply the radius of the circle.

Kepler published his discovery in 1619 in *The Harmony of the Worlds*. The relation is now known as his *third*, or *harmonic, law*.

> KEPLER'S THIRD LAW: *The squares of the sidereal periods of the planets are in direct proportion to the cubes of the semimajor axes of their orbits.*

It is simplest to express Kepler's third law with the algebraic equation:

$$P^2 = Ka^3,$$

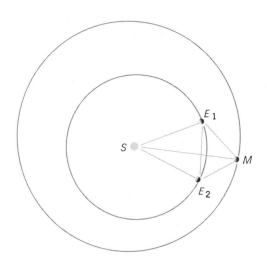

Figure 3.12 Kepler's method of triangulating the distance to Mars.

where P represents the sidereal period of the planet, a is the semimajor axis of its orbit, and K is a numerical constant whose value depends on the kinds of units chosen to measure time and distance. It is convenient to choose for the unit of time the earth's period—the year—and for the unit of distance the semimajor axis of the earth's orbit, the *astronomical unit* (AU). Note that the values found for the distances of the planets from the sun, both as determined by Copernicus and by Kepler, are in astronomical units. With this choice of units, $K = 1$, and Kepler's third law can be written

$$P^2 = a^3.$$

We see that to arrive at his third law it was not necessary for Kepler to know the *actual* distances of the planets from the sun (say, in kilometers), only the distance in units of the earth's distance, the astronomical unit. The length of the astronomical unit in kilometers was not determined accurately until later (see Chapter 22).

As an example of Kepler's third law, consider Mars. The semimajor axis, a, of Mars' orbit is 1.524 AU. The cube of 1.524 is 3.54. According to the above formula, the period of Mars, in years, should be the square root of 3.54, or 1.88 years, a result that is in agreement with observations. Table 3.2 gives for each of the six planets known to Kepler the modern values of a, P, a^3, and P^2. To the limit of accuracy of the data given, we see that Kepler's law holds exactly, except for Jupiter and Saturn, for which there are very slight discrepancies. Decades later, Newton gave an explanation for the discrepancies, but within the limit of accuracy of the observational data available in 1619, Kepler was justified in considering his formula to be exact.

Table 3.2 **Observational Test of Kepler's Third Law**

PLANET	SEMIMAJOR AXIS OF ORBIT, a (AU)	SIDEREAL PERIOD, P (Years)	a^3	P^2
Mercury	0.387	0.241	0.058	0.058
Venus	0.723	0.615	0.378	0.378
Earth	1.000	1.000	1.000	1.000
Mars	1.524	1.881	3.537	3.537
Jupiter	5.203	11.862	140.8	140.7
Saturn	9.534	29.456	867.9	867.7

Much of the rest of *Harmony of the Worlds* deals with Kepler's attempts to associate numerical relations in the solar system with music; indeed, he tried to derive notes of music played by the planets as they move harmoniously in their orbits. The earth, for example, plays the notes *mi, fa, mi,* which he took to symbolize the *"miseria* (misery), *fames* (famine), *miseria"* of our planet.

(c) The *Epitome*

In 1618, 1620, and 1621, Kepler published (in installments) his text on astronomy, the *Epitome of the Copernican Astronomy*. The book includes accounts of discoveries, both by himself and by Galileo (see Section 3.4), and firmly supports the Copernican view. Here, for the first time, Kepler implies that his first two laws had been tested and found valid for the other planets besides Mars (including the earth) and for the moon. Also, he states that the harmonic (third) law applies to the motions of the four newly discovered satellites of Jupiter as well as to the motions of the planets about the sun.

Kepler's Discussion of the Distance to the Sun

We recall (Section 2.3) that both Hipparchus and Ptolemy had rather accurately measured the moon's distance to be about 60 times the radius of the earth. Earlier, Aristarchus (Section 2.3a) had found that the sun was 18 to 20 times as far away as the moon, which placed the sun at a distance of at most 1200 earth radii. This figure survived until the seventeenth century, when Kepler revised it upward. Kepler was not able to detect any diurnal parallax

of Mars, that is, any apparent shift in direction caused by the rotation of the earth carrying the observer from one side of the earth to the other. Now the distance to Mars was known only in terms of the earth's distance from the sun—the astronomical unit; the actual assumed distance to Mars would be proportional to whatever value was assumed for the astronomical unit. If the latter were only 1200 earth radii, Mars should be near enough to allow observation of such a daily parallax. Kepler concluded that the astronomical unit must be at least three times the accepted figure—still a value seven times too small but nevertheless an improvement.

3.4 GALILEO

Galileo Galilei (1564–1642), the great Italian contemporary of Kepler, was born in Pisa. Galileo,

Figure 3.13 Detail from Kepler's *Harmony of the Worlds.* (Crawford Collection, Royal Observatory Edinburgh)

like Copernicus, began training for a medical career, but he had little interest in the subject and later switched to mathematics. In school he incurred the wrath of his professors by refusing to accept on faith dogmatic statements based solely on the authority of great writers of the past. From his classmates he gained the nickname "Wrangler."

For financial reasons, Galileo was never able to complete his formal university training. Nevertheless, his exceptional ability as a mathematician gained him the post, in 1589, of professor of mathematics and astronomy at the university at Pisa. In 1592 he obtained a far better position at the university at Padua, where he remained until 1610, when he left to become mathematician to the Grand Duke of Tuscany. While at Padua he became famous throughout Europe as a brilliant lecturer and as a foremost scientific investigator.

(a) Galileo's Experiments in Mechanics

Galileo's greatest contributions were in the field of *mechanics* (the study of motion and the actions of forces on bodies). The principles of mechanics outlined by Aristotle had still not been completely discarded. Although the seeds of experimental science had been sown by certain of the later Greek scholars, notably Archimedes, the practice of performing experiments to learn physical laws was not standard procedure even in Galileo's time.

Galileo experimented with pendulums, with balls rolling down inclined planes, with light and mirrors, with falling bodies, and many other objects. Aristotle had said that heavy objects fall faster than lighter ones. Galileo argued that if a heavy and light object were dropped together, even from a great height, both would hit the ground at practically the same time. What little difference there was could easily be accounted for by the resistance of the air.

Laws of Motion

In the course of his experiments, Galileo discovered laws that invariably described the behavior of physical objects. The most far-reaching of these is the *law of inertia* (now known universally as Newton's first law). The inertia of a body is that property of the body that resists any change of motion. It was familiar to all persons then as it is to us now that if a body is at rest it tends to remain at rest, and re-

quires some outside influence to start it in motion. Rest was thus generally regarded as the *natural state of matter*. Galileo showed, however, that rest was no more natural than motion. If an object is slid along a rough horizontal floor, it soon comes to rest, because friction between it and the floor acts as a retarding force. However, if the floor and object are both highly polished, the body, given the same initial speed, will slide farther before coming to rest. On a smooth layer of ice, it will slide farther still. Galileo noted that the less the retarding force, the less the body's tendency to slow down, and he reasoned that if all resisting effects could be removed (for example, the friction of the floor or ground, and of the air) the body would continue in a steady state of motion indefinitely. In fact, he argued, not only is a force required to start an object moving from rest, but a force is also required to slow down, stop, speed up, or change the direction of a moving object.

Galileo also studied the way bodies accelerated, that is, changed their speed, as they fell freely, or rolled down inclined planes. He found that such bodies accelerate uniformly, that is, in equal intervals of time they gain equal increments in speed. Galileo formulated these newly found laws in precise mathematical terms that enabled one to predict, in future experiments, how far and how fast bodies would move in various lengths of time. It remained for Newton to incorporate and generalize Galileo's principles into a few simple laws so fundamental that they have become the basis of a great part of our modern technology (Chapter 4).

(b) Galileo's Astronomical Contributions

Sometime in the 1590s Galileo accepted the Copernican hypothesis of the solar system. In Roman Catholic Italy, this was not a popular philosophy, for the Church authorities still upheld the ideas of Aristotle and Ptolemy. It was primarily because of Galileo that in 1616 the Church issued a prohibition decree which stated that the Copernican doctrine was "false and absurd" and was not to be held or defended.

The prevailing notion of the time was that the celestial bodies belonged to the realm of the heavens where all is perfect, unchanging, and incorruptible. Perpetual circular motion, being the "perfect" kind of motion, was regarded as the natural state of affairs for those heavenly bodies. Once Galileo had established the principle of inertia—that on the earth bodies in undisturbed motion remain in motion—it was no longer necessary to ascribe any special status to the fact that the planets remain perpetually in orbit. By the same token, even the earth could continue to move, one started. What *does* need to be explained is why the planets move in curved paths around the sun rather than in straight lines. Evidently, Galileo was sufficiently imbued with Aristotelian concepts that he accepted uniform circular celestial motion without subjecting the planets to the same objective scrutiny that he applied in his terrestrial experiments.

In answer to the common objection that objects could not remain on the earth if it were in motion, Galileo noted that if a stone is dropped from the masthead of a moving ship it does not fall behind the ship and land in the water beyond its stern, but rather lands at the foot of the mast, for the stone already has a forward inertia gained from its common motion with the ship before it is dropped. In an analogous way, objects on the earth would not be swept off and left behind if the earth were moving, for they share the earth's forward motion.

Galileo's Telescopes

It is not certain when the principle was first conceived of combining two or more pieces of glass to produce an instrument that enlarged distant objects, making them appear nearer. Claims for the discovery exist as early as the time of Roger Bacon (13th century). At any rate, the first telescopes that attracted much notice were made by the Dutch spectacle-maker Hans Lippershey in 1608. Galileo heard of the discovery in 1609, and without ever having seen an assembled telescope, he constructed one of his own with a three-power magnification, that is, it made distant objects appear three times nearer and larger. He quickly built other instruments, his best with a magnification of about 30.

Sidereal Messenger

It was a fairly obvious step to apply the newly invented telescope to celestial observations. The idea may have occurred to others about the same time as it did to Galileo, or possibly sooner. Galileo, however, rightly deserves the honor of having been the first to make significant astronomical telescopic observations, for he realized the importance of careful

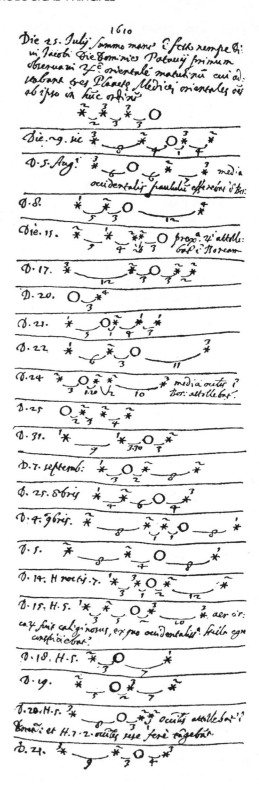

Figure 3.14 Galileo's drawings of Jupiter and its satellites.

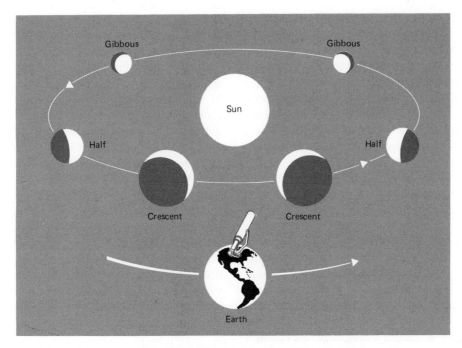

Figure 3.15 *Phases of an inferior planet.*

and persistent study of the objects he viewed. In 1610 he startled the world by publishing a list of his remarkable discoveries in a small book, *Sidereal Messenger (Sidereus Nuncius).*

Galileo found that many stars too faint to be seen with the naked eye became visible with his telescope. In particular, he found that some nebulous blurs resolved into many stars (for example, the Praesepe in Cancer) and that the Milky Way was made up of multitudes of individual stars. He found that Jupiter had four satellites or moons revolving about it with periods ranging from just under 2 days to about 17 days (12 other satellites of Jupiter have been found since). This discovery was particularly important because it showed that there could be centers of motion that in turn are in motion. It had been argued that if the earth were in motion the moon would be left behind, because it could hardly keep up with a rapidly moving planet. Yet here were Jupiter's satellites doing exactly that!

Phases of Venus

Another important telescopic discovery that strongly supported the Copernican view was the fact that Venus goes through phases like the moon. In the Ptolemaic system, Venus is always closer to the earth than is the sun, and thus, because Venus

never has more than about 45° elongation, it would never be able to turn its fully illuminated surface to our view—it would always appear as a crescent. Galileo, however, saw that Venus went through both crescent and gibbous phases, and concluded that it must travel around the sun, passing at times behind and beyond it, rather than revolving directly around the earth (Figure 3.15). Mercury also goes through all phases.

Irregularities in the Heavens

Galileo's observations revealed much about our nearest neighbor, the moon. He saw craters, mountain ranges, valleys, and flat dark areas that he guessed might be water (the dark *maria*, or "seas," on the moon were thought to be water until long after Galileo's time). Not only did these discoveries show that the heavenly bodies, regarded as perfect, smooth, and incorruptible, do indeed have irregularities, as does the earth, but they showed the moon to be not so dissimilar to the earth, which suggested that the earth, too, could belong to the realm of celestial bodies.

Of course, Galileo's conclusions did not go uncriticized. Ludovico delle Colombe, one of his critics, argued that the valleys and mountains on the moon were actually submerged beneath an invisible sea of a crystalline material whose outer surface was perfectly smooth. Galileo answered that if so he could argue that there are mountains of this same invisible substance rising ten times higher than any mountains that could be seen on the moon!

Galileo also found that Saturn seemed to appear strange, although his telescope was not good enough to show the true nature of the planet. It was not until 1655 that Huygens described the magnificent system of rings about Saturn (Chapter 18).

One of Galileo's most disturbing observations, to his contemporaries, was of spots on the sun, showing that this body also had "blemishes." Sunspots are now known to be large, comparatively cool areas on the sun that appear dark because of their contrast with the brighter and hotter solar surface (Chapter 31). Sunspots are temporary, lasting usually only a few weeks to a few months. Large sunspots actually had been observed before, with the unaided eye, but were generally regarded either as something in the earth's atmosphere or as planets between the earth and the sun silhouetting themselves against the sun's disk in the sky. In fact, some of Galileo's critics attempted to explain the spots as satellites revolving about the sun.

Galileo observed the spots to move, day by day, across the disk of the sun. He also noted that they moved most rapidly when near the center of the sun's disk and increasingly slowly as they approached the limb (the sun's limb is its apparent "edge" as we see it in the sky). Often, after about two weeks, the same spots would reappear on the opposite limb, and move slowly at first, then more rapidly toward the center of the disk. Galileo explained that the spots must be either on the surface of the sun or very close to it and that they were carried around the sun by its own rotation. Their variable speed, he showed, is an effect of foreshortening; when near the center of the sun's disk they are being carried directly across our line of sight, but near the limb most of their motion is either toward or away from us. He determined the sun's period of rotation to be a little under a month.

(c) Dialogue on the Two Great World Systems

As we have seen, Galileo had accumulated a great deal of evidence to support the Copernican system. By the decree of 1616 he was forbidden to "hold or defend" the odious hypothesis, but he still hoped to convert his countrymen to the heliocentric view. He finally prevailed upon his long-time acquaintance, Pope Urban VIII, to allow him to publish a book that explained fully all arguments for and against the Copernican system, not for the purpose of extolling it, but merely to examine it, and to show those of other nationalities that Italians were not ignorant of new theories.

The book appeared in 1632 under the title *Dialogue on the Two Great World Systems (Dialogo dei Due Massimi Sistemi)*. The *Dialogue* is written in Italian (not Latin) to reach a large audience and is a magnificent and unanswerable argument for Copernican astronomy. It is in the form of a conversation, lasting four days, among three philosophers: Salviati, the most brilliant and the one through whom Galileo generally expresses his own views; Sagredo, who is usually quick to see the truth of Salviati's arguments; and Simplicio, an Aristotelian philosopher who brings up all the usual

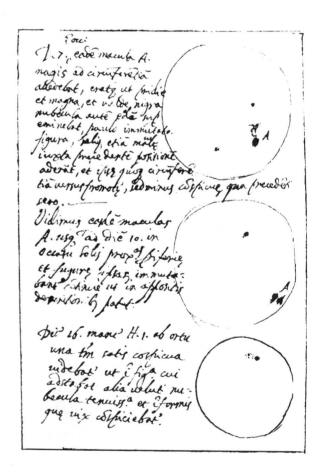

Figure 3.16 Galileo's drawings of sunspots. *(Yerkes Observatory)*

objections to the Copernican system, which Salviati promptly shows to be absurd.

It is pointed out in the preface to the *Dialogue* that the arguments to follow are merely a mathematical fantasy, and that divine knowledge assures us of the immobility of the earth. This was thinly cloaked irony, however, and Galileo's enemies acted quickly to build a case against him. He was called before the Roman Inquisition on the charge of believing and holding doctrines that are false and contrary to Divine Scriptures. Galileo was forced to plead guilty and deny his own doctrines. His life sentence was commuted to confinement in his own home at Arcetri, near Florence for the last ten years of his life. At the time of his inquisition, he was nearly 70.

The *Dialogue* joined Copernicus' *De Revolutionibus* and Kepler's *Epitome* on the *Index of Prohibited Books*. It was removed from the *Index*, however, in 1835. Moreover, in late 1980, Pope John Paul II ordered a reexamination of the evidence against Galileo. The result of the new investigation has not been announced at this writing, but it is expected that Galileo will be exonerated.

Figure 3.17 Villa Galileo, in Arcetri (near Florence), where Galileo spent the last years of his life in home imprisonment. The plaque under the bust between the front windows invites passersby to stop and contemplate, with respect, the great observer of the skies. *(Courtesy Professor G. Godoli, Arcetri Observatory)*

EXERCISES

1. A friend tells you that the earth cannot be rotating, as the scientists claim, because if a point on the equator were being carried eastward at about 1000 mi/hr, as claimed, a baseball pitcher could throw a ball straight up in the air, and the earth and pitcher would move to the east out from under the ball, which would land some distance behind (or to the west) of the pitcher. How might you straighten out this friend of yours?

2. Which if any of the following can never appear at opposition? at conjunction? at quadrature?

   (a) Jupiter   (d) Venus   (g) Mercury
   (b) Earth   (e) Saturn   (h) Moon
   (c) Sun   (f) Mars

3. Does full moon occur at intervals of the sidereal or synodic revolution of the moon about the earth? Why?

4. What is the phase of the moon when it is at quadrature?

5. If a superior planet had a synodic period equal to its sidereal period, what would this period be? Which planet most closely approximates this condition?

6. What would be the sidereal period of an inferior planet that appeared at greatest western elongation exactly once a year?

7. The synodic period of Saturn is 1.03513 sidereal years. What is its sidereal period?
   Answer: 29.5 years

8. What would be the distance from the sun, in astronomical units, of an inferior planet that had a greatest elongation of 30°? Assume circular orbits for the planet and the earth.

9. Draw an ellipse by the procedure described, using a string and two tacks. Arrange the tacks so that they are separated by one tenth the length of the string. Comment on the appearance of your ellipse. This (if you have been careful in your construction) is approximately the shape of the orbit of Mars.

10. Suppose a superior planet has a synodic period of two years. It is at opposition on January 1 and is next at quadrature on May 1. What is its distance from the sun in astronomical units? Assume circular orbits. Could such a planet actually exist? Why, or why not?

11. At best the positional accuracy attainable with the naked eye approaches one minute of arc. Nevertheless Tycho was able to determine the length of the year to within about one second of time, during which time the sun moves along the ecliptic through an angle of only 0.04 seconds of arc. How could he have done this?

12. (a) What is the major axis of a circle?
    (b) Where is the second focus of a parabola?
    (c) Which conic section could have an eccentricity of 0.3?
    (d) Which conic section could have an eccentricity of 3.2?

13. What is the eccentricity of the orbit of a planet whose distance from the sun varies from 180 million to 220 million km?

14. The earth's distance from the sun varies from 147.2 million to 152.1 million km. What is the eccentricity of its orbit?
    Answer: 0.016

15. Consider Kepler's third law as given in the equation in Section 3.3. Carefully explain why $K = 1$ when a is measured in astronomical units and P in years.

16. What would be the period of a planet whose orbit has a semimajor axis of 4 AU?

17. What would be the distance from the sun of a planet whose period is 45.66 days?
    Answer: 0.25 AU

18. Suppose Kepler's laws apply to the motion of Jupiter's satellites around that planet, and that one of the satellites has a period 5.196 times as long as another one. What would be the ratio of the semimajor axes of their orbits?
    Answer: 3:1

19. Draw a diagram showing Galileo's argument that sunspots are on or very near the solar surface.

20. Galileo's observations of the phases of Venus ruled out Ptolemy's system of cosmology. Did they also rule out Tycho Brahe's system? Why, or why not?

Sir Isaac Newton (1643–1727) had the insight to realize that the force that makes planets fall around the sun and the force that makes apples fall to the ground are different manifestations of the same thing: gravitation.

GRAVITATION: ACTION AT A DISTANCE

Kepler deduced that a force from the sun pulled on the planets. He did not determine the mathematical nature of this force, but as we have seen, he did discover some of the rules of planetary motion that result from it. Meanwhile, Galileo discovered some of the laws that describe the behavior of falling bodies, for example, that a freely falling body near the surface of the earth accelerates uniformly. Although he was not able to measure the precise value of that acceleration, it is now known that every second that a body falls (in a vacuum) its speed increases by 980 cm/s. Newton unified these insights by showing that the force of gravitation that accelerates falling bodies near the earth is the same force that keeps the moon in its orbit around the earth and the planets in their orbits about the sun.

Indeed, Newton's principles of mechanics and law of gravitation are so general and powerful that in the century following his death they strongly influenced the prevailing philosophy. Many thinkers held that the basic rules of nature were finally known, that all that remained was to fill in minor details. The philosophy of determinism reigned: namely, that every action in the universe follows by mechanistic laws from conditions immediately pre-

ceding the action. We shall see in the coming chapters, however, that this view of the ultimate success of science was overoptimistic. Newton's laws are not absolute; they fail to describe the motions of electrons in atoms, for example, for which the quantum theory is needed, and they fail when speeds are involved that are not small compared to that of light, in which case relativity provides a better model.

But the fact that Newton's laws are limited does not invalidate them; within the realm in which they work, they provide a magnificent description of the behavior of material objects. The success of our space program and of our other technology attests to the validity of Newtonian mechanics. Moreover, the Newtonian description of nature is so beautifully simple that, at least in essence, it can be understood and appreciated by those not specializing in science.

4.1 NEWTON'S PRINCIPLES OF MECHANICS

Isaac Newton (1643–1727) was born at Woolsthorpe, in Lincolnshire, England, only four days

49

less than one year after the death of Galileo. (Newton was born on Christmas Day, 1642, according to the calendar in use at his time, but by the modern Gregorian calendar his birth date was January 4, 1643.)

Newton entered Trinity College at Cambridge in 1661 and eight years later was appointed Lucasian Professor of Mathematics, a post that he held during most of his productive career. As a young man in college, he became interested in natural philosophy, as science was called then. The University was closed during the plague years of 1665 and 1666, during which Newton returned to Woolsthorpe. He wrote later that it was in those years that he worked out the main outline of his ideas on mechanics and gravitation. But on his later return to Cambridge, his research was mainly in mathematics and optics, and it was to be nearly two decades before he turned his attention again to gravitation.

Newton's return to gravitation was almost fortuitous. Physicist Robert Hooke, architect Christopher Wren, and astronomer Edmund Halley had all come independently to some notion of the law of gravitation and had realized that a force of attraction toward the sun must become weaker in proportion to the square of the distance from the sun. None, however, was able to solve the problem of how a planet should move under the influence of such a force. In 1684, Halley chanced to consult Newton on the matter. He was astonished to hear that Newton had solved the problem years previously and had found that the orbit of a planet should be an ellipse. Although Newton was unable to find his original notes containing the mathematical proof, he was able to re-solve the problem, and a short time later sent the demonstration of the proof to Halley. Early the following year Newton submitted a formal paper on the subject to the Royal Society. This treatise, consisting of four theorems and seven problems, was to become the nucleus of his great work on mechanics and gravitation, The *Mathematical Principles of Natural Philosophy*, usually known by the abbreviated form of its Latin title, *Principia*. Newton worked on the *Principia* for a year and a half during 1685 and 1686. It was published under the imprimatur of the Royal Society of London, and supervision of its publication was in the hands of Halley. As it turned out, the Society at that time was in financial difficulties, and Halley himself covered the cost of publication from his own personal funds.

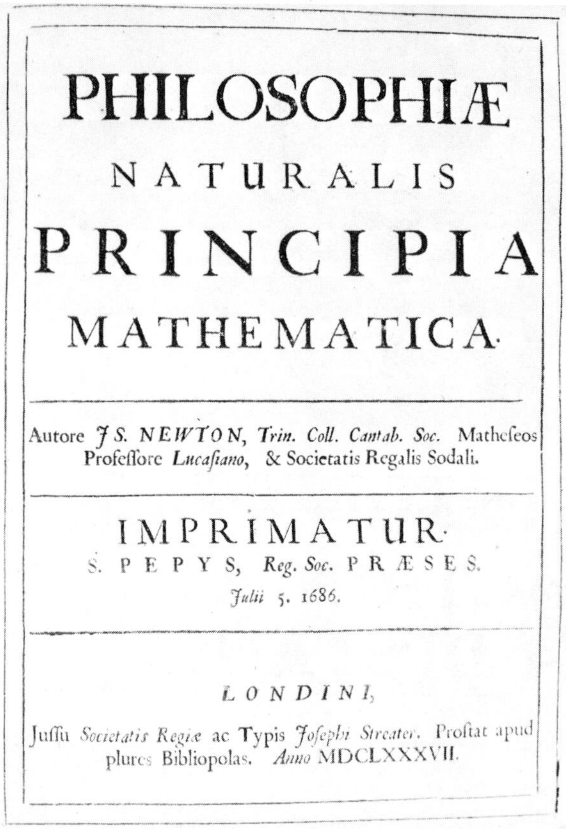
PHILOSOPHIÆ
NATURALIS
PRINCIPIA
MATHEMATICA.

Autore *J S. NEWTON*, *Trin. Coll. Cantab. Soc.* Matheseos Professore *Lucasiano*, & Societatis Regalis Sodali.

IMPRIMATUR.
S. PEPYS, *Reg. Soc.* PRÆSES.
Julii 5. 1686.

LONDINI,
Jussu *Societatis Regiæ* ac Typis *Josephi Streater*. Prostat apud plures Bibliopolas. *Anno* MDCLXXXVII.

Figure 4.1 Title page of the first edition of Newton's *Principia*. (*History of Science Collections, University of Oklahoma Libraries*)

In the *Principia* Newton gives his three laws of motion:

> *I. Every body continues in a state of rest, or of uniform motion in a straight line, unless it is compelled to change that state by forces impressed upon it.*
>
> *II. The change of motion is proportional to the force impressed; and is made in the direction of the straight line in which that force is impressed.*
>
> *III. To every action there is always an equal and opposite reaction: or, the mutual actions of two bodies upon each other are always equal, and act in opposite directions.*

Galileo had arrived at the first two laws, although he did not state them as precisely as Newton did. They are, however, deeper than Galileo could have realized. To appreciate Newtonian mechanics, we must understand thoroughly the meanings of certain terms.

(a) Some Basic Concepts: Length

In science, terms must be defined *operationally;* that is, we must supply a recipe or procedure for defining a quantity, so that someone else repeating our observation or experiment will obtain the same result. Consider, for example, *length.* It obviously will not do to define length as "how long something is." Suppose we wished to measure the length of a desk top. We might find a stick and define it as a unit of length; we could call that unit, say, a "thinga." Then by laying out the stick along the desk top we might learn that it has a length of 3.5 thingas.

The international standard unit of length is the *meter (m).* The meter was originally intended to be 10^{-7} times the distance from the equator of the earth to the North Pole. For years, however, the actual standard was a metal bar stored in a Paris vault. The modern standard was defined in 1960 at a General Conference sponsored by the International Bureau of Weights and Measures; it is equal in length to 1,650,763.73 times the wavelength (in a vacuum) of the orange light emitted by a certain isotope of krypton (^{86}Kr) under specified conditions of pressure and temperature. One hundredth of a meter is a *centimeter (cm).* Also by international agreement, although the system in common use in the United States, the archaic British imperial system of weights and measures, is being phased out, 1 inch = 2.5400 cm.

Areas of surfaces are defined as *square* measures. The area of our desk top, for example, might be the number of squares each one thinga on a side (and fraction thereof) it would take to cover its surface. The area of a rectangular surface is its length times its width.

Volumes of solids are *cubic* measures; for example, the volume of the desk might be the number of cubic containers of water, each one thinga on a side, it would take to fill a box the exact size and shape of the desk. The volume of a rectangular solid is its length times its width times its height.

Suppose we have two solids of identical shape but of different size—say, spheres 1 cm and 10 cm in radius, or a large man 2 m tall and an identical small one only 1 m tall. The surface area of one of the spheres (for example, the amount of paint needed to cover it) is proportional to the *square* of any of its linear dimensions, such as its radius. Thus the area of the larger sphere is 100 times that of the smaller (for 100 is the square of 10). Simi-larly, since 4 is the square of 2, it takes four times as much skin to cover the large man as it does the small man with half the large one's linear dimensions.

The volumes of solids are proportional to the *cubes* of their linear dimensions. Thus the sphere 10 cm in radius can hold 1000 times as much water as the one with a radius of 1 cm. Similarly, the large man has eight times as much flesh and bones as the small one. It is convenient to remember that areas are proportional to the squares, and volumes to the cubes of the linear dimensions of similarly shaped objects, whatever their shape may be.

(b) Time

Another basic concept is *time.* It is said that Galileo once used the beat of his pulse as a unit of time to measure the swing of a chandelier in church. He found that the time for one swing stayed the same, even though the length of the swing died down. The story of the swinging chandelier may be apocryphal, but Galileo did discover the law that determines the period of a pendulum—that the period of its oscillation depends only on the pendulum's length and not on the amount of arc of swing. Galileo later suggested that the pendulum would be a good device to regulate a clock, thus inventing the principle of the pendulum clock.

We could, of course, choose any convenient period for a unit of time. We could hold one end of that stick we used to define the thinga, and start the other end vibrating, like a tuning fork. The time of a single vibration might be called a "majig." We could count how many majigs it took for water to leak out of a can with a hole in the bottom.

A more universal unit of time is the *second(s).* It was originally meant to be $^1/_{60}$ of a minute, which is $^1/_{60}$ of a hour, which is $^1/_{24}$ of a day. But the earth does not rotate quite regularly enough to serve as an accurate enough standard for modern measurements. The 1967 General Conference turned to the *atomic clock* (Chapter 8) and defined the second as the duration of 9,192,631,770 periods of one of the radiations from a certain isotope of the cesium atom (^{133}Cs).

(c) Speed, Velocity, and Acceleration

With ways to measure length and time we can define *speed,* for example, how many thingas a hare can run in one majig. Now if we measured the distance the hare ran in 1000 majigs, that distance divided by 1000 is the *average* or *mean speed* of the hare. But if the hare had sped up part of the time, or had stopped to eat some grass along the way, his speed would be changing. At any point along his journey, the *instantaneous speed* of the hare is the number of thingas he would run in a majig if he kept that identical speed for one full majig. An automobile speedometer measures instantaneous speed, although we might have some trouble finding one calibrated in thingas per majig.

Speed and *velocity* are often confused. Velocity conveys more information than speed; it is a description of both the instantaneous speed and the *direction* of motion. Velocity is an example of a *vector* (Figure 4.2), a quantity that has both *magnitude* (speed in this case) and direction.

Any change in velocity requires *acceleration*. Acceleration, therefore, involves a change of speed, or of direction, or both. Starting, stopping, speeding up, slowing down, or changing direction are all accelerations. Acceleration is also a vector. The magnitude of acceleration is the rate at which the velocity changes, and its direction is the direction of that change. The acceleration produced by gravity at the surface of the earth, for example, is 980 cm/s per second (often written 980 cm/s^2), in a direction toward the center of the earth.

(d) Newton's First Law— Momentum

A moving body tends to keep moving and a stationary body tends to remain at rest. *Momentum* is a measure of this state of motion. Momentum depends on velocity, for clearly a body moving at 50 km/hr certainly has more "motion" than one moving 10 km/hr. But momentum also depends on the amount of matter in the moving object; an automobile going 30 km/hr certainly has more "motion," and is harder to speed up, stop, or turn than, say, a bicycle moving with the same speed.

Thus Newton defined momentum as proportional to velocity, and defined the constant of proportionality as *mass*. Mass is a quantity that characterizes the total amount of material in the body and is the property that gives the body its *inertia*, that is, that makes it resist acceleration. We have not yet defined mass operationally, but we shall see that Newton's third law provides a means of doing so. For now it is enough to say that our intuitive idea of what is usually meant by *weight* is actually mass. Weight, technically, is a measure of the gravitational pull upon an object. If one goes to the moon, he weighs only one-sixth of what he weighs on earth, or if he goes to a remote place in space he weighs essentially nothing, for there the pull of gravitation is almost zero. Still, that person's bulk or *mass* is unchanged; he is as obese as ever.

Thus Newton's first law says that the product of a body's mass and velocity is constant if no outside force is applied to it. This means that motion is as natural a state as is rest. With no force on it, a moving body would go in a straight line at a constant speed forever.

(e) Newton's Second Law— Force

The second law of motion deals with changes in momentum. It states that if a force acts on a body it produces a *change* in the momentum of the body that is in the direction of the applied force. The second law, then, defines *force*. The magnitude or strength of a force is defined as the *rate of which it produces a change* in the momentum of the body on which it acts.

Some familiar examples of forces are the pull of the earth, the friction of air slowing down objects moving through it, the friction of the ground or a floor similarly slowing bodies, the impact of a bat on a baseball, the pressure exerted by air, and the thrust of a rocket engine.

Note that Newton's first law of motion is consistent with his second; when there is no force, the change in momentum is zero.

There are three ways in which the momentum of a body can change. Its velocity can change, or its mass, or both. Most often the mass of a body does not change when a force acts upon it; a change in momentum usually results from a change in ve-

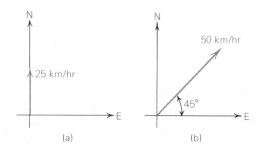

Figure 4.2 Two vectors, representing velocities of 25 km/hr to the north and 50 km/hr to the northeast.

locity. Thus, in the vast majority of examples, the second law can be written as the simple formula

force = mass × acceleration,

because acceleration is the rate at which velocity changes. If the acceleration occurs in the same direction as the velocity, the body simply speeds up; if the acceleration occurs in the opposite direction to the velocity, the body slows down. If acceleration occurs exactly at *right angles* to the velocity, only the direction of motion of the body, and not its speed, changes.

The acceleration of falling bodies is downward (in the direction toward which the gravitational pull of the earth is acting). Gravity accelerates a body in the direction it is already moving, and so simply speeds it up.

If a body is slid along a rough horizontal surface, it slows down uniformly in time. It is therefore accelerated in a direction opposite to its velocity. The acceleration is produced by the force of friction between the moving body and the rough surface.

In general, both the speed *and* direction of a body may change. Suppose, for example, its velocity is the vector **v** lying in direction *AB* in Figure 4.3. Now suppose that for a short time, Δt, a force with direction *BC* acts on the body. The force produces a change in velocity $\Delta \mathbf{v}$, in direction *BC*, and the new velocity **v′** is in direction *AC*. (The Greek letter Δ—capital delta—is commonly used to denote a small change in a quantity.) Acceleration is the *rate* at which the velocity changes, so we reason that if the velocity of a body changes by an amount $\Delta \mathbf{v}$ in time Δt, the rate at which it changes, that is, its acceleration, is $\Delta \mathbf{v}/\Delta t$. If the force (and hence acceleration) were not constant in magnitude and direction, $\Delta \mathbf{v}/\Delta t$ would be only the average acceleration over the time Δt. However, if the inter-

val Δt, and hence $\Delta \mathbf{v}$, is very small, $\Delta \mathbf{v}/\Delta t$ becomes a good approximation to the *instantaneous acceleration*. (Instantaneous rates of change are the subject of *differential calculus*, invented by Newton to handle such problems in mechanics.)

A body accelerates, that is, its momentum changes, only while a force acts upon it. The force of gravity is constantly pulling falling bodies downward; thus, they continually accelerate. On the other hand, once a baseball ceases to be in actual contact with the bat, it continues forward in a straight line at a constant speed and would do so forever if air friction did not slow it, the earth pull it down, or an eager outfielder intercept it.

Whenever the mass of an object is subject to change, we cannot assume that force = mass × acceleration. The variation of mass must be taken into account in calculating the acceleration produced by a force. An important and familiar example is a rocket. Exhaust gases are ejected from the back to accelerate the rocket forward, with its remaining fuel. Thus, as the rocket accelerates, its mass continually diminishes.

Two equal accelerations may correspond to entirely different forces. Consider the forces required to accelerate an automobile and a bicycle each to a speed of 30 km/hr in 20 seconds. Clearly, because of the car's greater mass, a proportionately greater force will be required to produce the necessary acceleration. Similarly, once the bodies are both moving at that speed, a far greater force is needed to *stop* the automobile as quickly as the bicycle.

Combinations of Forces

Newton showed how to compute the effect of two or more forces acting at once on an object. We shall illustrate the procedure when there are two forces, f_1 and f_2, acting together on a body at *O*. We can represent the forces by vectors just as we can velocities (Figure 4.4). The directions of the vectors show the directions of the forces, and the lengths show the magnitudes of the forces (say, so many dynes to the cm). The arrows representing the vectors both start from the object on which the two forces are acting. We note now that the two arrows define two sides of a parallelogram, Of_1ff_2. The diagonal of the parallelogram, *Of*, is the resultant force acting on the body; the body accelerates as if acted upon by just that one force. The figure shown is called a *parallelogram of forces*.

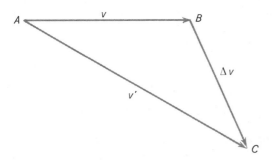

Figure 4.3   Change in velocity.

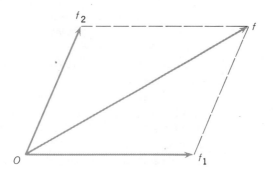

Figure 4.4 Parallelogram of forces.

Static Equilibrium

Of special interest is the situation in which two or more forces act upon a body in such a way as to cancel each other. Then the body is said to be in a state of *static equilibrium* and does not accelerate. An example is a ball hanging from a rubber band—the rubber stretches until its tension exactly balances the force of gravity pulling the ball downward. Another example is the column of mercury in a barometer; the mercury stands in static equilibrium between its own weight and the pressure of the air.

(f) The Third Law—Reaction

Newton's third law of motion was a new idea. It states that all forces occur as *pairs* of forces that are mutually equal to and opposite each other. If a force is exerted on an object, it must be exerted by something else, and the object will exert an equal and opposite force back upon that something. All forces, in other words, must be mutual forces acting *between* two objects or things.

If a man pushes against his car, the car pushes back against him with an equal and opposite force, but if the man has his feet firmly implanted on the ground, the reaction force is transmitted through him to the earth. Because of its enormously greater mass, the earth accelerates far less than the car. Suppose a boy jumps off a table down to the ground. The force pulling him down is a mutual gravitational force between him and the earth. Both he and the earth suffer the same total change of momentum because of the influence of this mutual force. Of course, the boy does most of the moving; because of the greater mass of the earth, it can experience the same change of momentum by accelerating only a negligible amount.

A more obvious manifestation of the mutual nature of forces between objects is familiar to all who have played baseball. The recoil of the bat shows that the ball exerts a force on the bat during the impact, just as the bat does on the ball. The momentum imparted to the bat by the ball is transmitted through the batter to the earth, so the acceleration produced is far less than that suffered by the ball. Similarly, when a rifle is discharged, the force pushing the bullet out the muzzle is equal to that pushing backward upon the gun and marksman.

Here, in fact, is the principle of rockets—the force that discharges the exhaust gases from the rear of the rocket is accompanied by a force that shoves the rocket forward. The exhaust gases need not push against air or the earth; a rocket operates best of all in a vacuum. Incidentally, passengers on a rocket ship would have to take care how they disposed of waste material. For example, if garbage were discharged through a port in the missile, the reaction force would accelerate the rocket slightly off course.

In all the cases considered above, a mutual force acts upon the two objects concerned; each object always experiences the same total change of momentum, but in opposite directions. Because momentum is the product of velocity and mass, the object of lesser mass will end up with proportionately greater velocity.

Generality of the Third Law

In fact, a system may be far more complex than just a pair of bodies exerting mutual forces on each other. Suppose, for example, a rocket in space were to explode into thousands of pieces. Each of the remnants of the explosion has exerted a force on each other in such a way that all the forces balance each other. Each particle has suffered a change in momentum, but the sum total of all changes in momentum of particles accelerated one way will be equal and opposite to the total change in momentum associated with all the particles going the opposite direction. All these changes in momentum balance each other, so that the momentum of the entire system, as long as it has not been acted on by *external* forces, is the same as before the explosion.

In other words, Newton's third law can be thought of as a generalization of his first. The first law states that in the absence of a force, a body's momentum is conserved. The third law means that if we isolate an entire system from outside forces, the total momentum of the system is conserved. *In-*

ternal forces in the system may result in changes of momentum within it, but these are always accompanied by equal and opposite changes. The total momentum of a rocket, for example, does not change, so long as we always include the momentum of its exhaust gases and any other object (such as the atmosphere of the earth) that may have come in contact with these gases.

Measuring Mass

We are now in a position to see how we can measure mass. Initially, some object must be adopted as a standard and said to have *unit mass*, for example, one cubic centimeter of water. The latter is defined as one *gram (g)*. Then the mass of any other object can be compared to it by measuring the relative accelerations produced when the same force acts on each of the two. We assure that each is subjected to the same force by isolating them from other objects and producing equal and opposite changes in their momenta with an *internal* force— say, by separating them with a compressed spring placed between them and released by an internal triggering device or by detonating a small charge between them. Thus, the third laws permits an *operational definition* of mass. This is called the *inertial* definition of mass.

Having found a way to measure mass, we can now express the value of a force numerically. The most common unit of force is the *dyne*, which is the force needed to give a mass of 1 g an acceleration of 1 cm/s^2.

(g)   Density

It is important not to confuse mass, volume, and *density*. Volume is simply a measure of the physical space occupied by a body, say in cubic centimeters or liters. In short, the volume is the "size" of an object—it has nothing to do with its mass. A lady's wrist watch and an inflated balloon may both have the same mass, but they have very different volumes.

The watch and balloon are also very different in *density*, which is a measure of how much mass is contained within a given volume. Specifically, it is the ratio of mass to volume:

$$density = \frac{mass}{volume}.$$

The units of density are usually expressed in grams per cubic centimeter (g/cm^3). Sometimes density is given in terms of the density of water (1 g/cm^3), in which case it is called *specific gravity*. Iron has a specific gravity of 7.9 or a density of 7.9 g/cm^3; gold has a specific gravity of 19.3. To sum up, then, *mass* is "how much," *volume* is "how big," and *density* is "how tightly packed."

(h)   Angular Momentum

Another useful concept is *angular momentum*, which measures the momentum of an object about some fixed point or origin. Angular momentum is defined in such a way as to involve the mass and velocity of the moving object and also its distance from the origin or reference point. It is not necessary to understand the precise definition of angular momentum to appreciate its consequences; thus we take up some of the details in the section that follows:

Consider (Figure 4.5) an object moving past point *P* along line *AB* with constant momentum, indicated in the figure by the vector *MN*. We denote the distance, *MP*, of the object from *P* at some arbitrary time by the distance *r*. Now, we can resolve the momentum *MN* into two components, one of which, *MD*, lies along the line of sight from *P*, and the other, *DN*, at right angles to the line of sight from *P*. The *angular momentum* of the object about *P* is defined as the product of its distance *r* from *P* and that perpendicular component, *DN*, of its momentum. Note that half of the numerical value of the angular momentum (*r* × *DN*) is the area of the triangle *MNP*.

The triangle *MDN* is similar to the triangle *MCP*, so we have the proportion

$$\frac{DN}{MN} = \frac{PC}{r},$$

from which we obtain

$$angular\ momentum = r \times DN = MN \times PC.$$

Now *PC* is the fixed distance of *P* from the line *AB*. If the momentum does not change (that is, if no force acts on the body), *MN* is always the same. Thus we see that if the momentum of the body is constant, so is its angular momentum. Moreover, the area of the triangle *MNP* is constant as well, that is, the line from the body to *P* sweeps out equal areas in equal times.

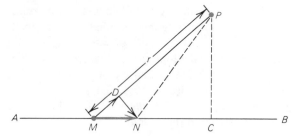

Figure 4.5 Angular momentum.

In the foregoing, it was shown that if the momentum of a body is constant, its angular momentum about any point or origin is also constant, that is, it is *conserved*. Moreover, the angular momentum may be conserved even if the body's momentum is not conserved, *provided that the only force on the body is along the direction toward or away from the origin point,* for a force toward or away from that point cannot affect the component of momentum perpendicular to that direction. In particular, a force of attraction between a planet and the sun cannot change the angular momentum of a planet around the sun. Since, as was shown above, the angular momentum is proportional to the area swept out by a line connecting the body and the origin, Kepler's law of areas (his second law) shows that the only force acting on a planet must be directed toward the sun.

Consider a rigid body rotating about an internal axis. Each part of that body is constrained to revolve about a point on that axis in a circular path. The forces causing this constraint are the cohesive forces binding the atoms and molecules of the body together. These forces are all radial, provided that there is no external force acting in such a way as to make a part of the body try to move faster or slower about its axis of rotation. Consequently, in the absence of any outside force tending to change its rotation (called a *torque*) a rotating body will continue to rotate indefinitely with constant angular momentum. Thus planets, such as the earth, keep spinning.

The rotation rate of a body, again assuming that there are no torques, can only change if that body can rearrange its parts in such a way as to move some of them farther away from or closer to the axis of rotation (the origin of angular momentum). This is exactly what is accomplished by figure skaters; suppose a skater is spinning on the tip of her (or his) skate with arms outstretched. If she holds her body rigidly until friction with the air and

ice produce enough torque to slow her down, she rotates at a constant rate. However, if she pulls her arms in to her body, some parts of her mass are closer to their rotation axes, and hence decrease in angular momentum unless she compensates by spinning faster. Thus, a figure skater can start a spin with her arms out, and then pull them in, thereby spinning faster so that her angular momentum is conserved. She can slow down again by pushing her arms (or a free leg) out from her body.

You can perform the same experiment with an old-fashioned rotating piano stool. Purchase another copy of this book and sit on the stool while holding both books out from your body at arm's length. Have a friend start you spinning—but not too fast, or you may not have the opportunity to finish reading even your first copy. Now pull your arms in so that the books are next to your body and you will experience a dramatic demonstration of the conservation of angular momentum.

The conservation of angular momentum is an important concept to an understanding of the formation of the solar system with its planets and their satellites, including the rings of Saturn, and even the formation of galaxies. We shall refer to it again in coming chapters.

4.2 ACCELERATION IN A CIRCULAR ORBIT

It might be assumed that some force or power is required to keep the planets in motion. However, Galileo argued from the principle of inertia (Newton's first law of motion) that once started, the planets would remain in motion—that the state of motion for planets was as natural as for terrestrial objects. What does require explanation, however, is why the planets move in nearly circular orbits rather than in straight lines (the latter motion would eventually carry them away from the vicinity of the solar system). Galileo had not considered this problem.

By Newton's time, a number of investigators had considered the problem of circular motion. The correct solution to the problem was first published (in 1673) by the Dutch physicist Christian Huygens (1629–1695). However, Newton had found the solution independently in 1666.

For a body to move in a circular path rather than in a straight line, it must continually suffer an acceleration toward the center of the circle. Such an acceleration is called *centripetal acceleration*. The central force that produces the centripetal acceleration *(centripetal force)* is, for a planet, an attraction between the planet and the sun. For a stone whirled about at the end of a string, the centripetal force is the tension in the string. With the help of Newton's laws of motion and some elementary mathematics we can calculate how great that central force has to be. We find that if a particle of mass m moves with a speed v on the circumference of a circle of radius r, the centripetal force is given by the formula

$$F = \frac{mv^2}{r}.$$

(a) DERIVATION OF THE CENTRIPETAL FORCE

Suppose the particle moves on the circumference of a circle centered at O (Figure 4.6). At some instant it is at position D and has a velocity of magnitude v and direction DE. By Newton's first law, if there were no forces acting upon the particle, it would continue to move in direction DE. After a brief interval of time Δt, it is at G, a distance along the circumference of the circle from D equal to its speed times Δt (for distance = rate × time); that is, G is a distance $v \times \Delta t$ along the arc from D. The particle's velocity still has magnitude v but is now in the direction GH. If the interval Δt is very short, the distance $v\, \Delta t$ along the arc is essentially the straight-line distance DG along the chord of the arc. (The angle α in Figure 4.6 is greatly exaggerated.)

The particle has accelerated, for it did not continue to move along direction DE. Because the particle is closer to the center of the circle than it would have been at E, the force that accelerated it must have acted in the direction toward the center. Since acceleration is the change of velocity per second, the acceleration the particle suffered during the time Δt is just the change of its velocity from direction DE to direction GH, divided by Δt. The angle through which the velocity changed must be the same as the angle α at the center of the circle subtended by the arc DG along which the particle moved, because its velocity at any point is always in a direction tangent to the circle, and the tangent to a circle at any point is always perpendicular to a line from that point to the center of the circle.

We can represent a change in velocity when that change is only in direction by a diagram such as in Figure 4.7. The original velocity has magnitude v and direction OA. After time Δt, the velocity has magnitude v and direction OB. The line representing the velocity has turned, in time Δt, through an angle α at O. The magnitude of the acceleration of the particle, its change in velocity during one second, is thus the change in velocity Δv, that is, the vector AB, divided by Δt.

Since the triangle OAB in Figure 4.7 is exactly similar to the triangle ODG in Figure 4.6 we have the simple proportion

$$\frac{\Delta v}{v} = \frac{v\, \Delta t}{r}.$$

The acceleration of the particle is then

$$a = \frac{\Delta v}{\Delta t} = \frac{v^2}{r}.$$

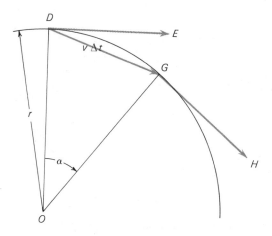

Figure 4.6   Motion in a circular orbit.

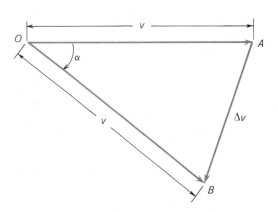

Figure 4.7   Centripetal acceleration.

If the particle has mass m, the central force required to produce this acceleration is $m \times a$, or

$$F = \frac{mv^2}{r},$$

as stated above.

Clearly, the above analysis is precise only if the angle α and the quantities Δv and Δt are extremely small. Here is another example in which Newton employed differential calculus (which he called *fluxions*); he obtained the result we derived, but rigorously, by allowing Δv and Δt to become infinitesimal; that is, he found the instantaneous rate of change of v.

4.3 UNIVERSAL GRAVITATION

It is obvious that the earth exerts a force of attraction upon all objects at its surface. This is a mutual force; a falling apple and the earth are pulling on each other. Newton reasoned that this force of attraction between the earth and objects on or near its surface might extend as far as the moon and produce the centripetal acceleration required to keep the moon in its orbit. He further speculated that there is a general force of attraction between *all* material bodies. If so, the attractive force between the sun and each of the planets could provide the centripetal acceleration necessary to keep each in its respective orbit.

Thus Newton hypothesized that there is a universal attraction between all bodies everywhere in space. Next he had to determine the mathematical nature of the attraction and test the hypothesis by using it to predict *new* phenomena. We shall now see how Newton formulated his law of universal gravitation.

(a) THE MATHEMATICAL DESCRIPTION OF GRAVITATION

For mathematical simplification we make the assumption that planets revolve around the sun in perfectly circular orbits. A more complicated analysis can be made to apply to the actual elliptical orbits. Using the results of Section 4.2 we find that the centripetal force that the sun must exert upon a planet of mass m_p, moving with speed v in a circular orbit of radius r, is

$$\text{force} = \frac{m_p v^2}{r}.$$

Now the period P of the planet, that is, the time required for the planet to go completely around the sun, is the circumference of its orbit ($2\pi r$) divided by its speed, or

$$P = \frac{2\pi r}{v}.$$

Solving the above equation for v, we find

$$v = \frac{2\pi r}{P}.$$

On the other hand, from Kepler's third law we know that the squares of the periods of planets are in proportion to the cubes of their distances from the sun. Because the sun is observed to be almost at the center of a planet's orbit, that distance is very nearly the radius of the orbit, r, and we have

$$P^2 = Ar^3,$$

where A is a constant of proportionality whose value depends on the units used to measure time and distance. Combining the last two equations, we find

$$v^2 = \frac{4\pi^2 r^2}{P^2} = \frac{4\pi^2 r^2}{Ar^3} = \frac{4\pi^2}{Ar},$$

that is,

$$v^2 \propto \frac{1}{r},$$

where the symbol $\propto$ means "proportional to."

If we substitute the above formula for v^2 into the one expressing the sun's centripetal force on the planet, we obtain

$$\text{force} \propto \frac{m_p}{r^2}.$$

The centripetal force exerted on the planet by the sun must therefore be in proportion to the planet's mass and in inverse proportion to the square of the planet's distance from the sun. According to Newton's third law, however, the planet must exert an equal and opposite attractive force on the sun. If the gravitational attraction of the planet on the sun is to be given by the same mathematical formula as that for the attraction of the sun on the planet, the planet's force on the sun must be

$$\text{force} \propto \frac{m_s}{r^2},$$

where m_s is the sun's mass. Since this is a mutual force of attraction between the sun and planet, it must be proportional to both the mass of the sun and the mass of the planet; therefore, the attractive force between the two has the mathematical form

$$\text{force} \propto \frac{m_s m_p}{r^2}.$$

Both the sun and a planet revolving around it experience the same change of momentum as a result of this mutual force between them. The fact that the sun is observed to remain more or less at the center of the solar system while the planets revolve around it is evidence that the sun's mass must be enormously greater than that of the planets. Therefore, its acceleration is relatively small, but it is actually observable.

(b)   The Law of Universal Gravitation

For Newton's hypothesis of universal attraction to be correct, there must be an attractive force between all pairs of objects everywhere whose value is given by the same mathematical formula as that above for the force between the sun and a planet. Thus the force F between two bodies of masses m_1 and m_2, and separated by a distance d, is

$$F = G\frac{m_1 m_2}{d^2}.$$

Here G, the constant of proportionality in the equation, is a number called the *constant of gravitation*, whose value depends on the units of mass, distance, and force used. The actual value of G has to be determined by laboratory measurement of the attractive force between two material bodies (see Section 4.3e). If metric units are used (grams for mass, centimeters for distance, and dynes for force), G has the numerical value 6.67×10^{-8}.★

The above equation expresses Newton's law of universal gravitation, which is stated as follows:

> *Between any two objects anywhere in space there exists a force of attraction that is in proportion to the product of the masses of the objects and in inverse proportion to the square of the distance between them.*

Not only is there a force between the sun and each planet, but also between any two planets. Because of the sun's far greater mass, the dominant force felt by any planet is that between it and the sun. The attractive forces between the planets have relatively little influence. Similarly, there is a gravitational attraction between any two objects on earth (for example, between two flying airplanes, or between the kitchen sink and a tree outside the house), but this force is insignificant compared to the force between each of them and the very massive earth.

Before we see how Newton tested his law of gravitation, let us investigate some of its other consequences.

(c)   Weight

Newton hypothesized a force of attraction between all pairs of bodies. The earth is a large spherical mass, however, that can be thought of as being composed of a large number of component parts. An object, say a man, on the surface of the earth feels the simultaneous attractions of the many parts of the earth pulling on him from many different directions. Exactly what is the resultant gravitational effect of the many parts of a sphere, each pulling independently upon a mass outside the surface of the sphere? Here was a difficult problem to which Newton had to find a solution before he could test his law of gravitation.

To solve the problem, Newton had to calculate the force between an object on the surface of the earth and each infinitesimal piece of the earth, and then calculate how all of these forces combined. It was necessary for him to invent and use a new method of mathematics which he called *inverse fluxions* (today we call it *integral calculus*). Fortunately, the solution to the problem gives a beautifully simple result. A spherical body acts gravitationally as though all its mass were concentrated at a point at its center (Figure 4.8).† This means that we can consider the earth, the moon, the sun, and the planets as geometrical points as far as their gravitational influences are concerned.

The gravitational force, then, between an object of mass m on the earth and the earth itself, of mass M, is equal to the constant of gravitation times the produce of the masses m and M, divided by the square of the distance from the object to the center of the earth. The latter distance is just the

---

★The notation 6.67×10^{-8} means 0.0000000667; see Appendix 3.

†Strictly, the statement is correct only if the density distribution within the body is spherically symmetrical.

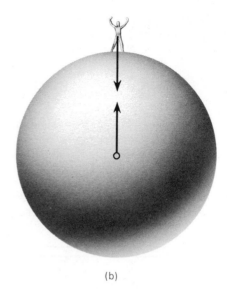

(a)   (b)

Figure 4.8   *Attraction of a sphere is as though all its mass were concentrated at its center.*

radius of the earth, R. This gravitational force between the earth and a body on its surface is the body's *weight*. Algebraically, the weight W of a body is given by

$$W = G\frac{mM}{R^2}.$$

We see that the weight of a body is proportional to its mass. This circumstance gives us another method of measuring mass, namely, by measuring the weight of a body. Whenever the mass of an object is determined by its gravitational influence, as by measuring its weight, the mass so determined is defined as a *gravitational mass*. So far, the results of all experiments indicate that gravitational mass is exactly equivalent to *inertial mass*, measured as described in Section 4.1f.

Because the gravitational attraction between two bodies decreases as the square of their separation, an object weighs less if it is lifted above the surface of the earth. A person actually weighs less at the top of a step ladder than on the ground. Careful laboratory experiments are able to detect such subtle changes in gravitational force. At 6400 km above the earth's surface, a body is twice as far from the earth's center as it would be at the surface, so an object there would weigh $\frac{1}{2}^2$, or $\frac{1}{4}$ times as much as at the surface of the earth. Far out in space, a person's weight drops virtually to zero. In the vicinity of another gravitating body, his weight is determined by the attraction between him and that body.

If an object is dropped from a height, the downward acceleration is equal to the force acting on it, that is, its weight, divided by its mass:

$$\text{acceleration} = \frac{W}{m} = G\frac{M}{R^2}.$$

Thus, although the gravitational force between the earth and a massive object is greater than between the earth and a less massive one, all objects experience the same acceleration and fall at the same rate, as Galileo had found.

(d)   Test of Gravitation: The Apple and the Moon

Suppose an apple is dropped from a height above the surface of the earth. We have seen that it accelerates 980 cm/s^2. If the hypothesis of gravitation is correct, the accelerations toward the earth of the apple (at the earth's surface) and of the moon should both be given by the equation

$$\text{acceleration} = G\frac{M}{D^2}.$$

where M is the mass of the earth and D is the distance of the object in question from the center of the earth. The acceleration, in other words, should be inversely proportional to the square of the distance from the earth's center. The apple's distance is about 6400 km, and the moon's distance is 384,403 km, about 60 times as far. Thus the acceleration of the moon should be $1/60^2$, or $1/3600$, as much as the apple's.

If we assume that the moon's orbit about the earth is a perfect circle (its orbit is, indeed, very nearly circular), we can use the formula for centripetal acceleration found in Section 4.2 to calculate the moon's acceleration:

$$\text{acceleration} = \frac{v^2}{D}.$$

The moon's distance (the radius of its orbit) is 3.844×10^{10} cm. Its average orbital speed is 1.023×10^5 cm/s. If we substitute these numbers in the above formula we find

$$\text{moon's acceleration} = \frac{(1.023 \times 10^5)^2}{3.844 \times 10^{10}}$$
$$= 0.272 \text{ cm/s}^2.$$

The acceleration predicted by the law of gravitation is

$$\text{moon's acceleration} = \frac{980}{(60)^2}$$
$$= 0.272 \text{ cm/s}^2.$$

The law of gravitation predicts that because the moon is 60 times as far from the center of the earth as an apple is, its acceleration should be $1/60^2$ as much; that is exactly the acceleration that we observe for the moon. The test gives results that are consistent with Newton's law of universal gravitation. It works! We can imagine the thrill that Newton must have felt in discovering that the same simple algebraic formula describes the law of gravitation that operates on the earth, on the moon and planets, and so far as we know, throughout the entire universe!

(e) The Mass of the Earth and the Determination of G

Note that in the formula for the weight of an object at the surface of the earth, or for its acceleration toward the earth, the constant of gravitation, G, always occurs multiplied by the mass of the earth. If the latter were known, G could be evaluated at once from the known acceleration of gravity:

$$G = \frac{R^2 g}{M_E},$$

where R and M_E are the radius and mass of the earth and g is the known acceleration of gravity at the earth's surface—980 cm/s^2. Conversely, if G were known, the mass of the earth could be found:

$$M_E = \frac{R^2 g}{G}.$$

Hence the determination of G is equivalent to the determination of the mass of the earth.

There are various methods of finding the mass of the earth. All depend on a comparison of the gravitational attraction between some object of known mass and the earth, with the attraction between two objects of known mass.

The first attempts to evaluate G (or, equivalently, the mass of the earth) were made by comparing the pull of the earth to that of large natural features, such as mountains. The most famous such experiment was carried out in 1774 by Nevil Maskelyne, Astronomer Royal, and Director of the Royal Greenwich Observatory. Overlooking Scotland's Loch Rannoch is Mount Schiehallion, a shallow conical peak so symmetrical that it is feasible to estimate its total mass. Now such irregularities in the earth's crust affect the "down" direction; a plumb bob suspended near a large mountain is pulled on dominantly by the massive earth itself, but the mass of the mountain is enough to attract the bob slightly, and hence to draw it a bit off vertical (Figure 4.9).

Maskelyne measured the angle by which Mount Schiehallion deflected a plumb bob suspended at two different places, one on the north slope, and the other on the south slope, by sighting the direction of its supporting cord against distant stars. He found that the mountain deflected the bob by 5″.63 from the vertical, and thus determined the pull of Schiehallion on the bob compared with the pull of the earth.

The strength of the pull on the bob by the mountain depends on the mountain's distance and mass, and that of the earth depends on its mass and the distance from the surface to the center of the earth (that is, the earth's radius). The distances involved are known, so by estimating the mass of the mountain one obtains enough information to calculate the mass of the earth (or G). Surveys of Schiehallion and samplings of its surface rocks were carried out and refined at various times, yielding fair estimates of the earth's mass, but the difficulty of knowing the precise density of the mountain's inte-

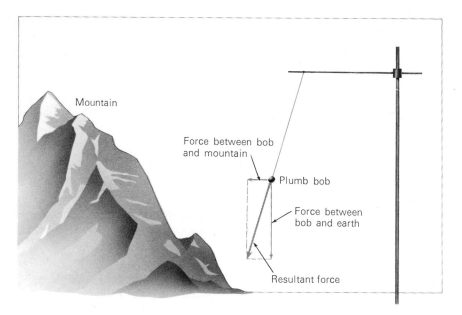

Figure 4.9 Deflection of a plumb bob from the vertical by a nearby mountain.

rior makes the method necessarily crude. Far more accurate were later techniques that could be carried out entirely within the laboratory.

One such procedure is with the use of a delicate balance. This method was applied by P. von Jolly in Munich in 1881. The principle of the balance method is shown in Figure 4.10.

Two equal weights, A and A', are placed in the two pans of the balance. Both weights have the same mass and are the same distance from the center of the earth, so the earth has the same attraction for both, and the system is in perfect balance. A very large mass B is then placed under the pan with the mass A, with its center only a small distance d from A. Now, in addition to the force between A and the earth, there is a small gravitational force between A and B. However, B is far enough from the weight A' in the other pan so that the attraction between B and A' is negligible. Thus A feels a greater downward force than A', and the balance is upset.

To restore the balance, a very small weight, C, is placed in the pan with A'. Now, the attraction between the earth and C is just enough to counteract the attraction between weight A and the large ball, B, and the pans hang evenly again. Since the system is again in perfect balance, the downward force on both pans must be the same. We let M_E and R be the mass and radius of the earth and $M_A = M_{A'}$, M_B, and M_C be the masses of the weights A, B, and C. The total downward

force on the left pan, the force between A and the earth *plus* the force between A and B, is

$$G\frac{M_E M_A}{R^2} + G\frac{M_A M_B}{d^2},$$

and the total downward force on the right pan, the force between A' and the earth *plus* the force between C and the earth, is

$$G\frac{M_E M_{A'}}{R^2} + G\frac{M_E M_C}{R^2}.$$

If we now equate these two forces (and remember that $M_A = M_{A'}$), we obtain

$$G\frac{M_E M_A}{R^2} + G\frac{M_A M_B}{d^2} = G\frac{M_E M_A}{R^2} + G\frac{M_E M_C}{R^2}.$$

After canceling out the common terms, and solving for M_E, we find

$$M_E = \frac{R^2}{d^2}\frac{M_A M_B}{M_C}.$$

Since all the masses and distances on the right side of the last equation can be measured directly, the mass of the earth can be calculated.

A more accurate modern method of "weighing" the earth is with a torsion balance. In a *torsion*

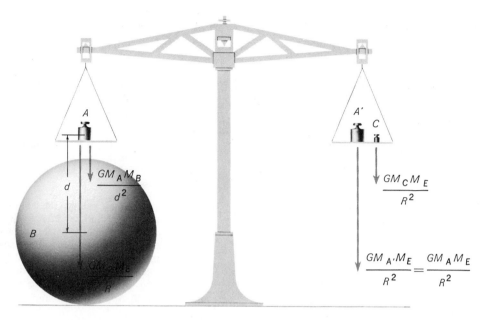

Figure 4.10 "Weighing" the earth with a balance.

balance two spheres are connected by a horizontal light rod that is suspended in the middle by a thin quartz fiber. Then another large mass is placed near one of the spheres. The gravitational force between the large mass and the sphere causes the sphere to move toward it, twisting the quartz fiber. The amount of twist, or torsion, in the fiber is an ac-

curate measure of the force between the two gravitating masses. This force can then be compared to that between an object and the earth to obtain the earth's mass.

The results of the best determinations of the mass of the earth give 5.98×10^{27} g, or about 6×10^{21} tons.

EXERCISES

1. What is the momentum of a body whose velocity is zero? Does the first law of motion (in the absence of a force the momentum of a body is constant) include the case of a body at rest?

2. Explain how a quantity of lead could be less massive than a quantity of feathers. How could a certain mass of lead be less dense than an equal mass of feathers?

3. If 24 g of material fills a cube 2 cm on a side, what is the density of the material?

4. How many accelerators are there in a standard passenger car? Explain.

5. A force of three dynes acts in a direction perpendicular to another force of four dynes. What is the magnitude of the resulting force?

6. Why is it nonsense to speak of the "force of forward motion"?

7. Suppose two billiard balls collide. Ignoring frictional forces, describe their velocities and accelerations.

8. Sometimes a cue ball hits another billiard ball and after the impact the other ball moves quickly forward while the cue ball stands motionless or moves only slightly. How can this be explained in terms of Newton's third law?

9. A body moves in a perfectly circular path at constant speed. What can be said about the presence or absence of forces in such a system?

10. Suppose you tie a string to a stone and whirl it around over you head at the end of the string. Now you hold up a razor blade directly in front of your face so that you cut the string just as the stone is out at the end of the string directly in front of you. Ignore the pull of the earth's gravity and the friction of the air.

(a) Just after the string is cut, what forces are acting on the stone?

(b) Which way does the stone move after the string is cut? Why?

11. What does a pan balance measure? A spring balance?

12. Suppose a 10-g ball and a 1000-g ball are simultaneously dropped from the top of a building. (a) What is the ratio of the weights of the two objects? (b) Explain why they would strike the ground at the same time. (Neglect air resistance.)

13. Did Newton need to know the size of the earth in kilometers to test his theory of gravitation with the apple and moon? Why or why not?

*14. Calculate the gravitational attraction between a man weighing 100 kg and a woman weighing 50 kg 10 m away from him. Compare this to the attraction between the woman and the earth. (For the purpose of this calculation, assume the man and woman to be perfect spheres.)

15. How much would a 100 kg man weigh at 6400 km (about one earth radius) above the surface of the earth? How much would he weigh at 25,600 km above the surface of the earth?
Answer to second part: about 4 kg

16. By what factor would a man's weight at the surface of the earth be reduced if (a) the earth had its present mass, but eight times its present volume? (b) the earth had its present size but only one-third its present mass?

17. At some point along a line connecting the earth and the moon a mass would experience equal but opposite forces due to the gravitational attractions of those two bodies. Verify that the distance of this point from the earth is about 346,000 km by calculating the acceleration at this point due to the earth and showing that it is the same as that due to the moon. Consult Appendices 10 and 11.

Pierre Simon, Marquis de Laplace (1749–1827), French mathematician and astronomer, developed many of the fundamental techniques in celestial mechanics and is responsible for the famous nebular hypothesis of the origin of the solar system—close to our modern view. (*Yerkes Observatory*)

THE TWO-BODY PROBLEM

Newton's laws of motion and gravitation enable us to predict the motions of bodies under the influence of their mutual gravitation. In this chapter we shall consider the interactions of two bodies, both of which are either point masses or spherically symmetrical, so that they act (gravitationally) as point masses. The subject is called the *two-body problem*.

5.1 CENTER OF MASS

According to Newton's third law, the total momentum of an isolated system is conserved; that is, all changes of momentum within it are balanced. We can, therefore, define a point within the system that remains fixed (or moves uniformly) as if the entire mass of the system were concentrated at that point. It is called the *center of mass*. It can be shown that the center of mass of a complex body (which is a collection of point masses joined rigidly together) is that point at which the body balances when placed near a gravitating body; thus it is also often called the *center of gravity*.

The center of mass (or gravity) for two bodies is given a special name: the *barycenter*. It must lie on a line connecting the centers of the bodies. We

now derive the location of the barycenter relative to the bodies. For simplicity we shall assume that each body revolves about it in a circular orbit. With somewhat more advanced mathematics, we would find that the result we derive is correct for any kind of motion of the two bodies—as long as they are acted on only by mutual forces between them.

Let the two bodies, of masses m_1 and m_2, revolve about and on opposite sides of the point O on the line between their centers (Figure 5.1). The distances of the two bodies from O are r_1 and r_2. To accelerate body 1 into a circular orbit, the force upon it must be $m_1 v_1^2/r_1$. The force on body 2 is $m_2 v_2^2/r_2$. As in Section 4.3, we can write the orbital speed of each body in terms of its period of revolution and its distance from the center of its orbit, that is,

$$v_1 = \frac{2\pi r_1}{P} \quad \text{and} \quad v_2 = \frac{2\pi r_2}{P}.$$

Of course, both bodies have the same period—the period of their mutual revolution.

The same force—the gravitational attraction between the bodies—produces the centripetal accelera-

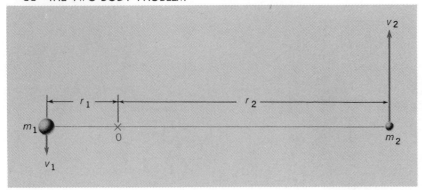

Figure 5.1 Center of mass; the two bodies, 1 and 2, mutually revolve about point O.

tion of each. Therefore, the centripetal force acting on each of them must be the same. Thus we can write

$$\frac{m_1 v_1^2}{r_1} = \frac{m_2 v_2^2}{r_2},$$

or

$$\frac{m_1 4\pi^2 r_1^2}{P^2 r_1} = \frac{m_2 4\pi^2 r_2^2}{P^2 r_2}.$$

After cancellation of common factors, the above equation becomes

$$\frac{r_1}{r_2} = \frac{m_2}{m_1}.$$

We see that the point of mutual revolution of two bodies on a line between them is located so that the distance of each body from that point is in inverse proportion to its mass.

The concept of center of mass reminds us of two boys on a seesaw. For the seesaw to balance properly, the fulcrum (support point) must be located proportionately closer to the boy of greater mass. Similarly, the barycenter of two revolving bodies lies proportionately closer to the body of greater mass.

Because of the sun's far greater mass, the center of mass of a system of revolving bodies consisting of the sun and a planet lies very close to the center of the sun—in most cases, within the sun's surface.

(a) Relative Orbits

As two bodies revolve mutually about their barycenter, they must always maintain the same relative distances from it; if one body doubles its distance so must the other, and if one comes closer, the other must lessen its distance by a proportional amount. We see, then, that the orbits of the two bodies are similar to each other (that is, are the same shape), the sizes of the orbits being in inverse proportion to the masses of the bodies.

The similarity of the two orbits makes it easy to refer to the orbit of one object relative to the other. It is conventional to choose the center of the more massive body as reference. Of course neither body is actually fixed—both move about the barycenter—but it is often convenient to *regard* one (the more massive) as origin, and consider the motion of the other with respect to it. This is called the *relative orbit*. The earth and moon, for example, revolve about their barycenter on a line between their centers. The earth is *not* at the center of the moon's orbit, but we can and often do speak of the relative orbit of the moon about the earth's center.

The size of the relative orbit is the sum of the sizes of the individual orbits about the barycenter. If the bodies move in circular paths (Figure 5.2), the relative orbit is a circle with a radius equal to the sum of the distances of the two bodies from the barycenter (it is also equal to the distance between the bodies). If the orbits are ellipses, the relative orbit is an ellipse of major axis equal to the sum of the major axes of the elliptical orbits of the two bodies; all three ellipses have the same eccentricity. The speed of the less massive object in the relative orbit (its speed with respect to the more massive body) is the sum of the speeds of the two individual bodies about their barycenter.

We now turn to the motions of the planets, satellites, and other celestial objects.

5.2 ORBITAL MOTION EXPLAINED

If an object near the earth's surface is dropped toward the ground, at the end of one second it has

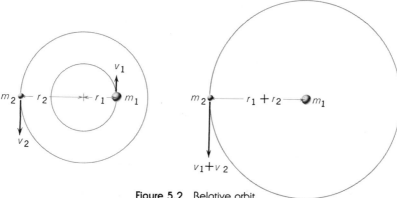

Figure 5.2 Relative orbit.

accelerated to a speed of 980 cm/s. Its average speed during that second is 490 cm/s. Thus it drops, in one second, through a distance of 490 cm. The moon, which accelerates only 1/3600 as much, drops toward the earth only about 490/3600 cm in one second, or about 0.14 cm. In other words, as the moon moves forward in its orbit for one second and travels about 1 km it falls 0.14 cm toward the earth. However, because of the earth's curvature, the ground has fallen away under the moon by that same distance of 0.14 cm, so the moon is still the same distance from the earth. In this way the moon literally "falls around the earth." In the period of about one month, it has "fallen" through one complete circuit of the earth and is back to its starting point (Figure 5.3).

Orbital motion is thus easily understood in terms of Newton's laws of motion and gravitation. At any given instant, the orbital speed of the moon would tend to carry it off in space in a straight line tangent to its orbit. The gravitational attraction between the earth and the moon provides the proper centripetal force to accelerate the moon into its nearly circular path. Consequently, the moon continually falls toward the earth without getting any closer to it. The orbital motions of the planets are similarly explained.

(a) The Fable of "Centrifugal Force"

Orbital motion is often incorrectly described as being a balance between two forces, the force of gravitation pulling the moon toward the earth and an outward "centrifugal force" that keeps the moon from falling into the earth. But if the moon were in a balance between two forces, it either would not

move at all or would move in a straight line, not in a nearly circular orbit.

If you whirl a rock around over your head at the end of a string, you feel a "tug" in the string. This tug is the tension in the string. It is the result of the mutual pull between your hand and the stone. You provide, through the string, the centripetal force needed to accelerate the stone into a circular path. The stone tugs with a mutual force on your hand. However, there is no outside force pulling the stone away from you. If there were, and the string were to break, removing the central or centripetal force, the stone would move radially away from you. Actually, if the string broke, the stone would fly off in a straight line in a direction *tangent* to its former circular path, continuing in the direction it was moving at the instant the string broke, in accord with Newton's first law of motion.

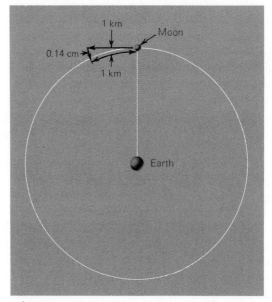

Figure 5.3 Velocity and acceleration of the moon.

As a further example, if you round a corner in a rapidly moving car you feel your body being shoved against the side of the car on the outside of the curve. The car has pushed against you to accelerate you into a curved path. The force you feel is the mutual force between you and the car that provides your centripetal acceleration. If the side of the car were to dissolve suddenly, you would continue moving *forward* in the direction you were going at that instant, and would not fly out at right angles to the side of the car.

In short, there is no such thing as centrifugal (or outward) force in orbital motion. No force is pulling outward on the moon to balance the attraction between the moon and earth. It is, in fact, just the *unbalanced* gravitational force that causes the moon to move in its nearly circular path around the earth. Of course, this discussion of the moon's motion applies equally to the planets' motions about the sun. (A "fictitious" centrifugal force, however, which arises from the introduction of a rotating coordinate system, is sometimes useful in the solution of more advanced problems in mechanics.)

(b) The Mutual Forces and the Moon's Mass

Some authors explain that "centrifugal force" is a reaction force to gravity, predicted by Newton's third law. This too is incorrect. The reaction to the earth's gravitational force on the moon is the moon's equal but opposite gravitational attraction for the earth. In other words, both pull mutually and centrally *toward* each other; there is no *outward* "reaction" force. The earth and moon suffer the same change of momentum as they revolve around each other, but the earth, being much more massive than the moon, accelerates far less. The center of mass about which they mutually revolve can be located by making careful observations of the other planets—especially of Mars—and, more recently, by tracking space probes. Small monthly periodic variations in the apparent motion of Mars result from the earth's monthly revolution about the barycenter. From the size of those variations we find that the barycenter is about 4700 km from the center of the earth, or about 1700 km below its surface. The moon is about 81 times as far from the barycenter and so is correspondingly less massive than the earth (Chapter 9).

5.3 NEWTON'S DERIVATION OF KEPLER'S LAWS

Kepler's laws of planetary motion are *empirical* laws, that is, they describe the way the planets are *observed* to behave. Kepler himself did not know of the more fundamental laws or relationships from which his three laws of planetary motion follow. On the other hand, Newton's laws of motion and gravitation were proposed by him as the basis of all mechanics. Thus it should be possible to derive Kepler's laws from them. Newton did so. In fact, as we saw in the previous chapter, it was Newton's derivation of the shape of the orbit of an object moving under the influence of an inverse-square force that had so astonished Halley. Today, the principal diagram Newton used in the proof (from the *Principia*) is reproduced on the British one-pound note (Fig. 5.4).

(a) KEPLER'S FIRST LAW

Consider a planet of mass m_p at a distance r from the sun moving with a speed v in a direction at right angles to the line from the planet to the sun. The centripetal force needed to keep the planet in a *circular* orbit, that is, at constant distance from the sun, is

$$\text{force} = \frac{m_p v^2}{r}.$$

Now suppose the gravitational force between the planet and the sun happens to be greater than the force given by the above equation. Then the planet will receive more acceleration than is necessary to keep it in a circular orbit, and it will move in somewhat closer to the sun. As it does so, its speed will increase, just as the speed of a falling stone increases as it approaches the ground.

Because of the planet's increased speed and decreased distance from the sun, a greater centripetal force is required to keep it at a constant distance from the sun. Eventually, as the planet continues to sweep in closer to the sun at higher and higher speed, a point will be reached at which the gravitational force between the two is no longer sufficient to produce enough centripetal acceleration to keep the planet from moving out away from the sun. Thus the planet will move outward as it rounds the sun until it has

Figure 5.4 Back side of a British one-pound note.

reached a position where the gravitational acceleration is again greater than the circular centripetal acceleration, and the process is repeated.

If the situation were reversed, and the planet were moving fast enough for the centripetal force required for circular motion to be greater than the gravitational attraction, the planet would move outward and consequently slow down until the gravitational force could pull it back again.

Thus we see, qualitatively, how a planet may follow an elliptical orbit. If, however, a planet had a high enough speed, the gravitational force between it and the sun might never be enough to provide sufficient centripetal force to hold the planet in the solar system, and the planet would move off into space. Its orbit would then be a *hyperbola* rather than a closed, elliptical path (Figure 5.5). There is a certain critical speed, which depends on the planet's distance from the sun, at which the planet can just barely escape the solar system along a *parabolic* orbit. This critical speed

is called the *parabolic velocity*, or the *velocity of escape*.

To prove rigorously that the gravitational force between the sun and a planet must result in an orbit for the planet that is either a circle, ellipse, parabola, or hyperbola is beyond the power of elementary algebra. Newton, in solving the problem, made use of his new *fluxions*, which we now know as differential calculus. He showed, in fact, that the gravitational interaction between *any* two bodies would result in an orbital motion of each body about the other that is some form of a *conic section* (see Section 3.3).

Circular and parabolic orbits require theoretically precise speeds that would not be expected to occur in nature; thus we would not expect to find a planet (or other object) with *exactly* a circular or parabolic orbit. The latter divides the family of elliptical (closed) from the family of hyperbolic (open) orbits that actually do occur. We shall see what conditions determine the type of orbit in Section 5.4. The planets, of course, do not have hyperbolic orbits or they would long since have receded into interstellar space; their orbits, then, must be elliptical, as found by Kepler.

(b) KEPLER'S SECOND LAW

In the preceding paragraphs we saw how a planet speeds up as it approaches the sun and slows down as it pulls away, in qualitative agreement with Kepler's second law. Newton derived the second law rigorously with a simple geometrical proof of this law of equal areas, which we repeat here.

Consider a planet at *A* revolving about the sun at *S* (Figure 5.6). In a short interval of time, the planet's

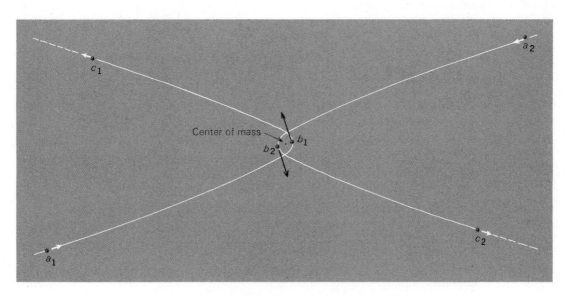

Figure 5.5 Relative hyperbolic orbits.

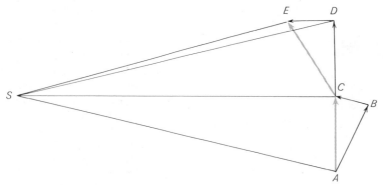

Figure 5.6 Geometrical proof of the law of areas.

forward velocity would ordinarily carry it to *B*. However, the gravitational pull between it and the sun accelerates it to *C*. Since we are considering a very brief interval of time, we can regard the acceleration of the planet as being along a direction *BC*, parallel to *AS*, the direction from the planet to the sun at the beginning of the instant. The planet now has a velocity along the direction *AC*. In the next brief interval of time, equal in length to the first interval, the planet would ordinarily continue moving in a straight line at a constant speed and would end up at *D*, along the extension of *AC*, so that the distance *CD* was equal to the distance *AC*. However, again the sun accelerates the planet toward it (now in direction *CS*), so the planet actually moves along *CE*.

Consider *AC* and *CD* to be the bases of the triangles *ASC* and *CSD*, respectively. Since *AC = CD*, the two triangles have equal bases. They also have the same altitude—the perpendicular distance of *S* from *AD* or its extension. Thus triangles *ASC* and *CSD*, having equal bases and altitudes, have equal areas.

Now *SC* is a common base of triangles *SEC* and *SDC*. Those triangles also have equal altitudes, the distance between the parallel lines *SC* and *ED*. Thus triangles *SEC* and *SDC* are equal in area.

Because triangles *ASC* and *CSE* are both equal in area to triangle *SCD*, they are equal in area to each other. These are the areas swept out by a line from the planet to the sun in two successive equal intervals of time. Many such brief intervals of time can be combined to show that the areas swept out in *any* two equal intervals of time are equal. Thus Kepler's second law is verified.

The argument given is rigorous only if the brief time intervals are infinitesimal. But we can imagine triangles *ASC* and *CSE* to be as small in area as we like, and the proof is still valid. Here is another example in which Newton employed differential calculus, letting time intervals approach zero.

In fact, we did not need to restrict ourselves to a planet moving around the sun. If any two objects revolve about each other under the influence of a central force, the law of equal areas will apply. The law of areas, as we have seen (Section 4.1h), is the geometrical manifestation of the conservation of angular momentum.

(c) KEPLER'S THIRD LAW

We may now use the ideas we have developed to demonstrate a simple derivation of Kepler's third law, but again only for circular orbits. Newton derived the same form of the law for the more general elliptical orbits.

For each of two mutually revolving bodies, the gravitational attraction between the two provides the centripetal acceleration to keep them in circular orbits. If the bodies have masses m_1 and m_2 and distances r_1 and r_2 from their common center of mass, they are separated by a distance $r_1 + r_2$ and we can equate the gravitational force to the centripetal force for each body. Equating the formulas derived in Sections 4.2 and 5.1 for centripetal force, we obtain for body 1,

$$\frac{Gm_1m_2}{(r_1 + r_2)^2} = \frac{m_1 4\pi^2 r_1}{P^2},$$

and for body 2,

$$\frac{Gm_1m_2}{(r_1 + r_2)^2} = \frac{m_2 4\pi^2 r_2}{P^2}.$$

If we cancel out the masses common to each side of each equation and add the two equations, we obtain

$$\frac{G(m_1 + m_2)}{(r_1 + r_2)^2} = \frac{4\pi^2}{P^2}(r_1 + r_2),$$

or

$$(m_1 + m_2)P^2 = \frac{4\pi^2}{G}(r_1 + r_2)^3.$$

Since $r_1 + r_2$ is the distance between the two bodies, we recognize it, in the case of a planet going around the sun in a circular orbit, as the semimajor axis, a, of the relative orbit. Then the above equation looks the same as the formula we gave in Chapter 3 for Kepler's third law, except for the factor $m_1 + m_2$ and the factor $4\pi^2/G$. The latter is simply a constant of proportionality. If the proper units are chosen for distance and time, G will take on such a value that $4\pi^2/G$ will equal unity. We shall see in a moment why Kepler was unaware of the factor $(m_1 + m_2)$.

First, let us illustrate two systems of units such that the constant of proportionality $4\pi^2/G = 1$. One such system is to measure the sum of the masses of the revolving bodies, $m_1 + m_2$, in units of the combined mass of the sun and earth, the period in years, and the separation of the bodies in astronomical units. Then if the equation is applied to the mutual revolution of the earth and sun, everything in the equation other than the factor $4\pi^2/G$ is equal to unity, so it must be unity also. It is easy to see that another suitable choice of units is the combined mass of the earth and the moon for the sum of the masses, the sidereal month (the moon's period of revolution about the earth) for the period, and the moon's mean distance from earth for the separation of the bodies.

Newton derived his equation not only for the planets moving about the sun but also for any pair of mutually revolving bodies—two stars, a planet and a satellite, or even a plate and a spoon revolving about each other in space.

Newton's version of Kepler's third law differs from the original in that it contains as a factor the sum of the masses of the two revolving bodies. It will become clear why Kepler was not aware of that term if we note that we can consider the sun and earth to be a pair of mutually revolving bodies. We shall see in Chapter 31 that the sun has a mass of about 330,000 times that of the earth. Thus the combined mass of the sun and the earth is, to all intents and purposes, the mass of the sun itself, the earth's mass being negligible in comparison. Suppose we choose the mass of the sun for our unit of mass. Then, in the earth-sun system, $m_1 + m_2 = 1$. Furthermore, the sum of the masses of the sun and any other planet is also very nearly unity. Even Jupiter, the most massive planet, has only 1/1000 of the mass of the sun; for the sun and Jupiter, $m_1 + m_2 = 1.001$, a number so nearly equal to 1.000 that Kepler was unable to detect the difference from Tycho's observations. (However, the fact that the masses of Jupiter and Saturn are not completely negligible compared to the sun accounts, in part, for the slight discrepancies in Kepler's version of his third law as applied to Jupiter and Saturn; see Table 3.2.) Thus, if we apply the equation Newton derived to the mutual revolution of the sun and a planet, and choose years and astronomical units for the units of time and distance, and the solar mass for the unit of mass, Newton's equation reduces to

$$(m_1 + m_2)P^2 = (1)P^2 = P^2 = a^3,$$

in agreement with Kepler's formulation of the law.

(d) Kepler's Laws Restated— Summary

We now restate Kepler's three laws of planetary motion in their more general form, as they were derived by Newton:

KEPLER'S FIRST LAW: If two bodies interact gravitationally, each will describe an orbit that is a conic section about the common center of mass of the pair. In particular, if the bodies are permanently associated, their orbits will be ellipses. If they are not permanently associated, their orbits will be hyperbolas.

KEPLER'S SECOND LAW: If two bodies revolve about each other under the influence of a central force (whether or not in a closed elliptical orbit), a line joining them sweeps out equal areas in the orbital plane in equal intervals of time.

KEPLER'S THIRD LAW: If two bodies revolve mutually about each other, the sum of their masses times the square of their period of mutual revolution is in proportion to the cube of the semi-major axis of the relative orbit of one about the other.

In metric units, the algebraic formulation of Newton's version of Kepler's third law is

$$(m_1 + m_2)P^2 = \frac{4\pi^2}{G}a^3,$$

where G is the constant of gravitation. (If the units of length, mass, and time are centimeters, grams, and seconds, respectively, G has the value 6.67×10^{-8}.) If either of the sets of units shown in Table 5.1 is used, the law becomes

$$(m_1 + m_2)P^2 = a^3.$$

Table 5.1 **Examples of Systems of Units for Which $4\pi^2/G = 1$**

	I	II
Units of $(m_1 + m_2)$	Sun's mass + earth's mass	Earth's mass + moon's mass
Units of P	Sidereal year	Sidereal month
Units of a	Astronomical unit	Mean distance of moon from earth

5.4 ENERGY OF A TWO-BODY SYSTEM—THE *VIS VIVA* EQUATION

It can be shown that for any two mutually revolving bodies of masses m_1 and m_2 with a relative orbit of semimajor axis a, the magnitude of the velocity, v, of one body with respect to the other at an instant when the bodies are at a distance r apart is given by the equation

$$v^2 = G(m_1 + m_2)\left(\frac{2}{r} - \frac{1}{a}\right).$$

This equation is called the *vis viva equation,* or, because it expresses the conservation of energy of the system, the *energy equation.* If the relative orbit of one body around the other is a circle, $r = a$, and the equation gives for the *circular velocity*

$$v^2 = G(m_1 + m_2)\frac{1}{r}.$$

If the relative orbit is a parabola, the bodies escape from each other, and the *vis viva* equation gives for the *parabolic* or *escape velocity,*

$$v^2 = G(m_1 + m_2)\frac{2}{r},$$

because a parabolic orbit can be considered an ellipse of eccentricity 1, with $a = \infty$ (infinity). Note that the velocity of escape is equal to the circular velocity times the square root of 2.

   If two bodies have the most minute sideways motion with respect to each other, they cannot fall straight toward each other but will move in elliptical orbits about each other. If that sideways velocity is just great enough so that the centripetal force required for a circular orbit is exactly equal to the bodies' mutual gravitational attractive force, they will move about each other in circular orbits. This critical speed is the *circular velocity* given by the second to last equation above. A still higher sideways motion will produce elliptical orbits of larger major axes than the diameters of the circular orbits. A sideways velocity equal to the velocity of escape of one body with respect to the other will result in parabolic orbits, and still higher velocities give hyperbolic orbits.

   The *vis viva* equation is quite general; the motion of one body with respect to the other need not be sideways. If v is the speed of one body with respect to the other, whatever the direction, the equation gives the corresponding value of a, the semimajor axis of the relative orbit. For any closed orbit (circle or ellipse),

a must be positive and finite. A value of v greater than the parabolic velocity results in open or hyperbolic orbits. The semimajor axis of a hyperbolic orbit is taken as negative.

   It is important to note that if two objects approach each other from a great distance in space they can never "capture" each other into elliptical orbits. Their mutual attraction will speed them up so that they pass each other with a relative speed greater than their mutual velocity of escape, and they will swing out away from each other again, moving in orbits that are hyperbolas (Figure 5.5). As an example, it is impossible to send a rocket to the moon, and to cause it to move on an elliptical orbit about the moon, without slowing it down when it is in the lunar vicinity. That is why a rocket intended for lunar orbit carries a *retro-rocket* designed to reduce its speed at an appropriate time so that it can enter an elliptical orbit about the moon. Otherwise, it would bypass the moon on a hyperbolic orbit.

5.5 MASSES OF PLANETS AND STARS

Our only means of measuring the masses of astronomical bodies is to study the way in which they react gravitationally with other bodies. Newton's derivation of Kepler's third law, which includes a term involving the sum of the masses of the revolving bodies, is most useful for this purpose.

   Consider a planet, such as Jupiter, that has one or more satellites revolving about it. We can select one of those satellites and regard it and its parent planet as a pair of mutually revolving bodies. We measure the period of revolution of the satellite (say, in sidereal months) and the distance of the satellite from the planet (in terms of the distance of the moon from the earth), and insert those values into the equation

$$m_1 + m_2 = \frac{a^3}{P^2}.$$

Since both a and P are observed, we can immediately calculate the combined mass of the planet and its satellite. Obviously most of this mass belongs to the planet, its satellites all being very small compared to it. Thus $m_1 + m_2$ is, essentially, the mass of the planet in terms of the mass of the earth.

As a numerical example, Deimos, the outermost satellite of Mars, has a sidereal period of 1.262 days and a mean distance from the center of Mars of 23,500 km. In sidereal months, the period of the satellite is 1.262/27.3 = 0.0462. In units of the distance of the moon from the earth, Deimos has a distance from the center of Mars of 23,500/384,404 = 0.0611. Thus the mass of Mars plus the mass of Deimos is given by

$$m_{\text{Mars}} + m_{\text{Deimos}} = \frac{(0.0611)^3}{(0.0462)^2} = \frac{2.28 \times 10^{-4}}{2.13 \times 10^{-3}}$$

$$= 0.11 \text{ earth mass.}$$

Since Deimos is a very tiny satellite (only about 13 km across), its mass can be neglected compared with that of Mars, and we find that Mars has a mass of just over one tenth that of the earth.

In Chapter 26 we shall see that we use the same mathematical technique to determine the masses of stars that are members of binary-star systems (a binary star is a pair of stars that revolve around each other). In fact, we can use Newton's version of Kepler's third law to estimate the mass of our entire Galaxy (Chapter 29) or even of other galaxies (Chapter 37).

5.6 ORBITS OF PLANETS

A classic problem in celestial mechanics is to compute the orbit of a planet (or minor planet) from observations of its directions at various times as seen from the earth. We have seen (Section 3.3a) how Kepler determined the orbit of Mars geometrically. With the additional knowledge of Newton's laws of motion and gravitation, it is possible to find the orbit of a planet with far fewer observational data. In 1801 Karl Friedrich Gauss (1777–1855) invented a method of determining the orbit of an object moving around the sun from observations of it extending over only a few weeks (Chapter 19). Since Gauss's time various mathematical techniques have been developed to handle the same problem. The complication is that we observe the other planets from a moving earth. Accurate positions in the sky of the object on each of at least three different dates, preferably separated by more than a week, must be known to calculate its detailed path around the sun. The orbit, when finally determined, is usually designated by a set of quantities known as its elements.

(a) ELEMENTS OF ORBITS

The orbit of a planet or some other body moving about the sun can be specified uniquely by six items of information, or *elements*.

Two elements are needed to describe the size and shape of an orbit. We have already seen, for example (Section 3.3a), that the size and shape of an elliptical orbit can be specified by the semimajor axis and eccentricity of the ellipse. In fact, the same two quantities serve to specify the size and shape of *any* conic section, and hence of any orbit. Three other data are required to specify the orientation of the orbit with respect to some reference system, say, one defined by the earth's orbit. A final element is needed to specify where the object is in its orbit at some particular time, so that its location at other times can be computed. A total of six such orbital elements are sufficient if the object is in orbit around the sun and if it has a mass that can be neglected in comparison with the sun's mass. If, however, the sum of the masses of two mutually revolving bodies is not known, a seventh datum is needed to specify their orbit completely. If the relative orbit is an ellipse, the period of mutual revolution of the bodies suffices for this seventh element. If it is a hyperbola, the *areal velocity* replaces the period. The areal velocity is the rate at which an imaginary line between the two bodies sweeps out an area in space with respect to one of the bodies.

The six (or seven) orbital elements can be specified in a multitude of ways. In Table 5.2 is summarized the set of elements that is most conventional for describing the orbit of an object revolving about the sun. It must be emphasized, however, that other sets of data can be used for the elements of an orbit and, indeed, often are used in modern practice. The elements described in Table 5.2 are illustrated in Figure 5.7.

If the set of elements given in Table 5.2 is used, the inclination and longitude of the ascending node, i and Ω, describe the orientation of the orbital plane. The argument of perihelion, ω, gives the orientation of the orbit in its plane. The semimajor axis a and eccentricity e give the size and form of the orbit. The time of perihelion passage T and period P are the data required to calculate the position of the object in its orbit. If the object is one of small mass circling the sun, the period is superfluous, for it can be obtained from the semimajor axis with the use of Kepler's third law. If the orbit is not an ellipse, the areal velocity rather than the period can be used.

Some of the elements of the orbits of the planets in the solar system are given in Appendix 9.

Table 5.2 Elements of an Orbit

NAME	SYMBOL	DEFINITION
Semimajor axis	a	Half of the distance between the points nearest the foci on the conic that represents the orbit (usually measured in astronomical units).
Eccentricity	e	Distance between the foci of the conic divided by the major axis.
Inclination	i	Angle of intersection between the orbital planes of the object and of the earth.
Longitude of the ascending node	Ω	Angle from the vernal equinox (where the ecliptic and celestial equator intersect with the sun crossing the equator from south to north), measured to the east along the ecliptic plane, to the point where the object crosses the ecliptic traveling from south to north (the ascending node).
Argument of perihelion	ω	Angle from the ascending node, measured in the plane of the object's orbit and in the direction of its motion, to the perihelion point (its closest approach to the sun).
Time of perihelion passage	T	One of the precise times that the object passed the perihelion point.
Period	P	The sidereal period of revolution of the object about the sun.

5.7 ARTIFICIAL SATELLITES

An artificial satellite is a manmade object that is in orbit around the earth or some other astronomical body. It is an astronomical body in its own right. If most of the artificial satellites that have been launched are temporary astronomical objects, it is because they dip into the atmosphere of the earth during some portions of their revolutions. The friction of the air causes a satellite to lose energy so that eventually it spirals into the denser part of the atmosphere where friction heats it until it burns up completely. If an artificial satellite is launched so that its entire orbit is outside the earth's atmosphere, it will remain in orbit indefinitely as an astronomical body.

To illustrate how a satellite is launched, imagine a man on top of a high mountain, firing a rifle in a direction exactly parallel to the surface of the earth (Figure 5.8—adapted from a similar diagram by Newton—Figure 5.9). Imagine, further, that the friction of the air could be removed, and that all hindering objects, such as other mountains, buildings, and so on, are absent. Then the only force that acts on the bullet after it leaves the muzzle of the rifle is the gravitational force between the bullet and earth.

If the bullet is fired with muzzle velocity v_a, it will continue to have that forward speed, but meanwhile the gravitational force acting upon it will ac-

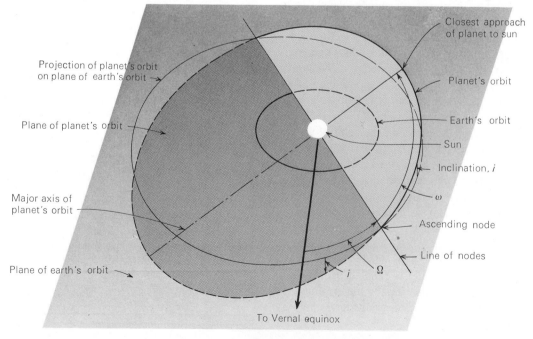

Figure 5.7 Elements of an orbit.

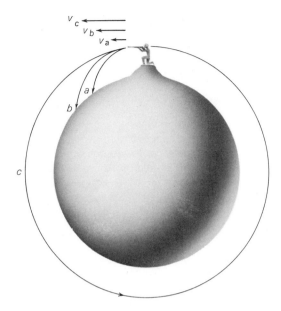

Figure 5.8 Firing a bullet into a satellite orbit.

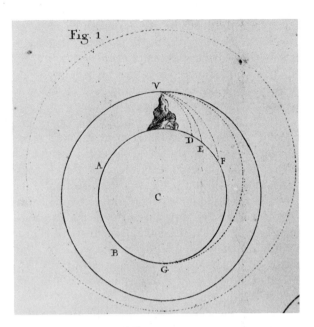

Figure 5.9 A diagram by Newton in his *De mundi system-atic*, 1731 Edition. (*Crawford Collection, Royal Observatory Edinburgh*)

celerate it downward so that it strikes the ground at *a*. However, if it is given a higher muzzle velocity v_b, its higher forward speed will carry it farther before it hits the ground, for, regardless of its forward speed, its downward gravitational acceleration is the same. Thus this faster-moving bullet will strike the ground at *b*. If the bullet is given a high enough muzzle velocity, v_c, as it accelerates toward the ground, the curved surface of the earth will cause the ground to tip out from under it so that it remains the same distance above the ground, and "falls around" the earth in a complete circle. This is another way of saying that at a critical speed v_c the gravitational force between the bullet and earth is just sufficient to produce the centripetal acceleration needed for a circular orbit about the earth. The speed v_c, the *circular satellite velocity* at the surface of the earth, is about 8 km/s.

Novelist Jules Verne anticipated earth satellites long ago. In one of his stories an enemy force was planning to bomb a city with a gigantic cannon ball. However, the cannon ball was propelled with too great a speed—in fact, the circular satellite velocity—so it passed harmlessly over the city and on into a circular orbit around the earth.

(a) Possible Satellite Orbits

Suppose that a missile is sent up to an altitude of a few hundred miles, then turned so that it is moving horizontally, and finally given a forward horizontal

thrust. It will proceed in an orbit the size and shape of which depend critically on the exact direction and speed of the missile at the instant of its "burnout," that is, the instant when the thrust supplied by its fuel is shut off. First, suppose that it is moving exactly horizontally, or parallel to the ground, at burnout. The possible kinds of orbits it can enter are shown in Figure 5.10.

If the missile's burnout speed is less than the circular satellite velocity, its orbit will be an el-

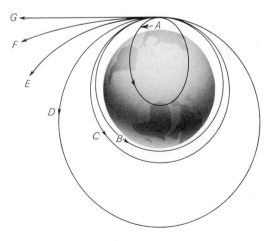

Figure 5.10 Various satellite orbits that result from different burnout velocities but that are all parallel to the earth's surface.

lipse, with the center of the earth at one focus of the ellipse. The *apogee* point of the orbit, that point that is *farthest* from the center of the earth, will be the point of burnout; the *perigee* point (closest approach to the center of the earth) will be halfway around the orbit from burnout.

If the burnout speed is substantially below the circular satellite velocity, most of its elliptical orbit will lie beneath the surface of the earth (orbit *A*), where, of course, the satellite cannot travel; consequently, it will traverse only a small section of its orbit before colliding with the surface of the earth (or more likely, burning up in the dense lower atmosphere of the earth). If the burnout speed is just slightly below the circular satellite velocity, the missile may clear the surface of the earth (orbit *B*), although its orbit will probably lie too low in the atmosphere for the satellite to be longlived.

If the burnout speed were exactly the circular satellite velocity, a circular orbit centered on the center of the earth would result (orbit *C*). It is extraordinarily unlikely that a missile could be given so accurate a direction and speed that a perfectly circular orbit could be achieved. A slightly greater burnout speed will produce an elliptical orbit with *perigee* at burnout point and apogee halfway around the orbit (orbit *D*).

A burnout speed equal to the velocity of escape from the earth's surface, that is, the parabolic velocity (about 11 km/s), will put the missile into a parabolic orbit that will just enable the vehicle to escape from the earth into space (orbit *E*). A still higher burnout speed will produce a hyperbolic orbit in which the missile escapes the earth with energy to spare (orbit *F*). The higher the burnout speed, the nearer will the orbit be to a straight line (orbit *G*).

We can apply the *vis viva* equation to the orbit of a satellite moving about the earth. Let us measure speed in terms of the circular-satellite velocity at the earth's surface, the masses in terms of the earth's mass, and r and a in units of the radius of the earth. In these units, the constant G is equal to unity, and the equation simplifies to

$$v^2 = \left(\frac{2}{r} - \frac{1}{a}\right).$$

Suppose a satellite is launched from a point near the earth's surface (say at an altitude of 300 km); r is

1.047 and v at that point is the burnout speed. Then the semimajor axis of the orbit, a (a measure of the size of the orbit), can easily be calculated if the burnout speed is known:

$$\frac{1}{a} = \frac{2}{r} - v^2.$$

Negative values of a correspond to hyperbolic orbits.

As an example, suppose the burnout speed is 10 km/s, or about 1.263 in units of the circular-satellite velocity. Then, we find for a

$$\frac{1}{a} = \frac{2}{1.047} - (1.263)^2 = 0.315,$$

or

$$a = 3.17 \text{ earth radii.}$$

Such a satellite would have an apogee distance of about 33,760 km from the center of the earth, or about 27,381 km above the surface.

The *vis viva* equation holds regardless of the direction the two bodies are moving with respect to each other. Note that there is no term in the equation for an earth satellite that involves the direction in which a missile is moving at burnout. Thus, even if the missile were not moving parallel to the ground at burnout, the major axis of its orbit would depend only on its burnout speed (see Figure 5.11). However, the *eccentricity*, or shape, of the orbit does depend on the direction of

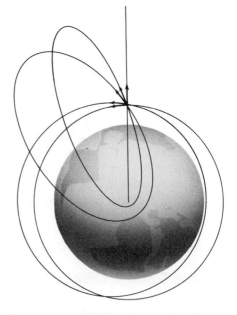

Figure 5.11 Various satellite orbits that result from the same burnout speed but in different directions. All these orbits have the same major axis.

motion of the missile. We see in Figure 5.11 that for a missile launched into a satellite orbit near the surface of the earth, unless the burnout direction is nearly parallel to the ground, the resulting orbit will be too eccentric to clear the surface of the earth.

Figure 5.12   Orbit of a ballistic missile.

(b)   Ballistic Missiles

A ballistic missile is a rocket or missile that is given an initial thrust and then allowed to coast in an orbit to its target. Such missiles can be considered earth satellites and temporary astronomical bodies.

A ballistic missile travels in an elliptical orbit with the center of the earth at one focus. Most of the orbit, however, lies beneath the surface of the earth, as in orbit A in Figure 5.10. One of the two intersections of the orbit with the earth's surface is at the launching point. The missile must have the correct burnout speed and direction so that the other intersection of the orbit with the surface of the earth will occur at the target. The missile, then, travels along that part of its orbit that lies outside the earth's surface until it collides with the earth at the calculated point (Figure 5.12).

The calculation of precise trajectories for ballistic missiles, as with regular earth satellites, is a complicated task. Account must be taken of various perturbations introduced by the slight asphericity of the earth (Chapter 6), the drag of the earth's atmosphere, and complicating effects due to the earth's rotation. However, the basic principles are quite simple, as we have seen, and were completely described by Newton.

(c)   First Artificial Satellites

For the International Geophysical Year 1957–1958, it was proposed to launch small earth satellites that would carry instruments to investigate the conditions just outside the earth's atmosphere.

The first successful launching of an artificial earth satellite, an important milestone in the history of human technology, was accomplished by the Soviet Union on October 4, 1957. This first Soviet satellite, called *Sputnik I* (the Russian word for "satellite"), had an overall weight of about four tons, and a scientific instrumentation package weighing about 80 kg. The instrument package and launching rocket traveled about the earth in a period of 96 minutes in an elliptical orbit that ranged

from 230 to 950 km above the earth's surface. Since Sputnik I was never completely outside the earth's atmosphere, it gradually lost energy, and finally burned up in the denser lower atmosphere on January 4, 1958. On November 3, almost a month after Sputnik I was put in orbit, the Russians launched a second satellite that survived in orbit until April 14, 1958. This second satellite was very similar to the first except that it had a heavier instrument payload and included a live dog.

The first successful American satellite was launched by an Army Jupiter C missile on January 31, 1958. By this time, the "space age" was well under way. Satellites carry instruments that radio back to earth data about our immediate space environment, make routine weather observations, carry out all manner of reconnaissance for military and other purposes, and, of course, are the means of what have now become routine communications. Ballistic rockets (which are satellites in a sense, as we have explained) carry instruments 150 or more kilometers above the earth's surface. One advantage of rockets that return to the ground is that instrument payloads can sometimes be recovered intact, so that photographs can be obtained of the earth, the sun, the solar spectrum, and so on.

5.8   INTERPLANETARY PROBES

We have now learned the principles of space travel. Rockets, once they have left the earth, are astronomical bodies. They obey the same laws of celestial mechanics as the planets and natural and artificial satellites. In other words, rockets or space probes travel in orbits. If the space vehicles carry auxiliary rocket engines and extra fuel, it may be possible to alter their orbits at will, but the principles remain the same.

We shall illustrate one particular kind of space trajectory by showing one of the many possible ways to reach each of the planets Mars and Venus. The orbits to Mars and Venus we show are those that require the expenditure of the least energy as the rocket leaves the earth and are thus the most economical of fuel. The orbits of the successful United States Mariner and Pioneer Venus probes, of the Mariner and Viking Mars probes, and of the similar Soviet probes, were all nearly of this type.

Suppose, for simplicity, that the orbits of Venus, Earth, and Mars are circles centered on the sun (when the slight ellipticity of planetary orbits is taken into account, the problem is similar but slightly more complicated). The least-energy orbit that will take us to Mars is an ellipse tangent to the earth's orbit at the space vehicle's *perihelion* (closest approach to the sun) and tangent to the orbit of Mars at the vehicle's *aphelion* (farthest point from the sun) (Figure 5.13).

The earth is traveling around the sun at the right speed for a circular orbit. For us on the earth to enter the elliptical orbit to Mars, we must achieve a speed, in the same direction as the earth is moving, that is slightly greater than the earth's circular velocity (which is about 29.8 km/s). To calculate this speed, we employ the *vis viva* equation. The major axis of the elliptical orbit we want to achieve is the sum of the radii of the orbits of the earth and Mars. Half of this major axis is the value a. The appropriate value of a is 1.26 AU. The value r is, of course, the earth's distance from the

sun, and $m_1 + m_2$ is the combined mass of the sun and the spaceship (the latter is negligible). The required speed turns out to be slightly under 33 km/s. Since the earth is already moving 29.8 km/s, we need to leave the earth with the proper speed and direction so that when we are far enough from it that its gravitational influence on us is negligible compared to the sun's, we are still moving in the same direction as the earth with a speed relative to it of about 3 km/s.

We have now entered an orbit that will carry us out to the orbit of Mars. The time required for the trip can be found from Kepler's third law, because our spaceship is a planet. The period required to traverse the entire orbit is $a^{3/2}$ years if a is measured in astronomical units. The entire period of the orbit is thus $(1.26)^{3/2} = 1.41$ years. The time required to reach the aphelion point (Mars' orbit) is half of this, or about 8½ months. The trip will have to be planned very carefully so that when we reach the aphelion point of the least-energy orbit, Mars will be there at the same time. Space probes generally carry rocket engines that can be activated by radio command from the earth, so that minor corrections in their trajectories can be made, as necessary, for them to achieve their missions.

The return trip from Mars to the earth is of the same type as the trip from the earth to Venus, to which we now turn our attention. The orbit to Venus is very similar to the orbit to Mars, except now it is at the *aphelion* point that the trajectory ellipse is tangent to the earth's orbit, and at the *perihelion* point that it is tangent to the orbit of Venus (Figure 5.14). The semimajor axis of this orbit is half the sum of the radii of the orbits of the earth and Venus, which is 0.86 AU. From the *vis viva* equation we find that the speed at the aphelion point in the orbit is 27.3 km/s, about 2.5 km/s *less* (rather than more) than the earth's speed.

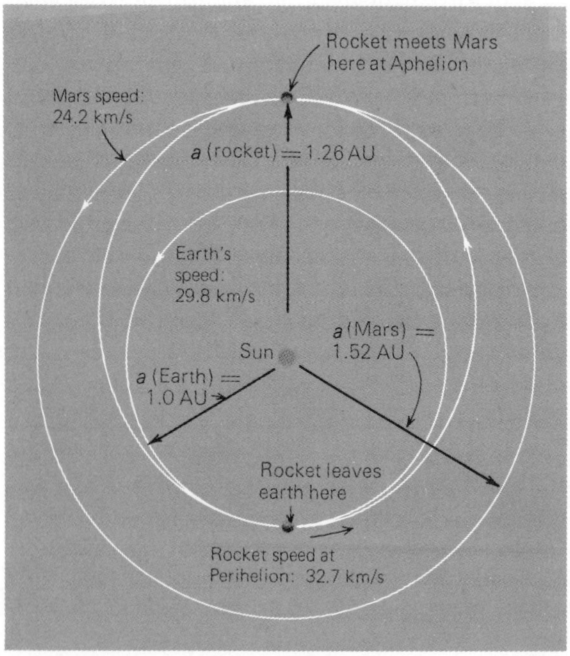

Figure 5.13 Least-energy orbit to Mars.

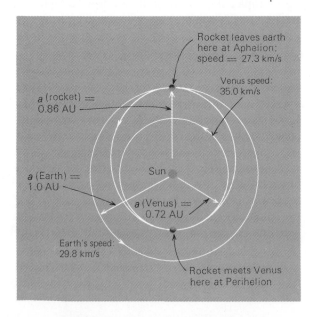

Figure 5.14 Least-energy orbit to Venus.

The space vehicle would have to leave the earth, as before, with enough speed so that when it has left the earth's vicinity it has a speed with respect to the earth of 2.5 km/s but in a direction *opposite* that of the earth's motion. Then, relative to the sun, the vehicle is moving at the required 27.3 km/s and will reach the orbit of Venus along the desired elliptical orbit. The travel time to Venus, found as before from Kepler's third law, is about five months. Returning from Venus to earth is similar to traveling from earth to Mars.

From the foregoing discussion it is obvious that the earth and the planet to be visited must be at a critical configuration at the time of launch, in order that the space vehicle meet the planet at the other end of the vehicle's heliocentric orbit. These critical configurations occur at intervals equal to the synodic period of the planet. In practice it is not necessary, and is seldom feasible, to launch the rocket at exactly the proper instant to achieve the least-energy orbit. However, there is a short range of time (typically a few weeks) during which a *nearly* least-energy orbit can be achieved. The length of this time period, called a "window" in space jargon, depends on the thrust capabilities of the available rockets (that is, on how much energy, above the least possible needed, can be supplied by the rocket). "Windows" for Mars journeys occur at intervals of about 780 days; those for Venus trips at intervals of about 584 days.

Most of our space probes have been flybys, which means they have made the relevant observations of the planets they visited during the brief periods during which they passed near those planets and then flew on into space. Closeup observations of the visited planets thus must be made in a matter of hours. The orbit of such a flyby probe relative to the center of the planet it visits is a hyperbola; the planet gravitationally deflects the probe into a different direction from that which it had when it approached the planet. By careful planning, we have sometimes been able to take advantage of this "gravitational whip" action of the planet on a flyby, by timing the launch and choosing the flyby distance just right so that the probe's post-encounter trajectory will direct it to another planet in the solar system. We did this in 1974 with Mariner 10, which was directed to Mercury by the gravitational action of Venus, and of course with Pioneer 11 in 1974 and with Voyagers 1 and 2 in 1979, which were sent on to Saturn by the gravitational deflection of Jupiter. In 1981 Voyager 2 was similarly redirected by Saturn into a trajectory that will take it close to Uranus.

On the other hand, if we want a prolonged visit, or a landing, we must slow the probe down with a retrorocket while it is near its destination planet so that it can enter an elliptical orbit about that planet. We did this with the Mariner 9 Mars probe in 1971 and the Pioneer Venus probes in 1978. The Viking landers sent to the surface of Mars in 1976 were launched from similar orbiters. We shall describe these probes more fully in Chapters 17 and 18.

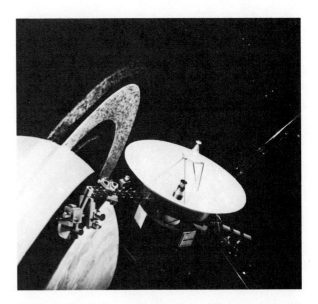

Figure 5.15 Artist's rendering of the Voyager spacecraft flying behind the rings of Saturn, using cameras and radios to measure how sunlight is affected as it shines between the ring particles. *(NASA)*

EXERCISES

1. Compare and contrast the earth's motion about the barycenter with the idea of epicycles and deferents (Chapter 2).

2. How far is the barycenter from a star of three times the mass of the sun in a double-star system in which the other star has a mass equal to the sun's and a distance of 4 AU from the first star?

3. What is the period of mutual revolution of the two stars described in Exercise 2?

4. Why does Newton's version of Kepler's third law have the form

$$(m_1 + m_2) P^2 = a^3,$$

with the constant of proportionality equal to unity, if $m_1 + m_2$ is in units of the combined mass of the earth and moon, P in sidereal months, and a in units of the moon's distance? Find another such set of units, other than those given in the text, for which the constant of proportionality is unity.

5. A cow attempted to jump over the moon but ended in an orbit around the moon instead. Describe how the cow could be used to determine the mass of the moon.

6. What would be the period of an artificial satellite in a circular orbit around the earth with a radius equal to 96,000 km? (Assume that the moon's distance and period are 384,000 km and $27\frac{1}{3}$ days, respectively.)
   Answer: About $\frac{1}{2}$ week

7. Why is it easier to get a space probe to escape the earth than to put a satellite into a perfectly circular orbit?

8. As air friction causes a satellite to spiral inward closer to the earth, its orbital speed *increases.* Why?

9. If a lunar probe is to be launched from the earth's surface into an elliptical orbit whose apogee point is at the moon, why must the eccentricity of the orbit be nearly one?

10. How could you calculate the period of an artificial satellite if its perigee and apogee altitudes above the earth's surface were known?

*11. Verify the periods given in the text for the times required to reach the planets Venus and Mars along least-energy orbits.

*12. Show why the times at which a space vehicle can be sent to a planet on a least-energy orbit occur at intervals of the synodic period of the planet.

*13. Describe how a space vehicle must be launched if it is to fall into the sun.

*14. If a satellite has a nearly circular orbit at a critical distance from the earth's center, it will have a period or revolution equal to one day and thus can appear stationary in the sky above a particular place on earth. Calculate the radius of the orbit of such a *synchronous* satellite.
   Answer: About 42,400 km

15. If the sun had eight times its present mass and the earth's orbit were twice its present size, what would then be its orbital period?

*16. Show that the circular satellite velocity at the surface of the earth is 7.9 km/s.

*17. Calculate the velocity of escape from the sun at the earth's distance, and compare this with the earth's actual orbital speed.

18. If an artificial earth satellite has an orbital eccentricity of 0.5, what is the ratio of apogee to perigee? If the semimajor axis of its orbit is twice the radius of the earth, what is its distance from the surface of the earth at apogee?

John Couch Adams
(1819–1892)

John Couch Adams
(1819–1892), *(top)*, and
Urbain Jean Joseph Leverrier
(1811–1877), *(bottom)*, English
and French mathematicians,
independently predicted the
direction in the sky of an
unknown planet that was
perturbing the motion of
Uranus. Their prediction led
to the discovery of Neptune.
(Yerkes Observatory)

The *n*-BODY PROBLEM

Until now we have considered the sun and a planet, or a planet and one of its satellites, as a pair of mutually revolving bodies. Actually, the planets (and different satellites of a planet) exert gravitational forces upon each other as well. These interplanetary attractions cause slight variations from the orbits that would be expected if the gravitational forces between planets were neglected. Unfortunately, the problem of treating the motion of a body that is under the gravitational influence of two or more other bodies, called the *n-body problem*, is very complicated.

6.1 THE *n*-BODY PROBLEM— PERTURBATION THEORY

The *n*-body problem is, in general, that of describing the motion of any body in a collection (or cluster) of many objects all interacting under the influence of their mutual gravitation.

If the exact position of each body is specified at any given instant, we can calculate the combined gravitational effect of the entire ensemble on any one member of the group—it is merely an extension of the application of the parallelogram of forces (Section 4.1e). Knowing the force on the body in question, we can find how it will accelerate; a knowledge of its initial velocity, therefore, is enough to calculate how it will move in the next instant of time, and thus to follow its motion. However, the problem is complicated by the fact that the gravitational acceleration of one body depends on the positions of all the other bodies in the system. Since they, in turn, are accelerated by all the members of the cluster, we must simultaneously calculate the acceleration of each particle produced by the combination of the gravitational attractions of all the others to follow the motions of all of them, and hence of any one. Such extremely complex calculations have been carried out, with electronic computers, to follow the dynamical evolution of hypothetical clusters of up to thousands of members.

Although computations of the type just described can be carried out, in principle, to study the motion of any one member of a group or cluster of bodies, it is not possible to write an equation that will describe the trajectory (or orbit) of that body for all time, as it is in the two-body problem (in which the orbits are always conic sections). Consequently, the *n*-body problem is often said to have no solution. Actually, by numerical calculation many problems can be solved to rather high precision, although for some problems of importance, such as the evolution of the solar system, even the biggest electronic computer is not adequate. In principle, however, the *n*-body problem is not solvable only in the sense that a single equation does not describe the motion.

Calculations have been performed to follow the evolution of hypothetical clusters, but they are not feasible for the study of all problems. It is, however, possible to derive certain properties of a cluster of particles interacting gravitationally in order to study it statistically, that is, to specify the average behavior of its members. We shall discuss some of these applications in Chapter 30, which deals with star clusters.

(a) PERTURBATION THEORY

Fortunately, the many-body problem can be solved rather accurately when a given body feels predominantly the gravitational force of one other mass. The motion of a planet around the sun, for example, is determined mainly by the gravitational force between it and the sun, the force between it and any of the other planets being very small in comparison. Thus the influences of the other planets can be regarded as small corrections to be applied to the two-body solution; these corrections are called *perturbations*.

There are two approaches to perturbation theory. One is to calculate directly the actual gravitational force on a planet (or minor planet or satellite) due to the combined attraction of all perturbing bodies, and knowing how each other planet is moving, to calculate in detail how the object in question will move. This is called the method of *special perturbations*. The procedure is especially tractable with modern computers and is the one most often used to calculate the orbit of a lunar or planetary probe moving under the combined influences of the various members of the solar system that it passes near.

The other approach is the method of *general perturbations*. Here the position at any moment of a planet (or minor planet or satellite) with respect to the sun (or earth), in combination with its velocity at that instant, is used to calculate the elliptical orbit it would follow if there were no perturbations. In other words, its orbit is computed on the basis of the two-body theory. This orbit is only a temporary one for the object because it will gradually be perturbed by other bodies than the sun (although it may well represent the body's actual motion for a considerable time); it is called an *osculating orbit*. Algebraic formulas can then be derived which express, with tolerable accuracy, how the elements of the osculating orbit will change with time, owing to the perturbations of the other planets. This procedure was usually employed before high-speed computers were developed, which facilitated the calculation of special perturbations. General perturbations are still widely used in many applications.

We shall see in Chapter 18 how perturbation theory led to the discovery of the planet Neptune.

(b) SPECIAL SOLUTIONS—THE RESTRICTED THREE-BODY PROBLEM

As stated above, the *n*-body problem can be solved in general only by laborious numerical calculation. There are, however, some special circumstances in which there exist solutions, or partial solutions, in the form of algebraic equations. Usually, these solutions apply only when the collection of particles has a very particular (and unlikely) configuration. One case, however, has partial solutions with very interesting applications. This is the *restricted three-body problem*, first considered by the French mathematician and astronomer Louis Lagrange (1736–1813).

Lagrange investigated the behavior of a small particle moving in the gravitational field of two objects revolving about each other in circular orbits. The restriction is that the particle must have too small a mass to have any gravitational influence on the other two bodies. Lagrange found that there are five positions relative to the two objects in mutual circular revolution (Figure 6.1) where the small mass, once placed, will move on a circular orbit always maintaining a fixed orientation with respect to the two greater masses. The three points marked *A* are unstable, in the sense that if the small body is displaced slightly from one of them it will leave its circular orbit. Because small perturbations are always likely to occur, we would not expect to find many examples in nature in which three bodies revolve exactly in those configurations.

The points marked *B* in Figure 6.1, however, are stable; the small object at one of those positions would not be forced away by slight perturbations. Note that the two bodies of larger mass and either of the *B* points are at the corners of an equilateral triangle. We do, in fact, find natural examples of this kind of motion. The best known is the equilateral configuration defined by the sun, the planet Jupiter, and the two groups of *Trojan* asteroids (Chapter 19). (The sun and Jupiter move in nearly circular paths around each other, and the minor planets have negligible mass in comparison, so the conditions of the problem are approximately met.) It has been suggested long ago that small particles could similarly revolve about the earth 60° ahead of and/or behind the moon. Some observers have even reported sighting faint patches of light

that could be clouds of such particles reflecting sunlight to us, but these sightings have never been confirmed.

Lagrange's solution to the restricted three-body problem also specifies the regions of space within which the small particle can move relative to the two larger ones. We shall see (Chapter 33) that in at least one stage in the evolution of a typical star, it becomes a giant, greatly distending its outer layers. Now there are many double-star systems in which the two stars revolve about each other in nearly circular orbits. If the two stars are relatively close together, and if one evolves to a large enough size, the atoms of its outer distended layers, having negligible mass, move about in the system in the manner predicted by Lagrange. We find, thus, that during the evolution of stars in binary systems, matter can flow from one star to another, or can flow in an orbit around one or both stars, or can even flow into space, escaping the two stars altogether. Examples will be encountered in Chapter 34.

6.2 THE GRAVITATIONAL EFFECTS OF NONSPHERICAL BODIES

Bodies with spherical symmetry act, gravitationally, as point masses, for which the gravitational influences are easily calculated. In nature, however, most bodies are not exactly spherical, and the simple two-body theory does not give precise results. If the shape of a body deviates only slightly from a sphere, we usually approximate its gravitational influence by that produced by a point mass and treat the small effects of its asphericity as perturbations. A common cause of the deformation of a star or planet from a perfect sphere is its rotation. Rapidly rotating planets, such as Jupiter, are noticeably flattened. The rotational flattening of the earth is slight but is important.

(a) The Shape of the Earth

Because of the earth's rotation, the inertia of its constituent parts tends to make them fly off tangentially into space. Therefore, each particle on and in the earth must be undergoing a constant *centripetal* acceleration to keep it in place. Of course, it is the earth's gravitation that provides this acceleration. Only a small part of this gravitational force on an object at the earth's surface is required to provide the centripetal acceleration that keeps it on the ground; the remainder accounts for the object's

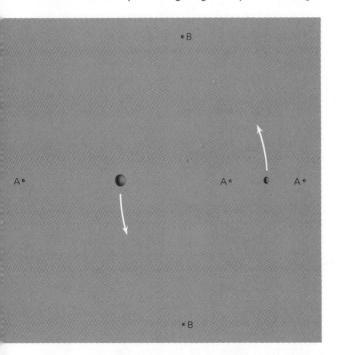

Figure 6.1 Lagrangian points, at each of which a body of infinitesimal mass moves in a circular orbit, maintaining a fixed orientation with respect to two bodies mutually revolving in circular orbits.

weight. In other words, the rotation of the earth slightly reduces the weights of objects on the earth's surface (and the interior parts of the earth itself as well), because some part of the gravitational force is used up in providing the centripetal acceleration.

On the other hand, a body exactly on the axis of the earth has zero speed and suffers no centripetal acceleration. Therefore, at either pole of the earth (at either end of its rotational axis) the full force of gravity goes into the weight of an object. As a person travels away from a pole, his weight gradually diminishes and is least at the equator, where the centripetal acceleration is greatest. The effect is slight; one would lose only about 1/2 percent of his weight this way.

Newton showed that the result of this effect should be a distortion of the earth from a purely spherical shape. Suppose the earth were spherical (Figure 6.2). The direction of the gravitational pull upon a constituent rock in the earth near its surface is toward the center of the earth. This force must produce the combined effects of the centripetal force on the rock and its weight. Now the centripetal force is not directed toward the center of the earth (except for objects at its equator) but rather is toward the center of the path on which the rock is moving. Thus the centripetal force is directed perpendicularly toward the earth's axis of rotation. We once again apply the principle of the parallelogram of forces. The gravitational force can be resolved into two components: the centripetal force and the weight of the body. In other words, the weight of the rock in question is *not directed toward the earth's center* and is not in a direction perpendicular to the surface the earth would have if it maintained a perfectly spherical shape. The weight *is* in the direction a freely hanging plumb bob would point, and we see that it is *not*, in general, toward the earth's center.

If a spherical earth were infinitely rigid, that would be the end of the story. But it would mean that the weights of the crustal rocks would not be perpendicular to the mean surface of the earth, so that there would be a component of force lying along the surface, constantly pulling on those rocks in a direction toward the equator. Now, the earth is not infinitely rigid, and if it is subjected to such a force over a very long time, like heavy molasses it eventually flows, until it takes on such a shape that its mean surface is everywhere perpendicular to the direction of the weight of the surface (technically, along what is called an *equipotential surface*). Superficial crustal features, however, such as mountains, are rigid enough to hold their shapes, like grains of gravel floating on the molasses.

Analysis shows that the resulting shape of the earth should be nearly that of an *oblate spheroid*, a slightly flattened sphere that has an elliptical cross section. The earth should be flattened at the poles, whereas the equatorial regions should be slightly bulged out. Consequently, a person at one of the poles would be nearer the earth's center than he would be at the equator. Here is an additional effect that causes the weight of an object to be less at the equator. Both effects together, the rotation of the earth and the earth's oblate shape, reduce the weight of an object at the equator compared to its weight at a pole by 1 part in 190.

The gravitational acceleration, in other words, varies slightly over the surface of the earth. The variation was first measured by John Richer, who was sent on a scientific expedition from Paris to Cayenne (in French Guiana) in the years 1671–1673. Richer found that at Cayenne (5° north latitude) a pendulum beat slower than it did at Paris. In 1673 Huygens showed that the period, or swinging rate, of a pendulum is proportional to the square root of the ratio of its length to the gravitational acceleration. Thus, Richer's measurements showed the variation of the acceleration of gravity between Paris and Cayenne. Later these results were used in a theoretical analysis to compute the shape of the earth.

Another consequence of the earth's oblateness is to make the distance along the surface of the

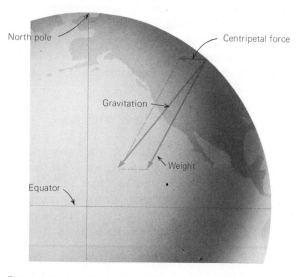

Figure 6.2 Direction of the weight of a body on a spherical earth.

earth corresponding to one degree of latitude longer in the vicinity of the poles than near the equator (Figure 6.3). The effect is easily observable, and the length of a degree at various latitudes has been measured. It varies from 110.6 km at the equator to 111.7 km at the poles.

The actual oblateness of the earth is small. Its diameter from pole to pole is only 43 km less than through the equator, about 1 part in 298.

(b) PERTURBATIONS ON EARTH SATELLITES

The equatorial bulge of the earth is responsible for a deformation in the earth's gravitational field from that which would be produced by a point mass or spherically symmetrical earth. These deformations are especially important near the surface of the earth and produce conspicuous perturbations in the orbits of low-altitude earth satellites.

As an example, we illustrate the perturbation on the orientation of the *line of nodes*—the line of intersection between the plane of the earth's equator and that of a satellite orbit. A rigorous discussion of this perturbation will not be attempted here; however, a physical feeling for how the asphericity of the earth can produce such an effect can be gained through the following oversimplified description.

Consider a satellite moving eastward and crossing the equatorial plane from south to north (Figure 6.4). As it approaches the equatorial plane, the equatorial bulge pulls it slightly northward. As it passes north of the equatorial plane, the bulge pulls it slightly southward. The orbit of the satellite has thus been displaced a little westward. On the other side of the earth, as the satellite passes from north to south it is pulled first to the south and then to the north; again its orbit is displaced slightly westward.

Actually, the changes are not abrupt, as suggested here and in Figure 6.4. In the nonspherical gravitational field of the oblate earth the satellite is continually being accelerated from the two-body orbit. Gradually, then, as it revolves about the earth, its orbit slides in a direction opposite to that of its motion, and the line of nodes slowly rotates to the west, or *regresses* (if the motion of the satellite is to the east). For a satellite with a period of two hours, an orbital eccentricity of 0.2, and an orbit inclined at 45° to the equatorial plane, the line of nodes rotates 3.4° per day.

Many other perturbations of satellite orbits exist. The exact nature of these effects, many of them subtle, depends critically on the precise shape of the earth, or, more precisely, on the distribution of mass within the earth.

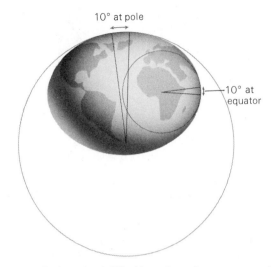

Figure 6.3 Length of 10° of latitude at the equator and at the North Pole (the difference is greatly exaggerated).

(c) SATELLITE INVESTIGATIONS OF THE EARTH'S SHAPE

Since the motions of earth satellites depend on the precise shape of the earth, careful studies of perturbations on satellite orbits should enable us to derive the earth's shape rather accurately.

The deformations of the earth's shape from that of a perfect sphere are usually represented by a series of what are called *spherical harmonics*. The most important deformation is the equatorial bulge, which is represented by the *second zonal harmonic* (Figure 6.5), a slightly oblate figure superimposed on the spherical earth. The kind of deformations represented by the third, fourth, and fifth zonal harmonics are also shown schematically in Figure 6.5. All these shapes are really three-dimensional and are symmetrical about the axis of the earth's rotation; only cross sections are shown in the figure. In addition, there are harmonics that are

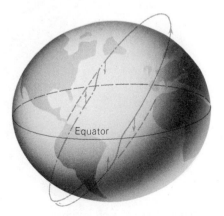

Figure 6.4 Effect of the earth's equatorial bulge on an earth satellite.

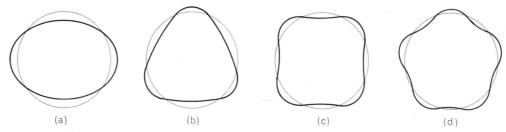

Figure 6.5 From (a) to (d), respectively: The form of the second, third, fourth, and fifth spherical harmonics—part of a series of surfaces superimposed to represent the mean shape of the earth.

not symmetric about the rotation axis. The general forms of the deformations represented by the various harmonics are correctly indicated, but the actual deformations of the earth are exceedingly small. The relative importance of the various harmonics is revealed by satellite observations. When a large number of harmonics, properly scaled, are superimposed on one another, the shape of the earth is obtained.

For example, observations of the earliest Navy Vanguard satellite first revealed a deformation represented by the third harmonic. This showed that there is a very slight amount of "pear shape" superimposed on the oblate earth, as shown in Figure 6.5(b). A complete representation of the gravity field requires an infinite set of harmonics. Hundreds of harmonics have now been measured with satellite radar altimetry.

Analyses of satellite perturbations, combined with sensitive measures of surface gravity over the earth, lead to an accurate knowledge of the form of the earth and permit the preparation of maps like the one in Figure 6.6, by William Kaula at UCLA. The earth is first represented by an oblate spheroid (a flattened sphere with elliptical cross section) in which the equatorial diameter exceeds the polar diameter by only 1

part in 298.25. The figures on the map are the heights, in meters, of the geoid above (positive numbers) or below (negative numbers) the surface of that oblate spheroid. The *geoid* is the same as mean sea level over the oceans; it is a surface everywhere perpendicular to the direction of gravity. The *topography*—the ups and downs of continents, mountains, and other crustal features—affects the geoid only indirectly through slight gravitational attractions.

6.3 DIFFERENTIAL GRAVITATIONAL FORCES

Two perfectly spherical bodies, as has been explained, attract each other as if they were point masses, located at their own centers. If however, a body deviates from perfect sphericity, even slightly, it no longer acts as a point mass, and we have seen in Section 6.2 how satellite orbits are consequently perturbed. The shape of a planet can be

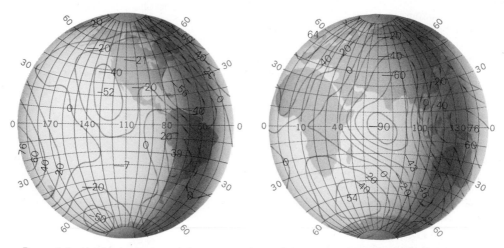

Figure 6.6 Height, in meters, of the mean surface of the earth above (positive numbers) and below (negative numbers) the surface of an oblate spheroid with a flattening of 1 part in 298.25. (*Courtesy of William Kaula, UCLA*)

Figure 6.7 Attraction of a large mass and two smaller ones.

affected by rotation. Another contribution to asphericity is supplied by the *differential gravitational forces* that two neighboring bodies exert on each other. These forces, in turn, result in such phenomena as *tides* and *precession*.

(a) One Body's Attraction on Two Others

A *differential gravitational force* is the *difference* between the gravitational forces exerted on two neighboring particles by a third more distant body. As an example, consider Figure 6.7, in which three bodies are shown in a line. These are either point masses or perfectly spherical objects whose gravitational effect on external objects is the same as that produced by point masses. To the left is a large body of mass M. To the right are two bodies, each of whose masses we shall assume, for ease of calculation, to be unity—say, each has a mass of 1 g. The first of the small bodies, body 1, is at a distance R from the large one; the other, body 2, is at a distance $R + d$.

The force of attraction between the large mass and body 1 is

$$F_1 = \frac{GM}{R^2},$$

and that between the large mass and body 2 is

$$F_2 = \frac{GM}{(R + d)^2}.$$

Note that F_2 is slightly *smaller* than F_1 because of the greater distance between the large mass and body 2. The difference $F_1 - F_2$ is the differential gravitational force of the large mass on the two smaller masses.

In Figure 6.8 the forces F_1 and F_2 are shown as vectors pointing toward the large mass to the left. Because the force on body 1 is greater than on body

2, the *differential* force tends to separate the two bodies.

Now the center of mass of two small bodies is halfway between them. If either of the two unit masses were at that point, the attraction it would feel toward the large body, M, would be

$$F_{CM} = \frac{GM}{(R + \frac{1}{2}d)^2}.$$

This force is intermediate between the force on body 1 and that on body 2. *With respect to the center of mass*, therefore, both body 1 and body 2 feel themselves pulled *outward*. If the bodies are free to move, they will separate unless their mutual gravitational attraction (not shown in Figure 6.8) is great enough to hold them together.

(b) CALCULATION OF DIFFERENTIAL GRAVITATIONAL FORCE

In the example described in the preceding paragraphs, the differential gravitational force ΔF was found to be

$$\Delta F = F_1 - F_2 = \frac{GM}{R^2} - \frac{GM}{(R + d)^2}.$$

Combining the two terms of ΔF, we find, with simple algebra,

$$\Delta F = GM\frac{d(2R + d)}{R^2(R + d)^2}.$$

Now let us suppose that the distance R is very much greater than the distance d. In this case, $R + d$ is so nearly equal to R that we can write

$$R + d \approx R.$$

Similarly,

$$2R + d \approx 2R.$$

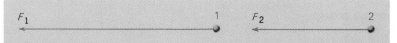

Figure 6.8 Forces on the smaller masses shown as vectors.

With this approximation, our equation for ΔF becomes

$$\Delta F = 2GM\frac{Rd}{R^4} = 2GM\frac{d}{R^3}.$$

Now let us denote by δF the differential force corresponding to a unit separation of the two small bodies, that is, for the case where $d = 1$. Then*

$$\delta F = \frac{2GM}{R^3},$$

and the total differential force is

$$\Delta F = d \times \delta F.$$

In the foregoing calculations it was assumed that the three bodies are in a line. In general, the bodies are not lined up, and the differential gravitational force between two of them is not simply the arithmetic difference between the forces exerted on each by the third body. Since a force is a *vector* (for it has both magnitude and direction), the difference must be calculated according to the rules of vector subtraction (see Section 4.1e). In Figure 6.9 a mass M (to the left) is attracting each of two masses, A and B. Mass B is shown in various orientations with respect to the line between A and M. In each case the *vector difference* between the force of attraction of M on B and on A is shown. Usually, this differential gravitational force acts in such a way as to tend to separate A and B. However, when B is nearly at right angles to the line joining M and A, M's force on the two is in slightly different directions and tends to pull them closer together; then the differential force on B is directed more or less toward A.

(c) Differential Forces and Perturbations

Many perturbations can be looked upon as an effect of differential gravitational forces. A third body perturbs the orbital motion of two bodies because its gravitational force on both bodies is not the same. If it were, the two bodies would be acceler-

---

*This result could have been obtained immediately with differential calculus by differentiating the gravitational force;

$$\frac{dF}{dR} = \frac{d}{dR}\left(\frac{-GM}{R^2}\right) = \frac{2GM}{R^3}.$$

The minus sign in the gravitational force denotes that the force acts in such a direction as to decrease R.

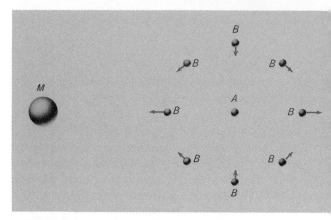

Figure 6.9 Vector differences between forces of attraction of mass M on each of masses A and B.

ated by the third body the same amount, and in the same direction, and their relative motion would be unchanged.

As an example, consider the influence of the sun on the mutual revolution of the earth and the moon. Both are about the same distance from the sun, so both are accelerated approximately the same amount and follow a nearly common orbit about the sun. The relative motion of the earth and the moon, then, depends mainly on their mutual attraction and not on the sun. For the sun to play no role at all in the orbital revolution of the earth and the moon about each other, however, its differential force upon them would have to be zero. Actually, at new moon the moon is accelerated more strongly toward the sun, and at full moon the earth feels the stronger acceleration. Therefore, the gravitational force between earth and moon is slightly altered by the differential force of the sun on the earth-moon system. This is the most important cause of those irregularities in the motion of the moon described in Chapter 9.

6.4 TIDES

Early in history it was realized that tides were related to the moon, because the semi-daily delay in the time of high tide ("high water") is half the daily delay in successive transits of the moon across the local meridian. A satisfactory explanation of the tides, however, awaited the theory of gravitation, supplied by Newton.

(a) Earth Tides

First, we shall consider the effects of the moon's attraction on the solid earth. For the moment, we

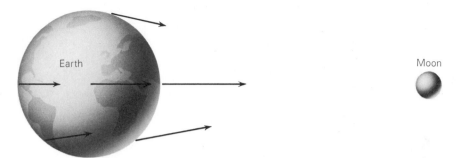

Figure 6.10 The moon's attraction on different parts of the earth.

ignore the flattening of the earth due to its rotation. Our planet can be regarded as being composed of a large number of particles, each of unit mass, all bound together by their mutual gravitational attraction and cohesive forces. The gravitational forces exerted by the moon at several arbitrarily selected places in the earth are illustrated in Figure 6.10. These forces differ slightly from each other because of the earth's finite size; all parts are not equally distant from the moon, nor are they all in exactly the same direction from the moon. If the earth retained a perfectly spherical shape, the resultant of all these forces would be that of the force on a point mass, equal to the mass of the earth, and located at the earth's center. Such is approximately true, because the earth is nearly spherical, and it is this resultant force on the earth that causes it to accelerate each month in an elliptical orbit about the barycenter of the earth-moon system.

The earth, however, is not *perfectly* rigid. Consequently, the differential force of the moon's attraction on different parts of the earth causes the earth to distort slightly. The side of the earth nearest the moon is attracted toward the moon more strongly than is the center of the earth, which, in turn, is attracted more strongly than is the side of the earth opposite the moon. Thus, the differential force tends to "stretch" the earth slightly into a *prolate spheroid* with its major axis pointed toward the moon. That is, the earth takes on a shape such that a cross section whose plane contains the line between the centers of the earth and moon is an ellipse with its major axis in the earth-moon direction.

Figure 6.11 shows the forces (as vectors) that are acting at several points on the surface of the earth. In each case, the forces are shown with respect to the earth's center. The dashed vectors represent the forces due to the earth's gravity, that is, the weights of various parts of the earth. The solid

vectors (much exaggerated in length) represent the differential gravitational forces due to the varying attraction of the moon on different parts of the earth. They are called the *tidal forces*. Those parts of the earth closer to the moon than the earth's center are attracted more strongly toward the moon than parts of the earth near its center. Thus the tidal forces are directed *toward* the moon. Those parts on the opposite side of the earth are attracted less strongly than are parts at the earth's center. The tidal forces there are directed *away* from the moon.

In each case, the vector representing the force can be broken into two components, one in the *vertical* direction, that is, toward or away from the direction of the earth's gravity, and one in the *horizontal* direction, along the surface of the earth. The effect of the vertical component of the tidal force is to change slightly the weight of the surface rocks of the earth. The effect of the horizontal component is to attempt to cause the surface regions of the earth to flow horizontally.

If the earth were perfectly spherical, its gravitational attraction for objects on its surface would be in a *vertical* direction, toward the center of the earth. The actual earth, however, distorts under

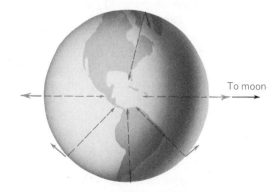

Figure 6.11 Gravitational and tidal forces at various places on the earth's surface.

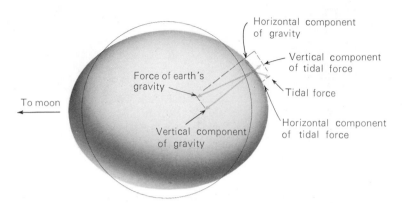

Figure 6.12 Deformation of the solid earth under the influence of tidal forces (much exaggerated).

the influence of the tidal forces and is not quite spherical (we are still ignoring distortion due to rotation). Consequently, the earth's gravitational pull upon objects on its surface is not exactly in a direction perpendicular to the surface; there is a slight *horizontal* component in the gravitational pull of the earth upon its surface regions (Figure 6.12).

If the earth were fluid, like water, it would distort until all the horizontal components of the tidal forces were exactly balanced by the horizontal pull of the earth at all points throughout it. Then the inward force on an object at the earth's surface would be in a vertical direction. It would depend on two factors, those components of the earth's gravitational attraction and of the tidal force that are normal to the surface of the earth at that point.

Measures have been made to investigate the actual deformation of the earth. It is found that the solid earth does distort, as would a liquid, but only about one third as much, because of the high rigidity of the earth's interior. In fact, the rigidity of the earth must exceed that of steel to account for the small degree of its tidal distortion, a result in agreement with seismic studies (Chapter 15). The maximum tidal distortion of the solid earth amounts at its greatest to only about 20 cm.

As the earth rotates, different parts of it are continually being carried under the moon, so the direction and magnitude of the tidal force acting at any given place on the earth's surface are constantly changing. If the earth were viscous, its distortional adjustment would lag somewhat as the tidal forces on it change. However, direct observation shows that the earth readjusts its shape under the influence of the changing tidal force almost instantaneously. This circumstance implies that the earth is not only almost perfectly rigid, but also highly elastic on the time scale of tides.

In summary, tidal forces on the earth, that is, the differential gravitational forces of the moon's attraction on different parts of the earth with respect to its center, cause the solid earth to distort continually from a spherical shape, rising up and down and tilting as a fluid surface would do. but by only about one third the amount. Furthermore, these deformations are nearly instantaneous, changing just as quickly as the tidal forces change due to the earth's rotation. These facts show the earth to be more rigid than steel and to be highly elastic.

Rotation, of course, also distorts the earth's shape. The slight elongation of the earth that results from the tidal distortions described above is superimposed on the equatorial bulge due to its rotation. The latter is a very much greater distortion than the distortion due to tides.

(b) A Comment About "Weight"

In Section 4.1(d) we defined weight as the gravitational pull of the earth on an object at its surface. Most authors define weight this way, but we have seen in this chapter that the earth's rotation and tides also affect, slightly, an object's weight. For that matter, we often speak of one's weight on the moon or on some other planet, or his lack of weight off in space remote from all gravitating bodies. Moreover, within a space station in near earth orbit, where the earth's gravity is almost as strong as it is on the ground, astronauts feel no effects of gravity at all. In one sense they cannot be truly weightless, for they are falling freely about the earth; yet they are often said to be experiencing "weightlessness." Clearly a more specific definition of weight would be welcome.

Some physicists prefer to define one's weight in terms of the force on the soles of his feet (I am indebted to Professor Roy L. Bishop for this suggestion). Then an object in free fall (or in orbit) would be weightless by definition, and weight would be the force with which such free fall is impeded or prevented (such as the supporting force of the ground upon one standing on the earth). The latter would be equal to the force of the earth's gravitation adjusted for rotational and tidal effects. But until some such rigorous convention is adopted, the reader had best be warned that the term "weight" is often used ambiguously.

(c) Ideal Ocean Tides

In Figure 6.13 the vectors represent tidal forces (relative to the earth's center) at various points on the earth's surface. These forces are directed generally toward the moon on the side of the earth facing the moon and away from the moon on the opposite side. Within a zone around the earth that is roughly the same distance from the moon as the earth's center, the tidal forces are directed more or less toward the center of the earth. At these points, the attraction toward the moon is the same in magnitude as it is at the center of the earth, but because of the relatively small distance to the moon, it is in a direction that tends to pull those points closer to the earth's center. Each of the vectors (solid arrows) representing a tidal force in Figure 6.13 is resolved into components perpendicular to and parallel to the earth's surface (dashed arrows). We have seen

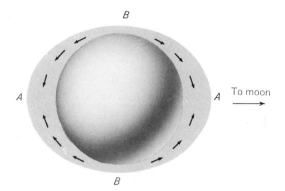

Figure 6.14 Tidal bulges in the "ideal" oceans.

that if the earth were fluid it would take on a shape such that points on its surface would feel no horizontal component of force. However, the earth is sufficiently rigid to be distorted from a sphere by only about one third of the amount required to remove these horizontal forces. Consequently, objects at the surface of the earth that are not restrained from horizontal motion, for example, the waters in the oceans, are free to flow in the direction of the horizontal components of the tidal forces. We shall assume, first, that the earth is covered uniformly by a deep ocean, and investigate the nature of the tides produced in it.

The actual accelerations of the ocean waters caused by the horizontal components of the tidal force are very small. These forces, acting over a number of hours, however, produce motions of the water that result in measurable tidal bulges in the oceans. Water on the lunar side of the earth is drawn toward the sublunar point (the point on the earth where the moon appears in the zenith), piling up water to greater depths on that side of the earth, with the greatest depths at the sublunar point. On the opposite side of the earth, water moves in the *opposite* direction, producing a tidal bulge on the side of the earth opposite the moon (Figure 6.14).

It is important to understand that it is the horizontal components of the tidal forces that produce the tidal bulges in the oceans. At the two opposite points on the earth where the moon is at the zenith and at the nadir, the tidal forces are exactly radial, that is, directed away from the earth's center. At those points the horizontal components are zero, and there is no acceleration causing the water to flow along the surface of the earth. The tidal forces serve only to reduce very slightly the weight of the water, but because of the low compressibility of water, its physical expansion (because of its re-

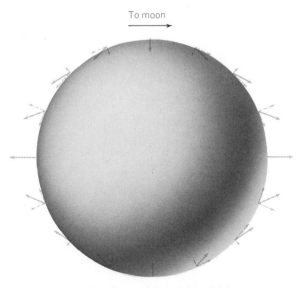

Figure 6.13 Components of the tidal forces.

duced weight) is completely negligible. Thus, at these points, the tidal forces play virtually no role at all, even though that is where the water is piled up the most.

In a beltlike zone around the earth from which the moon appears on the horizon, the direction of the tidal force is *inward*, toward the center of the earth, and again there are no horizontal components. The weight of the water is increased very slightly, but it is not appreciably compressed. Here, also, the tidal forces have no effect, although it is in this belt that the ocean level is lowest.

The tidal bulges in the oceans, then, do not result from the moon compressing or expanding the water, nor from the moon lifting the water "away from the earth." Rather, the tidal bulges result from an actual flow of water over the earth's surface, toward the regions below and opposite the moon, causing the water to pile up to greater depths at those places. It is the horizontal components of the tidal forces that produce this flow; those components, or "tide-raising forces," are greatest in regions of the earth intermediate between those from which the moon appears at the zenith or the nadir (points *A* in Figure 6.14) and on the horizon (points *B*).

The tidal bulge on the side of the earth *opposite* the moon often seems mysterious to students who picture the tides as being formed by the moon "lifting the water away from the earth." What actually happens, as we have seen, is that the differential gravitational force of the moon on the earth tends to stretch the earth, elongating it slightly toward the moon. The solid earth distorts slightly, but, because of its high rigidity, not enough to reach complete equilibrium with the tidal forces. Consequently, the ocean, moving freely over the earth's surface, flows in such a way as to increase the elongation and piles up at points under and opposite the moon.

In this section we have regarded the earth as though its ocean waters were distributed uniformly over its surface. In this idealized picture, not actually realized even in the largest oceans, the tides would cause the depths of the ocean to range through only a few feet. The rotation of the earth would carry an observer at any given place alternately into regions of deeper and shallower water. As he was being carried toward the regions under or opposite the moon where the water was deepest, he would say, "the tide is coming in"; when carried away from those regions, he would say, "the tide is going out." During a day, he would be carried through two tidal bulges (one on each side of the earth) and so would experience two "high tides" and two "low tides."

The two high tides during a day need not be equally "high," however. For example, in northern or southern temperate latitudes, the axis of the tidal bulges is periodically inclined to the equator. The observer in the northern hemisphere (shown in Figure 6.15) would find the high tide on the side of the earth facing the moon much higher than the high tide half a day later. An observer in the southern hemisphere would find the opposite effect. In extreme cases there may appear to be only one "high tide" a day.

(d) Tides Produced by the Sun

The sun also produces tides on the earth, although the sun is less than half as effective a tide-raising agent as the moon. Actually, the gravitational attraction between the sun and the earth is about 180 times as great as that between the earth and the moon. We recall, however, that the tidal force is the differential gravitational force of a body on the earth. The sun's attraction for the earth is much greater than the moon's, but the sun is so distant that it attracts all parts of the earth with almost equal strength. The moon, on the other hand, is close enough for its attraction on the near side of the earth to be substantially greater than its attraction on the far side. In other words, its *differential gravitational pull* on the earth is greater than the sun's, even though its total gravitational attraction is less.

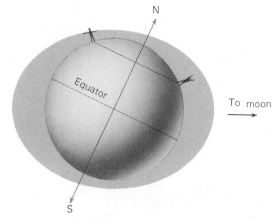

Figure 6.15 Inequality of the two "high tides" during a day.

If there were no moon, the tides produced by the sun would be all we would experience, and the tides would be less than half as great as those we now have. The moon's tides, therefore, dominate. On the other hand, when the sun and moon are lined up, that is, at new moon or full moon, the tides produced by the sun and moon reinforce each other and are greater than normal. These are called *spring tides*. Spring tides (which have nothing to do with spring) are approximately the same, whether at new moon or full moon, because tidal bulges occur on both sides of the earth—the side *toward* the moon (or sun) and the side away from the moon (or sun).

In contrast, when the moon is at first quarter or last quarter (at quadrature), the tides produced by the sun partially cancel out the tides of the moon, and the tides are lower than usual. These are *neap tides*. Spring and neap tides are illustrated in Figure 6.16.

Although spring tides are the highest type of tides, they are not all equally high, because the distances between the earth and sun and the earth and moon (and hence the tide-raising effectiveness of these bodies) both vary. The moon's distance varies by about 10 percent, and its tide-raising effectiveness varies by about 30 percent. The highest spring tides occur at those times when the moon is also at perigee.

(e) The Complicated Nature of Actual Tides

The "simple" theory of tides, described in the preceding paragraphs, would be sufficient if the earth were completely surrounded by very deep oceans, and if it rotated very slowly. However, the presence of land masses stopping the flow of water, the friction in the oceans and between oceans and the ocean floors, the rotation of the earth, the variable depth of the ocean, winds, and so on, all complicate the picture.

Both the times and the heights of high tide vary considerably from place to place on the earth. The earth's rapid rotation causes the tide-raising forces within a given mass of water to vary too rapidly for the water to adjust completely to them. These forces, however, recurring periodically, set up forced oscillations in the ocean surfaces, so that the water over a large area rises and lowers in step. Consequently, the highest water does not necessarily occur when the moon is highest in the sky (or lowest below the horizon), but rather when the oscillations of the ocean, produced by the tidal forces acting upon it, pile up the water to its greatest depth at that location. The latter depends critically upon the shape and depth of the adjacent ocean basin. The time, in hours and minutes, by which high tide lags behind the time when the moon is highest (or lowest below the horizon) is called the *establishment of the port*. The establishment of the port is different for different places but is very nearly constant for a given place. The United States Coast and Geodetic Survey prepares and publishes each year the *Tide Tables*, which give the times and heights of tides at principal ports throughout the world.

Sometimes shallow coastal seas have such shapes and sizes that the natural frequency of oscillation of water sloshing back and forth in the sea

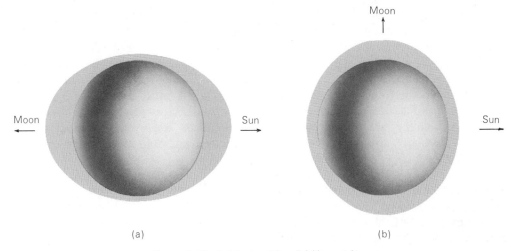

Figure 6.16 (a) Spring tides. (b) Neap tides.

Figure 6.17 A 16-meter (52-foot) tidal range at Hantsport, Nova Scotia, at the head of the Bay of Fundy. These photographs were taken by Prof. Roy L. Bishop, Head of the Department of Physics at Acadia University, on January 29, 1979, a date when several factors were favorable for an exceptionally high tide: The moon was just past new and was near perigee (nearest the earth), the earth was near perihelion (nearest the sun), and the barometric pressure was low. The low tide photograph *(left)* was taken at 8:30 AM; the high tide photo *(right)* was taken five hours later. The objects floating on the water are pieces of ice. *(Courtesy Roy L. Bishop)*

basins is very nearly the same as that of the tidal rise and fall of the water in the adjacent ocean. Then the ocean tides can set the water in these seas into strong resonance, like wind blowing into an organ pipe. The most famous such place is the Bay of Fundy between New Brunswick and Nova Scotia. The Bay of Fundy and Gulf of Maine act as a single oscillating system. The outer boundary of this system is the edge of the continental shelf with its approximately 50-fold increase in depth. The highest tides on earth occur at the head of the Bay of Fundy, in Minas Basin. Under favorable circumstances the tidal range here can exceed 50 feet (Figure 6.17).

Tides also occur in the atmosphere. These atmospheric tides are complicated by weather phenomena, but in principle they are the same as earth and ocean tides. They are, obviously, of importance to meteorologists.

(f) Measurement of Ideal Tides

In view of the complexity of tides, the problem of measuring the tidal forces produced by the sun and moon might seem very difficult indeed. However,

they were measured simply and directly by the American physicist Albert Michelson in 1913. He laid out horizontal pipe, in the north-south and east-west directions, 150 m long, half filled with water. Tides were produced naturally in the water in the pipes, causing periodic fluctuations in the water level that could be measured through sealed windows at both ends of the pipe. The range in depth of the water was less than 0.025 mm, but it was measured with a microscope to within 1 percent.

The tides produced in the pipes were only 69 percent as great as would be expected from calculations based on the assumption that the earth is completely rigid. It is from this experiment that the extent of the tides produced in the solid earth could be inferred. Today earth tides are measured more accurately and conveniently with delicate gravimeters.

(g) Other Tides

The tides produced by the moon and the sun upon the earth are not the only tides in nature. The earth exerts a tidal force upon the moon that is stronger than the one the moon exerts upon the earth. The earth's tidal force has acted as a brake on the lunar rotation, so that today the moon keeps the same face turned toward the earth.

In fact, all bodies in the universe exert a tidal force on all other bodies, just as they exert a general gravitational attraction. In most cases these tidal forces are too small to produce observable effects. For example, the tides produced by planets on each other, and on the sun, are entirely negligible, despite claims made in 1974 in a sensational book, *The Jupiter Effect*, by John Gribbin and Stephen Plageman (New York: Walker and Company). The thesis (subsequently retracted by one of the authors) was that a fortuitous configuration of the planets in 1982 would produce tides on the sun strong enough to influence solar activity, and indirectly, the earth's rotation, causing catastrophic earthquakes. The circuitous reasoning behind the theory was completely without sound basis.

(h) ROCHE'S LIMIT

There is a minimum distance a satellite can be from its planet. At smaller distances a large satellite could not withstand the differential, or tidal, forces exerted on it by the planet and would be torn apart. E. Roche investigated the problem in 1850 and found that if the constituent parts of a satellite are held together only by their mutual gravitation, as, for example, in a liquid body, and if the satellite has the same density as its planet, the critical distance is 2.44 times the planet's radius. At a greater distance, the satellite suffers only tidal distortion, but holds together. At a smaller distance it is torn apart by the tidal forces, for they are greater than the gravitational forces holding the satellite together. If the satellite has a higher density, or a high rigidity, so that cohesive forces add to gravitational ones in binding it together, it could survive at a somewhat smaller distance from the planet. The critical distance at which a large satellite can survive tidal destruction is called *Roche's limit*. The rings of Saturn are particles that are closer to the planet than the distance at which a large solid body with the same low density as Saturn has can survive—that is, the rings are within Roche's limit.

6.5 PRECESSION

The earth, because of its rapid rotation, is not perfectly spherical but has taken on the approximate shape of an oblate spheroid; its equatorial diameter is 43 km greater than its polar diameter. As we have seen, the plane of the earth's equator, and thus of its equatorial bulge, is inclined at about $23\frac{1}{2}°$ to the plane of the ecliptic, which, in turn, is inclined at 5° to the plane of the moon's orbit. The differential gravitational forces of the sun and moon upon the earth not only cause the tides but also attempt to pull the equatorial bulge of the earth into coincidence with the ecliptic.

The latter pull is illustrated in Figure 6.18. The solid arrows are vectors that represent the attraction of the moon on representative parts of the

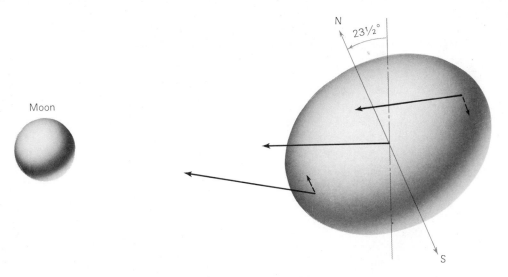

Moon

Figure 6.18 Differential force of the moon on the oblate earth tends to "erect" its axis.

earth. The part of the earth's equatorial bulge nearest the moon is pulled more strongly than the part farthest from the moon, and the earth's center is pulled with an intermediate force. The dashed arrows show the differential forces with respect to the earth's center. Note how they tend not only to "stretch" the earth toward the moon, but also to pull the equatorial bulge into the plane of the ecliptic. The differential force of the sun, although less than half as effective, does the same thing. Thus, the gravitational attractions of the sun and the moon on the earth act in such a way as to attempt to *change the direction of the earth's axis of rotation*, so that it would stand perpendicular to the orbital plane of the earth. To understand what actually takes place, we must digress for a moment to consider what happens when a similar force acts upon a top or gyroscope.

(a) Precession of a Gyroscope

Consider the top (a simple form of gyroscope) pictured in Figure 6.19. If the top's axis is not perfectly vertical, its weight (the force of gravity between it and the earth) tends to topple it over. The actual force that acts to change the orientation of the axis of rotation of the top is that component of the top's weight that is perpendicular to its axis. We know from watching a top spin that the axis of the top does not fall toward the horizontal, but rather moves off in a direction *perpendicular to the plane defined by the axis and the force tending to change its orientation*. Until the spin of the top is slowed down by friction, the axis does not change its angle of inclination to the vertical (or to the floor), but rather describes a conical motion (a cone about the vertical line passing through the pivot point of the

top). This conical motion of the top's axis is called *precession*.

(b) QUALITATIVE EXPLANATION OF PRECESSION

The surprising phenomenon of precession can be understood in terms of Newton's laws of motion. Consider, for simplicity, the jack-shaped gyroscope in Figure 6.20(a), consisting of four masses supported at the ends of rigid light rods perpendicular to each other and to the axis of rotation. As the gyroscope spins, the masses move in the plane indicated. Suppose now that a force F is applied to the axis in a direction perpendicular to the plane defined by the axis of the jack and the line between masses 2 and 4. The force is transmitted through the rods to each of the four masses. Mass 1 feels a force tending to raise it (in the orientation of the diagram), and mass 3 feels a force tending to lower it; only masses 2 and 4 do not feel forces in the vertical direction. Masses 2 and 4 tend to continue moving in the same plane as before the force was applied. Mass 1 accelerates upward, but because of its forward motion it moves along the path ab. Similarly, mass 3 accelerates downward, but because of its forward motion follows path cd. Thus, after a part of a revolution, the masses are in the positions shown in Figure 6.20(b). The axis of rotation has changed, not in the direction of the applied force, but at right angles to it.

The above discussion is not a very rigorous description of precession; it is intended only to give some feeling for the fact that the axis of a spinning top does not yield in the direction of a force acting on it. When we consider how each of the constituent parts of the top should behave under the influence of the applied force we can understand the apparently strange motion of the axis of the whole spinning body in terms of

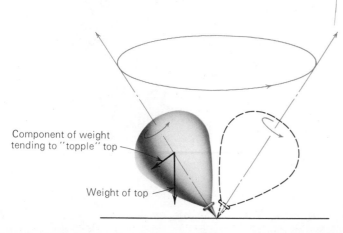

Component of weight tending to "topple" top

Weight of top

Figure 6.19 Precession of a top.

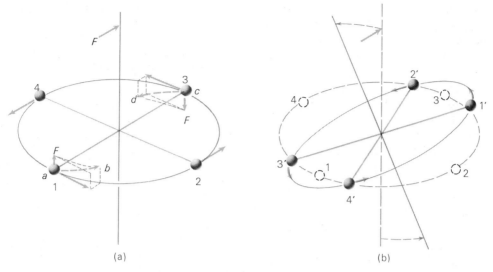

Figure 6.20 (a) Force applied to the axis of a simple gyroscope. (b) The new orientation taken by the gyroscope.

Newton's laws. It can be shown, however, by a rigorous mathematical treatment, that if a force is applied to the axis of any spinning body, the axis itself will move in a plane perpendicular to that defined by the force and the instantaneous axis of rotation.

(c) Precession of the Earth

The differential gravitational force of the sun on the earth tends to pull the earth's equatorial bulge into the plane of the ecliptic, and that of the moon tends to pull the bulge into the plane of the moon's orbit, which is nearly in the ecliptic. These forces, in other words, tend to pull the earth's axis into a direction approximately perpendicular to the ecliptic plane. Like a top, however, the earth's axis does not yield in the direction of these forces, but precesses. The obliquity of the ecliptic remains approximately $23^{1}/_{2}°$. The earth's axis slides along the surface of an imaginary cone, perpendicular to the ecliptic, and with a half-angle at its apex of $23^{1}/_{2}°$ (see Figure 6.21). The precessional motion is exceedingly slow; one complete cycle of the axis about the cone requires about 26,000 years.

Precession is this motion of the axis of the earth. It must not be confused with *wobble* or *variation in latitude* (Chapter 7), which is caused by a slight wandering of the terrestrial poles with respect to the earth's surface. Precession does not affect the cardinal directions on the earth nor the positions of geographical places that are measured with respect to the earth's rotational axis, but only the orientation of the axis with respect to the celestial sphere.

Precession does, however, affect the positions among the stars of the celestial poles, those points where extensions of the earth's axis intersect the celestial sphere. In the 20th century, for example, the north celestial pole is very near Polaris. This was not always so. In the course of 26,000 years, the north celestial pole will move on the celestial sphere along an approximate circle of about $23^{1}/_{2}°$

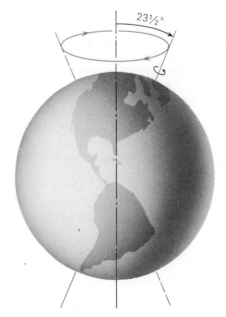

Figure 6.21 Precession of the earth.

radius, centered on the pole of the ecliptic (where the perpendicular to the earth's orbit intersects the celestial sphere). This motion of the pole is shown in Figure 6.22. In about 12,000 years, the celestial pole will be fairly close to the bright star Vega.

As the positions of the poles change on the celestial sphere, so do the regions of the sky that are circumpolar, that is, that are perpetually above (or below) the horizon for an observer at any particular place on earth. The Little Dipper, for example, will not always be circumpolar as seen from north temperate latitudes. Moreover, 2000 years ago, the Southern Cross was sometimes visible from parts of the continental United States. It was by noting the very gradual changes in the positions of stars with respect to the celestial poles that Hipparchus discovered precession in the second century B.C. (Section 2.3c).

(d) NUTATION

If the differential gravitational attractions of the sun and moon upon the earth's equatorial bulge were always exactly the same, precession of the earth's axis would be the smooth conical motion we have described in the preceding sections. However, the effect of the differential forces on the orientation of the earth's axis depends on the directions of the sun and moon with respect to the direction of its 23½° tilt. These directions change as the earth and moon move in their respective orbits. Moreover, the moon's orbit is inclined at about 5° to the ecliptic. Not only is that 5° inclination slightly variable itself, but the intersections of the moon's orbit with the ecliptic slide around the ecliptic in 18.6-year intervals (the regression of the nodes).

The average effect of the sun and moon on the earth's equatorial bulge is to produce the relatively smooth precession we have described. We define the *mean pole* of rotation of the celestial sphere as a fictitious one that describes this smooth precessional motion. The motion of the actual celestial pole varies slightly around the motion of the mean pole. These variations, which are quite small, can be fairly well represented by an elliptical orbit of the actual pole about the mean pole with a semimajor axis of 9".2, and a period of about 19 years. In other words, the

motion of the celestial pole about the ecliptic pole is not quite a perfect circle, but a slightly wavy circle, with the "waves" having amplitudes of about 9 seconds of arc—small compared to the 23½° radius of the precessional orbit of the pole in the sky. This slight "nodding" of the pole about a smooth circle is called *nutation*.

(e) PLANETARY PRECESSION

Up to now we have implied that the plane of the earth's orbit is fixed in space. The earth's orbit, however, is constantly being perturbed by the gravitational attractions of the other planets upon the earth. These perturbations are very slight, but they do measurably alter the plane of the earth's orbit and hence the position of the pole of the ecliptic on the celestial sphere. This motion of the pole of the ecliptic, only a fraction of a second of arc per year, adds to the complications of precession.

The motion of the mean celestial pole with respect to the ecliptic pole is called *lunisolar precession*. The motion of the ecliptic pole, because of planetary perturbations of the earth's orbital motion, is called *planetary precession*; the ecliptic pole moves only about one fortieth as fast as the celestial pole. The two kinds of motion combined give *general precession*.

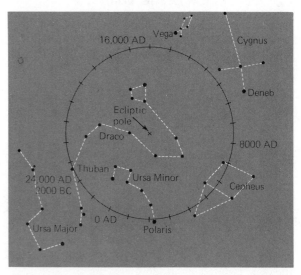

Figure 6.22 Precessional path of the north celestial pole among the northern stars.

EXERCISES

1. Find the separation d between two small bodies, each of unit mass, lined up with a large body of mass M, at a distance R from the nearest of the small bodies, such that the gravitational attraction between the small bodies is just equal to the differential gravitational force between them caused by their attraction to the large body. The answer should be in terms of G, M, and R.

2. If the three bodies described in the last exercise are free to move and no other bodies or forces are present, how may their motion be described for various separations of the small bodies? How do the forces change as the bodies move?

3. If the earth had its present size and rotation rate, but were far more massive than it actually is, would its shape be more or less oblate? Why?

4. The radius of curvature of the earth at a particular point is the radius of a sphere whose surface matches the curvature of the earth at that point (see Figure 6.3). How much greater is the radius of curvature of the earth at the poles than at the equator?

5. Strictly speaking, should it be a 24-hour period during which there are two "high tides"? If not, what should the interval be?

6. Compute the relative tide-raising effectiveness of the sun and the moon. For this approximate calcula-

tion, assume that the earth is 80 times as massive as the moon, that the sun is 300,000 times as massive as the earth, and that the sun is 400 times as distant as the moon.
Answer: Moon is $8/3$ times as effective

7. Explain why the north celestial pole moves in the sky along a circle centered on the pole of the ecliptic, rather than some other point.

8. What will be the principal north circumpolar constellations as seen from Los Angeles (latitude 34° north) in the year 18,000?

9. In the year 13,000, will Orion be circumpolar as seen from the North Pole? Explain.

10. What would be the annual motion of the equinoxes along the ecliptic if the entire precessional cycle required only 360 years?

11. Describe how perturbations of the earth's motion by Mars can be considered as due to differential gravitational force.

12. Does a bicycle offer another example of precession? Explain. (*Hint:* Consider how a rider can steer by leaning to one side.)

13. If the precessional rate is about 50" per year, show that the complete cycle is about 26,000 years.

Tycho Brahe (1546–1601), Danish astronomer whose extensive observations of the planets led to Kepler's discovery of their laws of motion. Brahe made meticulous measurements of the lengths of the seasons, precession of the equinoxes, length of the year, and nearly every other astronomical constant known at the time. (*Yerkes Observatory*)

7

EARTH AND SKY: THE CELESTIAL CLOCKWORK

In the preceding chapters we were concerned with the mechanics that dictate the motions of celestial bodies. Now we turn our attention to the motion of the earth, and the relation between earth and sky. We shall consider the physical properties of the earth as a planet in Chapter 15.

7.1 ROTATION OF THE EARTH

We have seen that the apparent rotation of the celestial sphere could be accounted for either by a daily rotation of the sky around the earth or by the rotation of the earth itself. Copernicus, Kepler, Galileo, and Newton had piled up convincing circumstantial evidence in favor of a rotating earth, but it was not until the nineteenth century that simple direct proofs were devised to show that Newton's laws require that the earth be in rotation.

(a) The Foucault Pendulum

In 1851 the French physicist Jean Foucault suspended a 60 m pendulum weighing about 25 kg from the domed ceiling of the Pantheon in Paris. He started the pendulum swinging evenly by drawing it to one side with a cord and then burning the cord. The direction of swing of the pendulum was recorded on a ring of sand placed on a table beneath its point of suspension. At the end of each swing a pointed stylus attached to the bottom of the bob cut a notch in the sand. Foucault had taken great care to avoid air currents and other influences that would disturb the direction of swing of the pendulum. Yet, after a few moments it became apparent that the plane of oscillation of the pendulum was slowly changing with respect to the ring of sand, and hence with respect to the earth.

The only force acting upon the pendulum was that of gravity between it and the earth, and, of course, this force was in a downward direction. If the earth were stationary, there would be no force that could cause the plane of oscillation of the pendulum to alter, and, in accord with Newton's first law, the pendulum should continue to swing in the same direction. The fact that the pendulum slowly changed its direction of swing with respect to the earth is proof that the earth rotates.

It is comparatively easy to visualize a Foucault pendulum experiment at the North Pole. Here we can imagine the plane of swing of the pendulum maintaining a fixed direction in space with respect to the stars, while the earth turns under it every day. Thus, at the North (or South) Pole, a pendu-

Figure 7.1   A Foucault pendulum. *(Griffith Observatory)*

lum would *appear* to rotate its plane of oscillation once completely in 24 hours (actually, 23 hours 56 minutes—see Chapter 8). At other places than the poles, the problem is complicated because the pendulum must always swing in a vertical plane that passes through the center of the earth. That plane of oscillation obviously must change with respect to the stars.

We must think of the pendulum as measuring the rate at which the earth turns around directly *beneath* it—that is, the rate of rotation of the earth about an imaginary line from the center of the earth out through the point of the pendulum's suspension.* About this line the plane of swing of the pen-

dulum does not rotate. If we imagine ourselves looking down upon the earth's North Pole, we can "see" the earth spinning beneath us like a phonograph record. On the other hand, if we imagine ourselves looking down on the earth's equator, we do not see a rotation of the earth beneath us, only a west-east translational motion. At intermediate latitudes we see beneath us a combination of west-east motion and a certain degree of rotation. At the equator, therefore, a pendulum would not appear to change direction of swing, while at latitudes intermediate between the equator and the poles its period—the time required for it to change its apparent plane of oscillation through 360°—would have a value somewhere between 24 hours and infinity, depending on the exact latitude. For example, at a latitude of 34° (the latitude of Los Angeles), the Foucault pendulum has a period of just under 43 hours. This is the time required for the earth to turn around a line from its center through Los Angeles, or the time required for a spectator to be carried completely around the pendulum by the turning earth.

---

*The following is a brief derivation of the period of a Foucault pendulum: Let the angular velocity of the earth about its polar axis be ω. Then the component of this angular velocity about a radius vector through a point on the earth's surface at latitude ϕ is $\omega \sin \phi$. The time required for one full rotation about this radius vector is $360°/(\omega \sin \phi)$. Since $\omega = 360°/(23$ hours 56 minutes), the period of a pendulum at latitude ϕ is

$$P = \frac{23^{\text{h}}56^{\text{m}}}{\sin \phi}.$$

It should be noted that the turning earth also turns the support system for the pendulum, and consequently the wire and bob of the pendulum itself. However, the rotation of the wire and bob of the pendulum does not alter the direction of swing. Try the following simple experiment. Improvise a small pendulum, say a watch and watch chain. Swing the watch to and fro, holding the end of the chain in your fingers. Now twist the chain in your fingers; the watch will twist with the chain, but will *not* change its direction of swing.

(b) The Coriolis Effect

The apparent rotation of the plane of oscillation of the Foucault pendulum is a demonstration of the rotation of the earth underneath a freely moving body. Any such apparent deflection in the motion of a body, resulting from the earth's rotation, is called the *coriolis effect*. The moving body need not be the bob of a pendulum. Any object moving freely over the surface of the earth appears to be deflected to the right in the Northern Hemisphere (to the left in the Southern Hemisphere) because of the rotation of the earth beneath it. As an example of the effect, consider a projectile fired to the north from the equator.

The projectile starts its northward trip with an *eastward* velocity that it shares with the turning earth just before it is fired (Figure 7.2); at the equator this eastward velocity is about 1700 km/hr.

There is no westward force on the projectile to slow it down, so it continues to move eastward after being fired. Proceeding northward over the curved surface of the earth, however, it comes closer to the axis of the earth's rotation. To conserve its angular momentum (Section 4.1h), the projectile's linear speed to the east must increase if its distance from the axis of rotation decreases. Meanwhile the ground beneath the northbound projectile moves eastward progressively slower, because that ground, closer to the earth's axis, has less far to move in its daily rotation. We see, then, that the eastward speed of the projectile increases and that of the ground beneath it decreases. Thus, relative to the ground, the missile veers off to the east, that is, to the right for one looking in the direction of its motion.

A similar analysis would show that no matter in what direction a projectile moves, in the northern hemisphere it veers off to the *right*, and in the southern hemisphere to the *left* of its target. This

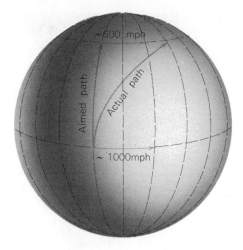

Figure 7.2 Coriolis effect.

effect must be corrected for in the firing of long-range artillery and of course, in the launching of missiles.

Winds blowing toward a low-pressure area similarly veer off to the right of this area (left in the southern hemisphere). However, the force continually trying to equalize the pressure of the air accelerates the wind toward the low-pressure area. The wind, rather than "falling" directly into the low center, is caused to circle *around* the low center by the inertia of the forward moving air (Figure 7.3). If it were not for the earth's rotation, winds would blow directly into low-pressure regions, but because the winds veer off and miss the lows, they end up with a *cyclonic* motion. In the northern

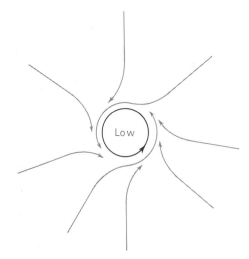

Figure 7.3 Circulation of winds about a low-pressure area in the northern hemisphere.

hemisphere, the winds always blow around storm centers in a *counterclockwise direction,* whether they be hurricanes, tornadoes, or those great general cyclonic storms that blow into our west coast from the Pacific. In the southern hemisphere, the winds are *reversed,* that is, they move around storm centers in a *clockwise* direction.

Falling bodies display a similar phenomenon caused by the rotation of the earth. Suppose a body is dropped down a deep vertical well located, say, at the equator. The eastward velocity of the body increases as it falls toward the axis of rotation. Progressively deeper down the well, on the other hand, the rocks, being closer to the axis of rotation, are moving to the east more and more slowly. Thus, the body is deflected toward the east wall of the well. The principle dictates that if the well were deep enough, the body would strike the east wall.

7.2 RELATION OF EARTH AND SKY

(a) Positions on the Earth

To denote positions of places on the earth, we must set up a system of coordinates on the earth's surface. The earth's axis of rotation (that is, the locations of its North and South Poles) is the basis for such a system.

A *great circle* is any circle on the surface of a sphere whose center is at the center of the sphere. The earth's *equator* is a great circle on the earth's surface halfway between the North and South Poles. We can also imagine a series of great circles that pass *through* the North and South Poles. These circles are called *meridians;* they intersect the equator at right angles.

A meridian can be imagined passing through an arbitrary point on the surface of the earth (see Figure 7.4). This meridian specifies the east-west location of that place. The *longitude* of the place is the number of degrees, minutes, and seconds of arc along the equator between the meridian passing through the place and the one passing through Greenwich, England, the site of the old Royal Observatory. Longitudes are measured either to the east or west of the Greenwich meridian from 0 to 180°. The convention of referring longitudes to the Greenwich meridian is of course completely arbitrary. As an example, the longitude of the bench mark in the clock house of the Naval Observatory

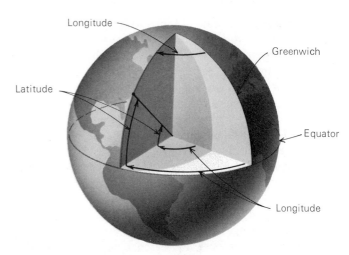

Figure 7.4 Latitude and longitude of Washington, D.C.

in Washington, D.C., is 77°03'56".7 W. Note in Figure 7.4 that the number of degrees along the equator between the meridians of Greenwich and Washington is also the angle at which the planes of those two meridians intersect at the earth's axis.

The *latitude* of a place is the number of degrees, minutes, and seconds of arc measured along its meridian to the place from the equator. Latitudes are measured either to the north or south of the equator from 0 to 90°. As an example, the latitude of the above-mentioned bench mark is 38°55'14".0 N. Note that the latitude of Washington is also the angular distance between it and the equator as seen from the center of the earth.†

(b) Variation of Latitude or Wobble

With astronomical observations it is possible to measure the latitude of a place to within about 0".01, or about 0.3 m. It is found that the latitude of any given place on earth shows a periodic variation of several hundredths of a second of arc or several meters. Apparently, this variation, usually called *wobble* by modern geophysicists, is caused

---

†Strictly, this is *geocentric* latitude. Because of the earth's oblate shape, there are several ways to define latitude. The *geodetic* (or *geographical*) latitude commonly used is defined by the angle between the equatorial plane and the perpendicular to the mean "sea-level" surface of the earth at the place in question. It may differ by several minutes from geocentric latitude. *Astronomical* latitude and longitude, obtained directly from astronomical observations, may differ from geodetic latitude and longitude by a few seconds of arc, owing to the deflection of the plumb bob by mountains or other crustal irregularities.

by a slight shifting of the solid earth with respect to its axis of rotation.

Systematic observations over the earth show that the exact positions of the poles—that is, the ends of the axis of the earth's rotation—wander about with respect to the ground. These wanderings seem to be composed of two independent motions. The first is a motion of each pole along the circumference of an approximate circle about 6 m in diameter, in a period of one year. The second motion has a period of about 14 months and is also a nearly circular wandering of the poles, but the diameter of this circle has varied from 3 to 15 m. The causes of these changes may be natural oscillations of the earth in response to seasonal and also irregular variations in atmospheric density and motions and to crustal shifts of the earth.

The variation of latitude can be thought of as a motion of the terrestrial poles over the ground. However, it is actually the earth itself that does the shifting, while the direction of its axis of rotation remains fixed relative to the stars; that is, the *celestial* poles are unaffected. It should be emphasized that the variation in latitude is very slight—only a few meters. The annual and 14-month wobbles do not seem to be connected in any way with such phenomena as the ice ages.

[Variation of latitude should not be confused with precession (Section 6.5c), which is a slow shifting of the earth's rotation axis with respect to the stars.]

(c) Positions in the Sky

In denoting positions of objects in the sky, it is often convenient to make use of the fictitious *celestial sphere,* a concept, we recall, that many early peoples accepted literally (see Chapter 2). We can think of the celestial sphere as being a hollow shell of extremely large radius, centered on the observer. The celestial objects appear to be set in the inner surface of this sphere, so we can speak of their positions *on* the celestial sphere. We have devised coordinate systems, analogous to latitude and longitude, to designate these positions. Of course we are really only denoting their *directions* in the sky.

The point on the celestial sphere directly above an observer (defined as opposite to the direction of a plumb bob) is his *zenith.* Straight down, 180° from his zenith, is the observer's *nadir.* Halfway between, and 90° from each, is his *horizon.* (This is

the *celestial* horizon and will not necessarily coincide with the apparent horizon, which may be interrupted with such things as mountains, buildings, and trees.) Note that observers at different places have different zeniths, nadirs, and horizons.

Celestial Equator and Poles

The apparent rotation of the sky takes place about an extension of the earth's axis of rotation. That is, the sky appears to rotate about points directly in line with the North and South Poles of the earth— the *north celestial pole* and the *south celestial pole.* Halfway between the celestial poles, and thus 90° from each, is the *celestial equator,* a great circle on the celestial sphere that is in the same plane as the earth's equator; it would appear to pass directly through the zenith of a person on the equator of the earth. Great circles passing through the celestial poles and intersecting the celestial equator at right angles (analogous to meridians on the earth) are called *hour circles.*

The Celestial Meridian

The great circle passing through the celestial poles and the zenith (and also through the nadir) is called the observer's *celestial meridian.* It coincides with the projection of his terrestrial meridian, as seen from the earth's center, onto the celestial sphere. The celestial meridian intercepts the horizon at the *north* and *south* points. Halfway between these north and south points on the horizon are the *east* and *west* points.

As the earth turns, the observer's terrestrial meridian moves under the celestial sphere, sweeping continually eastward around the sky. An equivalent way of putting it is to say that as the sky turns around the earth, the stars pass by the observer's stationary celestial meridian.

It helps to visualize these circles in the sky if we imagine that the earth is a hollow transparent spherical shell with the terrestrial coordinates (latitude and longitude) painted on it. Then if we imagine ourselves at the center of the earth, looking out through its transparent surface to the sky, the terrestrial poles, equator, and meridians will be superimposed upon the celestial ones.

Altitude and Azimuth

The most obvious coordinate system is based on the horizon and zenith of the observer. Great circles

passing through the zenith (*vertical circles*) intersect
the horizon at right angles. Imagine a vertical circle
through a particular star (Figure 7.5). The *altitude*
of that star is the number of degrees along this circle from the horizon up to the star. It is also the
angular "height" of the star as seen by the observer.

The *azimuth* is the number of degrees along the
horizon to the vertical circle of the star from some
reference point on the horizon. In astronomical tradition, azimuth formerly was measured from the
south point on the observer's horizon, but in modern practice azimuth is measured from the north
point, in conformity with the convention of navigators and engineers. In any case, azimuth is measured to the east (clockwise to one looking down
from the sky) along the horizon from 0 to 360°.

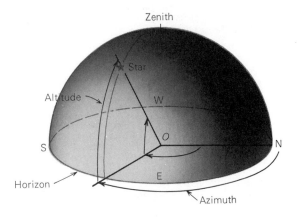

Figure 7.5 Altitude and azimuth.

Right Ascension and Declination

The principal disadvantage of the altitude and azimuth system (the *horizon* system) is that as the
earth turns the coordinates of the celestial objects
are constantly changing. It is desirable, therefore,
to devise a coordinate system that is attached to the
celestial sphere itself, just as the system of latitude
and longitude is permanently attached to the earth.
Then the positions of the stars remain fixed rather
than changing rapidly as the earth's rotation makes
the sky seem to rotate. A system that comes close
to meeting these requirements is *right ascension* and
declination, or the *equator system*.

Right ascension and *declination* bear the same
relation to the celestial equator and poles that longitude and latitude do to the terrestrial equator and
poles. *Declination* gives the arc distance of a star (or
other point on the celestial sphere) along an hour
circle north or south of the celestial equator. *Right
ascension* gives the arc distance measured eastward
along the celestial equator to the hour circle of the
star from a reference point on the celestial equator.
That reference point is the *vernal equinox*, one of
the two points on the celestial sphere where the celestial equator and the ecliptic intersect. Because of
precession, both the celestial equator and the vernal
equinox slowly move with respect to the stars; thus
the right ascension and declination of a star continually change, but the changes are so gradual as not
to be important, for most purposes, over a period
of one year or so. The lack of constancy of right
ascension and declination makes the system less
than ideal, but it is still the most convenient one

available, for it is based on the celestial equator and
is thus symmetrical with respect to the earth's axis
of rotation. Right ascension and declination are
therefore very useful for pointing telescopes and
moving them to follow the daily motions of the stars
(Chapter 11).

Several celestial coordinate systems are in common use. Each has its advantages for special purposes and is important to astronomers. These systems are defined in Appendix 7.

(d) The Orientation of the Celestial Sphere

The next step is to determine the orientation of the
celestial sphere with respect to the zenith and horizon of a particular observer on the earth. At the
North (or South) Pole, the problem is very simple
indeed. The north celestial pole, directly over the
earth's North Pole, appears at the zenith. The celestial equator, 90° from the celestial poles, lies
along the horizon. An observer at one of the terrestrial poles would never see more than half the sky.

At the equator the problem is almost as simple.
The celestial equator, in the same plane as the
earth's equator, passes through the zenith, and
since it runs east and west, it intersects the horizon
at the east and west points. The celestial poles,
being 90° from the celestial equator, must be at the
north and south points on the horizon. Evidently at
points on the earth between the equator and poles,
one of the celestial poles must be a certain distance
above the horizon.

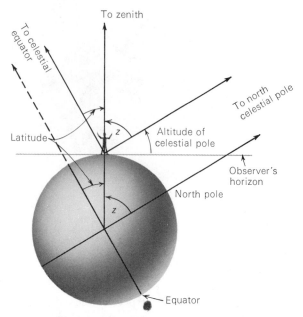

Figure 7.6 The altitude of the celestial pole equals the observer's latitude.

Figure 7.6 shows an observer at an arbitrary latitude north of the equator; for a southern latitude the case would be exactly analogous. Since the terrestrial North Pole is on the observer's terrestrial meridian, the north celestial pole will have to be on his celestial meridian, at some altitude above the north point on the horizon. Suppose the angle from the observer's zenith down to the north celestial pole is *z*. Being on the celestial sphere, the north

celestial pole is so distant that it is seen in the same direction as by an observer at the center of the earth. Thus the angle at the center of the earth between the observer and the north celestial pole is also *z*. (See Figure 7.6; the angles of intersection between each of two parallel lines and a third line are equal.)

We recognize that *z* is just 90° *minus* the observer's latitude. But also *z* is 90° (the altitude of the zenith) *minus* the altitude of the north celestial pole. Thus we see that *the altitude of the north (or south) celestial pole is equal to the observer's north (or south) latitude.*

Finally, since the celestial equator is 90° from the celestial poles, it must cut through the east and west points on the horizon, tilt southward as it extends up above the horizon, and cross the celestial meridian a distance south of the zenith that is also equal to the observer's latitude.

(e) The Motion of the Sky as Seen from Different Places on Earth

Imagine an observer at the earth's North Pole. The celestial north pole is at his zenith and the celestial equator along his horizon. As the earth rotates, the sky turns about a point directly overhead. The stars neither rise nor set; they circle parallel to the horizon. Only that half of the sky that is north of the celestial equator is ever visible to this observer. Similarly, an observer at the South Pole would only see the southern half of the sky (Figure 7.7).

The situation is very different for an observer at the equator (Figure 7.8). There the celestial

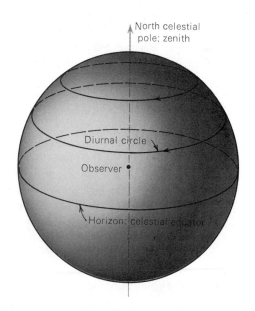

Figure 7.7 Sky from the North Pole.

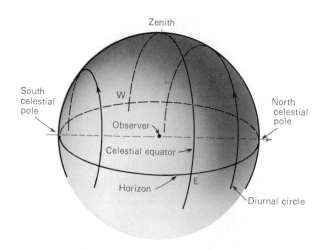

Figure 7.8 Sky from the equator.

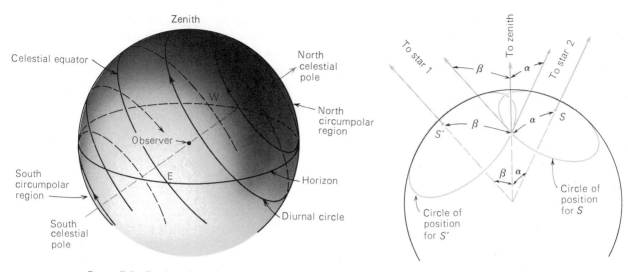

Figure 7.9 Sky from latitude 34° N.

Figure 7.10 Circles of position.

poles, the points about which the sky turns, lie at the north and south points on the horizon. All stars rise and set; they move straight up from the east side of the horizon and set straight down on the west side. During a 24-hour period, all stars are above the horizon exactly half the time.

For an observer between the equator and North Pole, say at 34° north latitude, the situation is as depicted in Figure 7.9. Here the north celestial pole is 34° above the observer's northern horizon. The south celestial pole is 34° *below* the southern horizon. As the earth turns, the whole sky seems to pivot about the north celestial pole, and the stars appear to circle around parallel to the celestial equator. For this observer, stars within 34° of the north celestial pole can never set. They are always above the horizon, day and night. This part of the sky is called the *north circumpolar zone* for the latitude 34° N. To observers in the United States, the Big and Little Dippers and Cassiopeia are examples of star groups that are in the north circumpolar zone. On the other hand, stars within 34° of the south celestial pole never rise. That part of the sky is the *south circumpolar zone*. To most U.S. observers, the Southern Cross is in that zone. At the North or South Pole the entire sky is circumpolar, half of it above the horizon and half below.

Stars north of the celestial equator, but outside the north circumpolar zone, in the greater parts of their daily paths, or *diurnal circles*, lie above the horizon; hence they are up more than half the time. Stars *on* the celestial equator are up exactly half the time, for their diurnal circle is the celestial equator; because it is a great circle, exactly half of it must be above the horizon. Stars south of the celestial equator, but outside the south circumpolar zone, are up less than half the time.

(f) CELESTIAL NAVIGATION

The concepts that we have just dealt with have been applied by navigators for centuries to determine position at sea. The celestial coordinates of the sun, the moon, the planets, and the brighter stars are computed for years in advance (in this country at the U.S. Naval Observatory) and are tabulated day by day in such publications as the *Nautical Almanac*. To find his position, a navigator must measure the altitudes of at least two celestial objects with a sextant (or of one object at least two different times), have available the celestial coordinates of those objects, say, from the *Almanac*, and for each of his observations know the exact time at some known place on earth—usually Greenwich, England. He obtains the time from a *chronometer*, an accurate clock set to keep Greenwich time. In modern practice, the chronometer is checked regularly with radio broadcasts of the accurate Greenwich or Washington time.

The basic principle of *celestial navigation* is that of Eratosthenes' method of determining the size of the earth (Chapter 2). The celestial bodies are so distant that the direction to any of them from all places on earth must be the same. One can easily calculate where on earth a particular star or planet will appear at the local zenith. Let that place be denoted S (Figure 7.10). Now the angle in the sky between the navigator's zenith and that star must be the same as the angle at the center of the earth between him and the *substellar point, S.* Thus, the navigator knows that he is somewhere along a circle on the surface of the earth whose radius, in degrees, is equal to the zenith angle of the star. The circle is called a *circle of position.* By observing a *second* star, which must be at the zenith of a place denoted *S',* the navigator finds that in addition he must be somewhere along a second circle of position, centered at *S'.* Since there are only two points of intersection of the two circles of position, there are only two places on earth where the ship can be. Usually, the navigator has a clear enough idea of where he is to eliminate one of those places (it might, for example, lie in a different ocean, or on land). If not, observation of a third star settles the matter.

In present-day practice, the procedure of celestial navigation is roughly as follows: First, the navigator measures the altitudes of several stars, and notes the times of his observations. Then he assumes a position (latitude and longitude) for his ship (or airplane) for the times of those observations. He can usually make an intelligent guess of his position by having kept track of the ship's speed and direction since his last *fix* (determination of his location). Thus, with the aid of the data in the *Almanac,* and from a knowledge of Greenwich time, he calculates what the altitude should be for each of the stars he observed, assuming that his guess of his position was accurate.

If he has actually guessed his position correctly, the observed and computed altitudes for the stars should agree. Generally, there are small discrepancies that reflect the error in his assumed position. The amount of the discrepancies tells him just how far off his guess was and how to correct it to obtain his actual position.

(g)   Nomenclature of Stars

While we are discussing the celestial sphere, it is in order to describe briefly the system for naming the stars and constellations. As has been stated (Section 2.2a), the ancients designated certain apparent groupings of stars in honor of characters or animals in their mythology. We retain most of these *constellations* today, although their number has been augmented to 88 (they are listed in Appendix 20). By action of the International Astronomical Union in 1928, the boundaries between constellations were established as east-west lines of constant declination and north-south segments of hour circles. Because of precession, over the years the constellation boundaries have gradually tilted slightly from precisely north-south and east-west. Although all constellation boundaries ran north-south and east-west,

they nevertheless jogged about considerably, so that the modern constellations still contain most or all of the brighter stars assigned to them by the ancients. Consequently, the boundaries often delineate highly irregular regions of the sky, reminding one of the boundaries between Congressional districts that result from the gerrymandering practices of many state legislatures. At any rate, because every position on the celestial sphere (or every direction in the sky) lies in one or another constellation, we commonly use constellations today to designate the places of stars or other celestial objects.

Many of the brighter stars have proper names. Often these are Arabic names that describe the positions of the stars in the imagined figures that the Greek constellations represented. For example, *Deneb,* is Arabic for "tail" and is the star that marks the tail of *Cygnus* (the Swan), and *Denebola* is the star at the tail of *Leo* (the Lion). Some star names, however, are of modern origin, for example, *Polaris,* the "pole star."

A superior designation of stars was introduced by the Bavarian, J. Bayer, in his *Atlas* of the constellations, published in 1603. He assigned successive letters of the Greek alphabet to the more conspicuous stars in each constellation, in approximate order of decreasing brightness. The full star designation is the Greek letter, followed by the genitive form of the constellation name. Thus Deneb, the brightest star in Cygnus, is α *Cygni,* and Denebola, the second brightest in Leo, is β *Leonis.* Bayer's ordering of stars by brightness was not always correct, and on occasion he deviated from the scheme altogether and assigned letters to stars of comparable brightness according to their geometrical arrangement in the constellation figure, for example, in the Big Dipper.

Fainter stars were subsequently given number designations, with the numbers increasing in order of the stars' right ascensions. The majority of stars, however, too faint to see without a telescope, are designated, if at all, only by their numbers in various catalogues. The most famous and extensive catalogue, which contains one third of a million stars, was compiled in the years following 1837 by F. W. Argelander at the Bonn Observatory. Stars in this Bonn Catalogue, or *Bonner Durchmusterung,* are known by their *BD numbers.* The Bonner Durchmusterung was later extended to the part of the sky too far south to observe at Bonn and was eventually supplemented with a catalogue of the southernmost stars, made at Cordoba in Argentina.

Many other catalogues, with ever-increasing accuracy of the star positions they record, have been and are being compiled. Many star catalogues are prepared for special purposes, or list only certain types of stars, or stars in certain regions of the sky. A commonly used catalogue produced by the Harvard College Observatory gives the spectral types of the stars (Chapter 25); stars in this *Henry Draper Catalogue* are denoted by their *HD numbers*. Data for the nearest stars, listed in Appendix 13, are seen, from the variety of nomenclature, to have been selected from several different catalogues. Many stars, of course, are listed in more than one catalogue and thus bear various names—a circumstance that has sometimes confused even astronomers.‡ The vast majority of stars, however, too faint and numerous to measure and catalogue, remain nameless.

7.3 THE REVOLUTION OF THE EARTH

We have seen (Chapter 2) that the earth's revolution about the sun produces an apparent annual motion of the sun along the ecliptic. However, we have also seen that the sun's apparent revolution about the earth can be explained either by a motion of the sun or of the earth. What evidence is there that it is the earth, not the sun, that moves?

(a) Proofs of the Earth's Revolution

If we adopt Newton's laws of motion and gravitation, it follows simply and directly that. the earth must revolve about the sun and not vice versa. It is obvious that *either* the earth goes around the sun or the sun around the earth. Thus we have a system of *two mutually revolving bodies*. The problem is simply to determine where the common center of revolution is. In Chapter 31 we shall see that the sun is about 330,000 times as massive as the earth. Thus the common center of mass (Section 5.1) of the earth-sun system must be less than 1/300,000 of the

‡A theoretical astronomer in France published, some years ago, a paper in which he reported calculations he had performed to derive some of the internal properties of a number of stars. His calculations were based on observational data selected by him from various catalogues. One of the stars he gave results for is Sirius; another is α Canis Majoris. He did not know that these were the same star!

distance from the center of the sun to the center of the earth. This puts it well inside the surface of the sun. Essentially, then, the earth revolves around the sun.

There are also some geometrical consequences of the earth's revolution that would be very difficult to explain if the earth were assumed to be stationary. These are *stellar parallax* and *aberration of starlight*. We shall discuss stellar parallax in Chapter 22.

To understand aberration, consider the analogy of walking in the rain, holding a straight drainpipe (Figure 7.11). If the drainpipe is held vertically, and if the raindrops are assumed to fall vertically, they will fall through the length of the pipe only if you are standing still. If you walk forward, you must tilt the pipe slightly forward, so that drops entering the top will fall out the bottom without being swept up by the approaching inside wall of the pipe. If the raindrops fall with a speed V, and if you walk with a speed v, the distance by which the top of the pipe precedes the bottom, divided by the vertical distance between the top and bottom of the pipe, must be in the ratio v/V.

Similarly, because of the earth's orbital motion, if starlight is to pass through the length of a telescope, the telescope must be tilted slightly forward in the direction of the earth's motion. In other words, the apparent direction of a star is displaced slightly from its geometrical direction, and the displacement is in the direction of the earth's orbital motion. Analogous to the tilt of the drainpipe, this forward tilt of the telescope is in the ratio of the speed of the earth to the speed of light. The speed

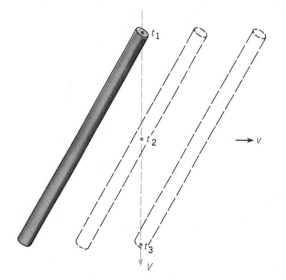

Figure 7.11 Raindrops falling through a moving drainpipe.

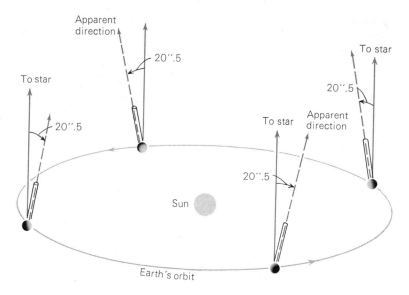

Figure 7.12 Aberration of starlight.

of light is about 10,000 times that of the earth in its orbit, so the angle through which a telescope must be tilted forward is 1 part in 10,000 or about 20″.5. The effect is greatest when the earth is moving at right angles to the direction of the star, and disappears when the earth moves directly toward or away from the star. A star that is on the ecliptic appears to shift back and forth in a straight line during the year, for through part of the year the earth is moving in one direction compared to the star and during the rest of the year the earth is moving in the opposite direction. A star in a direction perpendicular to the earth's orbit appears to describe a small circle in the sky, for its apparent direction is constantly displaced in the direction of the earth's orbital motion from the direction it would have as seen from the sun. Stars between these extremes appear to shift their apparent directions along tiny elliptical paths of semimajor axis 20″.5 (Figure 7.12).

7.4 THE SEASONS

The earth's orbit around the sun is an ellipse, its distance from the sun varying by about 3 percent. However, the changing distance of the earth from the sun is *not* the cause of the seasons. The seasons result because the plane in which the earth revolves is not coincident with the plane of the earth's equator. The planes of the equator and ecliptic are inclined to each other by about $23\frac{1}{2}°$. This angle of $23\frac{1}{2}°$ is called the *obliquity of the ecliptic*.

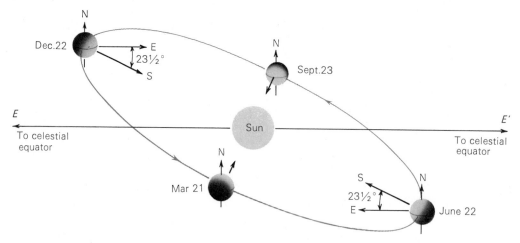

Figure 7.13 The seasons are caused by the inclination of the plane of the earth's orbit to the plane of the equator.

Globes of the earth are usually mounted with the earth's axis tilted from the vertical. This tilt is the same angle of $23\frac{1}{2}°$, for that is the angle the earth's axis must make with the perpendicular to the plane of its orbit around the sun. The result of the obliquity of the ecliptic is that the northern hemisphere is inclined *toward* the sun in June and away from it in December.

(a) The Seasons and Sunshine

Figure 7.13 shows the earth's path around the sun. The line EE' is in the plane of the celestial equator. In the figure the earth appears to pass alternately above and below this plane, but the celestial sphere is so large, and the celestial equator so far away, that a line from the center of the earth through the earth's equator always points to the celestial equator.

We see in the figure that on about June 22 (the date of the *summer solstice*), the sun shines down most directly upon the northern hemisphere of the earth. It appears $23\frac{1}{2}°$ *north* of the equator and thus on that date passes through the zenith of places on the earth that are at $23\frac{1}{2}°$ north latitude. The situation is shown in detail in Figure 7.14. To an observer on the equator, the sun appears $23\frac{1}{2}°$ north of the zenith at noon. To a person at a latitude $23\frac{1}{2}°$ N, the sun is overhead at noon. This latitude on the earth, at which the sun can appear at the zenith at noon on the first day of summer, is called the *Tropic of Cancer*. We see also in Figure 7.14

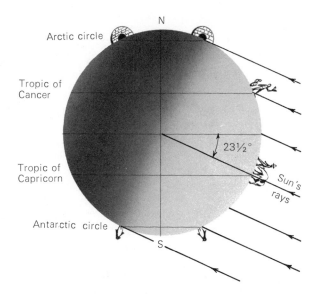

Figure 7.15 The earth on December 22.

that the sun's rays shine down past the North Pole; in fact, all places within $23\frac{1}{2}°$ of the pole, that is, at a latitude greater than $66\frac{1}{2}°$ N, have sunshine for 24 hours on the first day of summer. The sun is as far north on this date as it can get; thus, $66\frac{1}{2}°$ is the southernmost latitude where sun can ever be seen for a full 24-hour period (the *midnight sun);* that circle of latitude is called the *Arctic Circle*.

During this time, the sun's rays shine very obliquely on the southern hemisphere. In fact, all places within $23\frac{1}{2}°$ of the South Pole—that is, south of latitude $66\frac{1}{2}°$ S (the *Antarctic Circle*)— have no sight of the sun for the entire 24-hour period.

The situation is reversed six months later, about December 22 (the date of the *winter solstice*), as is shown in Figure 7.15. Now it is the Arctic Circle that has a 24-hour night and the Antarctic Circle that has the midnight sun. At latitude $23\frac{1}{2}°$ S, the *Tropic of Capricorn*, the sun passes through the zenith at noon. It is winter in the northern hemisphere, summer in the southern.

Finally, we see in Figure 7.13 that on about March 21 and September 23 the sun appears to be in the direction of the celestial equator, and, on these dates, the equator itself is the diurnal circle for the sun. Every place on the earth then receives exactly 12 hours of sunshine and 12 hours of night. These points, where the sun crosses the celestial equator, are called the *vernal* (spring) *equinox* and *autumnal* (fall) *equinox*. *Equinox* means "equal night."

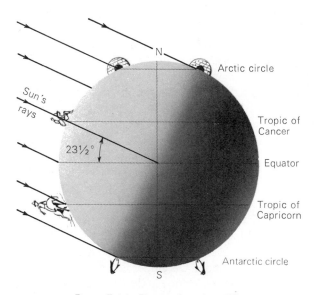

Figure 7.14 The earth on June 22.

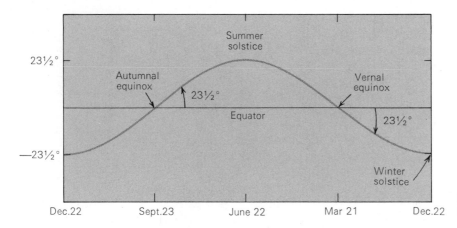

Figure 7.16 Plot of the ecliptic around the celestial equator.

Figure 7.16 is a map in which the sky is shown flattened out, as in a Mercator projection of the earth. The equator runs along the middle of the map, and the ecliptic is shown as a wavy line crossing the equator at the two equinoxes. Both equator and ecliptic are, of course, great circles, but they cannot both be shown as straight lines on a flat surface. Notice that the ecliptic intersects the equator at an angle of $23^{1}/_{2}°$, and that its northernmost extent is $23^{1}/_{2}°$ north of the equator (the summer solstice) and its southernmost extent is $23^{1}/_{2}°$ south of the equator (the winter solstice).

Figure 7.17 shows the aspect of the sky at a typical latitude in the United States. During the spring and summer, the sun is north of the equator and is thus up more than half the time. A typical spot in the United States, on the first day of summer (about June 22), receives about 14 or 15 hours of sunshine. Also, notice that the sun appears *high*

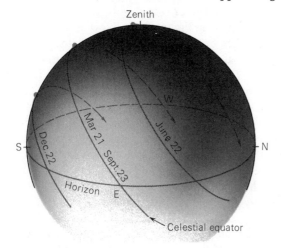

Figure 7.17 Diurnal paths of the sun for various dates at a typical place in the U.S.

in the sky, and so in these seasons the sunlight is more direct, and thus more effective in heating than in the fall and winter when the sun appears at a lower altitude in the sky.

In the fall and winter the sun is south of the equator, where most of its diurnal circle is below the horizon, and so it is up less than half the time. On about December 22, a typical city at, say 30° to 40° north latitude receives only nine or ten hours of sunshine. Also, the sun is low in the sky; a bundle of its rays is spread out over a larger area on the ground (Figure 7.18) than in summer; because the energy is spread out over a larger area, there is less for each square meter, and so the sun at low altitudes is less effective in heating the ground.

(b) The Seasons at Different Latitudes

At the equator all seasons are much the same. Every day of the year, the sun is up half the time, so there are always 12 hours of sunshine at the equator. About June 22, the sun crosses the meridian $23^{1}/_{2}°$ north of the zenith, and about December 22, $23^{1}/_{2}°$ south of the zenith.

The seasons become more pronounced as one travels north or south of the equator. At the Tropic of Cancer, on the date of the summer solstice, the sun is at the zenith at noon. On the date of the winter solstice, the sun crosses the meridian 47° south of the zenith. At the Arctic Circle, on the first day of summer the sun never sets, but at midnight can be seen just skimming the north point on the horizon. About December 22, the sun does not quite rise at the Arctic Circle, but just gets up to the south point on the horizon at noon. Between

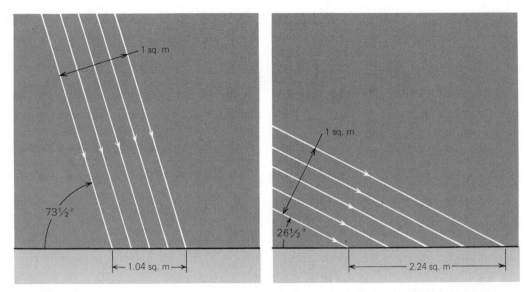

Figure 7.18 Effect of the sun's altitude. When the sun is low in the sky its rays are more oblique to the ground and are spread over a larger area than when the sun is high.

the Tropic of Cancer and the Arctic Circle, the number of hours of sunshine and the noon altitude of the sun range between these two extremes.

We recall that at the North Pole, all celestial objects that are north of the celestial equator are always above the horizon and, as the earth turns, circle around parallel to it. The sun is north of the celestial equator from about March 21 to September 23, and so at the North Pole the sun rises when it reaches the vernal equinox and sets when it reaches the autumnal equinox. There are six months of sunshine at the Pole. The sun reaches its maximum altitude of 23½° about June 22; before that date it climbs gradually higher each day, and after that it drops gradually lower. A navigator can easily tell when he is at the North Pole, for there the sun circles around the sky parallel to the horizon, getting no higher or lower (except gradually as the days go by).§

In the southern hemisphere, the seasons are reversed from those in the north. While we are having summer in the United States, in Australia it is winter. Furthermore, in the southern hemisphere, the sun crosses the meridian generally to the *north*

of the zenith. In Buenos Aires, you would want a house with a good *northern* exposure.

The earth, in its elliptical orbit, reaches its closest approach to the sun about January 4. It is then said to be at *perihelion*. It is farthest from the sun, at *aphelion*, about July 5. We see, then, that the earth is closest to the sun when it is winter in the north. However, it is summer in the southern hemisphere when the earth is at perihelion, and the earth is farthest from the sun during the southern hemisphere's winter. Therefore, we might expect the seasons to be somewhat more severe in the southern hemisphere than in the northern. However, there is more ocean area in the southern hemisphere; this and other topographical factors are more important in their influence on climate than is the earth's changing distance from the sun. We shall see that for Mars, whose orbit is considerably more eccentric than the earth's, the same kind of situation does have a pronounced effect upon the seasons.

(c) Precession of the Equinoxes

As the earth's axis precesses in its conical motion (Section 6.5c), the equatorial plane retains (approximately) its 23½° inclination to the ecliptic plane; the obliquity of the ecliptic remains constant. However, the intersections of the celestial equator and the ecliptic (the equinoxes) must always be 90° from the celestial poles (because all points on the celestial

§It is said that one botanist considered the North Pole to be an excellent place to raise sunflowers, because there were so many hours of sunshine during the summer months. He accordingly planted some there, and they did quite well for a while. However, sunflowers like to *face* the sun, and as they followed the sun around and around the sky, they ended by wringing their own necks!

equator are 90° from the celestial poles). Thus, as the poles move because of precession, the equinoxes slide around the sky, moving westward along the ecliptic. This motion is called the *precession of the equinoxes*. The angle through which the equinoxes move each year, the *annual precession*, is 1/26,000 of 360°, or about 50″. Each year as the sun completes its *eastward* revolution about the sky with respect to, say, the vernal equinox, that equinox has moved *westward*, to meet the sun, about 50″. Since it takes the sun about 20 minutes to move 50″ along the ecliptic (or, more accurately, because it takes the earth that long to move through an angle of 50″ in its orbit about the sun), a *tropical year*, measured with respect to the equinoxes, is 20 minutes shorter than a *sidereal* year, measured with respect to the stars.

Precession has no important effect on the seasons. The earth's axis retains its inclination to the ecliptic, so the northern hemisphere is still tipped toward the sun during one part of the year and away from it during the other. Our calendar year is based on the beginnings of the seasons (the times when the sun reaches the equinoxes and solstices), so spring in the northern hemisphere still begins in March, summer in June, and so on. The only effect is that as the precessional cycle goes on, a given season will occur when the earth is in gradually different places in its orbit with respect to the stars. In the 20th century, for example, Orion is a *winter constellation*; we look out at night, away from the sun, and see Orion in the sky during the winter months. In 13,000 years, half a precessional cycle later, it will be summertime when we look out in the same direction, away from the sun, and see Orion. Similarly, Scorpio is a summer constellation now, whereas in the year 15,000 it will be a winter constellation (see Figure 7.19).

The vernal equinox is sometimes called the *first point of Aries*, because about 2000 years ago, when it received that name, it lay in the constellation of Aries. Now, because of precession, the vernal equinox has slid westward into the constellation of Pisces.

(d) Twilight

We all know that the sky does not immediately darken when the sun sets. Even after the sun is no longer visible from the ground, the upper atmosphere of the earth can catch some of the rays of the setting sun and scatter them helter-skelter, il-luminating the sky. Gases in the earth's atmosphere are dense enough to scatter appreciable sunlight up to altitudes of about 300 km. The sun must be at least 18° below the horizon for all traces of this postsunset or presunrise sky light (*twilight*) to be absent. At latitudes near the equator, where the sun rises and sets nearly vertically to the horizon, twilight lasts only a little over one hour. However, at far northern and southern latitudes, the sun rises and sets in a much more oblique direction and takes correspondingly longer to reach a point 18° below the horizon, so twilight may last for two hours or more. In the far northern countries twilight lasts all night in the summertime. At the North Pole there are six weeks of twilight in the late winter before sunrise and again in the early fall after sunset.

(e) Lag of the Seasons

Places in the northern hemisphere receive the most sunlight about June 22, and the least about December 22. Yet these are not the dates of the hottest and coldest weather of the year, respectively. The hottest and coldest weather generally lags behind the opening of summer and winter by several weeks.

The actual amount of heat received from the sun by a given area on the earth is called its *insolation*. In the northern hemisphere, the insolation is greatest on the first day of summer and decreases thereafter. During the preceding winter, however, the hemisphere had cooled considerably, and large deposits of ice and snow had formed. During the spring, as the insolation increases, these snow deposits slowly melt, and the hemisphere gradually warms up. This warming-up process continues past the date of the summer solstice. In other words, during the early weeks of summer, a good portion of the insolation is going into the melting of ice and the heating of the land and oceans. The temperatures do not reach their maxima until the portions of the earth that contribute to the climate of a particular place have warmed as much as they are going to. This usually occurs in August in the northern hemisphere, although the date varies from place to place with the local topography.

Similarly, the coldest time of year is not at the winter solstice, even though that is the time of the least insolation, because the land and ocean, having been warmed up in the previous summer, are still cooling down. They reach their maximum chill sometime in midwinter.

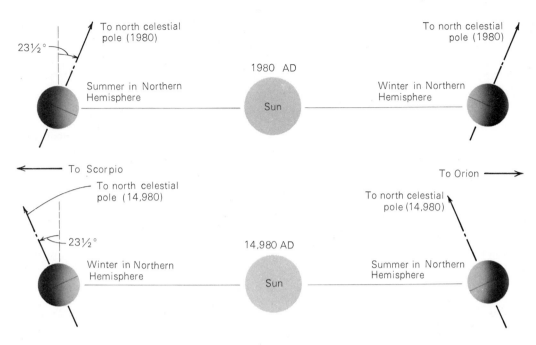

Figure 7.19 Summer and winter constellations change because of precession.

7.5 THE MANY MOTIONS OF THE EARTH

In this chapter we have discussed in detail two of the earth's motions: rotation and revolution. However, there are many other motions of the earth, dealt with in other sections. Here, for completeness, these motions are summarized:

1. The earth *rotates* daily on its axis.
2. The earth periodically shifts slightly with respect to its axis of rotation (*variation in latitude*) (Section 7.2b).
3. The earth *revolves* about the sun.
4. The gravitational pull of the sun and moon on the earth's equatorial bulge causes a very slow change in orientation of the axis of the earth called *precession* (Section 6.5).
5. Because the moon's orbit is not quite in the plane of the ecliptic, and because of a slow change in orientation of the moon's orbit, there is a small periodic motion of the earth's axis superimposed upon precession, called *nutation* (Section 6.5d).
6. Actually, it is the center of mass of the earth-moon system, or *barycenter*, that revolves about the sun in an elliptical orbit. Each month the center of the earth revolves about the barycenter (Section 5.1).

7. The earth shares the motion of the sun and the entire solar system among its neighboring stars. This *solar motion* is about 20 km/s (Chapter 23).
8. The sun, with its neighboring stars, shares in the general *rotation of the Galaxy*. Our motion about the center of the Galaxy is about 250 km/s (Chapter 29).
9. All other galaxies are observed to be in motion. Therefore, our *Galaxy is in motion* with respect to other galaxies in the universe (Chapter 39).
10. In view of the many motions of the earth, one might wonder what the absolute speed of the earth is in the universe. To determine the absolute velocity of the earth was the object of the Michelson-Morley experiment, which we shall see (Chapter 13) gave null results. A fundamental postulate of Einstein's special theory of relativity is that it is not possible to define an absolute coordinate system with respect to which the absolute speed of an object in space can be determined. We shall see in Chapter 39, however, that we now believe we have measured the speed of our Galaxy with respect to the average distribution of the matter in the universe around us. That speed is about 500 km/s.

EXERCISES

1. Show that the apparent deflection in the direction of swing of a Foucault pendulum in the southern hemisphere is to the *left*, rather than to the *right* as in the northern hemisphere.

2. What is the latitude of (a) the North Pole? (b) the South Pole?

3. Why has longitude no meaning at the North or South Pole?

4. Draw a diagram to show that for an observer south of the equator the altitude of the south celestial pole is equal to his latitude south.

5. Why is exactly half of any great circle in the sky above the horizon at once?

6. Prove that the celestial equator must pass through the east and west points on the horizon.

7. Suppose you observe a star 30° from the south celestial pole pass through your zenith. Which of the following will be above your horizon sometime during a 24-hour period? (a) Big Dipper; (b) south celestial pole; (c) Orion; (d) Southern Cross; (e) north ecliptic pole; (f) autumnal equinox; (g) nadir; (h) Ursa Major.

8. If a star rises in the northeast, in what direction does it set?

9. Prove that if vertically falling raindrops dropping with a speed of V are to fall through a drainpipe, the pipe must be tilted forward so that its top precedes its bottom by a distance which, when divided by the vertical extent of the pipe, is in the ratio v/V, where v is the speed of the drainpipe.

10. How can we tell that stars are displaced in the direction of the earth's motion (aberration), since all stars in a given part of the sky appear shifted by the same amount?

11. Where on earth is it possible for the ecliptic to lie, momentarily, along the horizon?

12. Explain why New York has more hours of daylight on the first day of summer than does Los Angeles.

13. Suppose the obliquity of the ecliptic were only $16\frac{1}{2}°$. What then would be the difference in latitude between the Arctic Circle and the Tropic of Cancer?

14. If the obliquity were only $16\frac{1}{2}°$, what would be the effect on the seasons as compared to the actual obliquity of $23\frac{1}{2}°$?

15. What are the approximate dates of sunrise and sunset at the South Pole? Would a lunar eclipse occurring in January be visible from there? Why or why not?

16. In far northern countries such as Canada and Scotland, the winter months are so cloudy that astronomical observations are nearly impossible. Why is it that good observations cannot be made in those places during summer nights?

Simon Newcomb (1835–1909) was the great American astronomer who laid a foundation of precise positional astronomy based on his measurements of the motions of the earth. (*U.S. Naval Observatory*)

TIME AND DATE

One of the most ancient uses of astronomy was the keeping of time and the calendar. From the earliest history in virtually every center of civilization —China, India, Mesopotamia, Egypt, Greece, the Mayan and Aztec civilizations in the western hemisphere, and evidently even in Bronze-Age England when Stonehenge was built—man followed the motions in the heavens to calculate the time and the date.

8.1 TIME OF DAY

The measurement of time is based on the rotation of the earth. As the earth turns, objects in the sky appear to move around us, crossing the meridian each day. Time is determined by the position in the sky, with respect to the local meridian, of some reference object on the celestial sphere. The interval between successive meridian crossings or *transits* of that object is defined as a *day*. The actual length of a day depends on the reference object chosen; several different kinds of days, corresponding to different reference objects, are defined. Each kind of day is divided into 24 equal parts, called *hours*.

(a) The Passage of Time; Hour Angle

Time is reckoned by the angular distance around the sky that the reference object has moved since it last crossed the meridian. The motion of that point around the sky is like the motion of the hour hand on a 24-hour clock. The angle measured to the west along the celestial equator from the local meridian to the hour circle passing through any object (for example, a star) is that object's *hour angle*. (An hour circle is a great circle on the celestial sphere running north and south through the celestial poles—see Chapter 7.) *Time* can be defined as the *hour angle of the reference object*.

As an example, suppose that the star *Rigel* is chosen as the reference for time. Then when Rigel is on the meridian it is $0^h0^m0^s$, "Rigel time." Twelve *Rigel hours* later, Rigel is halfway around the sky, at an hour angle of 180°, and the Rigel time is $12^h0^m0^s$, When Rigel is only 1° east of the meridian, and one *Rigel day* is nearly gone, the star is at an hour angle of 359°, and the Rigel time is $23^h56^m0^s$.

Time can be represented graphically by means of a *time diagram*, as in Figure 8.1. Here we imagine ourselves looking straight down on the north celestial pole from *outside* the celestial sphere. The pole appears as a point in the middle of the diagram, and the celestial equator appears as a circle centered on the pole. As the earth turns to the east, the local meridian of an observer sweeps around the sky, so that its intersection with the celestial equator would move *counterclockwise* around the circle in the time diagram. However, it is customary to represent the observer's meridian as fixed, intersecting

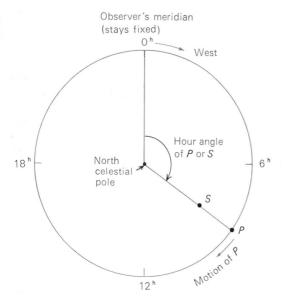

Observer's meridian (stays fixed)

Figure 8.1 Time diagram.

the equator, say, at the top of the diagram. Then the celestial sphere must be regarded as rotating *clockwise* with respect to the meridian. Let the reference object be denoted by S. Its hour circle intersects the equator at P, and as the celestial sphere rotates, the point P moves clockwise around the circle, like the hour hand of a clock. The hour angle of S (or P) increases uniformly with the rotation of the celestial sphere. In the diagram it is shown as about 120° (the time would then be about 8^h. Since the celestial equator intersects the horizon at the east and west points, P is below the observer's horizon in the time interval from 6 to 18 hours; the same is not true of S unless it happens to lie on the equator.

Because of the relation between hour angle and time, it is often convenient to measure angles in time units. In this notation, 24 hours corresponds to a full circle of 360°, 12 hours to 180°, 6 hours to 90°, and so on. One hour equals 15°, and 1° is four minutes of time.

Here we must distinguish between minutes and seconds of *time* (subdivisions of an hour), denoted by ^{m} and ^{s}, respectively, and minutes and seconds of *arc* (subdivisions of a degree), denoted by ′ and ″, respectively. The conversion between units of time and arc is given in Table 8.1.

Because we base time on the rotation of the earth, it might seem that the earth's rotation with respect to the stars would be the nearest we could come to defining the "true rotation rate" of the earth. We could then measure the passage of time by the hour angle of some fixed point on the celestial sphere. The interval between two successive meridian transits of this point would define such a "stellar day." This kind of rotation period of the earth would be measured, say, by the Foucault pendulum. (Actually, of course, the stars are in motion and do not define a perfectly fixed reference system; however, they are so distant that over moderate periods of time they appear fixed on the celestial sphere.)

We could measure time by such a scheme, but it turns out that other kinds of time are more convenient. The most common kinds of time in use are *solar* (sun) time and *sidereal* (star) time, which are based on the solar day and the sidereal day.

(b) The Solar and Sidereal Day

The solar day is the period of the earth's rotation with respect to the sun. The sidereal day is, instead, the time required for the earth to make a complete rotation with respect to a point in space. The point chosen is the *vernal equinox*, defined as the point on the celestial sphere where the sun in its apparent path around the sky (the ecliptic) crosses the celestial equator from south to north, marking the first day of spring (Section 7.4). Technically, the term "sidereal day" is a misnomer, because the vernal equinox slowly shifts its position in the sky as a result of precession (Section 6.5). This movement is so slow, however, that a sidereal day is within 0.01 s of the true period of rotation of the earth with respect to the stars.

A solar day is slightly longer than a sidereal day, as a study of Figure 8.2 will show. Suppose we start a day when the earth is at A, with the sun on the meridian of an observer at point O on the earth. The direction from the earth to the sun, AS, if extended, points in the direction C among stars on the celestial sphere. After the earth has made one rotation with respect to the stars, the same

Table 8.1 Conversion Between Units of Time and Arc

TIME UNITS	ARC UNITS
24^h	360°
1^h	15°
4^m	1°
1^m	15′
4^s	1′
1^s	15″

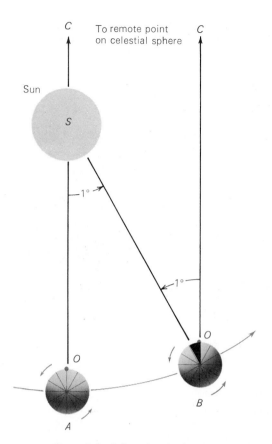

To remote point on celestial sphere

Sun

Figure 8.2 Sidereal and solar day.

stars in direction C will again be on the local meridian to the observer at O. However, because the earth has moved from A to B in its orbit about the sun during its rotation, the sun has not yet returned to the meridian of the observer but is still slightly to the east. The vernal equinox is so nearly fixed among the stars that the earth has completed, essentially, one *sidereal* day, but to complete a *solar* day it must turn a little more to bring the sun back to the meridian.

In other words, a solar day is slightly *longer* than a sidereal day, or one complete rotation of the earth. There are about 365 days in a year and 360° in a circle; thus the daily motion of the earth in its orbit is about 1°. This 1° angle, ASB, is nearly the same as the additional angle over and above 360° through which the earth must turn to complete a solar day. It takes the earth about 4 minutes to turn through 1°. A solar day, therefore, is about 4 minutes longer than a sidereal day.

Each kind of day is subdivided into hours, minutes, and seconds. A unit of solar time (hour, minute, or second) is longer than the corresponding unit of sidereal time by about 1 part in 365. In units

of solar time, one sidereal day is $23^h56^m4^s.091$. The period of the earth's rotation with respect to the stars is $23^h56^m4^s.099$ in solar time units.

(c) Sidereal Time

Sidereal time is based on the sidereal day with its subdivisions of sidereal hours, minutes, and seconds. It is defined as the *hour angle of the vernal equinox*. The sidereal day begins ($0^h0^m0^s$) when the vernal equinox is on the meridian.

Sidereal time is useful in astronomy and navigation. The common coordinate system used to denote positions of stars and planets on the celestial sphere (right ascension and declination) is referred to the celestial equator and the vernal equinox, much as latitude and longitude on the earth are referred to the earth's equator and the meridian of Greenwich, England (see Appendix 7). Therefore, the position of a star in the sky with respect to the observer's meridian is directly related to the sidereal time. Every observatory maintains clocks that keep accurate sidereal time.

We regulate our everyday lives, however, by the sun, not the vernal equinox; for example, the working day is usually determined according to the daylight hours. It is far more desirable to use solar time for ordinary purposes.

(d) Apparent Solar Time

Just as sidereal time is reckoned by the hour angle of the vernal equinox, so *apparent solar time* is determined by the hour angle of the sun. At midday, apparent solar time, the sun is on the meridian. The hour angle of the sun is the time *past midday* (*post meridiem,*, or P.M.). It is convenient to start the day not at noon, but at midnight. Therefore the elapsed apparent solar time since the beginning of a day is the hour angle of the sun *plus* 12 hours. During the first half of the day, the sun has not yet reached the meridian. We designate those hours as *before midday* (*ante meridiem*, or A.M.). We customarily start numbering the hours after noon over again, and designate them by P.M. to distinguish them from the morning hours (A.M.). On the other hand, it is often useful to number the hours from 0 to 24, starting from the beginning of the day at midnight. For example, in various conventions, 7:46 P.M. may be written as 19^h46^m, 19:46, or simply 1946.

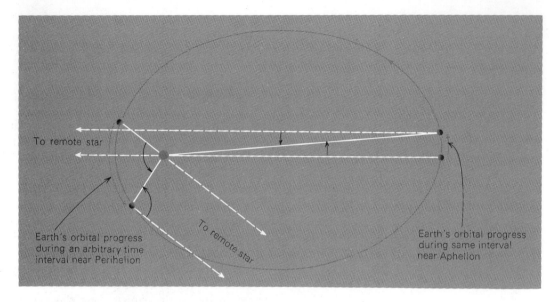

Figure 8.3 Variation in the length of an apparent solar day because of the earth's variable orbital speed. (The effect, of course, is greatly exaggerated.)

On about September 23, the sun passes through the autumnal equinox, halfway around the sky from the vernal equinox. On that date, at midnight, when the day begins, the vernal equinox is on the meridian, and so solar time and sidereal time are in agreement. With each succeeding day, however, sidereal time gains 3^m56^s on solar time, and the two kinds of time do not agree again until the daily difference between them accumulates to a full 24 hours—one year later.

Apparent solar time, defined as the hour angle of the sun plus 12 hours, is the most obvious and direct kind of solar time. It is the time that is kept by a sundial. In a sundial, a raised marker, or *gnomon*, casts a shadow whose direction indicates the hour angle of the sun. Apparent solar time was the time kept by man through many centuries.

The exact length of an apparent solar day, however, varies slightly during the year. Recall that the difference between an apparent solar day and a sidereal day, if time is counted from noon on one day, is the extra time required, after one rotation of the earth with respect to the vernal equinox, to bring the sun back to the meridian. The length of this extra time depends on how far *east* of the meridian the sun is after the completion of one sidereal day. The earth rotates to the east at a nearly constant rate of 1° every 4 sidereal minutes. Thus, if the sun were exactly 1° east of the meridian, about 4 sidereal minutes would be needed to bring it the rest of the way to the meridian. (Actually, it is just over four sidereal minutes, because the earth is still ad-

vancing in its orbit during that period, which moves the sun another 10″ to the east along the ecliptic.) If the sun were more or less than 1° east of the meridian, the extra time required would be a little more or less than 4 sidereal minutes.

The length of the apparent solar day would be constant if the eastward progress of the sun, in its apparent annual journey around the sky, were precisely constant. However, there are two reasons why the amount by which the sun shifts to the east is not the same every day of the year.

The first reason is that the earth's orbital speed varies. In accord with Kepler's second law—the law of areas—the earth moves fastest when it is nearest the sun (perihelion) in early January and slowest when it is farthest from the sun (aphelion) in July. However, its rate of rotation is nearly constant. Consequently, it moves *farther* in its orbit during a sidereal day in January than in July (Figure 8.3). The sun's apparent motion along the ecliptic is just the result of the earth's revolution, so the sun's daily progress to the east reflects the inequalities of the eath's daily progress in its orbit. We see, then, that the extra amount by which the earth must turn after a sidereal day to complete a rotation with respect to the sun is not always exactly the same.

The second reason for the variation in the rate of the sun's eastward progress, and the consequent nonuniformity in the length of the apparent solar day, is that the sun's path—the ecliptic—does not run exactly east and west in the sky, along the celestial equator, but is inclined to the equator by

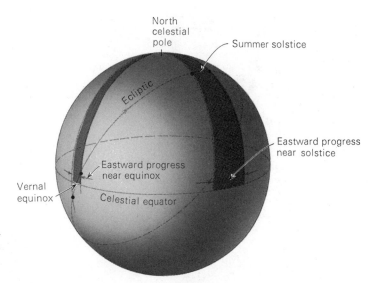

North celestial pole

Summer solstice

Ecliptic

Eastward progress near solstice

Eastward progress near equinox

Vernal equinox

Celestial equator

Figure 8.4 The sun's apparent eastward daily progress varies because of the obliquity of the ecliptic. (The obliquity, and hence the effect, have been greatly exaggerated.)

$23\frac{1}{2}°$. Even if the earth's orbit were circular, so that the sun moved uniformly along the *ecliptic*, the amount by which it moved to the east would vary slightly throughout the year. The situation is illustrated in Figure 8.4, which shows the celestial sphere, the celestial equator, and the ecliptic. To make the effect more obvious, the obliquity of the ecliptic is grossly exaggerated. Now suppose the sun moved equal distances along the ecliptic near March 21 and June 22; such equal distances are marked off on the ecliptic in the figure. Near the equinox, part of the sun's motion is northward, and it progresses less far to the *east* than it does along the ecliptic. At the solstice, on the other hand, not only is the sun moving due east but it is also north of the equator where the hour circles converge, so that a 1° advance on the ecliptic is *more* than a 1° advance to the east. A similar analysis shows the sun would also make more eastward progress near the winter solstice than near the autumnal equinox. With the actual $23\frac{1}{2}°$ obliquity, it turns out that a 1° advance on the ecliptic corresponds to 0.°92 advance to the east at the equinoxes and 1.°08 advance to the east at the solstices. Thus, even if the sun did move uniformly on the ecliptic, its eastward progress would be variable.

The apparent solar day is always *about* 4 minutes longer than a sidereal day, but because of the sun's variable progress to the east, the precise interval varies by up to one half minute one way or the other. The variation can accumulate after a number of days to several minutes. After the invention of clocks that could run at a uniform rate, it became necessary to abandon the apparent solar day

as the fundamental unit of time. Otherwise, all clocks would have to be adjusted to run at a different rate each day.

(e) Mean Solar Time

Mean solar time is based on the *mean solar day*, which has a duration equal to the *average* length of an apparent solar day. Mean solar time is defined as the hour angle plus 12 hours, of a fictitious point in the sky that moves uniformly to the east along the *celestial equator*, with approximately the same average eastern rate as the true sun. In other words, mean solar time is just apparent solar time averaged uniformly. Originally that fictitious point was called the *mean sun*, but in the modern definition of time standards, it is no longer related to the sun.

The irregular rate of apparent solar time causes it to run alternately ahead of and behind mean solar time. The difference between the two kinds of time can accumulate to about 17 minutes. The difference between apparent solar time and mean solar time is called the *equation of time*, shown graphically in Figure 8.5. One can read from the plot, for any date of the year, the correction to apply to mean solar time to obtain apparent solar time. When the equation of time is positive, apparent time is *ahead* of mean time. Often the equation of time is plotted on globes of the earth as a nomogram, shaped like the figure eight and placed in the region of the South Pacific Ocean.

Although mean solar time has the advantage of progressing at a uniform rate, it is still inconvenient for practical use. Recall that it is defined by the

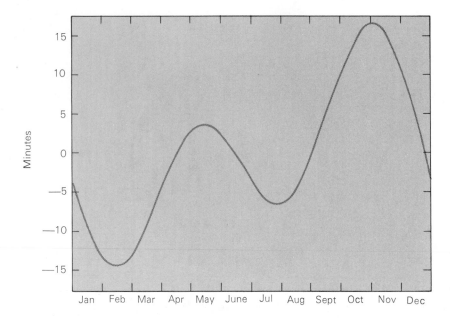

Figure 8.5 Equation of time (apparent minus mean solar time).

hour angle of the fictitious point. But hour angle refers to the local celestial meridian, which is different for every longitude on earth. Thus, observers on different north-south lines on the earth have a different hour angle of the point and hence a different mean solar time. If mean solar time were strictly observed, a person traveling east or west would have to reset his watch continually as his longitude changed, if it were always to read the local mean time correctly. For instance, a commuter traveling from Oyster Bay to New York City would have to adjust his watch as he rode through the East River tunnel, because Oyster Bay time is actually about 1.6 min more advanced than that of Manhattan.

(f) Standard and Zone Time

Until near the end of the last century, every city and town in the United States kept its own local mean time. With the development of railroads and the telegraph, however, the need for some kind of standardization became evident. In 1883 the nation was divided into four time zones. Within each zone, all places keep the same time, the local mean solar time of a standard meridian running more or less through the middle of each zone. Now a traveler resets his watch only when the time change has amounted to a full hour. For local convenience, the boundaries between the four time zones are chosen to correspond, as much as possible, to divisions between states. Mean solar time, so standardized, is called *standard time*. The standard time zones in the United States (not including Alaska and Hawaii) are Eastern Standard Time (EST), Central Standard Time (CST), Mountain Standard Time (MST), and Pacific Standard Time (PST), which respectively keep the mean times of the meridians of 75°, 90°, 105°, and 120° west longitude. Hawaii and Alaska both keep the time of the meridian 150° west longitude, two hours less advanced than Pacific Standard Time.

In 1884, largely under the impetus of two papers by Sanford Fleming, an international conference was held in Washington, D.C., in which 26 nations were represented. At that conference it was agreed to establish a system of 24 international time zones around the world. Each time zone, on the average, is 15° wide in longitude, although the zone divisions are usually irregular over land areas to follow international boundaries. At sea, the zone time of any place is the mean time of the standard meridian running through the center of the zone of that place. The zones are numbered consecutively from the Greenwich meridian; those west of Greenwich are denoted (+) and those east are denoted (−). The Eastern Standard time zone is zone number +5.

The procedure for determining standard time from apparent solar time, as read, say, from a sundial, is illustrated in the following example: At Los Angeles (118° west longitude) the apparent solar (sundial) time on March 16 was 11:30 A.M. From the equation of time we note that on March 16 apparent solar time is nine minutes behind mean solar time. Thus the local mean time is 11:39 A.M. Now, Los Angeles is in the Pacific Standard Time zone, which keeps the time of the meridian at 120° west longitude. Los Angeles is 2° east of that meridian, so its local time is eight minutes *more advanced* than that of the 120° meridian. Pacific Standard Time is thus 11:31 A.M.

(g) Daylight Saving Time

To take advantage of the maximum amount of sunlight during waking hours, most states in the United States as well as many other nations, keep what is called *daylight saving time* (or *summer time*) during the spring and summer, and sometimes during fall and winter as well. Daylight saving time is simply the local standard or zone time of the place *plus* one hour. Thus on a summer evening when it would ordinarily grow dark about 8:00 P.M. standard time it is light until 9:00 P.M. daylight saving time.

It might seem that daylight saving time is needed more in the winter than in the summer. It is not always popular in the winter, however, for when clocks are set an hour ahead in December, in many areas it is still dark at 7:30 A.M. while people are on their way to work.

(h) Time Around the World— The International Date Line

The direction of the earth's rotation is to the east. Therefore, places to the east of us must always have a time *more advanced* than ours. The celestial meridian of New York sweeps under the sun on the celestial sphere about three hours earlier than does the celestial meridian of San Francisco. Thus in New York the hour angle of the sun, the local time there, is three hours ahead of San Francisco time. The times of places halfway around the world from each other differ by 12 hours.

In general, the difference in the local time of any two places on earth is equal to their difference in longitude, if longitude is measured in time units rather than degrees. The more easterly place always has the more advanced sidereal, apparent solar, or mean solar time, because any kind of local time is defined as the hour angle of some reference object. The difference in the hour angles of the same object seen at two different places is just the difference in the directions of the local meridians of the two places, that is, the difference in their longitudes. Thus, the determination of longitude is equivalent to finding the difference between local time and the time at some known place on earth—say, the time at Greenwich (Figure 8.6).

The local mean solar time of the meridian running through Greenwich, England, is called *universal time*. Data are usually given in navigational tables, such as the *Nautical Almanac*, for various intervals of universal time.

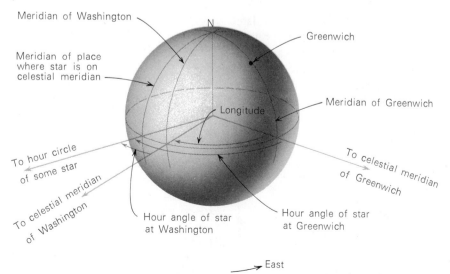

Figure 8.6 A difference in hour angle equals the difference in longitude.

The fact that time is always more advanced to the east presents a problem. Suppose a man moving eastward travels around the world. He passes into a new time zone, on the average, for about every 15° of longitude he travels and, each time, he dutifully sets his watch ahead an hour. By the time he has completed his trip, he has set his watch ahead through a full 24 hours, and has thus *gained a day* over those who stayed home. Let us look at the problem another way. It is 3 hours later in New York than in Berkeley, California. In London, it is 8 hours later than in Berkeley; in Tokyo, still farther to the east, it is 17 later. Because San Francisco is about 100° east of Tokyo, it follows that in San Francisco, a few kilometers to the west across the San Francisco Bay from Berkeley, it is 7 hours later than at Tokyo, or 24 hours—one day—later than at Berkeley! One might suppose that by going around the world to the east or west enough times, we could go into the future or past as far as we desired, creating the equivalent of a time machine.

The solution to the dilemma is the *international date line*, set by international agreement to run along the 180° meridian of longitude. The date line runs about down the middle of the Pacific Ocean, although it jogs a bit in a few places to avoid cutting through groups of islands and through Alaska. By convention, at the date line, the date of the calendar is changed by one day. If a person crosses the date line from west to east, so that he is advancing his time, he compensates by decreasing his date; if he crosses from east to west, he increases his date by one day.

If an ocean liner crosses the international date line from west to east just after its passengers have been served a Christmas dinner, the crew can begin the preparation of a second feast, for the next day will be Christmas again. On the other hand, one could skip Christmas altogether by crossing the date line from east to west.

(i) Summary of Time

We may briefly summarize the story of time as follows:

1. The ordinary day is based on a rotation of the earth with respect to the sun (not the stars).
2. The length of the apparent solar day is slightly variable because of the earth's variable orbital speed and the obliquity of the ecliptic.
3. Therefore, the average length of an apparent solar day is defined as the mean solar day.
4. Time based on the mean solar day is mean solar time. It is defined as the local hour angle of a fictitious point revolving annually on the celestial equator at a perfectly uniform eastward rate.
5. The mean solar time at any one place, called *local mean time*, varies continuously with longitude, so that two places a few kilometers east and west of each other have slightly different times.
6. Therefore time is standardized, so that each place in a certain region or zone keeps the same time—the local mean time of the standard meridian in that zone.
7. In many localities, standard time is advanced by one hour during part or all of the year to take advantage of the maximum number of hours of sunshine during the waking hours. This is daylight saving time.

8.2 TIME STANDARDS

At this point we must distinguish between the *time of day* and the *passage* of time. The former is defined in terms of the hour angle of some reference object in the sky, and thus is based on the orientation of the celestial sphere. The latter depends on a precise definition of a time standard, that is, the definition of the *second*.

At the United States Naval Observatory in Washington, D.C., and in certain other observatories throughout the world, time is measured as a routine procedure. Specially designed telescopes measure the exact instants when stars cross the meridian. Since the position on the celestial sphere of the fictitious point defining mean solar time is known with respect to the stars, the time of the meridian transit of a star indirectly gives the mean solar time. Time can be determined by astronomical observations to within about 0.001 s.

If the earth's rotation were absolutely constant the mean second (1/86,400 of a mean solar day) could serve as an appropriate standard for measuring the passage of time. For a long time the earth's rotation was the most stable time standard available. Mechanical clocks were in use in the thirteenth century, but they were not very accurate. A great improvement

Figure 8.7 A photographic zenith tube, which automatically takes pictures of objects as they cross the meridian. *(U.S. Naval Observatory)*

came after Galileo suggested the use of a pendulum for keeping time (Section 4.1b). In 1656, Christian Huygens published a comprehensive treatise, *The Pendulum Clock*, that outlined many of the properties of a swinging pendulum. The period of oscillation of a pendulum is regular and depends only on the length of the pendulum and the acceleration of gravity. These circumstances make it possible to construct a pendulum clock of high accuracy by fabricating the pendulum itself of a special alloy that expands only very slightly with changes in temperature and by keeping the clock in a temperature-controlled chamber. After Huygens' time pendulum clocks were made with increasing accuracy, but still they served only as secondary standards, and were routinely checked with astronomical observations.

(A) IRREGULARITIES IN THE EARTH'S ROTATION AND MODERN TIME STANDARDS

During the past century gravitational theory has been developed to the point where the motions of the sun,

the moon, and the planets can be predicted with great confidence. However, the precise times when they reach particular positions in the sky sometimes differ slightly from the times predicted from a clock regulated by the earth's rotation. Moreover, all of them are slightly ahead of or behind schedule in unison. By the 1920s it was realized that the causes of these discrepancies are small variations in the rate at which the earth rotates. Because the earth must conserve its angular momentum (Section 4.1h), its rotation rate can vary because of slight expansions or contractions in its crust, or because of seasonal shiftings of air masses and ice and snow deposits. The variations in the length of a single day are slight but measurable.

Some of the irregularities in the length of the day are *periodic*, while others are random, and unpredictable over extended periods of time. Superimposed on these periodic changes in the length of the day, however, is a *secular* increase; the day continues to lengthen as time goes on, and the changes accumulate until they become appreciable. Suppose a clock keeping accurate mean solar time had been set 2000 years ago. Today that clock would be out of step with the position of the sun in the sky by about 3 hours. This "clock error" was discovered by noting where historical records show that ancient eclipses actually occurred compared with the locations where they should have occurred if the earth's rotation had not changed. The only explanation consistent with gravitational theory is that the length of the day is slowly increasing. In historical times the average amount by which the day has lengthened over one century is estimated at a little under two thousandths of a second. The cause is believed to be the dissipation of energy in tidal friction. In this case, the angular momentum associated with the earth's rotation is not conserved, but it is slowly being transferred to the moon, so that the angular momentum of the earth-moon system is conserved. The results of this tidal evolution are discussed briefly in the next chapter.

The short-term, periodic variations in the earth's rotation rate have been measured independently of gravitational theory since the invention of quartz-crystal clocks about 1930. In such a clock, the natural frequency of oscillation of a vibrating quartz crystal is used to regulate the frequency of a current that runs an electric clock. The vibration of a quartz crystal is far more regular than that of a pendulum, or of the usual 60-cycle alternating current produced by commercial power-generating stations. However, the vibration even of a quartz-crystal clock changes very gradually with age.

More accurate still are modern atomic clocks, which use the frequency of electromagnetic radiation absorbed (or emitted) by atoms or molecules. This frequency standard is used to regulate the frequency of oscillation of a less accurate standard, such as a quartz crystal. In the ammonia clock, for example, a source of radio waves is set to a frequency of 2.4×10^{10} vibrations per second by a vibrating quartz crystal. This radio radiation is then passed into a tube containing molecules of ammonia gas, which absorb exactly that frequency, so no radiation gets through. However, if the quartz crystal gets off frequency slightly, the beam of radiation is not absorbed by the ammonia molecules, but passes through the tube. This transmitted radiation then activates a servomechanism that corrects the frequency of oscillation of the quartz crystal. Atoms of cesium gas also absorb at a certain radio frequency and can be used similarly to control the frequency of a vibrating crystal. Cesium clocks have an accuracy of up to a few parts in 10^{13}. Even more accurate timing may be potentially available. The hydrogen maser, for example, promises accuracy up to 100 times that of the cesium clock.

(B) ATOMIC TIME

In 1967 the General Conference of Weights and Measures adopted an atomic definition of the second, based on a natural resonance of atoms of a certain isotope of cesium. An atom of cesium 133 has two particular energy levels that depend on the way its nucleus spins (see Chapter 10 for a discussion of energy levels). It can transform itself from the lower of these energies to the higher by the absorption of radio waves with a frequency of 9,192,631,770 cycles per second. The *atomic second* is thus defined as the time required for 9,192,631,770 cycles of this radiation (the radiation has a wavelength of about 3.26 cm).

A cesium clock has at one end an oven in which a sample of the cesium[133] isotope is heated so that some of the atoms are boiled off into a vacuum chamber. Those atoms that are in the lower of the two energy states of interest are then selected by a magnetic field and passed through a beam of radio radiation of frequency near 9,192,631,770. If the frequency is just right, the radiation is absorbed by the atoms, which are thus raised to the higher of the two energy states. Those atoms of the higher energy are then detected at the far end of the clock. If, however, the frequency of the radio beam drifts off slightly, the radiation is not absorbed, and the number of cesium atoms of higher energy reaching the detector drops sharply, thereby activating a servomechanism that corrects the frequency of the radio beam. A cesium clock thus accurately maintains the frequency of the radiation. The overall accuracy of such a clock is claimed to be within one second in 6000 years. The National Bureau of Standards uses a cesium clock at its Boulder,

Colorado, laboratory to regulate the frequency of the time signals broadcast by the Bureau at its several shortwave stations.

A scale of International Atomic Time or TAI (French abbreviation) has been formed by coordinating data from about 30 participating institutions scattered about the world that operate cesium-beam atomic clocks. Included are the National Bureau of Standards and the United States Naval Observatory. Responsibility for this coordination lies with the Bureau International de l'Heure (BIH) in Paris. By definition, TAI agreed with the time of the Greenwich meridian (Universal Time—UT) at 0^h0^m on January 1, 1958. With the help of communications satellites, international synchronization of time is maintained to about one millionth of a second.

TAI provides the standard for the one-second interval between the beeps that are broadcast throughout the world. Time of day, however, is still based on the apparent rotation of the celestial sphere. That rotational time, now denoted UT1, and TAI slowly drift apart because of the irregularities in the earth's rotation. Consequently, a new system of international time, called Coordinated Universal Time (UTC), has been developed that will approximate UT1 to within one second, but always differ from TAI by precisely an integral number of seconds. The BIH decides when a full atomic second must be added to or subtracted from UTC to bring it into near accord with UT1. Recently such adjustments have been necessary about once per year and have generally been made at midnight at the end of either June 30 or December 31. The first such "leap second" was introduced on June 30, 1972, giving that day a total length of 86,401 atomic seconds, rather than 86,400.

It remains to be determined whether gravitational theory is as precise as our best modern time standards. It will be important in the future to make the most critical observations to see whether there is any evidence that the constant of gravitation or the properties of atoms (or both) change with time.

(C) EPHEMERIS TIME

Since 1960, for all calculations and tables of positions of astronomical objects, a kind of time called *ephemeris* time has been used. Ephemeris time is a uniformly flowing time in terms of which it is assumed that gravitational theory predicts the correct positions of celestial bodies.

An *ephemeris second* is equal to the length of one mean solar second at the beginning of the year 1900 (strictly, an ephemeris second is 1/31,556,925.97474 of the length of the *tropical year* at that date). Ephemeris and universal times were in

8.3 THE DATE OF THE YEAR

The natural units of the calendar are the day, based on the period of rotation of the earth; the month, based on the period of revolution of the moon about the earth, and the year, based on the period of revolution of the earth about the sun. The difficulties in the calendar have resulted from the fact that these three periods are not commensurable, that is, one does not divide evenly into any of the others.

The period of revolution of the moon with respect to the stars, the *sidereal month*, is about $27\frac{1}{3}$ days. However, the interval between corresponding phases of the moon, the more obvious kind of month, is the moon's period of revolution with respect to the sun, the *synodic month*, which has about $29\frac{1}{2}$ days (Chapter 9).

There are at least three kinds of year. The period of revolution of the earth about the sun with respect to the stars is called the *sidereal year*. Its length is 365.2564 mean solar days, or $365^{d}6^{h}9^{m}10^{s}$.

The period of revolution of the earth with respect to the vernal equinox, that is, with respect to the beginnings of the various seasons, is the *tropical year*. Its length is 365.242199 mean solar days, or $365^{d}5^{h}48^{m}46^{s}$. Our calendar, to keep in step with the seasons, is based on the tropical year. Because of precession (Section 6.5), the tropical year is slightly shorter than the sidereal year.

The third kind of year is the *anomalistic year*, the interval between two successive perihelion passages of the earth. Its length is 365.2596 mean solar days, or $365^{d}6^{h}13^{m}53^{s}$. It differs from the sidereal year, because the major axis of the earth's orbit slowly shifts in the plane of the earth's orbital revolution. Perturbations by the other planets cause this shift.

(a) THE WEEK

The week is an independent unit arbitrarily invented by man, although its length may have been based on the interval between the quarter phases of the moon. The seven days of the week are named for the seven planets (including the sun and moon) recognized by the ancients. In order of supposed decreasing distance from the earth, these seven objects are Saturn, Jupiter, Mars, the sun, Venus, Mercury, and the moon. (It was believed that the faster-moving objects were the nearest. The assumption is not necessarily correct—Mercury,

Figure 8.8 A cesium clock, one of the primary frequency standard devices maintained by the National Bureau of Standards in Boulder, Colorado. This device, made operational in 1972, is accurate to one part in 10 million million (10^{13}). *(National Bureau of Standards)*

agreement near the beginning of the century.* Changes in the earth's rotation rate, however, have now caused the two kinds of time to differ (in 1980) by more than one half minute.

Ephemeris time was originally determined from observations of the sun, the moon, or the planets, by calculating when, according to the rate of passage of ephemeris time, one of these bodies should reach its observed position among the stars on the celestial sphere. More recently, however, the ephemeris second has been defined in terms of the frequency of the cesium-beam atomic clock, and ephemeris time is now obtained directly from atomic time.

By action of the International Astronomical Union, on January 1, 1984, ephemeris time will be superseded by new time scales known collectively as *dynamical time*. One system, *terrestrial dynamical time (TDT)*, is based on the directions of solar system objects as viewed from the center of the earth. It is defined as TAI + 32.184 s for 1977. The other, *barycentric dynamical time (TDB)*, is based on the directions of those objects as seen from the center of mass of the solar system. The two differ only in subtle ways and only through periodic terms, so far as is known.

*For convenience, ephemeris time was originally adjusted to agree with the times for which the positions of the sun as given in Simon Newcomb's tables (prepared near the start of the twentieth century) were correct. For example, suppose the sun at some instant was observed at point P on the celestial sphere, and according to Newcomb's tables the sun should have reached P at time t; t, then, was chosen as the ephemeris time for that instant.

the fastest-moving planet, does not come as close to the earth as Venus.)

Each hour of the day was believed to be ruled by one of the planets in the order named. A particular day of the week was named for the planet that ruled it during the first hour of that day. The first day, Saturday, was ruled by Saturn during the first hour, Jupiter during the second, Mars during the third, and so on. Saturn thus ruled, in addition to the first hour, the eighth, fifteenth, and twenty-second. Jupiter and Mars were allotted the twenty-third and twenty-fourth hours of Saturday, leaving the first hour of the second day, Sunday, for the sun. If the scheme is continued, it is found that the moon rules the third day (Monday), and Mars, Mercury, Jupiter, and Venus the fourth, fifth, sixth, and seventh days. Our Anglo-Saxon names for the fourth through the seventh days come from the Teutonic equivalents of the Roman gods for which the planets Mars, Mercury, Jupiter, and Venus were named. The connection between those planets and the days of the week is even more obvious if we look, for example, at the Italian names for Tuesday, Wednesday, Thursday, and Friday: *Martedi*, *Mercoledi*, *Giovedi*, and *Venerdi*.

(b) The Roman Republican Calendar

The roots of our modern calendar go back to the Roman republican calendar, which derives from earlier Roman and Greek calendars dating from at least the eighth century B.C. The earliest Roman calendar probably had ten months, the last four of which have given us the names of our months, September, October, November, and December. But by the first century B.C., two additional months, January and February, had been added.

The original Roman calendar was lunar. The months were based on the moon's synodic period; each month began with a new moon. To give the months an average length of $29\frac{1}{2}$ days (the lunar synodic period) the months had 29 and 30 days alternately. The difficulty with the lunar calendar is that 12 lunar months add up to only 354 days, whereas the tropical year has about $365\frac{1}{4}$ days. After about three years, the difference accumulates to a whole month. To keep their year in step with the year of the seasons, the ancients adopted the policy of intercalation, that is, they simply inserted a 13th month every third year or so. The normal 12-month years were "empty years," and the 13-month years were "full years."

The Roman republican calendar, in use by about 70 B.C., had 12 months. These months, and their duration in days, were Martius (31), Aprilis (29), Maius (31), Iunius (29), Quintilis (31), Sextilis (29), September (29), October (31), November (29), December (29), Ianuarius (29), and Februarius (28). The year thus had 355 days. From the middle of the second century B.C. January (Ianuarius) 1 officially marked the beginning of the year, although in the popular view the year ended with February 23. When an extra month had to be intercalated every two to four years to bring the average length of the year to $365\frac{1}{4}$ days, it was added immediately after February 23. Then followed the last five days of February, and March.

Unfortunately, the management of the calendar was left to the discretion of the priests, who greatly abused their authority by declaring as full years those in which their friends were in public office. The intercalation process became such a political football that by the time of the reign of Julius Caesar, a Roman traveler going from town to town could find himself going from year to year! Thus, in 46 B.C., Julius Caesar instigated a calendar reform.

(c) The Julian Calendar

At the advice of the Alexandrian astronomer Sosigenes, Caesar adopted a new calendar, which had 12 more or less equal months averaging about $30\frac{1}{2}$ days in length rather than $29\frac{1}{2}$. The features of the Julian calendar reform of 46 B.C. were as follows.

1. The lunar synodic month was abandoned as a basic unit in the calendar. Instead, each year contained 12 months, which contained a total of 365 days. Caesar distributed the ten extra days among the 12 months, but there appears to be a difference of opinion among historians as to which months originally had how many days.

2. The calendar was to be based on the tropical year, whose length had at that time been determined to be $365\frac{1}{4}$ days. Of course, one fourth of a day could not be "tacked on" to the end of the calendar year. Therefore, *common years* were to contain only 365 days. However, after four years, this quarter-day per year adds up to one full day. Thus every fourth year

was to have 366 days (a *leap year*), the extra day being added to February. The *average* length of the year would then be $365\frac{1}{4}$ days. Note that the process of leap year is analogous to the intercalation of extra months in "empty years" to make them "full years" of 13 months.

3. To bring the date of the vernal equinox, which had fallen badly out of place in the Roman republican calendar, back to its traditional date of March 25, Caesar intercalated three extra months in the year 46 B.C., bringing its length to 445 days. Forty-six B.C. was known as the "year of confusion." The Julian calendar was introduced, then, on January 1, 45 B.C.

After Caesar's death in 44 B.C., the month Quintilis (the fifth month in the original Roman calendar) was renamed in his honor (thus our name, July). Later, the Roman senate did some further juggling with the Julian calendar. Sextilis (originally the sixth month) was renamed in honor of Augustus Caesar, successor to Julius, and the present number of days for each month resulted.

(d) The Council of Nicaea

The dates of observance of Easter and certain other religious holidays were fixed by order of the Council of Nicaea (Nice) in 325 A.D. Easter, according to the rule adopted, falls on the first Sunday after the 14th day of the moon (almost full moon) that occurs on or after March 21. At that time March 21 was the date of the vernal equinox. (The Sunday *after* full moon was specified intentionally to avoid the possibility of an occasional coincidence with the Jewish Passover.)

Note that between 45 B.C. and 325 A.D., the date of the vernal equinox had slipped back from March 25 to March 21. This was because the Julian year, with an average length of $365\frac{1}{4}$ days, is 11^m14^s longer than the tropical year of $365^d5^h48^m46^s$. The slight discrepancy had accumulated to just over three days in those four centuries.

(e) The Gregorian Calendar

By 1582, that 11 minutes and 14 seconds per year had added up to another ten days, so that the first day of spring was occurring on March 11. If the trend were allowed to continue, eventually Easter and the related days of observance would be occurring in early winter. Therefore, Pope Gregory XIII instituted a further calendar reform.

The Gregorian calendar reform consisted of two steps. First, ten days had to be dropped out of the calendar to bring the vernal equinox back to March 21, where it was at the time of the Council of Nicaea. This step was expediently accomplished. By proclamation the day following October 4, 1582, became October 15.

The second feature of the new Gregorian calendar was that the rule for leap year was changed so that the average length of the year would more closely approximate the tropical year. In the Julian calendar, every year divisible by four was a leap year, so that the average year was 365.250000 mean solar days in length. The error between this and the tropical year of 365.242199 mean solar days accumulates to a full day every 128 years. Ideally, therefore, one leap year should be made a common year, thus dropping one day, every 128 years. Such a rule, however, is cumbersome.

Instead, Gregory decreed that three out of every four century years, all leap years under the Julian calendar, would be common years henceforth. The rule was that only century years divisible by 400 should be leap years. Thus, 1700, 1800, and 1900, all divisible by four, and thus leap years in the old Julian calendar, were *not* leap years in the Gregorian calendar. On the other hand, the years 1600, and 2000, both divisible by 400, are leap years under both systems. The average length of this Gregorian year was 365.2425 mean solar days, and was correct to about one day in 3300 years.

The Catholic countries immediately put the Gregorian reform into effect, but countries under control of the Eastern Church and most Protestant countries did not adopt it until much later. It was 1752 when England and the American colonies finally made the change. The year 1700 had been a leap year in the Julian calendar but not the Gregorian; thus the discrepancy between the two systems had become 11 days. By parliamentary decree, September 2, 1752, was followed by September 14. Although special laws were passed to prevent such breaches of justice as landlords collecting a full month's rent for September, there were still riots, and people demanded their 11 days back. To make matters worse, in England it had been customary to follow the ancient practice of starting the year on March 25, originally the date of the vernal equinox.

In 1752, however, the start of the year was moved back to January 1, so in England and the colonies 1751 had no months of January and February and had lost 24 days of March! We mark George Washington's Birthday on February 22, 1732, but at the time of his birth, a calendar would have read February 11, 1731. Russia did not abandon the Julian calendar until the time of the Bolshevik revolution. The Russians then had to omit 13 days to come into step with the rest of the world.

The Gregorian calendar has now been modified slightly to come into better conformity with the tropical year: the years 4000, 8000, 12,000, and so on, all leap years in the original Gregorian calendar, are now common years. The calendar is thus accurate to one day in about 20,000 years.

At a meeting of the Congress of the Orthodox Oriental Churches at Constantinople in 1923, a slightly improved version of the Gregorian calendar was adopted for the Eastern churches. This Eastern calendar is shorter than the Julian by seven days every 900 years, rather than three days every 400 years. The rule for leap year is that century years, when divided by 900, will be leap years only if the remainder is either 200 or 600. The years 2000 and 2400 will be leap years in both the Gregorian and Eastern Orthodox calendars. The years 2100, 2200, 2300, 2500, 2600, and 2700 will not be leap years in either. The two calendars will not diverge until 2800, which will be a leap year in the Gregorian calendar but not in the Eastern Orthodox. The Eastern Orthodox calendar year has an average length of 365.2422 mean solar days, very nearly that of the tropical year. Its error is only one day in 44,000 years.

(f) The Julian Day

In 1582 (the same year as the Gregorian reform), J. J. Scaliger introduced the 7980-year *Julian period*, the least common multiple of three different calendar cycles that began in phase in 4713 B.C. The *Julian date* is the count of days that have elapsed since January 1, 4713 B.C. Thus, Julian Day 2,444,606 began at noon, universal time, on January 1, 1981. The Julian date is commonly used to refer to astronomical events, because simple subtraction gives the interval between dates in different months and years.

(g) THE MAYAN CALENDAR

Of the various calendar systems of other ancient civilizations, one of the most interesting was that of the Maya, whose civilization, flourishing in the Yucatan area in Central America, was contemporary with the early European civilizations. The Mayan calendar was later adopted, at least in part, by the conquering Aztecs.

The Mayan calendar was more sophisticated and complicated than either the Roman or Julian calendar. Apparently, the Maya did not attempt to correlate their calendar accurately with the length of the year or lunar month. Rather, their calendar was a system for keeping track of the passage of days and for counting time far into the past or future. Among other purposes, their calendar was useful for predicting astronomical events, for example, the positions of Venus in the sky.

The Mayan calendar consisted of three simultaneous systems for counting days. The first was the sacred almanac, called the Tzolkin by modern archaeologists, which was somewhat analogous to our week of seven named days that recur in specified order perpetually. However, the Tzolkin had 20 named days; moreover, each day's name was accompanied by a number. The numbers ran from 1 to 13 and were then repeated. The first day of the sacred almanac was 1 *Imix*, the second 2 *Ik*, and so forth, up to the 13th day, which was 13 *Ben*. The next day, *Ix*, was accompanied by the number 1 again, that is, 1 *Ix*. The twentieth day was 7 *Ahau*. Then the day names started over with *Imix* again, but this time *Imix* appeared with the number 8: 8 *Imix*. In other words, the numbers were always out of phase with the day names. After 13 × 20, or 260 days, 1 *Imix* appeared again, after which the whole series was repeated, and so on, indefinitely. Thus, the Tzolkin was a counting system containing 260 combinations of numbers and names.

The second counting system of the Mayan calendar was a 365-day period that is approximately equal to the year. However, there was no intercalation of extra days, or "leap-year" scheme, so the 365-day period did not remain fixed with respect to the seasons. This 365-day period was divided into 18 *uinals* (analogous to months, but not equal to the moon's period) of 20 days each, with five "unlucky" days tacked on as a 19th uinal. Each day was numbered according to its position in its uinal; thus the Maya would speak of 17 *Yaxkin*, much as we would say July 23. (Of course, there is no simple correspondence between the dates in our calendar and theirs.) To give both the Tzolkin day name and uinal date, the Maya might say, for example, 7 *Ik* 15 *Yaxkin*; analogously, we might say Thursday, January 11. In our calendar,

Figure 8.9 The famous Aztec calendar (based on the Mayan calendar) on display in the Anthropological Museum in Mexico City. *(Photograph by the author)*

January 11 can fall on a Thursday every several years (it would be every seven years if it were not for leap year), but a date in the Mayan calendar, specified like 7 *Ik* 15 *Yaxkin*, occurred exactly once every 18,980 days, or about every 52 years.

Finally, to specify completely a particular date, the Maya made use of what is called the *long count*, a perpetual tally of the days that had elapsed since a particular date about 3000 years in the past. This system is analogous to that of the Julian day, described in Section 8.3f. The starting date, however, was not meant to be that of "the beginning"——it was merely an arbitrary starting point, from which days could be counted. The significant feature of the long count is that it employed a vigesimal number system, that is, one based on 20 (rather than 10, as is our decimal system). The useful property of our decimal system is not, particularly, that it is based on the number 10 but that it employs the *zero*, without which arithmetic would be extremely tedious. (If the reader questions this statement, he should try multiplying 53,498 by 627 in Roman numerals.) The Maya made use of the zero in counting and arithmetic and employed a method of place value, analogous to our own method of writing numbers, many centuries before the Arabs introduced the concept to Europe.

EXERCISES

1. If there are 365¼ solar days per year what is the daily motion of the earth in its orbit in degrees per solar day?

2. Referring to Exercise 1, calculate how long it takes the earth to turn (rotate) through this angle.

3. Show that the difference between the rotation period of the earth with respect to the stars and with respect to the moving equinoxes is about 0.008 s per day.

4. If the sidereal month is 27⅓ days, show that the rotation period of the earth with respect to the moon is about 53 minutes longer than with respect to the stars.

5. What is the sidereal time when the vernal equinox rises?

6. If it is 3:00 P.M. local apparent time, what is the hour angle of the sun?

7. How many more sidereal days per year are there than solar days? Why?

8. (a) On what date, approximately, do sidereal and solar time agree? (b) On what date, approximately, is sidereal time 4 hours ahead of solar time? *Answer to b:* about November 22 or 23

9. If a star rises at 8:30 P.M. tonight, at approximately what time will it rise two months from now?

10. At New Orleans, Louisiana (longitude 90° W), on February 1, a sundial reads 10:25 A.M. What is the approximate Central Standard Time? (Use Figure 8.5) *Answer:* 10:38 A.M.

11. At Boston, Massachusetts (longitude 71° W), on November 1, a sundial reads 3:20 P.M. What is the approximate Eastern Standard Time? *Answer:* 2:47 P.M.

12. Point A is at latitude 10° N and longitude 47° W. Point B is at latitude 24° S and longitude 118° W. What is the difference in apparent solar time for these two locations? What is the difference in sidereal time? What is the difference in zone time?

13. If it is 1:00 A.M., July 17, at longitude 165° W, what are the time and date at longitude 165° E?

14. What is the greatest number of Sundays possible in February for the crew of a vessel making weekly sailings from Siberia to Alaska?
Answer: 10

15. If the local mean time is 2:30 P.M. and the Universal Time is 10:30 A.M., what is the longitude?

16. If the sun is on the meridian on October 10, and the Universal Time is 15:30, what is the longitude? (Ignore the equation of time.)
Answer: 52½° W

17. Show that the Julian calendar was in error by about 0.78 days per century.

18. Show that the Gregorian calendar will be in error by one day in about 3300 years.

19. If the earth were to speed up in its orbit slightly, so that a tropical year were completed in exactly $365^d 3^h$, how often would we need to have a leap year? What rule, if any, would we need for century years?

20. You wish to telephone a friend in London, but do not want to bother him during the night hours when he would normally be sleeping. During what local times (for you) should you avoid placing the call?

Theodor Egon von Oppolzer (1841–1915), Viennese (Bohemian-born) astronomer, was a specialist in orbital mechanics who compiled *Canon der Finsternisse*, a momentous volume in which are presented the results of detailed calculations of the circumstances of all lunar and solar eclipses from 1207 BC to AD 2163. *(Yerkes Observatory)*

9

THE MOON IN THE SKY: ASPECTS AND ECLIPSES

Having examined, in the last two chapters, the motions of the earth and how those motions affect the appearance of the sky, we now turn to the apparent and real motions of the moon and certain of its other aspects. The physical nature of the moon, however, is discussed in Chapter 16.

9.1 ASPECTS OF THE MOON

The moon, because of its proximity, appears to move more rapidly in the sky than any other natural astronomical object, except meteors, which are within the earth's atmosphere. The moon appears the same size as the sun in the sky, subtending $1/2°$ of arc. Its larger surface features are easily visible to the unaided eye and form the facial markings of the "man in the moon." As it travels about the earth each month, it displays different parts of its daylight hemisphere to our view and progresses through its cycle of phases.

(a) Moonlight

The most conspicuous property of the moon is its light. The amount of light we receive from the moon varies immensely with its phase. When the

moon is full, its light is nearly bright enough to read by; we receive only about 10 percent as much light from the moon at first and last quarter, and only one thousandth of the light of the full moon when the moon appears as a thin crescent 20° from the sun in the sky.

Despite the brilliance of the full moon, it shines with less than 1/400,000 the light of the sun. Even if the entire visible hemisphere of the sky were packed with full moons, the illumination would be only about one fifth or less of that in bright sunlight.

Because the moon shines by reflected sunlight, we can calculate the moon's reflecting power from its apparent brightness. The moon and earth are at about the same distance from the sun; consequently, the moon receives as much sunlight per unit area of its surface as does the earth. Calculation shows that if all this light were reflected back into space, the full moon would appear about 14 times as bright as it actually is. The fraction of incident light that is reflected by a body is called its *albedo*. The average albedo of the moon is thus about 0.07. The moon absorbs most of the sunlight that falls upon it; its surface is quite dull. The absorbed energy heats up the surface of the moon until the energy is radiated away again as infrared radiation.

When the bright part of the moon appears as only a thin crescent, the "night" side of the moon often appears faintly illuminated. Leonardo da Vinci (1452–1519) first explained this illumination as *earthshine*, light reflected by the earth back to the night side of the moon, just as moonlight often illuminates the night side of the earth.

(b) Sidereal and Synodic Months

The moon's sidereal period of revolution about the earth, that is, the period of its revolution with respect to the stars, is $27^d7^h43^m11^s5$ (27.32166 days). However, during this period of the moon's sidereal revolution, the earth and the moon together revolve about $1/13$ the way around the sun, or about 27°. The sun, therefore, appears to move 27° to the east on the celestial sphere during the period of the moon's sidereal revolution. In other words, the moon would not, in its sidereal period, have completed a revolution about the sky with respect to the sun, and consequently would not have completed a cycle of phases. To complete a revolution with respect to the sun, the moon requires, on the average, $29^d12^h44^m2^s8$ (29.530588 days). We have, then, two kinds of months: the *sidereal month*, the period of revolution of the moon with respect to the stars, and the *synodic month*, the period with respect to the sun (Figure 9.1). (Compare with the concept of sidereal and solar days, Chapter 8.)

(c) The Moon's Apparent Path in the Sky

If the moon's position among the stars on the celestial sphere is carefully noted night after night, it is seen that the moon changes its position rather rapidly, moving, on the average, about 13° to the east per day. In fact, during a single evening the moon creeps visibly eastward among the stars. The moon's apparent path around the celestial sphere is a great circle (or very nearly so) that intersects the sun's path, the ecliptic, at an angle of about 5°.

The moon's path intersects the ecliptic at two points on opposite sides of the celestial sphere. These points are called the *nodes* of the moon's orbit. The node at which the moon crosses the ecliptic while moving northward is called the *ascending node*, and the node at which the moon crosses the ecliptic moving southward is the *descending node*.

The moon's orbit is constantly and gradually changing because of perturbations, just as the or-

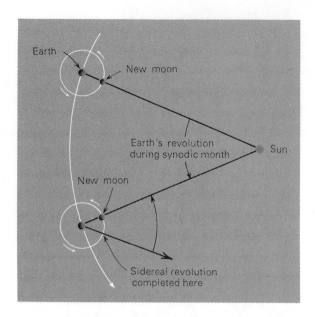

Figure 9.1 Sidereal and synodic months.

bits of artificial satellites change. The most important perturbations are caused by the gravitational attraction of the sun. One of the effects of the perturbations on the moon's orbit is that the nodes slide westward along the ecliptic, completing one trip around the celestial sphere in about 18.6 years. This motion is called the moon's *regression of the nodes*. Perturbations also cause the inclination of the moon's orbit to the ecliptic to vary from 4°57′ to 5°20′; the average inclination is 5°9′.

If it were not for the regression of the nodes, the moon's orbit would maintain a nearly fixed angle to the celestial equator. It maintains a nearly fixed angle of about 5° to the ecliptic, but because the nodes constantly shift, its angle of inclination to the equator varies from $23\frac{1}{2} + 5°$, or about $28\frac{1}{2}°$, to $23\frac{1}{2} - 5°$, or about $18\frac{1}{2}°$ ($23\frac{1}{2}°$ is the inclination of the ecliptic to the celestial equator).

(d) Delay in Moonrise from Day to Day

We have seen that the moon's average eastward motion with respect to the stars is about 13° per day. The sun, on the other hand, appears to move to the east about 1° per day. With respect to the sun, therefore, the moon moves eastward about 12° per day. As the earth turns on its axis, the moon, like other celestial objects, appears to rise in the east, move across the sky, and set in the west. But because of its daily eastward motion on the celestial sphere, it crosses the local meridian each day about 50 minutes later, on the average, than on the previous day. We could define this interval of 24^h50^m (see also Section 8.1) as the average length of an apparent lunar day.

Conditions similar to those that cause the length of an apparent solar day to vary also cause apparent lunar days to vary in length. The moon's true orbit is an ellipse; the moon's orbital speed consequently varies, in accordance with Kepler's law of areas. The moon's eastward progress in its orbit is therefore not constant. Moreover, since the moon's orbit is inclined to the celestial equator, the moon's *eastward* progress in the celestial sphere (that is, the projection of the lunar motion on the celestial equator) would not always be uniform even if the moon did move uniformly in its orbit. The daily retardation in successive transits of the moon across the local meridian ranges from 38 to 66 minutes.

Moonrise (and moonset) is similarly retarded from day to day. If the moon did not move with respect to the sun, it would rise at nearly the same time from one day to the next. At moonrise, the moon occupies some particular place on the celestial sphere. Approximately 24 hours later, the same place on the celestial sphere rises again, but the moon in the meantime has moved off to the east, so moonrise does not occur until a little later. At the equator, the daily delay is the same as the moon's delay in crossing the meridian. However, at other latitudes the moon and stars rise obliquely to the horizon, rather than in a direction perpendicular to it. Consequently, the time required for the earth to turn the sky westward through the angle representing the moon's eastward motion is not necessarily the same as the daily delay in moonrise. The phenomenon of the harvest moon, discussed below, provides an excellent example. In the northern parts of the United States, the daily delay in moonrise can vary from a few minutes to well over an hour.

(e) The "Harvest Moon"

The *harvest moon* is the full moon that occurs nearest the date of the autumnal equinox. Because the moon, when full, is opposite the sun in the sky, it must rise as the sun sets. When the sun is near the autumnal equinox, the full moon is near the vernal equinox, so at the time of the harvest moon the vernal equinox is rising with the full moon. When the vernal equinox is rising, the ecliptic makes its minimum angle with the horizon for an observer in intermediate northern latitudes.

Since the moon's orbit lies within 5° of the ecliptic, it is evident that at the same hour on successive nights the moon's apparent motion, being nearly parallel to the horizon, will not change the moon's relation to the eastern horizon appreciably. In Figure 9.2 the full moon (position 1) is shown rising at sunset. At the same time on the next night it has moved about 12° along its orbit; but it is not very far below the horizon and will rise by moving along the line *AB*, parallel to the celestial equator. The earth will not have to turn far to bring up the moon on this second night. Thus, for several nights near full moon in late September or early October there will be bright moonlight in the early evening—a traditional aid to harvesters. The phenomenon of the harvest moon is most striking in northern latitudes.

(f) The Progression of the Moon's Phases

The phases of the moon (explained in Section 2.2d) were thoroughly understood by the ancients. The relation between any phase of the moon and the moon's corresponding position in the sky at any time of day should now be clear. Except at extreme northern or southern latitudes, one can easily tell

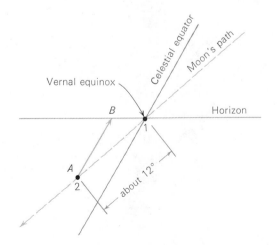

Figure 9.2 The harvest moon.

where to look for the moon in the sky, if he knows its phase, from a consideration of Figure 9.3.

In Figure 9.3 we imagine ourselves looking down upon the earth and the moon's orbit from the north. The moon is shown in eight positions in its monthly circuit of the earth. The sun is off to the right of the figure at a distance so great that its rays approach the earth and all parts of the moon's orbit along essentially parallel paths. The daylight sides of the earth and moon—the sides of those bodies turned toward the sun—are indicated. For each position of the moon, its phase, that is, its appearance *as viewed from the earth*, is shown just outside its orbit. Several observers are at various places on the earth, *A*, *B*, *C*, and so on. The time of day, indicated for each observer, depends on the position of the sun in the sky with respect to his local meridian or, equivalently, on his position on the earth with respect to the meridian where it is noon.

For person *A* it is 3:00 P.M. If he sees the moon on the meridian, it must be in the waxing crescent phase (the moon can be seen easily at noon when in this phase). If the moon is in the waning crescent phase it is setting, for it lies on his western horizon. West is the direction away from which the turning earth carries the observer, and his horizon lies in a plane tangent to the surface of the earth at the point where he is standing. If the moon is new it is in about the same direction as the sun in the western sky, and if it is at first quarter it is in the

eastern sky. If the moon is rising it must be in the waxing gibbous phase.

For person *B* it is 6:00 P.M. (Person *B* could be person *A* three hours later. During a period of even a full day, the moon does not move enough for its phase to change appreciably.) For *B* the moon is setting if new, and rising if full. If it is a waxing crescent or gibbous it is in the western or eastern sky, respectively. If it is in the first quarter phase, it appears on the meridian.

For person *D* it is midnight. If the moon is full, it rose at sunset, and is now on the meridian. The first or third quarter moon is just setting or rising, respectively. The waxing or waning gibbous moons must appear in the western or eastern sky.

By studying Figure 9.3 we can tell where the moon is in the sky at any time of day or night if we know its phase. For example, the full moon rises at sunset and sets at sunrise. The first quarter moon rises at noon and sets at midnight, and so on. We have ignored here the dependence of the moon's position on the latitude of the observer. However, for most places on earth the figure is a good enough approximation.

(g) Configurations of the Moon

The configurations of the moon are named like those of the planets. The first or third quarter moon

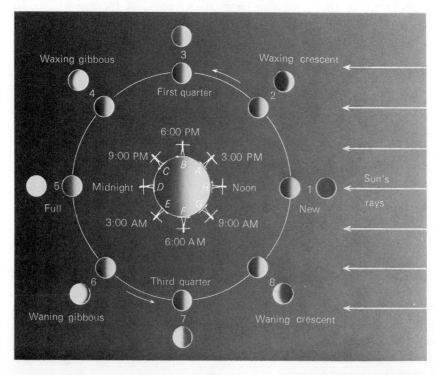

Figure 9.3 Phases of the moon and the time of day. (The outer series of figures shows the moon at various phases as seen in the sky from the earth's surface.)

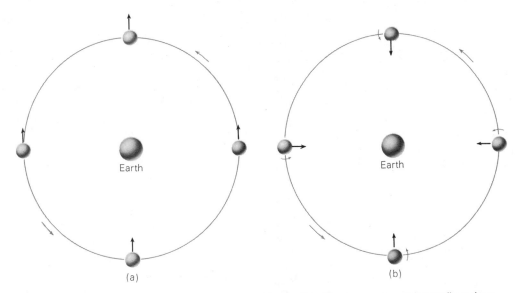

Figure 9.4 (a) If the moon did not rotate, it would turn all its sides to our view. (b) Actually, it does rotate in the same period as it revolves, so we always see the same side.

is at *quadrature*. The full moon is at *opposition*, and the new moon is a *conjunction*.*

(h) The Rotation of the Moon

Even naked-eye observation is sufficient to determine that as the moon goes about the earth it keeps the same side toward the earth. The same facial characteristics of the "man in the moon" are always turned to our view. Although the moon always presents the same side to us, it must not mislead us into thinking that it does not rotate on its axis. In Figure 9.4 the arrow on the moon represents some lunar feature. If the moon did not rotate, as in (a), we would see that feature part of the time, and part of the time we would see the other side of the moon. Actually the moon rotates on its axis with respect to the stars in the same period as it revolves about the earth, and so always turns the same side toward us (b).

The coincidence of the periods of the moon's rotation and revolution can hardly be accidental. It is believed to have come about as a result of the earth's tidal forces on the moon (Section 6.4).

We often hear the back side of the moon (the side we do not see) called the "dark side." Of course, the back side is dark no more frequently than the front side. Since the moon rotates, the sun rises and sets on all sides of the moon. The back side of the moon is receiving full daylight at new moon; the dark side is then turned toward the earth.

(i) LIBRATIONS

Actually, for several reasons, the moon does not always keep exactly the same face turned toward a particular terrestrial observer, but presents a little more than half its surface to our view over a period of time. The motions of the moon and earth that allow us to see slightly different parts of its surface at different times are called *librations*. Librations were first observed and explained by Galileo.

Three librations are geometrical. These *geometrical librations* are libration in longitude, libration in latitude, and diurnal libration.

Libration in longitude is a consequence of the fact that while the moon *rotates* at a uniform rate its angular velocity in its elliptical orbit is variable. Therefore, although the moon rotates and revolves at the same *average* rate, the two motions get slightly out of step with each other. As a consequence, we on earth are able to see, as the moon moves, first a little around its west side and then a little around its east side. The extra amount of the moon that we can observe ranges up to nearly 8° either way.

Libration in latitude occurs because the moon's axis of rotation, like the earth's, is not exactly perpen-

---

*Either full or new moon is also called *syzygy*. It is a term seldom used by astronomers, but it is sometimes encountered in crossword puzzles.

dicular to the plane of its orbit, but is tilted at an angle of about 6½°. The orientation of the moon's axis remains constant during its revolution about the earth, just as the direction of the earth's axis remains constant during the earth's revolution about the sun. This results in something quite like the way in which the sun's rays alternately fall upon the earth's north and south poles during the course of the year. In the case of the moon, we on earth see as much as 6½° beyond the moon's north pole when it is on one side of its orbit, and the same amount beyond the moon's south pole two weeks later.

Diurnal libration is caused by the rotation of the earth carrying the observer back and forth over a base line of several thousand kilometers. When we see the moon rising we are looking at it from a point some 6400 km to one side of the earth's center, and can see farther around the moon's western "edge" than when the moon is on the meridian. At moonset, our vantage point has shifted 6400 km in the opposite direction from the earth's center, and we see a little around the moon's eastern "edge." This libration amounts to about 1° in either direction.

Because of the geometrical librations, only 41 percent of the moon's surface is never presented to the earth; 41 percent is always presented, and 18 percent is alternately visible and invisible.

The moon also undergoes slight *physical librations*, which are actual irregularities in its rotation. These irregularities arise because the moon is not perfectly spherical but is technically a *triaxial ellipsoid*, that is, it has three different diameters in three mutually perpendicular directions. The longest diameter of the moon is directed toward the earth, and the earth's tidal attraction on this "bulge" perturbs the moon's rotation, causing it to oscillate slightly. Despite the moon's triaxial shape, it is more nearly a perfect sphere than the earth; its maximum and minimum diameters differ by less than 3 km. The physical librations, therefore, displace features on the moon's surface from their mean positions by only a kilometer or so.

(j) Alleged "Lunar Effects" on Humans

The sun has an obvious and profound effect on our lives, and the moon at least provides light and raises tides. Are there other, more subtle effects of the moon on humans? Legends of lunar effects abound—from lycanthropy (the "curse of the werewolf") to insanity to fertility. Most legends attribute special powers to the moon when it is at the full phase. It is easy to understand that if a person is emotionally unstable and impressionable and sincerely believes that the full moon can affect his behavior, he might be motivated to commit irrational acts when he observes a full moon. Indeed, many police officers believe that more violent crimes occur at full moon, and very many maternity-ward nurses believe that more babies are born then as well. Countless articles and books have been written promoting various theories of lunar effects. Even occasional statistical studies are published that seem to suggest evidence for a connection between the phase of the moon and human affairs.

However, more careful studies, using much larger samples and far better controls, fail to confirm evidence for such lunar effects. Every year thousands of reports are published describing results of experiments, many of which are based on poor experimental techniques, improper controls, and biased or selected data, and other results are even "fudged." If one combs the literature for bizarre results, one often finds them. But exceptional claims require exceptional proof. It would be very interesting indeed if the phases of the moon could have the effects claimed for them, but it would also be very unexpected, and such claims need to be subjected to exceptionally thorough analysis. Those thorough and competent studies do *not* confirm lunar effects beyond those already well understood, such as tides.

The moon is quite bright for several days before and after the full phase. Most casual observers call any bright moon "full," which means that the moon can be seen as "full" 25 to 30 percent of the time. If during a night of unusual activity (for instance, a bizarre crime or a high number of births at a particular hospital), one notices a bright moon, one is likely to associate the moon with the event and remember it, but this association may not be made at other times. It is like the tendency to notice and remember those dreams that seem to "come true" and to forget the overwhelming majority that do not.

I have personally participated in a study of the birth rate at various phases of the moon. In the 51 synodic months between March 17, 1974 and April 30, 1978, there were 11,691 live births at the UCLA Hospital. They showed no correlation whatsoever with the phase of the moon! We obtained the same negative results when we analyzed the 8142 of the 11,691 births that were completely natural (not induced by drugs or Caesarean section). Other studies have also failed to show a significant variation in the numbers of babies born according to the phase of the moon, despite widespread belief to the contrary.

There is not even convincing evidence that the human menstrual cycle has anything to do with the moon. The cycle varies greatly from woman to woman and also from time to time for a given individual, but it averages 28 days, not even a very close coincidence with the 29.53-day synodic month. The corresponding estrous cycles of other mammals are very different: 11 days for guinea pigs, 16 to 17 days for sheep, 20 to 22 days for sows, 21 days for cows and mares, 24 to 26 days for macaque monkeys, 37 days for chimpanzees, and only 5 days for rats and mice. For the opossum, as for the human, it is 28 days. If humans are affected differently from animals by the moon, the 28-day period of the opossum must be a coincidence, but if it is a coincidence for the opossum, why not also for humans?

One cannot rule out that some individuals may be psychologically affected by the phase of the moon, but the evidence is either unconfirmed or not convincing. Other alleged relationships are totally unsupported by "hard" (scientifically acceptable) evidence. So far as we know, the moon is important in raising tides, and it gives us moonlight, but its traditionally attributed influences are entirely mythical.

9.2 THE MOON'S SIZE AND DISTANCE FROM EARTH

The moon's distance from earth is only about 30 times the diameter of the earth; consequently, the direction of the moon differs slightly as seen from various places on the earth. We have seen (Section 2.3d) how Ptolemy made use of this principle to determine geometrically the distance of the moon. Today, however, the moon's distance can be determined more accurately by other methods.

(a) Modern Measures of the Distance to the Moon

One accurate way of finding the distance to the moon is by means of radar. The first successful radar contact with the moon was achieved by the United States Army Signal Corps in 1946. In this technique, radio waves, focused into a beam by a powerful broadcasting antenna, are transmitted to the moon; some of this energy is reflected back to the earth. Since radio waves are a form of electromagnetic energy, they travel with the speed of light (Chapters 10 and 11); hence half the time between when the waves are transmitted and the echo is received, multiplied by the speed of light, gives the distance to the moon. The time intervals involved can now be measured electronically to better than one millionth of a second. By 1957 the Naval Research Laboratory had determined the moon's distance by radar to within about 1 km.

More precise still is ranging by *laser* (Chapter 10). A laser is a device in which atoms are induced to emit light pulses that leave the device in a highly focused beam, in a very narrow range of frequency, and with the waves in each pulse accurately aligned with each other. One advantage of the laser beam is that it can be transmitted over large distances with high efficiency, and if it is reflected from an object, the round-trip travel time can be measured to one nanosecond (10^{-9}s).

Since late 1969 lasers have been attached to telescopes to send such light pulses to the moon. Mirrors left on the moon by Apollo astronauts and Soviet lunar probes reflect an extremely small fraction of this light back to earth, where it is observed with the telescopes. In principle the laser technique can determine the distance to the moon to within a few centimeters. At present this accuracy is not realized because the speed of light is known to a precision of only a few parts in 10 million (Chapter 13). Laser observations, however, can detect relative changes in the moon's distance as it moves in its elliptical orbit, and can also check the effects of perturbations on the moon.

The best determination to date gives for the distance from the center of the earth to the center of the moon the value 384,404 km, with an uncertainty of about 0.5 km.

(b) The Moon's Diameter

The mean angular diameter of the moon is 31' 5''. From its angular size and its distance, the linear size of the moon can easily be found. Because the method is the same as that applied to measure the diameters of planets and other astronomical objects that subtend a measurable angle, we shall explain the procedure in detail.

Notice in Figure 9.5 that because the moon's angular size is relatively small its linear diameter is essentially a small arc of a circle, with the observer as center and with a radius equal to the moon's distance. Obviously, the moon's diameter is the same fraction of a complete circle as the angle subtended by the moon is of 360°. A complete circle contains 1,296,000'' (there are 60'' per minute, 60' per degree, and 360° in a circle). As seen from the center of the earth, the moon's mean angular diameter of 31'5'', or 1865'', is thus ¹/₆₉₅ of a circle. The moon's diameter, therefore, is ¹/₆₉₅ of the circumference of a circle of radius 384,404 km. Since the circumference of a circle is 2π times its radius, we have

$$\text{diameter of moon} = \frac{2\pi(384404)}{695} = 3475 \text{ km}.$$

This type of calculation can be generalized if we note that the distance along the arc of a circle of radius R subtended by 1'' is $2\pi R/1,296,000 = R/206,265$. Thus, if an object, say a planet, subtends an angle of α seconds, and has a distance D, its linear diameter d is given by

$$d = \frac{\alpha D}{206,265}.$$

Accurate calculations of the moon's diameter give as a result 3475.9 km, with an uncertainty of a few hundredths of a kilometer. This is the moon's equatorial diameter in a direction perpendicular to its direction from the earth. We have already seen (Section 9.li) that the diameter differs slightly in different directions. The moon's diameter is a little over one quarter of the earth's, and its volume is about ¹/₄₉ of the earth's.

9.3 THE TRUE ORBIT OF THE MOON

When the distance to the moon is measured during different times of the month, it is found to vary by more than 10 percent. In obedience to Kepler's first law, the moon's orbit, basically, is an ellipse with the earth at one focus. The sun, however, produces strong perturbations on the moon, and the elements of its orbit can be stated only in an average sense. Indeed, the moon's orbit changes so rapidly that if the moon's positions over a month, relative to the center of the earth, are plotted carefully, even on a sheet of standard typing paper, the orbit is seen not to close on itself. The celestial mechanics of the moon's motion (lunar theory) is very complex; here we can only mention briefly a few of these complexities.

It is easy to visualize one of the effects of the sun's perturbations. The sun's tidal force on the earth-moon system tends, on the average, to separate the two. At new moon, the sun pulls on the moon more strongly than on the earth, tending to pull them apart, and at full moon the sun's greater pull on the earth has the same result. At first and third quarters the sun tends to pull the earth and the moon together slightly (because the sun is in slightly different directions as seen from the earth and the moon), but this effect is weaker than the tendency to separate the two at full and new moon. Consequently the month is about 53 minutes longer than it would be in the sun's absence.

The tidal force of the sun also acts to speed up and slow down the moon in different parts of its orbit. The net result of these accelerations is to cause the *line of apsides* (the major axis of the moon's orbit) to rotate toward the east in the orbital plane in a period of 8.85 years. Thus the position in the sky of the moon's *perigee* (closest approach of

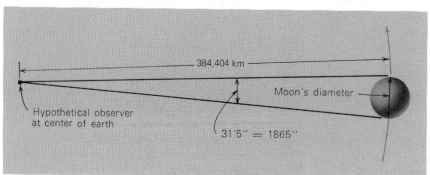

384,404 km

Moon's diameter

Hypothetical observer at center of earth

31'5'' = 1865''

Figure 9.5 Measuring the moon's diameter.

the moon to the earth) advances to the east about the sky in this period.

The plane of the moon's orbit is inclined at about 5°9′ to the ecliptic plane, and one component of the sun's force on the moon is toward the ecliptic plane, trying to pull the moon into it. Analogous to precession, however, the moon's orbit does not tilt back into the plane of the ecliptic, but rather the line of nodes regresses (that is, the nodes slide westward), moving around the ecliptic in 18.6 years.

The solar perturbations on the moon's motion depend critically on the average distance of the earth and moon from the sun. Because of planetary perturbations, the eccentricity of the earth's orbit slowly oscillates in value. In the present era the eccentricity is decreasing, resulting in a slight and slow increase in the mean distance to the sun. One effect is to shorten slightly the length of the month. The effect was discovered by Edmond Halley in the course of his investigation of ancient eclipse records. The explanation was later supplied by Laplace.

Because of such perturbations on the moon, the eccentricity of its orbit changes even during one month. The mean value, over many years, is 0.0549. The moon, however, may come as close as 356,334 km to the earth's center and pass as far from it as 406,610 km. At its farthest, however, the moon is still only about $1/370$ as far as the sun, and its orbital speed of about 1 km/s is only about one thirtieth of the earth's orbital speed about the sun. Thus, if the moon's path with respect to the sun is plotted, it is seen to vary only minutely from the earth's orbit; in fact, the moon's orbit is always concave to the sun.

The rapid changes in the moon's orbit make it useful to define two additional kinds of months, which are especially helpful in the prediction of eclipses. They are the *nodical* or *draconic month* of 27.212220 days, the time between two successive passages of the same node, and the *anomalistic month* of 27.55455 days, the time between two successive perigee passages.

(a) Mutual Revolution of Earth-Moon System

We saw in Section 5.1 that one body does not strictly revolve about another, but that the two bodies mutually revolve about their center of mass, or the *barycenter*. The barycenter of the earth-moon system revolves annually in an elliptical orbit about the sun, while the earth and moon simultaneously revolve about the barycenter in a shorter period—the sidereal month.

The elliptical orbit of the center of the earth about the barycenter constitutes an independent motion of the earth. The motion can be detected by careful observations of the nearer planets, or better yet, of near-approaching minor planets (Chapter 16). The motion of Mars, for example, shows monthly oscillations, carrying it a little ahead and then a little behind its regular orbital motion. This oscillation is only apparent and results from the motion of the earth, carrying us first to one side and then to the other side of the barycenter. When Mars is at its closest, the apparent displacements caused by the orbital motion of the earth around the barycenter amount to about 17″. The corresponding mean distance of the center of the earth from the barycenter is 4672 km. Thus, the earth and moon jointly revolve about a point approximately 1707 km below the surface of the earth.

(b) The Mass of the Moon

In Section 5.1 we saw that the distances of two bodies from their barycenter are inversely proportional to their masses. The 4672 km distance of the earth from the barycenter is 1/82.3 of the distance from the earth to the moon; hence the moon is 81.3 times as far from the barycenter as the earth and so is only 1/81.3 as massive. Because the earth's mass is 6.0×10^{21} tons (Section 4.3e), the moon's mass is $(6.0 \times 10^{21})/81.3 = 7.35 \times 10^{19}$ metric tons.

The best modern determinations of the moon's mass are obtained by measuring the accelerations the moon produces on space probes—either those sent to the moon itself or those which pass it by on interplanetary missions. Such analyses give for the earth-moon mass ratio the value 81.3008 ± 0.0003.

(c) TIDAL FRICTION AND EVOLUTION

One force that is believed to affect, over long periods, the mutual revolution of the earth-moon system is friction in the tides. As the earth rotates, the tidal waters continually flow back and forth over each other, across ocean floors, and in and out over coastal shallows. The resulting friction within the water and between the water and the solid earth draws a considerable amount of energy from the kinetic energy of rotation of the earth and expends this energy in the form of heat. Even friction in the tidal distortions of the solid earth and in atmospheric tides may play a role. The earth, consequently, is slowing in its rate of rotation, and the day is gradually lengthening at the rate of about 0.002 s per day each century (Section 8.2a). The continuous dissipation of the rotational energy of the earth is calculated to be approximately 2000 million horsepower.

Whereas the earth slows down in its axial spin as it loses kinetic energy, the angular momentum of the earth-moon system must be conserved. According to a theory of tidal evolution worked out by Sir George Darwin (son of the naturalist), the moon must slowly spiral outward away from the earth, thus maintaining the total angular momentum of the earth and the moon. (But this is superimposed on a temporary *decrease* of the moon's distance caused by a lowering of the eccentricity of the earth's orbit.) As the moon's distance from the earth increases, its period of revolution must also increase and its orbital speed must decrease, in accord with Kepler's third law. The day and the month, in other words, will both lengthen. Eventually the day will catch up with the month in length, when both require about 47 of our present days.

9.4   SHADOWS AND ECLIPSES

Eclipses occur whenever any part of either the earth or the moon enters the shadow of the other. When the moon's shadow strikes the earth, people on earth within that shadow see the sun covered at least partially by the moon, that is, they witness a *solar eclipse*. When the moon passes into the shadow of the earth, people on the night side of the earth see the moon darken—a *lunar eclipse*.

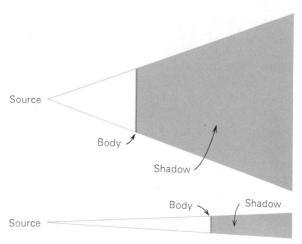

Figure 9.6   Shadow in light from a point source.

A shadow is a region of space within which rays from a source of light are obstructed by an opaque body. Ordinarily, a shadow is not visible. Only when some opaque material, which will show the contrast between lighted and unlighted areas, intersects the shadow and the surrounding area does the shadow become visible.

(a)   Shadow from a Point Source

If a source of illumination is a point source (Figure 9.6), the shadow cast by an opaque body has sharp boundaries. From a point inside the shadow, the light source is not visible; outside the shadow it is. The boundaries of the shadow diverge radially from the source. The more distant the source, the smaller is the angle of divergence. If the source is infinitely distant, the boundaries of the shadow are parallel.

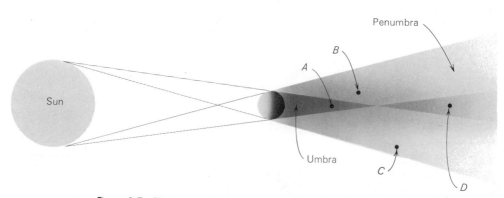

Figure 9.7   Shadow cast by an opaque spherical body in sunlight.

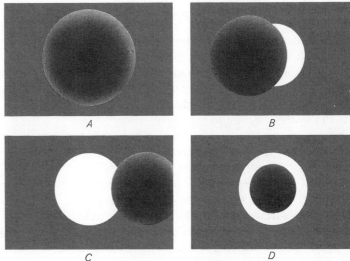

Figure 9.8 Appearance of the sun from points A to D in the shadow shown in Figure 9.7

(b) Shadow from an Extended Source

If a light source is not a point source but presents a finite angular size as seen from the opaque body, as is almost always the case, the shadow cast by the body is not limited to the inner part of the shadow, the *umbra*, where complete darkness prevails. From within the umbra, of course, no part of the light source is visible. At any point completely outside the shadow, there is no obscuration of light, and from such places the entire light source is visible. Between the umbra and the region of full light lies a space of partial illumination, within which the illumination ranges from complete darkness at the boundary of the umbra to full illumination at the outer boundary of the entire shadow. This transition region of the shadow is called the *penumbra*. From any point within it, a part, but not all, of the light source is visible.

As an illustration, consider the shadow cast by a spherical body, such as the earth, the moon, or a planet, in sunlight (Figure 9.7). It is obvious from the figure that because the umbra includes that region of space from which no part of the sun is visible, it must have the shape of a cone pointing away from the sun. The umbra of the moon or a planet is sometimes called the *shadow cone* of that body. Everything on the night side of the earth is within the umbra, or shadow cone, of the earth.

The penumbra, on the other hand, is that region from any point within which only part of the sun is covered by the eclipsing body. It is clear from the figure that the penumbra has the shape of a truncated cone pointed *toward* the sun, and that it includes the umbra, as a reversed cone symmetrical about the same axis. The appearances of the sun from points *A*, *B*, *C*, and *D* in the shadow are shown in Figure 9.8.

The size of the umbra cast by an opaque sphere in sunlight depends on the size of the sphere and its distance from the sun. The length of the umbra is directly proportional to the distance of the sphere from the sun. Figure 9.9 illustrates this relation.

(c) Eclipse Seasons

For the moon to appear to cover the sun and thus to produce a solar eclipse, it must be in the same direction as the sun in the sky, that is, it must be at the *new* phase. For the moon to enter the earth's shadow and produce a lunar eclipse, it must be opposite the sun; that is, it must be at the *full* phase. Eclipses occur, therefore, only at new moon and at full moon. If the orbit of the moon about the earth

Figure 9.9 Shadow lengths at various distances from the sun.

lay exactly in the plane of the earth's orbit about the sun—in the ecliptic—an eclipse of the sun would occur at every new moon and a lunar eclipse at every full moon. However, because the moon's orbit is inclined at about 5° to the ecliptic, the new moon, in most cases, is not *exactly* in line with the sun, but is a little to the north or to the south of the sun in the sky. Similarly, the full moon usually passes a little south or north of the earth's shadow.

However, if full or new moon occurs when the moon is at or near one of the *nodes* of its orbit (where its orbit intercepts the ecliptic), an eclipse can occur. The line through the center of the earth that connects the nodes of the moon's orbit is called the *line of nodes*. If the direction of the sun lies along, or nearly along, the line of nodes, new or full moon occurs when the moon is near a node, and an eclipse results. The situation is illustrated in Figure 9.10. The orientation of the moon's orbit, and the line of nodes, *nn'*, remains relatively fixed during a revolution of the earth about the sun. There are, therefore, just two places in the earth's orbit, points *A* and *B*, where the sun's direction lies along the line of nodes. It is only during the times in the year, roughly six months apart, when

the earth-sun line is approximately along the line of nodes that eclipses can occur. These times are called *eclipse seasons*.

We have seen (Section 9.3) that because of perturbations of the moon's orbit, the line of nodes is gradually moving westward on the ecliptic, making one complete circuit in 18.6 years. Therefore, the eclipse seasons occur earlier each year by about 20 days. In 1981 the eclipse seasons are near January and July; in 1986 they are April and October.

9.5 ECLIPSES OF THE SUN

One of the most fortunate coincidences of nature is that the two most prominent astronomical objects, the sun and the moon, have so nearly the same apparent size in the sky. Although the sun is about 400 times as large in diameter as the moon, it is also about 400 times as far away, so both the sun and moon subtend about the same angle—about 1/2°.

The apparent or angular sizes of both sun and moon vary slightly from time to time, as their respective distances from the earth vary. The average angular diameter of the sun (as seen from the center

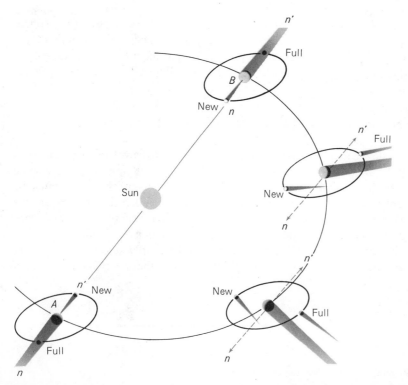

Figure 9.10 Eclipses occur only when the sun is along, or nearly along, the line of nodes.

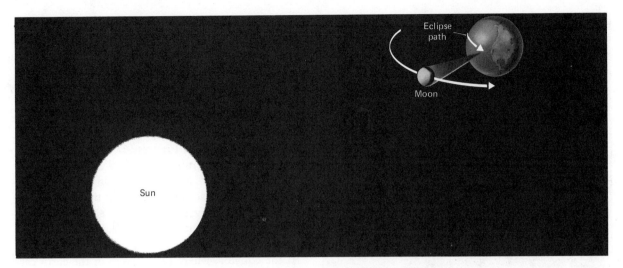

Figure 9.11 Geometry of a total solar eclipse (not to scale).

of the earth) is 31′59″, and the average angular diameter of the moon is slightly less, 31′5″. However, the sun's apparent size can vary from the mean by about 1.7 percent and the moon's by 7 percent. The maximum apparent size of the moon is 33′16″, which is larger than the sun's apparent size, even at its largest. Therefore, if an eclipse of the sun occurs when the moon is somewhat nearer than its average distance, the moon can completely hide the sun, producing a *total solar eclipse*. In other words, a total eclipse of the sun occurs when-

ever the umbra of the moon's shadow reaches the surface of the earth.

(a) Geometry of a Total Solar Eclipse

The geometry of a total solar eclipse is illustrated in Figure 9.11. The earth must be at a position in its orbit such that the direction of the sun is nearly along the line of nodes of the moon's orbit. Furthermore, the moon must be at a distance from the

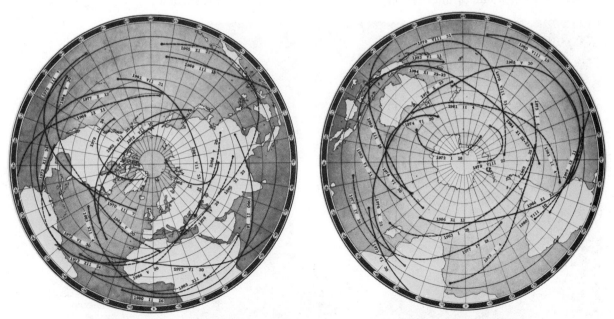

Figure 9.12 Some total eclipse paths in the twentieth century: (a) Northern hemisphere. (b) Southern hemisphere. (From *Canon of Solar Eclipses*, by J. Meeus, C. Grosjean, and W. Vanderleen; Pergamon Press, 1966)

Figure 9.13 Time-lapse photograph showing the moon passing in front of the sun during a total solar eclipse. (*American Museum of Natural History*)

a point on the earth's surface. Anyone on the earth within the small area covered by the moon's umbra will not see the sun and will witness a total eclipse. The moon's penumbra, on the other hand, covers a larger area of the earth's surface. Any person within the penumbra will see part but not all of the sun eclipsed by the moon—a partial solar eclipse. The regions of total and partial eclipse correspond to points *A* and *B* in Figures 9.7 and 9.8.

As the moon moves eastward in its orbit with respect to the sun at about 3400 km/hr, its shadow sweeps eastward across the earth at the same speed. The earth, however, is rotating eastward at the same time, so the speed of the shadow with respect to a particular place on earth is less that 3400 km/hr. At the equator, where the rotation of the earth carries places eastward at about 1670 km/hr, the shadow moves relative to the earth with a speed of about 1730 km/hr. In higher latitudes the speed is greater. In any case, the tip of the truncated cone of the umbra of the moon's shadow sweeps along a thin band across the surface of the earth, and the total solar eclipse is observed successively along this band (refer to Figure 9.11). This path across the earth within which a total solar eclipse is visible (weather permitting) is called the *path of totality*. Within a zone about 3000 km on either side of the eclipse path, a partial solar eclipse is visible—the observer, inside this limit, being located in the penumbra of the shadow.

surface of the earth that is less than the length of the umbra of the moon's shadow. Then, at new moon, the moon's umbra intersects the ground at

Figure 9.14 Total solar eclipse of 7 March 1970, near a time of maximum sunspot activity. A radially symmetric, neutral-density filter in the focal plane of the camera was used to compensate for the increase in brightness of the corona near the limb of the sun. Photograph by Gordon Newkirk, Jr. (*High Altitude Observatory, Boulder Colorado, a Division of the National Center for Atmospheric Research. NCAR is operated by the University Corporation for Atmospheric Research under sponsorship of the National Science Foundation.*)

Because the moon's umbra just barely reaches the earth, the width of the eclipse path, within which a total eclipse can be seen, is very small. Under the most favorable conditions, the diameter of the shadow cone is 269 km wide at the earth's surface. This would be the width of the path if the shadow struck the earth perpendicularly near the earth's equator. At far northern or southern latitudes, because the moon's shadow falls obliquely on the ground, it can cover a path somewhat more than 269 km wide.

It does not take long for the moon's umbra to sweep past a given point on earth. The duration of totality may be only a brief instant. It can never exceed about $7\frac{1}{2}$ minutes.

(b) Appearance of a Total Eclipse

A total solar eclipse is one of the most spectacular of natural phenomena. If you are anywhere near the path of totality of solar eclipse, it is well worth your while to travel into the eclipse path so that you may witness this rare and impressive event.

The very beginning of a solar eclipse is the *first contact*, when the moon just begins to silhouette itself against the edge of the sun's disk. The *partial phase* follows during which more and more of the sun is covered by the moon. *Second contact* usually occurs more than an hour after first contact, at the instant when the sun becomes completely hidden behind the moon. In the few minutes immediately before second contact (the beginning of totality) the sky noticeably darkens; some flowers close up, and wildlife, especially birds, exhibit nocturnal behavior. I saw bats appear during totality of the eclipse of March 7, 1970 as I watched it from a Mexican desert. Because the diminished light that reaches the earth must come solely from the edge of the sun's disk, and consequently from the higher layers in its atmosphere (see Chapter 31), the sky and landscape take on strange colors. In the last instant before totality, the only parts of the sun that are visible are those that shine through the lower valleys in the moon's irregular profile and line up along the periphery of the advancing edge of the moon—a phenomenon called *Baily's beads*. The final flash of sunlight through a lunar valley produces a brilliant flare on the disappearing crescent of the sun—the *diamond ring* effect. During totality, the sky is quite dark (like late twilight). The brighter planets are visible, and sometimes the brighter stars as well.

As Baily's beads disappear and the bright disk of the sun becomes entirely hidden behind the moon, the *corona* flashes into view. The corona is the sun's outer tenuous atmosphere, consisting of sparse gases that extend for millions of miles in all

Figure 9.15 Total solar eclipse of 30 June 1973, near a time of minimum sunspot activity. A radially graded filter was used, as in the photograph in Figure 9.14, with a technique developed by Gordon Newkirk, Jr. *(High Altitude Observatory, a Division of the National Center for Atmospheric Research. NCAR is operated by the University Corporation for Atmospheric Research under sponsorship of the National Science Foundation.)*

directions from the apparent surface of the sun. It is ordinarily not visible because the light of the corona is feeble compared to that from the underlying layers of the sun that radiate most of the solar energy into space. Only when the brilliant glare from the sun's visible disk is blotted out by the moon during the total eclipse is the pearly white corona, the sun's outer extension, visible. (It is, however, possible to photograph the inner, brighter, part of the corona with an instrument called a *coronagraph*, a telescope in which a black disk in the telescope's focal plane produces an artificial eclipse, enabling at least the inner brighter part of the corona to be studied at any time.)

Also, during a total solar eclipse, the chromosphere can be observed—the layer of gases just above the sun's visible surface. Prominences, great jets of gas extending above the sun's surface, are often seen. These outer parts of the sun's atmosphere are discussed more completely in Chapter 31.

The total phase of the eclipse ends, as abruptly as it began, with *third contact*, when the moon begins to uncover the sun. Gradually the partial phases of the eclipse repeat themselves, in reverse order. At *last contact* the moon has completely uncovered the sun.

In addition to being inspiring to watch, total eclipses of the sun have considerable astronomical value. Many data are obtained during eclipses that are otherwise not accessible. For example, during an eclipse we can determine the exact relative positions of the sun and moon by timing the instants of the four contacts. We can take direct photographs and make spectrographic observations of the sun's outer atmosphere and prominences. We can measure the light and heat emitted by the corona. We can determine how meteorological conditions are affected by solar eclipses and can learn something about the light-scattering properties of the earth's atmosphere. Sometimes, especially by being near the edge of the path of totality, even a rank amateur, or group of amateurs, can make scientifically useful observations with simple timing, as has been done several times in the past since 1715.

(c) Annular Eclipses of the Sun

More than half the time the moon does not appear large enough in the sky to cover the sun completely, which means that the umbra of its shadow does not reach all the way to the surface of the earth. The geometry of the situation is illustrated in Figure 9.16. When the moon's shadow cone does not reach the earth, a total eclipse is not possible. However, if the boundaries of the umbra of the shadow are extended until they intersect the earth's surface, they define a region on the ground within which the moon can be seen completely silhouetted against the sun's disk, with a ring of sunlight showing around the moon. This kind of eclipse is called an *annular eclipse*, from the Latin word *annulus*, meaning "ring." The extension of the moon's umbra, within which an annular eclipse is visible, sweeps across the ground in a path much like the path of totality of total eclipse. As is true for total eclipses, a partial eclipse is visible within a region of 3000 km or more on either side of the annular

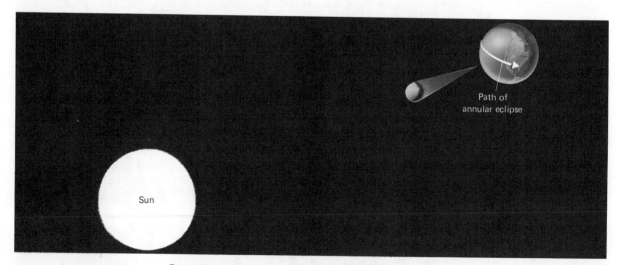

Figure 9.16 Geometry of an annular eclipse (not to scale).

Figure 9.17 An annular eclipse of the sun. Bright spots are sunlight streaming through lunar valleys. *(Lick Observatory)*

eclipse path. In this case, the regions of annular and partial eclipse correspond to positions *D* and *C*, respectively, in Figures 9.7 and 9.8. The maximum duration of an annular eclipse is about 12.5 minutes.

An annular eclipse begins and ends like a total eclipse. However, because the sun is never completely covered by the moon, the corona is not visible, and although the sky may darken somewhat, it does not get dark enough for stars to be seen. An annular eclipse is not so spectacular, nor has it the scientific value of total eclipse.

Sometimes, because of the earth's finite size and sphericity, the umbra of the moon's shadow may be long enough to reach the surface of the

earth only near the time when the moon is most nearly in a line between the centers of the earth and sun. Then it falls short of the earth's surface over the beginning and end of the eclipse path (see Figure 9.18). These mixed annular and total eclipses are relatively rare.

(d) Partial Eclipses of the Sun

A *partial eclipse* of the sun is one in which only the penumbra of the moon's shadow strikes the earth. During such an eclipse, the moon's umbra passes north or south of the earth, and from nowhere can the sun appear to be covered completely by the moon. Also, a total or annular eclipse appears partial from regions outside the eclipse path but within the zone of the earth that is intercepted by the moon's penumbra. Few people, therefore, have seen total or annular solar eclipses, whereas most have had the opportunity to see the sun partially eclipsed. The moon seems to "skim" across the northern or southern part of the sun. How much of the sun can appear covered depends on how close the observer is to the path of totality or of annularity. Partial eclipses are interesting but not spectacular. Only within a few hundred kilometers of the eclipse path does the sky darken appreciably.

(e) How to Observe an Eclipse

The progress of an eclipse can be observed safely by holding a card with a small (1 mm) hole punched in it several feet above a white surface, such as a concrete sidewalk. The hole in the cardboard pro-

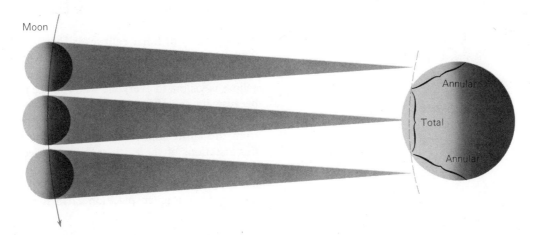

Figure 9.18 Mixed annular and total eclipse.

duces a pinhole camera image of the sun (Figure 9.19).

Although there are safe filters through which one can safely look at the sun directly, people are reported to have suffered permanent eye damage by looking at the sun through improper filters (or no filter at all!). In particular, neutral-density photographic filters are not safe, for they transmit infrared radiation that can cause severe damage to the retina.

Common sense (and pain) prevents most of us from looking at the sun directly on an ordinary day for more than a brief glance. Of course there is nothing about an eclipse that makes the radiation from the sun more dangerous than it is any other time; on the contrary, we receive less radiation from the sun when it is partly hidden by the moon. It is *never* safe, however, to look steadily at the sun when it is still in *partial* eclipse; even the thin crescent of sunlight visible a few minutes before totality has a surface brightness great enough to burn and permanently destroy part of the retina. Unless you have a filter prepared especially for viewing the sun, it is best to watch the partial phases with a pinhole camera device, as described above.

It is *perfectly safe*, however, to look at the sun directly when it is *totally eclipsed*, even through binoculars or telescopes. Unfortunately, unnecessary panic has often been created by well-meaning but uninformed public officials acting with the best intentions. I have witnessed two marvelous total eclipses in Australia, during which townspeople held newspapers over their heads for protection, and school children cowered indoors, with their heads under their desks. What a cheat to those people to have missed what would have been one of the most memorable experiences of their lifetimes! During totality, by all means look at the sun.

Nor should you be terrified of accidently catching a glimpse of the sun outside totality. How many times have you glanced at the sun on ordinary days while driving a car or playing ball or tennis? Common sense made you look away at once. Do the same if you inadvertantly glimpse the sun directly while it is partially eclipsed.

9.6 ECLIPSES OF THE MOON

A lunar eclipse occurs when the moon, at the full phase, enters the shadow of the earth. There are three kinds of lunar eclipses: *total, partial,* and *penumbral.*

(a) Geometry of a Lunar Eclipse

The geometry of lunar eclipses is shown in Figures 9.20 and 9.21. Unlike a solar eclipse, which is visible only in certain local areas on the earth, a lunar eclipse is visible to everyone who can see the moon, because its source of illumination is blocked off. Weather permitting, a lunar eclipse can be seen from the entire night side of the earth, including those sections of the earth that are carried into the earth's umbra while the eclipse is in progress. Lunar eclipses, therefore, are observed far more frequently from a given place on earth than are solar eclipses.

Figure 9.21 shows the cross section of the earth's shadow at the moon's distance from earth.

Figure 9.19 How to watch a solar eclipse safely.

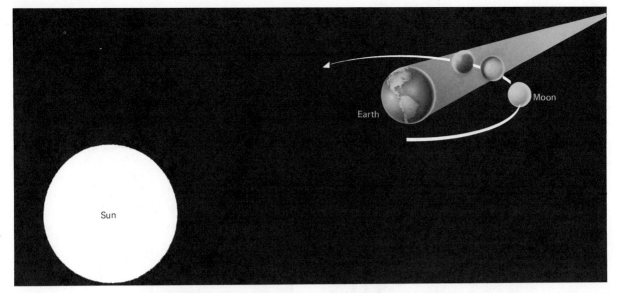

Figure 9.20 Geometry of a lunar eclipse (not to scale).

Since the cross-sectional diameter of a cone is proportional to the distance form the apex of the cone, the cross section of the earth's shadow cone at the moon's distance is in the same proportion to the size of the earth as the moon's distance from the end of the shadow is to the total length of the shadow. The umbra is thus found to be 9200 km in diameter at the moon's distance. The value varies slightly from one eclipse to another because the earth-moon distance varies, and because the diameter of the umbra at the place where the moon enters the shadow depends on the moon's and the sun's distances at the time of the eclipse. The penumbra of the earth's shadow averages about 16,000 km across at the moon's distance.

In Figure 9.21 four of the many possible paths of the moon through the earth's shadow are shown. A total lunar eclipse occurs when the moon passes completely into the umbra (path *A*). A *partial eclipse* occurs if only part of the moon skims through the umbra (path *B*), and a *penumbral eclipse* occurs if the moon passes through the penumbra, or partially through the penumbra, but does not come into contact with the umbra (paths *C* and *D*).

(b) Appearance of Lunar Eclipses

Penumbral eclipses usually go unnoticed even by astronomers. Only within about 1100 km of the umbra is the penumbra dark enough to produce a noticeable darkening on the moon. However, the dimished illumination on the moon's surface can be detected photometrically.

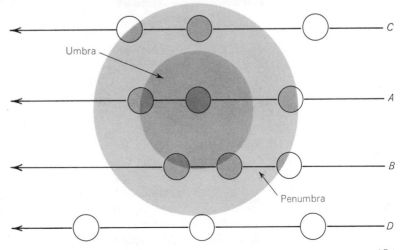

Figure 9.21 Different kinds of lunar eclipses. The penumbra pales from the full light of the outer boundary to the full dark of the umbra at the inner boundary.

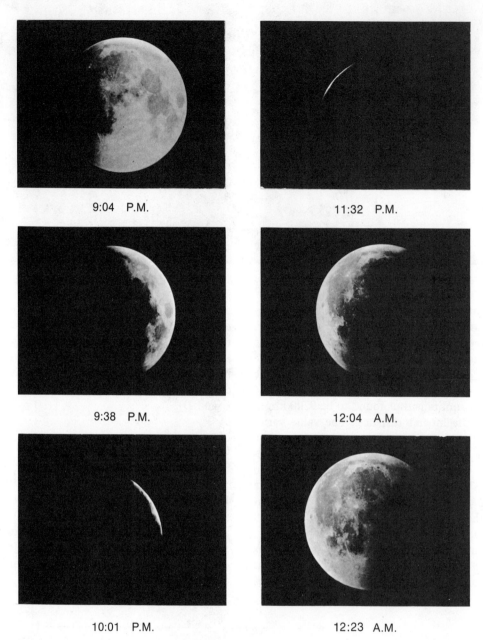

9:04 P.M.

11:32 P.M.

9:38 P.M.

12:04 A.M.

10:01 P.M.

12:23 A.M.

Figure 9.22 Sequence of photographs of the total lunar eclipse of November 17/18, 1956. (*Photographs by Paul Roques, Griffith Observatory*)

Every total or partial lunar eclipse must begin with a penumbral phase. About 20 minutes or so before the moon reaches the shadow cone of the earth, the side nearest the umbra begins to darken somewhat. At the moment called *first contact*, the limb of the moon (the "edge" of its apparent disk in the sky) begins to dip into the umbra of the earth. As the moon moves farther and farther into the umbra, the curved shape of the earth's shadow

upon it is very apparent. In fact, Aristotle listed the round shape of the earth's shadow as one of the earliest proofs of the fact that the earth is spherical (Section 2.2e).

If the eclipse is a partial one, the moon never gets completely into the umbra of the earth's shadow but passes on by, part of it remaining in the penumbra, where it still receives some sunlight. At *last contact* the moon emerges from the umbra.

On the other hand, if the eclipse is a total one, at the instant of *second contact* the moon is completely inside the umbra, and the total phase of the eclipse begins. Even when totally eclipsed, the moon is still faintly visible, usually appearing a dull coppery red. Kepler explained this phenomenon in his treatise *Epitome*. The illumination on the eclipsed moon is sunlight that has passed through the earth's atmosphere and has been refracted by the air into the earth's shadow (Figure 9.23).

The eclipse is darkest if the center of the lunar disk passes near the center of the umbra. The darkness of the lunar eclipse depends also upon weather conditions around the terminator of the earth. It is only here, on the line between day and night on the earth, that sunlight passing through the atmosphere can be refracted into the shadow. Heavy cloudiness in that critical region will allow less light to pass through. The light striking the eclipsed moon inside the umbra is reddish because red light of longer wavelengths penetrates through the long path of the earth's atmosphere most easily.

Totality ends at *third contact,* when the moon begins to leave the umbra. It passes through its partial phases to *last contact,* and finally emerges completely from the penumbra. The total duration of the eclipse depends on how closely the moon's path approaches the axis of the shadow during the eclipse. The moon's velocity with respect to the shadow is about 3400 km/hr; if it passes through the center of the shadow, therefore, about six

hours will elapse from the time the moon starts to enter the penumbra until it finally leaves it. The penumbral phases at the beginning and end of the eclipse last about one hour each, and each partial phase consumes at least one hour. The total phase can last as long as 1 hour 40 minutes if the eclipse is central.

9.7 ECLIPTIC LIMITS

We have seen that solar eclipses occur only when the moon is new and lunar eclipses when the moon is full. Furthermore, eclipses occur only when the new or full moon is near a node. If the sun, the earth, and the moon were geometrical points, the new or full moon would have to occur *exactly* at a node if there were to be an eclipse. Because of the finite sizes of the three bodies, however, part of the sun may appear covered by part of the moon for observers at certain places on earth even though at new moon the moon is not located *exactly* at the node. Similarly, in a lunar opposition, it is possible, because of the large size of the earth's shadow, that the full moon can pass partially through it or even pass completely into it, even if the full moon is not exactly at the node. Thus, the requirement that the moon be at a node can be relaxed slightly; there is some leeway within which eclipses can occur. The limits of this leeway are called *ecliptic limits.*

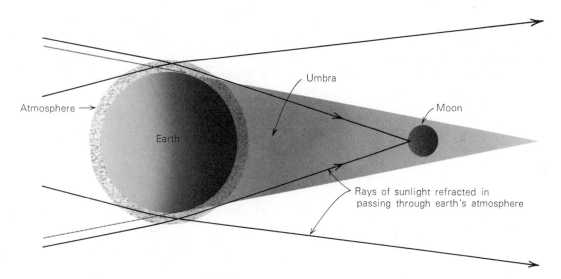

Figure 9.23 Illumination of the moon during a total eclipse by sunlight refracted by the earth's atmosphere into the earth's shadow.

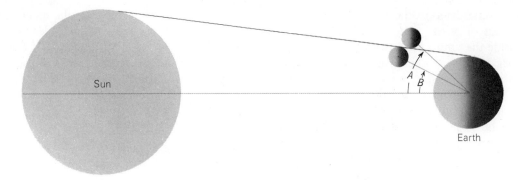

Figure 9.24 Condition for a solar eclipse.

(a) SOLAR ECLIPTIC LIMITS

Note in Figure 9.24 that a partial eclipse of the sun is just barely visible somewhere on earth if the new moon encroaches on the conical surface enveloping the earth and sun. Under these conditions, it is a simple problem in geometry to find the angle A at the center of the earth between the centers of the sun and the moon. Angle A turns out to be about $1\frac{1}{2}°$. For an eclipse to be seen as *central*, that is, annular or total, the moon's distance of the cone tangent to the sun and earth. The centers of these circles, Q, P, P', and earth's center, between the centers of the moon and sun must be less than or equal to the angle B, about $1°$.

Figure 9.25 shows the ecliptic and the moon's apparent path in the celestial sphere in the region of one of the nodes (the ascending node), which is indicated by O. The moon, M, is shown in four places in its path. The large circles, marked C, represent cross sections at the moon's distance of the cone tangent to the sun and earth. The centers of these circles, Q, P, P', and Q', are the positions of the sun's center at four different places along the ecliptic. Note that if new moon occurs at R, with the sun at Q, the centers of the sun and moon are separated by just the critical angle A for

which some kind of eclipse is visible at some place on earth. Similarly, an eclipse can just barely occur if the sun and moon are at Q' and R'. If new moon occurs when the sun is outside the part of the ecliptic lying between Q and Q', an eclipse cannot occur. If the new moon is at S or S', and the sun at P or P', respectively, the two bodies are separated by the critical angle B required for a central eclipse. Central eclipses cannot take place if the new moon occurs when the sun is not within the region on the ecliptic between P and P'. The angular distance OQ or OQ' is the *solar ecliptic limit*. The sun must be within this distance of the node at new moon to be eclipsed. The angular distance OP or OP' is called the *central ecliptic limit*; the sun must be within the central ecliptic limit of the node at new moon for an annular or total eclipse to occur. Given the angle of intersection between the moon's path and the ecliptic (about $5°$), the actual values of the ecliptic limit and central ecliptic limit can be found easily by trigonometric calculation or geometrical construction. They are about $17°$ and $10°$, respectively.

The ecliptic limits can vary considerably, for they depend on the exact angular sizes of the sun and moon in the sky (and hence on their exact distances from the earth) and on the exact inclination of the moon's orbit to the ecliptic; all these are variable quantities. If we consider the extreme range of their

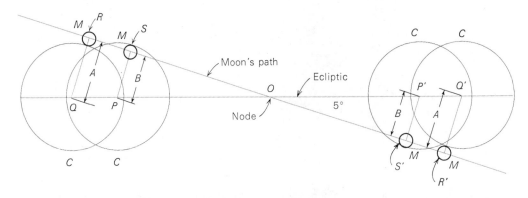

Figure 9.25 Solar ecliptic limits.

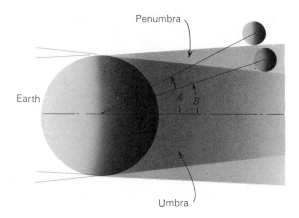

Figure 9.26 Condition for a lunar eclipse.

The penumbra pales gradually from the darkness of the umbra to full light outside; consequently, there is no observable phenomenon at the penumbra's outer boundary. The critical angle, *A*, for a penumbral eclipse, although well defined geometrically, has little real significance. Usually, we speak only of the limits of an umbral lunar eclipse (total or partial), for which the angle *B* applies. Angle *B* is a little less than 1°.

The *lunar ecliptic limit* is the maximum angle between the axis of the umbra and the direction of the node (or between the center of the sun and the opposite node) for which the moon can pass within an angular distance *B* of the umbral axis. The geometry of the situation, very similar to that of the solar ecliptic limit, is illustrated in Figure 9.27. If full moon occurs exactly at the node, *O*, the lunar eclipse is central. There will be no umbral eclipse if the moon passes the earth's shadow when it is farther from the node than at *Q* or *Q'*. The angle *OQ* or *OQ'* is thus the ecliptic limit for an umbral lunar eclipse. The angle varies somewhat, as does the solar ecliptic limit, because of the variations in the distances of the sun and the moon and of the inclination of the moon's orbit. The *major* and *minor lunar ecliptic limits* are 12°15' and 9°30', respectively.

variability, however, we can find the largest and smallest possible values for the ecliptic limits. These are called the *major* and *minor ecliptic limits,* respectively. The *major* and *minor solar ecliptic limits* are 18°31' and 15°21'. The *major* and *minor central ecliptic limits* are 11°50' and 9°55'. At the time of new moon, if the sun is farther from one of the moon's nodes than the major ecliptic limit, no eclipse can occur. If it is between the major and minor limits, an eclipse may occur; it is contingent upon the relative distances of the earth, the sun, and the moon at that instant, and upon the exact inclination of the moon's orbit. If the sun is within the minor ecliptic limit of a node at new moon, an eclipse must occur.

(b) LUNAR ECLIPTIC LIMITS

The *axis,* or center line, of the umbra of the earth's shadow is directed toward a point on the ecliptic exactly opposite the sun in the sky. If the moon is full while it is on or near the ecliptic (that is, near one of its nodes), it can enter the earth's shadow. In Figure 9.26 the angles *A* and *B*, at the center of the earth, lie between the direction to the center of the moon and the axis of the umbra at which the moon can encroach, respectively, on the penumbra and umbra.

(c) FREQUENCY OF ECLIPSES

The angle through which the sun moves along the ecliptic during a synodic month averages 29°6', and varies from this value only slightly. Between two successive new moons, therefore, the sun moves less than twice the minor solar ecliptic limit of 30°42' (2 × 15°21'). Consequently, it is impossible for the sun to pass through a node without at least one eclipse during that eclipse season. There must, then, be at least two solar eclipses during any one calendar year. If the sun is eclipsed within a few days after it reaches the western ecliptic limit, a second eclipse can follow at the next new moon, just before the end of the eclipse season. If the first season falls in January, five solar eclipses are possible in a single calendar year, for

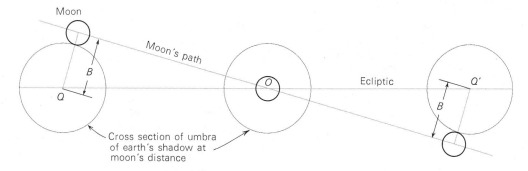

Figure 9.27 Lunar ecliptic limits.

that same eclipse season will begin again in the following December. This happened in 1935. Twice the *central* solar ecliptic limit is narrower (at most 23°40′), and no more than one total or annular eclipse of the sun is possible during an eclipse season; the number during a calendar year varies from none to three.

Eclipses of the moon are about as frequent as those of the sun if penumbral eclipses are counted, the number per year ranging from two to five. Umbral eclipses, on the other hand, occur about as often as central solar eclipses; there are from none to three in any one year.

Penumbral eclipses are inconspicuous, so the total number of observable solar eclipses over a period of time outnumbers the observable lunar eclipses by nearly three to two. Lunar eclipses are more common at any one station, however, for they can be viewed from more than half of the globe, whereas solar eclipses, even partial ones, are visible only in limited areas. Total solar eclipses occur on an average about once every 1½ years, but they are visible only within narrow eclipse paths; their average frequency at any one place is about once every 360 years.

The maximum number of all kinds of eclipses (solar and lunar) in any one calendar year is seven.

9.8 RECURRENCE OF ECLIPSES

Thousands of years ago the ancients noticed that similar eclipses occurred at regular intervals. This recurrence of eclipses made it possible for early astronomers to predict eclipses with fair accuracy.

(a) CIRCUMSTANCES FOR TWO SIMILAR ECLIPSES

In order that an eclipse may be followed, after a lapse of time, by another eclipse similar to the first one, the following conditions must be met: (1) the moon must be at the same phase again (new for a solar, full for a lunar eclipse); (2) the moon, when at that phase, must be in the same place in its orbit with respect to the node; and (3) the sun and moon must have the same distances from the earth again. If, in addition, the two eclipses are to have similar eclipse paths on the earth, they must occur at about the same time of the year.

Let us consider requirements (1) and (2) first. For the moon to return to exactly the same phase again, an integral (or whole) number of synodic months must have elapsed. To return to the same place in its orbit again with respect to the node, the moon must have made an integral number of revolutions about its orbit with respect to the nodes; that is, there must be an integral number of *nodical* or *draconic months*. (The

term "draconic" is derived from the ancient Chinese superstition that eclipses were caused by dragons swallowing the sun.)

Suppose the synodic and nodical months were exactly 30 and 27 days long, respectively. Then eclipses nearly identical to each other would occur every 270 days, because both nine synodic months and ten nodical months would add up to 270 days. However, the synodic month is actually 29.5306 days and the nodical month is 27.2122 days, and there exists no integral least common multiple of these two numbers.

There are, on the other hand, some periods of time that are very nearly integral multiples of both the synodic and the nodical months. For example, the interval of 47 synodic months is almost the same as the interval of 51 nodical months:

$$47 \text{ synodic months} = 1387.94 \text{ days;}$$
$$51 \text{ nodical months} = 1387.82 \text{ days.}$$

As an illustration of the significance of this coincidence, suppose there is a solar eclipse. On the forty-seventh new moon following that eclipse, there will be another eclipse in which the moon is only about one tenth of a day's journey beyond the original place in its orbit relative to the node. Since the moon moves about 13° with respect to the node during one full day, this second eclipse will occur at a time when the sun and the moon are situated only a little over 1° from where they were, relative to the node, at the time of the eclipse 47 months earlier. Suppose the first eclipse had occurred just inside the western ecliptic limit. The second eclipse will take place a little farther inside the limit; the eclipse path on the earth will lie at latitudes slightly different from those of the first path. After another 47 synodic months, a third eclipse will occur, again about 1° further in from the western end of the ecliptic limit, and its path on the earth will again be displaced only slightly from that of the second eclipse. There will be a series of similar eclipses, occurring at intervals of 47 months, or about 1388 days. The series of eclipses will continue until the position of the sun and moon during eclipse, shifting a little over one degree eastward from one eclipse to the next, has progressed along the ecliptic from the western end of the limit to the eastern end thereof, through a range of about 35°. There will be about 30 eclipses in the series.

Any two of these successive eclipses will have similar geometry, but they will not be identical; some will be total and others annular. For identical eclipses, the relative distances of the earth, the sun, and the moon must be the same in both cases. Because the eccentricity of the earth's orbit about the sun is rather small, we may, as a first approximation, regard the moon's changing distance from the earth as the factor that determines the type of eclipse. The moon's closest approach to the earth (perigee) occurs at intervals of the anomalistic month of 27.55455 days. The anomalistic month differs from the sidereal and nodical months because the major axis of the moon's elliptical orbit (the line of *apsides*) gradually changes orientation in the plane of the moon's orbit (Section 9.3). The successive eclipses in an eclipse series are of about the same type if the integral number of synodic months is not only nearly equal to an integral number of nodical months but also is nearly equal to an integral number of anomalistic months. In Table 9.1 the various intervals are given for four eclipse cycles.

Table 9.1   **Intervals of Eclipses in Four Representative Cycles**

1	47 synodic months	=	1387.94 days
	51 nodical months	=	1387.82
	50 anomalistic months	=	1377.73
2	223 synodic months	=	6585.32
	242 nodical months	=	6585.36
	239 anomalistic months	=	6585.54
3	3803 synodic months	=	112,304.8
	4127 nodical months	=	112,304.8
	4076 anomalistic months	=	112,312.4
4	4519 synodic months	=	133,448.7
	4904 nodical months	=	133,448.7
	4843 anomalistic months	=	133,446.7

(b) ECLIPSE SERIES; THE SAROS

Any of the above cycles can be used to predict eclipses; it is simplest to use the synodic month as the unit of time. Note that the second and fourth cycles contain nearly integral numbers of nodical, anomalistic, and synodic months. In both of those cycles the successive eclipses are of about the same type. In the fourth cycle, however, the interval is so long that changes in the orbit of the moon become significant. Most useful for the prediction of eclipses, therefore, is the second cycle of 223 synodic months, just 11 days in excess of 18 years. This cycle is called the *saros*.

Nearly identical eclipses, both solar and lunar, recur at intervals of the saros. Successive eclipses, following each other at these 18-year intervals, are said to belong to the same series. Of course, many eclipses occur during any 18-year period; thus, many series of eclipses are in progress at once. Because the saros cycle contains so nearly an integral number of tropical years, at each successive eclipse in the cycle the earth's axis has almost the same orientation relative to the ecliptic plane (or to the earth-sun line); thus successive eclipses even occur at approximately the same latitudes and have similar paths over the earth's surface.

The first solar eclipse in any series belonging to the saros cycle is a partial eclipse that occurs when the sun and the moon are at an eastern ecliptic limit. If it is the

descending node, the eclipse is just barely visible near the south pole of the earth. If the eclipse is at the eastern limit of the ascending node, it is just visible at the north pole. After 223 synodic months the next eclipse in the series occurs, with the moon just about one thirtieth of a day's journey, or slightly under $1/2°$, west of the same place relative to the node. Thus the eclipse occurs with the sun and moon about $1/2°$ farther inside the ecliptic limit. After about 70 successive eclipses separated by intervals of the saros (about 1200 years), the positions of the sun and moon at the times of the eclipses have shifted through the node to the western ecliptic limit, and the series ends. The first dozen eclipses in the series are only partial ones visible near one of the polar regions of the earth (North Pole for the sun and moon at the ascending node, South Pole for the sun and moon at the descending node). The last dozen eclipses are partial ones visible at the other polar region. The middle 45 eclipses of the series are total or annular, the eclipse path gradually shifting in latitude from one pole to the opposite one during the series.

If the saros interval contained an exactly integral number of days, successive eclipses in a series would occur at the same time of day for each place on earth, and the eclipses would all be visible at nearly the same longitudes as well as at the same latitudes. As it is, however, the saros interval is 6585.32 days; the approximate one third day remainder causes successive eclipses in the cycle to occur about one third of the way around the world from each other. However, every third eclipse in a series does follow a path lying in nearly the same terrestrial longitudes.

Series of lunar eclipses, each one similar to the preceding one, also occur at intervals of the saros. However, because the limits for umbral lunar eclipses are smaller than for solar eclipses, a series of lunar eclipses runs through only about 50 saroses, which requires about 870 years.

The saros interval contains a nearly integral number of anomalistic months, so any two successive eclipses in the same series will be of nearly the same type, that is, both total, both annular, and so on. Of course, the type of eclipse changes as the series progresses. Especially notable is the series of total solar eclipses to which belong the eclipses of 1937, 1955, 1973, and 1991, because the duration of totality of these eclipses is near the maximum possible.

Appendix 12 lists total solar eclipses of appreciable duration of totality that are visible from inhabited places on the earth during the interval 1950 to 2021 A.D. This interval contains nearly three saros cycles, and the similarity of successive eclipses in the same cycle is evident. All eclipses belonging to a given saros cycle are identified with the same letter in the table.

Figure 9.28   Series of photographs (left to right) showing the emergence of Jupiter and three of its satellites after their occultation by the moon. (*Photographs by Paul Roques, Griffith Observatory*)

9.9   PHENOMENA RELATED TO ECLIPSES

We have considered so far only eclipses that involve the sun, the moon, and the earth. However, there are phenomena with similar geometrical properties that involve other celestial bodies. Examples are *occultations* and *transits*.

(a)   Occultations

The moon often passes between the earth and a star; the phenomenon is called an *occultation*. The stars are so remote that the shadow of the moon cast in the light of a star is extremely long and sensibly cylindrical. Because a star is virtually a point source, there is no penumbra. During an occultation, a star suddenly disappears as the eastern limb of the moon crosses the line between the star and observer. If the moon is at a phase between new and full, the eastern limb will not be illuminated and the star may appear to vanish mysteriously as the dark edge of the moon covers it. Because the moon moves through an angle about equal to its own diameter every hour, the longest time that an occultation can last is about one hour. It can have a much shorter duration if the occultation is not central. Geometrically, occultations are equivalent to total solar eclipses, except that they are total eclipses of stars other than the sun.

The sudden disappearance of a star behind the limb of the moon during an occultation is evidence that the moon has no appreciable atmosphere. If there were one, the star would fade gradually as the moon's limb approached it, because the starlight would traverse a long path of the lunar atmosphere. Occultations also demonstrate the extremely small angular sizes of the stars (owing to their great distances). If a star had an appreciable angular size it would require a perceptible time to disappear behind the moon, as is true during the partial phases of a total solar eclipse. Actually, the partial phases of occultations have been measured photoelectrically, but they are extremely brief, less than a few hundredths of a second. The angular sizes of stars cannot be observed directly in a telescope, but they can often be determined by various techniques (see Chapter 26). It has been possible to observe that stars of large computed angular size require longer to disappear behind the moon than those of small angular size. Also observations of occultations of celestial radio sources have been useful in detecting the accurate positions and angular sizes of those objects.

Occultations of the brighter stars are listed in advance in various astronomical publications. The times and durations of the occultations and the places on earth from which they are visible are given. Also listed are the comparatively rare occultations of planets by the moon, and of stars by planets.

(b)   Transits

A *transit* is a passage of an inferior planet (Mercury or Venus) across the front of the sun's disk at the time of inferior conjunction. Usually each of these planets appears to pass north or south of the sun, but it can pass in front of the sun if inferior conjunction occurs when the planet is near one of the nodes of its orbit—the points where its orbit crosses the ecliptic.

The sun passes the nodes of Mercury's orbit in May and November; these are the seasons for transits of Mercury. However, transit "limits," analogous to ecliptic limits, are only a few degrees, and Mercury is seldom near enough to a node during the brief periods in May and November when transits of Mercury are possible. There are, on the average, about 13 transits of Mercury per century.

Transits of Venus are more rare than those of Mercury because the limits are narrower and because inferior conjunctions occur less frequently. Recurrences of transits occur just as recurrences of eclipses occur. Between two transits there must be an integral number of synodic periods of the planet, and there must be an integral number, or very nearly an integral number, of periods in which the planet has returned to the node. At present, transits of Venus occur in pairs separated by intervals of about eight years. The last pair of Venerian transits were those of 1874 and 1882. The next pair will be on June 8, 2004, and June 6, 2012.

Transits are analogous to annular eclipses of the sun. The planets Venus and Mercury, seen from earth, are too small to cover the sun completely; their shadow cones fall far short of reaching the surface of the earth. The appearance of a transit is that of a black dot slowly crossing the disk of the sun from east to west. The silhouette of Mercury against the sun is too small to see without a telescope. That of Venus can be barely observed, without optical aid, if the sun is properly viewed by projection, or through proper filters to protect the eye.

(c) Eclipses on Other Planets

Eclipses would be visible from planets with satellites, other than the earth. For example, total solar eclipses would be common to observers living on the outside of the dense cloud layers that shroud the planet Jupiter. Several of Jupiter's many satellites regularly cast their shadows on the atmosphere of the planet. The dark spots on Jupiter where these shadows fall on its atmosphere are easily visible through telescopes on earth. Only on the planets Mercury, and Venus, which have no known satellites, are eclipses impossible.

(d) Stellar Eclipses

The most important kind of eclipses, to astronomers, are those of *eclipsing binary stars*. A binary system consists of two stars that revolve about each other. Such double stars are very common—many naked-eye stars are actually binary. When the plane of mutual revolution of such a double star happens to be so oriented that we see it edge on, each star periodically passes partially or entirely behind the other, thus being eclipsed. Then some or all of the light from the eclipsed star is prevented from reaching the earth, and the combined light received from the system is diminished. The most famous eclipsing binary star is the bright star *Algol* in the constellation of Perseus. Algol is really two stars revolving around each other, a bright one and a faint one. About every 2 days and 21 hours, the bright star passes partially behind the faint one and the apparent brightness of Algol drops down to less than half of normal—a change readily observable by the naked eye. A study of the light variation of such binary systems gives much information about the sizes and masses of their member stars. Eclipsing binary stars are described in Chapter 26.

EXERCISES

1. When earthshine is brightest on the moon, what must be the phase of the earth, as seen from the moon?

2. How long would a synodic month be if the moon's sidereal month were four of our present calendar months? (*Hint:* See Section 3.1a.)

3. How many more sidereal months than synodic months must there be in a year? Why?

4. If the moon revolved from east to west rather than from west to east, would a synodic month be longer or shorter than a sidereal month? Why?

5. Describe the daily delay in the rising of the full moon, both in the northern and southern hemispheres, nearest the time when the sun is at the vernal equinox.

6. What is the phase of the moon if (a) it rises at 3:00 P.M.? (b) it is on the meridian at 7:00 A.M.? (c) it sets at 10:00 A.M.?

7. What time does (a) the first quarter moon cross the meridian? (b) the third quarter moon set? (c) the new moon rise?

8. Describe the phases of the earth as seen from the moon. At what phase of the moon (as seen from the earth) would the earth (as seen from the moon) be a waning gibbous?

9. Suppose the moon is at its maximum distance north of the equator. If the moon just grazes the northern horizon as it crosses an observer's meridian, at about what latitude would this observer be?

10. A waxing crescent moon is observed to lie in the direction of the vernal equinox. What season is it?

11. Suppose you lived in the crater Copernicus on the moon. (a) How often would the sun rise? (b) How often would the earth set? (c) Over what fraction of the time would you be able to see the stars?

12. The mean distance from the earth to the moon is 384,404 km, and light travels at 299,793 km/s. How long do radar waves take to make a round trip to the moon?
Answer: 2.56 seconds

13. When Mars is at a distance of 56,000,000 km, its angular diameter is 24″.9. What is its linear diameter?
Answer: About 6760 km.

14. Suppose the moon were exactly 386,000 km from the earth and that the earth's distance from the barycenter turned out to be 96,500 km. How much more massive than the moon would the earth be?

15. Sketch the shadow of a pencil cast by an electric lamp. Indicate the umbra and penumbra.

16. What is the angle at the apex of the shadow cone of the earth? (*Hint:* What is the angular size of the sun?)

17. Jupiter, having a diameter of about 140,000 km, is more than twice the diameter of Neptune, with a diameter of only about 48,000 km. Yet the umbra of Neptune's shadow is about twice as long as the umbra of Jupiter's shadow. Explain.

18. What are the relative lengths of the shadow cones of the earth and the moon? Assume that both objects are exactly the same distance from the sun. (Illustrate and show the method of your calculation, rather than looking up the figures in the chapter.)

19. What will be the eclipse seasons in 1990?

20. If a solar eclipse were annular over part of the eclipse path and total over part, would the total eclipse be of long or short duration? Why?

21. Does the longest duration of a solar eclipse occur when
    a) the sun is at its nearest and the moon at its nearest?
    b) the sun is at its farthest and the moon at its nearest?
    c) the sun is at its nearest and the moon at its farthest?
    d) the sun is at its farthest and the moon at its farthest?

22. Describe what an observer at the crater Copernicus on the moon would observe during what would be a total solar eclipse as viewed from the earth.

23. Compare the actual ratio of the size of the earth's shadow to the moon's diameter to the value 2 estimated by Aristarchus (Chapter 2).

24. Does the moon enter the shadow of the earth from the east or west side? Explain why.

25. Describe the phenomenon observed by a spectator on the moon while the moon is eclipsed.

26. If the penumbra of the earth's shadow is 16,000 km across, and if the moon moves 3400 km/hr with respect to the shadow, why does it take six hours instead of only five hours to get completely through the penumbra?

★27. Suppose there were five lunar and two solar eclipses during a calendar year. What would be the nature of the lunar eclipses? Explain.

★28. Can there be four solar and three lunar eclipses during a calendar year? If so, what are the natures of the lunar eclipses? Explain your answer.

29. Which planets can the moon never occult while at the full phase? Why?

30. If the earth and the moon had their present distance from each other and their present period of mutual revolution, but if they were removed to the distance of Jupiter from the sun, would total solar eclipses be more common or less common at any one place? Explain.

★31. Compare the annual motion (in degrees per year) of the moon's line of nodes and that of its line of apsides.

★32. Calculate the shadow length of the moon and compare it with the moon's distance from the earth. Discuss this result in terms of annular and total solar eclipses.

Max Planck (1858–1947) established the quantized nature of light and other electromagnetic radiation. The multicentered *Max-Planck Institut* of West Germany is named in tribute to his contributions. (*American Institute of Physics, Niels Bohr Library*)

10

ENERGY FROM SPACE: THE ELECTRO-MAGNETIC SPECTRUM

The earth is constantly exposed to energy of various forms from space. In this chapter we discuss the nature and properties of some of this cosmic energy and how it interacts with matter; in the next two chapters we shall describe the techniques and instruments we use to observe it. Among these various forms of energy are:

1. *Electromagnetic radiation*, if we include that which we receive from the sun, is by far the most important kind of energy reaching the earth; most of this chapter and the next two are concerned with it.

2. *Cosmic rays* are charged particles, mostly the nuclei of atoms, that strike the molecules of the earth's upper atmosphere, producing tremendous numbers of secondary subatomic particles that rain down to the earth's surface. The total energy we receive in the form of cosmic rays exceeds that which we get from starlight. We discuss these particles near the end of this chapter.

3. The *solar wind* is a constant flow of atomic nuclei (mostly of hydrogen) and electrons

emitted into interplanetary space by the sun. Most solar wind particles do not move fast enough to be called cosmic rays. Their effects can sometimes be observed, however, in the form of the light (especially auroras) emitted by upper atmospheric molecules ionized or excited by them. The solar wind is also regularly observed with instruments carried by space vehicles and artificial satellites. We take up the solar wind in Chapters 21 and 31. ·

4. *Neutrinos* are produced by the decay of certain subatomic particles and also in many of the nuclear reactions that keep stars shining. Until 1980 they were generally thought to have no mass, in which case they must travel with the speed of light (Chapter 13). Several recent experiments, however, have yielded some evidence that suggests neutrinos may have very small masses, and if so they cannot have speeds quite as great as that of light. At this writing (1981) the interpretation of those experiments is a matter of debate, and the question of whether or not neutri-

nos are truly massless is far from settled. In any case, they pass through great amounts of matter without influencing it in the slightest and so are very difficult to detect. Nevertheless, the universe must be completely bathed in the radiation of neutrinos. We discuss them further in Chapter 32.

5. Einstein's general relativity theory predicts that motions of matter must generate gravitational waves. They must be everywhere around us, but gravitational waves are extraordinarily weak and enormously difficult to detect. We shall describe them, and our efforts to detect them, in Chapter 35.

6. Finally, the earth, like Mars and the moon, is constantly bombarded by particles in orbit about the sun—meteoroids and micrometeoroids. They are particularly important to us, for we can subject them to chemical analysis and dating. We take up these objects in Chapter 21.

10.1 ELECTROMAGNETIC RADIATION

We expect that astronomers will continue to investigate astronomical objects by means of electromagnetic radiation for many decades to come. However, the overwhelming majority of all the radiation from space that will be observed throughout our lifetimes is already in space on its way to us. What is the nature of this energy approaching us with the speed of light from all directions, waiting to be sampled by our telescopes for the century to come—and beyond? What are the secrets it holds, and what are the revelations it will give us about those objects it left years, centuries, even thousands of millions of years ago?

Light is the most familiar form of electromagnetic energy. Actually, visible light constitutes only a small part of a wide range of different kinds of electromagnetic radiation, which differ from each other only in *wavelength* but are called by different names: gamma rays, X rays, ultraviolet, light, infrared, and radio waves. By its omnipresence, light is familiar to us all. Yet to describe its nature, or to represent it with pictorial conceptions, is extremely difficult. The best we can do is to describe its properties with mathematical models. Most of the char-

acteristics of electromagnetic radiation can be described adequately if it is represented as energy propagated in waves, although no medium is required to transmit the waves. Electromagnetic radiation is so named because it is the propagation of a disturbance in an electromagnetic field, which could be caused, say, by the oscillatory motion of a charge. All forms of electromagnetic radiation must travel (in a vacuum) with the same speed, denoted c. We shall have more to say of this in Chapter 13.

Any wave motion can be characterized by a *wavelength*. Ocean waves provide an analogy. The wavelength is simply the distance separating successive wave crests. Various forms of electromagnetic energy differ from each other only in their wavelengths. Those with the longest waves, ranging up to many kilometers in length, are called *radio waves*. Forms of electromagnetic energy of successively shorter wavelengths are called, respectively, infrared radiation, light, ultraviolet radiation, X rays, and finally the very short-wave gamma rays. Long infrared waves are sometimes referred to as *radiant heat*. All these forms of radiation are the same basic kind of energy and could be thought of as different kinds of light. Here, however, we shall reserve the term "light" to describe those wavelengths of electromagnetic radiation that, by their action upon the organs of vision, stimulate sight. Light is also called *luminous energy*.

(a) Wave Motion: Interference and Diffraction

If you stand in the water at the beach, just out beyond the breakers, you can experience the waves passing by you, with the water getting deeper as each crest, and shallower as each trough, moves in. Similarly, as light passes a point, it varies in intensity (as the ocean water does in depth); the waves of light pass by so rapidly, however—some thousand million million per second—that you are unaware of the wavelike nature of light.

To explain one of the many experiments that demonstrate the wave properties of light, it will be useful to begin with an analogy. Imagine ocean waves approaching a wall, labeled "front wall" in Figure 10.1. Let us suppose there are two openings, or gates, in that front wall. The waves of water are stopped, or reflected back, at every point along that front wall, except where it can flow through the open gates. Now, as the waves of water

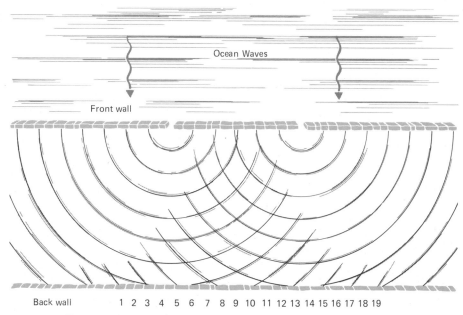

Ocean Waves

Front wall

Back wall   1 2 3 4 5 6 7 8 9 10 11 12 13 14 15 16 17 18 19

Figure 10.1 Interference of water waves passing through two openings.

flow through the gates, they spread out into semi-circular waves, radiating from each gate just as the ripples in a pond radiate from the place where a pebble is dropped. But the waves flowing through the two different openings intermingle; where two crests happen to come together, the water is especially deep, and where two troughs happen to combine, it is especially shallow.

Now consider the heights of the water on a second wall, behind the first one, the "back wall" in Figure 10.1. At points 1,3,5,7,9, . . . , along the wall, wave crests from the two gates always arrive at the same time. Between the successive arrivals of two crests, wave troughs from the two gates arrive together. Thus at those odd-numbered points the water alternately is very high and very low, as the successive crests and troughs of the waves from the two gates reinforce each other. At intermediate points, however, 2,4,6,8, . . . , the wave crest from one gate always meets a trough from the other, so the two sets of waves cancel each other.

If you have followed the preceding discussion, you will realize that if you were standing at the odd-numbered points (in the figure) you would experience large-amplitude waves striking you (*amplitude* simply means the height of the waves compared to the midpoint between crest and trough). In between, at the even-numbered points, the waves would disappear. Well, exactly the same experiment can be done with light and with the same result.

It works best if we select light of a very pure color, and let that light, from a distant source, approach an opaque surface with two narrow slits, exactly analogous to the gates in the front wall in Figure 10.1 Now suppose you put a white card or screen some distance behind the slits. Light passing through the two slits spreads out, just as the ocean waves do, and strikes the screen behind the slits. However, you will not see the screen uniformly illuminated by the light, but rather crossed by a series of bright and dark fringes, of the same color as

Figure 10.2 Interference fringes of light.

that of the light source (as in Figure 10.2). The reason is that light consists of waves, and where the wave crests and troughs of the light passing through the two slits reinforce each other at the screen, you see a bright fringe of light, just as you see large-amplitude waves at the odd-numbered points on the back wall in Figure 10.1. In between, though, the troughs and crests of the light from the two slits cancel each other, and you see no light; those places are the dark fringes in Figure 10.2 and correspond to the even-numbered places on the back wall in Figure 10.1.

The exact spacing of the fringes on the screen, or the distance between the odd-numbered points on the back wall in Figure 10.1, depends on how far apart the crests of the wave are; that distance between wave crests is called the *wavelength* of the waves. Typically ocean waves have wavelengths of tens of meters, but visible light typically has wavelengths of 0.00005 cm. This phenomenon, whereby light waves reinforce each other or cancel out each other and produce fringes, is called *interference*. The fact that light can display such interference is evidence that it propagates as waves.

It is easy to see interference in light with a very simple experiment. Place in front of one of your eyes two straight-edged cards, with the edges close together, so that they form a narrow gap or slit between them. Now look at a bright source of light through the slit. As you narrow the gap between the cards, you will see a succession of bright and dark lines running along the gap. These are interference fringes, evidence of interference of light waves alternately canceling and reinforcing each other to produce dark and bright bands.

The interference bands are formed in this experiment because of *diffraction*. The phenomenon of diffraction is a deflection of light when it passes the edge of an opaque object; light spreads itself out, slightly, into the shadow of the object it passes. The phenomenon is familiar also in sound: out of sight is *not* out of earshot! In the above experiment, the light passing the edges of the cards on either side of the slit fans out. The image of fringes you see as you look through the slit is produced by waves from different places across the width of the slit, which must travel different distances, resulting in alternate reinforcement and cancellation of the waves.

Diffraction is important in many astronomical applications; among other things it provides a theoretical limitation to the resolving power of a tele-scope. It is also central to the operation of many spectrographs (Chapter 11). Other phenomena whereby the direction of light is changed are *reflection* and *refraction*, both discussed later in this chapter.

(b) Propagation of Light: The Inverse-Square Law

Light propagates through empty space in a straight line (but in Chapter 35 we consider more deeply the meaning of "straight line"). An important property of the propagation of electromagnetic energy is the *inverse-square law*, a property that also belongs to the propagation of other kinds of energy, for example, sound. The amount of energy that would be picked up by a telescope, or an eye, or any other detecting device of fixed area, located at any given distance d from a light source O, is proportional to the amount of energy crossing each square centimeter of the surface of an imaginary sphere of radius d (Figure 10.3). A certain finite amount of energy is emitted by the source in a given time interval. That energy spreads out over a larger and larger area as it moves away from the source. When it has moved out a distance d, it is spread over a sphere of area $4\pi d^2$ ($4\pi d^2$ is the surface area of a sphere of radius d). If E is the amount of energy in question, an amount $E/4\pi d^2$ passes through each square centimeter (if d is measured in centimeters).

We see, then, that the amount of energy passing through a unit area *decreases* with the *square of*

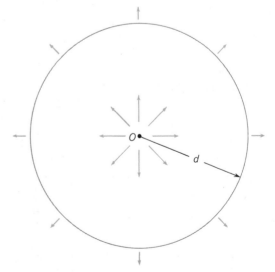

Figure 10.3 Inverse-square law of light.

the distance from the source. At distances d_1 and d_2 from a light source, the amounts of energy received by a telescope (or other detecting device), ℓ_1 and ℓ_2, are in the proportion:

$$\frac{\ell_1}{\ell_2} = \frac{4\pi d_2^2}{4\pi d_1^2} = \left(\frac{d_2}{d_1}\right)^2.$$

The above relation is known as the inverse-square law of light propagation. In this respect, the propagation of radiation is similar to the effectiveness of gravity, because the force of gravitation between two attracting masses is inversely proportional to the square of their separation. Figure 10.4 also illustrates the inverse-square law of propagation.

(c) The Electromagnetic Spectrum

We have seen that all forms of electromagnetic energy have certain characteristics in common: their wavelike method of propagation and their speed. Nevertheless, the various kinds of radiant energy, differing from each other in wavelength, are detected by very different means.

Radio waves have the longest wavelength, ranging up to several kilometers. Those used in short-wave communication and in television have wavelengths ranging from centimeters to meters. When these waves pass a conductor, such as a radio antenna, they induce in it a feeble current of electricity, which can be amplified until it can drive a loudspeaker or recording apparatus.

The shortest wavelengths of radio radiation, about one millimeter, merge into infrared radiation. Infrared radiation of wavelength less than 0.0015 mm can be photographed with special emulsions. There are also certain substances such as lead sulfide and indium antimonide whose conductivity increases when they absorb infrared radiation. Cells containing these materials as detectors measure the intensity of infrared radiation striking them by changes in their resistance to a current.

Another infrared detector is the *thermocouple*— a union of two metals through which a weak current is induced by the absorption of radiation. In the past, stacks of thermocouples called *thermopiles* were commonly used at the focus of a telescope to measure the infrared radiation from planets.

Far more sensitive modern devices utilize germanium or silicon semiconductors that change resistance when they absorb infrared radiation. The change in the current conducted by the semiconductor measures the energy it absorbs. With these devices infrared radiation of wavelength up to several millimeters can be detected.

A very simple device for detecting infrared radiation is a container of water. As the water absorbs the radiation, its temperature rises.

Electromagnetic radiation with a wavelength in the range 0.0004 to 0.0007 mm comprises visible light. It is more convenient to express the wavelengths of visible light in *angstroms*. One angstrom (abbreviated Å) is one hundred-millionth of a centimeter. Visible light, then, has wavelengths that range from about 4000 to 7000 Å. The exact wavelength of visible light determines its *color*. Radiation with a wavelength in the range of 4000 to 4500 Å gives the visual impression of the color violet. Radiations of successively longer wavelengths give the impression of the colors blue, green, yellow, orange, and red, respectively. The array of colors of visible light is called the *spectrum*. A mixture of light of all wavelengths, in about the same relative proportions as are found in the light emitted into space by the sun, gives the impression of white light. (Actually it is the frequency of the radiation rather than the wavelength that determines which color it appears. As we shall see below, radiations of different wavelengths also have different frequencies, so in a vacuum the distinction is moot. When radiation passes through a transparent medium,

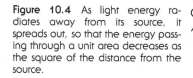
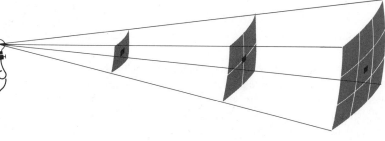

Figure 10.4 As light energy radiates away from its source, it spreads out, so that the energy passing through a unit area decreases as the square of the distance from the source.

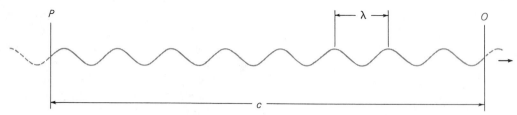

Figure 10.5 Relation between wavelength, frequency, and the speed of radiation.

however, such as glass or water, its wavelength is decreased, but the frequency is unchanged; thus, for example, a red light still looks red under water.)

Light not only affects the organs of vision but can also be detected photographically and photoelectrically. A photographic emulsion contains silver halides that undergo chemical changes when they absorb light. Upon development the molecules so altered release in the emulsion the grains of silver that make up the photographic image. A photoelectric detector contains a photoemissive substance that ejects an electron when it absorbs electromagnetic radiation. Hertz discovered this *photoelectric effect* in 1887.

Radiation of wavelengths too short to be visible to the eye is called *ultraviolet*. Radiation with wavelengths less than about 200 Å comprise X rays. Both ultraviolet radiation and X rays, like visible light, can be detected photographically and photoelectrically.

Electromagnetic radiation of the shortest wavelength, to less than 0.1 Å (10^{-8} mm), is called *gamma radiation*. Gamma rays are often emitted in the course of nuclear reactions and by radioactive elements. Gamma radiation is generated in the deep interiors of stars; it is gradually degraded into visible light by repeated absorption and reemission by the gases that comprise stars.

The array of radiation of all wavelengths, from radio waves to gamma rays, is called the *electromagnetic spectrum*.

(d) Relation Between Wavelength, Frequency, and Speed

Let us once again compare the propagation of light to the propagation of ocean waves. While an ocean wave travels forward, the water itself is displaced only in a vertical direction. A stick of wood floating in the water merely bobs up and down as the waves move along the surface of the water. Waves that propagate with this kind of motion are called *transverse waves*.

Light also propagates with a transverse wave motion, and travels with its highest possible speed through a perfect vacuum. In this respect light differs markedly from *sound*, which is a physical vibration of matter. Sound does not travel at all through a vacuum. The displacements of the matter that carry a sound impulse are in a *longitudinal* direction, that is, in the direction of the propagation, rather than at right angles to it. Sound is actually a traveling wave of alternate compressions and rarefactions of the matter through which it moves. Of course, sound also travels far more slowly than electromagnetic radiation—only about ⅓ km/s through air at sea level.

For any kind of wave motion, sound or light, we can derive a simple relation between wavelength and *frequency*. The frequency of light (or sound) is the rate at which wave crests pass a given point, that is, the number of wave crests that pass per second. Imagine a long train of waves moving to the right, past point O (Figure 10.5), at a speed c. If c is measured, say, in centimeters per second, we can measure back to a distance of c centimeters to the left of O and find the point P along the wave train that will just reach the point O after a period of one second. The frequency f of the wave train— the number of waves that pass O during that second—is obviously the number of waves between P and O. That number of waves, times the length of each, λ, is equal to the distance c. Thus we see that for any wave motion, the speed of propagation equals the frequency times the wavelength, or symbolically,

$$c = f \times \lambda.$$

(It is traditional to denote the length of a wave of electromagnetic radiation by λ, the Greek letter for "ℓ".)

10.2 HEAT AND RADIATION

A warm or hot solid is composed of molecules and atoms that are in continuous vibration. A gas consists of molecules that are flying about freely at high speed, continually bumping into each other, and bombarding against the surrounding matter. That energy of motion is called *heat*. The hotter the solid or gas, the more rapid is the motion of those molecules, and its *temperature* is just a measure of the average energy of those particles. One of the principles of *thermodynamics* (the *second law*) is that heat always tends to transfer from a hot object to a cooler one. Thus a solid or gas that is at a higher temperature than its surroundings radiates some of its heat energy into those surroundings, thereby cooling.

A century ago physicists were interested in the properties of this emitted radiation—that is, how much of which kind of radiation is radiated by warm and hot bodies? The situation is quite complicated, because bodies absorb some radiation, reflect some, and transmit some. A blue sweater, for example, reflects more of the relatively short-wave blue light than it does other colors; that is why it looks blue. A piece of black coal reflects relatively little of any visible light, and a window pane transmits most light through without either absorbing or reflecting it.

All these bodies, though, absorb some radiation, especially when the entire electromagnetic spectrum is considered; moreover, all bodies will eventually come into equilibrium with their surroundings, until they reemit energy at a rate which, averaged over time, is exactly the rate at which they absorb it.

When confronted with a complex problem, as this one is, the scientist usually tries to get a start in his analysis by finding a circumstance in which the problem is simplified. To this end, physicists invented the *ideal* or *perfect radiator*, a hypothetical body that completely absorbs every kind of electromagnetic radiation incident on it, reaches some equilibrium temperature, and then reradiates that energy as rapidly as it absorbs it. Because the perfect radiator absorbs everything and reflects and transmits nothing, it is also called a *black body*.

A black body, however, is not so called because it necessarily *looks* black; since it is radiating energy, it might be very bright indeed. It happens that a piece of coal, which is a crude approximation to a black body, does look black, but that is because (at least at room temperature) its reradiated energy is mostly in the invisible infrared; if we could see infrared radiation, we would find that coal to be glowing brightly. Stars happen to be good black bodies, because the gases they are made of are very opaque to virtually all electromagnetic radiation, and only from a star's outermost layers can energy escape into space. A star like the sun, of course, hardly looks black.

(a) Black-Body Radiation

Laboratory devices that are close approximations to perfect radiators were invented long ago. One, for example, is an enclosed chamber with well-insulated walls that are painted black on the inside. The internal energy, nearly all absorbed and reemitted

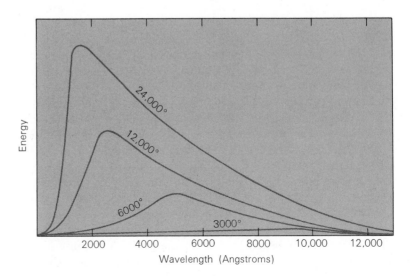

Figure 10.6 Energy emitted at different wavelengths for black bodies at several temperatures. (The curves are schematic only and are not plotted accurately to scale.)

by the interior walls, can be observed through a tiny hole in the chamber. Such a device was used to learn experimentally the nature of black-body radiation and to test the derived radiation laws. It was found that the distribution of energy at different wavelengths (*spectral energy distribution*) emitted by a given unit area of such a black body depends only on the body's temperature; two different black bodies of the same temperature always radiate in exactly the same way. Figure 10.6 shows, schematically, how much electromagnetic energy at various wavelengths black bodies of several different temperatures radiate from each square centimeter of their surfaces.

We note three interesting things about the graph in the figure. First, every black body emits some radiation at *every* wavelength. Second, at *every* wavelength, a hotter black body emits *more* energy than a cooler one. Third, for a black body at each temperature there is a certain wavelength at which it radiates a maximum amount of energy, and the *higher* the temperature, the shorter that wavelength of peak radiation.

(b) Wien's Law

The third point is expressed precisely by *Wien's law*, which states that the wavelength of peak emission is inversely proportional to the temperature (or, equivalently, the product of its temperature and that wavelength of maximum radiation is always the same number); if we denote by λ_{max} that wavelength of maximum radiation, we have

$$\lambda_{max} = \frac{\text{constant}}{T}.$$

When the wavelength is expressed in centimeters and the temperature in Kelvins, the constant has the value 0.2897 cm · K. The sun, for example, has a temperature of about 6000 K and emits its maximum light at about 5000 Å. Wien's law means that a star with twice the sun's temperature emits its maximum light at 2500 Å. Such a star appears blue to us, because most of its radiation is at short wavelengths. Some stars are so hot that most of their energy is emitted in the invisible ultraviolet. On the other hand, a star with a temperature of only 3000 K, half the sun's temperature, emits its maximum light at 10,000 Å—in the infrared. Note that since all black bodies emit some energy at all wavelengths, even the cool stars emit a little ultra-

violet light. Moreover, not only do the very hot stars emit some infrared and visible light, but they emit *more* of it (per unit area) than the cool stars do (remember the second point in the foregoing paragraph).

So a consequence of Wien's law is that we can calculate the temperature of a star from observing the wavelength at which it emits the most radiation. Since that also is what determines the color of a star, its color tells us its temperature. In practice we do not even have to observe the wavelength of maximum light; whenever we observe the color index of a star (Chapter 24), we are comparing the intensity of its radiation in two different wavelength regions. We can determine at what temperature a black body would have the same ratio of intensity in the same two wavelength regions and thereby find the star's temperature. A more accurate comparison with a black body can be obtained if more than two spectral regions are observed. A stellar temperature determined by comparison of the spectral distribution of the star's radiation with that of a black body is called a *color temperature*.

(c) The Color of the Sun

Now the sun's wavelength of maximum radiation is that of green light; yet the sun appears *white*, not green. The reason is that white light is the particular admixture of different wavelengths we receive from the sun (or, approximately, from a black body with a temperature of 6000 K). A piece of white paper looks white because it reflects all wavelengths about equally to our eye, so the reflected light from a white paper has the same distribution of wavelengths as sunlight—or any other source of white light. Stars that have higher temperatures than the sun look blue or violet to the eye, and those of lower temperature look yellow, orange, or red, because the peak of their light is at those shorter or longer wavelengths, respectively; a star you might expect to look green has about the same temperature as the sun and is white, more or less by definition. We would have to filter out most of the other colors to make sunlight look green.

Most people (even some astronomers, but *not* solar astronomers) describe the sun as *yellow*. It is not. Snow is a pretty good reflector of all visible wavelengths; the sun is roughly the color of snow. Then why do people think it is yellow? Perhaps because it is hard (indeed, dangerous!) to look at the sun when it is high in the sky. We can and often do look at the sun comfortably when it is low in the sky, and we see it through a long path of air, which scatters the short wavelengths out of the beam, allowing only the long waves to pass through. Consequently the sun, when seen low or setting, generally appears yellow, orange, or red. Also, in grade school when we drew and colored the sun we used yellow crayons, because white crayon doesn't show up well on white paper. Maybe it's just traditional to regard the sun as yellow. Still, it's white.

(d) The Stefan-Boltzmann Law

Since a hot black body emits more energy at all wavelengths than a cool one, the combined amount of electromagnetic radiation over all wavelengths emitted by a black body is extremely sensitive to its temperature. If we sum up the contributions from all parts of the spectrum, we obtain the total energy emitted by a black body over all wavelengths. The total energy emitted per second per square centimeter by a black body at a temperature T is given by the equation known as the *Stefan-Boltzmann law:*

$$E(T) = \sigma T^4.$$

The constant σ, called the *Stefan-Boltzmann constant,* has the value 5.672×10^{-5}, if T is in K and $E(T)$ is in ergs/cm^2/sec.

(e) Some Applications of Radiation Laws

Now imagine two stars of the same size and distance, one of which is blue and the other red. The blue star, being hotter, radiates very much more efficiently and so appears far, far brighter than the red one. But what if a red and a blue star, both the same distance, appeared *equally* bright? Knowing that the red star is an inefficient radiator (per unit area), you realize that it must be very much larger to appear as bright as the blue one.

So at once you see how we can use the Stefan-Boltzmann law to figure out approximately how big a star is. First, from its color, we find the wavelength of the star's maximum radiation; that tells us the temperature. The temperature, with the Stefan-Boltzmann law, tells us how much energy the star radiates from every square centimeter of its surface. Now if we know the star's distance (which we take up in Chapters 22 to 24) we can observe how much

energy it emits altogether from its entire surface. Knowing how much it radiates from each square centimeter and knowing how much it radiates altogether tell us how many square centimeters there are in its surface, that is, its size.

The earth, moon, and planets are not very good black bodies. Typically, they reflect about half of the sunlight incident on them, although it's not the same for all of them; Venus, for example, reflects more than half, and the moon less than 10 percent. But they all absorb some of the sunlight hitting them, and they must reradiate this energy. We see the planets in the sky by the sunlight they reflect. It's easy to calculate how much sunlight strikes them, so from their brightnesses we find out how much solar radiation each planet must absorb, which is, of course, how much it also emits. Now the Stefan-Boltzmann law enables us to calculate the temperature of a black body that emits the same amount of energy a particular planet does. Although a planet is not really a perfect radiator, its surface temperature will be in the general neighborhood of that black body's temperature.

Thus we find that the earth's temperature should average somewhere around 300 K—about room temperature. (Actually, 293 K is 20°C, or 68°F.) Mars, being farther from the sun and thus receiving less energy from it than we do, is a little cooler. Mercury, being closer than we, is quite a bit hotter.

Wien's law tells us the wavelength at which a black body of the same temperature as a planet would emit its maximum radiation; roughly, this is the wavelength of the peak of the energy radiated by the planet, too. The earth's temperature of 300 K is about $^1/_{20}$ the sun's 6000 K; if the sun radiates most strongly in green light of wavelength 5000 Å, the earth must radiate most strongly at 100,000 Å (or about 0.001 cm)—way out in the infrared. Everything about us, therefore, including ourselves, is radiating infrared energy. If we had infrared vision, we would easily see each other in the dark, in fact, we would find that our world had continuous illumination. "Night" is night only to visible sunlight; there is no night on earth to its own radiation.

Similarly the other planets radiate in the infrared. We can measure the radiation they emit with infrared detectors attached to our telescopes. By the way, by now you can appreciate one of the problems of the astronomer who specializes in infrared observations. The telescope dome and everything in it, including the telescope itself, and the astronomer and his assistants are all shining brightly in the very wavelengths he is trying to detect from celestial objects. It's rather like trying to watch a motion picture show with the theater lights on. The astronomer must take very special pains to shield his detector from all the radiation except that reflected from the telescope mirror.

10.3 THE QUANTIZATION OF LIGHT; PHOTONS

We have seen that quite a bit was known a century ago about the way things radiate. At that time, however, there was not yet a physical theory that accounted for the properties of radiation from black bodies, that is, those particular graphs shown in Figure 10.6. We did not understand radiation in the same sense that we understand the motions of the planets in terms of the law of gravitation. Now it was known that currents moving in a wire emit electromagnetic waves (for example, radio waves), so it was thought that charged particles (electrons) within the atoms of the radiating substances were oscillating back and forth to emit all kinds of electromagnetic energy. But if those electrons could oscillate in every way, there was no reason why a plot of the energy emitted at various wavelengths should look like the graphs in Figure 10.6.

The answer to that riddle was part of the start of the quantum theory—one of the great new branches of physics of the 20th century. In 1900 the German physicist Max Planck succeeded in deriving a theoretical formula for those graphs in Figure 10.6.★ To do so, he had only to assume that the tiny atomic oscillators in the radiating bodies can emit energy at any given wavelength or frequency only in certain discrete amounts. In other words, at each frequency the energy leaving a radiating particle is always some multiple of a minute unit of energy. That unit of energy is the frequency of the radiation multiplied by a tiny number—always the *same* number, symbolized *h*. That number, *h*, is now called *Planck's constant*. If energy is measured in ergs and frequency in cycles or waves per second, Planck's constant has the value $h = 6.62 \times 10^{-27}$ erg · s.

Since 1887 (Section 10.1c) it has been known that any metal will emit electrons when struck by electromagnetic radiation if that radiation is of short enough wavelength (or high enough frequency). In other words, the radiation must have a certain critical or threshold frequency. Radiation of *just* the threshold frequency barely dislodges electrons from the metal, but that of *higher* frequency dislodges them and gives them some kinetic energy also.

The subject was clarified by Albert Einstein in 1905. Einstein proposed that electromagnetic radiation itself travels in units of energy, each unit having energy equal to *h* (Planck's constant) times the frequency. Moreover, each electron emitted results from the metal absorbing just *one* of those units of radiant energy. Now it takes a certain minimum energy to remove an electron from the metal. Thus, of the units of electromagnetic energy that are absorbed, the only ones that can cause the ejection of electrons are those which have high enough frequencies to possess that requisite energy. On the other hand, any radiation unit of *higher* energy (that is, higher frequency) can be absorbed, and the excess energy absorbed goes into kinetic energy (energy of motion) of the dislodged electron.

In short, Einstein showed that the phenomenon of radiation causing the ejection of electrons—called the *photoelectric effect*—is evidence that electromagnetic energy is itself *quantized*, consisting of little packets of energy. These packets of radiant energy are now called *photons*. Einstein received the Nobel prize for this contribution to our understanding of the nature of radiation—not for his better known relativity theory!

(a) Particles and Waves

The photoelectric effect and many other experiments show that electromagnetic energy acts as though it were made up of little energetic particles—photons. But we have already seen that electromagnetic energy also propagates as *waves*. So these photons themselves must be waves, but always containing discrete amounts of energy. Thus photons of the shortest wavelength and highest fre-

---

★If *T* is the temperature in Kelvins, λ the wavelength in centimeters, and *k* Boltzmann's constant (1.37×10^{-16}), $E(\lambda, T)$, the energy in ergs emitted per unit wavelength interval per second per square centimeter and into unit solid angle, is

$$E(\lambda, T) = \frac{2hc^2}{\lambda^5} \frac{1}{e^{hc/\lambda kT} - 1}.$$

Wien's law can be derived from Planck's formula by finding the wavelength at which the derivative of Planck's formula equals zero, and Stefan's law by integrating Planck's formula.

quency—gamma rays—have the highest energy, and those of radio waves have the lowest energy; visible light is intermediate.

How can we reconcile our concept of photons of light and other electromagnetic radiation as having the properties of both particles and waves? I suggest two ways: First, we can think of each photon itself as waves propagating in all directions through space, like ripples in a pond. The photon can do all the things waves can do—it can pass through two slits at once and interfere with itself, for example. But as soon as we *observe* it, say by its absorption in the retina of the eye, in a photographic emulsion, or by means of the photoelectric effect, those waves immediately disappear, and the photon itself is known to have been at the place where it was observed.

Second, we can think of those waves as just a mathematical formula, which tells us the probability of the photon—itself a localized particle—being in any particular place. Thus the wave crests are the places where the photon is more likely to be found, and the wave troughs places where it is less likely to be. We never, of course, know where it *really* is until we observe it.

Perhaps the second way of thinking of photons makes physical sense—one that is conceptually credible. But actually, both ways are exactly equivalent as far as predicting results of experiments are concerned, and science, after all, doesn't concern itself with what a photon "really" is, but only *how it behaves.* Unless we can devise an experiment to distinguish between these viewpoints, it does not even make sense—within the realm of science, anyway—to ask which is *correct.*

In fact, it turns out that material particles—like protons and electrons, and yes, even billiard balls—can similarly be thought of as waves. We shall see that we cannot say just where an electron (for example) is located at any given instant. The best we can do is write down an equation that gives us the probability of its having various locations. The equation turns out to be the equation for a system of waves; we are, in other words, exactly where we are with the photon and may as well think of the electron itself as a wave.

These concepts may seem less satisfying than the simple, mechanical, predictable system of Newtonian theory, but it is just as successful in accounting for the behavior of matter on the atomic level as Newton's laws are on the scale of the solar system. And the two theories are not contradictory,

either. If you flip an honest coin in a random way one million times, you can safely predict that very close to 50 percent of the flips will come up heads. The gentlemen who run the casinos at Las Vegas make their fortunes by knowing exactly how statistics of large numbers work out. Similarly, planets and billiard balls have absolutely enormous numbers of atomic particles. Just as we cannot predict whether an individual gambler will win a particular hand of blackjack, so the behavior of each particle in a billiard ball is uncertain. But the behavior of the total ensemble is highly predictable. As the casino operator knows that on the average the customers will lose (and for that matter, so will each individual if he plays long enough), so the physicist knows how the billiard ball will carom, behaving according to the average of the motions of its tremendous numbers of atoms.

On the other hand, submicroscopic chance events can lead to chains of actions with macroscopic consequences. An individual gambler may win some money at roulette and purchase a new car, thus resulting in the automobile salesman's being promoted, and building a new home, and . . . you can continue the tale. Analogously, the chance absorption of a photon from a remote galaxy in a photomultiplier on a telescope can eject an individual electron, which can be detected, and the signal could be used, if we wished, to turn on the lights of a city.

10.4 LAWS OF GEOMETRICAL OPTICS

We now turn our attention to some properties of visible light. Those that are most important to the design and construction of telescopes and other astronomical instruments can be summarized simply in three laws of *geometrical optics*—the laws of *reflection, refraction,* and *dispersion.*

(a) Reflection

The law of reflection describes the manner in which light is reflected from a smooth, shiny surface. The *normal* to a surface at some point is simply a line or direction perpendicular to that surface at that point. If light strikes a shiny surface, its direction must make a certain angle with the normal to the surface at the point where it strikes. That angle is the *angle of incidence.* The angle that the reflected beam of

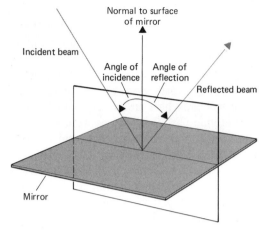

Figure 10.7 Law of reflection.

light makes with the normal is called the *angle of reflection*. The law of reflection states that the angle of reflection is equal to the angle of incidence and that the reflected beam lies in the plane formed by the normal and the incident beam (see Figure 10.7).

(b) Refraction

The law of refraction deals with the deflection of light when it passes from one kind of transparent medium into another. The *index of refraction* of a transparent substance is a measure of the degree to which the speed of light is diminished while passing through it; specifically, it is the ratio of the speed of light in a vacuum to that in the substance. Usually, media of higher densities have higher indices of refraction. The index of refraction of a vacuum is, by definition, exactly 1.0. That of air at sea level is about 1.00029. Water has an index of about 1.3; crown glass and flint glass have indices of about 1.5 and 1.6, respectively. Diamond has the high index of refraction of 2.4.

The law of refraction states that if light passes from one medium into a second one of a different index of refraction, the angle the light beam makes with the normal to the interface between the two substances is always *less* in the medium of higher index. Thus, if light goes from air into glass or water, it is bent *toward* the normal to the interface, while if it goes from water or glass into air, it is bent *away* from the normal. The law of refraction is expressed mathematically by Snell's law.†

It is the refraction of light when it passes from the water into the air that makes the handle of a spoon appear bent if a spoon is immersed in a glass of water. Similarly, light entering the earth's atmosphere from space is slightly bent. The light from stars, planets, the sun, and the moon is bent, upon entering the earth's atmosphere, in such a way as to make the object appear to be at a greater altitude above the horizon than it actually is. Atmospheric refraction is greatest for objects near the horizon. It raises the apparent altitude of objects on the horizon by about $1/2°$. The refraction of light

---

†*Snell's law* is

$$n \sin \alpha = n' \sin \alpha',$$

where α and α' are the angles between a light beam and the normal to the interface between two media of indices of refraction n and n', respectively.

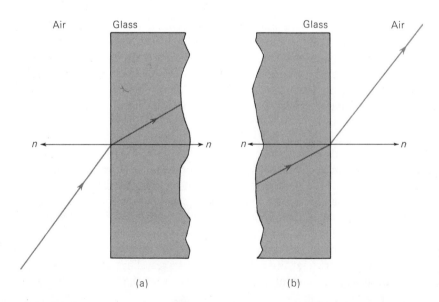

Figure 10.8 Refraction of light between glass and air. (a) When light passes from air to glass. (b) When light passes from glass to air.

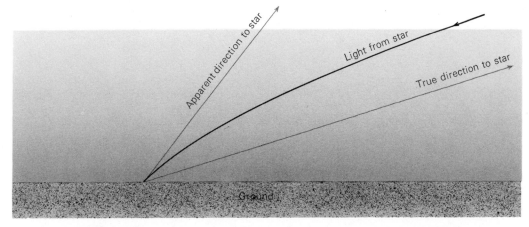

Figure 10.9 Atmospheric refraction. Light is bent upon entering the earth's atmosphere in such a way as to make stars appear at a higher altitude than they actually are. (The effect is grossly exaggerated in this figure.)

passing through glass and the earth's atmosphere is illustrated in Figures 10.8 and 10.9, respectively.

The illusion that the moon (or sun) looks larger near the horizon than when it is high in the sky is *not* due to refraction. Actually, refraction raises the lower limb of the moon more than the upper, so that the moon really looks smaller and oval near the horizon, not larger. The apparent enlargement of the moon or sun when seen near the horizon is a purely psychological effect that has been the subject of much discussion and investigation by psychologists.

(c) Dispersion

Dispersion of light is the manner in which white light, a mixture of all wavelengths of visible light, can be decomposed into its constituent wavelengths or colors when it passes from one medium into another. The phenomenon of dispersion occurs because the index of refraction of a transparent medium is greater for light of shorter wavelengths. Thus, whenever light is refracted in passing from one medium into another, the violet and blue light of shorter wavelengths is bent more than the orange and red light of longer wavelengths.

Figure 10.10 shows the way in which light of different wavelengths is bent different amounts in passing from air into glass. Figure 10.11 shows how light can be separated into different colors with a prism, a triangular piece of glass. Upon entering one face of the prism, light is refracted once, the violet light more than the red, and upon leaving the

opposite face, the light is bent again, and so is further dispersed. Even greater dispersion can be obtained by passing the light through a series of prisms. If the light leaving a prism is focused upon a screen, the different wavelengths or colors that compose white light are lined up side by side. The array of colors is a spectrum.

It has long been known that colors are produced when white light passes through glass, but before Newton's time it was believed that the glass itself produced the colors. Newton produced a spectrum with a glass prism and thereby isolated individual rays of nearly pure color. He found that

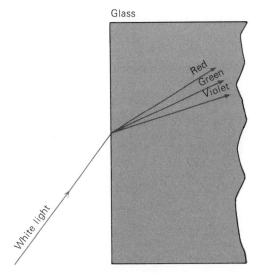

Figure 10.10 Dispersion at the interface of two media.

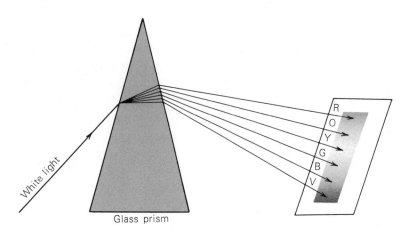

Figure 10.11 Dispersion by a prism and formation of a spectrum. The letters R, O, Y, G, B, and V stand for red, orange, yellow, green, blue, and violet.

these monochromatic rays were not further dispersed with a second prism, but merely refracted, which indicated that the glass prism had not produced the colors but had merely separated them out from white light.

Atmospheric refraction occurs in varying amounts for different colors, also. This atmospheric dispersion combined with turbulence in the air often produces colorful effects when a bright star is seen low above the horizon, and its different colors are bent by the air in differing amounts. When astronomers measure the exact positions of stars in the sky, they must not only take account of the distortion of the position of the star by atmospheric refraction but also take care to note the color of the light with which the star is observed, because the correction for refraction is different for different colors.

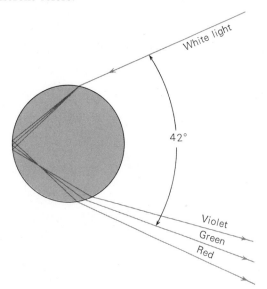

Figure 10.12 Dispersion in a raindrop.

(d) WEATHER OPTICS

Nature provides an excellent example of the dispersion of light in the production of a rainbow. Raindrops, tiny spherical droplets of water in the air, act like prisms. Light from the sun entering a raindrop is bent, the blue and violet light being bent the most. This bent light strikes the inside rear surface of the drop, and some of it is reflected back toward the front surface. This reflected light leaves the raindrop by passing through the same side that it enters. But when it leaves the drop, it is again refracted, and again dispersed, just as when light leaves a glass prism. Thus sunlight is spread into the rainbow of colors—the rainbow is nothing more than the spectrum of sunlight. Figure 10.12 shows how a raindrop produces a spectrum.

The light that emerges from a raindrop is most intense at an angle of about 42° from the direction at which it enters. Thus (Figure 10.13), to see a rainbow, the observer must have the sun behind him, at an altitude of less than 42° above the horizon. The rainbow then appears as an arc, with an angular radius of 42°, centered about a point exactly opposite the sun. Since the sun must be above the horizon to illuminate the raindrops, the center of the rainbow must be below the horizon; an observer on the ground can never see a rainbow as more than half a complete circle (observers in airplanes or on mountains may occasionally see more).

Although the arc of a rainbow has a radius of approximately 42°, the different colors of sunlight are refracted and reflected by the droplets back to the observer in slightly different directions, so the band of color has a finite width. It may be seen in Figure 10.13 that from the upper drops it is the red light, bent the least, that enters the observer's eye, and from the lower drops it is the violet light, bent the most, that the observer sees. Thus, the top, or outside, of the arc of the rainbow appears red, and the bottom, or inside,

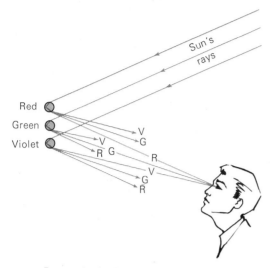

Figure 10.13 Formation of a rainbow.

of the arc appears violet. The other colors come from the drops in between.

A little of the light is reflected a second time before leaving the drops. This light emerges at an angle of about 51° from that of the incoming light and produces a fainter secondary rainbow in an arc of 51° radius, thus lying above the primary rainbow. In the secondary rainbow (less frequently seen), the red is on the inside of the arc and the violet on the outside, or upper part.

Another natural phenomenon that involves the refraction and dispersion of light is a *halo* about the sun or moon. A halo is a faint ring of light, of angular radius 22°, caused by the bending of light as it passes through the tiny ice crystals that form cirrus clouds at altitudes of more than 7000 m. As in the rainbow, the violet light is refracted most in passing through the crystals. Consequently, the outer edge of a solar or lunar halo appears violet, the inner edge red (Figure 10.14).

10.5 SPECTROSCOPY IN ASTRONOMY

The light from a star or other luminous astronomical body, as well as from the sun, can be decomposed into its constituent wavelengths, producing a spectrum. A device used to observe visually the spectrum of a light source is a *spectroscope;* one used to photograph a spectrum is a *spectrograph.* It will be explained in Chapter 11 how a spectroscope (or spectrograph) is constructed and attached to a tele-

scope so that we may study the spectra of astronomical objects.

(a) The Value of Stellar Spectra

If the spectrum of the white light from the sun and stars were simply a continuous rainbow of colors, astronomers would not have such intensive interest in the study of stellar spectra. To Newton, the solar spectrum did appear as just a continuous band of colors. However, in 1802, William Wollaston (1766–1828) observed several dark lines running across the solar spectrum. He attributed these lines to natural boundaries between the colors. Later, in 1814 and 1815, the German physicist Joseph Fraunhofer (1787–1826), upon a more careful examination of the solar spectrum, found about 600 such dark lines. He noted the specific positions in the spectrum, or the wavelengths, of 324 of these lines. To the more conspicuous lines he assigned letters of the alphabet, with the letters increasing from the red to the violet end of the spectrum. Today we still refer to several of these lines in the solar spectrum by the letters assigned to them by Fraunhofer.

Subsequently, it was found that such dark spectral lines could be produced in the spectra of artificial light sources by passing their light through various transparent substances or gases. On the other hand, the spectra of the light emitted by certain glowing gases were observed to consist of several separate bright lines. A preliminary explanation

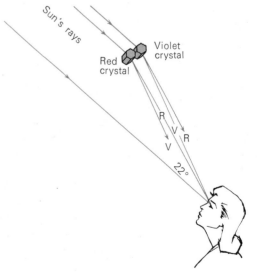

Figure 10.14 Formation of a halo around the sun or moon by ice crystals in cirrus clouds.

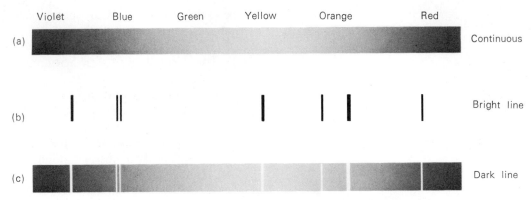

Figure 10.15 Three kinds of spectra: (a) continuous, (b) bright line, (c) dark line. The spectra are shown as they appear on photographic negatives, from which astronomers generally work.

of these phenomena (although grossly oversimplified in light of modern knowledge) was provided in 1859 by Gustav Kirchhoff (1824–1887) of Heidelberg. Kirchhoff's explanation is often given in the form of *three laws of spectral analysis:*

> *FIRST LAW: A luminous solid or liquid emits light of all wavelengths, thus producing a continuous spectrum.*

It is often stated that a highly compressed gas also emits a continuous spectrum, but the statement is an oversimplification. It is usually true if the gas is opaque.

> *SECOND LAW: A rarefied luminous gas emits light whose spectrum shows bright lines, and sometimes a faint superimposed continuous spectrum.*

> *THIRD LAW: If the white light from a luminous source is passed through a gas, the gas may abstract certain wavelengths from the continuous spectrum so that those wavelengths will be missing or diminished in its spectrum, thus producing dark lines.*

We distinguish, then, among three types of spectra (Figure 10.15). A *continuous* spectrum is an array of all wavelengths or colors of the rainbow. A *bright line* or *emission* spectrum appears as a pattern or series of bright lines; it is formed from light in which only certain discrete wavelengths are present. A *dark line* or *absorption spectrum* consists of a series or pattern of dark lines—missing wavelengths—superimposed upon the continuous spectrum of a source of white light.

The great significance of Kirchhoff's laws is that each particular chemical element or com-

pound, when in the *gaseous form,* produces its own characteristic pattern of dark or bright lines. In other words, each particular gas can absorb or emit only certain wavelengths of light, peculiar to that gas. The presence of a particular pattern of dark (or bright) lines characteristic of a certain element is evidence of the presence of that element somewhere along the path of the light whose spectrum has been analyzed.

Thus the dark lines (Fraunhofer lines) in the solar spectrum give evidence of certain chemical elements between us and the sun, absorbing those wavelengths of light. It is easy to show that most of the lines must originate from gases in the outer part of the sun itself.

The wavelengths of the lines produced by various elements are determined by laboratory experiment. Most of the thousands of Fraunhofer lines in the sun's spectrum have now been identified with most of the known chemical elements.

Dark lines are also found in the spectra of stars and in stellar systems. Much can be learned from their spectra, in addition to evidence of the chemical elements present in a star. A detailed study of its spectral lines indicates the temperature, pressure, turbulence, and physical state of the gases in that star; whether or not magnetic and electric fields are present, and the strengths of those fields; how fast the star is approaching or receding from us; and many other data. Other information about a star can be obtained by studying its continuous spectrum.

The study of the spectra of celestial objects is the most powerful means at the astronomer's disposal for obtaining data about the universe.

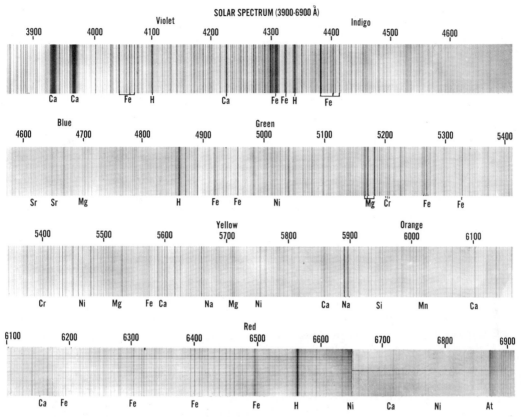

Figure 10.16 The solar spectrum. Labels indicate the elements in the sun's photosphere that cause some of the dark lines. The wavelengths in angstroms and the colors of the different parts of the spectrum are also labeled. *(Mount Wilson and Las Campañas Observatories)*

(b) The Doppler Effect

In 1842 Christian Doppler (1803–1853) pointed out that if a light source is approaching or receding from the observer, the light waves will be, respectively, crowded closer together or spread out. The principle, known as the *Doppler principle* or *Doppler effect,* is illustrated in Figure 10.17. In (a) the light source is stationary with respect to the observer. As successive wave crests 1, 2, 3, and 4 are emitted, they spread out evenly in all directions, like the ripples from a splash in a pond. They approach the observer at a distance λ behind each other, where λ is the wavelength of the ripple. On the other hand, if the source is moving with respect to the observer, as in (b), the successive wave crests are emitted with the source at different positions, S_1, S_2, S_3, and S_4, respectively. Thus, to observer *A,* the waves seem to follow each other more closely, at a decreased wavelength and increased frequency, whereas to observer *C* they are spread out and ar-

rive at an increased wavelength and decreased frequency. To observer *B,* in a direction at right angles to the motion of the source, no effect is observed. The effect is produced only by a motion *toward* or *away* from the observer, a motion called *radial velocity.* Observers between *A* and *B* and between *B* and *C* would observe some shortening or lengthening of the light waves (increase or decrease in frequency), respectively, for a component of the motion of the source is in their line of sight.

The Doppler effect is also observed in sound. We have all heard the higher than normal pitch of the whistle of an approaching train and the lower than normal pitch of a receding one. In the case of sound, the precise formula for the amount of shortening or lengthening of the sound waves depends on whether the source or observer, or both, are in motion. For light, however, it follows from the special theory of relativity that there is no way of determining which is in motion, and the formula for the increase or decrease in the wavelength, the

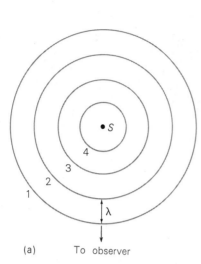
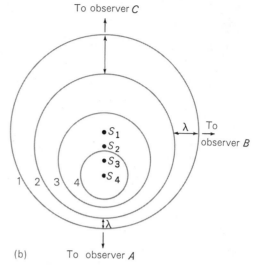

Figure 10.17 The Doppler effect.

Doppler shift, is identical for each case. The wavelengths are shortened if the distance between the source and observer is decreasing, and they are lengthened if the distance is increasing.

If the relative motion is entirely in the line of sight, the exact formula for the Doppler shift is

$$\frac{\Delta\lambda}{\lambda} = \frac{\sqrt{1 + v/c}}{\sqrt{1 - v/c}} - 1,$$

where λ is the wavelength emitted by the source, $\Delta\lambda$ is the difference between λ and the wavelength measured by the observer, c is the speed of light, and v is the relative line of sight velocity of the observer and source, which is counted as positive if the velocity is one of recession and negative if it is one of approach. If the relative velocity of the source and observer is small compared to the speed of light, however, the formula reduces to the simple form, which is usually used:

$$\frac{\Delta\lambda}{\lambda} = \frac{v}{c}.$$

Solving this last equation for the velocity, we find

$$v = c\frac{\Delta\lambda}{\lambda}.$$

If a star approaches or recedes from us, the wavelengths of light in its continuous spectrum appear shortened or lengthened, as well as those of the dark lines. However, unless its speed is tens of thousands of kilometers per second, the star does not appear noticeably bluer or redder than normal.

The Doppler shift is thus not easily detected in a continuous spectrum (except for very remote galaxies—see Chapter 39) and cannot be measured accurately in such a spectrum. On the other hand, the wavelengths of the absorption lines can be measured accurately, and their Doppler shift is relatively simple to detect. Generally, when the spectrum of a star or other object is photographed at the telescope, sometime during the exposure the light from an iron arc or some other emission-line source is allowed to pass into the same spectrograph, and the spectrum of the arc is then photographed just beside that of the star. The known wavelengths of the bright lines in the spectrum of the arc (or other laboratory source) serve as standards against which the wavelengths of the dark lines in the star's spectrum can be accurately measured. Further illustrations of the Doppler effect are given in later chapters.

10.6 ABSORPTION AND EMISSION OF LIGHT BY ATOMS

As explained in Chapter 1, atoms are the smallest particles into which a chemical element can be subdivided and still retain its chemical identity. An atom consists of two parts: a nucleus of protons and neutrons, each proton of which carries a small unit of electric charge‡, and a system of electrons, each

‡The amount of charge is 4.8×10^{-10} electrostatic units.

Figure 10.18 *Series of Balmer lines in the spectrum of hydrogen.*

with a negative charge equal to the positive one on a proton. In its ordinary neutral state, an atom has as many electrons as it has protons in its nucleus. The neutral hydrogen atom, for example, has one proton and one electron. A helium atom has a nucleus of two protons (and two neutrons) and two electrons. The various other elements have larger numbers of protons in their nuclei (and electrons outside): 8 for oxygen, 16 for iron, and 92 for uranium.

(a) Spectrum of Hydrogen

A clue to the structure of atoms came from the study of the spectrum of hydrogen, whose atoms are the simplest in nature. The dark lines of hydrogen that can be observed in the spectra of many stars occur in an orderly spaced series of wavelengths. The bright lines of hydrogen that are observed in the laboratory spectrum of glowing hydrogen are observed in the same series of wavelengths. The Swiss physicist Balmer found that these wavelengths could be represented by the formula

$$\frac{1}{\lambda} = \frac{1}{911.8}\left(\frac{1}{2^2} - \frac{1}{n^2}\right),$$

where λ is the wavelength in angstroms and n is an integer that can take any value from 3 on. If $n = 3$, the wavelength of the first line in the *Balmer series* of hydrogen is obtained (at 6563 Å in the red).

For $n = 4$, 5, and so on, the wavelengths of the second, third, and higher Balmer lines are obtained. As n approaches larger and larger values, the wavelengths of the successive Balmer lines become more and more nearly equal. The lines of hydrogen in stellar spectra are observed to do just this; they approach a limit at about 3650 Å (Figure 10.18), corresponding to a value of $n = \infty$.

After Balmer's work, other series of hydrogen lines were found. The *Lyman series*, in the ultraviolet, approaches a limit at about 912 Å. The *Paschen series*, in the infrared, approaches a limit at about 8200 Å. Still farther in the infrared are found the *Brackett series*, the *Pfund series*, and so on. All these series (including the Balmer series) can be predicted by the more general formula, known as the *Rydberg formula*:

$$\frac{1}{\lambda} = R\left(\frac{1}{m^2} - \frac{1}{n^2}\right),$$

where m is an integer and n is any integer greater than m. The *Rydberg constant*, R, has the value 109,678 if λ is measured in centimeters. For the Lyman series, $m = 1$; $m = 2, 3, 4, \ldots$ for the Balmer, Paschen, Brackett, and other series.

(b) The Bohr Atom

The Danish physicist Niels Bohr (1885–1962) suggested that the hydrogen spectrum can be explained

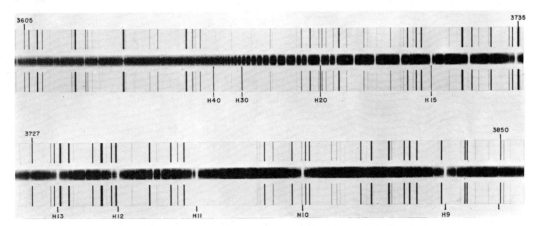

Figure 10.19 Photograph of part of the spectrum of the star HD 193182, showing many lines of the Balmer series. Several of the lines are identified by number. The spectrum has been split and is shown in two segments for convenience in reproduction on the page. *(Mount Wilson and Las Companas Observatories)*

if the assumption is made that the electrons in the hydrogen atom revolve about the nucleus in orbits, but that only orbits of certain sizes are possible. By specifying those permissible sizes for the electron orbits, Bohr was able to compute the values of energy, corresponding to the orbital motion of the electron, that are possible for an individual atom. He assumed that an atom can change from one allowed state of energy to another state of higher energy it its electron moves from a smaller to a larger allowed orbit. Conversely, according to the hypothesis, if the electron moves from a larger to a smaller orbit, the atom changes from a higher to a lower state of energy. One way in which an atom can gain or lose energy is by absorbing or emitting light. Since light is composed of *photons* whose energies depend on their wavelengths (the energy of a photon is hc/λ), the only wavelengths of light that could be absorbed or emitted are those corresponding to photons possessing energies equal to differences between various allowed energy states of the hydrogen atom. Those energy states are given by a simple formula that Bohr derived from the orderly progression of wavelengths in each series of lines in the hydrogen spectrum. In other words, by assuming that the allowed sizes of the electron orbits and thus the possible energies of the hydrogen atom are *quantized*, Bohr was able to account for the spectrum of hydrogen.

For example, suppose a beam of white light (which consists of photons of all wavelengths) is passed through a gas of atomic hydrogen. A photon of wavelength 6563 Å has the right energy to raise an electron in a hydrogen atom from the second to the third orbit, and can be absorbed by those hydrogen atoms that are in their second to lowest energy states. Since the energy of a photon is inversely proportional to its wavelength, the shorter the wavelength of a photon, the higher its energy. Photons with higher energies corresponding to the other successively shorter wavelengths in the Balmer series, therefore, have the right energies to raise an electron from the second orbit to the fourth, fifth, sixth and larger orbits, and can also be absorbed. Photons with intermediate wavelengths (or energies) cannot be absorbed. Thus, the hydrogen atoms absorb light only at certain wavelengths, and produce the spectral *lines*. Conversely, hydrogen atoms in which electrons move from larger to smaller orbits emit light—but again only light of those energies or wavelengths that correspond to the energy differences between permissible orbits. The transfer of electrons giving rise to spectral lines is shown in Figure 10.20.

A similar picture can be drawn for kinds of atoms other than hydrogen. However, since they ordinarily have more than one electron each, the energies of the orbits of their electrons are much

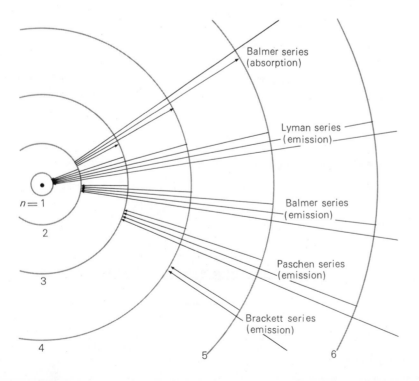

Figure 10.20 Emission and absorption of light by the hydrogen atom according to the Bohr model. Several different series of spectral lines are shown, corresponding to transitions of electrons from or to certain allowed energy levels. These series are named for the various physicists who studied them.

more complicated, and the problem of their spectra is much more difficult to handle theoretically.

(c) Energy Levels of Atoms and Excitation

Bohr's model of the hydrogen atom was one of the beginnings of quantum theory and was a great step forward in the development of modern physics and in our understanding of the atom. However, we know today that atoms cannot be represented by quite so simple a picture as the Bohr model. Even the concept of discrete orbits of electrons must be abandoned, because according to the modern quantum theory it is impossible at any instant to state the exact position and the exact velocity of an electron in an atom simultaneously. Nevertheless, we still retain the concept that only certain discrete energies are allowable for an atom. These energies, called *energy levels*, can be thought of as representing certain mean or average distances of an electron from the atomic nucleus.

Ordinarily, the atom is in the state of lowest possible energy, its *ground state*, which, in the Bohr model, would correspond to the electron being in the innermost orbit. However, an atom can absorb energy which raises it to a higher energy level (corresponding, in the Bohr picture, to the movement of an electron to a larger orbit). The atom is then said to be in an *excited state*. Generally, an atom remains excited only for a very brief time; after a short interval, typically a hundred-millionth of a second or so, it drops back down to its ground state, with the simultaneous emission of light, unless it chances to absorb another photon first and go to a still higher state. (In the Bohr model, this corresponds to a jump by the electron back to the innermost orbit.) The atom may return to its lowest state in one jump, or it may make the transition in steps of two or more jumps, stopping at intermediate levels on the way down. With each jump, it emits a photon of the wavelength that corresponds to the energy difference between the levels at the beginning and end of that jump. An energy-level diagram for a hydrogen atom and several possible *atomic transitions* are shown in Figure 10.21; compare this figure with the Bohr model, shown in Figure 10.20.

Energy differences in atoms are usually expressed in *electron volts*. An electron volt (abbreviated eV) is the small amount of energy acquired by an electron after being accelerated through a potential difference of one volt. The energy needed to raise the hydrogen atom from its ground state to its first excited state is about 10.2 eV. One eV = 1.602 × 10^{-12} ergs. One erg is very roughly the energy required by a fly to do a pushup. Ten million ergs of energy expended in 1 s is a *watt* of power. Typical electric light bulbs use about 100 watts.

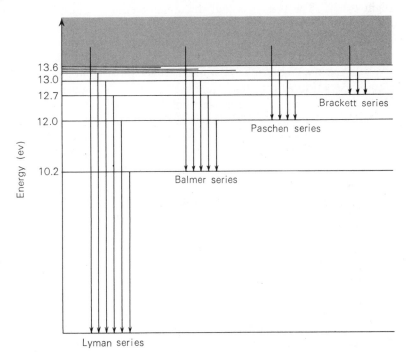

Figure 10.21 Energy-level diagram for hydrogen. The shaded region represents energies at which the atom is ionized (see Section 10.6d).

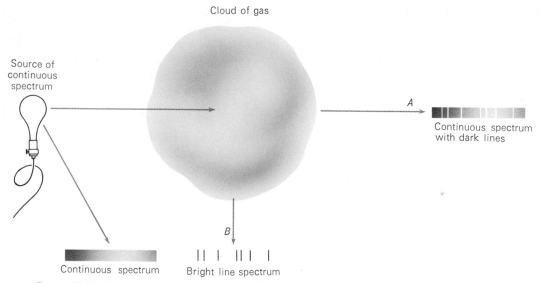

Figure 10.22 The atoms in the gas cloud produce absorption lines in the continuous spectrum of the white light source when viewed from direction A, but they produce emission lines (of the light they reemit) when viewed from direction B. The spectra are shown as they appear on photographic negatives.

Because atoms that have absorbed light and have thus become excited generally deexcite themselves and emit that light again, we might wonder why dark lines are ever produced in stellar spectra. In other words, why doesn't this reemitted light "fill in" the absorption lines? Some of the reemitted light actually *is* received by us and this light does partially fill in the absorption lines, but only to a slight extent. The reason is that the atoms reemit the light they absorb in random directions. Now the absorption of light we would otherwise observe is of that light "coming our way" from deeper, hotter levels in the star's atmosphere, while the light they reemit goes off in all directions, and not toward us. To be sure, other atoms can absorb light *not* coming our way and then reemit it in our direction, but many of these atoms are in a higher, cooler layer of the star's atmosphere, so the light that they thus scatter toward our telescope is not intense enough to replace that from the hot gas that is absorbed. Therefore, the dark lines persist. We can observe the reemitted light as emission lines only if we can view the absorbing atoms from a direction from which no light with a continuous spectrum is coming—as we do, for example, when we look at gaseous nebulae (Chapter 28). Figure 10.22 illustrates the situation.

We can calculate from theory the allowable energies of the simplest kinds of atoms and thus the wavelengths of the absorption lines that can be pro-

duced by those atoms. For the more complicated atoms, however, the wavelengths of the spectral lines are determined empirically by laboratory experiment.

Atoms in a gas are moving at high speeds and continually colliding with each other and with electrons. They can be excited and deexcited, therefore, by these collisions as well as by absorbing and emitting light. The mean energy of atoms in a gas depends upon its temperature (actually, *defines* the temperature), and if we know the temperature of the gas, we are able to calculate what fraction of its atoms, at any given time, will be excited to any given energy level. In the photosphere of the sun, for example, where the temperature is in the neighborhood of 6000 K, only about one atom of hydrogen in 100 million is excited to its second energy level by the process of collision. The Balmer lines of the hydrogen spectrum arise when atoms in this second energy level absorb light and rise to higher levels. At any given time, therefore, most atoms of hydrogen in the sun cannot take part in the production of the Balmer lines.

(d) The Uncertainty Principle

Actually, a particular energy state is one of the more likely energies an atom can have, but there is almost as good a chance that it will have a very slightly different energy. So the energy levels themselves are fuzzy, and therefore so are the spectral lines. Those lines are not *absolutely* sharp, but just pretty sharp.

The vagueness of where an electron is in an atom arises from the inherent uncertainty and statistical nature of fundamental units of matter touched on in the foregoing scene. The German physicist Werner Heisenberg considered an idealized experiment in which one tries to measure with perfect precision the simultaneous position and momentum of a particle. He showed that even with theoretically optimum circumstances, there are certain inherent limitations (such as the finite speed of light) to the experimental procedures that prevent those two quantities from ever being known better than to a certain limiting accuracy. There is always an uncertainty in the position, and one in the momentum of the particle; the more accurately one of the two is known, the less accurately is the other. The product of those two uncertainties is ultimately never less than a certain small number (Planck's constant, h, divided by 2π). The result is called the *Heisenberg uncertainty principle*.

The uncertainty principle applies to everything, large and small. However, a large object like a bowling ball has enough mass that even a tiny uncertainty in its velocity results in the momentum being uncertain enough that its position (so far as this quantum-mechanical limitation is concerned) can be specified to a far higher accuracy than we could ever hope to measure. Suppose, for example, the speed of the bowling ball is known to within 0.1 mm/s—in other words, darned accurately. Then, the uncertainty principle says that theoretically the ball's position can never be known to better than 0.000000000000000000000000000015 cm (more easily written 1.5×10^{-29} cm)—not a very interesting limitation to the average bowler.

But an electron has a tiny mass, and if its velocity is uncertain by that same amount, its position is uncertain by more than one meter!

It may seem that we are just talking about our inability to make a precise measurement, and that the real momentum and real position *exist* to perfect precision. But how can we define something like perfect precision if there is no possible experiment, even theoretically, by which we can determine it? According to the rules of science something that cannot be defined by means of a measurement or experiment (that is, operationally) does not exist—at least not in the realm of science. Since the limitation is theoretical, we must regard the uncertainty principle as *inherent* to nature. The *electron* cannot simultaneously know its precise position and momentum—those precise quantities do not exist at all! There is, in fact, very ample evidence to prove that such is the case.

(e) Ionization

We have described how certain discrete amounts of energy can be absorbed by an atom, raising the atom to an excited state, and moving one of its electrons farther from its nucleus. If enough energy is absorbed, the electron can be removed completely from the atom. The atom is then said to be *ionized* (Figure 10.21). The minimum amount of energy required to ionize an atom from its ground state is called its *ionization energy* or *ionization potential*. The ionization potential for hydrogen is about 13.6 eV. Still greater amounts of energy must be absorbed by the ionized atom (called an *ion*) to remove a second electron. The minimum energy required to remove this second electron is called the *second ionization energy* or *potential*. The third, fourth, and fifth ionization potentials are the successively greater energies required to remove the third, fourth, and fifth electrons from the atom, and so on. If enough energy is available (in the form of very short wavelength photons or in the form of a collision with a very fast-moving electron or another atom) an atom can become *completely ionized*, losing all of its electrons. A hydrogen atom, having only one electron to lose, can be ionized only once; a helium atom can be ionized twice, and an oxygen atom, eight times. Any energy over and above the energy required to ionize an atom can be absorbed also, and appears as energy of motion (kinetic energy) of the freed electron. The electrons released from atoms that have been ionized move about in the gas as free particles, just as the atoms and ions do.

An atom that has become ionized has lost a negative charge—that carried away by the electron—and thus is left with a net positive charge. It has, therefore, a strong affinity for a free electron, and eventually will capture one and become neutral (or ionized to one less degree) again. During the capture process, the atom emits one or more photons, depending on whether the electron is captured at once to the state corresponding to the lowest energy level of the atom, or whether it stops at one or more intermediate levels on the "way in." Any energy that the electron possessed as kinetic energy before cap-

Figure 10.23 Series of closely spaced lines comprising bands in the spectrum of the compound made up of molecules of titanium oxide. Note how the bands coalesce to form band heads. *(Mount Wilson and Las Compañas Observatories)*

ture can be emitted. Absorption or emission of light over a continuum of wavelengths therefore accompanies the process of ionization or recapture, at least over those wavelengths corresponding to energies higher than the ionization energy of the atom. Atoms of various kinds, being ionized and deionized, and absorbing and emitting light at various wavelengths, account for much of the continuous opacity and continuous spectrum of the sun and stars.

Just as the excitation of an atom can result from a collision with another atom, ion, or electron (collisions with electrons are usually most important), so also can ionization. The rate at which such collisional ionizations occur depends on the atomic velocities and hence on the temperature of the gas. The rate of recombination of ions and electrons also depends on their relative velocities, that is, on the temperature, but in addition it depends on the density of the gas; the higher the density, the greater the chance for recapture, because the different kinds of particles are crowded closer together. From a knowledge of the temperature and density of a gas, it is possible to calculate the fraction of atoms that have been ionized once, ionized twice, and so on. In the photosphere of the sun, for example, we find that most of the hydrogen and helium atoms are neutral, whereas most of the atoms of calcium, as well as many other metals, are once ionized.

The energy levels of an ionized atom are entirely different from those of the same atom when it is neutral. In each degree of ionization, the energy levels of the ion, and thus the wavelengths of the spectral lines it can produce, have their own characteristic values. In the sun, therefore, we find lines of *neutral* hydrogen and helium, but of *ionized* calcium. Ionized hydrogen, of course, having no electron, can produce no absorption lines.

There is an additional mechanism by which ionized atoms can absorb and emit light in the continuous spectrum. When an electron passes near an ion, it is attracted by that ion's positive charge (be-

cause opposite electrical charges attract each other). If the electron is not captured, it will pass the ion in a hyperbolic orbit, much like two stars passing in interstellar space. While the electron is passing, however, the ion can absorb or emit a photon which accompanies a corresponding increase or decrease in the kinetic energy of the hyperbolic motion of the electron. The process is called *free-free* absorption or emission (because the electron is "free" of the ion both before and after the encounter), or *bremsstrahlung*. Photons of any wavelength can be absorbed or emitted in bremsstrahlung.

(f) MOLECULAR SPECTRA

Molecules, combinations of two or more atoms, can also absorb or emit light. In addition to undergoing electronic transitions, molecules can also rotate and vibrate, all of which involve energy. The quantum theory predicts that the energy of vibration and rotation of molecules is quantized, like the energy of atoms. The vibrational and rotational energies are generally low, but they add to or subtract from the energies corresponding to electronic transitions. Consequently, in place of each atomic energy level there is a series of closely spaced levels, each one corresponding to a different mode of vibration or rotation of the molecule. Many more different transitions between energy levels are possible, therefore, differing from each other only slightly in energy or wavelength. Molecules, in other words, produce series of closely spaced lines known as *molecular bands*. In spectra of those stars in which molecules exist, these many molecular lines within a band are often not resolved as separate, and only a single broad absorption feature is observed.

(g) Summary of Emission and Absorption Processes

Atoms are characterized by energy levels which correspond to various distances of their electrons from

The eclipsed moon. *(Celestron International)*

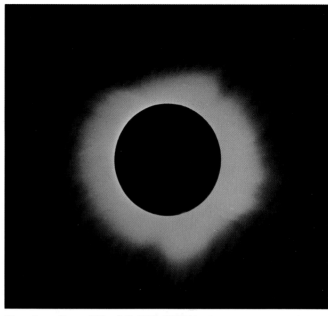

The solar eclipse of March 7, 1970. *(NASA)*

A sequence of photographs of the solar eclipse of June 20, 1974, taken by Stephen Schiller. The final frame is the totally eclipsed sun photographed with a special infrared film; the different colors represent different infrared wavelengths.

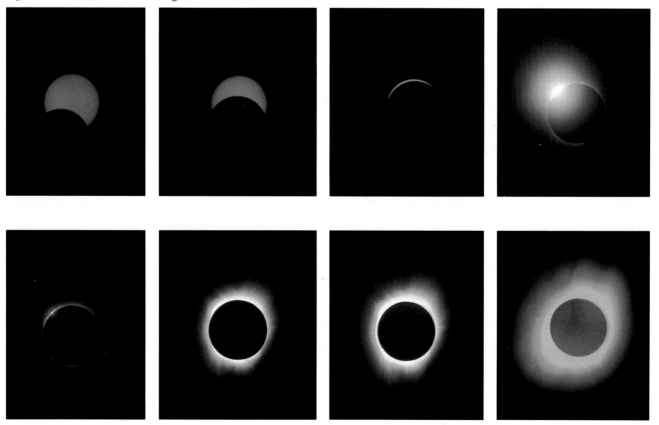

India and Sri Lanka, looking north with the Bay of Bengal on the right. Gemini XI photograph taken from an altitude of 410 nautical miles. *(NASA)*

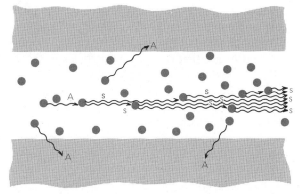

Figure 10.24 Principle of the maser and laser. Most of the atoms in an enclosure are excited from the ground state by a pumping process. The highly excited atoms transfer to lower energy levels, ending up in an excited state from which a few spontaneously jump to the ground state, emitting photons marked *A*. Some of these photons stimulate other excited atoms to emit photons, marked *S*. Because there are too few atoms in the ground state to reabsorb all of these photons from stimulated emission, they are able to induce still further emissions. Those that happen to travel along the length of the enclosure may be reflected back and forth many times (by mirrors in a laser), thus amplifying the radiation into an intense, coherent beam.

their nuclei. By absorbing or emitting radiant energy, an atom can move from one to another of these levels, thus raising or lowering its energy. Since only certain discrete energy levels exist for each kind of atom, the absorbed or emitted radiation occurs only at certain energies or wavelengths, producing dark or bright spectral lines.

An atom is said to be *excited* if it is in any but its lowest allowable energy level. It is said to be *ionized* if, by the absorption of energy, it has lost one or more of its electrons. It can be excited or deexcited by collisions as well as by the absorption and emission of radiation; it can be ionized by collision or by absorbing radiation. Any wavelength of light corresponding to an energy greater than the ionization energy of an atom can be absorbed or emitted in the process of ionization or of deionization when the ion captures an electron.

Ionized atoms (*ions*) can also absorb or emit light at continuous wavelengths, when free electrons pass near them. The relative energy of the ion and passing electron changes by the same amount as the energy absorbed or emitted.

(h) STIMULATED EMISSION, MASERS, AND LASERS

In our discussion of the emission and absorption of radiation, we deferred until now mention of stimulated emission. There are two ways in which excited atoms radiate photons in transitions to the ground state. Usually electrons jump down and emit photons spontaneously. However, if an atom in an excited level is passed by a photon of exactly the excitation energy, that photon can stimulate (or induce) the atom to emit an identical photon and transfer to the ground state. The two photons travel in the same direction and have the same phase as the one that was passing by (that is, the crests and troughs of their waves coincide). Emission by this process is called *stimulated* (or *induced*) *radiation*. The two photons that have identical direction of motion and phase are said to be *coherent*.

Photons of stimulated emission are usually quickly absorbed by other atoms in the ground state, exciting them to that same level. Thus we would normally not expect to see emission lines from a gas due to stimulated emission. However, in some circumstances a larger fraction of the atoms of a gas are in a particular excited state than in the ground state. Suppose, for example, that the atoms that are in the ground state are continually excited to high energy levels by, say, intense radiation or by an electric field. Now suppose that these excited atoms undergo downward transitions to a lower excited level from which there is a relatively low probability of a spontaneous emission of a photon with a transition the rest of the way to the ground state. Thus the atoms that linger in that lower excited state can outnumber those in the ground state. Then those photons emitted spontaneously from the few excited atoms that do jump to the ground state, passing other excited atoms, can induce them to emit. Now, because there are not enough ground-state atoms to absorb all of these photons of stimulated emission, those not absorbed that can induce still other atoms to emit. The radiation at that wavelength is thus amplified to a far greater intensity than would ordinarily occur.

Stimulated emission is the basis of the maser, at radio (microwave) wavelengths, and of the laser, at optical wavelengths. The word *maser* is an acronym for *microwave amplification by stimulated emission of radiation*, and *laser* for *light amplification by stimulated emission of radiation*. In these devices atoms in an enclosure, subjected to intense radiation or an electric field, are excited or ionized to energies from which they undergo downward transitions to levels from which stimulated emission can produce amplification. The energy supplied externally to produce the excitation or ionization is called *pumping*.

In the optical laser the light beam of stimulated radiation is reflected back and forth through the gas (sometimes a solid, as in the ruby laser) many times by mirrors at either end of the enclosure. With each pass the beam induces more and more radiation to reinforce or amplify itself. The mirror at one end is half silvered to reflect part of the beam back for further amplification, and to allow the rest to pass out. Because the emerging beam of photons is coherent, it is narrow and intense, and can be transmitted over long distances (see Section 9.2a).

Masers evidently occur naturally in some interstellar gas clouds. The radio emission lines from some molecules observed in interstellar space (Chapter 28), are far more intense than would be expected unless their energy were reinforced by the maser process. The pumping of the interstellar molecules to higher excited levels is believed to be due to infrared radiation from nearby stars.

10.6 COSMIC RAYS

Our bodies are constantly being subjected to a rain of invisible high-energy particles passing through them. These bullets of radiation, undetected by our senses, result from the entrance of some 10^{18} atomic nuclei into the earth's atmosphere each second at speeds near that of light. The total rate of influx of energy to the earth from these particles is comparable to the rate at which the earth receives energy that comes from starlight.

The physicist has learned, through the study of this phenomenon, of kinds of subatomic particles (for example, *muons* and *positrons*) hitherto unobserved. The incoming particles of extraterrestrial origin interest the astronomer because most are believed to come from beyond the solar system, and the discovery of the origin of these cosmic rays may provide us with new clues about the nature of the universe.

(a) Discovery of Cosmic Rays

For more than a century it has been known that the air has a slight electrical conductivity, because a charged body exposed to the air slowly loses its charge. Some of the atoms in the air, therefore, must be ionized. The electrons released from the atoms that have been ionized are attracted to a positively charged body, and the ions themselves are attracted to a negatively charged body; in either case, the charge on the body is reduced. The physicists J. Elster and H. Geitel investigated the conductivity of the air with an electroscope in 1899 and 1900 and found that the ionized particles in the air were continually being replenished.

An *electroscope* is a very simple device for detecting the presence of charged particles. An elementary homemade type of electroscope is illustrated in Figure 10.25. It operates on the principle that like charges repel. If the ball, A, at the top of the instrument is charged with electricity, part of that charge is conducted into the metal-foil leaves, B. Since the leaves have the same charge, they repel each other and separate; a measure of their separation is a measure of the total charge on the instrument. Now, if particles of the opposite charge are attracted to the ball, A, the charge on the electroscope is gradually neutralized, and the leaves come together. Thus, the rate of closing of the leaves measures the rate at which the electric charge leaks from the instrument as a result of ionization (that is, the presence of charged particles) in the air.

In August 1912 the Austrian physicist Victor Hess carried an electroscope aloft in a balloon. He found that except near the ground the conductivity of the air *increased* with altitude. In 1914 this surprising result was confirmed by D. Kolhörster, who showed that the increase in conductivity with altitude continued to more that 8000 m. Apparently

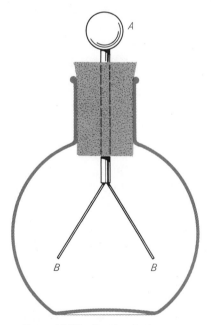

Figure 10.25 Simple electroscope.

the radiation that ionized the air came either from high in the atmosphere or from beyond the earth.

Further evidence for extraterrestrial origin of the mysterious radiation was provided in 1928, when R. A. Millikan and G. H. Cameron lowered sealed electroscopes into two freshwater California lakes. They found that at successively greater depths under water, the radiation decreased, a result which would not be expected if it originated from within the earth's crust or atmosphere. Millikan gave the radiation the name *cosmic rays.*

(b) The Charged Nature of Cosmic Rays

At first, cosmic rays were believed to be very high energy photons, that is, electromagnetic energy of wavelengths even less than those of gamma rays. In 1927, however, the Dutch physicist J. Clay found that the intensity of the ionizing radiation (cosmic rays) varies with latitude, being least near the geomagnetic equator (the circle halfway between the geomagnetic poles), and increasing as the geomagnetic poles are approached.§ Clay's observations have been confirmed with many subsequent experiments. It is difficult to understand why photons, which have no electrical charge, should be in any way affected by the magnetic field of the earth as they approach it.

It is well known, on the other hand, that charged particles move on curved paths through a magnetic field. Charged particles entering the earth's magnetic field, therefore, would be deflected by the field; if their deflections, that is, the curvatures of their paths, are great enough, they will never strike the earth's atmosphere. Near the geomagnetic equator, only those particles of very high energy can reach the atmosphere. At higher geomagnetic latitudes, however, the lines of force of the earth's field curve toward its surface, and particles of lower energy can penetrate the field. At the magnetic poles, where the lines of force are perpendicular to the surface of the earth, particles of all energies reach the upper atmosphere. Thus, if the cosmic rays were charged particles, rather than

§The *geomagnetic poles* are in line with the ends of a hypothetical ideal bar magnet whose magnetic field most nearly matches that of the earth. The actual field of the earth, however, is somewhat irregular, and the earth's *magnetic poles,* where the actual lines of force are perpendicular to the surface, deviate by some hundreds of kilometers from the idealized geomagnetic poles.

photons, it would be easily understood why cosmic-ray intensity is greater nearer the magnetic poles. Today, it is well established that this explanation is, indeed, the correct one.

The realization only a few decades ago that the ionization noticed in the earth's atmosphere is due to charged particles striking the earth from space was one of the important discoveries of the 20th century. Investigation of these particles became, in the 1940s, a major effort of modern physics.

(c) DETECTION OF COSMIC RAYS

The first evidence for cosmic rays came from observations with electroscopes. More sophisticated instruments, however, are used today. A *radiation counter* (for example, a *Gieger counter*) consists of a gas-filled chamber across which an electric field is provided by oppositely charged electrodes on either side. When a high-energy charged particle enters the chamber, it ionizes some of the gas, so that the gas becomes momentarily conducting. This pulse of current flowing through the tube is amplified and recorded on a meter, or detected by means of a loudspeaker. A counter, therefore, can detect and record those individual charged particles that pass through it with enough energy to ionize the gas within the chamber. (It can detect gamma rays as well, by the same process.)

A *cloud chamber* is a chamber filled with a gas that is saturated with the vapor of water or of some other liquid. The chamber is so designed that its volume can be enlarged suddenly, usually by moving a rubber diaphragm or a tightly fitting piston. As it is enlarged, the gas contained within it lowers in density and cools. If the chamber were enlarged sufficiently, the gas would cool enough for the liquid to precipitate. In actual use the cloud chamber is not enlarged quite enough for the precipitation to occur by itself. If, however, a charged particle should happen to pass through the chamber and ionize some of the atoms of the gas just as it is cooling, those ions will serve as condensation nuclei, and a line of droplets will form, marking the path of the ionizing particle. The track can be observed visually or it can be photographed.

A more useful device for many experiments is the *bubble chamber,* similar in function to the cloud chamber. The bubble chamber contains a liquid which is superheated when the chamber is suddenly expanded. A charged particle moving through the liquid when it is in this condition causes it to boil along the track of the particle, leaving a string of tiny bubbles that can be observed or photographed.

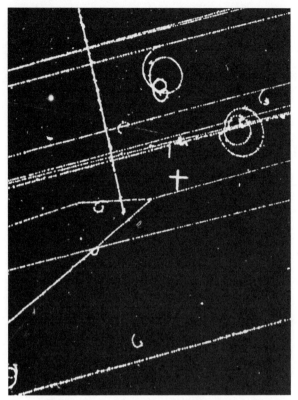

Figure 10.26 Bubble chamber photograph. (*Courtesy H. Ticho, UCLA*)

In either the cloud or bubble chamber, additional information is obtained by applying a magnetic field through the chamber, because the path of a charged particle in a magnetic field is curved. The amount of curvature depends on the field strength, on the charge on the particle, and on its momentum. Cloud and bubble chambers with magnetic fields, therefore, not only show the tracks of the impinging particles but indicate something about their charges and momenta.

Figure 10.26 shows tracks of charged particles through a bubble chamber containing liquid hydrogen. Most of the tracks are of the short-lived subatomic particles π-mesons, moving so fast that the magnetic field curves their trajectories only slightly. The branching near the middle of the picture is where a π-meson, coming from the right, struck a proton and produced two other charged nucleons. The upper one (a sigma-particle) quickly decayed into a neutral particle that left no track and another π-meson, whose track is visible in a still different direction. If all the particles, charged and neutral, are taken into account, the momentum is always conserved in such collisions and decays. The small spiral tracks are of low-energy electrons ejected from hydrogen atoms that are ionized by the high-speed mesons. The paths of these electrons are

curved more and more by the magnetic field as they lose energy moving through the liquid.

Tracks of charged particles can also be recorded in *photographic emulsions*. The grains in an emulsion are made capable of being developed not only by photons of light but also by ionization by charged particles. Ordinarily, a photographic emulsion is a very thin coating on a piece of celluloid or glass. For detecting particle radiation, however, many layers of emulsion are often piled on top of one another, forming a thick emulsion *stack*. A charged particle passing through the stack leaves a track of developable grains behind it. After development the emulsion layers are separated and the track of the particle through each of them is measured; its course is thus determined.

Certain solid substances (for example, zinc sulphide) when bombarded by subatomic particles emit flashes of light or *scintillations*. A *scintillation counter* utilizes such a scintillating phosphor to detect particles; the light flashes produced are amplified with a photomultiplier and recorded. Certain plastics fluoresce when hit by energetic particles, and so can also be used as scintillating materials. Large disks of such plastic spread out over a substantial area of the ground (for example, in the M.I.T. experiments at Volcano Ranch, New Mexico) have been used to detect great numbers of cosmic-ray particles.

One technique by which particles from a particular direction can be isolated is to arrange two or more counters in a line, and to record only pulses that occur almost simultaneously in all of them. Such coincidences are almost always the result of single particles that pass successively through each counter, and hence which approach along (or nearly along) the direction in which the counters are lined up. If sheets of lead (or other materials) of various thicknesses are placed between the counters, the penetrating power of a particle, and thus its energy, can be determined.

Electroscopes, cloud and bubble chambers, and geiger counters are no longer (or at best, rarely) used in cosmic-ray research, and emulsion stacks are used less and less in recent years, but these devices were important in the discovery and early investigation of cosmic rays. Space does not permit a discussion here of the many modern instruments which, in addition to scintillation counters, are in wide use today. At least, however, we have seen some ways in which fast-moving subatomic particles can be detected.

(d) Variation of Intensity with Altitude

The early observations of Hess and Kolhörster, as mentioned above, showed cosmic-ray intensity to increase with altitude. Modern investigations have shown, however, that the increase of intensity with altitude does not continue indefinitely. In southern California at a height of 3000 m the cosmic-ray intensity is four times its sea-level value. The increase over sea-level intensity is 30-fold at 8000 m and 100-fold at 20,000 m. At 30,000 m, however, where the atmospheric pressure is only about 1 percent of its sea-level value, the cosmic-ray intensity is *less* than at 20,000 m.

At first thought we might reason that the source of cosmic rays is near 20,000 m above the earth's surface. Actually, however, at 20,000 m and below, most cosmic-ray particles are *secondary* particles, produced by collisions between the *primary* extraterrestrial particles and molecules of air. At 30,000 m, on the other hand, the atmospheric density is low enough that such collisions are relatively rare and most of the observed cosmic rays are the original particles from outer space.

(e) Primary and Secondary Particles

Analysis of the primary cosmic-ray particles shows that most of them are high-speed protons (nuclei of hydrogen atoms), that most of the rest are alpha particles (nuclei of helium atoms), and that a few are nuclei of the still heavier atoms. A primary particle traverses, on the average, only about one tenth of the earth's atmospheric gases, however, before colliding with the nucleus of an air molecule.

When such a collision occurs, the nucleus in the air molecule breaks into several smaller subatomic particles. If the primary particle has high energy, each of these secondary particles is also given considerable energy. It, in turn, collides with still another nucleus in an air molecule, producing more secondary particles. In this way an original primary particle moving with high speed dissipates its energy in a great many secondary particles, producing many of the particles that are recorded at intermediate and low altitudes in the atmosphere. A large proliferation of particles by successive collisions following the impingement of a primary particle of very high energy is called a *shower*.

Perhaps it is well that we do not see these primary and secondary particles or feel them as they strike us. Otherwise, it might be discomforting to see repeated skyrocketlike bursts high above our heads, and then observe many of the burst particles passing into one side of our bodies and out through the other. Those few that are intercepted by the atoms of our bodies contribute to the natural radiation our bodies are constantly receiving, and probably are responsible for some mutations and even cancer.

From tracks in cloud chambers and through magnetic fields, physicists have learned the energies, charges, and masses of these secondary cosmic-ray objects. Most of them are charged particles with masses intermediate between those of protons and electrons. Such particles of intermediate mass are called *mesons*. At sea level the most common mesons in secondary cosmic rays are called *mu mesons* (or *muons*); each carries either a positive or negative charge that is numerically equal to the charge on the electron and has a mass equal to 207 electron masses.

Muons are very unstable particles and disintegrate in average periods of only about 1.5 microseconds. When a muon disintegrates (or *decays*) it forms an electron (if it had a negative charge) or a *positron* (if it had a positive charge). A positron is a particle equivalent to an electron but carrying a positive charge that is numerically equal to the negative charge on an ordinary electron. The excess mass of a muon, over that of an electron (or positron), is converted into energy and into *neutrinos*—particles that have energy but no mass. We shall return later (Chapter 32) to some of these unusual subatomic particles; it is worth noting, however, that a few of them were first observed in cosmic rays.

(f) Energies of Primary Cosmic-Ray Particles

The energies of most cosmic-ray primaries are near 10^9 eV (a thousand million electron volts, abbreviated GeV), but a small percentage have energies in excess of 10^{18} eV. The relative number of particles of various energies is called the cosmic-ray *energy spectrum*. The energy spectrum is now known fairly well for cosmic-ray particles with energies in the range 10^8 to 10^{19} eV. In that range the numbers of primaries of successively higher energy drops rapidly; the total number of particles, $N(E)$, with total energies greater than E is represented, approximately, by the empirical formula

$$N(E) = \frac{\text{constant}}{E^{1.6}}.$$

The constant in the above equation is different for the different kinds of atomic nuclei that make up the primary cosmic-ray particles. The conclusion is that particles of higher energy are much less common than those of lower energy.

For energies above 10^{19} eV, the energy spectrum is much less well established. It is known, however, that a few primaries of extremely high energy exist. Spectacular showers have been detected for several decades with many techniques. For example, with the large fluorescent plastic disks at their Volcano Ranch site, the M.I.T. physicists have observed very large showers of secondary particles since about 1960. Since primary particles of the highest energy produce the largest showers, the energy of a primary particle can be inferred from the number of shower particles it produces. The M.I.T. group has found that for every million showers of 1 million particles each, there are 1000 showers of 10 million particles each, and only one shower of 100 million particles.

Nevertheless, occasional showers of thousands of millions of particles are observed. A few are produced by primary particles with energy as great as 10^{20} eV; this is an amount of energy that could keep a one-watt electric light burning for one second.

The energies observed for some primary cosmic-ray particles are many times those that can be obtained in laboratory accelerators (cyclotrons, synchrotrons, betatrons, and so on). The total energy density of cosmic rays in space is estimated at about one eV/cm^3. Some of the possible origins of this energy will be considered in later chapters.

(g) Latitude Effects for Particles of Different Energies and the Isotropy of Cosmic Rays

Particles of low energy are deflected by the earth's magnetic field and can come close enough to the ground to interact with molecules of the earth's atmosphere only in the vicinity of the earth's magnetic poles. Particles that approach the earth near the geomagnetic equator cannot reach the denser parts of the atmosphere if they have energies much less than about 10 GeV (10^{10} eV). Particles of energy greater than about 60 GeV, on the other hand, always strike the atmosphere, even at the equator. Particles of intermediate energy may or may not reach the earth's dense atmosphere, depending on their direction of approach.

The *latitude effect,* that is, the variation of cosmic-ray intensity with latitude, should therefore be present only for cosmic-ray particles of relatively low energy. This prediction is confirmed by observation; for particles of energy above 60 GeV, the number striking the atmosphere is approximately the same at all latitudes.

Those high-energy particles from which the geomagnetic field does not successfully shield us are observed to approach the earth in approximately equal numbers from all directions. The high-energy cosmic-ray intensity, in other words, is *isotropic* about the earth. We presume that if it were not for the effect of the earth's magnetic field, the lower-energy particles would also appear to be isotropic. Further evidence for the isotropy of cosmic rays is that their diurnal (day to night) variation in intensity is less than $\frac{1}{2}$ percent.

(h) Composition of the Primary Particles

As has been stated, the primary cosmic-ray particles are mostly atomic nuclei. The majority are protons (nuclei of hydrogen atoms). Most of the rest—about 15 percent—are alpha particles (nuclei of helium). The remaining 1 percent or so are nuclei of heavier elements; those of elements as heavy as iron are moderately abundant, and heavier ones have been observed. Very roughly, the relative abundances of the various atomic species in cosmic rays resemble those of elements elsewhere in the universe. There are exceptions, however. The cosmic-ray abundance of nuclei heavier than helium seems to be several times higher than the general cosmic abundance of those elements. The nuclei of the elements lithium, beryllium, and boron, in fact, are thousands of times as abundant in cosmic rays as in stars. Lithium, beryllium, and boron, however, are unstable at temperatures of a few million Kelvins, and in stars they would undergo nuclear transformations to other elements; their comparatively low abundances in stars, therefore, is not surprising. The high abundance of these nuclei in cosmic rays may result from the breakup of heavier nuclei. Finally, there is a small percentage (about 1 percent) of primary particles that are electrons.

EXERCISES

1. How many times brighter or fainter would a star appear if it were moved to (a) twice its present distance? (b) ten times its present distance? (c) half its present distance?

2. "Tidal waves," or *tsunamis*, are waves of seismic origin that travel rapidly through the ocean. If tsunamis traveled at the speed of 600 km/hr, and approached a shore at the rate of one wave crest every 15 minutes, what would be the distance between those wave crests at sea?

3. Stars are fairly good approximations to black bodies. Explain why they do not look black.

4. Suppose the sun radiates like a black body. Explain how you could calculate the total amount of energy radiated into space by the sun each second. What solar data would you need to make this calculation?

5. If the reemitted infrared radiation from Pluto has a wavelength of maximum intensity at 0.005 cm, what is the temperature of Pluto?

6. What color would you expect light to appear that was equally intense at all visible wavelengths?

7. What is the temperature of a star with a wavelength of maximum light of 2.897×10^{-5} cm?

8. What is the energy of a proton of wavelength 3 cm? *Answer:* 6.62×10^{-17} erg

9. Because of refraction, the sun appears to rise before it is above the geometrical horizon and to set after it has dropped below the geometrical horizon. By how much does atmospheric refraction increase the hours of sunshine in a typical day? *Answer:* About four minutes or more.

10. Suppose that a spectral line of some element, normally at 5000 Å, is observed in the spectrum of a star to be at 5001 Å. How fast is the star moving toward or away from the earth? *Answer:* About 60 km/s away from the earth

11. How could you measure the rotation rate of the sun by photographing the spectrum of light coming from various parts of the sun's disk?

12. How could you measure the earth's orbital speed by photographing the spectrum of a star at various times throughout the year?

13. Refer to the Rydberg formula and indicate to what series each of the following transitions in a hydrogen atom corresponds. State whether the spectral line produced is one of absorption or emission. List the lines in order of *increasing wavelength*.
   a) $m = 3$ to $n = 8$ d) $m = 2$ to $n = 13$
   b) $m = 1$ to $n = 2$ e) $n = 17$ to $m = 2$
   c) $m = 1$ to $n = 4$ f) $n = 14$ to $m = 5$

14. Most hydrogen atoms in the sun are in their lowest state of energy, so what series of absorption lines would hydrogen produce most strongly in the sun?

15. The decay of naturally radioactive elements in the earth's crust would be expected to produce some ionization in the atmosphere. How did Hess's balloon experiment rule out the possibility that the total ionization of the air was so produced?

16. Describe how the observed cosmic-ray intensity at the surface of the earth would vary with the energy of the primary particle and with latitude if (a) the earth had no magnetic field, (b) the earth had a magnetic field millions of times stronger than its actual field.

17. Where on earth would you like to live to have the best chance of avoiding radiation resulting from primary cosmic-ray particles striking the earth at energies of 100 GeV? Why?

18. A cosmic ray primary has an energy of 2.08×10^{11} eV. What is its energy in ergs? How many such particles would be required each second to yield the power equivalent of a 1000-watt light bulb? *Answer to second part:* about 3×10^9 such particles

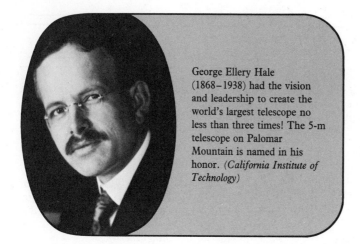

George Ellery Hale (1868–1938) had the vision and leadership to create the world's largest telescope no less than three times! The 5-m telescope on Palomar Mountain is named in his honor. (*California Institute of Technology*)

11

OPTICAL ASTRONOMY

Astronomical instruments have been used since ancient times. Most of the instruments used by Hipparchus and Ptolemy, and the later, more refined ones of Tycho Brahe, were essentially calibrated sighting devices for measuring the directions of celestial objects. Brahe, for example, designed and built accurate sextants and quadrants, in which sights for observing a star were carried on a long pivoted arm. The end of the arm moved along an arc of a circle with a graduated scale on which the precise angle of the arm and hence of the sights could be read, indicating in turn the angle of the star being observed above the horizon (that is, the star's altitude). Optical instruments were not em-

ployed in astronomy, however, until the time of Galileo.

Electromagnetic radiation from space comes at all wavelengths, from gamma rays to radio waves. Unfortunately for astronomical observations (but fortunately for biological organisms) much of the electromagnetic spectrum is filtered out by the terrestrial atmosphere. There are two spectral windows in the atmospheric filter through which we can observe cosmic radiation (Figure 11.1). One of these is the *optical window*, which includes the near ultraviolet (wavelengths longer than about 3000 Å), light, and portions of the infrared (with wavelengths up to about 0.03 mm). The other is the *ra-*

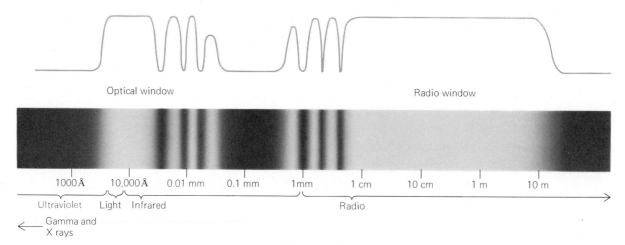

Figure 11.1 A portion of the electromagnetic spectrum, showing those regions (windows) to which the earth's atmosphere is transparent. The dark regions are those to which the atmosphere is opaque. The upper graph is a plot of the transparency of the atmosphere.

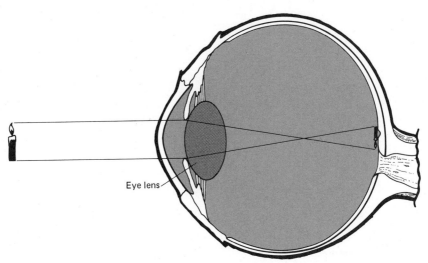

Figure 11.2 Production of an image on the retina of the eye.

dio window, which includes radio waves of length from about a millimeter to about 20 m (the long wavelength cutoff depends somewhat on the variable conditions of the ionosphere).

In this chapter, we consider the optical window.

11.1   OPTICAL INSTRUMENTS

The simplest kind of optical detector would do no more than reveal the presence of radiation striking it. But for most purposes it is more useful to form an *image* of an object of known direction in the sky. The image can then be detected, measured, reproduced, and analyzed in a host of ways. One image-forming device is the eye (Figure 11.2). It consists of a *lens,* which focuses light into an optical image on the *retina* at the back of the eye. There the image is detected by the light-sensitive retinal nerve endings, which transmit the information (through the optic nerve) to the brain for analysis. The retinal nerves can sense only that kind of electromagnetic energy we call light, but images can be made of other electromagnetic radiation as well. Ground-based optical astronomy is concerned with the forming of images in the near ultraviolet, light, and infrared, but the image-making principles are the same for even shorter and longer wavelengths.

(a)   Formation of an Image by a Lens or a Mirror

Having examined the laws of geometrical optics (Section 10.4), we are in a position to understand the formation of images by optical systems. Images were first produced by simple convex lenses. To illustrate the principle, let us imagine two triangular prisms, base to base, as in Figure 11.3. Now suppose we select two of the parallel rays of light from a distant object and allow one ray to enter each prism. The light rays are refracted by the prisms and meet at a point *F*.

This is the principle underlying the formation of an image by a lens. In Figure 11.4 a simple convex lens is shown. Parallel light from a distant star or other light source is incident upon the lens from the left. A *convex* lens is thicker in the middle than at the edges. In cross section it is no more than a series of segments of prisms piled one on another, with slightly different slopes to their sides. If the curvatures of the surfaces of the lens are correct, light passing through the lens will be refracted in such a way that it converges toward a point. Convex lenses whose surfaces are portions of spherical surfaces are easiest to manufacture. Such lenses will refract a parallel beam of light to a point as shown in Figure 11.4, if the curvature of the surfaces is slight.

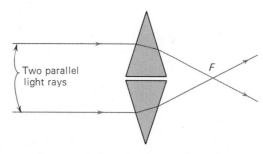

Figure 11.3   Principle of image formation.

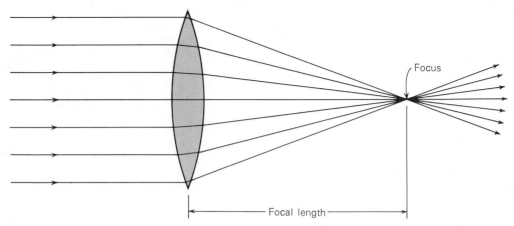

Figure 11.4   Formation of an image by a simple convex lens.

The point where light rays come together is called the *focus* of the lens. At the focus, an *image* of the light source appears. The distance of the focus, or image, behind the lens is called the *focal length* of the lens. A lens or other device that forms an image is called an *objective*.

We have seen how an image can be formed of a point source, say, a star. However, the image itself in that case is just a point of light. In Figure 11.5 we see how an image is formed of an extended source, for example, the moon. From each point on the moon, light rays approach the lens along parallel lines. However, from different parts of the moon, the parallel rays of light approach the lens from different directions. The light from each point on the moon strikes all parts of the lens (or *fills* the lens); these rays of light are focused at a point at a distance behind the lens equal to the focal length of the lens. If a screen, such as a white card, is placed at this distance behind the lens, a bright spot of

light appears, representing that point on the moon. Light from other points on the moon similarly focuses at other points, producing bright spots on the card in different places. Thus, an entire image of the moon is built up at the focus of the lens. The plane in which the image is formed is called the *focal plane*.

(We discuss here only the case where the object whose image is formed is so distant that light from any point on it can be regarded as approaching the lens along parallel rays. This is always true when any astronomical body is observed. Nearby terrestrial objects may be so close that the assumption is not valid. Then the image is formed at a point farther from the lens than the focal length.)

Note that if part of the lens is covered up, or if the middle is cut out, or if ink or mud is splattered over it, as long as part of the lens is still transparent to light, an entire image will be formed. All parts of the lens contribute to each part of the im-

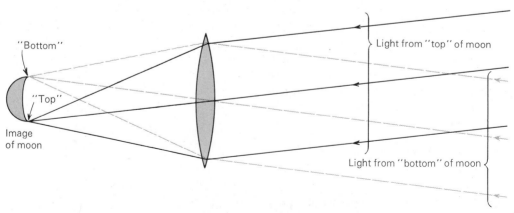

Figure 11.5   Formation of the image of an extended object.

age. Covering up part of the lens cuts down the total amount of light that can strike each portion of the image and thus makes the image fainter, but nevertheless the whole image is formed. An ordinary camera lens produces an image at the focal plane (where the film is placed) just as is shown in Figure 11.5. Every photographer knows that he can cover up part of his camera lens by "stopping it down" with an iris diaphragm. In so doing he will cut down the *brightness* of the image (and hence the effective exposure on the film), but the outer parts of the image will not be removed. The part of the lens that remains uncovered still produces the entire image.

Finally, we note that the image formed is always inverted and reversed (upside down and left to right) with respect to the object. The eye lens forms inverted images on the retina, but the brain interprets the image so that it appears upright. The eye is equipped with an adjustable iris diaphragm that automatically enlarges or contracts to allow more or less light to enter the eye, so that an image of the optimum brightness can be produced on the retina. An ordinary camera is a nearly complete analogy to the eye.

Rays of light can also be focused to form an image with a *concave* mirror—one hollowed out in the middle. Parallel rays of light, as from a star, fall upon the curved surface of the mirror (Figure 11.6), which is coated with silver or aluminum to make it highly reflecting. Each ray of light is reflected according to the law of reflection. If the mirror has the correct concave shape, all the rays are reflected back through the same point, the focus of the mirror. At the focus appears the image of the

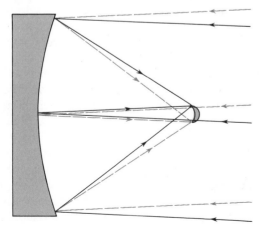

Figure 11.6 Formation of an image by a concave mirror. The dashed lines are from the "top" of the moon.

star. As in a lens, the distance from the mirror to the focus is called its focal length.

Rays of light from an extended object are focused by a mirror, exactly as they are by a lens, into an inverted image of the object. The principal difference between image formation by a lens and by a mirror is that the mirror reflects the light back into the general direction from which it came, so that the image forms in front of the mirror. The image can be inspected, as with the lens, by allowing the light to illuminate a screen, such as a white card, held at the focus of the mirror. The card, of course, will block off part of the incoming light, but since, as in the lens, all parts of the mirror contribute to the formation of all parts of the image, the presence of the card will not produce a "hole" in the image but will merely reduce its brightness.

If the shape of the mirror is part of a concave spherical surface, it will produce a fair-quality image, provided that the entire mirror constitutes only a very small part of a sphere, so that its size is small compared to its focal length. We shall see that the curves of telescope mirrors are usually parabolic in cross section.

(b) Properties of an Image

The most important properties of an image are its *scale* or size, its *brightness*, and its *resolution*. We shall consider these in turn. In this section, we ignore the limitations imposed by the earth's atmosphere (Section 11.4).

The *scale* of an image is a measure of its size. In all astronomical applications we are dealing with objects whose sizes can be expressed in angular units, and it is generally convenient to express the scale of an image as the linear distance in the image that corresponds to a certain angular distance in the sky. For example, suppose that an image of the moon were produced that is exactly 1 cm across. The moon has an apparent or angular size of $\frac{1}{2}°$, that is, it subtends $\frac{1}{2}°$ in the sky. The scale of the image is thus $\frac{1}{2}°$ per centimeter, or 2 cm per degree.

The scale of an image depends only on the focal length of the lens or mirror that produces it. Numerically, the distance s in an image corresponding to 1° in the sky is given by the equation

$$s = 0.01745f,$$

where f is the focal length of the lens or mirror. For

example, the 5-m (200-inch) mirror of the Hale telescope on Palomar Mountain has a focal length of 16.764 m. It produces images with a scale of 0.292 m per degree, which corresponds to 12″.3 per millimeter. Most astronomical telescopes have lenses or mirrors that give image scales ranging from 10″ to 200″ per millimeter.

The *brightness* of an image is a measure of the amount of light energy that is concentrated into a unit area of the image, such as a square millimeter. The brightness of an image determines whether it is above the threshold of visibility or, alternatively, how long a time would be required to record the image photographically.

The brightness of the image of an extended object, such as the moon, a planet, a nebula, a galaxy, or the faint illumination of the night sky, is greater the greater the amount of light flux that passes through the objective (lens or mirror) to form the image and is less the larger the image area over which that amount of flux must be spread. The amount of flux reaching the image is proportional to the area of the objective and hence to the square of its diameter or aperture (the area of a circular region of radius R or diameter D is πR^2 or $\frac{1}{4}\pi D^2$). The area the flux is spread over is proportional to the square of the focal length of the objective, for as we have seen the image diameter varies directly with the focal length. Hence the brightness B of an extended image is given by

$$ B = \text{constant} \times \left(\frac{a}{f}\right)^2 , $$

where a is the diameter or aperture and f is the focal length of the lens or mirror. The constant of proportionality is a number whose value depends on the units chosen to measure the various quantities, and also on the amount of light actually leaving each unit area of the object.

The quantity f/a is called the *focal ratio* or simply the *f ratio* of the lens. In common notation, if the focal length is, say, eight times the aperture of the lens, the focal ratio is written $f/8$, which should be interpreted as $a/f = \frac{1}{8}$. (Note that $f/8$ means $a = f/8$). Every photographer is familiar with the concept of focal ratio when applied to his camera lens. In all but the simplest cameras the clear aperture of the lens can be increased or decreased by adjusting an iris diaphragm, thereby changing the focal ratio of the lens. A typical 35-mm camera, for example, might have focal ratio adjustments varying from $f/1.7$ to $f/16$. Since the focal ratio determines the image brightness, it determines the effective exposure on the film for a given exposure time, that is, the time required to expose the image to a particular blackness. For this reason, the focal ratio is often called the "speed" of an optical system. In high-speed systems the focal length is short compared to the lens size. The focal ratio is thus smaller (that is, the ratio of aperture to focal length is a larger number) in high-speed lenses than in low-speed lenses.

Stars are in effect point sources, for even through the largest telescopes the stars still appear too small to show any apparent disks. Therefore a lens or mirror of good quality concentrates the starlight into a "point" image regardless of the focal length. For a point-source object such as a star, the amount of light in the image thus depends only on the amount of light gathered by the lens or mirror, and hence is proportional to the square of the aperture.

Resolution refers to the fineness of detail inherently present in the image. Even if the lens or mirror is of perfect optical quality, it cannot produce perfectly sharp and detailed images. Because of the phenomenon of diffraction and interference, a point source does not form an image as a true point but as a minute spot of light surrounded by faint, concentric, evenly spaced rings. The angular size of that central spot of light, called the central *diffraction disk* of the image, is inversely proportional to the aperture of the lens or mirror and directly proportional to the wavelength of the light observed. No detail can be resolved in the image if that detail is smaller than the diffraction disk. For example, the diameter predicted by geometrical optics for the image of a star produced by the lens of a telescope is much smaller than the size of the diffraction disk. Therefore, we do not see the geometrical image of a star itself but only a *diffraction pattern* that the telescope lens produces with the star's light. If a star is viewed telescopically under good conditions, the diffraction pattern, consisting of the bright central disk and faint surrounding rings, is clearly visible.

In the image of an extended source, the diffraction patterns of the various parts of the image, all overlapping each other, wash out the finest details. A feature on the surface of the moon or Mars that is smaller than the diffraction disk produced by the telescope lens may be visible, but its true size and shape will not be distinguishable. If two stars are so near each other that their images are closer together than the size of the diffraction disk of either, they will blend together to produce a single spot, or perhaps a slightly elongated diffraction pattern. The ability of an optical system (lens or mirror) to distinguish fine detail in an image it produces, or to produce separate images of two close stars, is called its *resolving power*. In astronomical practice, the resolving power of a lens or mirror is described in terms of the smallest angle between two stars for which separate recognizable images are produced. Two stars that lie less than 1′ from each other cannot be separated with the human eye. With a 15-cm lens of good quality, separate images are produced of two stars only 1″ apart. The 15-cm lens thus has higher resolving power than the eye.

We have said that the angular size of the central diffraction disk of a point source depends on the wavelength of light used and the diameter of the lens or mirror. If the wavelength and aperture are measured in the same units (for example, both in centimeters), the smallest angle (α) in seconds of arc that can be resolved by a lens or mirror of aperture d is given by the equation,

$$\alpha = 2.1 \times 10^5 \times \frac{\lambda}{d} \text{ seconds of arc,}$$

where λ is the wavelength. If λ is chosen as 5.5×10^{-5} cm (5500 Å)—near the middle of the visible spectrum—and if d is measured in centimeters, the formula becomes

$$\alpha = \frac{11.6}{d} \text{ seconds of arc.}$$

The above formula, called *Dawes' criterion*, represents a compromise or average of empirical studies of the ability of telescopes of various apertures to resolve double stars. It does not hold for the eye because the coarse structure of the retina limits the resolution ideally obtainable with the eye lens. Also, as we shall see, atmospheric turbulence usually further degrades the actual resolving power of large telescopes.

(c) Aberrations of Lenses and Mirrors

An image produced by any optical system always has imperfections. These are called *aberrations*. We describe briefly a few of the more important kinds

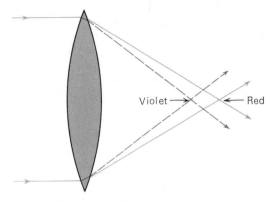

Figure 11.7   Chromatic aberration.

of aberrations. Aberrations are always most serious in telescopes of low focal ratios (high speeds).

Chromatic or *color aberration* is a consequence of dispersion. Consider the simple lens shown in Figure 11.7. Suppose parallel rays of white light from a remote star are incident upon the left side of the lens. Because the different wavelengths that comprise white light are refracted in different amounts upon entering and leaving the lens, they do not all focus at the same place. The shorter wavelengths (violet and blue light) are bent the most and focus nearest the lens, while the longer wavelengths of orange and red light focus farther from it. The effect of this chromatic aberration is to produce color fringes in the image.

Chromatic aberration is less serious in lenses of large focal ratio. Some telescopes constructed in the late 17th century, therefore, utilized lenses of extremely long focal length to obtain as large a focal ratio as possible for a given aperture. Telescopes more than 30 m long were in common use, and ones with lengths of up to 180 m are said to have been built. Such extremely long telescopes, however, were so difficult to use that they were, for the most part, impractical.

Chromatic aberration is now corrected by constructing a lens of two pieces of glass of different types (usually crown and flint glass) and thus of different indices of refraction. One of the pieces of glass, or *elements* of the lens, is a *concave* lens, that is, it is thinner in the middle than at the edges, so that it diverges the light rather than converging it. (The concave element of the lens diverges light only enough to correct for the chromatic aberration; the combined lenses still converge incident light to a focus.) The light of shorter wavelengths is diverged the most, just as it is converged the most in the

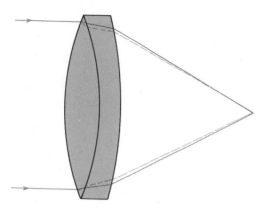

Figure 11.8 *Correction of chromatic aberration.*

usual convex lens. Dispersion, like refraction, is greater in the glass of higher index. It is possible to make the chromatic aberration completely cancel out at any two given wavelengths, if exactly the right curvatures are chosen for the surfaces of the two lenses (Figure 11.8). Lenses can be designed so that this aberration is nearly canceled out, or is greatly reduced, over a considerable range of wavelengths.

When a lens is designed, a choice must be made as to the range of wavelengths for which the chromatic aberration is to be corrected. Lenses designed for visual observation are corrected for the green and yellow spectral region, to which the eye is most sensitive. If one directly views an image produced by a lens corrected for photography in other spectral regions, disturbing color fringes may be observed. Use of more than two elements enables the correction to be extended over a wide range of wavelengths. Good camera lenses designed for color photography usually have several elements.

When an image is produced by a mirror, the light never has to pass through any glass, so it is not dispersed. Thus an image produced by a mirror does not suffer chromatic aberration.

Spherical surfaces are the most convenient to grind and polish, whether they be the concave or convex surfaces of a lens or mirror. Unfortunately, however, a simple lens or mirror with spherical surfaces suffers from the imperfection called *spherical aberration* (Figure 11.9). Light striking nearer the periphery focuses closer to the lens or mirror; light striking near the center focuses farther away.

In a lens, spherical aberration can be corrected or greatly reduced by constructing the lens of two pieces of glass or elements of different indices of refraction, just as in the correction of chromatic aberration. The elements are so designed that the spherical aberration introduced by one is canceled out by the other. Most lenses designed for astronomical purposes consist of two elements. The curvatures of the spherical surfaces of the elements are designed to reduce both the spherical and chromatic aberrations to a tolerable amount.

For a mirror, spherical aberration can be eliminated by grinding and polishing the surface not to a spherical shape, but to a *paraboloid of revolution,* that is, to a surface whose cross section is a parabola (Figure 11.10). A paraboloid has the property that parallel light rays striking all parts of the surface are reflected to the same focus. Similarly, light leaving the focus of a paraboloid is reflected at the surface in a parallel beam. An automobile headlight or searchlight, for example, has a parabolic reflector with the light source at the focus. Mirrors designed for astronomical telescopes usually have parabolic surfaces.

Coma is an aberration that distorts images formed by *off-axis* light rays that do not strike the lens or mirror "square on" (that is, those light rays from objects away from the center of the field of view). This off-axis distortion is particularly serious

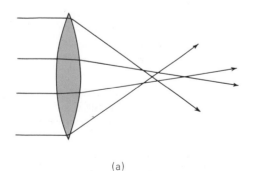

(a)

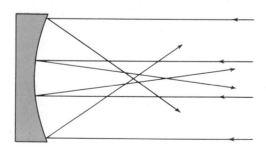

(b)

Figure 11.9 *Spherical aberration. (a) In a lens. (b) In a mirror.*

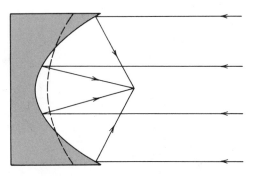

Figure 11.10 Correction of spherical aberration with a concave parabolic mirror. The dashed line is where the surface of a spherical mirror of the same focal length would lie.

in the images formed by the parabolic mirrors used in the largest existing telescopes. Photographs taken with such telescopes are reproduced in the latter part of this book. If you carefully inspect one of these (or Figure 11.11), you will find that the star images near the center of the picture appear as sharp round dots, whereas those near the corners, which are formed by light entering the telescope off axis, are distorted into tiny "tear drops" or "commas" pointing toward the center of the photograph. The cometlike shape of these images accounts for the name "coma" given to this particular aberration.

In modern telescopes, special *field-correcting lenses* are often placed in the light beam before it comes to a focus. Coma can be substantially reduced by these lenses, thereby greatly increasing the useful field of view of the telescope.

Astigmatism is an aberration produced in an optical system when rays of light approaching the lens in different planes do not focus at the same spot. In the example shown in Figure 11.12, the rays in a vertical plane focus farther from the lens than those in a horizontal plane. A geometrically perfect lens produces astigmatism only for off-axis rays.

Distortion is an aberration in which the image forms sharply but its shape is distorted. For example, straight lines in the object may appear as curved lines in the image.

Curvature of field is an aberration in which the image is sharp but parts of it are formed at different distances from the lens or mirror, so that the image cannot be formed on a flat screen or photographic plate.

(d) The Schmidt Optical System

An ingenious optical system that utilizes both a mirror and a lens was invented by the Estonian optician Bernhard Schmidt of the Hamburg Observatory. We noted in the last section that images formed by a mirror do not suffer from chromatic aberration. To avoid spherical aberration, parabolic rather than spherical mirrors are used. The principal disadvantage of the parabolic mirror is that it produces good images over only a relatively small field of view, that is, for light that approaches the mirror very nearly on axis.

On the other hand, for a spherical surface, any line reaching the surface through its center of curvature (that is, through the center of the sphere of which the surface is a part) is perpendicular, that is, "square on" to the surface, and hence is on axis. The Schmidt optical system, utilizing this principle, employs a spherical mirror that is allowed to receive light only through an opening located at its center of curvature (Figure 11.13). Thus there can be no off-axis aberration. The only trouble is that a spherical mirror, suffering as it does from spherical aberration, produces generally poor images for light coming from any direction. Schmidt solved the problem by introducing a thin correcting lens at the aperture at the center of curvature of the mirror. The lens is of the proper shape to correct the spherical aberration introduced by the spherical mirror

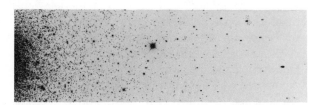

Figure 11.11 Photograph of the globular cluster M3 (at left) taken with a large reflecting telescope. The original photograph was centered on the cluster. Note how the coma distortion increases from center to edge. (*California Institute of Technology/Palomar Observatory*)

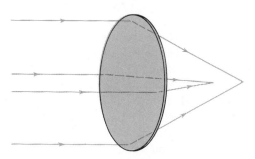

Figure 11.12 Astigmatism.

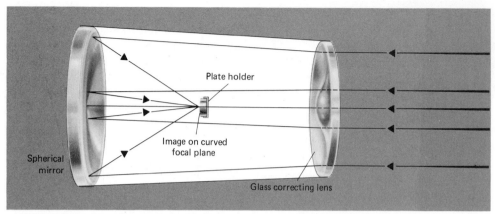

Figure 11.13 Schmidt optical system. The mirror is generally larger than the lens, so images of unreduced intensity can be produced over a wide field of view.

but not thick enough to introduce appreciable aberrations of its own. Thus, the Schmidt optical system produces excellent images over a large angular field. A disadvantage of the Schmidt system is that the focal surface is not a plane but a sphere concentric with the spherical mirror.

Today, the Schmidt optical system and modifications of it invented by Maksutov, Wright, and others are widely used in science and industry.

(e) The Complete Telescope

Now that we have seen how an image is produced, we are able to understand the operation of a telescope. Today there are two general kinds of optical astronomical telescopes in use. They are (1) *refracting* telescopes, which utilize lenses to produce images; and (2) *reflecting* telescopes, which utilize mirrors to produce images. As we have seen, Schmidt telescopes, and modifications thereof, are primarily reflecting telescopes but have refracting correcting lenses.

The refracting telescope is the most familiar. This is the kind of telescope that we can literally "look through." Ordinary binoculars are two refracting telescopes mounted side by side. A lens, generally consisting of two or more elements, is usually mounted at the front end of an enclosed tube. The tube is not really essential—its purpose is merely to block out scattered light; an open framework would suffice. In a refracting telescope, the objective, the optical part that produces the principal image, is the lens at the front of the tube. The image is formed at the rear of the tube, where various devices inspect, photograph, or otherwise utilize it. There is no particular problem in gaining access to the image. All that is required is a suitable ladder or platform to lift the observer to the point

from which he can conveniently reach the rear of the tube of the telescope.

The reflecting telescope was first conceived by James Gregory in 1663, and the first successful model was built by Newton in 1668. Here a concave mirror (usually a paraboloid) is used as an objective. The mirror is placed at the *bottom* of a tube or open framework. The mirror reflects the light back up the tube to form an image near the front end. Because of the difficulty of precisely producing the reflecting surface the reflecting telescope did not become an important astronomical tool until the time of William Herschel, a century later.

With a reflecting telescope the problem of image accessibility arises, because a concave mirror produces the image in front of the mirror, in the path of the incoming light. For some types of image analysis, bulky equipment must be placed at the telescope focus. There are, however, various arrangements for getting at the focus; which one is adopted depends on the type of telescope and on the purpose for which the image is to be used.

The place where the image is formed by the mirror, shown in Figure 11.14 (a), is the *prime focus*. If the image is to be photographed, a small plate or film holder can be suspended at the prime focus in the middle of the mouth of the telescope tube. The plate holder generally blocks out only a small fraction of the incoming light, so that the brightness of the image is only slightly dimmed. However, if the image is to be analyzed otherwise, the observer or the necessary apparatus may block out so much incoming light that the prime focus is impractical. Thus the light is usually diverted before it comes to a focus so that the image is formed in a more convenient location.

In the *Newtonian* reflecting telescope, the problem is solved by a flat mirror mounted diago-

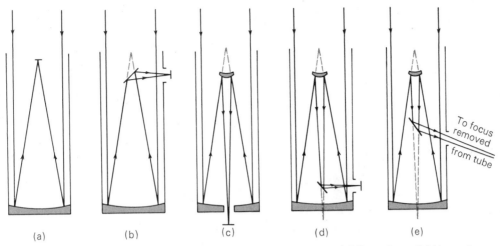

Figure 11.14 Various focus arrangements for reflecting telescopes: (a) Prime focus. (b) Newtonian focus. (c) and (d). Two types of Cassegrain focus. (e) Coudé focus.

nally in the middle of the tube so that it intercepts the light just before it reaches the focus. The mirror reflects the light to the side so that it comes to a focus just outside the tube, as in Figure 11.14 (b).

Another arrangement for a reflecting telescope is the *Cassegrain* system, in which a small convex mirror rather than a flat mirror is suspended in the telescope tube. The convex mirror intercepts the light before it reaches the prime focus and reflects it back down the tube of the telescope. In many reflecting telescopes (for example, the 5-m Hale telescope) a hole is provided in the center of the objective mirror so that the light reflected from the convex mirror can form an image behind the objective. If the objective has no hole (for example, the 2.5-m telescope on Mount Wilson), the image must be reflected out to the side of the tube with a third, flat mirror, mounted diagonally just in front of it. These arrangements are illustrated in Figure 11.14 (c) and (d).

Finally, in the *coudé* arrangement, Figure 11.14 (e), a convex mirror intercepts the light just before the prime focus is reached. The light is reflected back down the tube until it reaches one of the pivot points about which the telescope tube can be rotated to point to various parts of the sky. There it is intercepted by a flat mirror that reflects the light outside the tube to a stationary observing station. Because the station is not attached to the moving part of the telescope, heavy equipment can be used there.

Most large reflectors in use in astronomical observatories are equipped with auxiliary flat and con-

vex mirrors so that several of the possible focus arrangements can be used.

Schmidt telescopes are used for wide-angle photography in astronomy. The image is formed directly in front of the spherical mirror about halfway from the mirror surface to the correcting lens. Most often the entire system is enclosed in a light-tight tube, and the photographic emulsion, curved to fit the spherical focal surface, is inserted at the focus of the mirror in the center of the tube. The exposure can be started and stopped by uncovering and recovering the correcting lens.

For any type of telescope, the tube must be mounted in such a way that it can turn to any direction in the sky. Nearly all astronomical optical telescopes have *equatorial mounts*. An equatorial mount (see Figure 11.15) allows the telescope to turn to the north and south about one axis and to the east and west about another. The axis for the east-west motion of the telescope is parallel to the axis of the earth's rotation, and the other axis, about which the telescope can rotate to north or south, is perpendicular to this axis.

The two axes of motion of a telescope with an equatorial mount allow the telescope to be turned directly in right ascension and declination, the celestial coordinates in which astronomical positions are generally tabulated. Modern telescopes are equipped with setting circles or displays that indicate the direction or coordinates in the sky toward which the telescope is pointing. An important advantage of the equatorial mount is that a simple slow motion of the telescope about its axis parallel

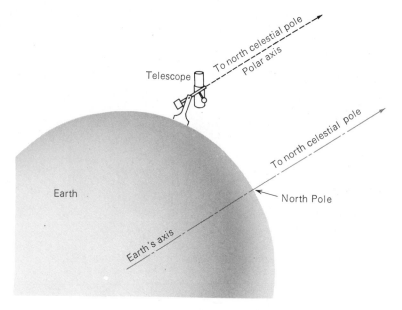

Figure 11.15 Equatorial mount.

to the earth's axis—the *polar axis* of the telescope— is sufficient to compensate for the apparent motions of the stars across the sky that result from the earth's rotation. A mechanical or electric mechanism called a *clock drive* is usually employed to drive the telescope automatically about its polar axis, so that it will follow the stars accurately as they move across the sky.

An equatorial mount, however, is a disadvantage for a very large telescope because it is difficult to adjust for the gravitational stress on its heavy mirror. Consequently, the world's largest optical telescope, the 6-m reflector in the Soviet Union, has an altitude-azimuth mount in which the telescope rotates simply about one vertical and one horizontal axis. Other even larger telescopes now under design will be similarly mounted, and so are large radio telescopes (next chapter). It is more complicated for them to compensate for the earth's rotation and track the stars, but today that problem can be handled with sophisticated computer-controlled driving mechanisms.

(f) Housings for Astronomical Telescopes

The conventional housing for an astronomical telescope is a hemispherical dome. The dome usually has an oblong window or slot on one side, extending from the spring line of the dome (its "base") to the top of the dome. Some kind of shutter can close over the window to protect the telescope during daylight and bad weather, when it is not in use.

The dome is generally mounted on rails so that the window can be turned to any direction. In a modern observatory, the dome usually turns automatically to keep the slot always oriented in the direction in which the telescope points. The domes of the largest telescopes are usually well insulated against heat, so that the interior can be maintained at nighttime temperatures, thus preventing rapid temperature changes from distorting the critical shape of the telescope mirrors or lenses.

(g) Advantages of Various Kinds of Telescopes

Refracting, reflecting, and Schmidt telescopes all have their special advantages. The choice of the kind of telescope to build or use depends on the type of project to be worked on.

Refracting telescopes are usually constructed with long focal lengths and relatively slow speeds ($f/12$ to $f/16$). In such instruments comatic aberration and distortion can be kept small, resulting in a larger field of view for these telescopes than for typical reflectors. Refractors are best suited for *astrometry*, the measurement of accurate positions and small angles in the sky. Also, because the image is formed at the back of the tube of a refractor, it is convenient for visual observations. On the other hand, refractors suffer from chromatic aberration, and can be constructed to perform well only for a limited spectral region. Because a refracting lens must contain at least two elements to produce good images, there are at least four surfaces of glass to be ground and polished. Furthermore, a lens can be supported only along its periphery; a very large lens sags of its own weight and distorts its shape.

A reflecting telescope, on the other hand, utilizes a mirror that has only one optical surface to be perfected. Because the light does not go through the mirror, this need not be made of optically perfect glass; furthermore, a mirror is easier to support, for it can be braced at all points along its back. Thus it is feasible to make much larger mirrors than lenses; the largest objective lens in use is 102 cm in diameter, whereas optical-quality mirrors of aperture up to 6 m have been fabricated. Because it is easier to construct reflecting telescopes, most homemade telescopes built by amateur astronomers are of this type.

Because the ordinary reflecting telescope is free of chromatic aberration, it is better suited to study of the spectra of stars. Reflecting telescopes of good optical quality can be made with relatively short focal lengths—that is, with high optical speeds—so they can photograph faint surfaces in shorter times than refracting telescopes can.

The Schmidt telescope is a compromise between reflectors and refractors. Schmidts can be made to produce excellent images over a large field of view. The nearly identical 124-cm (48-inch) Schmidts on Palomar Mountain and at Siding Spring, Australia, can record an area of the sky the size of the bowl of the Big Dipper on a single photograph, whereas with the 5-m reflector, also at Palomar, about 400 photographs are required to cover the same area. Schmidt telescopes are consequently especially useful for surveying. However, the correcting lens of a standard Schmidt system has nonspherical surfaces and is difficult to make.

(h) Some famous Optical Telescopes

The "size" of a telescope refers to the diameter of its lens or mirror, or of the correcting lens in the case of a Schmidt. Thus, a 15-cm telescope has a lens or mirror 15 cm in diameter.

The largest existing refractor is the 102-cm (40-inch) refracting telescope at the Yerkes Observatory of the University of Chicago. The Yerkes Observatory is located at Williams Bay, Wisconsin. The second largest refractor is the 91-cm (36-inch) telescope at the Lick Observatory of the University of California, on Mount Hamilton, California. These big refractors were built long ago, before the technique of building large reflectors was perfected.

The world's largest optical-quality telescope is the 6-m reflector at Mount Pastukhov in the Caucasus, U.S.S.R. (Figure 11.17). To support the 42-ton mirror, the telescope has an altitude-azimuth mounting, as described above. Until the 6-m telescope went into operation in the mid-1970s, the largest telescope was the 5-m (200-inch) Hale reflec-

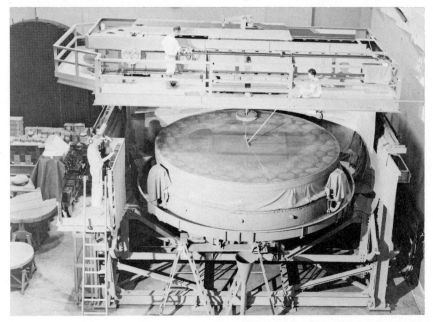

Figure 11.16 The 200-inch (5-m) mirror of the Hale telescope, shown on the polishing machine at the California Institute of Technology. *(California Institute of Technology/Palomar Observatory)*

Figure 11.17 The Special Astrophysical Observatory of the Academy of Sciences of the USSR, on Mount Pastukhov in the Caucasus, home of the 6-m telescope, the world's largest telescope. (*Fotokhronica TASS—Sovfoto*)

tor on Palomar Mountain,* California. At the prime focus of the Hale telescope is a cage 1.8 m across in which the observer can ride while he makes his observations. The Hale telescope is one of the few existing ones in which the astronomer is carried "inside" the telescope.

An interesting recent innovation is the multiple-mirror telescope (MMT) put into operation in the 1970s by the Smithsonian Astrophysical Observatory and the University of Arizona on Mount Hopkins, about 64 km south of Tucson. The MMT (Figure 11.18) utilizes six 1.8-m reflectors clustered about a common axis so that they point together to the same place in the sky. By multiple reflections, the images from the six telescopes are combined to

a single focus for analysis. It is easier to build six small telescopes than it is to build a single large one; yet the MMT has the light-gathering power equivalent to a single 4.5-m telescope. Although the multiple-mirror system does not achieve the resolution of a single large telescope, for many purposes it is an admirable substitute.

The largest single optical telescope currently under consideration is the ten-meter reflector being designed by the University of California. According to present plans, the mirror will consist of an array of more than 30 individual hexagonal segments, each 1.8 m in diameter, under continual adjustment by special activators to keep the overall surface of the mirror in the correct shape. Its unique features, at the forefront of modern telescope technology, make the 10-meter one of the most exciting projects ever contemplated in ground-based optical astronomy.

*Incidently, contrary to common misuse (even by astronomers) there is no such place as "Mount Palomar." The name of the place is Palomar Mountain; "Palomar" is an Indian word meaning "pigeon roost."

Figure 11.18 The Multiple Mirror telescope (MMT) on Mount Hopkins. The MMT is an array of six 1.8-meter mirrors, from which the light is brought to a common focus. *(Courtesy MMT Observatory, Smithsonian Institution/University of Arizona)*

One of the largest Schmidt telescopes is the 124-cm Schmidt of the Palomar Observatory† (Its correcting lens actually has an aperture of 49 in. rather than the 48 in. usually quoted.) It can photograph an area of the sky 6°6 square on a single 35-cm square photographic plate, and it is ideal for surveying the sky. The first main project assigned to the Schmidt after its completion in 1949 was to produce a photographic atlas of the sky. This sky survey, financed by the National Geographic Society, and called the National Geographic Society-Palomar Observatory Sky Survey, took seven years to complete. It provides the most comprehensive coverage of the sky available. The Palomar Schmidt, however, cannot reach the southern quarter of the sky.

In 1973, therefore, two other large Schmidt telescopes went into operation in the Southern Hemisphere. The 100-cm Schmidt of the European Southern Observatory (ESO) is at La Silla, Chile (600 km north of Santiago). The 122-cm Schmidt at Siding Spring Observatory, Australia, is a near copy of the Palomar Schmidt. These two southern Schmidts are cooperating in a venture to supplement the Palomar Sky Survey with a similar one for the Southern Hemisphere skies. Table 11.1 lists some of the more important optical astronomical observatories.

11.2 OPTICAL OBSERVATIONS

The popular view of the astronomer is of a person in a cold observatory peering through a telescope all night. Most astronomers do not live at observatories but near the universities or laboratories where they work. A typical astronomer might spend a total of only a few weeks a year observing at the telescope, and the rest of the time measuring or analyzing his or her data. Many astronomers work only with radio telescopes or with space experiments. Still others work at purely theoretical problems and never observe at a telescope of any kind. Even optical astronomers seldom inspect telescopic images visually except to center the telescope on a desired region of the sky or to make adjustments. On the contrary, the image is generally utilized in one of many other ways.

---

†The universal telescope of the Tautenburg Observatory can be used as a Schmidt by inserting a removable 135-cm correcting lens; it then becomes the world's largest Schmidt.

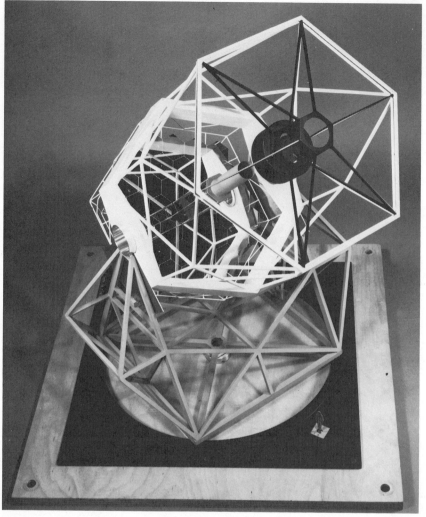

Figure 11.19 Scale model of the 10-meter telescope, now under design by the University of California. The mirror is a mosaic of 36 hexagonal segments. Although larger than any other existing optical telescope in aperture, its stubby construction will allow it to be housed in a dome smaller than that of the 5-meter telescope on Palomar Mountain. *(University of California)*

(a) Projection on a Screen

The most direct way to use the image is to place a screen, for example, a white card, at the focus of the telescope and allow the image to fall on it. This is the most convenient way to view the sun, for example. However, a visual image does not provide a permanent record that can be measured and studied later in the laboratory. Therefore, it is far more common to photograph the image instead of viewing it directly.

(b) Telescopic Photography

To photograph the image, the screen is replaced by a photographic plate or film. The image then forms on a light-sensitive coating which, when developed, provides a permanent record of the image—one that can be measured, studied, enlarged, published, and inspected by many individuals. When used for photography, a telescope becomes nothing more than a large camera; the lens or mirror of the telescope serves as the camera lens.

Table 11.1 **Some Major Optical Observatories**

LOCATION	INSTITUTION	LARGEST TELESCOPES
Mt. Pastukhov, Caucasus, U.S.S.R.	Soviet Special Astrophysical Observatory	6-m (236-inch) reflector
Palomar Mountain, California	Palomar Observatory (California Institute of Technology)	5-m (200-inch) Hale reflector
		1.5-m (60-inch) reflector
		1.2-m (48-inch) Schmidt
Canary Islands	La Palma Observatory (United Kindgom)	4.2-m (165-inch) reflector—under construction
		2.5-m (98-inch) Newton reflector
Kitt Peak, Arizona	U.S. National Observatory	4-m (158-inch) Mayall reflector
		2.1-m (84-inch) reflector
		1.5-m (60-inch) McMath solar telescope
	University of Arizona	2.3-m (90-inch) reflector
Cerro Tololo, Chile	Inter-American Observatory	4-m (157-inch) reflector
		1.5-m (60-inch) reflector
Siding Spring, Australia	Anglo-Australian Observatory	3.9-m (155-inch) reflector
		1.2-m (48-inch) Schmidt
Mount Hopkins, Arizona	U. of Arizona and Smithsonian Astrophysical Observatory	Six 1.8-m (70-inch) mirrors in the multiple-mirror telescope (MMT)
Mauna Kea, Hawaii	U.K. Science Research Council (operated by Royal Observatory, Edinburgh)	3.8-m (150-inch) infrared telescope
	Canada-France-Hawaii Telescope Corporation	3.6-m (142-inch) reflector
	NASA (operated by the U. of Hawaii)	3-m (120-inch) infrared telescope
	U. of Hawaii	2.2-m (88-inch) reflector
La Silla, Chile	European Southern Observatory	3.6-m (142-inch) reflector
		0.9-m (36-inch) Schmidt
Mount Hamilton, California	Lick Observatory	3-m (120-inch) Shane reflector
		1-m (40-inch) reflector
		0.9-m (36-inch) Crossley reflector
		0.9-m (36-inch) refractor
Mount Locke, Texas	McDonald Observatory (U. of Texas)	2.7-m (107-inch) reflector
		2.1-m (84-inch) reflector
Crimea	Crimean Astrophysical Observatory (U.S.S.R.)	2.6-m (102-inch) reflector
Armenia	Byurakan Observatory	2.6-m (102-inch) reflector
Las Campañas, Chile	Mount Wilson and Las Campañas Observatories	2.6-m (101-inch) du Pont reflector (L. C.)
Mount Wilson, California		2.5-m (100-inch) Hooker reflector (M. W..)
		1.5-m (60-inch) reflector (M. W.)
Wyoming	Wyoming Infrared Observatory	2.3-m (92-inch) infrared telescope
Williams Bay, Wisconsin	Yerkes Observatory (U. of Chicago)	1-m (40-inch) refractor

The most important advantage in using a telescope as a camera is that photographic emulsions can accumulate luminous energy and build up an image during a long exposure. Most astronomical objects of interest are remote, and hence the light we receive from them is feeble. However, long-time exposures can be made. Until electronic image intensifiers became available, astronomical exposures often would run hours in length, and occasionally over several successive nights. The longer the exposure, the more faint light gradually accumulates to help build up the photographic image. Objects can be photographed that are about a hundred times too faint to see by just looking through a telescope. The layman is often disappointed when he takes his first look through an astronomical telescope, for what he can see is nothing compared to the spectacular photographs, such as those reproduced in this book, that are the result of long time exposures.

A photographic emulsion consists of a thin layer of gelatin, usually mounted on a base of celluloid or glass. Within the emulsion are suspended silver compounds that undergo a chemical reaction when activated by light, resulting in the formation of metallic silver. When the emulsion is immersed in an appropriate chemical solution (developed), more silver is formed where light started the process and those molecules of silver compounds that were not activated by light are chemically removed, leaving grains of deposited silver in the parts of the emulsion that were exposed.

Astronomers, like all photographers, have their problems. They must determine the proper exposure to use and the best way to develop photographic plates. (For scientific purposes, photographic emulsions are usually spread on glass plates rather than on celluloid film, for the glass will not curl or stretch and is better for accurate measurements.) During a time exposure, the telescope must be carefully "guided" to keep it centered on the object being photographed. As mentioned earlier, telescopes are generally driven automatically to follow the gradual motion of the stars that results from the earth's rotation. However, minor irregularities in the driving mechanism of the telescope and motions of the image caused by atmospheric disturbances must be continually corrected for. Traditionally, astronomers would sight a star through an auxiliary telescope mounted securely alongside the main telescope tube, or sometimes they used a star image produced by the main telescope objective, but outside the field of view that they were photographing. They kept this star centered on illuminated cross hairs by moving the telescope slightly from time to time, thus assuring that the main telescope is tracking properly. In modern observatories, however, the human eye is usually replaced with an automatic photoelectric guiding mechanism.

(c) Visual Inspection with an Eyepiece

Occasionally it is desirable to inspect a telescopic image visually, doing what is commonly called "looking through" a telescope. To best inspect detail in the image, it is customary to view it with a magnifying lens. A common hand magnifier could be used for this purpose. A telescopic *eyepiece* or *ocular* is simply a high-quality magnifying lens that is used to view the image. Figure 11.20 shows how an eyepiece is used in a refracting telescope and in a Newtonian reflector. In practice, eyepieces usually have two or more elements to reduce aberrations.

When an extended celestial object is viewed through a telescope equipped with an eyepiece, it appears enlarged, that is, closer, than when viewed naturally. The factor by which an object appears larger (or nearer) is called the *magnifying power* of the telescope. For example, the moon appears to subtend an angle of $\frac{1}{2}°$ when viewed with the naked eye. If, when viewed through a particular telescope, the moon appears to subtend $10°$, the magnifying power of the telescope is 20.

Everyone knows that if he has to read fine print he must use a magnifying glass of higher power than when he reads the print in a newspaper. A higher power magnifying glass is one of shorter focal length. Similarly, eyepieces of different focal lengths, used in conjunction with the same telescope objective, produce different image magnifications. It is the purpose of the objective of a telescope to produce an image; it is the purpose of the eyepiece to magnify the image to the point where details in it can be viewed. In principle, any desired

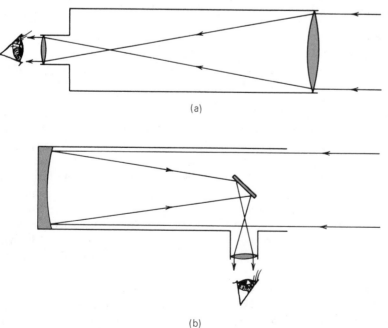

(a)

(b)

Figure 11.20 Use of an eyepiece. (a) With a refracting telescope. (b) With a reflecting telescope.

magnification can be obtained if an eyepiece of sufficiently short focal length is used. Therefore, it does not make sense to ask an astronomer what "power" his telescope is. The power can be changed at will by using different eyepieces. At every observatory there is a collection of eyepieces of different focal lengths that can be interchanged for various magnifications.

The exact value of the magnifying power can easily be calculated by dividing the focal length of the objective of the telescope (f_o) by the focal length of the eyepiece (f_e); that is,‡

$$\text{magnifying power} = \frac{f_o}{f_e}.$$

The highest useful magnification is that at which the finest details resolved in the image can be viewed. Additional magnification reveals no further detail but merely enlarges the blur of the diffraction pattern. Atmospheric disturbances further limit useful magnification, especially in large telescopes (Section 11.4).

(d) Measurement of an Image

One way of measuring the image produced by a telescope is with a *filar micrometer*. In this device, two

---

‡The formula for magnifying power can be verified by the accompanying figure. Light rays passing through the center of a lens are not deflected. Thus the angle θ subtended by the image at the objective is the same as the angle subtended by the object (presumed to be very distant) as seen from the objective (or to the naked eye). As viewed through the eyepiece, however, the image appears to have an angular diameter ϕ. From inspection,

$$\sin \phi = \frac{AB}{f_e} \quad \text{and} \quad \sin \theta = \frac{AB}{f_o}.$$

For small angles, $\sin \theta = \theta$, and $\sin \phi = \phi$; therefore,

$$\frac{\phi}{\theta} = \frac{f_o}{f_e}.$$

By definition, ϕ/θ is the magnifying power.

illuminated parallel cross hairs are placed at the focus of the telescope where the image is formed. The image with the cross hairs is viewed through a high-power magnifying eyepiece. The separation of the cross hairs can be varied by turning a precision screw, and the exact distance between them is indicated by the position of the screw. Thus the diameter of the image of a planet, or the distance between the images of two stars, can be measured with the instrument. The linear size of an image depends on its angular size in the sky and on the focal length of the telescope (see Section 11.1b). Therefore, the filar micrometer provides a means of measuring small angular sizes and separations in the sky. The instrument can also be used to measure the direction of one celestial object from another nearby one. The filar micrometer is seldom used in modern astronomy, but in the past it was particularly valuable for following the revolution of visual binary stars about each other (Chapter 26), and it is still an instructive class laboratory instrument.

(e) Spectroscopy

Spectroscopy occupies at least half of the available observing time of most large telescopes. The technique of spectroscopy is made possible by the phenomenon of dispersion of light into its constituent wavelengths. We saw in Chapter 10 that white light is a mixture of all wavelengths and that these can be separated by passing the white light through a prism, producing a spectrum.

A spectrum can also be formed by passing light through, or reflecting it from, a *diffraction grating*. These are of two kinds: A *transmission grating* is a transparent sheet or plate (such as glass) that is ruled with many fine, parallel, closely spaced scratches (typically, thousands per inch), so that light can pass through the material only between the scratches. A *reflection grating* consists of a shiny reflecting surface, say a mirror, upon which are ruled similar fine scratches or grooves, so that light can be reflected only from the places on the surface

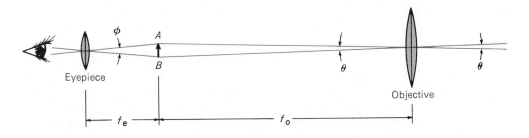

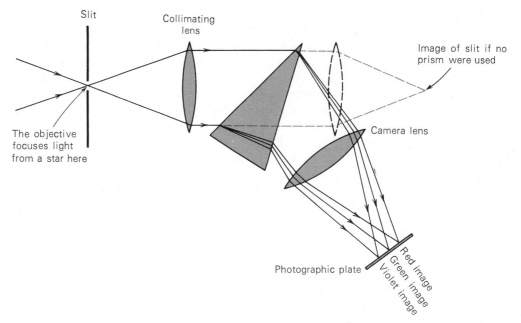

Figure 11.21 Construction of a simple prism spectrograph.

between the scratches. In either kind of grating, light emerges as if from a series of long, parallel, finely spaced slits. Because of diffraction, the light fans out in all directions as it passes through these "slits." The waves of light from different slits interfere with each other as described in Section 10.12. At any given point beyond the grating, the waves cancel each other out except at one specific wavelength. With an appropriate lens system, successive wavelengths of light can be focused at successive points along a line parallel to the surface of the grating, and in a direction perpendicular to the scratches in it. A spectrum can thus be produced by a grating as well as by a prism.

A *spectrograph* is a device with which the spectrum of a light source can be photographed or otherwise recorded. Attached to a telescope, the spectrograph can be used to record the spectrum of the light from a particular star. The construction of a simple spectrograph is illustrated in Figure 11.21. Light from the source enters the spectrograph through a narrow slit and is then collimated (made into a beam of parallel rays) by a lens. In Figure 11.21 the collimated light is shown entering a prism, but it can just as well pass through or be reflected from a grating. Different wavelengths of light leave the prism (or grating) in different directions, because of dispersion (or diffraction). A second lens placed behind the prism forms an image of the spectrum on the photographic plate. If the light

from the collimating lens had been passed directly into the second lens, rather than through the prism first, the second lens would simply produce an image of the slit through which the light entered the spectrograph (dashed lines in Figure 11.21). However, because of the dispersion introduced by the prism, light of different wavelengths enters the second lens from slightly different directions, and consequently the lens produces a different image of the slit for each different wavelength. The multiplicity of different slit images of different colors lined up at the photographic emulsion is the desired spectrum. In modern practice, the spectrum is often measured and recorded electronically rather than photographically, but the optical principle is the same. If the photographic plate (or other recording device) is removed and an eyepiece is used to inspect the spectrum, the whole instrument becomes a *spectroscope*.

All of the parts of a spectrograph—slit, two lenses, prism (or grating), and photographic plate—must be attached to a rigid framework so that they do not shift during long exposures. The whole spectrograph may weigh hundreds of pounds and require a lot of space behind the focus.

An image of the spectrum shows the star's light spread into a streak with short wavelengths (violet) at one end and long wavelengths (red) at the other. Absorption lines in the spectrum manifest themselves as narrow ranges of wavelength where the light intensity is low. Now the entrance slit of the spectrograph is much longer than the width of the star image. In fact, if the stellar image were only a point of light, its spectrum would be a thin line, and it would be hard to recognize the dark lines crossing it. It is common practice, therefore, to move the telescope back and forth very slightly during the exposure so that the image of the star will trail back and forth a short distance along the slit (Figure 11.22). This will cause the spectrum to be spread out perpendicular to the dispersion of wavelengths to a width that is great enough to facilitate measurement of details in the spectral image.

Now we need to know at just what wavelengths the dark lines and other spectral features occur. Thus it is also necessary to record some wavelength reference. Astronomers usually do this several times while recording the star's spectrum. They temporarily interrupt the exposure on the star by covering the part of the slit through which the star's light is allowed to enter the spectrograph; that part of the slit is shown as "*A*" in Figure 11.22. Then the light from an arc lamp or other laboratory source (whose spectrum consists only of bright emission lines) is passed briefly through the parts of the slit ("*B*" in Figure 11.22) on either side of the part used to admit light from the star. The spectrum of the laboratory reference source thus appears as a series of bright lines flanking the spectrum of the star. The wavelengths of the various parts of the star's spectrum match precisely those of the bright lines in the adjacent spectra of the reference source.

"Singling Out" the Star

The images of many stars appear in the focal plane of a telescope. However, it is no problem to single

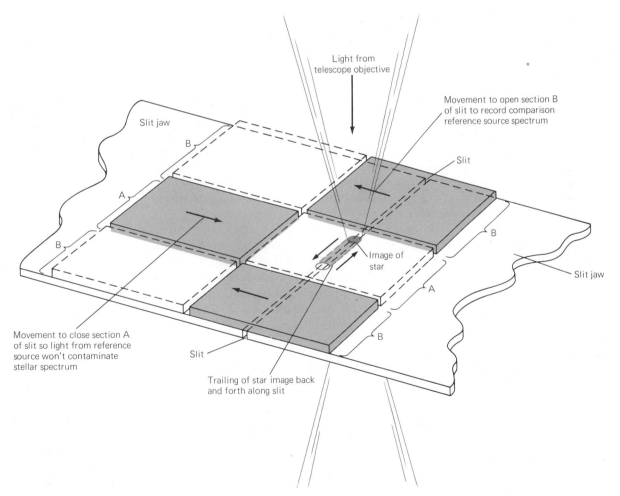

Figure 11.22 Different parts of the spectrograph slit are used to admit light from a star and from a comparison source.

out the star whose spectrum is desired. The entire spectrograph is attached to the telescope; a metal plate containing the small slit of the spectrograph is centered on axis at the focus. The desired star image must now lie on that slit so that its light will pass into the spectrograph. In front of the slit a flat mirror can usually be flipped into the telescope beam at 45°, so that the astronomer can examine the pattern of stars and pick out the one he wants. Then the whole telescope can be moved slightly so that star's image is centered on the slit.

(f) The Objective Prism

An objective prism is a special plate of glass that may be placed in front of the objective lens of a telescope. This plate is relatively thin; both of its faces are flat but at a slight angle to each other, so that, in effect, it is a thin prism. Starlight entering the telescope must first pass through the objective prism and is slightly dispersed. With the objective prism, if a photograph is taken of a field of stars, the image of each star is not a point but rather a small spectrum. In this way, the spectra of many stars can be photographed at once. The spectra are too small and of too low a dispersion (the colors are not widely spread out) to reveal much detail, and a comparison spectrum (a spectrum of a laboratory light source) cannot be photographed along side each stellar spectrum to provide a wavelength standard. Nevertheless, enough of the major features of the spectra show to enable the stars to be classified roughly according to their spectra.

(g) Photoelectric Photometry

An important datum about a star is the amount of light we receive from it, or its *magnitude*. The matter will be discussed in detail in Chapter 24. At present, the most accurate device for measuring stellar radiation received at the telescope is the *photomultiplier*, a light-sensitive electron tube placed at the focus of the telescope. Light from a star is gathered by the telescope and is focused on a light-sensitive surface just inside the glass envelope of the tube. A photon striking this *photocathode* dislodges an electron, which is attracted to a positively charged element of the tube called a *dynode*. When the speeding electron strikes the dynode, four or five secondary electrons are dislodged. These in turn are attracted to a second dynode, where each secondary electron again dislodges four or five more electrons. After passing through ten or eleven dy-

nodes, the original *photoelectron* has been amplified to a *pulse* of the order of a million electrons. This pulse is detected electronically and recorded as a *count*. The photomultiplier can thus be used to count the number of incident photons; the number recorded in a specific time is a measure of the star's apparent brightness. (Not all incident photons dislodge a photoelectron; hence the counts must be corrected for the *quantum efficiency* of the photomultiplier.)

Other radiation-sensitive devices in which incident photons induce or control electric currents are semiconductors whose resistance changes with changes in their temperature or because electrons are freed from their bound positions by absorption of photons. We discuss their use in the following section.

11.3 ADVANCES IN DETECTORS

The photographic plate is a superb device for collecting a large amount of information, but it does have limitations. The photographic emulsion, for example, is non-linear—that is, equal differences in exposure do not produce equal differences in the blackening of the emulsion. Moreover, the photographic plate saturates (turns black) with long enough exposure, and no additional data can be recorded. Consequently, it is far from ideal for photometry—the measuring of the brightnesses of astronomical objects.

Another limitation is resolution. The resolution of the photographic plate is limited by the graininess of the emulsion; on typical emulsions used in astronomy the grain size is 20 to 30 microns (0.02 to 0.03 mm). These smallest picture elements are called *pixels*. Finer-grain emulsions exist (that is, with smaller pixels), but they are less efficient and require much longer exposures, with consequent cost in valuable telescope time.

There are ways of increasing the sensitivity of photographic plates, for example, by baking them at 60°C or by soaking them in ammonia, nitrogen, or hydrogen. But at best the quantum efficiency of a photographic plate is only about one percent; it requires, typically, a hundred incident photons to make a pixel developable. It has long been desirable, therefore, to find a more ideal detector—one that is linear and does not saturate, one that can yield higher resolution, and one that has high quantum efficiency.

(a) Photocathodes

We have already mentioned the photoelectric effect (Section 10.3), whereby a photon of greater than a threshold energy can dislodge an electron from certain photoemissive surfaces. If that surface is given a negative electrical potential with respect to another surface (an anode), it becomes a photocathode, and the freed electrons are attracted to the more positive anode, as in a photocell (described in the previous section). A photocell (or photomultiplier), however, only records the amount of light energy and does not produce an image of the astronomical object.

In the type of television camera tube known as an SEC vidicon, on the other hand, as a free electron moves (through a vacuum) from the photocathode to the positive target (anode), it is magnetically focussed, so that the place where it strikes the target corresponds to the position at which the original photon struck the photocathode. The anode is a thin wafer of potassium chloride, which exhibits a gain mechanism called *secondary electron conduction* (hence the name SEC vidicon): For each impinging electron, about a hundred appear on the opposite side of the wafer. Thus a charge pattern builds up that corresponds to the optical image on the photocathode. These charges are read off by scanning the surface with an electron beam; where the charge is highly negative the electrons in the scanning beam are repulsed, but where the charge is low the electrons from the beam can flow onto the surface. The consequent variations in the current in the beam are the record of the stored image. In the U.S. television system, the scanning beam traces across the target in 525 parallel scan lines each 30th second (in the British system, there are 625 scan lines traced in every 25th second).

In commercial television cameras the image cannot be accumulated on the target for long exposure times, and the system is not very linear. In short, commercial television is not up to the exacting standards required in astronomical applications. Modifications of the principle, however, have led to very successful devices.

(b) Image Tubes

The earliest *image intensifier tubes* (or simply *image tubes*) used in astronomy were invented by Professor A. Lallemond in Paris. The photographic emulsion is far more sensitive to high-speed electrons than to photons, so Lallemond arranged that the electrons dislodged from the photocathode are electronically focussed directly on a photographic plate. This meant that the cathode and plate had to be encased in the same glass tube, which was then evacuated. Actually, a magazine of several plates could be enclosed in the vacuum, and the plates changed automatically from outside the tube; even so, to retrieve them the glass tube had to be broken, and this also destroyed the cathode. In the *electronographic camera*, a modification by G. Kron, the electrons pass out of the tube through a very thin mica window and expose the photographic plate that is placed in contact with the window. Although these early devices were successful, they were difficult to use.

In a newer type of image tube, much easier to use, the photographic plate is replaced by a phosphor screen that is set aglow by the impinging electrons (as in a television picture tube). The electrons from the photocathode are focussed onto the screen either electrostatically or magnetically. The picture is sometimes conducted from the glowing phosphor to a more convenient location by a bundle of fiber optics (optical fibers). It can then be photographed directly. Since a single incident photon can dislodge an electron at the photocathode, and that single electron can set a phosphor aglow, the system is far more sensitive than a photographic plate alone. Even higher sensitivity can be achieved by passing the picture from one image tube directly into another one and amplifying it again. Sometimes several such tubes are ganged in series.

(c) Use of Modern Detectors in Spectroscopy

So far the greatest application of modern detectors has been to record the spectra of faint sources with relatively short exposure times. One of the most successful systems was developed by L. B. Robinson and E. J. Wampler at the Lick Observatory. The image of the spectrum is first amplified by a series of three image tubes; the cathode of the first tube is placed directly at the focus of the telescope's spectrograph, replacing the photographic plate. The greatly enhanced image is then focussed on the photocathode of a device called an *image dissector*. The image dissector is similar to a photomultiplier, except that at any given time only electrons emitted from a small spot on the photocathode are selected

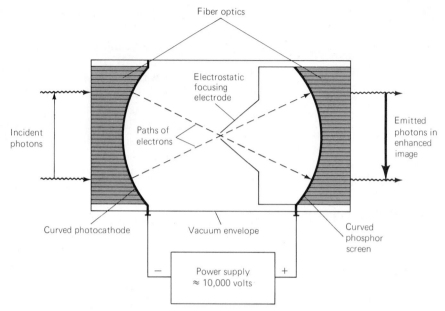

Figure 11.23 Sketch of a single-stage electrostatic image tube.

by a magnetic field for amplification by the dynodes and pulse amplifier. The magnetic field is varied periodically so as to scan along the spectrum. The time variation of the output from the pulse amplifier thus corresponds to the variation in intensity at different wavelengths along the spectrum. The intensity at each of 2048 different points along the spectrum is stored in a computer, and as the spectrum is continually rescanned the stored data gradually accumulate until a strong enough record of the spectrum is built up. The spectral intensity in the 2048 pixels can be printed out as a graph, displayed on a television screen, or stored on magnetic tape for later analysis.

Another widely used scheme was developed by S. Shectman at the Mount Wilson Observatory and W. A. Hiltner at the University of Michigan. Their system uses six stages of amplification and finally images the spectrum on an array of silicon diodes known as a *Reticon* array (for the company that manufactures it). Each silicon diode has the property that an impinging photon can give rise to an internal current flow of at most one electron. Every millisecond (0.001 s), under control of external "clocking" signals, the Reticon turns on and off successive switches on the array and connects each diode in turn to an amplifier that measures the charge necessary to restore the diode to its original voltage—hence the amount lost due to the previous

exposure to light. The diodes themselves and the electronic readout switches are all mounted on an integrated circuit produced on a small slice of material roughly the size of a postage stamp, called a *chip*. One Reticon chip much used in spectroscopy has 936 separate elements (diodes), each of which is a single pixel. The elements, 375 by 30 microns in size, are mounted side by side, producing a detecting surface 28 mm long, ideal for recording the light at various points along a spectrum imaged on it.

Still another detector that has been used in spectroscopy, called the *Digicon*, uses silicon diodes to detect directly the electrons emitted from the photocathode. Each diode conveys its signal current to a separate amplifier. A photoelectron is accelerated by a high voltage, as in an image intensifier, and produces an "avalanche" of about 5000 charges when it strikes a diode, thus giving rise to an easily detected pulse of electrical current. A variant of this device uses a self-scanned array such as a Reticon instead of the discrete diodes with individual wires, and is therefore known as a *self-scanned Digicon*.

(d) Two-Dimensional Arrays

With the exception of image-intensified photography and television camera tubes, so far we have described the application of low light level detectors only to spectroscopy. The natural question is whether they can also be applied to record directly the two-dimensional image produced by a telescope. There are currently two kinds of two-dimensional silicon photoarrays suitable for astronomical imaging in the visible and near-infrared parts of the spectrum. The *charge-injection device (CID)* and the *charge-coupled device (CCD)* are fabricated from single pieces of silicon. In operation, photosensitive regions, referred to as *field-induced photodiodes*, are created by application of appropriate voltages. The two arrays differ in the way the accumulated charge in each field-induced photodiode (corresponding to each pixel) is read out. In the CID the charge on any individual pixel can be read out at will without erasing the information stored there. In the CCD, during readout the charge in each pixel is passed on to the next one in the same row, and then to the next, and so on, like a bucket brigade. As each pixel's charge reaches the end of the row it is transferred to a low-noise on-chip amplifier, and the amplifier signals are converted to digital codes and stored in the memory unit of a computer. The CID, on the other hand, requires a sensitive low-noise amplifier external to the device. The user can select among various options for the reconstruction and display of the image, once the set of digital codes has been stored in the computer memory.

Of the two array types, CCDs are perhaps the most attractive for most astronomical applications, because of their low noise. But they are expensive to produce, and so far most of those that have been produced have been for the space program or for military uses. They are, however, beginning to become available to some observatories and are used for direct imaging at telescopes. CCDs are available on a limited basis as square matrices of 800 by 800 (or 640,000 in all) pixels, and cover an area somewhat more than one square centimeter. Slightly larger arrays have been made on an experimental basis, and it is hoped that these and still larger arrays will be available in the future.

We do not, on the other hand, expect to see CCDs displace the use of photography entirely in the foreseeable future. Photographic plates are inexpensive and easy to use, and those being used in current sky surveys (at Palomar and Siding Spring, Australia) are 35 cm square and contain more than 2×10^8 pixels. No other detector now on the horizon can approach this ability for collecting data at one time.

11.4 ATMOSPHERIC LIMITATIONS

The earth's atmosphere, so vital to life, is the biggest headache to the observational astronomer. In at least five ways the air imposes limitations upon the usefulness of telescopes.

1. The most obvious limitation is weather—clouds, wind, rain, and the like. Even on a clear, windless night, however, the atmospheric conditions may render a telescope virtually useless for many types of work.

2. Even for visible wavelengths the atmosphere filters out a certain amount of starlight and dims the stars slightly. For wavelengths longer than the near infrared out to the radio wavelength of about 1 cm, the air is almost opaque, and it is opaque also to all wavelengths shorter than 2900 Å. Thus, most of the electromagnetic spectrum remains invisible to ground-based observatories.

3. In the daytime, the air molecules scatter about so much sunlight that the resulting blue sky hides all celestial objects except the sun, moon, and Venus. At night the sky is darker, but never completely dark.

Figure 11.24 A CCD (charge-coupled device) wafer containing 169 CCDs. The wafer is three inches in diameter and is used in communication filter circuits. *(Courtesy of Bell Laboratories)*

Near cities the air scatters about the glare from city lights, producing an illumination that hides the faintest stars and limits the depths that can be probed by telescopes as well as by the naked eye. Even starlight scattered by the air contributes to the brightness of the night sky.

4. To make matters worse, the air emits light of its own. Charged particles from the sun and beyond, funneled into the atmosphere in the polar regions by the earth's magnetic field, set the air aglow and cause auroras. In addition, at night there is an airglow all over the earth, for atoms in the upper atmosphere are ionized by ultraviolet photons from the sun and fluoresce. The brightness of this airglow varies from time to time. During times of maximum solar activity, the night sky brightness may be two or three times normal (see Chapter 31). The faint illumination of the night sky limits the time that a telescopic photograph can be exposed without fogging and thus also limits the faintness of the stars that can be recorded.

5. Finally, when the air is unsteady, star images are blurred. In astronomical jargon, the measure of the atmospheric stability is the *seeing*. When the seeing is good, the stars appear as sharp points and fine detail can be seen on the moon and planets. When the seeing is bad, images of celestial objects are blurred and distorted by the constant twisting and turning of light rays by turbulent air. It is the variations in density of the upper atmosphere that causes stars to "twinkle."

(a) "Seeing"

Because stars present too small an angular size for existing telescopes to resolve their disks, their images should be geometrical points or rather tiny diffraction patterns. But in bad seeing, a star may appear as a "mothball" or as a dancing firefly. The star image usually looks "bigger" in bad seeing, but not because the telescope has magnified it, only because the air has distorted it. Use of higher magnification only magnifies these atmospheric disturbances. Under typical observing conditions, a star image appears to be from 1 to 3 inches in diameter. This is called the "seeing disk" of the star, and of

course has nothing to do with the star's actual size. Under the very best conditions obtainable, star images seldom have diameters less than about $1/4$ inch.

The cells of turbulence in the atmosphere that cause seeing may be as small as a few centimeters—occasionally even less. However, they are rarely, if ever, larger than about 75 cm. If a star is viewed through a telescope whose aperture is less than the size of the turbulent seeing cells, its image may appear sharp and display the diffraction pattern characteristic of a telescope of that size, but simply dance around in the field of view as the moving air currents refract the starlight first one way and then the other. This phenomenon is called "hard" seeing. If the seeing cells are smaller than the telescope aperture, on the other hand, the telescope must see the star in several directions at once, so that its image is smeared out into that "mothball" referred to above; starlight passing through different seeing cells produces different parts of the mothball image. This is called "soft" seeing. Because seeing cells are generally less than 75 cm across, large telescopes always have soft seeing, and the size of the disk of soft seeing is seldom less than $1''$ in diameter. The 5-m telescope at Palomar, which can theoretically resolve $0''.05$, thus usually falls short by 20 times of achieving its potential resolving power.

In view of the limitations set by the earth's atmosphere, the advantages of the space telescope (Chapter 12) are obvious. Already in use are telescopes on rockets and satellites. From space vehicles and rockets, important astronomical observations have been made, which could not have been obtained from the ground. Also, balloons have carried men and instruments, or instruments alone, to altitudes of 30,000 m for the purpose of obtaining astronomical observations. At those altitudes, most of the atmosphere is below the observer, and the seeing is much improved.

(b) Speckle Interferometry

One rather modern (since about 1970) method can circumvent the seeing limitation of a large telescope. The technique is *speckle interferometry*. Suppose a very short time exposure (0.01 s or less) is made of a highly magnified image of a star with a telescope of large aperture, and through a filter that admits only a very narrow range of wavelengths, so that the image is nearly monochromatic. In this case the photograph will reveal not one image but a mul-

The 1-m (40-inch) refracting telescope. *(Yerkes Observatory)*

The 122-cm (48-inch) Schmidt telescope at Palomar. *(California Institute of Technology/Palomar Observatory)*

The 5-m (200-inch) telescope. *(California Institute of Technology/Palomar Observatory)*

The observer's cage at the prime focus of the 5-m telescope. *(California Institute of Technology/Palomar Observatory)*

The 3-m (120-inch) telescope of the Lick Observatory. *(Lick Observatory)*

The 4-m telescope of the Inter-American Observatory at Cerro Tololo, Chile. (© *Association of Universities for Research in Astronomy, Inc. The Cerro Tololo Inter-American Observatory)*

The Kitt Peak National Observatory. *(Kitt Peak National Observatory)*

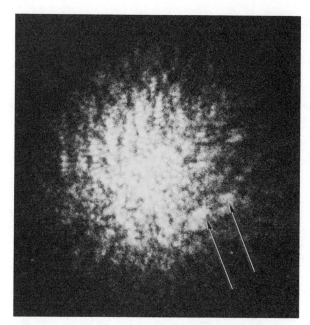

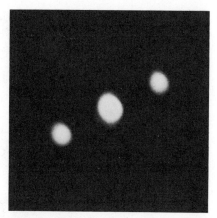

Figure 11.26 The autocorrelation image of the double star Kappa Ursae Majoris, reconstructed from 50 speckle photographs. In effect, each individual speckle image is placed at the position of the bright central spot, which serves as origin. Since each image is that of one of the two members of the binary system, the position of the image of its companion star must lie on one side or the other of the central spot (depending on which star happened to be selected as being at the origin). The companion star's position is thus represented as one of the two fainter spots flanking the central one. The separation of the two flanking spots (or the distance of either from the central spot) reveals the angular separation of the two stars in the sky, and their orientation gives the direction of one star from the other in the sky. (*Figures 11.25 and 11.26 courtesy of Harold McAlister, Georgia State University; the research was carried out at Kitt Peak National Observatory.*)

Figure 11.25 Photograph of a speckle pattern of the double star Kappa Ursae Majoris. There are really two nearly identical speckle patterns, one of each star, that are slightly displaced from each other and overlap. On a normal time exposure, the pattern would blur into the familiar-seeing disk. The eye can actually pick out the speckle pairs, one of which is marked with arrows. It is more accurate and convenient, however, to measure the system by using a composite transform made with an optical processor that passes a laser beam through the negatives of many such speckle photographs taken in rapid succession. The procedure yields what is called an *autocorrelation image* like that shown in Figure 11.26.

titude of images, arranged roughly in a circular pattern, with a greater concentration of images toward the center. The ensemble of images is called a *speckle pattern*. Each minute image in the pattern is that produced by the full aperture of the telescope of the starlight through a single seeing cell. That image may be too faint to reveal a useful picture, but in principle it contains information at the limit of resolution of the telescope. The technique of speckle interferometry involves the combining of many such short exposure photographs and a reconstruction of the true resolution-limited image by computer analysis.

With speckle interferometry, a considerable amount of information has been learned about the orientations of the pairs of stars in certain close binary star systems (Chapter 26). The stars in such a system have an angular separation great enough that they could theoretically be resolved by a large telescope, but that is still far smaller than the sizes of typical seeing disks. The technique has also provided information about a very few stars that are near enough and large enough that their disks would, in principle, be resolved by existing large telescopes but whose angular diameters are very much smaller than seeing disks. Already we have learned something about the distribution of light in a few giant stars, and as the speckle technique is perfected, we may be able to detect large irregularities in the surface brightnesses of some stars—such as gigantic starspots. In particular, speckle interferometry is a promising technique for determining the diameters of large nearby stars (Chapter 26).

(c) Site Selection

Despite the new observations from balloons, rockets, satellites, and space probes, ground-based ob-

Figure 11.27 A composite satellite photograph of the nighttime United States. Note the sky illumination in populated areas. *(Kitt Peak National Observatory)*

servatories will continue to be the workhorses of astronomy for the foreseeable future. But at the same time that we are improving our methods of observation, we are losing sites available for good observations, and existing observatories are losing effectiveness.

Observatory site selection must include consideration of many factors: freedom from cloudy weather much of the time, and also from winds; relatively high altitudes to reduce atmospheric interference and water vapor absorption; steady air (that is, good seeing); moderate latitudes where summer twilight and winter storms do not wipe out observing conditions; and accessibility.

Today we have the additional problem that the best sites are often rendered useless by pollution. Some of this is in the form of smog (already a major problem at the Mount Wilson Observatory), but the worst problem is light pollution. Figure 11.27 is a composite satellite photograph of the nighttime United States. The sky illumination in the populated areas is obvious. The actual situation is even worse. When the conditions for a good observatory site listed above are considered, it turns out that

nearly all satisfactory sites in the United States are in the southwest. Figure 11.27 shows that most of these sites are already gone, as far as astronomical observations are concerned. Mount Wilson has an enormously bright sky due to the lights of Los Angeles, and the sky at the Lick Observatory is illuminated by nearby San Jose and metropolitan San Francisco. At Palomar, once heralded as a magnificent site in the darkest part of the nation, the sky brightness has already increased (since construction of the Observatory) by 50 percent or more because of lights from both San Diego and Los Angeles.

The future of ground-based optical astronomy may well depend on the ability of astronomers to convince the rest of society to preserve some areas (and perhaps some spectral regions) free of light pollution. Some success has already been achieved at Tucson, Arizona, to the great credit of that city's legislature, where, at the urging of astronomers at the Steward and Kitt Peak National Observatories, ordinances were passed to limit the nature of street and commercial illumination. We must hope that this exemplary start will become a model for future development.

EXERCISES

1. Only the Balmer series of hydrogen lines is ordinarily observed in the spectra of stars. Can you suggest an explanation for why the Lyman series, for example, is hard to observe? What about the other series of hydrogen lines?

2. What would be the size of the image of the moon produced by the 5-m telescope (focal length = 16.76 m)? What size image of the moon would be formed by a telescope of the same focal length but with an aperture of only 2.5 m?

3. If the moon can be photographed in one second with a telescope of 50-cm aperture and 100-cm focal length, how long a time would be required to photograph the moon with a telescope of 100-cm aperture and 400-cm focal length?

4. Suppose that all stars emitted exactly the same total amount of light and that stars were distributed uniformly throughout space. Show that the depth in space to which stars could be observed would (ideally) be proportional to the aperture of the telescope used.

5. What is the smallest angle that could theoretically be resolved by the 5-m telescope at a wavelength of 5000 Å?

6. What kind of telescope would you use to take a color photograph entirely free of chromatic aberration? Why?

7. Suppose one wants the moon to appear 45° across when viewed through a telescope of 125-cm focal length. What focal-length eyepiece must be used?

8. Ordinary 7 × 50 binoculars magnify seven times (magnifying power = 7) and have objective lenses of 50-mm aperture (5 cm). For light of 5000 Å, what is the smallest angle that can be resolved by the lenses of binoculars? Could two stars separated by this angle actually be seen as separate stars when viewed through 7 × 50 binoculars? Why?

9. Because of the night airglow, the sky is a faint luminous surface. What determines how long it would take for this sky illumination to expose a telescopic photographic negative to a given blackness? What determines how intense or black the image of a star on the same photograph will be? After a photographic plate has been exposed to a certain blackness, further exposure merely fogs the plate, rendering faint star images invisible on the photograph. Show that the limiting faintness of stars that can be photographed with a given telescope depends only on the focal length of that telescope.

10. When Mars is at its closest, it subtends about 24″ in the sky. The diameter of Mars is about 6800 km. In principle, how close together could two features on Mars be and still be distinguished with the 6-m telescope? In practice, because of the limitations set by the earth's atmosphere, how close together can they be? With a 60-cm telescope, could a person be sure that "canals" on Mars were only 3 km wide? Why?

11. If the 6-m telescope were used as a radio telescope to observe radio waves of 20 cm wavelength, how accurately could it "pinpoint" the direction of a radio source?

12. The 3-m reflector at the Lick Observatory can be operated as an $f/5$ prime-focus telescope or as an $f/17.2$ Cassegrain telescope. (a) In which of these modes can an extended source be exposed fastest, and by what factor? (b) In which of these modes can a point source be exposed fastest and by what factor? (c) Which mode has a fainter limiting magnitude?

Grote Reber (1911–   ), an amateur astronomer and electronics expert, built in his own back yard, out of wooden two-by-fours and galvanized iron, the first radio telescope specifically designed to observe radio waves from space. From 1937 until after World War II, he was the world's *only* active radio astronomer. *(Ohio State University)*

12

NEW WINDOWS: FROM RADIO WAVES TO X- AND GAMMA RAYS

The first half of the 20th century was a very exciting one for science—especially so for physics. Einstein's special and general theories of relativity had provided new insight into the meaning of space and time; physicists had begun to make real progress in understanding the workings of atoms; and with the realization of the equivalence of mass and energy, we began to learn something of the forces that bind together the atomic nucleus itself. Great discoveries were made in astronomy as well—especially the structure of our own Galaxy, the knowledge that it is but one of many in the universe, and that the universe is expanding. We were, in other words, pushing the frontier of science to the realms of the very tiny and the very large. These frontiers are the "hot" fields in science today.

In the middle of the century, a fantastic change occurred that placed astronomy squarely at the forefront of science. Many physicists and scientists of other specialties changed their research interests to astronomy, and many are still doing so. What was that dramatic development? The opening of new windows in the electromagnetic spectrum. Before 1931 we recorded information from space only in optical wavelengths of electromagnetic radiation. Radio astronomy had its humble beginning

in that year but did not really develop until after World War II.

Consult again Figure 11.1. The two windows of wavelengths for which the atmosphere is transparent are the optical spectrum and part of the radio spectrum. The optical window is limited to wavelengths from about 3000 to about 7000 Angstroms—just over one octave. Radio waves can be observed from the ground over a range of wavelengths from a few millimeters to several tens of meters—more than a dozen octaves. But there's more—far more—radiation from space, ranging to the shortest gamma rays. That broad short-wave window in the electromagnetic spectrum did not open until rockets and space vehicles could carry our receptors above the atmosphere.

In less than three decades our vision, formerly responding only to light, had broadened to encompass the entire electromagnetic spectrum. It was predictable that there would be new discoveries, but no one could have foreseen just what those discoveries would be, or how far-reaching. Surely we do not yet know all of the laws of physics. Many scientists expect that some of those laws awaiting discovery could leave clues in the cosmos, and that our new eyes on the universe may be bringing us to the verge of new physics. The new astronomy of the latter half of the 20th century occupies substantial parts of the chapters to follow; here we summarize briefly how we observe through those new windows to hitherto invisible radiation.

12.1 THE RADIO WINDOW

In 1931 Karl G. Jansky (1905–1950), an American radio engineer at the Bell Telephone Laboratories, built a rotating radio antenna array designed to operate at a wavelength of 14.6 m (Figure 12.2). With this array he attempted to investigate the source of shortwave interference. In addition to temporary intermittent interference due to such phenomena as thunderstorms, he encountered a steady hiss-like static coming from an unknown source. He discovered that this radiation came in strongest about four minutes earlier on each successive day and correctly concluded that since the earth's sidereal rotation period is four minutes shorter than a solar day, the radiation must be originating from some region of the celestial sphere. Subsequent investigation showed that the source of the radiation was the Milky Way.

It was well over a decade before the astronomical community turned serious attention to Jansky's important discovery. Jansky has, however, received belated honors; today the standard unit of radio flux received from space is the *Jansky* (*Jy*).

Figure 12.1 Karl G. Jansky, the Bell Laboratories engineer who first detected radio radiation from space and correctly identified it as coming from the Milky Way. (*Bell Laboratories*)

(One Jansky is 10^{-26} watts striking a square meter of the earth's surface in one unit of frequency.)

Actually, the new radio astronomy did not go *totally* unnoticed, thanks to the American amateur

Figure 12.2 The rotating radio antenna used by Jansky in his serendipitous discovery of radio radiation from the Milky Way. (*Bell Laboratories*)

Figure 12.3 Grote Reber's original radio telescope. *(National Radio Astronomy Observatory)*

astronomer, electronics engineer, and radio ham, Grote Reber (1911–). In 1936 Reber built the first radio telescope—an antenna specifically designed to receive cosmic radio waves. He constructed it in his back yard in Wheaton, Illinois, of wooden two-by-fours and galvanized iron. Subsequently he built other improved antennas and remained active in the field for more than 30 years. During the first decade, Reber worked practically alone. By 1940 he confirmed Jansky's conclusion that the Milky Way is a source of radio radiation, and in 1944 he published in the *Astrophysical Journal* the first contour maps of the radio brightness of the Milky Way as it appears at a wavelength of 1.87 m. He also discov-

ered discrete sources of radio emission in the galactic center, Cygnus, and Cassiopeia, as well as radio waves from the sun. From 1937 until after World War II, Reber was the world's *only* active radio astronomer.

Meanwhile, in 1942 radio radiation from the sun was picked up by radar operators in England. After the war, the technique of making astronomical observations at radio wavelengths developed rapidly, especially in Australia, the Netherlands, England, and later in the United States. Radio waves have now been received from many astronomical objects—the sun, moon, some planets, gas clouds in our Galaxy, other galaxies, and many other objects. The technique of radio astronomy has become an integral and vastly important tool in observational astronomy.

(a) Detection of Radio Energy from Space

First, it is important to understand that radio waves are not "heard"; they have nothing whatever to do with sound. Although in commercial radio broadcasting, radio waves are modulated or "coded" to carry sound information, the sound itself is not transmitted. The radio waves merely carry the information that a radio receiver must "decode" and convert into sound by means of a loudspeaker or earphones. Sound is a physical vibration of matter; radio waves, like light, are a form of electromagnetic radiation. We can also code visible light to carry sound information, as is done, for example, by the sound track on a movie film.

Many astronomical objects emit all forms of electromagnetic radiation—radio waves as well as light, infrared and ultraviolet radiation, and so on. The radio waves we can receive from space are those that can penetrate through the ionized layers of the earth's atmosphere—those with wavelengths in the range from a few millimeters to about 20 m. The human eye and photographic emulsions are not sensitive to radio waves; we must detect this form of radiation by different means. Radio waves induce a current in conductors of electricity. An antenna is such a conductor; it intercepts radio waves which induce a feeble current in it. The current is then amplified in a radio receiver until it is strong enough to measure or record.

If we lay a photographic plate out on the ground in daylight, it will be exposed by sunlight, indicating the presence of a light source in the sky. We can place various color filters in front of such plates and detect the presence of various colors in the light that exposes them. Such an experiment, however, does not indicate the direction in the sky of the light source.

Similarly, a radio antenna can be strung up outside, and currents induced in it indicate the presence of a source of radio radiation. Electronic filters in the radio receiver can be "tuned" to amplify only a certain frequency at a time, and thus can determine what frequencies or wavelengths are present in the radio radiation. The earliest astronomical observations of radio energy were detected in this way. As in the case of a photographic plate

laid out in sunlight, however, a single antenna does not indicate the direction of the source.

(b) Radio Reflecting Telescopes

Radio waves are reflected by conducting surfaces just as light is reflected from an optically shiny surface according to the same law of reflection. A radio reflecting telescope consists of a parabolic reflector, analogous to a telescope mirror. The reflecting surface can be solid metal, or a fine mesh such as chicken wire. In the professional jargon, the reflecting paraboloid is called a "dish." Radio dishes are usually mounted so that they can be steered to point to any direction in the sky and gather up radio waves just as an optical reflecting telescope can be directed in any direction to gather up light. The radio waves collected by the dish are reflected to the focus of the paraboloid, where they form a radio image. In an optical telescope, a photographic plate, image tube, photomultiplier, spectrograph, or some other device is placed at the focus to utilize the image. In a radio telescope, an antenna or wave guide is placed at the focus of the dish. Radio waves focused on the antenna induce in it a current. This current is conducted to a receiver, not unlike ordinary home radio receivers in principle, where the current in amplified. As the optical astronomer chooses the type of photographic emulsion that is sensitive to the colors of light he wants to detect, the radio astronomer tunes his receiver to amplify the specific wavelengths he wants to receive from space.

One advantage of astronomical observations at radio wavelengths is that some of the important atmospheric effects discussed in Section 12.2j are not bothersome. In particular, radio observations are not affected by atmospheric seeing (although there is a similar *scintillation* effect due to clouds of ions in interplanetary space). They are less affected by weather and sky brightness, and at some wavelengths can even be made throughout the entire 24-hour day. Manmade radio interference, however, is a serious problem.

The ability of a radio telescope to gather radiation depends on its size. The radio energy received from most astronomical bodies is very small compared with the energy in the optical part of the electromagnetic spectrum. Hence radio dishes are usually built in large sizes; few are under 6 m across. At first thought, the problem of constructing a large parabolic reflecting surface to sufficient accu-

Figure 12.4 The 100-meter radio telescope at Effelsburg, near Bonn, W. Germany. *(Max-Planck Institut für Radioastronomie)*

racy might seem prohibitive. The 5-m mirror of the Hale telescope has a surface accurate to about five millionths of a centimeter, about one eighth of the wavelength of visible light. Ideally, radio reflectors should be built to similar accuracy. However, a radio dish designed to receive radio waves of a length of 25 cm, for example, need be accurate to only about 3 cm to achieve the same resolution. That is why an open wire mesh can be used as a reflecting surface.

(c) Resolving Power of a Radio Telescope

With a radio reflecting telescope, radio radiation can be detected from a particular direction, and the direction of the source in the sky can be determined. We have seen that optical telescopes can resolve images of very small angular size. A 15-cm telescope can determine the location of a star to within about 1″. The formula given in Section 11.1b for the smallest resolvable angle,

$$\alpha = 2.1 \times 10^5 \frac{\lambda}{d},$$

where d is the aperture of the telescope, holds for radio telescopes as well as optical telescopes. Thus we can easily compute the resolving power of a radio telescope.

The main difficulty with radio telescopes is that the wavelength of radio radiation, λ in the above formula, is far greater than for visible light, so the resolving power for a telescope of a given size is correspondingly less. Radio waves of 20 cm, for example, are some 400,000 times longer than waves of visible light, so to resolve the same angle, a radio telescope would have to be 400,000 times larger than an optical telescope. To resolve 1″ at 20 cm wavelength, a radio telescope would have to be nearly 40 km across. The largest steerable radio telescopes in use today are only about 100 m in diameter. At a wavelength of 20 cm they are capable of resolving two points about 800 km apart on the moon. If it were not for atmospheric "seeing," the 5-m telescope could resolve, in visible light, two points 40 m apart on the moon. The human eye can resolve points on the moon separated by about 100 km. Thus the largest radio telescopes have far poorer resolving power than even the human eye. Consequently, it is a special problem to determine accurately the positions of radio sources in the sky.

A radio astronomer commonly speaks of the *beam width* of his telescope. The *beam* is the cone-shaped bundle of radiation from space of such an angular size that within it features are smeared out and cannot be resolved. The beam width would correspond in optical astronomy to the angular size of the central diffraction disk of a star image (Section 11.1b). For a large optical telescope, that size

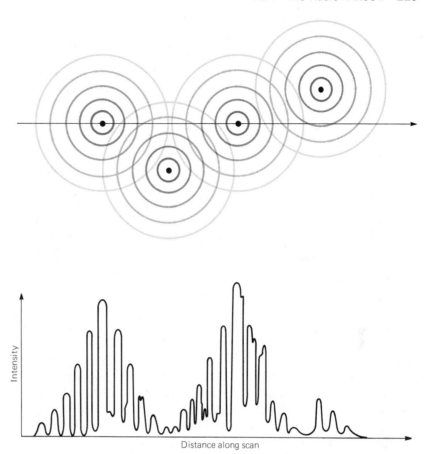

Figure 12.5 *(Above)* The overlapping diffraction patterns of five hypothetical point radio sources, as produced by a single antenna of circular aperture. *(Below)* The sort of signal that would be obtained by scanning through the overlapping images along the line shown.

might be a few hundredths of a second of arc, but for a typical radio telescope, the beam width is several minutes of arc. The beam width depends, of course, on the aperture of the telescope and on the wavelength of the radiation being detected. Now, the central disk in the optical diffraction pattern of a star is surrounded by concentric diffraction rings. Similarly, the central part of the image of a point radio source is surrounded by concentric rings of radio radiation. Suppose a radio telescope is slowly moved so that the field of view of the antenna at the focus of its dish (that is, the beam) slices across the image of such a point source of radio waves (or, equivalently, the telescope can remain stationary, and the turning earth can cause the source to move by). In that case, the beam scans across the diffraction pattern, detecting first a series of successively brighter pieces of the diffraction rings it crosses, then the central image of the source, and finally another series of successively fainter segments of the rings on the other side. Each image, in other words, is recorded as a bright central spot of size equal to the telescope beam width flanked by fainter

images on either side; these are called *side lobes*. With a large optical telescope the entire diffraction pattern of a star is usually small compared with the seeing disk caused by the earth's atmosphere (Section 11.4), so the star appears as a single image and many separate stars can be seen or photographed as separate images. With a radio telescope, however, the diffraction pattern is generally thousands of times larger, and the patterns of adjacent radio sources often overlap each other, creating general confusion (Figure 12.5). To make sense of his observations, the radio astronomer must sort out the mess. Fortunately, with modern computer technology, there are techniques for accomplishing this.

(d) Radio Interferometry

Angular resolution at radio wavelengths can be greatly enhanced with the technique of *interferometry*, which uses two or more separate radio telescopes. Suppose, for example, two radio dishes are placed some distance apart. Unless the source of radio radiation happens to lie along a perpendicular

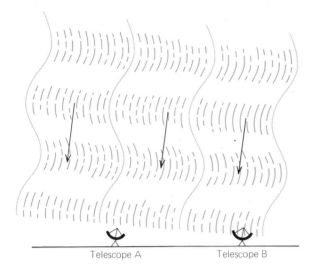

Figure 12.6 The principle of the radio interferometer. Two radio telescopes simultaneously observe the same source, which is not in a direction perpendicular to the line connecting the two telescopes. Thus the waves reach telescope *A* slightly out of phase with (behind) those reaching telescope *B*. The phase lag depends on the direction of the source, and hence determines it.

bisector of the line between the antennas, the radio waves will strike one antenna a brief instant before the other, so that the two antennas will receive the same waves at slightly different times and thus become "out of phase" with each other, that is, the antennas receive different parts of a given wave (Figure 12.6). The difference in phase between the waves detected at the two antennas can be measured electronically. Because this phase difference depends on the angle the direction to the source makes with the line between the antennas, that angle can be determined. If the two antennas are due east and west of each other, an observation of this

sort gives a much more accurate measure of the east-west position of the source in the sky than could be obtained with a single radio telescope. If the antennas are placed due north and south of each other, the other coordinate of the source can be found.

The farther apart the components of an interferometer are placed (the longer the *baseline*), the more sensitive is the phase discrimination of slightly off-axis sources, and the more accurately can we pinpoint the direction of the source. Most major radio astronomy observatories, therefore, have several telescopes so that pairs of them can be operated together as interferometers. The overall sensitivity to radio radiation is then the sum of the collecting areas of the telescopes so combined, and the resolution (or beam width) in a direction along the line between the telescopes is that of a single telescope of total aperture equal to that separation.

Caltech's observatory in Owens Valley near Big Pine (Figure 12.7) has two 27-m dishes and one 40-m dish; the telescopes are mounted on rails so that their separation can be varied to achieve a range of beam widths. At the National Radio Astronomy Observatory in Greenbank, West Virginia, any of three large telescopes can be used as an interferometer with a smaller one high in the Allegheny Mountains 35 km away.

Interferometry need not be limited to two telescopes. Entire arrays can be used together, and the radiation detected at each can be compared, by computer, with that from each other one in the array. In this way an image of the source can be reconstructed as it would appear under the resolution of a single telescope of aperture equal to the maximum separation in the array. The technique is

Figure 12.7 The twin 90-foot radio telescopes of the Radio Observatory of the California Institute of Technology. (*California Institute of Technology*)

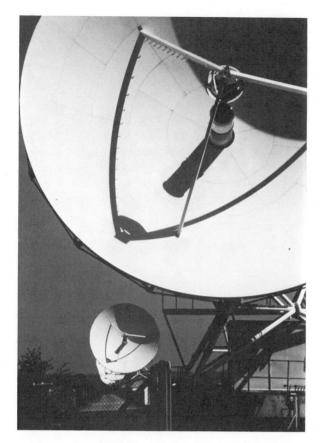

Figure 12.8 A portion of the Cambridge 5-km Array. *(Photograph by the author)*

called *aperture synthesis* and was largely pioneered by astronomer Martin Ryle at Cambridge, England. There are three arrays at the radio observatory at Cambridge, the largest consisting of eight dishes that can be separated to a maximum of 5 km by moving four of them along a section of the old Oxford/Cambridge railway track, on which they are mounted. When operating at a radio wavelength of 2 cm, the 5-km array can obtain a resolution of 1″, comparable to the usual resolution permitted by atmospheric seeing at optical wavelengths. Another large array at Westerbork, Netherlands, has 12 telescopes, each 25 m in diameter, spaced over a total baseline of 1.6 km.

By far the world's most impressive radio telescope array is the *Very Large Array (VLA)* near Socorro, New Mexico. The VLA, operated by the National Radio Astronomy Observatory (financed by the National Science Foundation), was essentially completed and dedicated in 1980. It has 27 telescopes, each of aperture 25 m, that can all be moved along rails laid out in a large "Y" configuration with a total span of about 36 km.

(e) Very Long Baseline Interferometry

Now that international time standards can be coordinated to high precision, we can extend the inter-

Figure 12.9 Part of the Y-shaped Very Large Array (VLA) near Socorro, New Mexico. *(National Radio Astronomy Observatory)*

ferometer principle to very long baseline interferometry (VLBI). Two different radio telescopes, thousands of kilometers apart, can simultaneously observe the radio waves from the same source and record them on tape, along with marks from a very accurate time standard (like the "ticks" of a very accurate clock). Later, these two tapes can be analyzed with a computer to find the phase difference between the radio radiation at the two stations, and hence the direction of the source. Baselines as long as from California to Parkes (in Australia) and from Greenbank, West Virginia, to Crimea have been used. The resulting angular resolution of the sources observed is as great as a few ten-thousandths of a second of arc—far surpassing the angular resolution of optical telescopes.

The principle of the radio interferometer is like that of the optical interferometer of the Michelson type, which has been used to measure small angles optically. The method of operation of the interferometer will not be described here, but some results of measures made with a Michelson interferometer are given in Chaper 26.

(f) Radar Astronomy

Radar is the technique of transmitting radio waves to an object and then detecting the radio radiation that the object reflects back to the transmitter. The time required for the radio waves to make the round trip can be measured electronically, and be-

Figure 12.10 One of the 27 25-m antennas of the VLA. *(Photograph by the author)*

cause they travel with the known speed of light, the distance of the object is determined. The value of radar in navigation, whereby surrounding objects

Figure 12.11 The 300-meter (1000-ft) dish of the Arecibo Observatory, Puerto Rico, operated by the National Astronomy and Ionosphere Center (Cornell University) and sponsored by the National Science Foundation. *(Cornell University)*

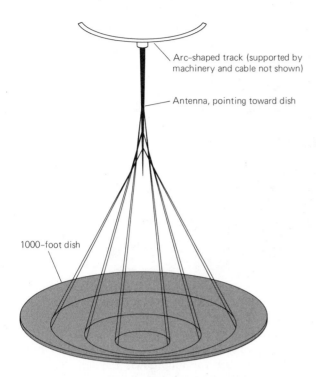

Arc-shaped track (supported by machinery and cable not shown)

Antenna, pointing toward dish

1000-foot dish

Figure 12.12 Because of spherical aberration produced by the spherical reflecting surface, different places along the needle-shaped antenna receive radiation reflected from different concentric zones of the 1000-foot dish.

can be detected and their presence displayed on a screen, is well known.

In recent decades the radar technique has been applied to the investigation of the solar system. Radar observations of the moon and most of the planets have yielded our best knowledge of the distances of those worlds. In addition, as will be discussed in later chapters, radar observations have determined the rotation periods of Venus and Mercury, provided information about their surfaces, and furnished rather detailed surface contour maps.

(g) Famous Radio Telescopes

The largest single radio telescope, in fact, the largest area focusing telescope of any kind in the world, is the 1000-foot (305-m) bowl near Arecibo, Puerto Rico, a facility of the National Astronomy and Ionosphere Center, funded by the National Science Foundation and operated by Cornell University. The telescope was first built in 1963, and in 1974 its reflecting surface was rebuilt with such high precision that it can be operated at radio wavelengths as short as 7 cm. Similar to several other much smaller radio "bowls" throughout the world, the

Arecibo telescope is not movable but consists of a fine wire mesh stretched with astonishing accuracy into a spherical surface over a natural valley in the Puerto Rican mountains. Because a spherical surface has no preferential axis, it produces comparable images over a wide field. A movable antenna is supported by a large arc-shaped track at the focus of the bowl, and the track itself can rotate about a vertical axis, allowing the antenna to move in any direction; it can thus record objects as far as about 20° away from the zenith. The rotation of the earth assures that all objects within a 40-degree belt around the sky come into view of the telescope at one time or another. The reflector has spherical aberration, of course, so a rather long antenna, pointing toward the bowl, is used. Different points along the antenna thus pick up radiation reflected from different ring-shaped zones of the bowl, and all these radiations are combined electronically (Figure 12.12). The 1000-foot dish is used not only to record extremely feeble radio signals from the remote universe but also to study the earth's atmosphere. Moreover, it is used as a very powerful radar telescope to explore the surfaces of planets (Chapter 17). Frank Drake, former director of the facility, has pointed out that the volume of the bowl could contain the world's annual consumption of beer. My own reaction is that the surface, as is, would be a mecca for skateboard enthusiasts.

Large movable dishes include the 100-m (328-ft) telescope at Bonn, the 91-m (300-ft) telescope of the National Radio Astronomy Observatory at

Figure 12.13 Closeup of the reflecting surface of the Arecibo radio telescope. The debris that collects on the wire mesh does not cause appreciable interference and is not removed frequently in order to avoid unnecessary risk of bending the dish out of shape. *(Photograph by the author)*

Figure 12.14 The Molonglo Radio Telescope, of the "Mills Cross" type, operated by the University of Sydney. (*Courtesy B. Y. Mills, University of Sydney, N.S.W.*)

Greenbank, the 76-m (250-ft) at Jodrell Bank, England, the 64-m (210-ft) dish at Goldstone, California (operated by the Jet Propulsion Laboratory in Pasadena), and the 64-m (210-ft) Parkes dish in Australia. The VLA and some other large arrays have already been mentioned.

Many arrays do not use separate parabolic dishes. For example, the Molonglo Radio Telescope, near Canberra, Australia, operated by the University of Sydney, consists of wire mesh stretched along two large troughs, each a mile long and crossing at right angles at their centers. De-

signed by B. Y. Mills, it is of the type known as the "Mills cross." Each trough is parabolic in cross section, so rather than concentrating the radiation at a single point focus, it reflects the radio waves to a long antenna running down the length of the trough at the focus of its cross-section parabola. The different signals striking different parts of the cross are used to reconstruct a radio image of the source observed by an aperture synthesis technique.

Of the many other important radio observatories, a few of the most famous are described in Table 12.1.

Figure 12.15 Closeup of one of the parabolic troughs of the Molonglo Radio Telescope. The long beam running along the focus contains the many small collecting antennas. (*Photograph by the author*)

Table 12.1 **Some Famous Radio Astronomy Observatories**

OBSERVATORY	DATE FOUNDED	LOCATION	IMPORTANT TELESCOPES
Arecibo	1963	Arecibo, Puerto Rico	305-m (1000-ft) bowl
Very Large Array	1980	Socorro, NM	27 25-m (82-ft) dishes in 36 km array
Max-Planck Institut für Radioastronomie	1971	Effelsburg, near Bonn, W. Germany	100-m (328-ft) steerable dish
National Radio Astronomy Observatory	1958	Greenbank, WV	91-m (300-ft) dish 43-m (140-ft) dish 26-m (85-ft) dish
Jodrell Bank	1949	near Manchester, England	76-m (250-ft) dish 38-m (125-ft) dish 66-m (218-ft) bowl
Parkes National Radio Observatory	1961	North Goobang, near Parkes, Australia	64-m (210-ft) dish
Goldstone Tracking Station (Jet Propulsion Laboratory)	1958	near Barstow, CA	64-m (210-ft) dish 26-m (85-ft) dish
Westerbork	1955	Netherlands	12 25-m (82-ft) dishes in 1.6-km array
Owens Valley (Caltech)	1958	near Big Pine, CA	2 27-m (90-ft) dishes 40-m (130-ft) dish
CSIRO Radioheliograph (Solar radio telescope)	1967	Culgoora, Australia	96 13.7-m (45-ft) dishes in 3-km diam. circular array
Molonglo Radio Observatory (U. of Sydney)	1967	near Canberra, Australia	2 crossed arms 1.6 km (1 mi) long and 12 m wide
Mullard Radio Astronomy Observatory	1945	Cambridge, England	5-km array (8 15-m dishes) 1.6-km array (3 18-m dishes) 0.8-km array (4 9-m dishes)
Haystack telescope	1956	Westford, MA	36-m (120-ft) dish 25-m (84-ft) dish

12.2 INFRARED ASTRONOMY

Infrared radiation with wavelengths less than about 2.5 μ, the *near* infrared, is adjacent to the optical spectrum and is emitted with fair intensity by the sun. (1 μ—one *micron*—is 10^{-6} m, or 10^{-3} mm; also, 1 μ = 10,000 Å.) The oxygen, nitrogen, and argon in the earth's atmosphere are transparent to all infrared radiation, but water vapor and carbon dioxide absorb strongly in certain spectral regions. The atmosphere is transparent to much of the near infrared, although there is absorption by water vapor at intermittent wavelengths longer than 0.8 μ. The near infrared can be detected photoelectrically, and its shorter wavelengths can be photographed with certain types of emulsions. The near infrared is thus sort of an extension of the optical window to longer wavelengths.

Infrared radiation of wavelengths from 2.5 μ to 50 μ is called the *intermediate* infrared. Most of the intermediate infrared does not penetrate the atmosphere to sea level; water vapor and carbon dioxide

combine to absorb strongly from 4 μ to 8 μ, for example, and over much of the spectrum beyond 13.5 μ. But there is a clear window from 8 μ to 13.5 μ. Ground-based observations in the intermediate infrared are best made at high mountain observatories, which lie at elevations above most of the absorbing water vapor.

The *far* infrared extends from 50 μ to about 1000 μ (1 mm), which overlaps the microwave radio spectrum. There are some clear atmospheric windows in the far infrared, but between 35 μ and 350 μ the atmosphere is completely opaque, making ground-based infrared astronomy impossible in those wavelengths. That gap, however, has been filled by carrying infrared detectors above the atmosphere with rockets and space probes, and even by Apollo astronauts. High-flying aircraft (including NASA's U-2 reconnaissance planes) and balloons have also been used to carry instruments above much of the atmosphere, facilitating infrared observations in certain spectral regions. One way or the other, the entire infrared spectrum has been observed.

(a) Infrared Telescopes

Infrared telescopes are essentially the same as optical telescopes. They are all reflecting telescopes, and since the far infrared cannot be focused with glass lenses, they are generally equipped with spectrographs using reflection optics and diffraction gratings, rather than lenses and prisms.

There are some marked differences between infrared and optical observations. Typical temperatures on the earth are from 290 to 300 K, which means that the earth and every warm thing on it—the telescope, the dome, and the astronomer—are radiating infrared radiation with a peak wavelength of about 10 μ. To infrared eyes, everything is brightly aglow! The problem is to detect faint cosmic infrared sources against this sea of light. The telescopic infrared image must be shielded as much as possible from this radiation from its surroundings. Telescopes intended especially for infrared observations are designed to greatly reduce the radiation from them. The detector itself is generally operated at a low temperature by cooling it with liquid nitrogen or some other agent. For observations in the far infrared, the detectors must be kept near liquid helium temperature (below 10 K). (Infrared detectors are described in Section 10.1c.)

Photons of infrared radiation have lower energy than do those of visible light, and cool bodies that radiate mostly in the infrared radiate less intensely than do hot ones at all wavelengths (see Section 10.2). Consequently, most infrared sources are relatively weak ones, and we need large telescopes to collect enough energy to record them. Infrared telescopes, therefore, are generally of large aperture. Recently completed infrared telescopes are the 3-m NASA Infrared Telescope Facility and the 3.8-m United Kingdom infrared telescope, both atop Mauna Kea in Hawaii. The Mauna Kea Observatory is especially suitable for infrared work because its high altitude (4205 m; 13,756 ft) places it above much absorbing water vapor.

Under the best conditions, infrared sources are faint compared with the background of radiation around the telescope. It is common practice, therefore, to observe an infrared source (or a suspected infrared source) directly and then to observe a region of blank sky next to it. The difference in the infrared radiation detected must be that due to the source. But the background level is so high, and often changeable in time, that it is necessary to look back and forth from source to sky many times and as rapidly as possible. In special infrared telescopes this is generally done by tilting a secondary mirror back and forth rapidly, but very slightly, so that the source's image produced by the primary mirror is alternately on and off the detector. The alternating current produced by this periodic change of radiation intensity on the detector is a measure of the infrared radiation from the source.

Infrared observations, especially near microwave frequencies, are hampered by radiation from the air. Consequently, it is desirable to get above as much of the atmosphere as possible. One remarkable infrared and microwave observatory is the Kuiper Airborne Observatory. It is a 0.9-m (36-inch) Cassegrain reflector carried aboard a Lockheed C-141 airplane, which operates out of NASA Ames Research Center at Moffett Field, in Mountain View, California. A sophisticated mechanism keeps the telescope accurately tracking while it is flown at an altitude of 12,500 meters (41,000 feet) through air at −50°C at a speed of 800 km/hr. It is better yet to carry the infrared telescope outside the atmosphere altogether. In 1982 NASA, in collaboration with the Netherlands and United Kingdom, plans to launch the Infrared Astronomy Satellite (IRAS), which will carry a 0.6-m telescope designed to map the sky with a resolution of 2 minutes of arc. Several years later NASA plans to use the Space Shuttle to launch the Shuttle Infrared Telescope Facility (SIRTF), with a 0.85-m infrared telescope, which will also share data with European astronomers.

Today, when very large optical astronomical telescopes are planned and built, their potential for work in the infrared is also considered. The 10-m telescope under design by the University of California, for example, will be intended for both optical and infrared astronomy.

12.3 THE SHORTWAVE WINDOW

The atmosphere is completely opaque to electromagnetic radiation of wavelength less than about 3000 Å. Consequently, ultraviolet, X-ray, and gamma ray observations must be made from space. Such observations first became possible in 1946, when the United States Naval Research Laboratory launched a captured German V2 rocket to observe the far-ultraviolet radiation from the sun. Subse-

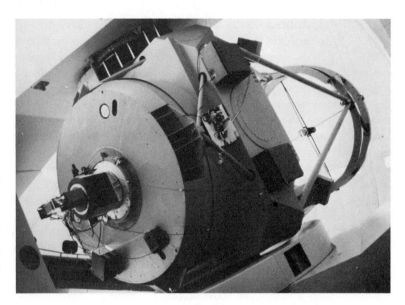

Figure 12.16 NASA's 3-m infrared telescope on Mauna Kea, Hawaii. The center section, supported by the large yoke mounting, contains the optics. The 3-meter-diameter primary mirror is inside the circular section at the lower end of the telescope, and the 10-inch-diameter secondary, at the front of the tube, reflects radiation gathered by the mirror to the detectors. (NASA/University of Hawaii)

quently many rockets have been launched to make X-ray and ultraviolet observations of the sun, and later of other celestial objects as well.

In the United States most of these sounding rockets have been launched from White Sands Missile Range in New Mexico, and the most extensively used of these rockets have been the Aerobee 150 and Aerobee 170; each of the latter can carry a 100-kg payload to an altitude of 250 km. The ultraviolet photographs were recovered from instrument packages that had to be parachuted back to the ground. Both ultraviolet and X-ray radiation produce images on film, but because of their high energy, X-rays cannot be focused by most conventional optical systems. The early X-ray detectors were just gas chambers in which the X-rays produced ionization.

Since the 1960s earth satellites have been launched to carry out astronomical observations. These have included the orbiting solar observatories (OSOs), the orbiting astronomical observatories (OAOs), and the high-energy astronomy observatories (HEAOs). Scientist astronauts also made astronomical observations at short wavelengths from Skylab, and the Apollo astronauts have operated ultraviolet cameras from the moon. Planetary space probes, of course, carry instruments to remote parts of the solar system.

Figure 12.17 The Infrared Telescope Facility, one of several large telescopes located at the summit of Mauna Kea, Hawaii, at an altitude of 13,796 feet (4200 m). (NASA/University of Hawaii)

(a) The Orbiting Ultraviolet Observatories

Ultraviolet telescopes are similar to optical telescopes, except that their optical surfaces need special coatings that have high ultraviolet reflectivity, and, of course, the telescopes must be taken outside the atmosphere. In addition to the small ultraviolet instruments flown in rockets and those operated on the moon and from Skylab, the United States has launched a highly successful series of satellites especially designed as ultraviolet observatories.

A series of eight satellites, launched during the years 1962 through 1975, were primarily for solar observations. These orbiting solar observatories (OSOs) have obtained thousands of ultraviolet spectra of the sun. The last and most successful observatory, OAO 8, was launched in June 1975 and carried eight instruments. Two of these (one American and one French) are for obtaining ultraviolet spectra of light from tiny regions of the sun (down to one arc second across) and six are X-ray and gamma ray detectors for exploring sources of this high-energy radiation from other directions in space. Many guest scientists have gone to the scientific headquarters at Boulder, Colorado, to participate in obtaining OSO observations of the sun.

The first successful orbiting astronomical observatory, OAO 2, was launched December 7, 1968, and carried instruments developed by the University of Wisconsin and by the Smithsonian Astrophysical Observatory at Harvard for obtaining ultraviolet spectra of astronomical objects in the spectral range 1200 to 4000 Å. During the useful life of the observatory (to February 1973), more than a thousand observations were made. The next important ultraviolet observatory was OAO Copernicus, launched in August 1972, which carried a 0.8-m ultraviolet telescope and three small X-ray telescopes. Hundreds of astronomers served as guest investigators to obtain observations with it.

The most successful orbiting ultraviolet observatory to date (still in operation at this writing— early 1981) is the International Ultraviolet Explorer (IUE) satellite, launched in January 1978 and carrying instruments developed by NASA and by the United Kingdom's Science Research Council. The IUE can obtain high-quality spectra in the range 1150 to 3200 Å and uses a vidicon image converter that can allow exposure times of up to 15 hours. The IUE is in a synchronous orbit (that is, its period of revolution about the earth is equal to the period of the earth's rotation), so that it is always in view of its control headquarters at NASA's Goddard Space Flight Center, Greenbelt, Maryland. More than 200 astronomers from 17 nations have already gone to Greenbelt as guest investigators. They must, of course, have their applications for observing time on the IUE approved, and the competition is great. Careful preparation must be made to be able to orient the satellite so that its telescope will point to the desired object. The guest observer at Greenbelt can actually see a TV-displayed image of the field of view through IUE's 0.45-m cassegrain telescope, as it is televised live to earth, and can compare what he sees with his star charts and make final adjustments as needed. When the exposure on his object is over, the observer sees his spectral data displayed on the TV screen; a printout on paper or magnetic tape is provided for the astronomer to take back to his home institution, where he can analyze it at his leisure.

(b) X-ray Observations

Although there is no special feature that distinguishes the ultraviolet from X-rays, we generally regard X-rays as electromagnetic radiation of wavelength less than about 100 or 120 Å. The shorter the wavelength, the higher the energy of the photons (Section 10.3), and X-ray and gamma ray astronomers often speak of the energies of the photons they observe rather than of their wavelengths. A photon of wavelength 12.4 Å has an energy of 1000 eV, abbreviated 1 keV. Those of lower energy are called *soft* X-rays, and those of higher energy (shorter wavelength), *hard* X-rays. Radio waves are characteristic of the energy emitted from cold bodies, and visible light and the ultraviolet are characteristic of that from hot bodies of temperature 10^3 to 10^5 K (like the surfaces of stars). X-rays are emitted from gas at very high temperatures—10^7 to 10^9 K. Thus in different spectral regions we preferentially observe parts of the universe that are at different temperatures.

X-rays from space were first observed with instruments flown on balloons and rockets. By 1967 about 30 discrete sources of X-rays had been discovered. X-ray astronomy made a sudden advance in December 1970, when the first orbiting X-ray observatory was launched from Kenya through an international program. That observatory was named *Uhuru*, the Swahili word for "freedom." Uhuru systematically scanned the sky for X-ray sources and charted more than 200 of them during its life-

Figure 12.18 Launch of a Paiute Tomahawk sounding rocket. *(U.S. Army photograph)*

time (to 1973). The convention for naming these sources has been numerically according to the constellation in which they are found; for example, the famous source Cygnus X-1 (Chapter 35) is the first X-ray source to be found in Cygnus.

In 1977 the United States launched the first of its series of high-energy astronomy observatories,

HEAO 1. At this writing HEAO 1 is still active and has recorded many more X-ray sources and measured their intensities. Now, up to 1978, X-ray telescopes did not produce images that could be inspected but could merely record the presence of X-rays of certain intensity from a particular direction. The detectors most commonly used were gas ionization detectors. By putting in front of the detector a shield or baffle of metal (for example, lead) that is opaque to the X-rays, the telescope could be made to record radiation only from such a direction as to pass through an aperture in the shield. One kind of shield consisted of many banks of parallel wires, one bank behind the other; the whole apparatus would have to be aligned in a critical way for the spacing between the wires to be lined up from bank to bank so that the radiation could pass through to the detector. With such arrangements it was possible to determine the directions of the X-rays' sources, but not their shapes or much about their sizes.

X-ray astronomy received a truly spectacular boost with the launching of HEAO 2, the *Einstein Observatory*, in November 1978. The Einstein X-ray telescope was the culmination of a nearly 20-year dream of Harvard-Smithsonian scientist Riccardo Giacconi, a pioneer in X-ray astronomy and the Director of the Einstein Observatory during its useful life (until 1981).

Although X-rays are easily absorbed in ordinary optical systems, they can be reflected from polished surfaces that they strike at a grazing angle—like stones skipping across water. The Einstein satellite has an X-ray telescope consisting of a complex set of concentric parabolic and hyperbolic

Figure 12.19 Recovery operations of a Tomahawk rocket on 24 September 1977. *(U. S. Army photograph)*

Figure 12.20 The Einstein X-ray Observatory (HEAO-2) before being launched with an Atlas Centaur rocket from Cape Kennedy in November 1978. *(NASA)*

cylindrical surfaces that use the grazing reflection principle to focus X-rays into an actual X-ray image that can be detected electronically and transmitted to earth. The size of the telescope aperture is 58 cm (23 inches), but because of the grazing angles of the reflecting surfaces to the incoming X-ray photons, the actual mirror surface is equivalent to that of the 2.5-m (100-inch) telescope on Mount Wilson. The Einstein telescope was designed to record X-rays of wavelength from 3 to 50 Å and had a field of view of about 1°. In the first few months of its operation, it was clear that it was an unqualified and spectacular success, with a sensitivity for detecting weak sources 1000 times as great as anything that preceded it. This is equivalent to changing from a

small amateur telescope to the 200-inch Hale reflector on Palomar. In later chapters we shall have more to say about the thousands of X-ray sources discovered with the Einstein telescope and what they mean.

The third satellite in the HEAO series, launched in September 1979, is used to detect high-energy particles from space (cosmic rays).

(c) Gamma-ray Astronomy

Gamma rays were first discovered among the radiation emitted during the decay of radioactive elements. We know today that atomic nuclei have excited energy states, analogous to those of an atom when it changes its configuration of electrons. The differences in energy between the nuclear states, however, are very much higher, and the photons that can be absorbed or emitted by nuclei changing states are gamma rays. Many physicists, in fact, *define* gamma rays as electromagnetic radiation involved in nuclear transitions and call all other high-energy photons X-rays. Although this gamma radiation is of far higher energy than that of common X-rays, there is no gap in the electromagnetic spectrum, and indeed photons can be produced by other mechanisms that have higher energy than the gamma rays emitted from many nuclear transitions. Physicists, in other words, sometimes call the same kind of photons by different names (gamma rays or X-rays) according to their origin. There are cosmic sources of electromagnetic radiation of all energies, however, so astronomers tend to ignore the nuclear definition and call *all* photons of high energy gamma rays. The boundary between X-rays and gamma rays is, of course, arbitrary, but is usually taken as about 0.1 Å (or the order of 100 keV).

Gamma rays of cosmic origin were first discovered by the *Vela* satellites. The Vela satellite system, under the direction of the Atomic Energy Commission, conducts worldwide surveillance for possible explosions of nuclear bombs, which would emit gamma rays. In 1967 they detected bursts of gamma radiation that investigation showed could not originate from within the solar system. Such bursts have also been detected by Uhuru and other orbiting observatories equipped with detectors of high-energy radiation, including some of the OSOs. A few dozen bursts have been observed altogether, sometimes single and sometimes complex and multiple. Each lasts anywhere from about a tenth of a second to several seconds. The sources of these

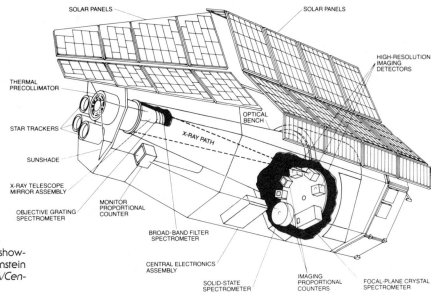

Figure 12.21 labels: SOLAR PANELS · SOLAR PANELS · HIGH-RESOLUTION IMAGING DETECTORS · THERMAL PRECOLLIMATOR · STAR TRACKERS · SUNSHADE · X-RAY TELESCOPE MIRROR ASSEMBLY · OBJECTIVE GRATING SPECTROMETER · MONITOR PROPORTIONAL COUNTER · OPTICAL BENCH · X-RAY PATH · BROAD-BAND FILTER SPECTROMETER · CENTRAL ELECTRONICS ASSEMBLY · SOLID-STATE SPECTROMETER · IMAGING PROPORTIONAL COUNTERS · FOCAL-PLANE CRYSTAL SPECTROMETER

Figure 12.21 Cutaway drawing showing the instrument array in the Einstein orbiting X-ray observatory. *(NASA/Center for Astrophysics)*

bursts, not yet identified, seem to be scattered more or less at random about the sky.

The bursts are of relatively low energy, as gamma rays go. Sources of far higher energy gamma rays—greater than 10 million eV (10 MeV)—have been found subsequently. The first certain detection of high-energy gamma rays of cosmic origin was with an experiment on OSO 3, in which the plane of our Galaxy was detected in the light of gamma rays of energy more than 50 MeV. Other satellites, especially the second Small Astronomy Satellite, SAS 2, have provided a far better picture of the gamma-ray sky and supplement that obtained from lower energy gamma rays detected from other satellites and from experiments flown in high balloons. In high-energy gamma radiation, the Milky Way appears somewhat as it does to the unaided eye, only narrower—a belt around the sky about 2° wide. There are also dis-

(a)

(b)

Figure 12.22 A balloon being launched (*a*) to carry instruments (*b*) to an altitude of more than 25,000 m to study microwave radiation from space. *(David T. Wilkinson, Princeton University)*

crete gamma-ray sources, mostly along the Milky Way. About 30 localized sources had been found by 1980, and two were definitely identified with old supernovae (Chapter 34). There are other probable identifications, at least one being a quasar (Chapter 37).

Among the devices that can be used to detect gamma rays are *scintillation counters* and *spark chambers*. Scintillation counters use a material that fluoresces—emits a flash of visible light—when it absorbs a gamma ray. The flash is generally recorded photoelectrically. A spark chamber consists of a stack of slightly separated conducting plates with high electrical voltages between them. When an incoming gamma ray is absorbed in one of the plates, its energy is converted to an electron-positron pair (see Chapter 13), and these charged particles are attracted by the voltage differences between the plates, creating a current that manifests itself as a series of sparks between the plates. The light from the sparks is recorded photoelectrically, and the direction of the sparks gives some indication of the incoming photon's direction.

(d) The Space Telescope

Encouraged by the phenomenal success of our observations from space, astronomers are now excited about the prospects of a large optical telescope in space—the *Space Telescope*, under construction and planned for launch from the Space Shuttle in late 1983.

The Space Telescope will be a cassegrain reflector with a mirror of aperture 2.4 m (94 inches). At an altitude of 500 km it will be above essentially all of the atmosphere. Unhindered by atmospheric seeing, it is expected to obtain a resolution of 0″.05, and because of the darkness of space, it is expected to detect objects 50 times fainter than is possible with any existing telescope on the earth's surface. Teams of astronomers from America and Europe are planning special experiments to be carried out with the Space Telescope, making use of its television cameras, spectrographs, and photometers, which operate not only in the visible spectrum but in the near ultraviolet and near infrared as well (from 1150 to 11,000 Å).

In the future, engineers carried aloft by the Space Shuttle will be able to make repairs on the Space Telescope, if need be, or even bring it back to earth for extensive overhaul, and then return it to space. The Space Telescope will not put ground-based astronomy out of business, but if fully successful should be the most effective optical observatory ever built.

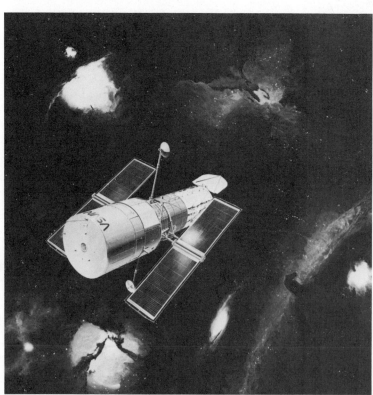

Figure 12.23 Artist's rendering of the Space Telescope. *(NASA)*

EXERCISES

1. When the 1000-foot radio telescope at Arecibo is used to observe an object that is not precisely overhead, less than the full aperture of the telescope is effective. Why?

2. Some early interferometers used only one radio antenna located on a cliff overlooking the ocean. Can you suggest how such a system can work? *Hint:* It involves the reflection of radio waves from the ocean surface.

3. Suppose there were a stationary array of radio telescopes lined up in an east-west direction and located at the equator of the earth. Would the system determine more accurately the right ascension or the declination of a cosmic radio source? Why?

4. Suppose at some northern latitude a straight-line array of radio telescopes is laid out in an arbitrary direction (not necessarily north-south or east-west, but not excluding these possibilities). Explain how the use of the array as an interferometer at different times during the day (or night) can yield two-dimensional information about the structure of the source and its direction in the sky.

5. Suppose the VLA to have a maximum baseline of 36 km. What is the angular resolution possible with the system operating at a wavelength of 6 cm?

6. If, with a baseline of 6000 km, a VLB interferometer just resolved an angle of $0''0042$, at what wavelength would the observations have been made?
Answer: 12 cm

7. If a radio telescope array has four separate antennas, how many different pairs can be selected among them for interferometry? Actually, the number of different pairs among n objects is $n(n - 1)/2$. How many pairs exist among the 27 antennas in the VLA?
Answer: 351

8. The IUE is a synchronous satellite, that is, it remains over the same meridian of longitude on the earth.
   a. What, to the nearest minute, is its period of revolution about the earth?
   b. If the moon has a sidereal period of 27.3 days, and the IUE is in a synchronous orbit, what is the ratio of its distance from the center of the earth to the moon's distance?
Answer to part b: 0.11

9. How long do signals transmitted from a synchronous satellite take to reach the earth? We regularly use synchronous satellites for long-distance and overseas telephone communications. Suppose you are in San Francisco and are talking by telephone to a friend in London. Do you think the two of you could be aware of the fact that you are conversing via a synchronous satellite? How.

10. Discuss the technology that had to be developed before satellite observatories were possible. Consider not only rocket and trajectory technology, but methods of getting data back to earth and analyzing it. Think of some practical devices, not connected with astronomy, that have been developed as a consequence of this space-age technology.

11. What is the wavelength of 50 MeV gamma-ray photons?
Answer: about 2.5×10^{-4} Å

James Clerk Maxwell (1831–1879), the great Scots physicist, unified electricity and magnetism into a coherent theory, much as Newton had unified celestial and terrestrial mechanics. Maxwell's theory was the cornerstone on which Einstein built his theory of special relativity. (*American Institute of Physics, Niels Bohr Library*)

13

SPECIAL RELATIVITY: A CONFLICT BROUGHT TO LIGHT

Newton unified science in 1687, and his theories of mechanics and gravitation dominated scientific thought for nearly two centuries. But by the middle of the 19th century, there was a change in the wind—a change set in motion by the Scottish physicist James Clerk Maxwell (1831–1879), who unified electricity and magnetism and showed that they are really different manifestations of the same thing. In the vicinity of an electric charge, another charge feels a force of attraction or repulsion, depending on whether the two charges have the opposite or the same sign, respectively. But in the vicinity of a *moving* charge another charge is acted on by a *magnetic* force. Maxwell spoke of an *electric field* around a static charge, and a *magnetic field* around a moving charge (as in an electric current).

Maxwell's theory of electromagnetism was codified in 1873 in his four famous equations, which describe electric and magnetic fields, and also how a change in one always induces a change in the other. If a charge moves back and forth, for example, as in an alternating current, the magnetic field set up is continually breaking down, and reforming with the opposite polarity. These changes in the magnetic field induce a constantly changing electric field, which in turn induces a changing magnetic field, and so on. These rapidly alternating fields are in the form of a disturbance that propagates as waves away from the moving charge. Maxwell's equations also predict that this disturbance should move with a very definite speed, equal to the ratio of the electromagnetic and electrostatic units of electricity. Maxwell recognized that this speed is remarkably similar to the speed that had been measured for light, and he suggested that light must be one form of this electromagnetic radiation. He even suggested that there were other forms of electromagnetic radiation, and of course he was right, but it was more than 20 years before radio waves were generated for the first time by the German physicist Heinrich Hertz.

13.1 THE SPEED OF LIGHT

Long before Maxwell's time it was known that light travels with a finite speed, and by the time of his prediction of electromagnetic radiation that speed had even been rather accurately measured. The speed of light is found by measuring the time required for it to travel an accurately known distance.

Galileo suggested a way to measure the speed of light with two experimenters, separated by a mile or more, and each equipped with a lantern that can be covered. The first opens his lantern, and the second, on seeing the light from the first, uncovers his. The time that elapses between the time that the first experimenter opens his lantern and the time that he sees the light of his associate's lantern, after correction for the human reaction time, is how long light spends making the round trip. It is not clear whether Galileo actually conducted this experiment, but he correctly concluded that the speed of light is too great to be measured by so crude a technique.

(a) Roemer's Demonstration of the Finite Speed of Light

The first demonstration that light travels at a finite speed was provided by observations of the Danish astronomer Olaus Roemer (1644–1710) in 1675. Jupiter's inner satellites are regularly eclipsed when they pass into the shadow of the planet. Thus a convenient way of timing the period of revolution of a satellite is to note the interval between successive eclipses. Roemer was doing just this when he noted that the period of a satellite seems longer when the earth, in its orbit, is moving away from Jupiter than it is six months later when the earth is approaching Jupiter. Roemer correctly attributed the effect to the time it takes light to travel through space.

Suppose the earth is at A (Figure 13.1) when we note an eclipse, indicating that a cycle of revolution of the satellite has begun. Before the revolution is completed the earth has moved somewhat further from Jupiter, and light from Jupiter has farther to travel to catch up with the earth, at B, to bring us news of the next eclipse. Six months later the situation is reversed; the earth then approaches Jupiter, and thus advances to meet the oncoming light. If the first eclipse is then observed with the earth at C, the second is observed at D a little ahead of the time it would have been seen had the earth remained at C. Observations like Roemer's indicate that light takes about $16\frac{1}{2}$ minutes to cross the orbit of the earth. Because the earth revolves about the circumference of its orbit in one year, it is easy to see that light must have a speed about 10,000 times that of the earth. Later, when the distance from the sun to the earth, and hence the speed of the earth, was well determined, the speed of light could be deduced in km/s.

(b) Measuring the Speed of Light

In 1849, the French physicist Hippolyte Fizeau (1819–1896) invented a laboratory method of measuring the speed of light. His procedure was to send a beam of light to a distant mirror. On the way, the light had to pass through the gap between two teeth in a toothed wheel. If the wheel was stationary, or rotating only slowly, the mirror would reflect the light beam back through the same gap. If, however, the wheel was set in rapid rotation, by the time the reflected light beam reached the wheel a tooth would have moved into the location occupied by the gap when the light first passed the wheel. Fizeau's procedure was to find at what speed he had to rotate the wheel in order that the reflected light beam would be so eclipsed. Knowing the speed of rotation of the wheel, he could calculate the time required for a tooth to move into the position occupied by its adjacent gap. This was the same time that the light spent traversing the distance from the wheel to the mirror and back. In Fizeau's experiment, the mirror was a little over 8 km from the toothed wheel, so that the light had to travel about 17.2 km. His result for the speed of light was accurate to within about 4 percent. Cornu later applied the same method with an improved apparatus and measured the speed of light to an accuracy of 1 percent.

A superior procedure was developed independently in 1850 by the French physicist Jean Foucault (1819–1868), also famous for his pendulum experiment (Section 7.1a). The principle of Fou-

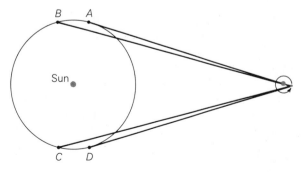

Figure 13.1 Roemer's method of demonstrating the finite speed of light.

cault's method is shown in Figure 13.2. Light from a source, S, is reflected from a rapidly rotating mirror, M. Each time that the mirror turns through the correct orientation, the reflected beam strikes a stationary mirror B some distance away. A second reflection sends the beam from B back to M, but while light is making that round trip, M, because of its rapid rotation, has turned a bit (dashed line in Figure 13.2), so that the light does not reflect directly back to S, but off to the side, to E, where it can be observed with an eyepiece. From the geometry of the setup, it is easy to calculate the angle through which the mirror M must rotate to send the light into the eyepiece at E. Thus by adjusting the rotation speed of the mirror until he could see the final reflected beam at E, Foucault could determine how long it took the mirror to turn through the requisite angle while light was making the round trip from M to B and back.

Foucault's experiment was performed entirely within the laboratory, and his total light path was only about 20 m; even so, he found the speed of light to an accuracy better than 1%. An adaptation of Foucault's method was later applied by Albert A. Michelson (1852–1931), the first American physicist to win the Nobel prize. Michelson measured the speed of light a number of times between 1878 and 1926. His most accurate determinations (1924 to 1926) were made by passing the light beam from a rotating eight-sided mirror on Mount Wilson, Southern California, to a stationary mirror on Mount San Antonio, 35 km away. Michelson arranged to have the distance between the two peaks surveyed with the help of the United States Navy and the National Bureau of Standards to a precision of better than 1 cm; it was, at that time, the most precise land survey ever attempted.

Michelson's best result at Mount Wilson gave, for the speed of light in air, the value 299,729 km/s. But light is slowed in passing through transparent media; in air it is retarded by nearly 70 km/s. Michelson began the construction of a mile-long vacuum tube on the Irvine Ranch in Southern California, through which he had hoped to measure the speed of light in empty space. Unfortunately he died before the experiment could be completed, and without his expert touch the results were not as good as had been hoped.

Today electronic timing techniques can measure the speed of light with high accuracy completely in the laboratory, eliminating the necessity of sending light over long distances. The modern

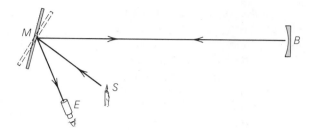

Figure 13.2 Principle of Foucault's method of measuring the speed of light.

accepted value for the speed of light in a vacuum, c, is 299,792.458 km/s.

(c) The Special Nature of the Speed of Light

The speed of light is very unlike other speeds that we normally encounter. First, of course, it is very great. Indeed, the speed of light is an absolute barrier. Nothing can go faster than light; in fact, no material body can ever travel quite at the speed of light. But in addition to these properties, the speed of light is always the same for all observers, irrespective of how they may be moving with respect to each other or to the source of the light. This absoluteness of the speed of light (and other electromagnetic radiation) is predicted by Maxwell's equations. But it is not something one would be likely to expect from "common sense." It is, indeed, a most remarkable thing.

Consider, for example, a material body, such as a bullet fired from a pistol. There is a certain muzzle velocity of the bullet, with respect to the gun, but the speed of the bullet with respect to an observer also depends on the speed with which the gun is moving when it is fired. If the bullet is discharged from a moving car, for example, its speed would certainly have the speed of the car added to its own speed. Alternately, one can, in principle, catch up with a speeding bullet, or even outrun it. (Astronauts in orbit, for example, travel much faster than bullets.) None of this is so for light! No matter how fast you approach or recede from the source, the speed of light, with respect to you, is always the same as if you and the source had no relative motion at all. Even if you could race from the earth in a spaceship at 99% the speed of light, and a colleague on earth were to send you a light signal, when that light caught up with your ship, and entered the rear port, its speed, with respect to the ship, would still be *c*. The speed of light depends in no way on the speed of the source.

But neither does the speed of sound, which is a compressional wave that travels through the atmosphere with a (rather slow) speed that depends on the temperature and other characteristics of the air. Similarly, ocean waves travel across the surface of the water as a transverse wave—the water level rising and falling vertically while the wave moves forward—and their speed depends on the wind and water conditions. But you can outrun sound waves, as is done in a supersonic airplane, and you can swim into ocean waves, increasing their speed with respect to you. Not so for light! No matter how fast you move, or in what direction, light waves approach you with that same speed—*c*. You can race forward to meet the waves of light, like the swimmer in the ocean, and, to be sure, that light will reach you sooner than if you were stationary, just as Roemer found for the times of eclipses of Jupiter's satellites, but the speed of that light when it reaches you is nevertheless the same, with respect to you, as if you were not moving; the *speed* of the light from Jupiter is the same all year round.

Light (and other electromagnetic radiation) behaving as predicted by Maxwell's theory, therefore, does not act the way one would expect from considering only the science of Newton. Does light point out a conflict between Newtonian mechanics and Maxwellian electrodynamics?

The existence of just such a contradiction was realized in 1895 by a 16-year-old schoolboy in the Luitpold Gymnasium in Munich. The boy was regarded as backward and indifferent by his teachers and was advised to leave the gymnasium without a diploma, because he "would never amount to anything and his indifference was demoralizing." But his thought experiment at that time directed his thinking along lines that 10 years later would lead him to his special theory of relativity.

Albert Einstein (1879–1955) reasoned that it should be possible to catch up with any uniformly moving object, after which the relative velocity of the two would be zero. But if you could catch up with a light beam—an electromagnetic wave—you would find it still oscillating back and forth in time, and varying in intensity in space, but not moving! But not only is no such electromagnetic field known, it is impossible according to Maxwell's equations, which say it must be moving with a speed *c*. Here is surely a contradiction; either Max-

well's equations must be wrong, or our fundamental Newtonian concepts must be wrong. Yet all the predictions of Maxwell's theory that could be tested turned out to be correct. The electronic technology available to us today certainly attests to the power and success of electromagnetic theory.

But to young Einstein there was also a strong philosophical reason for suspecting that Maxwell was right: the *principle of relativity*.

13.2 THE PRINCIPLE OF RELATIVITY

The principle of relativity states that there is no physical experiment by which one can detect his state of uniform relative motion. What this means is that if two observers, moving uniformly with respect to each other, perform the identical experiment in their own moving environment they will obtain identical results, so neither can say, from anything the experiment told him, that he was or was not moving, or how fast he was moving.

For example, two people standing in the aisle of an airliner going 1000 km/hr can play catch exactly as they would on the ground. On that same airplane you can drop a heavy and light object together, and they will hit the cabin floor at the same time, and they fall at the same rate as they would if you had dropped them on the ground (provided that the airplane is moving uniformly—in a straight line at a constant speed). You can play table tennis quite normally on a moving ship on a calm sea. You can swing pendulums in an automobile (so long as it is not turning or accelerating in some other way) and they will swing in the same way, with the same periods, obeying the same pendulum laws as do pendulums in the laboratory. When you are moving uniformly, you experience no physical sensation of speed, or any other sensation that will tell you that you are in motion. You can, of course, look out the window and see the ground moving by, but if you were stationary and the *ground* were moving you would feel the same and see the same thing. It is common to sit in a train in a station, and momentarily wonder whether it is your train or the one on the next track that starts to move. For that matter, none of us can feel the motion of the earth carrying us about the sun with its orbital speed of 30 km/s; so emphatically do we not feel it, that scarcely three centuries have elapsed since it has been generally accepted that, indeed, the earth *does* move.

But does this principle of relativity apply only to mechanics, or does it apply to electromagnetic phenomena as well? All experiments indicate that it does apply to electromagnetic experiments, and that the principle of relativity is quite general. Do not radios and tape recorders work the same on an airplane as in the house? You can pick up iron filings with a magnet just as easily in a moving automobile as in a classroom.

On the other hand, if the speed measured for light depended in any way on the velocity of the observer, then relatively simple experiments could be performed that would reveal the observer's motion. All such experiments invariably fail. Like Einstein, we are forced to the conclusion that you cannot catch up with a light beam; light will always have the same speed (in a vacuum)—*c*—and the prediction of Maxwell's electromagnetic theory is correct. The principle of relativity holds, and there is no experiment—whether mechanical, or involving electricity and magnetism, or light, or anything else—by which we can detect our state of uniform motion.

The principle is profound, for if it is impossible to detect uniform motion, the idea of *absolute motion*—motion with respect to absolute space, as envisioned by Newton—can have no meaning. There can be no absolute reference frame or coordinate system that is guaranteed to be at rest in the universe, and with respect to which other motion can be referred. All we can define is *relative* motion with respect to something else. Thus the earth moves 30 km/s with respect to the sun; the sun moves 20 km/s with respect to the average of its stellar neighbors; these nearby stars all move at about 250 km/s with respect to the center of our Galaxy; and so on. But there is no way to know how fast we are "really" moving, with respect to absolute space.

But it is a bizarre thing that two observers, one stationary with respect to a light source, and one moving rapidly away from (or toward) it will still measure the light from it to be approaching them with the same speed. All the seemingly strange results of special relativity come about because of that bizarre fact, and once we can swallow it, everything else in relativity makes perfect sense. Before proceeding, therefore, let's make sure we all understand that it is the real world we are talking about, and not fantasy. We describe briefly, there-

fore, a few (of very many) observations and experiments that demonstrate the absoluteness of the speed of light.

(a)   *c* Cannot Depend on the Motion of the Source

A simple experiment to show that the speed of light cannot depend on its source is provided by nature herself, and was pointed out by the Dutch astronomer Willem de Sitter (1872–1934) early in the century. Many stars are found in double-star systems—in which the two stars revolve about their common center of mass. Take one star, say, the brighter, in such a system whose orbit lies roughly edge on to our line of sight (Figure 13.3). Suppose the orbital speed of that star about the center of mass of the system is v, and consider what would happen if the speed of the light it emits included the speed of the star itself, just as the speed of a bullet fired from a moving automobile has the speed of that car added to it. Then when the star approaches us, at point A in Figure 13.3, light from it should be traveling toward the earth at a speed of $c + v$, and when the star is moving away from us in its orbit, at B in the figure, its light should approach earth with a speed of $c - v$. To be sure, the speeds of the stars in binary systems (v) are very small compared to the speed of light (c), but the stars are very far away and over the many years it takes their light to reach us, the faster beam, traveling at $c + v$, can gain considerably over the slower beam, traveling only at $c - v$. If the distance to the binary system were just right, we could

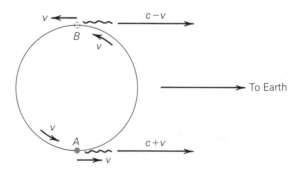

Figure 13.3 If the speed of light depended on the speed of the source, the light emitted by a binary star in a binary system would have a speed toward the earth that depended on the location of the star in its orbit.

be receiving light from the star at position A at the same time as the light sent to us at a slower speed at an earlier time, when the star was at position B. A little thought will show that under some circumstances we could be seeing the same star in a double-star system at many different places in its orbit at once, and analysis of the orbit would end in hopeless confusion. But we have actually analyzed the orbital motions of stars in thousands of double-star systems with distances ranging from a few light years to many hundreds of light years, and the orbital motions are all well-behaved, with the stars moving in accordance with Newton's laws. The speed of light from them therefore cannot include the speeds of the stars themselves.

Further proof that the speed of light is independent of the speed of its source comes from the nuclear physics laboratory. In nuclear accelerators, subatomic particles moving at nearly the speed of light are often observed to change form (*decay*) and emit photons, but these photons are always observed to move with the normal speed of light, c, with respect to the laboratory.

(b) c Cannot Depend on the Motion of the Observer

The most famous experiment showing that the speed of light does not depend on the motion of the observer was performed in 1887 in Ohio by A. A. Michelson and E. W. Morley, but its results have been confirmed by hundreds if not thousands of other experiments since then. Michelson and Morley, unaware at that early date of the principle of relativity, were attempting to measure the absolute speed of the earth through space by measuring the speed of light in two different directions.

To illustrate the idea of the Michelson-Morley experiment, consider three hypothetical astronauts, Able, Baker, and Charley, as shown in Figure 13.4(a), all stationary in space. Suppose further (contrary to experiment, as we shall see) that the speed of light is constant through absolute space. Baker and Charley are each 4 light years (LY) away from Able, but in directions at right angles to each other. Able sends radio signals (which travel with the speed of light) to Baker and Charley at the same time, and those signals reach their destinations 4 years later; immediately Baker and Charley respond, and Able receives their answers simultaneously, 8 years after his original transmission.

Now suppose the three astronauts maintain their relative positions, but all three are moving at 60% the speed of light, and in the direction from Able toward Charley. As before, Able sends out the two messages, and, by our supposition, those signals move at the same speed, c, but with respect to stationary absolute space. This time, however, Charley is moving away from the point where the signal was emitted at $0.6c$ [Figure 13.4(b)], so waves from the signal approach Charley at only $0.4c$, and take 2.5 times as long to reach him as before—that is, 10 years. On the other hand, Able is approaching the point where Charley sends back his response, at $0.6c$, so Able moves forward to meet Charley's transmission at a relative speed of $1.6c$; thus those waves take only 2.5 years to span the 4 LY, and reach Able 12.5 years after his original transmission.

But how about the message going from Able to Baker and back? The radio waves that reach Baker from Able must be directed *ahead* of Baker's position (B) as seen from Able (A) at transmission time, just as a hunter must "lead" his running prey [see Figure 13.4(c)]. Thus the message that reaches Baker from Able travels an oblique path, and since Baker's speed is 60% that of the radio waves, Baker has traveled 0.6 times as far as the message has when it catches him at P. The three points—A, B, and P—make a right triangle. Since BP and AP are in the ratio 3 to 5, the theorem of Pythagoras tells us that AB must be $4/5$ of AP; but AB is 4 LY, so the radio message, traveling at the speed of light, took 5 years to reach Baker. The same geometry holds for the return message, so Able receives Baker's reply 10 years after his original transmission, and 2.5 years ahead of hearing from Charley!

We selected a speed and separation for the astronauts of 60% c and 4 LY to make the arithmetic easy. What is the effect, in general, for a speed of v and separation L? The signal from Able to Charley has speed $c - v$, and the return one from Charley to Able has speed $c + v$. Thus the total round-trip light travel time in this direction parallel to the motion of the astronauts is

$$T_\parallel = \frac{L}{c - v} + \frac{L}{c + v} = \frac{2L}{c} \frac{1}{(1 - v^2/c^2)}.$$

Now the effective speed of the light from Able to Baker (and return) in a direction perpendicular to the motion is $\sqrt{1 - v^2/c^2}$, from the theorem of

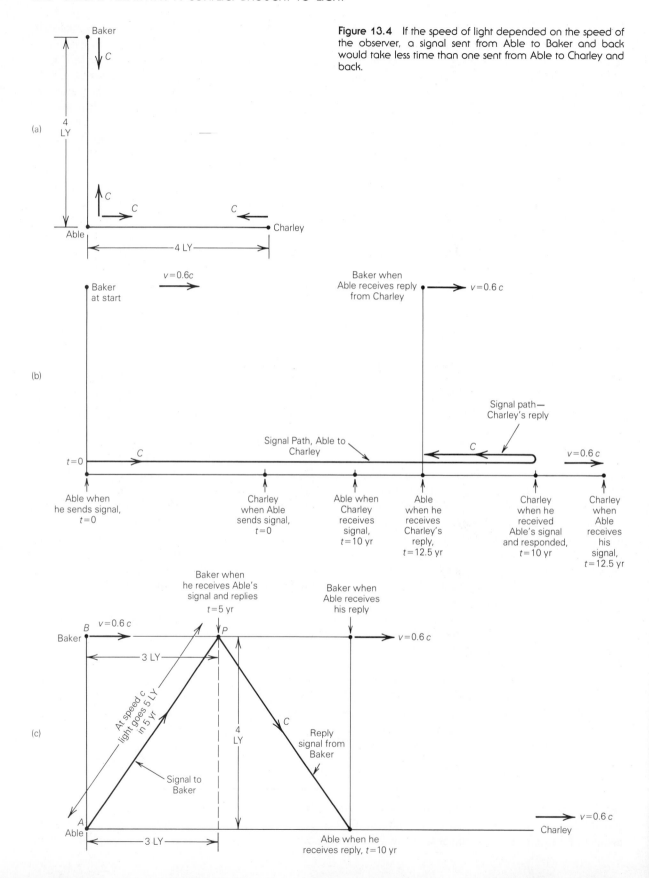

Figure 13.4 If the speed of light depended on the speed of the observer, a signal sent from Able to Baker and back would take less time than one sent from Able to Charley and back.

(a)

Baker

4 LY

Able   Charley

4 LY

(b)

v=0.6c

Baker at start

Baker when Able receives reply from Charley   v=0.6 c

Signal Path, Able to Charley

Signal path— Charley's reply

t=0   C   C   v=0.6 c

Able when he sends signal, t=0

Charley when Able sends signal, t=0

Able when Charley receives signal, t=10 yr

Able when he receives Charley's reply, t=12.5 yr

Charley when he received Able's signal and responded, t=10 yr

Charley when Able receives his signal, t=12.5 yr

(c)

Baker when he receives Able's signal and replies t=5 yr

Baker when Able receives his reply

B   v=0.6 c
Baker   P   v=0.6 c

3 LY

At speed c light goes 5 LY in 5 yr

4 LY

C

Reply signal from Baker

Signal to Baker

A
Able   v=0.6 c
Charley

3 LY

Able when he receives reply, t=10 yr

Pythagoras, so the round-trip time for that perpendicular signal is

$$T_\perp = \frac{L}{c\sqrt{(1 - v^2/c^2)}} + \frac{L}{c\sqrt{(1 - v^2/c^2}}$$

$$= \frac{2L}{c} \frac{1}{\sqrt{(1 - v^2/c^2)}}.$$

Hence,

$$T_\parallel = \frac{T_\perp}{\sqrt{(1 - v^2/c^2)}} = \gamma T_\perp,$$

where we define γ by $1/\sqrt{(1 - v^2/c^2)}$. Thus if Able had not known his speed through space, he could have deduced it from the ratio of $T_\parallel$ to $T_\perp$.

Michelson and Morley performed their experiment in a basement laboratory, but the principle is the same. They hoped to determine the absolute speed of the earth through space by measuring the difference in times required for light to travel across distances in the laboratory that were at right angles to each other. The two light paths were set up by multiple reflections between mirrors on the horizontal surface of a heavy sandstone slab. The total effective round-trip light path for each beam was about 22 m, and the slab floated on a layer of mercury to reduce vibration, and also so that the slab could be rotated easily.

At this point, we should consider what kind of difference we might expect. The highest speed for the earth that Michelson and Morley knew about for sure was its 30 km/s orbital velocity—only 10^{-4} that of light. The ratio of the travel times, γ, introduced by that relatively tiny speed of the earth, differs from unity by only 5 parts in a thousand million. The perpendicular beam should return ahead of the parallel beam by only about 10^{-5} cm, about a fifth of a wavelength of visible light! How can one determine so slight a lead of one beam over the other?

Here is where the ingenuity of Michelson came into play. He arranged that the two returning beams come together and interfere with each other, producing that familiar pattern of alternately dark and light fringes (Section 10.1a). Now, obviously no one could assume that the light paths of the two beams were precisely of the same length in the first place, so they will return with some unknown

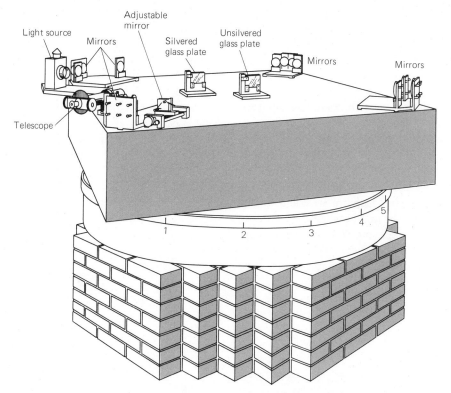

Figure 13.5 Schematic illustration of the Michelson-Morley apparatus. (From "The Michelson-Morley Experiment" by R. S. Shankland. Copyright © 1964 by Scientific American, Inc. All rights reserved.)

phase difference between them (that is, the crests of one beam will not coincide with those of the other), and that phase difference depends both on the slightly unequal lengths of the light paths and on the effect due to the earth's motion that was being sought. Consequently, the observation of the fringe pattern does not, in itself, reveal directly the difference in light travel time. That was the reason for being able to rotate the slab supporting the apparatus. Turning it through 90° interchanges the roles of the beams; if one happened, at the outset, to be parallel to the direction of motion of the apparatus due to the velocity of the earth and the other was perpendicular to that direction, after rotation the first beam would be perpendicular and the second parallel to the motion. This would change the difference in arrival times of the beams by exactly twice the amount Michelson and Morley were trying to measure. The pattern of fringes would thus shift, and the amount of shift would reveal the speed of the earth.

Michelson and Morley made their observations at noon and at 6:00 P.M. on several consecutive days beginning with July 8, 1887. During that six-hour interval the direction of the earth's motion must change by 90° relative to the laboratory. During each observation the slab was rotated slowly, and the fringe position was measured at 16 different positions around the circle. The earth's orbital velocity alone should cause a shift in the interference pattern of two fifths of a fringe width. And what did they observe?

Nothing! Rotating the slab made absolutely no difference in the light travel time along the two beams at right angles to each other. Their experiment was accurate enough to detect a small part of the earth's orbital motion, but there was no difference in light travel times. It was as if the earth were absolutely stationary, but Copernicus could not be wrong, for the success of gravitational theory shows that the earth has to be moving. Michelson and Morley thought their experimental setup was at fault, so they repeated the experiment with even greater accuracy. Still no result, and there has never been any in all the many, many times this and comparable experiments have been repeated. The only conclusion is that the speed of light does *not* depend on the motion of the observer; it is always *c* with respect to him.

Strange as this result seems, however, it is completely consistent with Maxwell's theory and with the principle of relativity.

13.3 THE SPECIAL THEORY OF RELATIVITY

How can we understand the bizarre properties of the propagation of light? Einstein gave the solution in his special theory of relativity. He showed that different observers in uniform relative motion (moving at constant velocity with respect to each other) perceive space and time differently. There are two assumptions on which the special theory is based: the principle of relativity and, embodied within it, the absolute constancy of the speed of light.

(a) Time Dilation

Let us imagine the construction of an ideal clock. Of many possible designs, we shall choose a clock consisting of two parallel mirrors, and a pulse of light reflecting back and forth perpendicularly between them. Each time the pulse passes from one mirror to the other, we shall count it as a "tick" of the clock. Because light travels at an absolutely constant rate, by carefully standardizing the spacing of the mirrors, we can agree that all such clocks should keep identical time.

On the other hand, what if an observer is moving very rapidly to the right with respect to us, carrying his two-mirror clock with him? Further, suppose his direction of motion with respect to us is parallel to the surfaces of the mirrors (see Figure 13.6). As far as he is concerned, his clock, in his own system, is at rest, for there is no experiment by which he can detect his own motion; consequently, as far as he is concerned his clock is op-

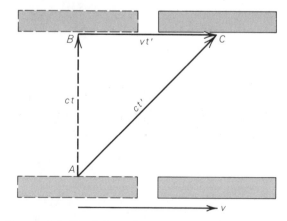

Figure 13.6 The light path in a moving observer's "ideal clock."

The 90-m (300-foot) radio telescope. *(National Radio Astronomy Observatory)*

The 100-meter radio telescope near Bonn, West Germany. *(Max-Planck Institut für Radioastronomie)*

Voyager spacecraft sent to Jupiter and Saturn. *(NASA)*

The earth and moon, photographed by Voyager 1 on its way to Jupiter. *(NASA)*

erating normally, with the light pulse reflecting perpendicularly back and forth between the mirrors. But as *we* see the situation, because of his clock's rapid motion to the right, the light pulse is not bouncing simply back and forth along a single line, but is following a slanting path. In other words, we see the moving observer's light pulse traveling farther between ticks than he sees it traveling. But he and we agree on the *speed* of the light pulse, so *we* must conclude that the interval between *his* pulses is longer than it is between ours; that is to say, his seconds are too long, and his clock is running slowly. On the other hand, he, aware of no motion on his part, argues that it is *we* who are moving to the left, and that it is in *our* clock that light travels obliquely, and that it is *our* clock that runs slowly. Each of us insists that the other's clock is slow.

By isolating a triangle in Figure 13.6 we can see by how much we disagree on the rate of passage of time. As far as our moving friend is concerned, he is stationary, and the light pulse has traveled vertically from A to B at a speed c, and the time it has taken to do so is t; since distance equals rate times time, the distance from A to B must be ct. But we see the light taking the slanting path AC, and requiring, at the same speed c, a longer time, t', to do so; thus we say that the pulse traveled a distance ct'. Meanwhile, our friend with his moving clock has gone from B to C, and if his speed relative to us is v, that distance must be vt'. The theorem of Pythagoras tells us that $c^2t^2 = c^2t'^2 - v^2t'^2$, from which we find, upon solving for t',

$$t' = \frac{t}{\sqrt{1 - v^2/c^2}} = \gamma t.$$

Thus what the moving observer thinks is an interval t, we see to be a longer interval t', and it is longer by the factor $1/\sqrt{1 - v^2/c^2}$. He, of course, regards his time intervals as normal and *ours* as too long by the same factor, γ.

Which of us is right? We both are. Time really *does* move at different rates in two different systems in uniform relative motion. We simply perceive time differently: Time is not absolute; each of us has his own private time.

(b) Reality of the Time Dilation

We must not think of this time dilation (the stretching out of time) between observers in uniform rela-

tive motion as some artifact of the clock we choose to construct. It is a very real thing. All processes slow down in moving systems; moving observers actually age more slowly than we do.

Nature provides a spectacular example of time dilation. The upper atmosphere of the earth is continually being bombarded by cosmic rays—atomic nuclei moving at very nearly the speed of light. When a cosmic ray particle strikes a molecule in the upper air, it breaks the molecule into a number of subatomic particles. Common among these are particles called *muons*. A muon is rather like an electron, but has about 200 times the mass. Muons spontaneously *decay*, turning into electrons, and emitting certain other radiation, in an average time of 1.5 millionths of a second. The muons formed by collisions of cosmic rays high in the atmosphere are moving at very high speeds, close to that of light. But even if they moved *at* the speed of light, they could travel, on the average, only about a half kilometer before decaying. Now the muons are formed at altitudes of 10 to 20 km; yet they rain down to the surface of the earth in enormous numbers. In fact, the muon is the principal kind of cosmic ray particle observed at ground level. If they are formed 15 km above the ground and decay before having time to travel as far as 1 km, how can we observe them at the bottom of the atmosphere?

As far as the muons are concerned, they *do* decay on the average in 1.5 millionths of a second, but because of their high speeds, as we observe them their time has slowed down, so that they have time to go very much farther than 0.5 km before decaying, and, in fact, most of them survive all the way to the earth's surface. Muons are also observed to live very much longer when they are accelerated to high speeds in the nuclear physics laboratory. In 1976 at CERN, the international laboratory of the European Council of Nuclear Research in Geneva, muons were accelerated to a speed of 0.9994c; the formula for time dilation predicts that their time should slow down by a factor of 30 at that speed. Indeed, the average lifetime of those high-speed muons was 44 millionths of a second, 30 times the 1.5 millionths of a second they survive at rest. Note that we are not speaking here of light pulses bouncing between mirrors, but of muons waiting to disintegrate; time dilation is not just a strange property of our light clock, but a fundamental property of time itself!

Would people live longer if they were rapidly moving? Not to their own way of thinking, of course, for they would have no sensation of moving. But relative to *us* they most certainly would age more slowly. In principle, long space trips could be made by astronauts if they were moving near enough to the speed of light. If we were to send a manned spaceship at a speed of $0.98c$ on a round trip to a star 100 LY away, the round trip would take just over 200 years of our time, but time for the astronauts would slow down by a factor of about 5, and on their return they would be only 40 years older than when they left. (Although such relativistic space travel is theoretically possible, the virtually prohibitive energy requirements make it unfeasible in practice—see Chapter 36.)

(c) Contraction of Length and Distance

Let us return to those astronauts traveling to a star 100 LY away at 98% the speed of light. If they make the trip in 20 years of their time (40 years round trip) does that mean that they have traveled at 5 times the speed of light? No, for lengths (and distances) as perceived by different observers in uniform relative motion are also different. As perceived by the astronauts moving $0.98c$ with respect to the earth, the earth and star are moving at that same speed in the opposite direction. The astronauts see the separation of the earth and star to be very much less than as perceived by earthlings; in fact, they find the distance from earth to star to be just under 20 LY. If a system is in uniform motion with respect to us, we see all dimensions in that system that lie along the direction of relative motion to be *shorter* than as perceived by an observer in that moving system. He, on the other hand, sees lengths in *our* system (that lie parallel to the direction of relative motion) to be shorter than we see them. All objects in a moving system, in other words, appear foreshortened in the direction of motion.

We can see how this foreshortening must come about by reconsidering the astronauts Able, Baker, and Charley, who are lined up at right angles to each other. Suppose, as before, they are moving with respect to us along the direction from Able to Charley. We have already seen that there is no way that they can detect their own motion; thus if Able sends signals to Baker and Charley and receives simultaneous replies, he must conclude that they are equidistant from him. But not so for us, for we see Charley moving away from Able's signal until it

catches up with him, and then Able rushing forward to meet the return signal from Charley. And we see the signal from Able to Baker, and Baker's return signal, traveling on slanting paths, as shown in Figure 13.4. As we found before, if Baker and Charley are equidistant from Able, we should see Able receive Baker's reply first. Thus if we see the two signals return to Able at the same time, we must conclude that Charley is *closer* to him than Baker is. As judged by Able, Baker and Charley are the same distance from him, but to us the moving system of astronauts is foreshortened in the direction of motion. Only a bit of algebra is needed to show that the factor of foreshortening is just the same factor by which time intervals in the moving system are too long. That is, a distance in the moving system, along the direction of motion, that the moving observer would say is D, we would say is only $D\sqrt{1 - v^2/c^2}$. Of course the moving observer sees *our* distances as foreshortened, not his own. *Length* is just as private a matter as time is!

═══════════════════════════

(d) THE "TWIN PARADOX"

There is a famous story, the so-called "twin paradox" (which is not really a paradox, as we shall see), describing a debate between two identical twins, Peter and Paul. Peter, the astronaut, travels to a remote star and returns, aging less than his brother, Paul, who stays on earth. But Peter claims that he was actually stationary all the time and that it was the earth that rushed away from him, and then came back again, carrying Paul. Thus, argues Peter, it is really Paul who has aged less. Who is right? It turns out to be Peter who ages less, for he had to change direction to return to earth and hence is *not* an equivalent observer. To see how this comes about, I have adapted the discussion by N. David Mermin in his book, *Space and Time in Special Relativity* (Chapter 16).

Imagine that Peter makes a trip to the star Vega, 25 LY distant, at a speed that is 99% that of light (0.99c). To keep track of what is happening, both Peter in the spacecraft and Paul on earth are equipped with a flasher that sends out a brilliant signal exactly every second.

What does Paul see when Peter is on his way to Vega? Since Peter is moving at 0.99c, his time, relative to Paul, is slow by a factor $\gamma = 1/\sqrt{(1 - v^2/c^2)} = 1/\sqrt{(1 - 0.99^2)} = 7.09$. This fact alone would slow the frequency of Peter's flashes, as Paul sees them, to one every 7.09 s. But in addition, since Peter is moving away at 0.99c, each successive flash has farther to go to reach Paul. In the 7.09 s between Peter's flashes (as

Paul sees them), Peter moves 7.09 s × 0.99c = 7.02 light seconds farther away, and it takes light 7.02 s to go that extra distance, so Paul sees Peter's signals coming every 7.09 + 7.02 = 14.11 s.

Peter, moving away from Paul, sees exactly the same effect, for Paul is moving relatively away from Peter, and Paul's flashes reach the spacecraft every 14.11 s.

Now Peter, moving at 0.99c, judges the distance to Vega to be 25/γ = 25/7.09 = 3.53 LY, and he arrives there in 3.53/0.99 = 3.56 years. During that time, Peter has received flashes from earth every 14.11 s, but he knows that those flashes were emitted one each second of earth time, so he reasons that when he reaches Vega only 3.56/14.11 = 0.25 years had passed on earth while those pulses were emitted.

Immediately upon reaching Vega, Peter turns around and starts his return to earth at 0.99c. The distance home is the same as the distance out, of course, so Peter arrives at earth after another 3.56 years, or after a total journey (of his time) of 7.12 years. During his return trip, how much time does he calculate to have passed on earth?

While Peter approaches earth he perceives, as before, that earth time is running slower than his by the factor γ = 7.09. Thus he would expect Paul to be emitting pulses every 7.09 s (of Peter's time), just as during the outward journey. But this time Peter *approaches* Paul at 0.99c, so during the interval of 7.09 s between the emission of those pulses, Peter has closed the gap by 7.09 s × 0.99c = 7.02 light seconds. Thus Peter's approach toward Paul saves 7.02 s of the 7.09 s that would otherwise separate the arrival of Paul's pulses at the spacecraft, so Peter receives the pulses every 7.09 s − 7.02 s = 0.07 s, or at the rate of 14.11 per second (what may appear to be a small arithmetic error here and elsewhere arises because I have rounded all of the figures to two decimals; I have, however, carried out the calculation to higher precision, so the answers given here are actually the correct ones). In other words, Peter's movement toward Paul at 0.99c more than makes up for the slowing of Paul's time as perceived by Peter, and he receives Paul's flashes about 14 times as frequently as he would if he and Paul were not moving with respect to each other.

During the 3.56 years of Peter's time that he is returning home, he receives signal flashes from Paul at the rate of 14.11 per second. But he knows that Paul emitted them, in Paul's own time, at the rate of one each second. Consequently, Peter realizes that the interval of time that has elapsed on earth must be 14.11 times as long as the 3.56 years of his time in the spacecraft. Peter realizes, therefore, that while he is returning home Paul has aged 14.11 × 3.56 = 50.25 years. The total amount that Paul has aged during Peter's round trip to Vega is therefore 50.25 + 0.25 =

50.50 years. Meanwhile, Peter has aged just 7.12 years.

Now let's explore things from the point of view of Paul, on earth. He received flashes from Peter every 14.11 s while Peter was en route to Vega, as we have seen above. However, Paul does not receive that last flash emitted by Peter at Vega, just before he began his return trip home, for 25 more years, because Vega is 25 LY away. On the other hand, Paul knows that Peter, traveling at 0.99c, takes 25/0.99 = 25.25 years to reach the star. Thus Paul easily calculates that by the time he has received that last flash emitted by Peter at Vega, Peter has been gone from earth for 25.25 + 25 = 50.25 years. In fact, by that time, Peter is already only 2 × 25.25 − 50.25 = 0.25 years from home and is closing fast! Now during those 50.25 years of Paul's time, Peter had been emitting a pulse of light every second of his own time, but those flashes were received by Paul at the rate of one every 14.11 s. Thus Paul finds that Peter, while en route to Vega, must have aged 50.25/14.11 = 3.56 years—exactly what Peter himself had calculated.

Finally, during those 0.25 years between the time Paul received Peter's last flash from Vega and the time of Peter's splashdown in the Gulf of Iran, Paul receives all of those flashes that Peter emitted on his way home. They arrive at earth at the rate of 14.11 per second—just the rate that Peter received flashes from Paul during his return to earth. Paul, of course, knows that those flashes were emitted at one per second by Peter's time, so Peter must have aged, on his way home, by Paul's reckoning, 0.25 times 14.11, or 3.56 years, making a total aging for Peter on his round trip of 7.12 years—again, exactly in agreement with what Peter himself found. Meanwhile, Paul ages 2 × 25/0.99 = 50.50 years, again in agreement with Peter's calculations.

Note that Peter and Paul each received all of the pulses emitted by the other—all flashes are accounted for. Moreover, Paul and Peter, fully understanding how relative motion affects time, are in complete agreement on how old each of them is. There is no paradox; Paul and Peter are not equivalent observers. Peter, on his outward trip, received flashes every 14.11 s, and then *changed direction*, henceforth receiving flashes at the rate of 14.11 per second. Paul, who never moved, received flashes from Peter at the rate of one every 14.11 s during all but the last quarter-year of Paul's 50½-year absence; during that last 3 months, Peter received 14.11 flashes per second.

Now, although the traveling astronaut came back having aged far less than his brother at home, it is not correct to say that he bought any additional longevity by taking the journey. To be sure, he would likely outlive his brother after returning to earth, but in his own time he didn't get any extra years of life; he only *lived* those 7 years he was gone. His brother at home had a full life of 50 fun-filled, adventurous (perhaps even amorous) years!

(e) Relativistic Mass Increase

If different observers in uniform relative motion disagree on length and time, they must also dis-

agree on velocity, which is distance covered in a given time. Thus they must, in turn, disagree on such things as momentum and energy, which depend on velocity. But they do agree on the laws of physics and the results of physical experiments— such as the conservation of momentum.

Suppose Jane and Mary are astronauts in space, moving together so that their relative velocity is zero [Figure 13.7(a)]. At a given instant each fires an elastic missile, such as a billiard ball, to-

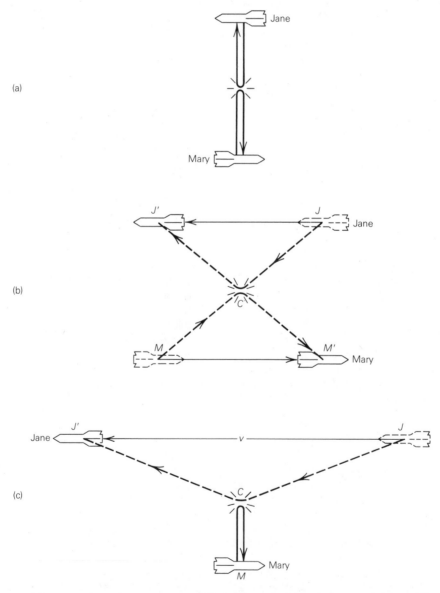

Figure 13.7 A hypothetical experiment involving collisions of elastic missiles, as seen from different perspectives. In all cases, momentum is conserved.

ward the other. The two balls are identical and are fired at identical speeds. They meet halfway between the spaceships at C, rebound, and return to the ships from which they were launched. The balls had equal but opposite momenta before the impact (since they were moving in opposite directions), and since each was turned about, they had equal and opposite momenta after the collision, so the total momentum is conserved, as it must be.

Now suppose that Jane and Mary are moving with equal speeds (with respect to us) but in opposite directions. As before, Jane and Mary discharge missiles toward each other, but because of their relative motion they fire the balls at $\mathcal{J}$ and M, respectively, and in directions perpendicular to their relative velocity. Because the balls move forward with the spaceships, they follow the dashed paths shown in Figure 13.7(b), meet at C, rebound, and return to their own ships at $\mathcal{J}'$ and M', respectively. Again, each is reversed in a symmetrical way and momentum is conserved.

But let's look at the last experiment from the point of view of Mary [Figure 13.7(c)]. Now Mary is stationary (in her own system), so her missile moves straight out perpendicular to the path of Jane's ship. But Jane's missile is released when she was way back at $\mathcal{J}$. As before, the two missiles meet at C, rebound, and Mary's missile returns to her ship, while Jane's returns to hers at $\mathcal{J}'$, as must happen, since it is the identical experiment we described in the last paragraph. Both Mary and Jane must agree that momentum is conserved (if not, one of them would be able to detect something about her own motion).

But now there is a problem, because Jane is moving rapidly with respect to the stationary Mary, hence Jane's time passes more slowly. Similarly, all physical processes in Jane's system must slow down, including the component of velocity with which Jane's missile is fired toward Mary, perpendicular to the direction of their relative motion. But if Jane's missile is moving more slowly than Mary's, we would expect it to have less momentum perpendicular to the direction of motion as well, since the balls are of the same mass. But then how can each turn the other around, conserving momentum? We would expect Mary's missile, with the greater momentum, to suffer less change in the impact, and not return to Mary's ship. But from our own vantage point [Figure 13.7(b)], we saw that it *did* return, and that momentum *is* conserved. The only explanation is that Jane's missile,

as observed by Mary, must have greater mass to compensate for its lower velocity in the direction perpendicular to that of their relative motion.

If two observers are in uniform relative motion, each will say that the masses of objects in the other's system are greater than they would be if they were at rest. The factor by which mass is increased is exactly the same as the factor by which time is slowed. If an object has a mass m_o when it is at rest, when it is moving with a speed v its effective mass is $m_o/\sqrt{1 - v^2/c^2}$. The quantity m_o is called the *rest mass* of the object.

The increase in mass of rapidly moving objects is not illusory; it is real. We observe it commonly in nuclear accelerators. As subatomic particles are sped up to nearly the speed of light, their masses increase manyfold, and enormously more power is required to provide them additional acceleration. Note that if the speed of a body were equal to the speed of light, v/c would be 1, and $\sqrt{1 - v^2/c^2}$ would be zero. Anything (except zero) divided by zero is infinite, so the mass of a particle moving with the speed of light would be infinite, which, of course, is impossible. Thus no material body (a body with nonzero rest mass) can ever travel at quite the speed of light. Here is the physical explanation of the fact that the speed of light is an absolute barrier that no body can cross. To accelerate a body of appreciable mass to a speed even very close to that of light would require absolutely tremendous amounts of energy; so far we have succeeded in making only objects of the mass of subatomic particles reach speeds close to that of light.

(f) Mass and Energy

All material bodies in motion possess energy of motion called *kinetic energy*. With a little algebra, it can be shown that the increase in mass of an object caused by its motion is its kinetic energy divided by the square of the speed of light, or equivalently, its kinetic energy equals its mass increase times c^2. Thus there is an equivalence between the mass and energy of a moving body. Einstein postulated that even when a body is at rest there is an energy equivalence to its rest mass, so that its total energy is equal to its total mass times c^2, a concept made famous by that equation that is the hallmark of special relativity,

$$E = mc^2.$$

With a little algebra, we can derive the above relation. We have

$$m = \frac{m_o}{\sqrt{(1 - v^2/c^2)}} = m_o(1 - v^2/c^2)^{-1/2}.$$

We can expand the factor $(1 - v^2/c^2)^{-1/2}$ with the binomial theorem, and obtain

$$m = m_o\left(1 + \tfrac{1}{2}\frac{v^2}{c^2} + \text{terms of order } \frac{v^4}{c^4}\right).$$

For v very much smaller than c, v^4/c^4 is so small that it can be ignored, and

$$m = m_o + \tfrac{1}{2} m_o v^2 \frac{1}{c^2},$$

or,

$$(m - m_o)c^2 = \tfrac{1}{2} m_o v^2.$$

We call $m - m_o$ the increase in mass, Δm, and note that $\tfrac{1}{2}m_o v^2$ is the kinetic energy due to the motion of the body. Thus Δmc^2 is the extra energy gained by motion, which is ΔE. But if kinetic energy has an equivalence to mass gained by virtue of its motion, what if the velocity is zero, and the kinetic energy is zero? Einstein postulated that there must be energy associated with mass even at rest, and proposed:

$$E = mc^2,$$

where E is the total energy, including that associated with the rest mass, and m is the total mass of a moving body. In particular, if $v = 0$, $m = m_o$, and the rest energy is

$$E_o = m_o c^2.$$

The equivalence of mass and energy stated in the above equation suggests that matter can be converted into energy and vice versa. Indeed, conversions in both directions are commonly observed in experiments with subatomic particles. For example, the electron has a twin called a *positron*, which has the opposite charge of the electron but identical mass. When a positron and electron come into contact, they annihilate each other, turning into photons of energy equal to the combined mass of the positron and electron times the square of the speed of light. Energetic photons can also combine to produce a positron and electron pair.

Because c^2 is a very large quantity, the conversion of even a small amount of mass results in a very great amount of energy. For example, the mutual annihilation of one gram of electrons and one gram of positrons (about $\frac{1}{14}$ ounce in all) would produce as much energy as 30,000 barrels of oil. Here is the source of nuclear energy. Commercial nuclear power plants do not, however, involve the complete conversion of the nuclear fuel, but only a small fraction of it. In the hoped-for hydrogen reactor of the future, hydrogen is converted to helium with the destruction of a little under half of one percent of the original hydrogen. Still, the conversion of only 15 kg (about 33 lb) of hydrogen into helium per hour annihilates enough matter to produce energy at the rate of the current United States oil consumption. We are still a long way from the technology to accomplish this, but the sun and stars derive their energy by a similar process, as we shall describe in Chapter 32.

The fact that mass can be converted into energy and vice versa means that the old concepts of conservation of mass and conservation of energy are not strictly correct. However, the total of mass and energy equivalence *is* conserved; that is, if we calculate the energy equivalence of all mass by multiplying that mass by c^2, the resulting figure, added to the total energy, is conserved. Of course, we could also divide the total energy by c^2 and add that to the total mass to obtain a quantity that is conserved.

(g) HOW VELOCITIES ADD

Let an object move with a speed U' in the system of coordinates of an observer moving at a speed v with respect to another observer, taken to be at rest. Let U_x' be the component of U' along the direction of relative motion of the two observers, and let U_y' be one of the components of U' perpendicular to the direction of relative motion. It can be shown that the corresponding components of U, U_x and U_y, in the system of the stationary observer, are

$$U_x = \frac{U_x' + v}{1 + \dfrac{v}{c^2}U_x'},$$

and

$$U_y = \frac{U_y'}{\gamma\left(1 + \dfrac{v}{c^2}U_x'\right)}.$$

With Newtonian reasoning, one would expect the velocity of the object in the moving system and the velocity of that system itself to simply add—that is, $U_x = U_x' + v$ —but we see that the velocity is "corrected" by the terms in the denominator.

As an example, suppose $U_x' = 0.9c$, and $v = 0.9c$. Newton would have said $U_x = 1.8c$, but relativity says:

$$U_x = \frac{1.8c}{1 + 0.81} = \frac{1.8c}{1.81} = 0.9944c.$$

If $U_x' = c$, we have

$$U_x = \frac{c + v}{1 + \dfrac{vc}{c^2}} = \frac{c+v}{(c+v)/c} = c;$$

that is, the speed of light is the same in both systems (as we have seen it must be!)

(h) Faster than Light?

It has been speculated that there could exist particles that always must move faster than light, and can never slow down to the speed c. Such hypothetical particles have been called tachyons, and experiments have been performed to search for them, to date with negative results. But as Nobel Laureate physicist Julian Schwinger (1918–   ) has pointed out, there is excellent reason for believing that tachyons cannot exist. If there *were* particles that could travel faster than light, then we could, at least in principle, use them to transmit signals at a faster rate than light can. But then we could build an ideal clock that ticks at more nearly the same rate for different observers in uniform relative motion, and all the special relativity effects we have described would not be correct; in fact if we could communicate with infinite speed (instantaneously), there would be no special relativity at all. Michelson and Morley's experiment would have given a positive result, muons would not arrive at the ground, electrons and protons would not gain mass in accel-

erators, and all of the many thousands of extremely accurate tests of relativity would not have turned out the way they did. In particular, E would *not* equal mc^2, and we would not have nuclear bombs and reactors. One could, of course, hypothesize that tachyons exist but are totally unobservable, but then we can never know of them nor detect their existence, directly or indirectly, and their existence would have no practical significance on the real world. Most physicists now discard the tachyon hypothesis.

But this does not stop many people from feeling that somehow science and technology will somehow find a way to "break the light barrier." Perhaps they read the wrong science fiction authors. Anyway, irrespective of Captain Kirk's taking the *Enterprise* to "warp II," it is impossible for a material body to ever reach the speed of light. It is not a technological problem, but a fundamental principle of nature. As we have seen, the mass of such a body would become infinite; it would become the entire universe itself.

Nor is there any need to travel faster than light, for, at least in principle, a person can travel at a speed as close to that of light as he wishes (given enough energy), and the closer his speed is to light's, the smaller all distances ahead of him become. Our hypothetical astronauts going only 98% the speed of light could reach a star 100 LY away in 20 years, but by going even closer to the speed c they could make the trip in a far shorter time. As one approaches c, his time slows and distances shrink so that he can go anywhere in as short a time as he likes.

If one *could* travel at the speed of light, riding on a photon as it were, his time would stand still and he would be everywhere in the universe at once. Of course only a photon or other body of no rest mass can really do that—a body of pure energy. Massless bodies—photons and neutrinos*— can only travel at the speed of light; material bodies may approach that speed but never attain it.

---

*Neutrinos are particles released in certain nuclear reactions. They carry energy and have other properties, and can (with difficulty) be detected, but until recently were thought to have no rest mass. Some recent (1980) experiments suggest that a neutrino may have a very tiny rest mass, in which case it could not move with the speed of light. At this writing, none of the experiments is considered very definitive, and the question of whether or not neutrinos have rest mass is still open.

Table 13.1 **Gamma** $(1/\sqrt{1 - v^2/c^2})$ **for Various Speeds**

MOVING OBJECT	v	v/c	GAMMA (γ)
Automobile	100 km/hr	0.00000009	1.000000000
Concorde SST	2000 km/hr	0.000002	1.000000000
Rifle bullet	1 km/s	0.000003	1.000000000
Earth escape speed	11 km/s	0.000037	1.000000001
Orbital speed of earth	30 km/s	0.0001	1.000000005
10% light's speed	30,000 km/s	0.1	1.005
		0.5	1.155
		0.9	2.294
		0.98	5.025
		0.99	7.089
		0.999	22.37
Muons in CERN experiment		0.9994	28.87
		0.9999	70.71
		0.999999	707.1
		0.999999999	22360.7

13.4 CONCLUSION

Special relativity can be understood with very little effort in mathematics. Yet the concepts are very difficult to grasp. The mathematics is easy enough, but the ideas are totally alien to our experience, and present no easy conceptual hurdle. Why is this so? Because we have all grown up in a world where speeds around us are very small compared to the speed of light. All of the relativistic effects we have discussed depend on that factor $1/\sqrt{1 - v^2/c^2}$, usually denoted by the Greek letter gamma (γ). Values of gamma corresponding to several values of v/c are given in Table 13.1. Values in the table show by how much masses increase, lengths shrink, and clocks slow in moving systems. Until v/c is a pretty good-sized fraction, gamma is essentially equal to 1. In such low-velocity systems, Newton's laws of motion apply with admirable precision. Even the earth's speed about the sun—30 km/s—is only $0.0001c$, and gamma is equal to unity within one part in a hundred million. We have become used to the low-velocity world, and it has prejudiced our ideas of "common sense."

On the other hand, how about a hypothetical civilization living on another world in an environment where speeds close to that of light are commonplace? Relativity would not seem strange to them. They, like us, given enough time, would discover the laws of physics, but not in the same order. As Schwinger has put it, "They would have their Maxwell and their Einstein, but alas, no Newton."

EXERCISES

1. Do you think that two observers on systems that are rotating with respect to each other would find that all physical laws are the same in their two systems? Could they tell which observer was rotating more rapidly? If so, how?

2. Suppose a ball is thrown forward at 60 km/hr from an automobile moving at 100 km/hr. How fast is the ball moving with respect to an observer on the roadside? What if the ball is thrown toward the rear of the car with the same speed?

3. Compare the ways in which different observers compare the speed of the ball in the last exercise with how they compare the speed of light.

4. From the fact that light takes $16\frac{1}{2}$ minutes to cross the orbit of the earth, show that the speed of light is about 10,000 times that of the earth.

5. Prepare a pendulum consisting of a small weight at the end of a string exactly 40 cm long. Start the pendulum swinging and time it for 10 complete oscillations (one oscillation is to and fro). What is your result? Now take the pendulum into an automobile, try to arrange that the car drive as smoothly and at as constant a speed as possible, and repeat the experiment. Now what is your result?

6. Suppose a star in a binary system revolves about its companion with an orbital speed of 90 km/s (three

times the earth's orbital speed). Now suppose (which, as we have seen, is incorrect) that the speed of light emitted by the star toward us *did* depend on how fast the star was approaching us or receding from us.

a. How much faster, in km/s, would light be approaching us that is emitted when the star approaches us in its orbit, than when it is moving away (assume that the orbit is edge-on to our line of sight)?

b. One year is approximately 3×10^7 s. If the double star-system were 200 LY away, by how many kilometers would the faster approaching light, by the time it reached the earth, have gained on the slower light, emitted when the star was on the opposite side of its orbit?

c. How long (approximately) would it take the slower light to make up this distance?

d. How far does the star move in its orbit in this time? If the star's orbit were the size of the earth's, what fraction of its orbit would the star have traversed?

7. Show that in the Michelson-Morley experiment the predicted lead of the faster light beam over the slower one (due to the earth's orbital revolution) is 10^{-5} cm.

8. Refer to Table 13.1. What would we measure for the mass of a 100-kg body moving past us with a speed of 90% that of light?

9. According to Newton's laws the ordinary kinetic energy of a body of mass m moving at a speed v is $^1\!/_2 mv^2$. Calculate the kinetic energy of a body of mass 1 g moving with a speed of 10^6 cm/s (about one third the orbital speed of the earth). Now calculate the energy associated with the rest mass of the same body. How do the two energies compare?

10. By what factor does time slow for an astronaut moving 99.99% the speed of light? How long would it take an astronaut going that fast to make a round trip to a star 100 LY away: (a) according to people who stayed behind on earth? (b) according to his own time?

11. By what factor is the mass of an object increased if it moves: (a) 3/5 the speed of light? (b) 99.99% the speed of light?

Sir James Hopwood Jeans (1877–1946), English mathematician and physicist, was a pioneer in the theory of the formation of stars and galaxies and was an active researcher in the theory and evolution of the solar system. His theory that planets were formed from material drawn from the sun by a passing star is now considered incorrect, but he played an important role in influencing future investigation. *(Yerkes Observatory)*

14

THE SOLAR SYSTEM: ITS STRUCTURE AND ORIGIN

The ancient observer, who considered the earth to be central and dominant in the universe, regarded the sun, moon, and planets as luminous orbs that moved about on the celestial sphere through the zodiac.

Our solar system is indeed dominated by one body, but it is the sun, not the earth. Our sun, so important to us, is merely an ordinary, "garden-variety" star. Only careful scrutiny at close range would reveal the tiny planets to an imaginary interstellar visitor. First Jupiter, the largest, would be seen; then Venus and Saturn; and perhaps only with the greatest difficulty, the earth and other planets. Almost 99.9 percent of the matter in the system *is* the sun itself; the planets comprise most of what is left—the earth scarcely counts among them. The countless millions of other objects in the solar system, mostly unknown to the ancients, would probably remain unnoticed by a casual traveler passing through the solar neighborhood.

We turn now to those worlds of the solar system. They will be considered individually in detail in later chapters. Here we take only a brief look at some of the general characteristics of the solar system and remark on a few of the properties that its constituent worlds have in common.

14.1 INVENTORY OF THE SOLAR SYSTEM

The solar system consists of the sun and a large number of smaller objects gravitationally associated with it. These other objects are the planets, their satellites, the comets, the minor planets or asteroids, the meteoroids, and an interplanetary medium of very sparse gas and microscopic solid particles. The relative prominence of the various kinds of members of the solar system can be seen from Table 14.1, which lists the approximate distribution of mass among the bodies of the solar system. The last four entries in the table are order of magnitude guesses only.

Table 14.1 **Distribution of Mass in the Solar System**

OBJECT	PERCENTAGE OF MASS
Sun	99.85
Planets	0.135
Satellites	0.00005
Comets	0.01 (?)
Minor Planets	0.0000002 (?)
Meteoroids	0.0000001 (?)
Interplanetary medium	<<0.0000001

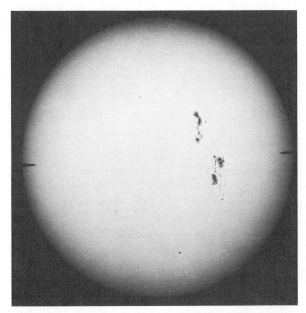

Figure 14.1 The sun. *(Mount Wilson and Las Compañas Observatories)*

(a) The Sun

The sun, practically, *is* the solar system. It is a typical star—a great sphere of luminous gas. It is composed of the same chemical elements that compose the earth and other objects of the universe, but in the sun (and other stars) these elements are heated to the gaseous state. Tremendous pressure is produced by the great weight of the sun's layers. The high temperature of its interior and the consequent thermonuclear reactions keep the entire sun gaseous. There is no distinct "surface" to the sun; the apparent surface we observe is optical only— the level in the sun at which its gases become opaque, preventing us from seeing deeper into its interior. The temperature of that region is about 6000 K. Relatively sparse outer gases of the sun extend for millions of kilometers into space in all directions. The visible part of the sun is 1,390,000 km across, which is 109 times the diameter of the earth. Its volume is $1^{1}/_{3}$ million times that of the earth. Its mass of 2×10^{33} g exceeds that of the earth by 333,000 times. The sun's energy output of 4×10^{33} ergs/s, or about 5×10^{23} hp, provides all the light and heat for the rest of the solar system. The sun derives this energy from thermonuclear reactions deep in its interior, where temperatures exceed 14 million K. We shall describe the sun, a typical star, in Chapter 32.

(b) The Planets

Most of the material of the solar system that is not part of the sun itself is concentrated in the planets. The nine known planets include the earth, the five other planets known to the ancients (Mercury, Venus, Mars, Jupiter, and Saturn), and the three discovered since the invention of the telescope (Uranus, Neptune, and Pluto). In contrast to the sun, the planets are small, relatively cool, and are solid or liquid. They are not self-luminous at visible wavelengths and shine by reflected sunlight.

We should say something of the nomenclature given certain groups of planets. We have seen (Chapter 3) that the two planets nearer the sun than the earth (Mercury and Venus) are called inferior planets, and the ones with orbits outside the earth's are called superior planets. The four innermost planets (Mercury through Mars) are also called *inner planets;* Jupiter, Saturn, Uranus, Neptune, and Pluto are then the *outer planets.* Finally, the four largest planets (Jupiter, Saturn, Uranus, and Neptune) are often called the *Jovian planets* (after Jupiter) or, occasionally, *major planets,* and the others, including Pluto, are the *terrestrial* (earth-like) *planets.*

The masses of the planets, in terms of the mass of the earth, range from 0.002 (Pluto) to 318 (Jupiter). The mass of Jupiter is greater than that of all the other planets combined. In diameter, the planets range from 3000 km (Pluto) to 143,000 km (Jupiter). Most, but not all, of the planets are surrounded by gaseous atmospheres. All but two of the planets are known to have natural satellites; Jupiter and Saturn lead with at least 16 known moons each (and new ones keep turning up in our explorations of those worlds with interplanetary probes).

All the planets revolve about the sun in the same direction—from west to east (counterclockwise as viewed from the north). Their mean distances from the sun range from 0.39 AU (58 million km) for Mercury to 39.46 AU (5900 million km) for Pluto. Their periods of orbital revolution range from 88 days for Mercury to 248 years for Pluto; the corresponding mean orbital velocities range from about 48 to about 5 km/sec. The orbits of the planets are all in nearly the same plane. Pluto is the planet whose orbit is inclined at the greatest angle (17°) to that of the earth; next is Mercury, whose orbit is inclined to the earth's by 7°. The orbits of the planets are also nearly circular; all have eccen-

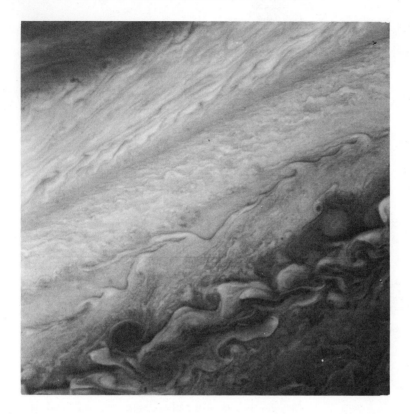

Figure 14.2 Mid latitudes of Jupiter, photographed March 2, 1979, by Voyager 1. *(NASA/JPL)*

tricities of less than 0.1 except for Pluto ($e = 0.25$) and Mercury ($e = 0.21$).

The planets all rotate as they revolve about the sun. Rotation is a turning of an object on an axis running through it, as distinguished from *revolution,* which is a motion of the object as a whole about another object or point. The Jovian planets are all rapid rotators; Jupiter rotates most rapidly, in a period of 9^h50^m. Mercury rotates $1\frac{1}{2}$ times during its 88-day revolution about the sun. Venus rotates still more slowly, in 243 days, but from east to west, reverse to the rotation of most of the planets (that is, *retrograde*). Some of the planets, especially Jupiter and Saturn, show marked oblateness or flattening because of their rapid rotation.

(c) Satellites

The satellites of the planets are the next most prominent members of the solar system. Only Mercury, and Venus do not have known satellites. Jupiter and Saturn have at least 16 each, Uranus has five, Neptune and Mars two each, and the earth and Pluto one each. Six of the 43 known natural satellites of the solar system are about as large as or larger than our moon. Only *Titan,* Saturn's largest satellite, is known to have an appreciable atmo-

sphere. Most of the satellites revolve about their planets from west to east, and most have orbits that are approximately in the equatorial planes of their primary planets. (The moon is an exception, with an orbit nearly in the ecliptic plane—Section 9.3.)

(d) Comets

Comets are chunks of frozen gases with solid particles (nuclei) that revolve about the sun in orbits of very high eccentricity, that is, very elongated elliptical orbits. Comets spend most of their time in those parts of their orbits that are very far from the sun, where they receive negligible radiant energy. However, as a comet moves in closer to the sun, it warms up and some of the volatile materials vaporize to form a cloud of gas and intermixed dust, or *coma,* around the solid nucleus. The particles and surrounding coma make up the *head* of a comet. When a comet is only a few astronomical units from the sun, the pressure of radiation from the sun and the electrical repulsion of ions in the solar wind (Chapter 10) sometimes force particles and gases away from the head to form a *tail*. The heads of comets are usually between 10,000 and 200,000 km across, and tails sometimes grow to lengths of many millions of kilometers. The entire mass of a typical

served. It is believed that there is a vast cloud of hundreds of billions of comets surrounding the sun at a distance of millions of millions of kilometers and that occasional perturbations from passing stars start some of those comets moving in toward the sun. Unlike the planets, which move in orbits that are all nearly coplanar, comets approach the sun along orbits that are inclined at all angles to the plane of the earth's orbit.

(e) Asteroids, or Minor Planets

The *asteroids*, or *minor planets*, are small planets, differing from the nine principal planets primarily in size. There are probably tens of thousands of them large enough to observe with existing telescopes. Ceres, the largest minor planet, has a diameter of 1025 km; about a thousand asteroids are more than 30 km across. Most are only a few kilometers or less in diameter. None of these objects is large enough to retain an atmosphere.

Like the principal planets, the asteroids move from west to east in elliptical orbits of small eccentricity and lie close to the plane of the earth's orbit. Typical minor planets have mean distances of about 2.5 to 3 AU from the sun and have periods in the range four to six years; thus, they occupy a region of the solar system that lies between the orbits of Mars and Jupiter. A few of the minor planets, however, have orbits that bring them close to the earth; several minor planets pass closer to the earth than any other known objects except the moon and meteoroids.

(f) Meteoroids

The number of small solid objects revolving about the sun that are too small to observe with telescopes is great indeed, and increases for objects of smaller and smaller size. These tiny astronomical bodies, too small to observe individually as each travels unhindered in its orbit, are called *meteoroids*. Their presence becomes known only when they collide

Figure 14.3 Saturn's small satellite, Mimas, with its heavily cratered surface, photographed by Voyager 1. *(NASA/JPL)*

comet, however, is less than one thousand millionth that of the earth. A comet is, therefore, a rather trivial entity.

More than a thousand comets have been observed, and usually some five to ten new ones are discovered telescopically each year. Most newly discovered comets have never before entered the inner part of the solar system, where they can be ob-

Figure 14.4 Comet Kohoutek, photographed December 7, 1973, at the Joint Observatory for Cometary Research, South Baldy Mountain, New Mexico. *(NASA)*

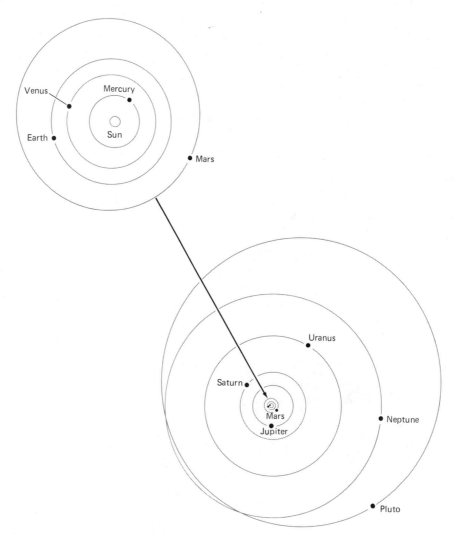

Figure 14.5 The relative sizes of the orbits of the planets. Two different scales are used to show the outer and inner planets.

with the earth, and plunging through the earth's atmosphere, heat with friction until they vaporize. The luminous vapors that are produced look like stars moving quickly across the sky and are popularly known as "shooting stars." Such phenomena are correctly called *meteors*. On a typical clear dark night, about half a dozen meteors can be seen per hour from any given place on earth. The total number of meteoroids that collide with the earth's atmosphere during a 24-hour period is estimated at 200 million. Meteoroids have also been recorded as they strike space vehicles.

Most meteoroids are associated with comets. Their orbits are very elongated (of high eccentric-

ity), and they approach the earth from any direction, at any angle to the plane of the earth's orbit. Frequently, they collide with the earth in groups or *showers,* and some such showers have been found to have orbits in common with known comets. Some meteoroids, on the other hand, revolve about the sun, as do the major planets, in orbits of relatively small eccentricity that lie near the principal plane of the solar system. These meteoroids are believed to be small fragments of asteroids. Rarely, one of them survives its flight through the earth's atmosphere and lands on the ground. It is then called a *meteorite*. A number of such fallen meteorites can be inspected in various museums. The largest known meteorites have masses of about 50 tons. Most are the size of pebbles. Chemical analysis reveals them to be formed of the same chemical elements that exist on earth and elsewhere in the cosmos.

The smallest meteoroids are slowed down by the atmosphere before they have a chance to heat up and vaporize. They are microscopic in size and are known as *micrometeorites*. They eventually settle to the ground as meteoritic dust. Micrometeorites are the most common kind of meteoroids and are the kind most frequently encountered by space rockets.

(g) The Interplanetary Medium

The interplanetary medium has two components: *interplanetary dust* and *interplanetary gas*.

Interplanetary dust can be thought of as a sparse distribution of micrometeorites throughout the solar system, or at least throughout the main disk that contains the orbits of the planets. Individual particles have been detected as they strike spacecraft. Collectively, the particles can be observed by the sunlight they reflect. On a dark clear night a faint band of light can be seen circling the sky along the ecliptic. This band of light is generally brightest near the sun and is best seen in the west within a few hours after sunset or in the east within a few hours before sunrise. Sometimes, however, it can be seen as a complete band across the sky. Because this light is confined to the region of the ecliptic or zodiac, it is called the *zodiacal light*. Spectrographic analysis of the zodiacal light shows it to be sunlight, presumably reflected from microscopic solid particles.

There is also a tenuous distribution of gas through the solar system. It is thought that the earth's atmosphere, rather than ending abruptly, must gradually thin out into the interplanetary gas. Evidence of interplanetary gas also comes from space probes, whose instruments have recorded rapidly moving atoms and charged atomic particles. High-altitude rockets, carrying cameras that photograph in the far ultraviolet, have recorded a faint illumination that is apparently light emitted by hydrogen gas, present either high in the earth's atmosphere or in interplanetary space. Practically all of the interplanetary gas consists of ions and electrons ejected into space from the sun. This flow of corpuscular radiation from the sun is called the *solar wind*. It is actually an outer extension of the solar corona (see Chapter 31).

We conclude that the region of interplanetary space contains minute, widely spread particles and very sparse gas. In the neighborhood of the earth, there are only a few ions per cubic centimeter. This is a far better vacuum than can be produced in any terrestrial laboratory.

14.2 THE PLANETS

The other planets of the solar system are of special interest, for they, like the earth, are worlds that revolve about the sun and derive their light and warmth from it. The planets are considered individually in Chapters 17 and 18. Here we shall summarize how certain information about them can be obtained.

(a) Some Basic Characteristics

The basic characteristics of a planet are its mass, its size, and its distance from the sun. From these data alone we can predict to some extent many of its other characteristics, for example, whether or not it can be expected to have an atmosphere, what kind of atmosphere, what its temperature is likely to be, and many other physical data.

The distances to most planets are found most accurately today by radar and from analyses of radio signals from our interplanetary probes. Before the development of these techniques (in the 1960s and 1970s), interplanetary distances had to be determined by surveying—specifically, by sighting the directions to the planets from different places on the earth's orbit, in principle the way Kepler mapped the orbit of Mars (Section 3.3a). We shall return to celestial surveying in Chapter 22.

The mass of a planet must be determined by measuring the gravitational acceleration that it produces on other objects. Four different methods have been used: (1) observing the acceleration a planet produces upon one of its satellites (Section 5.5), (2) observing the perturbations a planet produces upon the motions of other planets, (3) observing the perturbations a planet produces upon the motion of a close-approaching minor planet, and (4) observing the effects a planet produces on manmade space probes. The last technique has given the best results for all but the outermost planets.

As we have seen (Section 6.1), all of the planets in the solar system affect all other objects in it. Observations of planetary motions must be extended over long periods of time to detect the small changes brought about by these perturbations. More than a century ago gravitational theory had been refined to the point that just such observations were used to predict the existence of the planet Neptune.

With today's technology we can monitor accurately the motions of space probes, and with modern high-speed computers we routinely calculate the masses of all of the planets affecting them. When a space probe passes near a planet, it and the planet can be considered a two-body system, and they describe hyperbolic orbits about each other. Because of the planet's enormously greater mass, only the probe suffers a measurable change in its motion. A comparison of the orbital elements of the probe before and after the encounter reveals how much it was accelerated by the planet, and hence the planet's mass.

The diameter of a planet is found from its angular diameter and distance, as described in Section 9.2b.

(b)   Rotation of the Planets

All the planets are observed to rotate. At least four techniques have been employed to determine planetary rotation rates.

(1) The most direct method is to watch permanent surface features move across its disk. If a particular surface feature is observed from earth or a spacecraft for a large number of rotations, an accurate value for the time required for one rotation is obtained. (2) The rotation period of a rapidly rotating planet can be found from the Doppler effect in the lines in the spectrum of sunlight reflected from it. Since one side of the planet is turning toward us and the other side away, light reflected from the approaching limb is Doppler-shifted to slightly shorter wavelengths (relative to light reflected from the center of the planet's disk) and light from the opposite limb is shifted to longer wavelengths. Suppose the planet's image is focused across the slit of a spectrograph at the telescope and that the apparatus is oriented so that the slit runs along the equator of the planet's image. Then all the lines in the spectrum appear tilted (the top of the spectrum being light from one limb and the bottom being light from the other). A measure of the

amount of tilt reveals the rotation rate (Figure 14.6). The same technique is used to measure the rotation of galaxies (Chapter 37). (3) Sometimes a planet's rotation may be detected from variations in its brightness. Pluto, for example, seems to have surface irregularities so that different sides reflect light with different intensity. Thus as the planet rotates, it varies in brightness (as seen from the earth) with a period equal to the rotation period. (4) Radar (radio) waves beamed to a planet from earth or a space probe are reflected back, but are Doppler-shifted according to the line-of-sight velocity of the earth (or probe) and the planet. The waves reflected from the approaching limb of the planet thus are shifted to shorter wavelengths than those that are reflected from the center of the planet's disk, and radio waves reflected from the receding limb are shifted to longer wavelengths. Thus, although the transmitted beam may consist of waves of a very narrow wavelength range, those reflected back comprise a somewhat larger range of wavelengths. Measurement of this range of wavelengths (or *bandwidth*) gives the planet's rate of rotation.

(c)   Atmospheres

All of the planets except Mercury (and possibly Pluto) are surrounded by appreciable gaseous atmospheres. Some of the constituents of those atmospheres have been known for many years from earth-based spectrographic analysis of sunlight reflected from the planets. Sunlight reflected from the surface or a cloud layer of a planet must traverse part of its atmosphere, and the molecules of the gases in that atmosphere absorb certain wavelengths, leaving dark lines in the spectrum of the reflected light that are not present in the spectrum of direct sunlight. Proper identification of those lines reveals the identity of the atmospheric gases that produced them.

Except for the most remote planets, however, our modern knowledge of the planetary atmospheres comes from measures made with instruments carried on interplanetary space vehicles. Some of these instruments carry out spectroscopy from close range—in the ultraviolet and infrared as well as in visible light. The atmospheres of both Venus and Mars have also been penetrated with probes that have analyzed their gases directly. We shall return to this subject in Chapter 17.

Which gases are present in a planet's atmosphere must depend, at least in part, on how that

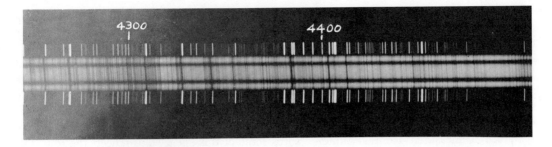

Figure 14.6 Spectra of Saturn and its rings, showing rotation. (*Lowell Observatory*)

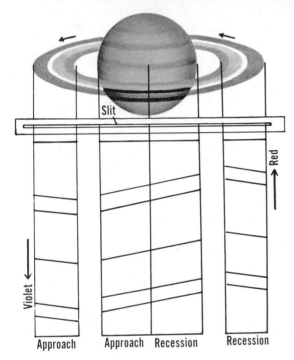

Kinetic Theory

Gases, like all matter, are composed of units of matter called *molecules*. Molecules are composed, of course, of the still smaller *atoms*. Molecules of pure elements consist of one or more atoms of that element. Molecules of a chemical compound consist of two or more atoms of two or more different kinds bound together. For example, a molecule of the compound *water* consists of one oxygen atom and two hydrogen atoms. The atomic structure of the molecules determines the identity of the substance they compose.

Molecules of a gas are in rapid motion, darting this way and that, frequently colliding with each other. At sea level on the earth there are some 10^{19} such molecules bouncing about in each cubic centimeter of air. The *kinetic energy* (or energy of motion) of a moving object is defined as $\frac{1}{2}mv^2$, where m is the mass of the object and v is its speed. Ordinary temperature, *kinetic temperature*, is a measure of the mean (or average) kinetic energy of molecules. Specifically, the absolute temperature of a gas, T, is related to the mean energy of its individual molecules by the formula

$$\overline{\tfrac{1}{2}mv^2} = \tfrac{3}{2}kT,$$

where the bar over the term on the left indicates that it is the average kinetic energy of the gas molecules, and k is Boltzmann's constant (in metric units, $k = 1.38 \times 10^{-16}$ erg/deg). If the above formula is solved for the mean speed of molecules in a given kind of gas, there results

$$\overline{v} = \sqrt{\frac{3kT}{m}}.$$

planet was formed and on its subsequent history. Most of the gases in the earth's atmosphere, for example, have escaped (that is, *outgassed*) from the earth's crustal rocks, and have changed subsequently due to photosynthesis and other chemical activity, whereas the constituents of Jupiter's atmosphere may reflect the original composition of the material from which Jupiter was formed.

On the other hand, the kind of atmosphere a planet has also depends on its ability to hold the various gases. The individual molecules of a gas are always in rapid motion; if their speeds exceed the velocity of escape of a planet (Section 5.4), that kind of gas can gradually "evaporate" from the planet into space. To investigate the conditions under which certain kinds of gases can be retained in a planetary atmosphere, we turn briefly to the kinetic theory of gases.

Thus we see that the mean speed of molecules in any particular gas is proportional to the square root of the temperature and inversely proportional to the square root of the mass of a single molecule of that gas. The higher the temperature, the faster the molecules move; at a given temperature, molecules of greater mass move slower, on the average, than those of smaller mass.

The temperature is measured from absolute zero; at $T = 0$, the mean energy of the molecules is zero. Here is the meaning of absolute zero: As gases are cooled, their molecules move more and more slowly, and at absolute zero all molecular motion ceases. This occurs at $-273°C$ ($-459°F$). Absolute, or Kelvin, temperature is thus the Celsius temperature $+273K$. (See Appendix 5.)

At any given time, some molecules move at less than the average speed, and others move faster. A few are moving at several times the average speed. An individual molecule, suffering frequent collisions, constantly exchanges energy with other molecules. Sometimes it moves relatively slowly; at other times it may get a good jolt in a collision and move far faster than the average. From kinetic theory we can calculate the relative numbers of molecules moving at various speeds if we know the kind of gas (and hence the mass of each molecule) and the temperature. We consider now what determines the temperature in a planetary atmosphere.

Effective Temperatures of Planets

The effective temperature of a planet is the temperature of a black body (Section 10.2), or perfect radiator, of the same size as the planet that receives the same amount of radiation from the sun that the planet absorbs. The effective temperature corresponds roughly to the kinetic temperature at the surface of the planet or at the level in its atmosphere from which it reradiates most of the energy it absorbs back into space. For Mars, whose atmosphere is nearly transparent, the effective temperature is approximately that of its solid surface. Venus and the Jovian planets, on the other hand, have opaque atmospheres, and their effective temperatures correspond to kinetic temperatures high in their atmospheres, from which is emitted the radiation we observe from them.

Just outside the earth's atmosphere, 1.37×10^6 ergs/s of solar radiation are incident on each square centimeter. This is called the *solar constant*. The earth's distance from the sun is 1 AU. Thus, from the inverse-square law of propagation of elec-

tromagnetic radiation (Section 10.1a), we see that a planet of distance r AU from the sun receives $1.37 \times 10^6/r^2$ ergs/s per square centimeter. The total solar energy intercepted by a planet each second is $1.37 \times 10^6/r^2$ times the cross-sectional area of the planet, which is just the area of a circle of radius R equal to the planet's, that is, πR^2. Thus the planet intercepts $\pi R^2 \times 1.37 \times 10^6/r^2$ ergs/s. Now, it absorbs some of this energy and reflects the rest. The fraction of incident energy reflected is called the planet's *albedo*, denoted by A. If it reflects a fraction A, it must absorb a fraction $1 - A$. Thus, finally, the total energy absorbed each second by a planet is:

$$\text{energy absorbed} = \frac{(1 - A)\pi R^2 \times 1.37 \times 10^6}{r^2}.$$

On the average, a planet must reradiate energy at exactly the rate it absorbs it; otherwise, it would heat up or cool off until this equilibrium is achieved.[*] Now the energy emitted by a black body at temperature T is σT^4 ergs/s from each square centimeter (Section 10.2d). Thus, for a spherical black body of radius R cm, the total energy radiated per second is $4\pi R^2 \sigma T^4$ ($4\pi R^2$ is the area of the surface of a sphere of radius R). Because the effective temperature of a planet is that of a black body that absorbs (and hence reradiates) as much energy as the planet, that effective temperature is found by equating the energy absorbed by the planet to that emitted by a similar black body; that is,

$$\frac{(1 - A)\pi R^2 \times 1.37 \times 10^6}{r^2} = 4\pi R^2 \sigma T^4,$$

from which we find

$$T = \left[\frac{(1 - A) \times 1.37 \times 10^6}{4\sigma r^2} \right]^{1/4}.$$

The albedo of a planet is found by comparing its observed brightness to that calculated for a perfectly reflecting body of the same size and distance from the sun.

As an example, consider Mars. Mars is 1.524 AU from the sun, so r^2 in the above formula is $(1.524)^2$, or 2.32. The Stefan-Boltzmann constant,

---

[*]Actually, Jupiter and Saturn emit appreciably more energy than they absorb. Part of the energy radiated by these planets must be from their store of internal heat.

σ, is 5.67 × 10⁻⁵. Mars absorbs about 85 percent of the incident energy on it and reflects the rest; thus A is 0.15. The above formula gives, for the effective temperature of Mars, 216 K, or −71°F.

Most of the energy emitted by planets is in the infrared. We can observe this energy directly at the telescope by using infrared detecting devices. This gives a separate check and is an independent way of finding the albedo of a planet.

Planetary temperatures derived in this way, of course, hold only in an average sense. A planet receives all its solar energy on its daylit hemisphere. If it does not rotate rapidly it may not reemit the energy it absorbs uniformly in all directions. Mercury and the moon, for example, are hot on their sunlit sides and cold on their night sides. Moreover, it must be remembered that the effective temperatures do not correspond to temperatures at the solid surfaces of planets with opaque atmospheres. Venus is much hotter at its surface than it is high in its atmosphere, from which most of its radiated energy escapes into space (Chapter 17).

Retention of Atmospheres

A molecule moving with a speed greater than the velocity of escape from a planet does not usually escape because collisions with molecules above it deflect it downward. Molecules at high enough levels in the atmosphere, on the other hand, which are moving in the right direction with enough speed do escape. Because some molecules are moving with much higher than average speeds, it is not necessary for the mean molecular speed in a gas to be equal to the escape velocity of a planet for that gas

to be lost to space. The high level in a planetary atmosphere from which molecules can escape is called the *exosphere*. Calculations show that if the mean molecular speed is as much as one third the velocity of escape, gases in the exosphere will escape in a few weeks. Molecules of a particular species in the lower atmosphere gradually diffuse upward into the exosphere, and if they find themselves in an environment of high enough temperature there, they will escape also. To be conservative, we shall suppose that for a planet to hold a particular gas in its atmosphere for several thousand million years, its velocity of escape must be at least six times the mean molecular speed for that gas.

In summary, to see whether a planet can hold a particular kind of gas in its atmosphere, we must (1) estimate a representative temperature appropriate to the planet's atmosphere (by the method described above), (2) calculate the mean speed of the molecules in the gas at that temperature from kinetic theory, and (3) compare six times that speed to the velocity of escape from the planet. The latter is $\sqrt{2GM/R}$, where G is the gravitational constant (6.67 × 10⁻⁸ dynes·cm²/g²) and M is the planet's mass. (Care must be taken to use the same units of speed in all calculations.) The condition that a gas, whose molecules have mass m each, be retained is thus

$$6\sqrt{\frac{3kT}{m}} < \sqrt{\frac{2GM}{R}}.$$

Figure 14.7 shows, roughly, the relative distributions of molecular speeds of oxygen and hydro-

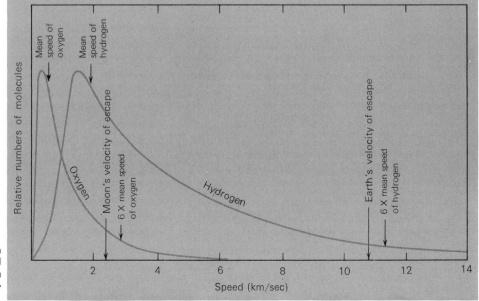

Figure 14.7 Distribution of speeds of oxygen and hydrogen molecules at a temperature appropriate for the earth and moon.

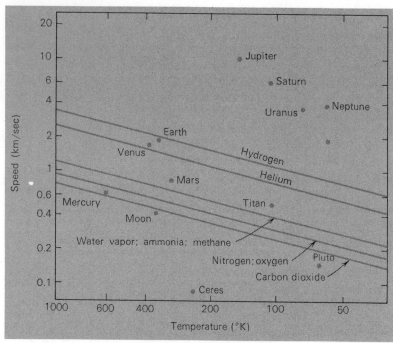

Figure 14.8 Molecular speeds of common gases at various temperatures (see text).

gen that would be expected at a temperature of about 300 K—which would be representative of the earth and moon if they both had appreciable atmospheres. The solid curves show the relative numbers of molecules of each gas that would be moving at various speeds. In each case, the mean speed and six times the mean speed are indicated. It is seen that neither earth nor moon should be able to hold hydrogen, for six times its mean molecular speed would exceed the escape velocity of both. Oxygen, however, would be retained by the earth but not by the moon.

In Figure 14.8 are shown plots (straight lines) of the mean molecular speeds of various common gases as a function of the absolute temperature in Kelvins. On the same plot are shown points representing the planets, the moon, the satellite Titan, and the minor planet Ceres. The position of the point for each body indicates one sixth of its velocity of escape and a temperature that is probably a minimum value of that appropriate to the levels where an atmospheric constituent can escape (generally higher than the effective temperature of the planet). A gas is not retained by a planet if the line representing its molecular speeds passes above the point for that planet. Moreover, it must be recognized that the escape of gases occurs at the uppermost levels of a planet's atmosphere, where the temperature (as a result of ionization by ultraviolet solar radiation) may be many times higher than the typical value—as is true for the earth. Thus the points representing the planets should probably be shifted to the left in the plot, to the high temperatures appropriate to the "escape layers" of their atmospheres. A particular atmospheric gas may well escape from a planet even if the line representing its molecular speed passes a little below the point for that planet.

It is apparent from the figure that Mercury, Ceres, and the moon would not be expected to retain any of the common gases, even if they ever did possess atmospheres. Mars could not hold hydrogen and helium but would be expected to retain the heavier gases. The Jovian planets could retain any gas. Hydrogen and helium are abundant in the atmospheres of all of them. One would not expect to find atmospheric oxygen on these planets, for the free hydrogen present would combine chemically with oxygen to form water, which in turn would freeze out of the atmospheres. Hydrogen is observed on those planets, both in molecular form and in combination with carbon in the form of methane (CH_4). Ammonia (NH_3) is also present in the atmosphere of Jupiter.

Hydrogen could not be retained for many millions of years in the atmospheres of Venus or Earth. Nor could appreciable helium be retained at the high temperatures expected for the escape layers. Moreover, substantial amounts of hydrogen and helium may never have been present in the gaseous state on the earth and Venus (see Section 13.3).

(d) Interiors of Planets

The study of the internal structure of planets is a difficult subject. Our knowledge of the interior of the sun is far more advanced than our knowledge of the cold worlds that revolve around it. Stars, like the sun, are completely gaseous; the physical laws that pertain to the behavior of gases are relatively simple and hence are better understood than those that pertain to the nature of solid or liquid matter under high pressure.

Yet some clues enable us to draw a few conclusions about the interiors of the planets. A knowledge of the mass and radius of a planet are sufficient for calculation of its mean density. Knowing the density, we can say something about the material that composes the planet. Clues to the distribution of mass in a planet can be gathered from the degree to which the planet flattens under the influence of its rotation, and from the perturbative effects that its flattening has on its satellites and on the planet's rate of precession. Additional valuable information is obtained for the earth and moon from the transmission and reflection of seismic waves at various levels in their interiors (Chapters 15 and 16).

Some of the conditions in the interior of a planet can be inferred from the fact that the planet is in *hydrostatic equilibrium;* that is, it is neither contracting nor expanding rapidly. The various layers of matter are all attracted toward the planet's center by the gravitational force exerted upon them by the deeper regions. These weights of the overlying layers must be supported by the pressures of the layers beneath. If we know how the mass of a planet is distributed throughout its interior, we can calculate how great these pressures must be at every point from the surface to the center. Even without information on the mass distribution we can calculate lower limits, below which the central pressures of planets cannot lie.

As far as their internal structures are concerned, the planets fall naturally into two groups: the terrestrial planets, Mercury, Venus, Earth, and Mars (sometimes the moon and Pluto are also included); and the Jovian planets, Jupiter, Saturn, Uranus, and Neptune. The terrestrial planets are relatively small and dense and are believed to be composed mostly of rocky and metallic material. The Jovian planets are relatively large and of low mean density (indeed, Saturn has a mean density less than that of water). The chemical composition of the Jovian planets is probably more nearly typical of the general cosmic abundances of the various chemical elements—for example, the relative abundances found in the sun. Hydrogen and helium, the most abundant elements in the universe, dominate in the Jovian planets. Jupiter and Saturn, at least,

Table 14.2. Some Approximate Planetary Data

	MERCURY	VENUS	EARTH	MARS	JUPITER	SATURN	URANUS	NEPTUNE	PLUTO
Distance from sun (AU)	$2/5$	$3/4$	1	$1\frac{1}{2}$	5	10	20	30	40
Mass in terms of earth's mass	$1/20$	$4/5$	1	$1/9$	318	95	15	17	0.002
Radius in terms of earth's radius	$2/5$	1	1	$1/2$	11	9	4	4	$1/4$
Rotation rate	59 days	234 days retrograde	24 hours	$24\frac{1}{2}$ hours	10 hours	10 hours	20 hours retrograde	16 hours	6 days
Velocity of escape (km/s)	4	10	11	5	60	35	20	22	1
Approx. effective temperatures* (K)	450	235	300	220	200	130	50	40	40
Important gases in atmosphere	—	CO_2, H_2O, N_2	N_2, O_2, A, CO_2, H_2O, etc.	CO_2, H_2O, N_2	H_2, CH_4, He, NH_3	H_2, CH_4, He	H_2, CH_4, He	H_2, CH_4, He	?
Number of known satellites	0	0	1	2	16	16	5	2	1

*Temperatures on the sunlit sides of planets, or at their surfaces, may be much higher. The sunlit side of Mercury, for example, has a temperature of over 700 K and the surface temperature of Venus is also near 700 K (Chapter 17).

are almost certainly composed mostly of hydrogen compressed to a liquid or solid state (see Chapter 18).

(e) Summary

The various planets of the solar system differ greatly, and in many ways. These differences can be largely understood, however, in terms of the respective distances of the planets from the sun and their respective sizes and masses. These three basic parameters are important to an understanding of Chapters 15 to 18, which deal with the planets individually.

In Appendices 9 and 10 are tabulated various data pertaining to the planets. However, many precise figures are often more confusing than helpful. Therefore, in Table 14.2 are summarized, very roughly, a few of the most important data concerning the planets. It is easier to remember that Jupiter is about 5 AU from the sun than to remember figure 778,730,000 km. The relative sizes of the planets and sun are shown in Figure 14.9.

14.3 ORIGIN OF THE SOLAR SYSTEM

Analysis of rocks in the earth and moon (Chapters 15 and 16) and of meteorites (Chapter 21) shows that the oldest of them all have ages of about 4500 million years. Theoretical studies of the early evolution of the sun suggest that its age may be about the same (Chapter 33). We conclude that the solar system must have formed slightly less than 5000 million years ago. On the other hand, our present theory of stellar evolution suggests that the oldest systems of stars in our Galaxy, and hence the Galaxy itself, have at least twice as great an age. Thus the formation of the sun and solar system took place long after that of some of the stars surrounding us in the sky; this should not surprise us, for as we shall see, we have strong evidence that stars are forming in the Galaxy even today.

We have noted that the organization of the planets in the solar system is quite orderly. They all lie in nearly the same plane, their orbits are nearly circular and are rather regularly spaced, they all revolve in the same direction—from west to east (the same as the direction of the sun's rotation)—and most planets and most of the satellites of planets rotate from west to east as well. These facts alone would appear to rule out the once-popular catastrophic theory of the origin of the solar system—that a passing star pulled out matter from the sun that later condensed into planets. Any matter so extracted, even if it were not lost to the solar system, could hardly have formed into planets with regularly spaced near-circular orbits, and the hot gas would have been dispersed, not formed into planets. Moreover, such close encounters of passing stars are exceedingly rare.

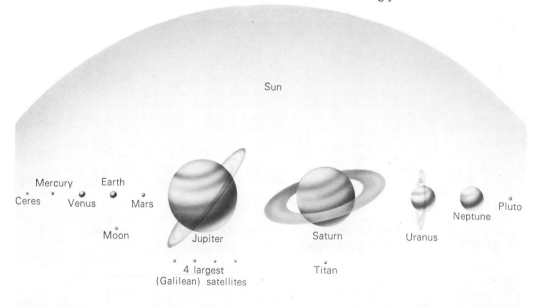

Figure 14.9 Relative sizes of the sun and planets.

Today it is generally accepted that the sun and planets formed together from the same original cloud of material. The idea appears to have been first suggested by the German philosopher Immanuel Kant, in the mid-18th century, and it was developed into a specific model by the French astronomer Marquis Pierre Simon de Laplace near the end of the same century. The Kant-Laplace idea is known as the *nebular hypothesis*. The Laplace model cannot be correct in detail, but modern versions are consistent with known laws of mechanics.

Whereas there is general agreement on the broad outline of the formation of the solar system, as described in the following paragraphs, the details are uncertain and still rather speculative—not because the subject is particularly mysterious but because we simply do not yet have enough information to choose among many possible models.

It is thought that about 5000 million years ago the solar system condensed from a tenuous cloud of interstellar gas and dust (Chapter 28). The original solar nebula would have to have had a diameter thousands of times that of the orbit of the most distant present planet—perhaps as much as a light year or so. It would also have had to have some original net rotation, probably due to the differential rotation of the Galaxy itself. The original cloud rotation could have been exceedingly slight— merely a slight net imbalance of the many random motions of the gases within it.

We shall see (Chapter 34) that some stars eject matter into space. Thus the interstellar medium (from which our solar system condensed) was not simply left-over matter that did not condense into stars when the Galaxy first formed, but also it contained a good deal of matter that was formerly parts of other stars. Most stars derive their energy by the thermonuclear conversion of light atoms into heavier ones, thereby slowly changing their chemical composition. The solar nebula therefore, contained many atoms that had been built up by nuclear reactions in the interiors of other stars. It is interesting to realize that our own bodies are made up of atoms, many or most of which were formed in earlier stars.

Perhaps the solar nebula represented a fluctuation of density, so that it was very slightly more dense than the gas in the interstellar medium surrounding it. In any case, it must have been gravitationally unstable, in that its own gravitation was enough to pull its parts together; that is, the random motions of its gaseous parts and dust particles

were not great enough for them to escape each other, dispersing the cloud. So it began to contract under its own gravity. As it contracted, however, it had to conserve its angular momentum (Section 4.1h), which forced it to rotate faster and faster as it drew itself together. After a time the rotation began to produce an orderly structure.

Eventually matter in the outer equatorial region of the rotating cloud, moving ever faster as it contracted, had a high enough speed to stay in a circular orbit about the center of mass of the cloud. Material in that part of the nebula could not come any closer to the center. To do so it would have had to speed up still more (to conserve its angular momentum); an increase in speed, however, would force it to move out *away* from the center into a larger orbit, and there was no energy available to move it out against the pull of gravity. (Recall how, in Section 5.8, we give a Mars space probe a larger orbital speed than the earth has in order to move it out from the earth so it can intercept Mars.) Thus, material in the equatorial region was simply left behind in a roughly circular orbit as the rest of the cloud continued to fall inward. As time went on, more and more material was left behind the shrinking cloud, moving in circular orbits and forming a disk of material. Matter in the disk could no longer contract toward its center, although matter on either side of the disk could fall toward it (falling toward the disk in a direction parallel to the axis of rotation). In this way the rotating solar nebula flattened itself into a disk.

The infalling atoms picked up speed as they fell, and when the gas density became high enough for them to collide with each other, the kinetic energy was distributed among the atoms, becoming heat. Most of this heat was radiated away from the disk, but in the central condensation—to become the sun—the density grew until the gases of the protosun became opaque. The opacity trapped the heat inside and the pressure produced by the heat slowed down the contraction. The shrinking nebula had become a great globe of hot gas that could contract only very gradually as it was able slowly to radiate away the heat trapped in its interior. Thus a star (the sun) was born at the center, containing perhaps half or more of the material of the original cloud. The rest of the nebula was in the form of a relatively cold rotating disk, from which the planets and their satellites formed.

In the inner part of the solar nebula the high luminosity of the young sun would have evaporated the grains that were composed of volatile substances. Thus particles of water ice and frozen carbon dioxide could exist only far out in the disk. Rocky and metallic grains, on the other hand, could survive throughout the disk. In all parts of that rotating disk, though, the orbiting particles were constantly colliding and often sticking together, and many began to grow by accretion. A few began to get big enough to gravitationally affect those which came near. Sometimes smaller particles would pass close enough to bigger ones to bump into them and stick. But if they didn't pass close enough to hit, they could be gravitationally deflected to another part of the disk, or even out of the solar system altogether.

In this way, a few large chunks gradually won out over their neighbors, either capturing them or getting rid of them, thereby sweeping out ring-shaped swaths in the solar nebula, all centered on the sun. They became the planets. In the final stages of this accretion the young planets swept up the last of the solid chunks remaining in the disk. There must have been many crater-producing explosions as these chunks smashed home. On the planets without dense atmospheres, we can still observe the heavy cratering produced, we think, in this period.

Those planets in the inner part of the solar system—Mercury, Venus, Earth, and Mars—were built up of rocky and metallic particles. They have a lot in common, and are called *terrestrial* (earth-like) planets. They could attract and hold on to none of the gases in the solar nebula. Their present atmospheres have outgassed from the rocks beneath their surfaces; Mercury, however, is too small to retain even this kind of atmosphere.

Far out in the nebula, on the other hand, it was cool enough for icy grains to exist, as well as rocky and metallic ones. The planets that accreted out there—Jupiter, Saturn, Uranus, and Neptune—thus formed out of lots of ices as well as of rocks and metals. They too have a lot in common, and are called *Jovian* (Jupiter-like) planets. Jupiter and Saturn were large enough to even attract and hold a large amount of the gas in the solar nebula. Jupiter, in particular, (as we have seen) has a present composition almost like the sun's; it is mostly hydrogen and helium.

The favored theory of the moon's origin is that it and the earth formed together, the moon accreting from material in orbit about the primordial earth. Some theorists, however, argue that the moon and earth could have formed independently in different parts of the solar nebula, but at about the same distance from the sun. In this case the earth and moon would have to have been trapped in each other's mutual gravitational fields at a later time. Most of the other planets have satellites, and many of them are believed to have formed by accretion from material in orbit about their parent planets. Some satellites, however, such as the outer ones of Jupiter, have eccentric orbits and even revolve from east to west (opposite to the revolution of the planets and most satellites); these were probably captured after their formation.

The asteroids may simply be objects that never accreted to a single large planet, perhaps because there was too little mass in that part of the solar system to begin with, or perhaps because of the tidal influence of Jupiter. They may, however, be still fragmenting by collisions, and have started from a much smaller original number of bodies. We discuss the possible origins of the comets and meteoroids in Chapters 20 and 21.

The planets and minor members of the solar system contribute only minutely to its total mass. The original mass of the disk must have been much greater, for it is doubtful if practically all of the solar nebula could have condensed into the sun itself. We shall see, however (Chapter 33), that when it first begins shining, a star is temporarily much more luminous than when it is fully developed. In that period of the sun's existence a large flux of energy from it, both in the form of photons and corpuscular radiation (atomic nuclei and electrons) may have interacted with the uncondensed gases and the tiny unaccreted particles and "blown" them from the solar system. Such an early solar wind of corpuscular radiation could also have carried away most of the sun's original angular momentum; today its angular momentum comprises only about 2 percent of that of the solar system, despite the fact that the sun has more than 99.8 percent of the system's present mass.

I think that the foregoing account of the solar system's beginnings is probably close to what actually took place, but future research can be expected to fill in many details and show others to be incorrect. We know, however, that all stars don't form in just this way. Often the original protostar fissions into two condensations, to become a double star, and the original angular momentum is conserved in the orbital motions of the two stars rather than in one star with its system of planets. Other

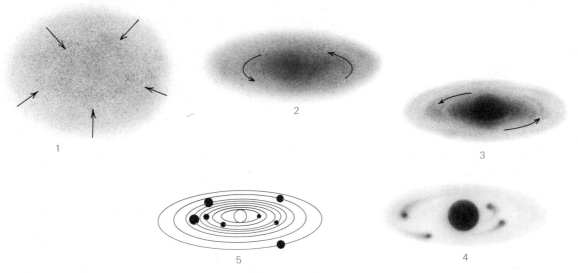

Figure 14.10 Schematic representation of the formation of the solar system. (1) The solar nebula condenses from the interstellar medium and contracts. (2) As the nebula shrinks, its rotation causes it to flatten, until (3) the nebula is a disk of matter with a concentration near the center, which (4) becomes the primordial sun. Meanwhile, solid particles condense to the inner solar nebula. These (5) accrete to form the terrestrial planets. The pressure of radiation and the wind of corpuscular radiation from the primordial sun blow the solar system clean of most of the matter in the disk that did not form into planets. The five drawings are not to the same scale; the original solar nebula had to contract greatly before its rotation produced appreciable flattening.

times the clouds break up into clusters of stars. We don't know how often planetary systems can form. At most it's only half the time, because about half the stars around us are members of binary-star systems. Some recent studies of the incidence of duplicity among stars have suggested that the formation of planetary systems might be relatively rare.

Alas, we have no direct evidence of a single other planetary system, but you must realize that as yet we know of no way to detect planets like the earth in revolution about other stars. At least it happened here, and most of us would be very surprised if it is so rare that it has not happened many millions of times in our Galaxy.

EXERCISES

1. From the data given in the chapter, show that Kepler's third law holds for Pluto.

2. What would be the period of a comet whose orbit has a semimajor axis of 10,000 AU?

3. It was once suggested that the asteroids and meteoroids may have originated from a planet that once broke up. In what way would that planet have had to differ from the other major planets?

4. A double planet has a period of mutual revolution of about $27\frac{1}{3}$ days, and the two bodies are separated by about 385,000 km. What is the combined mass of the two? If they are in the solar system, what are their names?

5. The velocity of escape of Mars is only a little greater than that of Mercury. Why then does Mars have an appreciable atmosphere while Mercury does not?

6. On the earth we have free nitrogen. Why then do we not find ammonia (NH_3) in our atmosphere as we do in the atmosphere of Jupiter?

7. Calculate the sidereal periods corresponding to each of the distances of the planets from the sun given in Table 14.2. Compare your results with the accurately observed values tabulated in Appendix 9.

8. Suppose there were a planet 100 AU from the sun, of mass and radius like the earth's. Would it be expected to have helium in its atmosphere? How about ammonia? Explain in each case.

9. Explain how observations of the infrared radiation emitted by a planet can help us learn its albedo.

10. Using the approximate data of Table 14.2, calculate the mean densities of both Mercury and Jupiter in terms of the mean density of the earth.

11. Calculate the mean density of the sun in terms of the earth's density, and compare your results with those of Exercise 10.

12. Can you think of any ways it might be possible to distinguish telluric lines (those due to absorption of light by the earth's atmosphere) from the absorption lines produced in the atmosphere of a planet being observed?

13. Use the approximate values for albedo, distances from the sun (in AUs), and radii (in units of the earth's radius) given in Appendix 10 to calculate the brightnesses of the planets, relative to that of Jupiter, as observed by a hypothetical interstellar traveler approaching the solar system from a distance that is large as compared with those of the planets from the sun. (This exercise was suggested by Prof. Clyde Zaidins.)

Alfred Wegener (1880–1930), German meteorologist, suggested the idea of "continental drift" in 1912. His arguments, although not accepted then, are now regarded as precursors of the theory of plate tectonics—the most important area of geophysics in the mid-20th century. (*Historical Pictures Service, Chicago*)

15

THE EARTH: ITS CHANGING FACE

In addition to the nine known planets in our solar system, there are at least 13 satellites large enough to dignify with the noun "world." Most of them have now been investigated at close range with interplanetary probes. Of these nearly two dozen worlds, only the earth has conditions even remotely suitable for life as we know it. But even on earth that situation may be temporary. Ours is a very fragile environment, and not only do we now have the power to alter it, but we have already begun to do so. We cannot say what the future of our world as a habitable planet may be, but if we care about it, it behooves us to look at it carefully and attempt to understand its workings.

Here we turn our attention to our own planet. Many of its gross properties have already been described, and some of these are summarized in Table 15.1.

15.1 APPEARANCE FROM SPACE

Satellite observations of the earth show that it reflects approximately 35 percent of all solar radiation incident on it. Its albedo (over all wavelengths) is thus about 0.35.

When the earth is at opposition as seen from Venus, it is at the full phase, its entire illuminated hemisphere being presented to that planet. At this point the apparent brightness of the earth, from Venus, would be from 10 to 20 times as great as that of Venus as we see it. Venus reflects more sunlight than the earth does, but we see Venus at its brightest only as a crescent. The moon would appear about as bright from Venus as Jupiter does to us, and, as the months go by, the moon would seem to swing back and forth around the earth,

Table 15.1. **Gross Properties of the Earth**

Mass	5.977×10^{27} g	$= 6.589 \times 10^{21}$ short tons
Diameter	12,756 km	$= 7926$ miles
Mean density	5.5 g/cm^3	
Surface gravity	980 cm/s^2	$= 32.15$ ft/s^2
Velocity of escape	11.19 km/s	$= 6.95$ mi/s
Rotation period	23 hours 56 min 4.099 sec	
Surface area	5.099×10^8 km^2	$= 1.969 \times 10^8$ mi^2
Land area	1.489×10^8 km^2	$= 5.751 \times 10^7$ mi^2 (29%)
Water area	3.610×10^8 km^2	$= 1.394 \times 10^8$ mi^2 (71%)

Figure 15.1 The first view of the earth taken by a spacecraft from the vicinity of the moon; Lunar Orbiter I photograph. *(NASA)*

reaching a maximum separation of just over $1/2°$. From Mars and the more distant planets, the earth would appear brightest when in the crescent phase; because of its greater distance from the sun, and a reflecting power lower than that of Venus, the earth would appear only about one fourth as bright as Venus.

15.2 THE EARTH'S ATMOSPHERE

We live at the bottom of the ocean of air that envelops our planet. The atmosphere, weighing down upon the surface of the earth under the force of gravitation, exerts a pressure, which at sea level amounts to that produced by the weight of 1.03 kg over each square centimeter. If the mass of the air over 1 cm^2 is 1.03 kg, the total mass of the atmosphere may be found by multiplying this figure by the surface area of the earth in square centimeters. We find, thus, that the total mass of the atmosphere is about 5×10^{15} tons, or about a millionth that of the earth.

At higher and higher altitudes the air thins out more and more until it disappears into the extremely sparse gases of interplanetary space. The real upper limit to the earth's atmosphere is the altitude at which the solar wind carries away atmospheric molecules. By observing auroras, we know that thin vestiges of the atmosphere extend to heights of at least 1000 km.

Auroras are discharges in the upper atmosphere caused when charged particles from the sun bombard the upper air. The height of an aurora can be found by triangulation. The same auroral display is photographed simultaneously with two different cameras some distance apart. The angle between the aurora and the horizon is measured on each photograph, and the distance between the cameras being known, the altitude of the display in kilometers above the ground can be calculated with simple trigonometry. Most auroras are found to occur at heights of from 80 to 160 km, but a few are as high as 1000 km.

Analysis of drag on earth satellites reveals evidence for some atmosphere at even greater heights—1600 to 2000 km.

(a) Composition and Origin of the Earth's Atmosphere

The chemical composition of the earth's atmosphere is determined by quantitative chemical analysis. At the earth's surface, the constituent gases are 78 percent nitrogen, 21 percent oxygen, and 1 percent argon, with traces of water (in the gaseous form), carbon dioxide, and other gases. At lower elevations, variable amounts of dust particles and water droplets are also found suspended in the air.

The gases nitrogen and argon are relatively inert chemically. It is oxygen that sustains most animal life on earth, by allowing animals to oxidize their food to produce energy. Oxygen is also required for all forms of combustion (rapid oxidation) and thus is necessary for most of our heat and power production. Green plants absorb carbon dioxide in the process of photosynthesis and release oxygen, which helps to replenish the oxygen consumed by man and other animals.

The earth's present atmosphere is not primordial. As explained in the last chapter, the earth is believed to have accreted from smaller solid particles. These particles contained hydrates, carbonates, and nitrogen compounds. Many of these compounds were evidently trapped in the outer layers of the earth, where they were subjected to heating by the decay of certain radioactive elements (such as uranium and thorium). The heating caused these compounds to lose water (in the gaseous state), carbon dioxide, and probably nitrogen. These gases escaped through the crustal layer to form the earth's atmosphere—a process called *outgassing*.

The principal means of outgassing is volcanism. Water was by far the most plentiful substance to outgas; when released from the earth's crust it cooled and condensed into the oceans. Carbon dioxide is next most important, and what did not dissolve into the oceans reacted chemically with crustal rocks. Chemical activity at the earth's surface is thought to have produced considerable concentrations of methane and ammonia and some hydrogen in the early atmosphere. These three gases, however, either escaped or dissociated relatively quickly. Nitrogen was far less abundant than carbon dioxide, but is rather inert chemically, so it has remained to be the principal component of the atmosphere today. At least some of the present nitrogen may have had a biological origin. The argon (the isotope argon-40) resulted from the decay of the radioactive isotope of potassium, ^{40}K. Among the planets the earth is unique in having an appreciable amount of free oxygen in its atmosphere. The earth's oxygen is believed to have resulted from photosynthesis by green plant life (mostly in the oceans) within the past thousand million years.

(b) Levels of the Atmosphere

The density of the air drops rapidly at higher and higher elevations. In fact, half of the total atmosphere is packed down within 6 km of the earth's surface. The gravitational attraction between the atmosphere and the earth crowds the air close to the ground. The pressure exerted by the rapid motions of the molecules that comprise air prevents it from collapsing completely and supports a small fraction of the atmosphere even to great heights.

In the lower part of the atmosphere, the temperature also drops rapidly with increasing elevation. It is this region, the *troposphere*, where all weather occurs. The upper extent of the troposphere varies from about 8 km at the poles to about 16 km at the equator.

Above the troposphere, and extending to a height of about 80 km, is the *stratosphere*. The temperature through most of the stratosphere is fairly constant—a little lower than 50° C below zero. Between 40 and 65 km up, however, there is a warm zone; in 1947, V2 rockets launched at White Sands, New Mexico, first recorded above freezing temperatures in this region.

The hot layer is due to the presence of ozone in that level of the atmosphere. Ozone is a heavy form of oxygen, having three atoms per molecule instead of the usual two. It has the property of being a good absorber of ultraviolet light. In absorbing the sun's short-wavelength ultraviolet light, the ozone is heated up and warms the parts of the atmosphere where it is present. Incidentally, the protective ozone layer helps to prevent some of the sun's dangerous ultraviolet radiation from penetrating to the earth's surface. The region of the stratosphere extending upward from the ozone layer is sometimes called the *mesosphere*.

From 65 to 80 km, the temperature drops to below −50° C again. Above 80 km the temperature rises rapidly through a region called the *thermosphere*, and at 400 to 500 km reaches values above 1000° C. The highest layer of the atmosphere, above about 400 km, is called the *exosphere*. In the upper atmosphere, the thermosphere and exosphere, molecules of oxygen and nitrogen break up into individual atoms of those elements. Ultraviolet radiation from the sun ionizes many of these atoms. Therefore, part of the upper part of the atmosphere (above 50 km) is also called the *ionosphere*.

Auroral discharges occur, and meteoroids burn up and produce meteors in the thermosphere and

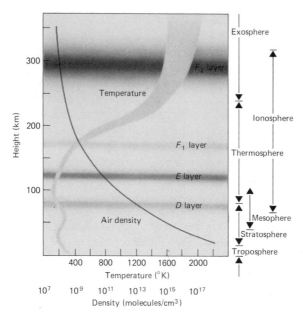

Figure 15.2 The structure of the earth's atmosphere. The broad dark band shows the ranges of temperature measured and estimated at various altitudes in the atmosphere. The single smooth line indicates the approximate density variation with altitude. The shaded regions indicate the zones in the atmosphere where the relative ionization of the atmospheric gases is highest.

exosphere. The density of the air there is extremely low. Studies of the effects of atmospheric drag on satellites indicate that at 500 km there are only about 10^7 atoms per cubic centimeter as compared to 10^{19} molecules per cubic centimeter at sea level. An air density of 10^8 atoms per cubic centimeter is considered a very good vacuum by laboratory standards.

Sharp increases in the densities of ionized or charged particles in the ionosphere occur at several levels, known as the D, E, F_1, and F_2 layers (see Figure 15.2). These layers reflect radio waves of wavelengths greater than about 17 m and make long-range radio communication possible. Radio waves can travel around the world by reflecting alternately between the ground or ocean and the ionized layers in the ionosphere. Wavelengths up to 17 m (sometimes up to 30 m), however, can penetrate the ionosphere and those from space are detected with radio telescopes.

15.3 THE EARTH'S MAGNETISM

The earth has a magnetic field similar to that produced by a bar magnet. Nearly everyone is familiar

with the way iron filings align themselves along the lines of force that extend between the north and south poles of a magnet. Between the magnetic poles of the earth stretch lines of force along which compass needles align. The positions of the magnetic poles, and the orientations of the lines of magnetic force, gradually shift. During periods of solar activity (Chapter 31) rapid short-period fluctuations in the earth's magnetic field also occur. The current position of the North Magnetic Pole is about 76° N latitude and 100° W longitude, which places it in Bathurst Island in northern Canada. The South Magnetic Pole is at latitude 66° S and longitude 139° E, just off the Adélie Coast of Antarctica.

The overall strength of the field is fairly weak, averaging about one half gauss at the surface of the earth. Its origin is believed to be in the earth's core. Fluid motions in the electrically-conducting core are thought to cause it to act like a dynamo. The rotation of the earth causes the magnetic field to be aligned approximately with the rotation axis. The energy source of this dynamo is still uncertain.

Not only are the locations of the magnetic poles constantly changing, but the polarity of the earth's field completely reverses itself from time to time. We detect these changes in polarity of the earth's field from the direction of magnetism in rocks. Molten rocks, as in volcanic lava, containing iron compounds are weakly magnetized by the earth's field. When the rocks harden their magnetic polarity is permanently frozen in them. We measure the strength and polarity of magnetism in rocks of different ages to trace the magnetic history of the earth, and find that its field has reversed polarity 171 times in the past 76 million years.

(a) The External Field and Van Allen Layer

Both U.S. and Soviet satellites and space probes have carried magnetometers into space to measure the strength of the earth's magnetic field far from its surface. The field strength is observed to decrease rapidly with distance from the surface and is roughly inversely proportional to the cube of the distance from the center of the earth, as predicted from theory. The strength and extent of the field is also observed to vary greatly. These fluctuations are probably associated with the ejection of charged particles from the sun during times of solar activity (Chapter 31).

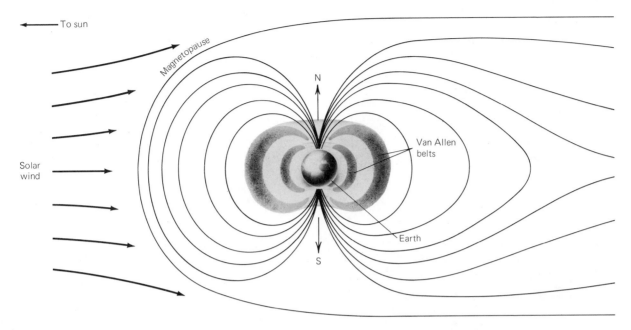

Figure 15.3 Cross section of the earth's magnetosphere and the Van Allen layer.

An important discovery made by the early artificial satellites was the presence of a zone of highly energetic charged particles above the earth. The inner part of this radiation zone was detected by instruments carried aboard the Army satellite Explorer I, launched on January 31, 1958. More information about the nature and extent of the belt of radiation was obtained with instruments on Explorer III, Explorer IV, and Pioneer I. In December 1958, the Army space probe Pioneer III revealed an additional region of radiation high above the one first discovered. The radiation-detecting experiments on Explorer I were under the direction of James A. Van Allen, physicist at the State University of Iowa; the radiation zone is called, in his honor, the Van Allen layer.

Figure 15.3, showing a cross section of the magnetic field surrounding the earth, displays the approximate location and extent of the most intense part of the Van Allen layer. It surrounds the earth, rather like a doughnut. The inner intense region is centered about 3000 km above the surface of the earth and has a thickness of 5000 km or more. The outer region of the Van Allen layer is about 15,000 to 20,000 km from the earth's surface and has a thickness of from 6000 to 10,000 km.

The Van Allen radiation zone consists of rapidly moving charged particles trapped in the earth's magnetic field. Most of the fastest particles are electrons, although the inner part of the Van Allen layer contains high-energy protons as well. The particles in the inner region are moving with speeds far greater than those in the outer one. Some are moving with energies as great as those possessed by electrons that have been accelerated through an electric field of over 10^9 volts. The radiation strengths, heights, and thicknesses of the regions in the Van Allen layer, especially of the outer one, vary considerably from time to time. These regions are not really sharply defined; trapped charged particles must exist throughout the earth's magnetic field. The inner and outer parts of the Van Allen belt may be thought of as regions where the density of the moving particles is much greater than elsewhere.

The origin of the trapped particles is not known with certainty. It seems likely, however, that the outer region consists mostly of particles captured from the "wind" of corpuscular radiation from the sun. It has been suggested that the protons and electrons in the inner region may be produced in interactions between molecules of the earth's atmosphere and cosmic rays.

The overall magnetic field surrounding the earth in space above the altitude of the E layer is called the earth's *magnetosphere*. The outer boundary of the magnetosphere—the magnetopause—marks the transition to the interplanetary field. The side of the field toward the sun has a rather blunt shape, where the solar wind impinges on it. The

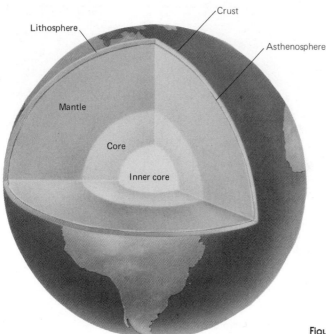

Lithosphere
Crust
Asthenosphere
Mantle
Core
Inner core

Figure 15.4 The earth's interior.

charged solar wind particles carry the field lines of the outer magnetosphere into a long tail pointing away from the sun, so the entire magnetosphere has a shape rather like that of a comet (see Figure 15.3). The geocentric distance of the magnetopause on the sunward side of the earth varies, according to the strength of the solar wind, between 6 and 14 times the radius of the earth, and is usually about 10 earth radii. But the magnetic tail extends away from the sun to a distance of from a few hundred to a few thousand earth radii.

15.4 THE EARTH'S INTERIOR

By sampling surface rocks, we find that the mean density of the outer crust of the earth is about 2.7 g/cm^3. On the other hand, the mean density of the earth as a whole is 5.5 g/cm^3. Because of the low surface density but high average density, we must conclude that the interior of the earth is very dense indeed.

The weight of the various layers of the earth bearing down upon its interior causes the pressure to increase inward. Calculations indicate that the pressure at the earth's center must be close to 4 million kg/cm^2. When subjected to this great pressure, matter is highly compressed and heated. We can expect the central regions of the earth, then, to be hot and dense.

(a) Regions of the Earth's Interior

Stresses that build up gradually in the crust of the earth are often released by slippages along fissures or *faults*. The energy released in these sudden movements in the earth's crust results in earthquakes. Vibrations are sent out and travel to all parts of the earth. Some types of the vibrations travel along the surface; others pass directly through the interior.

The study of these waves that originate in earthquakes is *seismology*. The seismic vibrations are picked up and recorded by delicate instruments stationed around the earth's surface. An analysis of the time required for seismic waves to reach these various stations provides information about the earth's interior.

The transmission of the waves through the earth's interior indicates that most of it is solid and rigid. This major part of the earth is called its *mantle*. The density in the mantle increases downward from about 3^1/$_2$ to 5^1/$_2$ g/cm^3. It is believed to be composed mostly of basic silicate rocks.

At the upper boundary of the mantle is the *crust*, a shell of lower density extending on the average about 35 km inward from the earth's surface under the continents. The crust consists of surface rocks such as granite and basalt, with overlying oceans and sedimentary rocks. Evidence from seis-

Astronaut Harrison Schmitt working beside a large boulder on the moon. The lunar roving vehicle is at the left. *(NASA)*

Mars, photographed from
Viking 1 spacecraft in 1976.
(*NASA*)

Mars, photographed
from the Viking 2
lander in 1976.
(NASA)

Looking south from the Viking 2 lander on September 6, 1976. (NASA)

Jupiter, photographed by Voyager 1 at a distance of 33,000,000 km. (NASA/JPL)

Closeup of Jupiter, photographed by Voyager 1 at a distance of 5 million km. The famous red spot is at the upper right. Note the turbulence to the left (west). (NASA/JPL)

mic studies suggests that the crust extends somewhat deeper under the continents than under the ocean floors, where it may be only a few kilometers thick. The boundary between the crust and mantle is called the *Mohorovicic discontinuity*.

At the inner boundary of the mantle, about 2900 km below the surface, is the *core* of the earth. The outer part of the core acts like a liquid, for it does not transmit certain kinds of seismic waves. The innermost part of the core (about 2400 km in diameter) is extremely dense and hot (in the thousands of degrees), and is probably solid.

The temperature increases downward in the crust of the earth about 1° C for every 30 m. As the core slowly cools, conduction gradually transmits heat from the earth's interior out through the mantle and crust. However, the conductivity of the interior is relatively low, and cooling of the hot core must occur at a very slow rate. Calculations have shown that most of the heat in the crust, perhaps over 80 percent, does not come from the deep interior but from the radioactive decay of uranium, thorium, and potassium—unstable radioactive elements in the crust itself.

(b) Age of the Earth

The age of the earth is determined from the degree of radioactive decay of certain unstable elements in its crust. Like muons (Chapter 13), many atomic nuclei are unstable and spontaneously convert to other nuclei with the emission of particles, such as neutrons, electrons, photons, or even the nuclei of light atoms, such as helium (helium nuclei are called *alpha particles*). For any given nucleus, this decay process, called *radioactivity*, is random, but for each kind of radioactive nucleus there is a specific time period, called the *half life*, during which chances are fifty-fifty that decay will occur. A particular nucleus may last a shorter or longer time than its half life, but in a large sample of nuclei, almost exactly half will have decayed in one half life, and half of those remaining (three quarters in all) will have decayed in two half lives. After three half lives, only one eighth of the original sample remains, and so on (Figure 15.5). Thus radioactive elements provide accurate nuclear clocks; by comparing the relative abundances of a remaining radioactive element and of the element it decays to, we can learn how long the process has been going on, and hence the age of the sample.

Most kinds of nuclei heavier than uranium have very short half lives—far, far less than one second; this is why they are not found in nature, although they can be produced for brief instants in the nuclear physics laboratory. (Even the relatively stable plutonium—atomic number 94—has a half life of only about 24,000 years, very short compared with the age of the earth.) On the other hand, the earth's crust contains radioactive elements that decay slowly. Among these are potassium-40, which decays to argon-40 with a half life of 1.3×10^9 years, rubidium-87, which decays to strontium-87 with a half life of 5×10^{10} years, and uranium-238, which decays through a series of elements (including radium) to lead-206 with a half life of 4.5×10^9 years. These nuclear clocks enable us

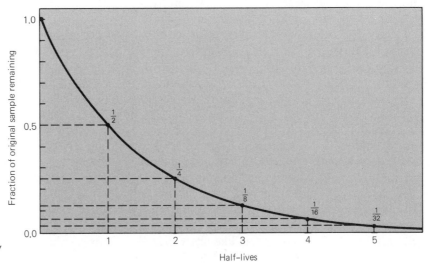

Figure 15.5 Radioactive decay and half-life.

to measure the ages of the rocks that these elements are found in.

There are rather few rocks on the earth with ages greater than 2.8×10^9 years; the oldest yet found (in west Greenland) have ages of 3.9×10^9 years. The ages of these rocks give a lower limit to the age of the earth itself. On the other hand, lunar rocks, returned by Apollo astronauts, and meteorites have been found with ages up to 4.5×10^9 years. Analysis of such evidence, along with information gleaned from our planetary probes on early accretion and melting at the surfaces of the moon and planets, leads us to believe that the bodies of the solar system, including the earth, formed during a 100-million-year period beginning 4.6×10^9 years ago.

15.5 PLATE TECTONICS

Lots of schoolchildren, in studying maps or globes of the earth, notice that North and South America, with a little juggling, look as though they could almost be nestled up against Europe and Africa; it is as if those great land masses were once together but somehow tore apart. The same idea also occurred to the German meteorologist Alfred L. Wegener early in the 20th century. However, he looked at the matter in considerable detail and made a pretty good case that the continents had, indeed, drifted apart. He based his arguments, in part, on mountain ranges and other geographical features that appeared on continents on both sides of the Atlantic and which looked as though they would be continuous if the continents were together. Nobody paid much attention to Wegener at the time, probably because too many schoolchildren had similar ideas.

Since the late 1950s, however, evidence accumulated that the continents really *are* drifting apart. Wegener had been right! The study of this phenomenon is called *plate tectonics,* and it has revolutionized geophysics just as the Copernican theory revolutionized astronomy.

The thin skin of surface rocks on the earth, the crust, extends to a depth of only 35 or so km, and even less under the oceans. The crust and upper mantle, to a depth of some 50 to 100 km, consists of a layer of the earth called the *lithosphere.* The lithosphere is not one solid piece, however, but consists of ten major (and some minor) separate *plates,* fitting together something like jigsaw-puzzle pieces. These plates of the lithosphere float on a hot, plastic, and possibly partly molten layer just beneath them called the *asthenosphere;* it is from 100 to 200 km thick.

There is a general circulation of the plates, which causes them to slide about over the asthenosphere. The continents drift because they are of lighter, more buoyant material, and float on top of the moving plates.

But how can a system of interlocking plates covering the entire surface of the earth move about? One might expect that they would not be free to circulate because of the presence of adjacent plates. The plates move about by jamming into their neighbors, and burrowing under them, right down into the mantle. Continents themselves don't actually burrow down under; when a continent reaches a plate boundary, it sort of stays there while the plate slides under it into the mantle. Far from a boundary a continent just rides along on top of its plate. It is rather like logs floating down a stream, and then jamming up against a low bridge over the water. When continents crash into each other, though, they can build mountains.

But if the plates go sliding around, bumping into each other and burying themselves, how can they cover the entire earth? Why don't they all dig under and disappear? Why isn't there a shortage of all that crockery? The answer is that the plates are being continually replenished, as quickly as they disappear. The earth seems to be a heat engine. Its internal heat, by mechanisms not yet thoroughly understood, evidently provides the energy to support a continual convection within the mantle of the earth that circulates material from the burying plates into emerging ones.

New material flows up from the mantle between separating plates and joins on to their trailing edges (or perhaps pushes them along). This is what happens at the *oceanic ridges* like the famous Mid-Atlantic Ridge. Sometimes all this activity occurs out of sight at the ocean floors, but sometimes some of the upflowing matter forms volcanic land masses. The whole nation of Iceland is sitting on the Mid-Atlantic Ridge and is currently being formed by new material from the mantle; that's why Iceland has so many volcanos and so much geothermal activity. Places where the plates burrow under are called *trenches,* a fine example being the Japan Trench, along the shore of Japan. Figure 15.6 shows, schematically, a cross section of the earth

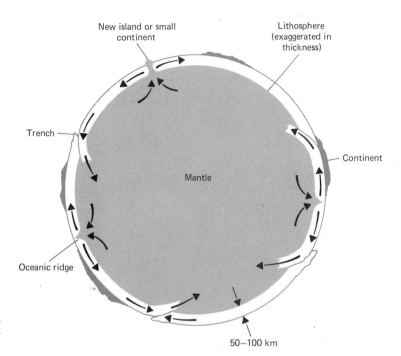

Figure 15.6 The lithosphere (of much exaggerated thickness), showing the emergence and immersion of plates.

(a) Evidence for Continental Drift

Part of the evidence for the motions of the plates, giving rise to continental drift, comes from a reexamination of the coastlines of existing continents. Especially if the continental shelves are taken into account, the fits between continents that were once together are quite good. Geologically speaking, this tectonic motion of the plates occurs very rapidly and has profoundly rearranged the continents over the earth's surface in the last couple hundred million years, as can be inferred, for example, from fossil remains of tropical plants and animals in what are now polar regions of the earth.

The clinching evidence, though, is in the record of the changing magnetic field of the earth. Consider what happens when molten rock flows into the floor of the Atlantic along the Mid-Atlantic Ridge. It spreads out in both directions, hardens, and becomes part of the separating plates. As it hardens, it freezes in the magnetism of the earth at the time of its emergence from the mantle. We can measure the magnetism of the rocks on the ocean floor with delicate magnetometers towed behind

ships, and see how their polarity switches back and forth many times as the ship moves away from the Mid-Atlantic Ridge. Since we have other records of the changing polarity of the earth's field, these Atlantic observations tell us just when the material in each part of the ocean floor oozed up from below and hardened, and how fast since then it has moved along with its plate.

Figure 15.7 is a map of the world showing the outlines of the major plates. The many tiny dots along those boundaries (and the few elsewhere) are the sites of moderate and major earthquakes in recent years. As you might expect, most earthquakes occur at plate boundaries where the major disturbances due to the motions of the plates are occurring. The Nazca Plate, for example, burrowing under the west coast of South America, has produced all kinds of havoc while building the Andes. The Indian Plate, shoving into Asia, has built the Himalayas. A very seismically active area is Japan, where the Pacific Plate pushes under the Eurasian Plate at the Japan Trench.

If we extrapolate the plates' motions backward, we arrive, some 200 million years ago, to the supercontinent of *Pangaea*. At that time the continents of the earth were all together. The original Pangaea is shown in Figure 15.8. The map may not be absolutely accurate but shows how geophys-

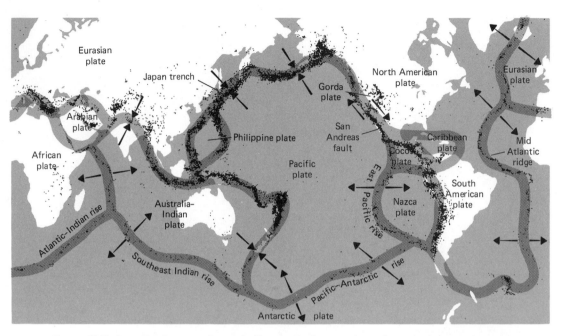

Figure 15.7 Tectonic plates on the earth. The dots indicate regions of seismic activity, where the boundaries of the plates generally lie. The major plates that have been identified are labeled, and the arrows indicate the direction of motion of the plates.

icists think the continents lined up about the time early dinosaurs roamed the earth.

About 200 million years ago Pangaea broke up. Ever since, the moving plates have been separating the continents. The plates that consist mostly of ocean areas tend to move the fastest—typically about 5 cm/yr. Those plates consisting mostly of continents are moving more slowly—typically 2 cm/yr, about the speed with which Europe and North America are separating. This may seem extremely insignificant, but consider that 2 cm/yr is 20 km/million years, or 2000 km/100 million years. Thus merely a few hundred million years separates Europe and North America.

(b) Earthquakes

Because I live in Los Angeles, I have a special interest in the boundary between the Pacific Plate and the North American Plate. Naturally, where the solid crustal parts of the earth are pushing into each other or (in the case of California) are sliding against each other, fissures (*faults*, or cracks) in the earth's crust will result. One of the most famous of these is the *San Andreas Fault,* running from the gulf of California in the south to the Pacific Ocean just west of San Francisco in the north, and sepa-

rating the Pacific Plate, containing southwestern California, from the North American Plate, containing the rest of the state. Displacements along it slowly carry Los Angeles northward; in several million years it will be on an island off the coast of San Francisco.

This creeping motion of the plates against each other builds up stresses in the ground around the fault, which are released in sudden slippages every 50 to 100 years, producing major earthquakes, including such famous ones as that in San Francisco in 1906. There is an impending one (at this writing) in the Los Angeles area.

Charles Richter (of the Richter scale of earthquake intensity) once said: "Some people have an irrational fear of cats, and others an irrational fear of earthquakes; the former should not have cats for pets, and the latter should not live in California." Dr. Richter was kidding about living in California—but only partly. It is a region, like many on the earth, that is near or on a boundary between plates, and can expect many interesting seismic events over the years. Some may even be very damaging.

On the other hand, I relish the opportunity to live where I do. My time on this earth is short, cosmically speaking, and it is exciting to witness

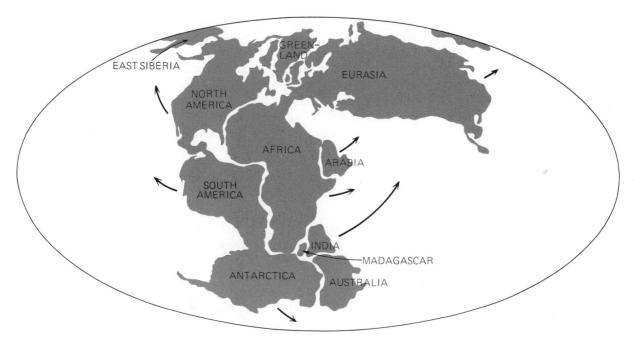

Figure 15.8 A map of the earth approximately as it appeared 150 million years ago. It is believed that at that time the continents were clustered together in a supercontinent that has been named Pangaea.

some of those grand changes that shape its face. If we could see a time-lapse motion picture of the earth's surface, with millions of years collapsed to a few seconds, we would find it a tremendously alive place, with its entire surface constantly rearranging itself; the motions of the plates would seem smooth and continuous. It is only on our tiny time scale, confined to barely more than a snapshot of the earth's cosmic existence, that things seem so abrupt and chaotic. In witnessing an earthquake, we are witnessing a tiny sliver of the evolution of the earth.

15.6 THE ICE AGES

Other gradual changes on earth are the glacial periods, or so-called ice ages. There is some hope now that we understand what brought on the previous ice ages. Because of the perturbative gravitational influences of the other planets, the earth's orbit is constantly changing, as is the tilt of its axis to the plane of its orbit (that is, the obliquity). Recent investigations show highly suggestive correlations between these periodic orbital changes in the earth's motion and glacial activity on earth over the past 450,000 years, as read from geological records. If the indications are correct, and investigators are not

at all sure yet that they are, it means that we should now be approaching a new ice age, but that the cooling trend extends over the next 20,000 years. Thus we should have plenty of time to check out the theory, correct for its shortcomings, and plan for the future. But there is a catch: All this ignores the *anthropogenic effects*, which could well have a short-range and immediate effect; that is, how man's own influence on his environment might dominate over other, natural effects.

15.7 ORIGIN OF LIFE

It seems to me to be almost a miracle that throughout most of the earth's history there has been life on it. Not very advanced life, to be sure; only in the past few hundred million years have there been complex animal forms; but at least single-cell, blue-green bacteria (formerly called blue-green algae) and similar organisms have been found in rocks as much as 3500 million years old. So far as we know, the earliest organisms required the existence of liquid water on earth to form, feed, and reproduce. Only on earth, of all known places, do we now find liquid water. On the other hand, under the present conditions on earth, we doubt that life could have ever begun.

Most biologists who have studied the origin of life believe that it required a *reducing* atmosphere, rather than the present *oxidizing* one. However, the earth almost certainly had no oxygen in its primordial atmosphere, and the best guess is that it had methane and ammonia, as well as water and carbon dioxide, to give it the right chemical makeup for synthesizing simple organic molecules necessary for the development of life. Biologists refer to the early conditions of an ocean of water and atmosphere of water vapor, carbon dioxide, methane, ammonia, and nitrogen as a "soup." Laboratory experiments show that if such a soup is irradiated with ultraviolet radiation, organic molecules such as sugars and amino acids are produced with large yield.

Three thousand to 4000 million years ago the earth, presumably, had no oxygen, and hence no ozone layer to filter out ultraviolet radiation from the young sun. Thus we would expect those prebiological organic molecules to be formed. That much seems almost inevitable. Somehow those amino acids and other molecules combined into proteins and eventually into the highly complex and long helical molecules we call DNA, those which form the chromosomes of living cells and which have the power of reproducing themselves. We cannot say whether their original formation was also almost inevitable or whether it was an incredibly long shot. Anyway, it happened; then began the long road of biological evolution that took more than 2500 million years to produce animal organisms with different cells serving different functions, and another 1000 million years to produce the earliest manlike creature.

There is an interesting thing about the organic molecules in living organisms on the earth. The basic molecules are helical—corkscrew shaped—and the more complex ones are built up of these simpler ones. They are all with the right-hand thread; that is to say, in all living organisms, the helical molecules twist the same way. It's easy enough to understand why this should be so. In the early development of life, some molecule with the ability to reproduce itself got formed; it had to be of either right-hand or left-hand thread—it couldn't be both at once. It just happened to have a right-hand thread; it then fed and reproduced, and fed and reproduced, and had mutations, and natural selection, and eventually ended up with something holding an astronomy book in its hand. If life had sprung up independently all over the earth, we might expect by chance to find some

forms with right-hand and others with left-hand helices in their chromosomes. But all of us, from the lowliest bacterium to the biggest elephant, are made of organic molecules twisting the same way, and all are based on the same genetic code. As Carl Sagan has put it, we are all closely related to the oak tree.

(a) Oxygen in the Atmosphere

Eventually organisms formed that filled the oceans and absorbed carbon dioxide; they used the carbon contained therein to build themselves, and released the oxygen into the atmosphere. Thus oxygen gradually began to build up in the atmosphere. It was at first very gradual, and 1000 million years ago, oxygen was probably a very minor constituent of the air. If fact, 90 percent or more of the earth's atmospheric oxygen has likely been produced in the past several hundred million years from this photosynthesis by green plants.

The production of oxygen, however, is always competing with its absorption. Oxygen is absorbed in respiration of animals, of course, but even today that is a fairly minor effect. Mostly, it is absorbed in the oxidation of dead vegetable and animal material. Biological matter rots when it dies—which means that it eventually oxidizes, the dead plants reabsorbing the oxygen they released during life. The only reason there is any net production of oxygen at all is that a small fraction of the plant material, probably one part in ten thousand to one part in one hundred thousand, escapes decay by being conserved as carbon, buried in the earth. This material is our fossil-fuel supply—coal and oil. The present atmosphere of the earth contains about 21 percent oxygen, built up gradually by the slight imbalance of photosynthesis over the decay of dead organic matter. If the only source of oxygen is photosynthesis (we think it is the only important source) and if we could burn up (that is, oxidize) all fossil fuel on earth, we should break even with the bank and reduce the oxygen in the atmosphere to nil. At the present rate of our burning of these fuels, this should take some thousands of years.

15.8 SUMMARY

Figure 15.9 is a figurative railway track, built as the earth has aged. Each tie in the track corre-

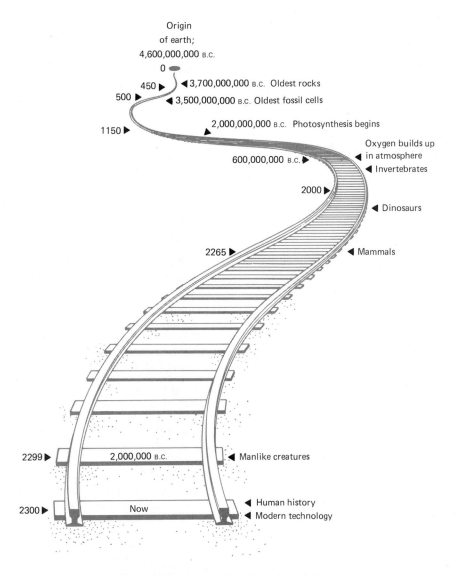

Figure 15.9 A symbolic railway through time.

sponds to 2 million years. There are 2300 ties so far, and events in the history of the earth are marked at various ties along the way. Suppose typical railway ties are about 1 m apart and that the last tie is being laid as I write these words. The second to last one, 1 m back, was laid when man-like creatures first stood upright on earth. Only a few centimeters from the front end of the last tie is where the history of man begins, as gleaned from his drawings and crude writing. Modern history begins a few millimeters back. Now bear in mind that the entire length of track is more than 2 km. Yet our technological society goes back only 200 microns—0.2 mm—just the thickness of that grain of dust on the front side of the tie that was just laid.

In the time represented by that grain of dust, man has drastically changed his environment. He has, of course, plundered our resources and polluted our waters. We could probably solve those problems, but in the past few years a few of us have begun to realize that we may have permanently altered, to the disastrous worse, that fragile, thin, atmospheric envelope that protects us from the outside universe and makes our own existence possible.

But to the earth itself, chugging along with its plates burrowing under each other, switching its magnetic field around chaotically, having ice ages come and go, and whose existence is pretty much assured for 5000 million years, what man does to himself is a matter of no consequence whatsoever.

EXERCISES

1. Planet X is the larger component of a double-planet system. It has twice the radius of its companion and reflects twelve times as much light. If the smaller companion absorbs 85 percent of the light incident upon it, what is the albedo of planet X?
Answer: 0.45

2. What fractions of the volume of the earth are occupied by the inner core, the entire core, and the mantle?

3. In what respects are the earth's oceans like an atmosphere and in what respects are they different? How would your answer have differed if the earth's orbit lay between those of Mercury and Venus? What if it were beyond the orbit of Jupiter?

4. Assume that the atmospheric pressure decreases by a factor of one-half with every increase of 5500 m in altitude above the earth's surface. At what altitude would the pressure be $^1/_{16}$ that at sea level?
Answer: 22,000 m

5. If the earth's crust were exactly 30 km thick, and if the temperature were to continue to increase by one °C for every 30 m of depth, how much higher would the temperature be at the base of the crust than at the surface?

6. If Europe and North America separated from each other at a constant rate of 2 cm/yr, how long a time would have been required for them to have moved from contact to their present separation of about 4000 km?

7. Identify on the map in Figure 15.7 other places besides Iceland where land masses may have been built up by the emergence of mantle material in an ocean ridge.

8. Identify on the map in Figure 15.7 several places that you would expect to be very active seismically and several that you would expect to be very quiet.

9. If you wish to move away from California to escape earthquakes, how about settling in Guatemala, Chile, Turkey, Italy, China, or Japan? Explain.

10. Consider what would happen in your town or city if snow fell at the rate of 1 m/week over the entire North American continent. Do you suppose the snow could be cleared fast enough to keep communications open? How? How do you imagine your community could survive?

11. If a light beam is passed through a solution of natural sugar (cane or beet) in water, the light is always circularly polarized because the sugar molecules all twist the same way. Now, sugar is a simple molecule that can be synthesized in the laboratory. Synthetic sugar, however, has molecules with right- and left-hand twists in equal numbers. Light passed through a solution of synthetic sugar, therefore, is not polarized. However, if bacteria are introduced into the solution of synthetic sugar, after a while the solution polarizes light, but in the opposite sense from a solution of natural sugar. Can you suggest an explanation?

12. Helical organic molecules have been found in meteorites, but such molecules are found in equal numbers that twist to the right and to the left. Why do we conclude that these molecules are probably not of biological origin?

Johannes Kepler (1571–1630) discovered the first quantitative laws that describe planetary motion. Kepler's laws made possible Newton's formulation of universal gravitation. He also wrote what may be the first science fiction story, the *Dream* (*Somnium seu astronomia lunari*), describing a trip to the moon, what that world is like, and how the earth would appear from the moon. (*The Bettmann Archive*)

16

THE MOON: A GIANT STEP

The moon is the nearest large celestial object to the earth and the only one to have been visited by humans. The cost of the Apollo Program of manned missions to the moon amounted to about $100 per American, which seems a rather trifling sum to pay for what was learned and for the incredible advances in technology that have so influenced our lives. I suspect that in my youth I spent that much money going to motion pictures about man going to the moon!

16.1 GENERAL PROPERTIES OF THE MOON

The gross properties of the moon have been described elsewhere in this book; some of the more important ones are summarized in Table 16.1.

(a) Lack of Lunar Atmosphere and Water

As we have seen (Section 14.2), the moon could retain only gases of rather heavy atoms; the common gases that would be expected to outgas would escape. Extensive Orbiter and Apollo studies of the moon confirm that the moon has no appreciable atmosphere, although trace concentrations of gases around the moon have been measured.

In the absence of air, the moon can have no liquid water on its surface. It is well known that water boils more easily (at a lower temperature) at high mountain altitudes than at sea level. If all air pressure is removed above water, it boils away rapidly. On the moon, therefore, if there ever were any liquid water, it would have evaporated and

Table 16.1 **Some Gross Properties of the Moon**

Mass	7.35×10^{25} g	$= 8.10 \times 10^{19}$ short tons $= 1/81.3$ earth masses
Diameter	3476 km	$= 2160$ miles
Mean density	3.34 g/cm³	$= 61\%$ that of the earth
Surface gravity	161 cm/s²	$= 0.165$ that of the earth
Velocity of escape	2.38 km/s	$= 1.48$ mi/s
Rotation period		
With respect to stars	$27^d7^h43^m11\overset{s}{.}5$	$= 27.32166$ days
With respect to sun	$29^d12^h44^m2\overset{s}{.}8$	$= 29.530588$ days
Inclination of polar axis to the perpendicular to its orbit $6\frac{1}{2}°$.		

then, as a gas, dispersed into space with the rest of the moon's atmosphere.

In the absence of air and water, there can be no weather on the moon. There are no clouds, winds, rain, snow, or even smog, which accounts for many of the differences between the moon and earth. Weather on the earth and the running of water over its surface have been major sources of erosion that have washed and worn away entire mountain ranges, and have reshaped the faces of continents many times over during geologic time. On the moon, where air and water cannot contribute to erosion, features are more nearly permanent. There are formations of all ages on the moon, standing side by side; features that were formed thousands of millions of years ago are often still intact, standing with those formed in the recent past. We find the history of the moon written upon its face.

Although there can be no surface water on the moon, we cannot absolutely rule out the possibility of underground water. Some astronomers have suggested that the moon could have a considerable supply of water trapped beneath its surface rocks. No trace of liquid water was found by the Apollo astronauts, nor was evidence found for underground "water tables." Some of the lunar rocks returned to earth did, however, contain chemically-bound water.

(b) Temperatures on the Moon

Because of the moon's lack of atmosphere and oceans, its temperature range between day and night is rather extreme. The earth's atmosphere and water cover serve to reflect and disperse much of the sun's radiation during the day and to blanket heat in near the surface at night.

The moon rotates with respect to the sun in about $29\frac{1}{2}$ days. Therefore, at a typical place on the moon day and night each lasts about two weeks. Without any atmosphere to absorb or reflect some of the sun's radiation, the day surface temperature rises to a high value, as more than 90 percent of the incident energy from the sun is absorbed. Eventually this energy is reradiated into space, most as infrared radiation, and some at radio wavelengths as well.

The rate at which a body like the moon radiates electromagnetic energy at various wavelengths depends on its surface temperature. The infrared radiation from the moon can be measured with a heat or infrared-sensitive detecting device, such as a lead sulfide cell or thermopile, placed at the focus of a telescope, and radio telescopes can measure the radio energy it emits. Thus, earth-based observations had determined the lunar temperatures long before they were measured directly by the Surveyor lunar probes and Apollo astronauts. The moon's temperature ranges from just above the boiling point of water where the sun is shining, down to about 100 K ($-173°C$) on its dark side.

The rate at which the moon's temperature drops when sunlight is blocked off has been measured during lunar eclipses. As the moon enters the earth's shadow, its surface temperature has been observed to drop more than 150° C in one hour. Certain craters, however, cool more slowly than their surroundings, giving eclipse "hot spots." This rapid temperature drop can be explained partially by the moon's lack of oceans and an atmosphere to act as a blanket for the heat. However, if the moon's surface were covered with solid rocks like those over much of the earth's surface, the temperature drop would occur more slowly. Such rocks, when exposed to sunlight, conduct heat well into their interiors and take a long time to cool. This evidence suggested that most of the moon's surface must be covered with fine dust particles or granules which conduct heat very poorly, heating up only at their surfaces and cooling rapidly. The prediction was later confirmed by the Surveyor and Apollo experiments.

(c) Lunar Magnetism

Magnetometers carried to the moon by Apollo astronauts, and those on earlier lunar probes and orbiters, have not detected a permanent general lunar magnetic field. There are some very weak local fields, indicating slight magnetism in some of the surface rocks. On the other hand, to be magnetized at all those rocks must have been exposed to a somewhat stronger field in the past. Possibly the rocks were magnetized in the earth's field at a time in the past when the moon was nearer the earth.

(d) Interior of the Moon

The moon's mass and mean density are both lower than the earth's. Because of the lower weight of lunar material, it should not compress as much as terrestrial matter, and internal pressures in the moon are expected to be far less than in earth.

Analysis of the perturbations on orbits of the lunar orbiting satellites gives some indication of the mass distribution within the moon. It is found that the moon is, indeed, much less centrally concentrated than the earth is. However, surface irregularities in the moon's mass distribution, called *mascons* (for "mass concentrations"), have been detected. These relatively dense subsurface regions are possibly submerged material that collided with the moon in the past. It is also found that the center of mass of the moon is not at its geometrical center but displaced about 2 km, roughly in the direction toward the earth.

One model of the moon that has been constructed to satisfy the gravitational data and also the offset center of mass predicts that the moon's crust is much thicker than the crust of the earth. Moreover, according to the model, it is very uneven in thickness. On the side of the moon turned toward the earth the crust extends to a depth of only about 60 km, while on the back side it is at least 150 km thick. The typical rocks of the lunar crust probably have higher melting points than those on earth, and

in addition the temperature increase with depth is less rapid than on earth.

Four seismometers left on the moon by Apollo astronauts have shown that compared to the earth, the moon is inactive seismically. In particular, there is no evidence at all for tectonic activity, as on the earth. However, some moonquakes, meteoritic impacts, and such artificial disturbances as grenade explosions and spacecraft booster impacts have been recorded. The natural seismic activity appears to be correlated with times when the earth exerts the greatest tidal stress on the lunar crust.

The surface material on the moon transmits seismic waves very slowly, consistent with the idea that it is loose rocks, like rubble. At deeper layers, however, the lunar crust transmits the waves very rapidly, suggesting that the material has undergone heating and fusing in the past. At depths greater than about 1000 km, the waves do not appear to be transmitted efficiently. Some investigators consider this as evidence for a hot or even molten or partially molten interior core in the moon. The case for a hot lunar interior, however, is still controversial. The moon appears to have cooled more rapidly than the earth, and much of its mantle is solid and rigid.

Figure 16.1 The moon in a gibbous phase as seen mostly from the far side. Mare Crisium is visible at the limb of the moon in the upper left. The photograph was taken by the Apollo 16 crew. (NASA)

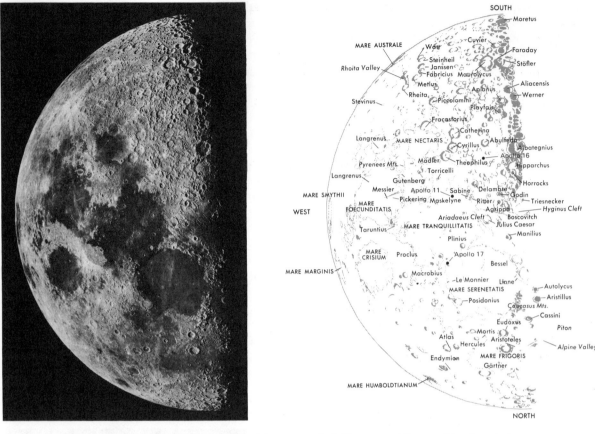

Figure 16.2 The western hemisphere of the moon. *(Lick Observatory)* The numbered symbols on the identification chart show the locations of Apollo landings.

16.2 THE SURFACE OF THE MOON

The most conspicuous of the moon's surface features are easily visible to the unaided eye. In medieval Europe these markings were generally regarded as colorations rather than actual irregularities in the surface of the moon. Galileo's telescope observations, however, revealed mountains, craters, valleys, and what appeared to be seas. The idea developed that the moon might be a world, perhaps not unlike our own. Many other lunar observers followed Galileo's lead. In 1647 John Hevel of Danzig (1611–1687) published his comprehensive treatise on the moon, *Selenographia*. In a series of carefully prepared plates, Hevel identified many of the moon's features and gave them names, in many cases in honor of similar features on the earth. The names he gave the mountain ranges on the moon survive today.

The early lunar observers regarded the moon as having continents and oceans and as being a possible abode of life. We know today, however, that the resemblance of lunar features to terrestrial ones is superficial. The moon's lack of air and water makes most of its features unlike anything we know on earth.

(a) Modern Observations

Many excellent photographs of the moon have been obtained with earth-based telescopes, but today most of the lunar photographs obtained from earth have been rendered obsolete by those obtained by the Apollo astronauts, and earlier, by the spectacular Ranger and Surveyor lunar probes and the lunar Orbiters. A Soviet space vehicle was the first to send to earth photographs of the moon's far side. The Soviet Luna 9 made the first soft landing on the moon on January 31, 1966, and transmitted close-up photographs to earth for three days.

Figure 16.3 The eastern hemisphere of the moon. *(Lick Observatory)* The numbered symbols on the identification chart show the locations of Apollo landings.

In the Ranger series in 1964 and 1965, Rangers 7, 8, and 9 made successful "hard" landings on the moon, that is, they crashed directly onto the moon as planned. On the way a great many views of the moon were televised directly back to earth, with resolution on the lunar surface as high as 0.3 m. The best observations from earth telescopes gave a resolution of about 1 km on the moon.

The first successful unmanned American soft landing on the moon was by Surveyor I, launched on May 30, 1966. Surveyor I alone sent over 11,000 photographs back to earth. In the next year and a half, Surveyors III, V, VI, and VII made successful landings. Each had television cameras that showed both the lunar landscape and close-up views of experiments done on the lunar "soil" by equipment carried on the probe. In addition to obtaining photographs, the Surveyors analyzed the chemical nature of lunar surface material and studied its physical and magnetic properties. Observations were

even made of part of the solar corona far from the sun's surface (Chapter 31).

The five lunar Orbiters were all highly successful. The first was launched on August 10, 1966, and the last on August 1, 1967. Each was sent near the moon and then put into a lunar orbit by firing a retrorocket. In addition to providing a highly detailed photographic survey of all sides of the moon, the Orbiters delineated several possible landing sites for astronauts in the Apollo project, photographed the earth, made measures of micrometeorites and the moon's magnetic field (not detected), and measured charged particles in the vicinity of the moon.

The most extensive studies of the moon have come from the six Apollo missions in which astronauts carried out experiments on the lunar surface and returned core samples and rocks to the earth. In each mission two astronauts landed on the moon in the lunar module, while a third remained in orbit about the moon in the command module. There have also been three unmanned Soviet missions that returned material from the moon's surface to earth: Lunas 16, 20, and 24, launched in 1970, 1972, and 1976, respectively. The Apollo missions are listed in Table 16.2.

(b) The Lunar Seas

The largest of the lunar features are the so-called seas, still called *maria* (Latin for "seas"). They form the features of the "man in the moon." The maria are of course dry land. They are great plains, with relatively smooth, flat floors that appear darker than the surrounding regions.

The largest of the 14 lunar "seas" on the moon's visible hemisphere is *Mare Imbrium* (the "Sea of Showers"), about 1100 km across. The other maria have equally fanciful names, such as Mare Nubium ("Sea of Clouds"), Mare Nectaris ("Sea of Nectar"), Mare Tranquillitatis ("Tranquil Sea"), Mare Serenitatis ("Serene Sea"), and so on. Most of the maria are roughly circular in shape, although many of them are interconnected or overlap slightly, and all have irregularities and baylike inlets, such as Sinus Iridum ("Bay of Rainbows") on the north "shore" of Mare Imbrium.

Inspection of good photographs shows that the maria do not have perfectly smooth floors, as they appear to have at first glance. They are speckled with thousands of tiny craters resembling potholes ranging in size down to less than 1 m across, and inside some maria there are large craters. Some of

the mare floors show wavelike ripples when the sunlight strikes them at a glancing angle. In a few places cliffs are found, suggesting slippages of the moon's crust along faults—fissures or lines of weakness in the crust. The best example is the "Straight Wall" in Mare Nubium, which is a cliff 180 m high and 130 km long.

Apollo and Surveyor landings in several different maria found them to be remarkably similar to each other. The mare floors are covered with a layer of material that analysis shows to be basaltic in origin, although severely eroded, probably by meteoritic bombardment. The material consists mostly of fine grains with a variety of sizes, typically about 0.02 mm in diameter. However, mixed in are numerous aggregates or "clots" of fine grains and some hard rocks. The material is, on the average, very porous and compressible, but of course, it is able to support easily the weight of an astronaut. Its density is betweeen 0.7 and 1.2 g/cm^3.

The far side of the moon (that turned away from the earth), in marked contrast to the near side, does not have large maria.

(c) The Lunar Craters

Even from earth there can be observed on the moon some 30,000 craters, circular depressions ranging in size from about 1 km to over 200 km across. Some of the largest are *Clavius* and *Grimaldi*, both nearly 240 km in diameter. Following the custom started by John Riccioli in 1651, craters are gener-

Table 16.2 **Apollo Lunar Landings**

APOLLO NUMBER	DATE	ASTRONAUTS	NUMBER OF ORBITS OF COMMAND MODULE	WHERE LANDED
11	1969 July 16–24	Neil A. Armstrong Michael Collins* Edwin W. Aldrin, Jr.	31	Mare Tranquillitatis
12	1969 Nov. 14–22	Charles Conrad, Jr. Richard F. Gordon, Jr.* Alan L. Bean	45	Mare Nubium
14	1971 Jan. 31–Feb. 9	Alan B. Shepard, Jr. Stuart A. Roosa* Edgar D. Mitchell	34	Mare Nubium
15	1971 July 26–Aug. 7	David R. Scott Alfred M. Worden* James B. Irwin	74	Hadley Rill in the Apennine Mountains
16	1972 Apr. 16–27	John W. Young Thomas K. Mattingly II* Charles M. Duke	64	Highlands east of Mare Nectaris, near the crater Descartes
17	1972 Dec. 7–19	Eugene A. Cernan Ronald E. Evans* Harrison H. Schmitt	75	Taurus Mountains near Mare Serenitatis

*Command module pilot.

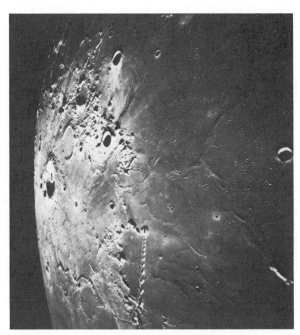

Figure 16.4 A closeup of a part of a lunar mare photographed by the Apollo 16 crew. (NASA)

ally named after famous scientists and philosophers. Other famous craters are Tycho, Copernicus, Kepler, Aristarchus, and Plato. Craters are still named after outstanding selenologists (astronomers who study the moon) today.

Although craters can be seen on the moon at almost any time, even with binoculars, the best time to view them is when the moon is near the first or last quarter phase. Then the terminator, dividing day and night on the moon, runs about down the middle of the apparent disk of the moon in the sky. The sun strikes craters near the terminator at a glancing angle and they cast long shadows, which makes them stand out in bold relief.

In general, craters are found over those regions of the moon that are not covered by the seas. There are comparatively few craters in the maria themselves, which suggests that the maria may be more recent features on the moon than most of the craters. The existence of crater tops barely visible in some of the mare floors supports this view. Most of the craters are found in the rugged regions of the

Figure 16.5 A bleak lunar landscape from the window of Apollo 11 landing module. (NASA)

Figure 16.6 Apollo 16 photograph of its landing site in the region of the lunar crater Descartes. (NASA)

moon. There they occur in all sizes and probably all ages as well. Frequently they overlap. An especially rugged region of overlapping craters is found in the vicinity of Tycho, near the moon's south pole.

The largest craters are often called *walled plains*. Examples are Clavius, Plato, and Ptolemaeus. A walled plain often appears to be a sunken region with little or no outside wall, sloping up to the crater rim. Some of the walled plains do not have circular walls, but are irregular in shape.

Most of the craters of moderate size are quite circular and have outside walls. In a few cases the crater floors are higher than the surrounding landscape, but in the majority of cases they are lower. Their inside walls are almost always steeper than the outside walls and rise to heights of as much as 3000 m above the crater floors. Many of these craters have mountain peaks in their centers. They somewhat resemble craters left by bomb explosions and may have had a violent origin. An excellent example is the crater Copernicus.

The most numerous craters are the smallest ones, those ranging in size down to craters less than a meter across. Often craters, especially the smaller ones, occur in clumps or clusters or in long lines or rows of craters. In some cases lines of craters seem to follow shallow cliffs or what appear to be faults.

(d) The Lunar Mountains

There are several mountain ranges on the moon. Most of them bear the names of terrestrial ranges— the *Alps, Apennines, Carpathians,* and so on. The similarity of mountains on the moon and the earth ends, however, with their names. Because of the absence of water, the lunar mountain ranges are devoid of the drainage features so characteristic of our own mountain ranges, and the lack of weather erosion on the moon results in a different appearance of mountains there.

Even from Earth the heights of lunar mountains above the surrounding plains can be determined by measuring the lengths of the shadows they cast. In Figure 16.7(a) is illustrated one of the

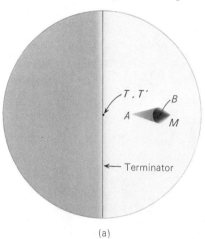

(a)

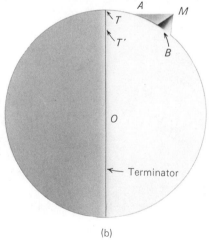

(b)

Figure 16.7 Measuring the height of a lunar mountain. (a) The apparent disk of the moon as it appears in the sky; (b) the geometry of the situation seen at right angles to the line of sight from the earth.

possible procedures for determining the elevation of a lunar feature. A mountain peak, *M*, near the terminator and near the center of the moon's disk, casts its shadow over the distance *AB* from its base toward the terminator. The length of the shadow *AM* and the projected distance of the mountain from the terminator, *T'B*, can be measured in seconds of arc, and these measures can be converted to kilometers.

The geometry of the problem is shown in Figure 16.7(b), in which the moon is seen from a direction at right angles to the line from the moon to the earth. The line *AM* is perpendicular to *OT*, and *AB* is perpendicular to *OM*. Therefore, the two right triangles *OT'B* and *ABM* are similar, and we have the proportion

$$\frac{BM}{AM} = \frac{T'B}{OB}.$$

Since *OB* is the known radius of the moon, and *T'B* and *AM* are measured quantities, the height of the mountain, *BM*, can be calculated.

Pains were taken that the photographs made with lunar Orbiters would be at such angles that the features on the moon would have shadows, enabling the determination of the heights of those features.

The highest lunar peaks range up to elevations of 8000 m or more, comparable to the highest peaks in the Himalayas on the earth. It is not cer-

Figure 16.8 A region of dense cratering on the moon, photographed by the Apollo 16 crew. *(NASA)*

tain, however, that any mountain on the moon is quite as high as Mount Everest. (Of course, on the earth mountain heights are measured from sea level rather than from the elevations of the surrounding plains.)

(e) Other Lunar Features

There are many other lunar features besides the seas, craters, and mountains. The *Alpine Valley*,

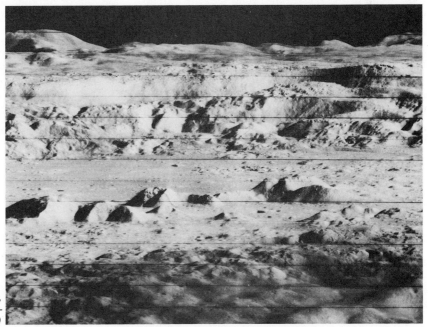

Figure 16.9 The crater Copernicus, viewed from the south, photographed by Lunar Orbiter II. *(NASA)*

Figure 16.10 An Apollo 16 photograph of a region of the moon showing several large craters and a large number of smaller secondary craters. *(NASA)*

for example (Figure 16.11), is a deep straight gorge cutting through the Alps. Also on the moon are many crevasses or clefts, half a mile or so across, called *rilles*.

Especially interesting are the *rays*—bright streaks that seem to radiate from certain of the craters that appear to be of explosive origin, notably Tycho and Copernicus (see Figure 16.13). The rays have widths of 8 to 20 km or more and in many cases extend for distances of hundreds of kilometers. Some of the rays from Tycho seem to extend completely around the visible hemisphere of the moon. They follow nearly great circles on the moon's surface, and sometimes are *tangent* to the craters from which they originate rather than coming from the exact centers of the craters. They cast no shadows and are seen best near full moon, when the sun shines down on them most directly. At full moon the ray system around Tycho gives the moon almost the appearance of a peeled orange. Ranger photographs show some of the rays to be associated with lines of craterlets. They were evidently formed by material thrown out in the explosion that caused the craters.

(f) Composition of the Lunar Surface

The chemical composition of the lunar surface material was first analyzed by alpha-particle scattering experiments carried out by the soft landers, Surveyors V, VI, and VII. More recently, of course, chemical analysis has been carried out directly on lunar rocks and soil samples brought back to earth by the Apollo astronauts, and by the unmanned Soviet Luna missions.

The results from widely separated places on the moon are strikingly similar. The most abundant element is oxygen, accounting for about 58 percent of the atoms present. Most of the oxygen atoms are chemically united with silicon, next most abundant, and accounting for about 20 percent of the atoms. Also abundant are aluminum, calcium, iron, magnesium, titanium-rich compounds, and other elements that are abundant in the earth's crust. The chemical composition of the lunar material, and also the mineral structure, is similar to that of volcanic basalts on earth but is lower in iron abundance and in volatile compounds than are earth rocks. The most common lunar minerals (which are also common in terrestrial volcanic rocks) are plagioclase and pyroxene. Some of the lunar minerals, however, are unknown on earth, so while the moon has had a similar chemical history to earth's, it is not identical.

16.3 THE ORIGIN OF THE MOON

Most of the various theories for the moon's origin are one or another of three general ideas: (1) that the moon separated from the earth by fission when the earth was just forming; (2) that the moon formed elsewhere in the solar system and was later captured; (3) that the moon accreted from particles in revolution about the young earth. Whereas no one is sure, most experts lean to the third of these ideas.

The idea of fission is that the forming earth had an extremely high rate of rotation. Under some circumstances a rapidly rotating body can be unstable and can divide into two bodies, which could then slowly separate as a result of their mutual tidal interactions. In fact, we have seen that the earth and moon do exert appreciable tides on each other, and that the friction of the tidal flow of water against itself and over coastal areas is gradually slowing the earth's rotation. In order that the earth-moon system conserve its angular momentum, the moon is, indeed, very slowly moving away from

Figure 16.11 The Alpine Valley, photographed by Lunar Orbiter V. *(NASA)*

particles in roughly the same region of the solar nebula, but in completely different orbits. Later, perturbations by other planets changed the moon's orbit, so the moon came close to the earth and was slowed down enough at the same time to be captured. The problem here is that such a capture is improbable, and even if it did happen, we would expect the moon to have a very eccentric orbit, rather than its present nearly circular one lying almost in the ecliptic plane.

The preferred hypothesis is that the moon accreted in roughly its present orbit from particles revolving about the earth. It is much easier for the earth to have captured a disk of small particles than one large, fully formed moon. While the earth was forming in the solar nebula, lots of particles were colliding with each other near it, thereby slowing each other down to the point where many can end up in an orbit around the forming earth. Drag on the particles by the gas in the solar nebula can also slow them down. It is estimated that an accretion disk of particles could so form in orbit around the earth in a time as short as 1000 years, but easily in a few million years, which is still a snap of the fingers in the geologic time scale.

(a) Origin of Lunar Craters

As the moon accreted, either in orbit around the earth or elsewhere in the solar system, it gradually

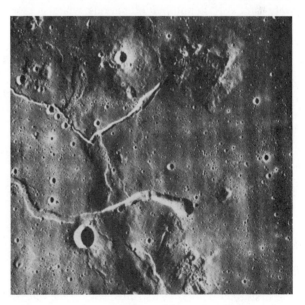

Figure 16.12 Lunar Orbiter V photograph of a lunar rille. *(NASA)*

the earth, increasing the size of its orbit. But it is not known how the earth could have a rapid enough spin to fission into two bodies, and if it did, it should have a much faster spin today. In other words, it is hard to conceive a satisfactory fission model that works in detail.

The second kind of theory—that the moon was formed elsewhere in the solar system and then captured—is more feasible than the fission model, but still presents problems. The idea of this theory is that the earth and moon both accreted from solid

Figure 16.13 The full moon, photographed with the 2.5-m telescope on Mount Wilson. The crater Tycho, with its many rays, is the bright feature near the bottom of the photograph.

collected matter and built up to its present size. In those very early years, though, there were still a great many particles left in the solar nebula, some of them very large. It must have taken some hundreds of millions of years to clean out the solar system of all that debris. During that time there were probably a number of periods when a new supply of rather big hunks became available, through collisions of large bodies—just as the asteroids seem to be colliding and fragmenting even today (Chapter 19). During the first 500 or 600 million years there were many rains of particles on the moon (and, of course, on the earth and other inner planets), producing the extensive cratering we still observe there. Those collisions may have made enough heat to keep the moon's outer layer molten for a while; we have seen evidence that the moon's crust may have once been completely liquid.

Gradually the chunks were swept out of the interplanetary space, and the impacts became less and less frequent. By 4000 million years ago the moon's crust had hardened and most of the cratering had ceased, although occasional collisions causing overlapping lava flows must have continued for the first 1000 million years, thus leaving the maria. By that time the cratering must have been at a rate at least 1000 times less than in the beginning, and perhaps only a few tens as great as it is today. By

2000 million years ago, impacts became very rare, and only once in a great while did large particles, probably collision fragments of asteroids chance to strike the moon and leave those craters of more recent origin. By now the moon is completely cooled off, to a depth of at least 1000 km. Seismic studies show that only an inner core is still soft and warm. To all intents and purposes that world, once seething with activity and molten rock, is now geologically dead.

If meteoritic collisions with the moon have formed so many craters there, one might wonder why we do not find such features on the earth. There must have been many thousands of large craters on the earth in the past, especially following the time of its original accretion, but they would have been washed away long since by the erosion of wind and rain; only those formed in the relatively recent past would still be intact. Still, many craters of collision origin are present on the earth today. The most famous is the Barringer (or Great Meteor) Crater, near Winslow, Arizona, which is more than 1 km across (Chapter 21). But there are many much larger craters, or suspected fossil craters, probably of impact origin, as well; for example, the 22-kilometer diameter Rieskessel structure in south Germany. In the first thousand million years of its existence, the earth must have been as heavily cratered as the moon.

Figure 16.14 A large boulder on the moon with many cracks. It is one of many lunar features described in considerable detail by Apollo 17 Astronaut Harrison H. Schmitt. (NASA)

EXERCISES

1. Does every place on the moon have daylight for exactly half of each period of rotation with respect to the sun? Why, or why not? (*Hint:* Think about the seasons on the earth.)

2. Astronaut David Scott dropped a hammer and a feather together on the moon, and both reached the ground at the same time. There are two reasons why this experiment on the moon has a distinct advantage over a similar one on the earth to show that light and heavy objects fall at the same rate. What are these advantages?

3. What would a 100-kg man weigh on the moon?

4. If the moon had its present radius but four times its present mass, what would be the velocity of escape from its surface?

5. At what phase of the moon could we most conveniently measure the temperature of the night side? Why?

6. Suppose a mountain is observed on the moon 100 km from the terminator. Its shadow is 40 km long. How high is the mountain?
   Answer: About 2.3 km, or 2300 m

*7. Suppose a mountain peak on the night side of the moon rises just high enough to catch some of the rays of the rising sun and shine like a bright spot of light. If the mountain is just 100 km from the terminator, what is its height?
   Answer: About 2.9 km

8. Consult an appropriate reference to learn the molecular weights of various gases and suggest one or two gases that the moon could hold for 5×10^9 years or more.

9. Describe a baseball game being played on the moon. Include discussions of the distance the ball might be batted, the strides of the base runners, the ease or difficulty of fielding the ball, the likelihood of rain checks and of calling the game because of darkness, and how the fans might boo the umpire.

10. Some people have noticed that the Pacific Ocean basin is roughly circular, and they have speculated that the Pacific Ocean is therefore a scar left when the moon pulled out of the earth. Criticize this hypothesis on as many relevant grounds as you can.

*11. Suppose that the moon has 30,000 craters more than 1 km across, and that the earth has only three. Assume the earth's radius to be four times the moon's. If craters have been formed on the earth and moon at a constant rate for the past 5×10^9 years, and if the moon's craters are never destroyed by erosion, how old, roughly, are those craters on earth?
   Answer: About 3×10^4 years or less

Percival Lowell (1855–1916) founded an observatory in Flagstaff, Arizona, for the primary purpose of studying the planets and searching for life in the universe. His vision provided inspiration that kept alive for decades the dream of exploring those other worlds— a dream realized in our present space age. (*Yerkes Observatory*)

17

THE TERRESTRIAL PLANETS

The ancients named the planets for their gods and associated them with those gods. Even in modern America tens of millions of people still believe in the Greek religion of astrology—that the planet-gods somehow control or at least influence our lives. In contrast to this 2000-year-old belief, with the technology made possible by modern science we have actually explored at close range all of the planets known in antiquity. What we have seen is fabulous indeed, and far more stimulating to the human intellect than the man-invented gods could ever have been.

The program of space exploration is less than two decades old, but in that short time its spectacular achievements have accomplished what hundreds of years of viewing from afar, and highly imperfect viewing from the bottom of our own ocean of air at that, could not even lead us to imagine. Our speculations and dreams have been replaced by hard data and close-up pictures. Yet, it was thousands of years of watching the planets move in the sky that led to the discovery of the gravitational theory that made our space explorations possible in the first place.

As we saw in Chapter 14, the planets can be sorted into two groups: the terrestrial (earth-like) and the Jovian (Jupiter-like) planets. In this chapter we describe the terrestrial planets (other than the

earth, which was considered in Chapter 15). Some of the more significant probes to the terrestrial planets are listed in Table 17.1 (the table is *not* a complete list of all U.S. and Soviet missions). All of these probes carried scientific instruments to obtain data on such things as magnetic fields, temperature, and atmospheric constituents. Many also carried television systems to transmit photographs back to earth. The Soviet Union has sent a long series of probes to each of Venus and Mars; their Venera missions to Venus have been the more successful. All of these probes, of course, were unmanned.

17.1 MERCURY

Mercury is one of the brightest objects in the sky. At its brightest it is inferior in brilliance only to the sun, the moon, the planets Venus, Mars, and Jupiter, and the star Sirius. Yet, most people—including even Copernicus, it is said—have never seen Mercury. The planet's elusiveness is due to its proximity to the sun. Its orbit is only about 40 percent the size of the earth's; it can never appear further from the sun than about 28°. It is visible to the unaided eye, for a period of only a week or so, at times when it is near eastern elongation, appearing

Table 17.1 **Some of the More Significant Planetary Probes (All Unmanned)**

PLANET	SPACECRAFT	ARRIVAL DATE	COMMENTS
Mercury	Mariner 10 (U.S.)	1974	Three flybys
Venus	Venera 7 (U.S.S.R.)	1970	Landed; broadcast for 23 min
	Venera 8 (U.S.S.R.)	1972	Landed; broadcast for 50 min
	Venera 9 (U.S.S.R.)	1975	Landed; sent photo
	Venera 10 (U.S.S.R.)	1975	Landed; sent photo
	Venera 11 (U.S.S.R.)	1978	Atmospheric probe
	Venera 12 (U.S.S.R.)	1978	Atmospheric probe
	Pioneer Venus 1 (U.S.)	1978	Orbiter
	Pioneer Venus 2 (U.S.)	1978	Multiprobe lander
Mars	Mariner 4 (U.S.)	1965	Flyby; 22 photos sent back
	Mariner 6 (U.S.)	1969	Flyby; many photos
	Mariner 7 (U.S.)	1969	Flyby; many photos
	Mariner 9 (U.S.)	1971	Orbiter; 7329 photos
	Viking 1 (U.S.)	1976	Orbiter and Lander; photos and life experiments
	Viking 2 (U.S.)	1976	Orbiter and Lander; photos and life experiments

above the western horizon just after sunset, and also when it is near western elongation, rising in the east shortly before sunrise. Because the synodic period of Mercury is 116 days, the intervals of its visibility as an "evening star" after sunset, and as a "morning star" before sunrise, occur about three times a year. However, Mercury sets so soon after the sun (or rises so shortly before the sun), that only rarely can it appear above the horizon when the sky is completely dark; generally, one must look for it in twilight.

Mercury was well known to the ancients of many lands. The earliest observers, however, did not recognize it as the same object when it appeared as an evening star and as a morning star. The early Greeks, for example, called it Apollo when it was seen in the morning twilight.

(a) Mercury's Orbit

Mercury is the nearest to the sun of the nine major planets and, in accordance with Kepler's third law, has the shortest period of revolution about the sun (88 of our days) and the highest mean orbital speed (48 km/s). It is appropriately named for the fleet-footed messenger god.

The semimajor axis of the orbit of Mercury, that is, the planet's median distance from the sun, is 58 million km or 0.39 AU. However, because Mercury's orbit has the high eccentricity of 0.206, its actual distance from the sun varies from 46 million km at perihelion to 70 million km at aphelion. Pluto is the only major planet with a more eccentric orbit. Furthermore, the 7° inclination of the orbit of Mercury to the plane of the ecliptic is also greater

than that of the orbit of any other planet except Pluto.

(b) General Properties and Structure

The mass of Mercury, which has no known satellite, is determined most accurately from the perturbations its gravitational influence produced on the Mariner 10 space probe. Mercury's mass is $1/18$ that of the earth; Pluto is the only planet with a smaller mass.

Mercury is also the second smallest of the planets, having a diameter of only about 4880 km, less than half that of the earth. Its mean density is about $5^{1}/_{2}$ times that of water—about the same as the mean density of the earth.

If Mercury were identical in chemical composition to the earth, it should not have compressed itself to so high a density. We believe it has a higher relative abundance of iron than the earth does, probably in the form of a sizable iron core. Heavier, more refractory elements must have contributed more to the material near the sun, where Mercury formed, than to that at the earth's position in the solar nebula.

(c) Telescopic Appearance

Mercury, like Venus, presents different portions of its illuminated hemisphere to us as it revolves about the sun and hence goes through phases similar to those of the moon (see Section 3.4b). The alternate crescent and gibbous shape of Mercury is its only conspicuous telescopic characteristic.

Figure 17.1 The planet Mercury in a mosaic of photographs made by Mariner 10 during its approach to the planet on March 29, 1974. *(NASA/JPL)*

Observations of Mercury are very difficult because of its nearness to the sun. At night it can be seen only close to the horizon, and then for but a short while. Its light must traverse a long path through the earth's atmosphere, which not only dims the planet, but subjects the rays to disturbances in the air ("seeing"), which are especially troublesome for all observations of objects near the horizon. Consequently, nearly all telescopic observations of the planet are carried out in daylight, when it is high in the sky. Unfortunately, daytime observational conditions are seldom good, and although some observers had been able to detect darker areas on Mercury, something like the maria

on the moon, earthbound studies never revealed the surface of the planet in detail.

(d) Rotation

Visual studies of Mercury's indistinct surface markings seemed to indicate that the planet kept one face to the sun, and for many years it was widely believed that Mercury's rotation period equaled its period of revolution about the sun of 88 days.

Radar observations of Mercury in the mid-1960s, however, showed conclusively that Mercury does rotate with respect to the sun. Its sidereal period of rotation (that is, with respect to the distant stars) is about 59 days. G. Colombo first pointed out that this is very nearly two thirds of the planet's period of revolution, and subsequently Goldreich and Peale showed that there are theoretical reasons for expecting that Mercury can rotate stably with a period *exactly* two thirds that of its revolution— 58.65 days. They argue that if Mercury were not perfectly spherical, but were deformed so that one dimension through the equator were longer than the others, the sun's force on that bulge should force the long axis of the planet to point to the sun when it is at perihelion where the sun's differential force on Mercury is strongest. This condition would be met if the planet rotated with its revolution period, but also with certain other rotation periods, the most likely being two thirds the revolution period, so that at successive perihelions alternate ends of the long axis of the planet are pointed toward the sun. [The actual deformation of the planet, of course, is at most very small; Mercury is spherical to a high approximation. Recall (Section 9.1) that the moon is also a triaxial ellipsoid.]

(e) Temperature

Although the temperature of Mercury had been measured from the earth, better data came from the Mariner 10 fly-by in 1974. The daylight temperature on the surface ranges up to about 700 K at noontime. Just after sunset, however, the temperature drops quickly to about 150 K, and then slowly descends to about 100 K at midnight. Extrapolation of the Mariner data leads to a temperature estimate of about 90 K just before dawn. The range in temperature on Mercury is thus over 600 K, more than on any other planet.

Figure 17.2 Cratered terrain of Mercury photographed by Mariner 10. The large flat-floored crater at the right is about 100 km in diameter—about the size of the lunar crater Copernicus. Many of the smaller craters are probably caused by particles thrown out by the impact that formed the large crater. *(NASA/JPL)*

(f) Lack of Atmosphere

Mercury was not expected to have an atmosphere because of its high daylight surface temperature and low velocity of escape. As expected, no evidence for a permanent atmosphere was found by Mariner 10. There is a very tenuous hydrogen cloud around the planet, but the gas pressure near the surface is completely negligible—less than 10^{-17} that of the earth's atmosphere.

The sparse hydrogen may be due to an interaction of the planet with the solar wind. Rather surprisingly, Mariner 10 found Mercury to have a weak magnetic field, with a strength at the surface of 1 to 2×10^{-3} gauss. This field is hundreds of times weaker than the earth's, but it can influence the motions of ions in the solar wind.

(g) The Surface of Mercury

The first close-up look at Mercury came on March 29, 1974, when the Jet Propulsion Laboratory's Mariner 10 passed 9500 km from the surface of the planet at a speed of about 11 km/s. On that date, and for several days before and after the encounter, Mariner 10 televised more than 2000 photographs to earth, revealing details with a resolution down to 150 m. Mariner 10, in a planned elliptical orbit

about the sun, passed Mercury again on September 21, 1974, and finally on March 16, 1975, obtaining over 2000 more pictures.

Mercury strongly resembles the moon in appearance. It is covered with thousands of craters up to hundreds of kilometers across, and larger basins up to 1300 km in diameter. There are also scarps (cliffs) over a kilometer high and hundreds of kilometers long, as well as ridges and plains. Some of the brighter craters are rayed, like Tycho and Copernicus on the moon, and many have central mountain peaks.

Some of the Mercurian features have been named in commemoration of artists, writers, composers, scientists, and the like. A crater 41 km across is named Kuiper, in honor of the late astronomer who contributed so much to our knowledge of the solar system. A large basin has been named Caloris (Figure 17.3), for the sun is at the zenith of the basin when Mercury is at perihelion, and Caloris is probably thus the place on the planet with the highest noon temperature.

The larger basins resemble the lunar maria, both in size and appearance. They show evidence of much flooding, as do the maria, from lava flows, evidently released during the impacts that produced the features. From the amount of flooding present around the Mercurian basins, extending in some

Figure 17.3 In the left half of this mosaic is the ring basin Caloris, 1300 km in diameter. It is the largest structural feature on Mercury seen by Mariner 10 in its first pass of the planet in late March 1974. *(NASA/JPL)*

gin. If so, Mercury, being so much nearer the sun, should show far less cratering than the moon and Mars. But Mercury is as cratered as the moon, which lends strong support to the opposing hypothesis that the craters of those worlds were formed near the ends of their accretion processes, as they swept up the remaining large objects in their regions of the solar system. Many of the smaller craters on Mercury are secondary ones, formed by debris thrown out by the impacts that caused the main craters.

17.2 VENUS

Venus, named for the goddess of love and beauty, is sometimes called the "earth's sister," for it is most like the earth in mass and size of all the planets. As we shall see, however, conditions on the two worlds could hardly be more different. Nor would visitors find Venus appropriately named, for its surface environment is certainly not conducive to romance.

Still, Venus is a beautiful object in the night sky; its brilliance is exceeded only by that of the sun and moon, save for rare comets and bright meteors. At night it can cast a shadow; at its brightest it can be seen easily in broad daylight if one knows exactly where to look for it. Indeed, when Venus

places for more than 1000 km, it can be concluded that the lava material must extend hundreds of kilometers beneath the surface. If this material has a density of about 3 g/cm³ (typical of silicate rocks) the presumed iron core of Mercury must be quite large in order for the planet to have a mean density of 5.4. Mercury is almost certainly a differentiated planet (rather than being of uniform composition).

Unlike Earth there is no evidence of faulting on Mercury; plate tectonics evidently does not occur there. Nor are there straight rills or grabens (as on the moon), which would indicate faulting. The long scarps, sometimes cutting across craters, seem to be due to crustal compression.

Some astronomers had once held the view that the craters on the moon and Mars had been formed by bombardment by material of minor planet ori-

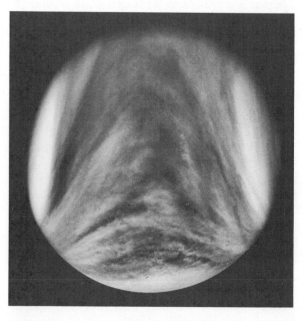

Figure 17.4 Venus, photographed by Pioneer Venus Orbiter at a distance of 58,000 km. *(NASA)*

happens to be conspicuous in the evening sky in December, people, surprised at its brilliance, often call observatories to inquire whether it is the star of Bethlehem.

(a) Venus' Orbit

Venus approaches the earth more closely than any other planet—at its nearest it is only 40 million km away. Like Mercury, Venus is an inferior planet (nearer the sun than the earth); consequently, it appears to swing back and forth in the sky, during its synodic period, from one side of the sun to the other. Like Mercury, therefore, it appears sometimes as an "evening star" and sometimes as a "morning star." The early Greeks, thinking it to be two objects, called it *Phosphorus* and *Hesperus* when it was seen in the morning and in the evening, respectively. Pythagoras, in the sixth century B.C., is credited with being the first to recognize that Phosphorus and Hesperus were one and the same planet.

Because Venus is farther than Mercury from the sun, it reaches much greater eastern and western elongations, and can appear as far as 47° from the sun. It can only be seen in the west in the early evening or in the east before sunrise; however, it is visible for a longer time than Mercury.

The median distance of Venus from the sun is 108 million km. Its orbit is the most nearly circular of any of the planetary orbits, having an eccentricity of only 0.007. Its distance from the sun, consequently, varies by only about 1.5 million km.

To complete its revolution about the sun in 225 days, Venus moves with a mean orbital speed of about 35 km/s. Its synodic period—the interval between successive conjunctions or oppositions—is 584 days. The inclination of its orbit to the ecliptic is 3°24'.

(b) Gross Characteristics

Like Mercury, Venus has no known natural satellite, but its mass is easily found from the perturbations it produces on space vehicles, and especially from the orbit of the Pioneer Venus Orbiter. Venus is 0.82 times the mass of the earth, and has a mean radius (corresponding to its solid surface) of 6051.4 km, only about 300 km less than the earth's. Its mean density is 5.2 times that of water.

It is possible to calculate what the approximate internal pressures and densities of Venus are if we assume that the compressibility of its material is the same as that of the material in the earth, known from seismic studies. At the center of Venus, the pressure exerted by the weight of the overlying rocks must be near 2.6×10^{12} dynes/cm^2, or about $2\frac{1}{2}$ million times the pressure of the earth's atmosphere. The central density must be about 11 times that of water. Most investigators believe that the cores of Venus and other terrestrial planets are largely metallic, probably mostly iron and nickel.

(c) Telescopic Appearance

Venus, like Mercury, goes through phases. Galileo was the first to report the phases of Venus, a discovery that had great importance in disproving the validity of the Ptolemaic cosmology (Section 3.4b).

Venus is not observable from earth at the full phase, because then it is at superior conjunction—on the other side of the sun from the earth—and is too nearly in line with the sun. Even if it could be seen, however, Venus would not show its greatest brilliance at full phase because at that time it is also at its greatest distance from the earth and subtends an angle of only 10″. Venus has its largest angular size of 64″ when it is nearest the earth, but then it is at the new phase (at inferior conjunction) and, again, is unobservable. Venus is at its greatest bril-

Figure 17.5 A Mariner 10 picture of Venus made from computer-enhanced and retouched ultraviolet photographs. *(NASA/JPL)*

liance when it is a crescent, with an elongation of about 39°; this occurs about 36 days before and after inferior conjunction.

The surface of Venus is not visible because it is shrouded by dense clouds. These clouds reflect sunlight very efficiently; measures of the albedo of the planet show that more than 70 percent of the incident visible sunlight is reflected, a circumstance that contributes greatly to its brightness. Some rather indistinct dark markings can be observed on Venus from earth; they show up best when the planet is photographed in violet light.

(d) Rotation

The rotation period of Venus, like that of Mercury, has been determined by radar. The first radar observations of the planet's rotation were made in the early 1960s. Surprisingly, they showed Venus to rotate from east to west—in the reverse direction from the rotation of most other planets—in a period of about 250 days. Subsequently, topographical surface features were identified on the planet that show up in the reflected radio signals. The rotation period of Venus, determined from the motion of such features across its disk, is 243 days retrograde (east to west).

Because Venus rotates so slowly, it was not expected to have a strong magnetic field, for there is no driving force to set up a current of moving charges in its interior. The prediction has been confirmed by the various space probes sent to the planet; no general magnetic field has been detected. Also because of its low rotation rate, Venus does not show appreciable polar flattening; its oblateness is 100 times less that is the earth's.

(e) Temperature

We might expect that Mercury, being nearest the sun, would be the hottest planet in the solar system. Not so: That honor goes to Venus, with a surface temperature of more than 700 K. Its high temperature is due to its dense atmosphere, largely composed of carbon dioxide. While solar radiation of visible light and near infrared can pass inward through Venus' atmosphere, the energy reradiated by that planet is farther in the infrared, which is blocked by the atmosphere, maintaining a high temperature near the surface.

Earth-based observations of the infrared radiation emerging from Venus revealed a temperature, roughly constant from day to night, of 230 to 240 K. This temperature corresponds to a level high in the Venerian atmosphere, where radiation emitted from gases above its opaque clouds can escape into space. Before any space probes had been sent to Venus, however, radio observations were made of Venus at wavelengths of several centimeters and longer. The Venerian atmosphere is transparent to this radiation, so it must be emitted by the surface of the planet. Those radio observations showed for the first time that Venus' surface is very hot—up to 730 K (450°C, or about 850°F). The high temperature of the surface of Venus has been verified by the several space probes sent to the planet.

(f) Space Probes to Venus

There have been several U.S. and Soviet probes to Venus (Table 17.1), among them the two successful Soviet landers that transmitted the only existing photographs made at the surface of the planet, as well as the Venera 11 and 12 probes of the Venerian atmosphere in 1978. A great deal of our information about Venus comes from the U.S. Pioneer Venus probes.

Pioneer Venus 1 arrived at its destination and went into a highly eccentric orbit about the planet on December 4, 1978. Pioneer Venus 2 was a multiple-probe vehicle that on December 9, 1978, sent five separate instrument packages into the Venerian atmosphere and down to its surface. These consisted of a large probe, three small ones known as North, Day, and Night (for the areas of the planet they explored), and the Bus—the vehicle on which the other four were mounted before deployment. The probes began collecting data from an altitude of 70 km. The large probe descended by parachute through the middle atmosphere, while the smaller probes descended without parachute and took 55 minutes to fall the 200 km from entry to the atmosphere to the surface. None of the probes was designed to survive impact; nevertheless, the Day probe transmitted data for 67 minutes after landing. Much of the following discussion of the Venerian atmosphere is based on data from Pioneer Venus 2.

(g) Atmosphere

From the earth (and even from the Pioneer orbiter) we see light reflected from upper cloud layers about 60 km above the surface. It is very cold up there, from 230 to 240 K. Those upper clouds form two

Figure 17.6 Pioneer Venus 2 Multiprobe *(foreground)* and Pioneer Venus 1 Orbiter. The entry probes can be seen mounted on the Bus, which itself carried instruments into the Venerian atmosphere. *(NASA)*

distinct layers, which together are about 12 km thick; they consist mostly of drops of sulfuric acid. Below them is a dark opaque cloud some 4 or 5 km thick, composed largely of solid and liquid sulfur with intermixed sulfur dioxide, which has shrouded the planet in mystery until the eyes of the Soviet Venera and U.S. Pioneer probes penetrated it. Below this main cloud a haze extends down to an altitude of about 32 km, but below that the atmosphere is clear—right down to the surface.

However, both the Soviet and U.S. probes detected frequent and almost constant displays of lightning in the middle atmosphere of Venus. These are not caused by thunderstorms of the earthly variety, but result from electrical or chemical activity not yet thoroughly understood.

Even the earlier space vehicles, such as Mariner 10, revealed high speed winds—100 to 200 km/hr—in the upper Venerian atmosphere, showing evidence of atmospheric circulation. The Pioneer probes found that at the north pole of the planet there is an area about 1100 km in diameter where the clouds are thinner. Evidently, the winds circulate poleward and down toward the surface at the poles, depressing the clouds to lower elevations, where they evaporate. In the lower atmosphere of Venus, however, there is no sign of weather whatsoever.

The first gas to be detected in the Venerian atmosphere, from earthbound observations in 1932, was carbon dioxide. Now that the chemical composition of the atmosphere has been measured accurately by the Pioneer probes, we can confirm that Venus' air is 96 percent carbon dioxide. Next in importance is nitrogen, at 3.4 percent. Water vapor accounts for 0.1 to 0.5 percent, and there are traces of argon, oxygen, neon, and sulfur dioxide.

The argon in the Venus atmosphere presents a puzzle. On the earth, the argon is mostly the isotope argon-40, which arises from the radioactive decay of potassium-40, and then outgases. But in Venus there is a fairly large proportion of argon-36—very rare on earth. We expect argon-36 to be primordial; that is, it was present in the solar nebula from which the planets formed. But then why is it not more plentiful on earth? Unless it is locked in the earth's interior, we have no immediate explanation; its presence on Venus was one of the unexpected findings by Pioneer.

The Venera, and later the Pioneer Venus probes verified what earlier space probes had indicated, and what had been calculated theoretically from the high surface temperature of Venus inferred by radio measures from earth: Venus has a very dense atmosphere—about 90 times the sea-level pressure of the earth's atmosphere.

We expect that the atmosphere of Venus, like that of the earth, originated by outgassing from the crust. The total amount of nitrogen on Venus is about the same as on earth, and so is the carbon dioxide, except that on Venus it is in the gaseous state in the atmosphere, while on the cooler earth it has reunited with surface rocks. But the major compound to outgas on earth was water, which exists only in small quantities on Venus. Where is the Venerian water? The favored guess is that it circulated to the top of the atmosphere, where solar ultraviolet radiation decomposed it to hydrogen, which would easily escape, and oxygen, which is chemically active enough to form compounds with other elements. It is possible, though, that hydrates (chemical compounds that contain water) were not common in the stuff from which Venus accreted, in which case there may never have been any water there.

At least our new knowledge of the composition of the Venerian atmosphere has completely settled the question of the high surface temperature of the planet. Earth-based radio observations of Venus provided the first evidence of the planet's high temperature, and the various space probes confirmed it, but some astronomers were uncertain whether the absorption of sunlight that penetrates to the lower atmosphere and surface could heat up the planet so much. They realized that the planet's re-radiated energy, in the form of infrared radiation, would be partly trapped by the carbon dioxide to keep the temperature high—the so-called *greenhouse effect*. But more information was needed to make sure the numbers work out right; that information was provided by the Pioneer probes. There is, indeed, enough carbon dioxide and water vapor, combined, to account for the observed temperature.

(h) The Surface of Venus

On October 22 and 25, 1975, the Soviet Venera 9 and 10 spacecraft landed at two different places on Venus, and each televised to earth a photograph clearly revealing surface rocks. At the landing site of Venera 9 the rocks were sharp-edged and angular, rather than showing the considerable erosion expected. Evidently they are of relatively recent origin—possibly volcanic. The rocks at the Venera 10 site, on the other hand, some 2000 km away, appeared flat, rounded and eroded. The Soviet scientists described the landscapes of Venera 9 and 10 as like young and old mountainscapes, respectively. The Veneras also measured wind speeds and found them to be very low—only 1 to 4 km/hr, indicative of a stagnant atmosphere.

The Veneras have provided our only optical observations of the ground beneath the opaque clouds. We can, however, see the surface of Venus with *radar*, whose radio waves penetrate the clouds. Out best radar data come from two complementary sources.

Figure 17.7 The surface of Venus, showing fairly rough rocks, photographed by the landing apparatus of Venera 9. *(Novoste from SOVFOTO)*

Figure 17.8 The surface of Venus, photographed by the landing apparatus of Venera 10. The vertical stripes in this figure and in Figure 17.7 occur where the transmission carried nonpicture information for other scientific experiments. *(TASS from SOVFOTO)*

The first is earth-based radar observations from the world's largest radio telescope, the 1000-foot dish at Arecibo. The Arecibo observations reveal features on Venus down to 10 to 20 km across. The earth-based radar data easily detect relief on the Venerian surface due to its changing slope, but do not give accurate information on the vertical extent (height) of the features. Radar observations from earth also have a severe handicap: we can only receive a strong enough reflected signal to obtain good resolution when Venus is at its closest to us, so observations are made only for a few months around the times of inferior conjunction, which occurs at intervals of Venus' 584-day synodic period. Unfortunately, it happens that the planet's 243-day retrograde rotation is just right so that at every such conjunction Venus displays the same side to earth. We therefore never see more than about half of the planet from Arecibo. (The coincidence is thought not to be exact; the best determination of the rotation period of Venus suggests that at successive conjuctions the hemisphere turned earthward changes by a small fraction of a degree, but at this rate it would take centuries to be able to map all sides of the planet by radar from earth.)

The other source of radar data is from the radar range experiment on the Pioneer Venus orbiter. The surface resolution possible from the Orbiter experiment (about 75 to 100 km) is poorer than that from Arecibo, but the range data give altitudes of features to better than 200 m. Moreover, the Orbiter can see all sides of Venus, and because its orbit has a high inclination (74°) to the ecliptic, it can even observe near the Venerian poles. Over a period of time the Orbiter can map 93% of Venus' surface by radar.

The most fundamental aspect learned about Venus' surface is that it is very flat. About 35% of the earth's surface in contained within the outer boundaries of the continental shelves. In contrast, 80% of the surface of Venus is smooth to within 2 km, and 60% is smooth to within only 1 km! It is with respect to this smooth spherical surface that the mean radius of Venus (6051.4 km) is defined. The extreme heights and depths on the planet are 6062 km and 6049.5 km from the planet's center, respectively.

Venus is not, however, without features, and some are spectacular. Except for a few early exceptions, the Venerian formations have been given feminine names, in keeping with the gender of the love goddess. One of the exceptions is Maxwell, named to honor the Scottish physicist. Maxwell is a rugged highland region at about 65° N latitude, and it contains the point of highest elevation on the planet—about 11 km. Although Maxwell is 2 km higher above the mean surface of Venus than Mt. Everest is above sea level on earth, much of the terrestrial ocean floor is at a depth of more than 5 km. Just west of Maxwell is a very large high flatland called the Lakshmi Plateau (for a Hindu goddess). The "continent" containing both Maxwell and the Lakshmi Plateau is Ishtar (the Babylonian love goddess). Some of the other highlands are the Beta region, at 25° N latitude, containing the mountains Rhea Mons and Theia Mons, and the Alpha region at 25° S latitude. The mountain peaks in these areas, especially in the Beta region, are thought to be shield volcanos. Perturbations on the Orbiter as it passes over the Venerian highlands indicate that they are probably of lighter material than the average in the planet's crust, and thus "float" higher on the underlying material, as do the continents on earth.

The surface of Venus also has valleys and canyons, some quite spectacular. Just north of the

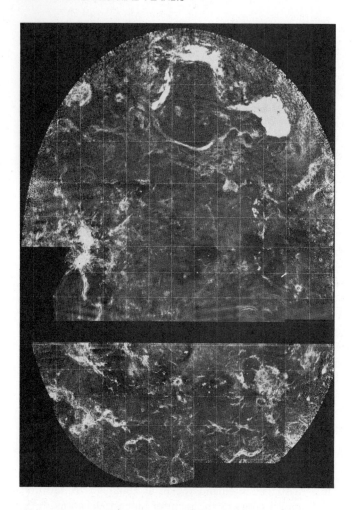

Figure 17.9   Radar image of Venus made with the 1000-ft telescope of the Arecibo Observatory. For identification of features, see Figure 17.10. (*Courtesy D. B. Campbell, Arecibo Observatory*)

equator, for example, is a furrowed arc about 1000 km long, consisting of two ridges, each 2 km high, straddling a valley 90 km wide. East of a continent known as Aphrodite, just south of the equator, is a rough region whose pattern suggests tectonic activity; a straight valley with high walls (up to 4 km) has a lateral offset, reminiscent of faults in terrestrial valleys.

Particularly interesting are the ring-like features, which are thought probably to be impact craters. An especially good one, Eve, is just southwest of Alpha highland region, at a latitude of 32°S. A bright spot at its center (mountain peak?) has been proposed as the origin of Venerian longitude. There are many other ring features. If they are impact craters, as is generally hypothesized, they provide information about the history of Venus' surface. The number and distribution with size of such features that are greater than 80 km across are consistent with what we would expect to have been formed due to bombardment of Venus with large particles

during the past six hundred million to thousand million years. Craters smaller than 20 km would have to be produced by chunks small enough that they would not be expected to survive flight through the Venerian atmosphere with explosive force. Craters of size between 20 and 80 km seem to be deficient on the planet's surface, but it may be because they are hard to recognize with existing radar resolution. Overall, the distribution of craters present suggests that the surface of Venus is intermediate in age between that of the earth and that of the moon or Mercury.

In summary, Venus is a world about the size of the earth, whose surface has been modified as has the earth's by all means except rain and running water. A significant surface radioactivity (measured by the Soviet Venera probes) and the probable presence of volcanos suggest that Venus has undergone considerable outgassing, as has the earth. Tectonic activity on Venus does not appear to have played a dominant role, but may be present. There are fea-

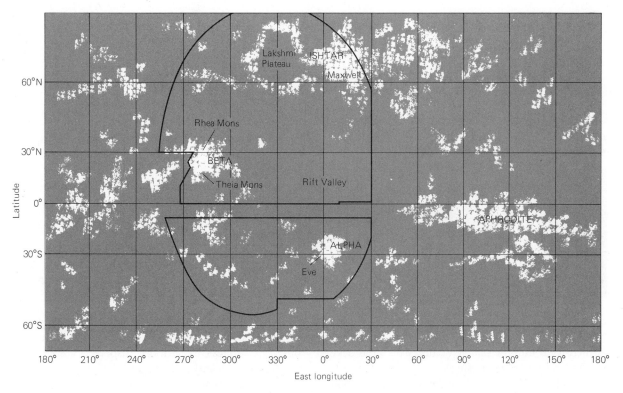

Figure 17.10 A map of Venus, based on radar observations from the Pioneer Venus Orbiter data. The longitudes are based on the proposed system whereby Eve's bright center defines 0°. Some of the conspicuous features described in the chapter are labeled. The large oval region is the area of Venus covered by radar from Arecibo (Figure 17.9).

tures analogous to continents and evidence of bombardment by meteoritic material, as on earth, but the scars of which have survived from an earlier time.

Still to be answered are such questions as whether Venus has a liquid metallic core, convection in its mantle, or active volcanos. And obviously, we have only begun to map the surface, so far with very crude resolution. There are plans for a Venus Orbiting Imaging Radar (VOIR), a satellite of the planet that if funded, could be launched late in the 1980s. VOIR, with aperture synthesis technology, will be able to map the surface of our sister planet with a resolution down to 100 meters; this would surely be a gigantic step toward our understanding of the nearest earth-like planet, and indirectly, the origin of our world as well.

(i) Velikovsky's Theory about Venus

It may seem inappropriate in a textbook on astronomy to discuss a theory that is demonstrably wrong.

But there exists a wrong theory that is so widely publicized and believed that I am asked about it with increasing frequency, and anyone who reads widely about Venus is almost sure to come across it.

The theory is Immanuel Velikovsky's (1895–1979), a Russian psychiatrist who immigrated to the United States in 1939. In the 1940s Velikovsky became interested in what he regarded as interesting correspondences between biblical stories and the mythologies of various people throughout the world, and he attempted to interpret these stories in terms of catastrophic encounters between the earth and other celestial bodies.

Briefly, Velikovsky's astronomical hypothesis is that a comet (or large body of planetary mass) was ejected from Jupiter and, about the middle of the second millennium B.C., the comet passed twice near the earth, causing great upheaval and raining hydrocarbons on the earth. At one passing the comet stopped the rotation of the earth temporarily, accounting for Joshua's making the sun "stand still upon Gibeon." The comet also passed near Mars, diverting that planet from its orbit, so that Mars also passed near the

Figure 17.11 Artist's conception of the Great Rift Valley on Venus. (*NASA*)

earth, causing additional tidal disruptions. Eventually, about the seventh century B.C., the comet transformed itself into the planet Venus, which has occupied its present orbit ever since.

Velikovsky's ideas were described in his book *Worlds in Collision*, published in 1950. The book and its sequels have sold millions of copies, and a number of other books have been written about the theory. There have also been several magazines devoted to a defense of Velikovsky. There is an active group of followers, including a few with some scientific training. These Velikovskyites have argued that certain discoveries about Venus that came as a surprise to most astronomers—for example, the high surface temperature, the retrograde rotation, and the high relative abundance of the isotope argon-36—are proofs of Velikovsky's hyothesis.

Astronomers and others who are active in planetary research, however, have never taken Velikovsky's theory seriously because it contradicts a great body of contrary evidence and well-understood laws of physics. To be sure, experts are sometimes later shown to be wrong, but generally in areas at the frontier of knowledge, not in matters that have been well established and are well understood. Moreover, the scientific heretics who occasionally turn out to be correct despite early skepticism of their ideas are almost always highly trained scientists themselves, not people untrained in the relevant disciplines.

There are many problems with Velikovsky's theory for the origin of Venus; I shall mention only a few. First, the composition of Jupiter (next chapter), like that of the sun, is mostly hydrogen and helium, rather than metallic and rocky material. We expect that there is *some* rocky material in Jupiter, but unless it is thinly dispersed among the light hydrogen and helium, it must be in a central core, from which it could not be ejected by any mechanism we can even imagine. Even if it could be, it would involve so much energy that the matter ejected would have to be vaporized and would disperse in space. Moreover, whereas a very critical speed and direction of ejection from Jupiter could send an object to the vicinity of the earth, it is thousands of times as likely that the body would be ejected in other directions, and probably escape the solar system altogether. A body of planetary mass coming close enough to Mars and the earth would perturb the orbits of these bodies, but then how could they return to their regularly spaced, nearly circular paths about the sun? And such a body could not stop the earth from rotating, let alone allow it to start up again with the same rotation period as before (we have ample evidence that the earth's rotation has been relatively stable over the past few tens of millennia). It is not clear how hydrocarbons would fall from the tail of the comet to the earth. Nor can we understand how the comet could lose an enormous amount of kinetic energy and settle into a nearly circular orbit about the sun.

There is much we do not yet know about Venus, and the more we do learn the more new questions can be expected to arise. But none of the present anomalies suggests a Jovian origin for the planet, involving impossible difficulties ranging from celestial mechanical to chemical. On the contrary, Venus, for the most part, appears to be a perfectly normal terrestrial planet, showing evidence of having accreted with its neighbors in its quite regular orbit about the sun. Its structure, composition, and atmosphere are typical of the terrestrial planets, and its high surface temperature is completely understood in terms of its atmospheric composition and pressure.

Moreover, there are historical records (cuneiform writing) of observations of Venus at least a millennium before its supposed origin according to Velikovsky.

At the time of antiquity, or perhaps even through the Middle Ages, Velikovsky's hypothesis might have been a viable one. Little was then known about the true nature of the other worlds or of the physical laws that govern their behavior. But we have come a long way since then, as our modern technology attests. Velikovsky's ideas are simply naive. They concern well-trodden ground with which we are thoroughly familiar and are simply incompatible with long-understood and extraordinarily well-document principles.

(j)   The Venerian Environment

Our modern exploration of Venus has shown it to be very different from the earth and hardly well named for the love goddess: Venus is an absolute hellhole! Imagine life at the surface of that world.

The massive stagnant atmosphere of carbon dioxide weighs down with a pressure averaging 91 times that of the earth's atmosphere. The temperature is between 721 and 732 K (near 850°F). Since Venus' upper clouds reflect about 70% of the incident sunlight back into space (which is what makes Venus appear so bright), not much passes through the clouds. In fact, only about 2 or 3% reaches the surface, and most of that is reddish light of longer wavelength. So the sky would be dark and red and gloomy, with a visibility of only a couple of kilometers. Moreover, it is dry and dusty. Except for the changing illumination, day and night on Venus are much the same, with virtually no change in temperature.

Despite its scenery, Venus would not be a nice place to live. It would not even be a nice place to visit!

17.3   MARS

Mars is the most favorably situated of the planets for observation from earth and has excited more interest and comment than any other. Only on Mars and Mercury is the solid surface visible with terrestrial optical telescopes, and Mercury is too near the sun to observe easily. Venus comes a little closer to the earth, but at its closest it is at the new phase

 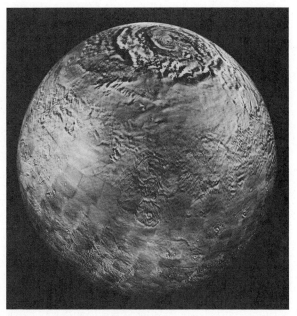

Figure 17.12   Mars, photographed from 472,000 km by Mariner 7 in August 1969. *(NASA/JPL)*

when it cannot be studied optically. Mars, on the other hand, is sometimes favorably situated for observation all night long. We have seen (Chapter 3) that Mars was the planet most extensively observed by Tycho Brahe, and the planet whose motion led Kepler to discover his laws of planetary motion.

(a)   Orbit of Mars

The median distance of Mars from the sun is 227 million km, but its orbit is somewhat eccentric (0.093) and its heliocentric distance varies by 42 million km. The sidereal period of revolution is 687 days. The orbit is inclined to the ecliptic plane by 1°51′.

At intervals of the synodic period of Mars, 780 days, the earth passes between it and the sun. Then Mars, at opposition, is above the horizon all night. It is also at its nearest to the earth and is most favorably disposed for observation.

(b)   Satellites

Centuries ago, Kepler, hearing of Galileo's discovery of four satellites of Jupiter, speculated that Mars should have two moons. The speculation was based on numerological, not scientific considerations. Again, in 1726, Jonathan Swift in his satire *Gulliver's Travels* described Gulliver's fictional visit to the land of Laputa, where he found scientists engaged in many interesting investigations. In one of them, Gulliver reported, Laputian astronomers had discovered

> . . . [*two*] *satellites, which revolve about Mars, whereof the innermost is distant from the centre of the primary planet exactly three of the diameters, and the outermost five; the former revolves in the space of ten hours, and the latter in twenty one and a half; so that the squares of their periodical times are very near in the same proportion with the cubes of their distance from the centre of Mars, which evidently shows them to be governed by the same law of gravitation, that influences the other heavenly bodies.*

It is an interesting coincidence that in 1877, 150 years later, Asaph Hall of the Naval Observatory actually discovered two small satellites of Mars that closely resemble those described by Swift. (More significant than Swift's "prediction" of the Martian moons, perhaps, is that he was aware of and understood Kepler's third law. Swift was not a

scientist. How many intellectuals of today are knowledgeable in natural philosophy?)

The moons are named *Phobos* and *Deimos*, meaning "fear" and "panic"—appropriate companions of the god of war, Mars. Phobos is 9380 km from the center of Mars, and revolves about it in 7^h39^m; Deimos has a distance of 23,500 km and a period of 30^h18^m. The "month" of Phobos is less than the rotation period of Mars; consequently, Phobos would appear to an observer on Mars to rise in the west.

Both satellites are too small to present disks that can be measured from earth. They have been photographed, however, by the Mariner and Viking interplanetary probes. They are slightly elongated and pitted by meteoritic impacts. The greatest diameters of Phobos and Deimos are about 25 and 13 km, respectively.

(c)   Physical Characteristics

The mass of Mars can be found by applying the distances and periods of its natural satellites, Mariner 9, and the Viking orbiters to Newton's formulation of Kepler's third law (Section 5.5). Mars is only 0.107 times as massive as the earth. The diameter of the planet is about 6800 km which gives it a mean density of approximately four times that of water and a surface gravity of 38 percent of the earth's.

The interior structure can be determined if, as in the case of Venus, we assume the same compressibility at a given pressure as is found for the material of the earth's interior. The central pressure of Mars is calculated to be about 400,000 times the earth's atmospheric pressure; the central density is about 8.6 times that of water.

(d)   Telescopic Appearance

Until Mariner 4, launched November 28, 1964, flew by Mars 228 days later and televised to earth photographs obtained at an altitude of 10,000 km, our only views of the Martian surface were through ground-based telescopes. The layman, having heard so much about Mars, is usually disappointed when he first sees it through a telescope. At its nearest, Mars is 150 times the moon's distance; a feature 1 km wide on the moon can be resolved about as well as a feature 150 km wide on Mars. At its nearest, the planet subtends only about 25″.

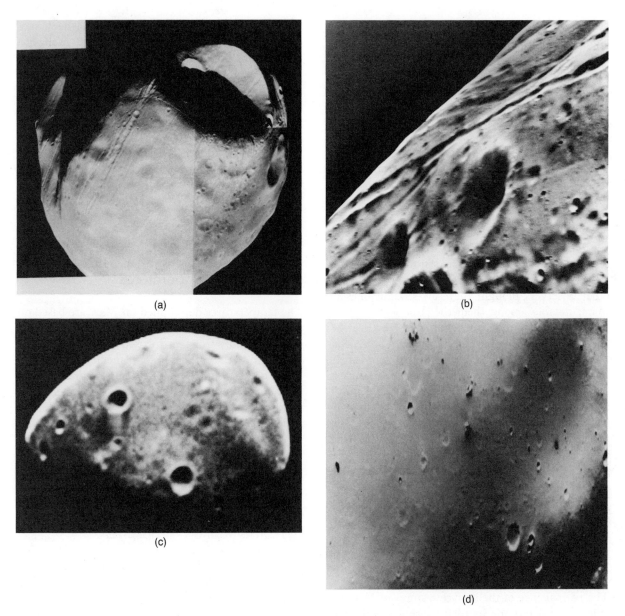

(a)

(b)

(c)

(d)

Figure 17.13 Mars' satellites, photographed with the Viking Orbiters 1 and 2. *(a)* Phobos, just before it entered the shadow of Mars. The large (10-km diameter) crater at the upper left (partly in shadow) is Stickney. *(b)* Phobos at a range of 120 km. The photograph covers an area of 3.6 × 3 km, and the largest crater visible is 1.2 km across. *(c)* Deimos, from a range of 3300 km. The satellite is seen in the gibbous phase; the illuminated portion facing the camera has an area of about 12 × 8 km. *(d)* Deimos, photographed by Viking Orbiter 2 at a range of only 30 km. The picture covers an area of 1.2 × 1.5 km and shows features as small as 3 m across, including strewn boulders of sizes 10 to 20 m. *(NASA)*

Since Mars is farther than the earth from the sun, it does not go through crescent phases. When it is near quadrature, however, it does appear distinctly gibbous. Except for its occasional gibbous shape, Mars usually resembles a shimmering or-

ange ball—the shimmering is caused, of course, by the earth's atmosphere. Most of the Martian surface appears yellowish orange or red. Often white areas, the *polar caps*, can be seen at one or both of the planets' poles. Under good conditions large dark areas can be seen as well. They have a grayish color and somewhat resemble oceans; they were once believed to be bodies of water and so were named *maria*, like the "seas" on the moon.

Under the best observing conditions, a large amount of detail on Mars is telescopically visible. The finest features, however, are seen best visually, during fleeting instants of steadiness of the earth's atmosphere, and are blurred during even brief time exposures needed for photographs. Consequently, the interpretation of Martian surface markings as seen from earth has always been controversial. The so-called "canals" provide an important example.

These straight-appearing streaks were first observed in 1877 by the Italian astronomer Schiaparelli, who called them *canali* ("channels"). Many other observers thought they saw straight lines on Mars, notable among them the American astronomer Percival Lowell, who charted more than 400 "canals," about 50 of which he thought were double, and many of which intersected at what he called "oases." Lowell believed these markings to be too nearly geometrically perfect to be natural, and he envisioned them to be actual canals built by intelligent beings to carry the melting waters from the polar caps across the desert to irrigate Martian crops. So grew the Martian dream.

The eye is easily deceived into interpreting poorly observed complex features as simpler ones. The Mars probes have shown that Mars does, indeed, abound in detailed surface features, but there are no canals.

(e) Rotation and Seasons

The permanence of the surface features of Mars enables us to determine its rotation period with great accuracy; its sidereal day is $24^h37^m23^s$, very near the rotation period of the earth. This high precision is not obtained by watching Mars for a single rotation, but by noting how many turns it makes in a long period of time. Good observations of Mars date back for more than 200 years, during which period tens of thousands of Martian days have passed. The value accepted today is accurate to within a few hundredths of a second.

The equator of Mars is inclined to its orbital plane by about 25°—very close to the $23\frac{1}{2}$° angle between the earth's orbit and its equator. Thus each of Mars' poles is alternately tipped toward and away from the sun, and the planet goes through seasons,

much like those on earth. Because of the longer Martian year, however, seasons there each last about six of our months. We recall that the earth is at perihelion in January (during summer in the southern hemisphere) and that, therefore, differences between the seasons south of the equator would be more extreme than those in the northern hemisphere, were it not for the unequal distribution of land and water over the earth. The same situation exists with Mars; however, because its orbit is more eccentric than the earth's, seasonal climatic variations in its southern hemisphere are considerably more pronounced than those in the northern hemisphere.

(f) Martian Temperatures

The temperatures determined for Mars are based on measures of infrared radiation from its surface, obtained not only from earth-based observatories, but also with detectors carried by planetary probes.

The maximum equatorial temperatures of Mars are about 30°C. The sunrise and sunset temperatures at the equator are much lower, around −40 and −20°C, respectively. At night, the temperature drops to nearly −75°C. The temperature at the south polar cap was first measured by Mariner 7 and was found to range down to about −150°C, which is near the frost point for carbon dioxide.

(g) The Martian Atmosphere

Mars was known for a long time to have an atmosphere. Vast yellow dust storms were sometimes seen to cover the planet. In fact, when Mariner 9 reached Mars and went into orbit around it in 1971, the entire planet was hidden by such blowing dust during the first weeks of observation. White clouds on Mars have long been observed from earth, too. We know now that some of these are quite high—about 45 km, and are probably carbon dioxide crystals. Other clouds are lower, 15 to 30 km, and are believed to be of ice crystals, like our cirrus clouds.

The Martian atmosphere, however, is now very thin; the surface pressure is somewhat less than 1 percent sea-level pressure on earth. The best determination of the abundances of the various gases present there is from data gathered by the Viking landers in 1976:

Carbon dioxide (CO_2)	95%
Nitrogen (N_2)	2 to 3%
Argon (A)	1 to 2%
Oxygen (O_2)	0.1 to 0.4%
Water (H_2O)	0.01 to 0.1%
Krypton (Kr)	less than 0.0001%
Xenon (Xe)	less than 0.0001%

The abundance of the argon is a good indication of how much total gas outgassed, and hence what Mars' atmosphere may have been like in the past. A best guess at the time of this writing is that the Martian atmosphere may have once had a pressure of about 10 percent that of the earth's.

(h) The Martian Surface

The first close-up of Mars came in 1965, when Mariner 4 flew past the planet and televised 22 pictures back to earth. The photographs showed a bleak and barren planet, covered with craters, more like the moon than the earth. Next Mariners 6 and 7 flew by in 1969, and sent back a couple of hundred more pictures, including 33 of the south polar cap. But no picture showed canals. The first extensive survey of Mars began in 1971, when Mariner 9 went into orbit about the planet and sent back 7329 photographs that have mapped Mars far better then we knew the face of the earth 200 years ago.

Much of Mars is covered with craters, which range in size up to hundreds of kilometers across. Their appearance reminds us of the craters on the moon, except that the Martian craters show greater evidence of erosion, because of winds and possibly other weather effects. Indeed, there is a much smaller percentage of small craters on Mars than on the moon; evidently most of the smaller craters have been eroded away entirely or perhaps buried in dust or lava. Many craters appear to have floors filled with dust, and surprisingly, there are large areas of the planet that are very smooth, almost devoid of craters altogether. The latter seem to be regions which have been flooded with lava relatively recently (less than 10^9 years ago). In the large Hellas basin, on the other hand, a layer of dust or other material has evidently been deposited. At least partly because of lava flows and drifting dust, Mars can be separated into cratered and uncratered terrains. For example, the southern hemisphere of the planet is more heavily cratered than the northern hemisphere.

On the floors of some of the larger craters there are rows of dunes, not unlike the dunes found on many terrestrial deserts, for example, in central Australia. We can estimate the speeds of the winds that have produced those dunes from their sizes and spacing, and we find that in the thin air on Mars, winds occasionally blow there with velocities of up to 100 m/s.

The culmination of the U.S. exploration of Mars came in 1976 when Viking 1 and 2 landers settled on the surface of Mars and proceeded to send back pictures and to analyze chemically the surface material. Each Viking lander was dispatched from a "mother" craft, which remained in orbit, made independent observations, and helped relay transmission from the landers to the earth.

The Martian polar caps are now well understood. They change rapidly with the seasons. The south cap, in particular, reaches halfway to the Martian equator in midwinter, and often disappears entirely in summer. The main stuff of those white caps is frozen carbon dioxide (like Dry Ice). But part of the caps is water. At least in the north, there is a permanent cap of water ice. It doesn't go away in summer, and it is probably hundreds of meters thick. If it could all be melted, it is estimated that it could cover the planet with water to a depth of one-half meter. Furthermore, a lot of water could be frozen as permafrost under the Martian surface; we don't know, of course, for sure.

There is also volcanism, and on a large scale. At least 12 huge volcanoes exist on Mars, and their eruptions have flooded a large part of the planet with newly formed lava rock. Most spectacular among them is Olympus Mons, 25 km high and 600 km across and displaying a giant caldera 70 km across at its summit. It is by far the largest volcanic pile ever seen by man.

Canyons also are present, especially a great canyon (Valles Marineris) 5000 km long, 75 km wide (on the average), and 6 km deep. Although there is what appears to have been lifting and subsiding, there is no evidence for *horizontal* plate tectonics, as on earth. This can explain why certain Martian volcanos, such as Olympus Mons, have grown so huge. On the earth, a volcano grows over a "hot spot" in the mantle beneath it. Now, if a crustal plate is slowly moving over that hot spot, the volcano emerges at different places on the surface of the crustal plate as time goes on. Thus island chains, like that of Hawaii, are built up over the

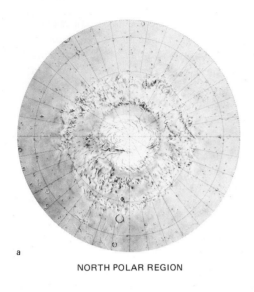

NORTH POLAR REGION

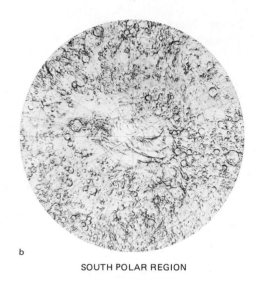

SOUTH POLAR REGION

NORTH

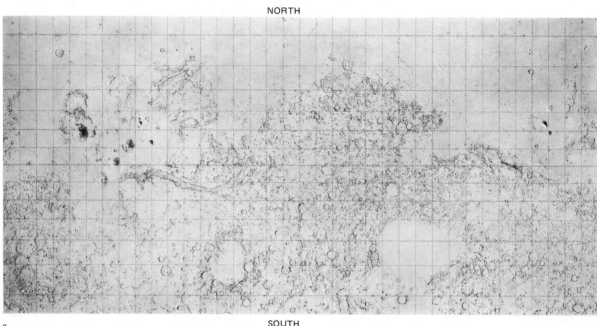

SOUTH

SHADED RELIEF MAP OF MARS

Figure 17.14 A topographical map of the planet Mars. (a) the north polar region, (b) the south polar region, (c) the equatorial and temperate regions. *(NASA/JPL)*

tens of millions of years, while a moving plate (in the case of Hawaii, the Pacific plate) carries new parts of the earth's crust over the exuding magma.

The most interesting things found on Mars are dry river beds. Almost all experts now agree that running water was once present on Mars. Braided streams have been found, and even ancient islands in dry stream beds. Certainly there has been erosion by some liquid, and water seems to be the only candidate.

But there is a problem: Geophysicists have pointed out that liquid water can simply not survive in the low-pressure atmosphere Mars has today. Various experts estimate that the atmospheric pressure would have to be from 5 to 50 times as great as at present to allow liquid water to flow over the surface of the planet, and not immediately evapo-

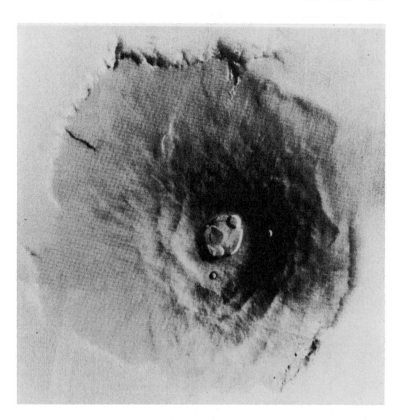

Figure 17.15 Olympus Mons, a gigantic volcanic mountain on Mars, photographed by Mariner 9. *(NASA/JPL)*

rate. Yet, the abundance of argon suggests that the requisite atmospheric pressure may have existed in the past, perhaps more than 1000 million years ago.

Does this mean that there was running water on Mars in the past, as on the earth, but not today? If so, what has happened to that water? Has it evaporated and dispersed into space, or frozen into a permafrost beneath the ground, where it can not be detected? Or perhaps some of each?

Mars could be in a glacial period at present. A 15% increase in the solar constant would probably be enough to melt such frozen water. We cannot rule out such a fluctuation in the solar luminosity over a period of hundreds of millions of years, but it can hardly occur in a few thousand years. On the other hand, if the obliquity of Mars' ecliptic (that 25° tilt in its axis of rotation) were to increase to just 30°, the seasons would become severe enough to melt the hypothetical ice in the polar caps. It is calculated that perturbations by Jupiter are great enough to periodically increase the Martian obliquity to, actually, 35°.

Counts of craters of various sizes in different parts of the Martian surface give some indication of the relative ages of those regions. Such counts have indicated that different areas on Mars have very different ages, and that, in particular, the features that appear to be river beds have formed at several different geological times in the past, rather than all at once. It would seem, therefore, that the disappearance of the rivers, at least in the past, has not been due to the depletion of the Martian source of water. Perhaps the rivers will flow again.

Of all known planets besides the earth, Mars has been the most popular candidate for one supporting life. Percival Lowell was one of the most ardent supporters of the hypothesis of not only life on Mars, but even intelligent life. He built the Lowell Observatory in Arizona primarily for the purpose of studying Mars. The supposed canal system on the planet and the changing colors of the maria seemed to give credence to the idea. Moreover, at one time there seemed to be spectroscopic evidence for the hydrocarbon radical—basic to organic molecules—in the reflected sunlight from the Martian maria. Subsequent observations, however, discredited the evidence for the hydrocarbon radical, and we have already seen how Mariner observations have failed to substantiate the canals on Mars. Furthermore, sophisticated observations of the surface of the planet by the Mars probes have found no evidence for growing vegetation on the

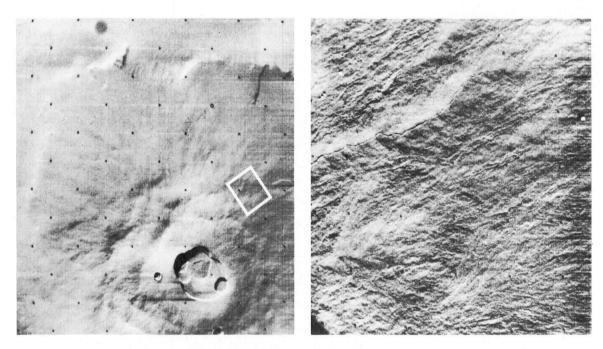

Figure 17.16 Olympus Mons region. *(left)* Summit crater of the volcanic mountain. *(right)* Detail of white rectangular region of left photograph showing lava flow. Mariner 9 photographs. *(NASA/JPL)*

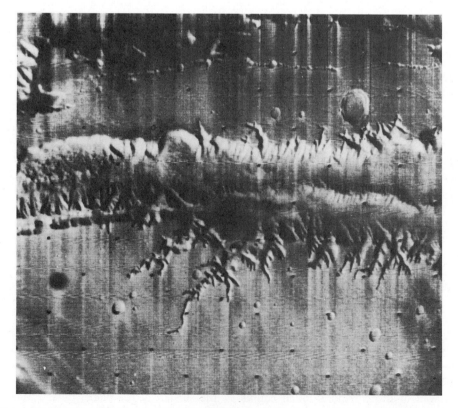

Figure 17.17 Huge chasm or canyon extending some 5000 km in an east-west direction starting about 1000 km southeast of Olympus Mons. Note the branching tributaries. *(NASA/JPL)*

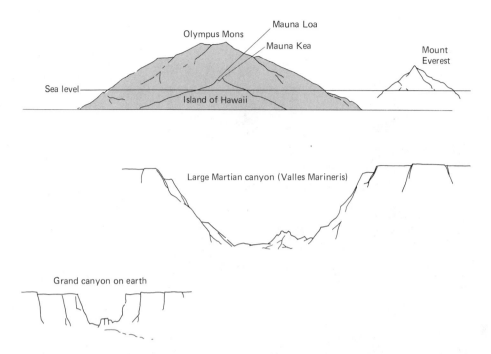

Figure 17.18 The relative sizes of Olympus Mons, Mauna Loa, Valles Marineris, and the Grand Canyon. The vertical scale is enlarged 5 times.

Figure 17.19 A Martian rille showing on its floor the braided pattern characteristic of many old terrestrial stream beds. (NASA/JPL)

Figure 17.20 The Martian landscape, photographed by Viking 1. *(NASA)*

planet's surface. Nor has any evidence been found for roads or cities or other artifacts of civilization.

All of this, however, is evidence neither for nor against life on Mars. The Viking landers—in addition to carrying experiments to measure weather, seismic events, and other things, as well as television cameras to see any Martians who might be walking up—had long arms that went out a few meters to scoop up soil and bring it back into the laboratory for analysis. There were four biological experiments. Three were to test for respiration of living animals, absorption of nutrients by any living organisms, and exchange of gases between the Martian soil and the laboratory environment for any reason whatsoever. In the various experiments the soil samples were isolated and incubated in contact with various gases, radioactive isotopes, and nutrients to see what would happen. The fourth experiment pulverized the soil sample, and analyzed it carefully to see what organic material it contains.

The Viking experiments were sensitive enough that had one of the probes landed anywhere on earth, with the possible exception of Antarctica, it would easily have detected life. Those experiments that tested for absorption of nutrients and gas exchange did show activity, but this was caused by the chemically active soil, not living or dead organisms. The organic chemistry experiments showed no trace whatsoever of any organic material. The possibility of biological organisms somewhere else on Mars has not been completely eliminated, but most experts consider the chance of any life on that planet to be negligible.

Compared to Venus, Mars would not be so bad a place to visit, but nobody seems to be home.

EXERCISES

1. Explain why Mercury is visible in the west after sunset when it is at eastern elongation, and in the east before sunrise when it is at western elongation. Draw a diagram.

2. At what seasons of the year would Mercury, when at eastern elongation, be above the horizon longest after sunset? Explain.

3. From the range of Mercury's distances from the sun, find the range of its orbital velocities. (*Hint:* Use Kepler's second law.)
   Answer: 38 to 58 km/s

*4. Show that the mean period from noon to noon at a place on Mercury is 176 days.

5. Give several reasons why Mercury would be a particularly unpleasant place to live.

6. Compare Mercury, Mars, and the moon. What do these worlds have in common? How do they differ? Explain in each case.

7. Venus requires 440 days to move from greatest western to greatest eastern elongation but only 144 to move from greatest eastern to greatest western elongation. Explain why. A diagram will help.

8. At its nearest, Venus comes within about 40 million km of the earth. How distant is it at its farthest?

9. Why isn't Venus always *exactly behind* the sun at superior conjunction?

10. On what occasion *can* Venus be observed at inferior conjunction? Explain.

11. Would astronomers be likely to learn more about the earth from observatories on Venus or Mars? Explain.

12. What problems would be encountered by scientists planning a station on the surface of Venus?

13. When might it be possible to observe both Mercury and Venus, at the same time, when they are at inferior conjunction? Why would this be an extraordinarily rare event?

14. Why does Mars have the longest synodic period of any planet but a sidereal period of only 687 days?

15. Show that the satellites of Mars obey Kepler's third law. Use the data in this chapter or in Appendix 11.

16. Why is it not easy for us to get a good look at the southern polar cap of Mars from the earth?

17. Describe at least three respects in which Mars resembles the earth, and three others in which it differs from earth.

Galileo Galilei (1564–1642) advocated that we perform experiments or make observations to ask Nature her ways, rather than deciding how things must be on the basis of preconceived notions. When Galileo turned the telescope to the sky, he discovered, among other things, the four largest, or *Galilean*, satellites of Jupiter—worlds which Voyagers 1 and 2 have shown to be of intense interest.

18

THE JOVIAN PLANETS (AND PLUTO)

The United States has now launched interplanetary probes that will, if all goes well, by 1990 have explored at close range all four giant planets. Only Pluto, among the nine principal planets, will have been missed. The missions to the Jovian planets are summarized in Table 18.1.

18.1 JUPITER

Next to the sun, Jupiter is the largest and most massive object in the solar system, so it is well named for the master of the gods. Jupiter has at least 16 satellites, four of which are the sizes of small planets. In a sense Jupiter is almost a miniature solar system in its own right.

The median distance of Jupiter from the sun is 778 million km, 5.2 times that of the earth. Its orbit has an eccentricity of 0.048, so its distance from the sun varies by 76 million km. Jupiter's mean orbital speed of 13.1 km/s carries it once around its orbit in 11.86 years. Its orbital plane is inclined 1°18′ to the ecliptic.

(a) Physical Properties

The mass of Jupiter has been calculated from the periods of its satellites and the sizes of their orbits, from the perturbations it produces on the orbits of passing minor planets, and most recently, from its acceleration of the Pioneer and Voyager spacecraft.

Table 18.1   U.S. Planetary Probes to the Jovian Planets

PLANET	SPACECRAFT	ARRIVAL DATE	COMMENTS
Jupiter	Pioneer 10	Dec. 1973	Flyby; photos
	Pioneer 11	Dec. 1974	Flyby; photos
	Voyager 1	Mar. 1979	Flyby; photos of Jupiter and satellites
	Voyager 2	July 1979	Flyby; photos of Jupiter and satellites
	Galileo	1988(?)	Jupiter orbiter and probe
Saturn	Pioneer 11	Sept. 1979	Flyby; photos
	Voyager 1	Nov. 1980	Flyby; photos of Saturn and satellites
	Voyager 2	Aug. 1981	Flyby; photos of Saturn and satellites
Uranus	Voyager 2	Jan. 1986	Deflected to Uranus by gravitational effect of Saturn
Neptune	Voyager 2	Aug. 1989	Deflection planned to Neptune by gravitational effect of Uranus, if the systems and communications with probe remain operative

Jupiter is found to be 318 times as massive as the earth, a value which is very close to ¹/₁₀₀₀ the mass of the sun. Jupiter is more massive than all the other objects in the solar system combined, except the sun itself.

Jupiter is also the largest planet, having a diameter of 143,000 km, or about 11 times the earth's diameter. If it were as near to the sun as Mars, it would be the brightest planet in the sky and would easily be visible in daylight.

Jupiter's surface gravity is higher than the earth's; a body at its surface would weigh 2.53 times what it would weigh on earth. Jupiter has the highest velocity of escape of any planet, 60 km/s; it can therefore retain all kinds of gases in its atmosphere. Its mean density, however, is much less than the earth's—only 1.33 times the density of water.

Because of the tremendous gravitational attraction of Jupiter for its constituent parts, it would most certainly be compressed to a far greater mean density unless it were composed almost entirely of hydrogen and helium—the lightest and most common elements in the universe. Even these elements are probably compressed to a liquid metallic state throughout most of Jupiter's interior, although there may possibly be a small rocky core at the planet's center. The largest uncertainties in the calculation of a model for Jupiter's internal structure are due to our incomplete knowledge of the compressibility of liquid hydrogen and helium at the various temperatures and pressures that exist inside Jupiter, and to the unknown abundances of other elements there. One model predicts a central pressure of over 100 million times the earth's sea-level atmospheric pressure, and a density of about 31 times that of water.

It is interesting to note that Jupiter has very nearly the maximum possible size for a body of "cold" hydrogen, that is, one that is not generating energy. Less massive bodies than Jupiter would occupy a smaller volume. More massive bodies, by virtue of their greater gravitation, would also be compressed to a smaller volume than Jupiter's. In this latter case, the extreme internal pressures would force the electrons to leave their atoms. The atoms and electrons then would comprise a gas of a sort, but it would not obey the laws of physics appropriate to ordinary gases (Chapter 34).

A remarkable thing about Jupiter is that it radiates about 2.5 times as much energy into space (at infrared wavelengths) as it receives from the sun. The source of most of this energy is probably stored heat trapped in the planet's interior at the time of its formation. Jupiter must be slowly radiating away this energy as it cools. It might also be deriving some energy from a slow gravitational contraction.

Because Jupiter is so far from the sun, it can appear only slightly gibbous, and usually presents a nearly full disk to the telescopic viewer. Its relatively high albedo (0.52) makes it quite bright, despite the small amount of sunlight that reaches it. Its four brightest satellites are easy to see, even with binoculars; the shadows of some of the satellites occasionally can be observed crossing the planet's disk.

(b) Rotation and Flattening

Distinct details in the cloud patterns on Jupiter allow us to determine its rotation rate; it is the most rapidly spinning of all the planets. However, it does not rotate quite as a solid body, or at least the outer clouds that we observe do not. For most of the planet, the rotation rate is fairly constant, averaging one rotation every 9^h55^m. A broad band about 16,000 to 25,000 km wide along the equator, however, moves about 300 km/hr faster than the surrounding regions, and completes a rotation in 9^h50^m. Jupiter's equator is inclined at only 3° to its orbit plane, so the planet has no appreciable seasons.

Jupiter's rapid rotation has caused it to become noticeably oblate, just as the earth is slightly flattened because of its rotation. When observed visually through a telescope, Jupiter's flattening is readily apparent; its polar diameter is less than its equatorial diameter by 1 part in 15. The polar and equatorial diameters of the earth differ by only 1 part in 298.

(c) Appearance and Atmosphere

The first close-up look at Jupiter was provided by Pioneers 10 and 11 in December 1973 and December 1974, respectively. But the really good look came in 1979 when Voyagers 1 and 2 reached Jupiter, and transmitted more than 33,000 pictures of the planet and its satellites back to earth. All four of these probes were flybys, but Pioneer 11 and both Voyagers were placed on trajectories such that Jupiter's gravitational deflection on them sent them on toward Saturn. Pioneer 11 passed Saturn in 1979, and the Voyagers in 1980 and 1981.

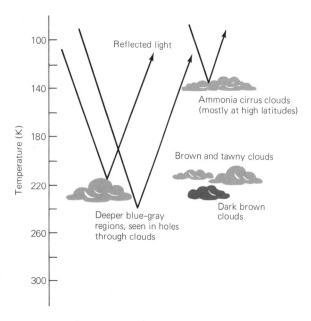

Figure 18.1 Schematic diagram of Jupiter's upper atmosphere. Incident light is reflected from a variety of levels, displaying high ammonia cirrus clouds at high latitudes, the varicolored clouds at other and deeper levels, appearing blue-gray through holes in the denser clouds.

Jupiter is crossed with alternate light and dark, brightly colored bands parallel to its equator. The bands abound in detail and as seen telescopically from earth exhibit gradual changes. Time-lapse photographs from the Voyagers, however, show them to be enormously active and complex features. Jupiter is the most rapidly rotating planet, turning once in just under ten hours. Because the different latitudes have slightly different rotational velocities, the cloud bands, especially at their boundaries, show almost every conceivable kind of current and flow.

Even more striking is the Great Red Spot. The spot was first seen telescopically from earth in 1831, although it may possibly have been the "spot" on Jupiter Cassini described in 1660 and used to determine the planet's rotation period. The Red Spot has changed slightly in size and shape since then, as well as in intensity of color, but on the whole it has persisted. It has been as large as 50,000 km across and has always been far bigger than the earth. The Pioneer data suggested that it was some kind of a long-lived storm system in the planet's atmosphere. The Voyagers showed it to be a massive eddy with enormously complex, and changing, small eddies in gas streams flowing around it.

When Voyager 1 visited Jupiter it sent us a surprise: Jupiter has a thin faint ring in its equatorial plane. The ring has an outer diameter of 256,000 km, and although it is most concentrated over a width of about 6000 km, it extends very faintly all the way in to the planet's surface. The ring is very thin, no more than 30 km thick, and is composed of very tiny particles. The famous rings on Saturn are not only far more substantial, but contain sizable chunks. Analysis of the scattering of light from the Jovian ring shows the particles making it up to have sizes of only 8 to 10 microns (a micron is one-millionth of a meter). It is doubtful that particles that small can remain in stable orbits in the presence of the various perturbing forces about Jupiter, which suggests that they are not permanently in that ring. The ring itself probably results from a continual flow of tiny particles swept

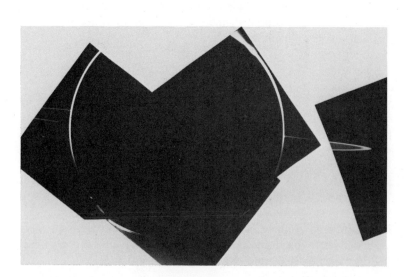

Figure 18.2 A mosaic of photographs obtained by Voyager 2, showing the night side of Jupiter, with the ring silhouetted against the dark sky. (NASA)

up by Jupiter from interplanetary space. After its discovery by Voyager, the Jovian ring has been confirmed by observations from earth.

Because of the high velocity of escape from Jupiter we would expect it to retain even the lightest gases; hydrogen and helium, therefore, should be abundant constituents of its atmosphere. Hydrogen has been observed in the Jovian atmosphere for some years, and Pioneer 10 confirmed that helium is present there as well. Methane and ammonia are easily identified in spectra of Jupiter obtained from earth and have long been known to exist in Jupiter's atmosphere; however, they comprise only about 1 percent of its total mass. Detailed spectrographic studies show that the relative abundances of such elements as hydrogen and carbon (in various chemical combinations) in Jupiter are very close to the abundance ratios of those same elements in the sun.

Various investigators, Sagan, Ponnamperuma, and others, have simulated Jovian atmospheric conditions in the laboratory. They find, for example, that a complex red polymer can be formed. Thus Sagan and his associates argue that substantial organic chemistry should occur on the Jovian planets. They suggest that the most apparent coloration on Jupiter may be due to organic chemistry.

Theoretical studies indicate that the total atmosphere cannot extend downward more than a small fraction of the planet's radius. Within a few hundred km below the visible clouds, the pressures are high enough to liquefy hydrogen. The atmosphere is pictured as gradually becoming thicker and "slushier" with depth, until it finally is condensed to an "ocean" of liquid hydrogen and other elements. The ocean probably gradually becomes liquid metallic hydrogen in another few hundred kilometers depth.

(d) Jupiter's Magnetism

In the late 1950s radio energy was observed from Jupiter that is more intense at *longer* than at shorter wavelengths—just the reverse of what is expected from a body radiating away its heat. It is typical, however, of the radiation emitted by electrons accelerated by a magnetic field (synchrotron radiation—see Chapter 29). Observations made at the Radio Observatory of the California Institute of Technology, near Bishop, California, showed that the radio energy originated from a region surrounding the planet whose diameter is several times that of Jupiter itself. Furthermore, the radio energy was found to be polarized, another characteristic of radiation from accelerated electrons. (Radio waves can be polarized, like visible light; the polarization is detected by changing the orientation of the receiving antenna at the focus of the parabolic reflecting dish of the radio telescope.) The evidence suggested, therefore, that there are a vast number of charged atomic particles circulating around Jupiter, spiraling through the lines of force of a magnetic field associated with the planet. The phenomenon is like the Van Allen belt around the earth (Section 15.3).

Subsequent investigations showed that at radio wavelengths Jupiter is very "noisy" indeed. Every now and then it emits powerful bursts of radio energy. Remarkably, these bursts seem to be correlated with the orbital motion of the satellite, Io. Evidently, Io interacts with the magnetic field of Jupiter in such a way as to trigger the radio bursts.

The Voyagers cast considerable light on the nature of Jupiter's magnetic field and radiation belts. They found Jupiter's surface magnetic field to be from 20 to 30 times as strong as the earth's field. Because of Jupiter's great size, moreover, its total magnetic energy is enormous compared with the earth's. The Pioneers and Voyagers encountered highly energetic charged particles (both electrons and atomic nuclei) in Jupiter's field. During Pioneer 10's passage through the field, for example, it was subjected to a total of some 250,000 roentgens from these particles; 500 roentgens are considered lethal to man.

The Jovian magnetic axis, like that of the earth, is not aligned exactly with the axis of rotation of the planet, but is inclined at some 15°. Jupiter's magnetic axis, moreover, does not pass through the planet's center, but is displaced by about 18,000 km.

The Jovian magnetosphere is actually much larger than the sun and completely envelops the innermost satellites. If we could see the magnetosphere with the eye from earth, it would appear the size of the full moon.

(e) Satellites

Jupiter's 16 known satellites (three of which were discovered with the Voyagers) fall into three groups. The outer satellites have eccentric orbits of large inclinations to Jupiter's equator, and revolve about the planet from east to west—backward with respect to most motions in the solar system. The

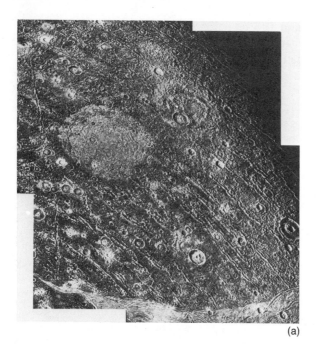

(a)

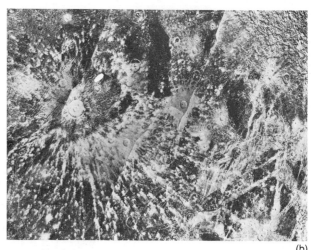

(b)

Figure 18.3 (a) and (b) Cratered terrain on Jupiter's largest satellite, Ganymede. (*NASA*)

most distant of these is about 24 million km from Jupiter. They are probably objects (asteroids?) captured by Jupiter long after its formation. The satellites in the middle group have direct orbital revolution, but their paths are also eccentric and highly inclined; they too may be captured asteroids. The inner satellites, including the four largest ones, on the other hand, all move in the usual west-east direction, and all have nearly circular orbits lying in Jupiter's equatorial plane; the best guess is that they were formed from a disk of matter revolving about the primordial Jupiter much as the planets are thought to have formed from the solar nebula.

Of the inner Jovian moons, the nearest to the planet are 1979 J1 and 1979 J3, discovered by Voyager. Next out is Almalthea, about the size of California, and the Voyager discovery 1979 J2. The next four, all larger than our moon, are the satellites discovered by Galileo, and are called the *Galilean* satellites. They are amazing little worlds. Callisto, the most remote of the four, is 4840 km in diameter and is completely covered with craters, so evidently it has suffered very little erosion since its accretion days. Ganymede, the next in, and the largest satellite in the solar system, is 5270 km across. It is also cratered, but less so than Callisto, and shows many mountain ridges and valleys. Europa, with a diameter of 3130 km, still further in, has relatively few craters, but many more cracks. Io, the innermost of the Galilean moons, has a diameter of 3640 km. Io has no sign of craters or

cracks, but a brightly colored dusty material covers its entire surface. That material is believed to be mostly sulfur.

Ganymede and Callisto have low densities (1.9 and 1.8, respectively), and are probably about 50 percent water ice, with the rest of their mass being made up of rocky material. Io and Europa have higher densities, and are at most only 20 percent ice. Io is now believed to be mostly molten inside a solid crust, its core kept hot and fluid by the strong tidal forces exerted on it by nearby Jupiter. Io also has hot spots on its surface—up to 400° warmer than the surrounding landscape.

But the most striking phenomena on Io, and perhaps the greatest surprise found by Voyagers, are its active volcanos. Nine volcanos were observed, but when account is taken of the fact that we cannot see all sides of Io at once, we predict that it may have a dozen or more. These volcanos are erupting almost continually, spewing clouds of sulfur and gases into graceful plumes hundreds of kilometers above its surface. The solid particles fall back, which is believed to result in the satellite being slowly buried in its own sulfur. Fresh sulfur must be depositing at an average rate of a millimeter or so per year, which over a period of millions of years is enough to hide all evidence of any old erosion.

The gases belched out by Io's volcanos, thought to be largely sulfur dioxide, would stay as an atmosphere were it not for the fact that the mol-

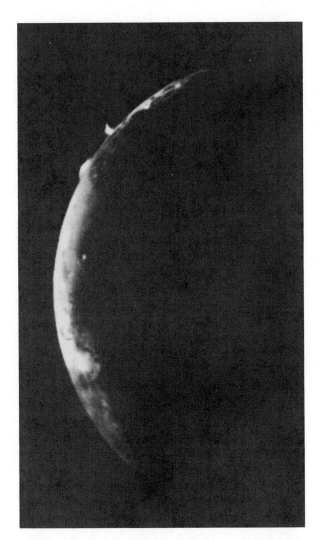

Figure 18.5 A portion of the limb of Io, showing a particularly striking plume from an erupting volcano. (NASA)

Figure 18.4 One of Jupiter's satellites, Io, showing two plumes from erupting volcanoes. Photographed from the night side by Voyager 2. (NASA)

ecules are dissociated and ionized by the charged particles in Jupiter's magnetosphere. Once ionized, they are swept up in Jupiter's magnetic field and drawn away from Io into a large doughnut or torus lying roughly in the orbit of the satellite. This *plasma torus* of Io is another discovery of Voyager 1.

None of the Jovian satellites has an appreciable permanent atmosphere.

The four Galilean satellites have played an important role in the history of science long before our space exploration of them. Their discovery provided strong support for the Copernican theory (Section 3.4), for their revolutions about Jupiter showed that a center of motion can itself be in motion. The timing of the intervals between their eclipses in Jupiter's shadow also led to the first dem-

onstration (by Roemer) of the finite speed of light (Section 13.1)

18.2 SATURN

Saturn is the second-largest planet in the solar system and is the most remote one known in antiquity. It is named for the Roman god of agriculture and the father of Jupiter. Its beautiful ring system makes it one of the most impressive telescopic objects, and it has long been a favorite object for public viewing. It also has at least 16 satellites, nearly half of which have been discovered in the past decade or so, several with the Voyager spacecraft.

The orbit of Saturn has an eccentricity of 0.056, about the same as that of Jupiter. The planet's distance from the sun varies from about 1347 to 1505 million km and has a median value of 1427 million km. The orbit is inclined $2\frac{1}{2}°$ to the ecliptic plane. At a mean orbital speed of 9.6 km/s, Saturn completes one sidereal revolution in $29\frac{1}{2}$ years.

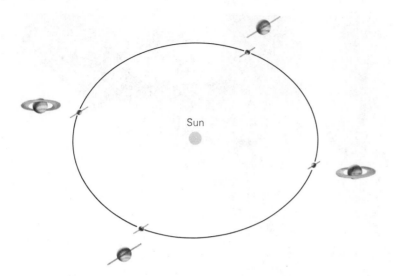

Figure 18.6 Orientation of the rings as seen from the sun with Saturn at different places in its orbit.

(a)   Physical Properties

From the orbits of its satellites, as well as from the perturbations of space probes, we find Saturn's mass to be 92 times that of the earth. Its mean diameter is 120,000 km. These figures give it a mean density of only 0.7—the lowest of any planet, except possibly Pluto. In fact, Saturn would be light enough to float, if an ocean existed large enough to launch it. Its surface gravity is only 7 percent greater than the earth's; nevertheless, it has a velocity of escape of 36 km/s, great enough for Saturn to hold the lightest gases.

The internal structure of Saturn is very similar to that of Jupiter. One model gives a central pressure of about 50 million times sea-level atmospheric pressure and a central density of about 16 times that of water. Also like Jupiter, Saturn probably has a rocky core, which may be twice as massive as the earth.

The rotation period of the planet, as determined both from the Doppler shift in its spectrum and from the apparent motions of the spots on its disk, is just over 10 hours at the equator. Like Jupiter, however, Saturn rotates more slowly at latitudes away from the equatorial regions. The mean rotation period is near 10^h38^m. Because of its rapid rotation, Saturn is the most oblate of all the planets; its equatorial diameter is about 10 percent greater than that through its poles.

Unlike Jupiter, Saturn does have seasons. Its axis of rotation is inclined at 27° to the perpendicular to its orbit. Because the planet's rings lie in its equatorial plane, they are tipped by this same angle to its orbit. Saturn is ten times as far from the sun

as the earth is, so we see the planet in almost the same aspect as it would appear from the sun. Thus during part of Saturn's orbital revolution we look obliquely at one face of the rings, and when the planet is on the opposite side of the sun we see the other face (Figure 18.6). At intermediate points we view the rings edge-on, at which times they vir-

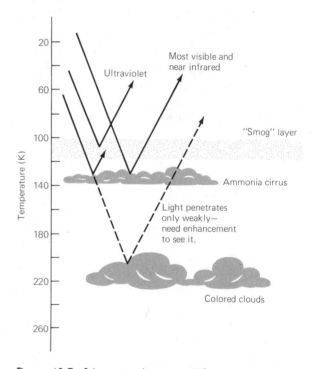

Figure 18.7 Schematic diagram of Saturn's atmosphere. Most of the light is reflected from high-level ammonia cirrus clouds; only very little light penetrates to the lower-lying clouds and is reflected out. Computer enhancement is required to bring up the contrast enough in the Voyager photographs to see these regions clearly.

tually disappear because the ring system is so thin—probably only 1 to 2 kilometers.

(b)   Appearance and Atmosphere

In 1979 Pioneer 11 passed close to Saturn and gave us our first closeup view of the planet. Far more spectacular, however, are the observations obtained with the much more sophisticated equipment of the Voyagers in 1980 and 1981.

Compared with Jupiter, Saturn has a bland appearance. The reason is that it is colder at Saturn's greater distance from the sun, and the chemical activity giving rise to Jupiter's spectacular surface features must take place relatively deeper in Saturn's atmosphere, where the temperature is comparable to that at a higher level in Jupiter's atmosphere. The Saturn features, in other words, are buried beneath a greater depth of haze.

Still, by electronically enhancing the contrast of the Voyagers' televised images, the features on Saturn are made visible, and they appear similar to the major markings on Jupiter. There are bands parallel to the equator, and colored bands. The winds are stronger than on Jupiter. Velocities are typically 150 m/s—well beyond hurricane intensity. The Voyagers also found darker spots with cyclonic (or anti-cyclonic) circulation about them, reminiscent of Jupiter's red spot, but far less spectacular. The largest such Saturnian feature, found at a high northern latitude, is an oval brown spot about 1000 km across (Figure 18.10).

Most of the visible and near infrared sunlight incident on the planet is reflected from ammonia cirrus (ice) clouds at a high cold level in the Saturnian atmosphere. The ultraviolet radiation is scattered still higher in a smog-like region (see Figure 18.7, and compare with Figure 18.1). In chemical composition the atmosphere of Saturn is very similar to that of Jupiter and to the sun.

Also like Jupiter, Saturn emits more radiation into space than it receives from the sun, so it too has an internal heat source, perhaps from a slow compression, or a sinking of heavier material. The internal structures of the two planets are thought to be similar as well; some hundreds of kilometers below the clouds Saturn's hydrogen must liquefy and at a somewhat greater depth become a highly-conducting metallic form of hydrogen.

Saturn has a magnetosphere, which contains an enormous amount of magnetic energy compared with the earth's magnetosphere, but it is nevertheless much weaker than Jupiter's. There are trapped ions and electrons in Saturn's magnetic field, but far fewer than in Jupiter's. It is thought that many of the ions that would otherwise be trapped are absorbed by collisions with particles in the ring system.

(c)   Rings

Telescopic observations from the earth show Saturn's rings to have three sections, labeled (from

(a)

(b)

Figure 18.8 (a) Saturn, photographed by Voyager 1 on October 30, 1980, from a distance of 18 million kilometers. (b) Image of the crescent Saturn transmitted from 1.5 million kilometers by Voyager 1 on November 13, 1980, as the spacecraft left the Saturn system. (NASA/JPL)

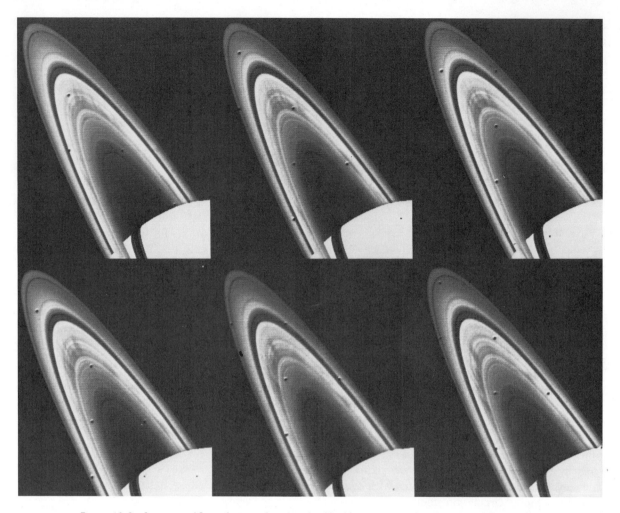

Figure 18.9 Six views of Saturn's rings photographed by Voyager 1 on October 25, 1980, at intervals of about 15 minutes. Note the motion of the spokelike features in the bright (B) ring. *(NASA/ JPL)*

outer to inner) the A, B, and C rings. The outer radius of the A ring is 137,400 km. The B ring is brightest, and the C or crepe ring is quite faint and has an inner radius of about 71,000 km. It has long been known that the rings are not solid, but consist of a myriad of tiny particles. First, they lie within Roche's limit (Section 6.4), where a large solid object would be torn apart by the tidal force of Saturn. Also, Doppler shifts in the spectrum of sunlight reflected from different parts of the rings (Section 14.2) show them to be revolving about the planet at lower speeds, the farther out they are, in compliance with Kepler's laws. Also stars can sometimes be seen shining with nearly full brilliance through the rings.

Between the A and B rings is a dark gap at least 4800 km wide, known as *Cassini's division.* This gap was long thought to be understood in terms of what is called *resonance* theory. A particle at that distance from Saturn would have a period of revolution of $11^h17\frac{1}{2}^m$, just half the period of the satellite Mimas. Such a particle would be nearest Mimas at the same place in its orbit every second time around and at that location would feel the maximum gravitational "tug" of the satellite. These perturbations, occurring repeatedly in the same direction, would gradually accumulate until they removed any particle from the region, leaving a gap. When one revolving body has a period that is a simple integral multiple of the period of another, those periods are said to be *commensurable* with each other. Many observers had reported other, less conspicuous divisions in the rings that would occur at places where particles would have periods commensurable with those of other satellites.

It is now known that the Saturnian rings are far more complex than was thought. First, the interplanetary probes found evidence for a very faint D ring, inside the C ring, and also for a very narrow F ring of radius 140,600 km (outside the A ring), and a suggestion of a G ring of radius 150,000 km, and a broad very faint extension of the system into an E ring out to more than 240,000 km.

But more surprising are the structures of the A, B, and C rings. Rather than appearing relatively smooth, as had been expected, each is sectored into many thousands of fine ringlets and gaps that fill the band. Even Cassini's division is not empty, but contains many hundreds of distinct ringlets, although there are far fewer particles there than in the surrounding A and B rings. Voyager 2 measured the changing light from a star that was occulted by the ring system; the fluctuations in the star's brightness revealed organized ring structure on a scale at least as small as one-fourth kilometer. In all, there are probably hundreds of thousands of separate rings within the classic three (A, B, and C), giving the entire ensemble the appearance of a grooved phonograph record.

No longer can Cassini's division be understood in terms of resonance with satellites, although such resonance may play a role in its structure. That theory, however, can certainly not account for the many thousands of thin divisions separating fine, sharply defined ringlets. After Voyager 1's pass, it was hypothesized that many of the divisions were due to very small satellites within the rings, cutting swaths along their paths through the ring material. A moonlet one kilometer across can clear a gap 10 km wide this way. Voyager scientists were, in fact, almost confident of finding somewhat larger satellites—a few that are 20 to 30 km across—accounting for some of the sharp boundaries within Cassini's division. Voyager 2, therefore, made a special search for those intra-ring moons. None was found! It is now almost certain that Cassini's division contains no satellites larger than about 5 km across.

The fine structures of the rings remains a mystery. Perhaps many small moons may account for the finer divisions, but at this writing the larger sharp gaps (some with even finer ringlets within them!) remain unexplained.

Voyager observations have, however, provided a great deal of information on the nature of the ring particles from the way they scatter light. From earth we see the rings by the light they reflect back to us. Voyager 1, however, also passed behind the rings to observe them in the light that the particles scatter mostly in a forward direction. Zones in the rings that appear bright from earth appear darker from behind, while the darker regions as seen from

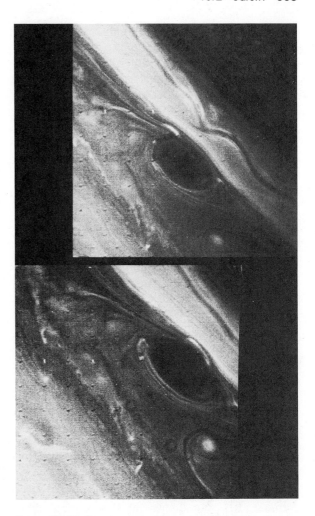

Figure 18.10 Two images of Saturn's atmosphere near the north polar region, taken August 23 and 24, 1981, by Voyager 2 from distances of 2.7 and 2.3 million kilometers, respectively. Note the circulation around the large brown spot. (NASA/JPL)

earth appear bright in forward-scattered light. The amount of forward-scattered light (that passes through the rings) depends on the numerical density of the particles (how many per given volume) and on their sizes. Particles of all sizes reflect or scatter some light in all directions, but large ones scatter more light backward (like a mirror) and small ones scatter more forward, analogous to the way large boulders and small pebbles interfere with the motion of ocean waves. Mostly small particles are found in the darker gaps and fainter rings, and large ones in the bright rings.

The amount of forward-scattered radiation shows that most of the particles in the F ring are very small—a few thousandths of a millimeter across. In the bright A, B, and C rings the particles range in size from very tiny up to centimeters and meters across. The attenuation of radio waves transmitted from Voyager to earth while the spacecraft was behind the rings provided a measure of the sizes of the larger particles. These have effective diameters of about 2 m in the C ring, 8 m in the Cassini division, and 10 m in the A and B rings. Most of the particles, however, probably have much smaller sizes. The composition of the rings is not entirely known, but studies from earth indicate that they are at least covered with ice, and Voyager data strongly suggest that they are largely ice. If all of the material of the entire ring system could be compressed to a uniform disk, it would be only about 1 m thick. The combined mass of the rings is estimated at 3×10^{-8} that of Saturn itself.

A remarkable feature of the B ring discovered by Voyager 1, and studied in detail by Voyager 2, is a series of radial "spokes" that appear from time to time, as if some giant laid his hand flat on the ring and left the impressions of his fingers. Analysis of the light reflected from and transmitted through the ring shows a larger abundance of very fine particles—a few thousandths of a millimeter across—in the spoke regions, because the spokes are very bright in forward-scattered light. The spokes are not permanent; after forming they begin to trail back, with the parts closer to the planet moving faster, in keeping with Kepler's laws. After a few hours the spokes dissipate, but new ones keep forming. It is not known (at this writing) what causes the phenomenon. One speculation is that the tiny particles are levitated above the ring plane by electrostatic charges generated on the particles by the photoelectric effect of sunlight striking them, or perhaps by ions in Saturn's magnetosphere.

Another surprise from the Voyagers was the structure of the F ring. It is not only very thin, but consists of several separate strands, as if it were a braid of hair (Figure 18.11b). The individual particles are moving in rapidly perturbed orbits. Although the details are not yet worked out, at least in a general way it is pretty clear what is going on, and in particular why the ring is narrowly confined. Two satellites discovered by Voyager 1, temporarily designated 1980 S 26 and 1980 S 27, have closely separated orbits straddling the ring. According to an idea first proposed by P. Goldreich at Caltech and S. Tremaine at Princeton, these little moons act like sheepdogs to keep the "flock" particles of the F ring in formation. The perturbative effects of these satellites are always such as to deflect particles to-ward the middle of the ring. Particle 1 (Figure 18.11a) approaching satellite S 26, moving faster on its inside track, is pulled back slightly, so it loses energy and falls in closer to Saturn. Particle 2, being passed by Satellite S 27, is pulled forward slightly, gaining energy so that it moves out into a higher orbit.

A similar, but much finer, "kinky" ring was found by Voyager 2 in another gap—the Encke division in the A ring; no shepherding satellites were found near it, however.

But how can satellites exist inside Roche's limit, where large bodies cannot? Roche's limit applies to bodies of the same mean density as the parent planet and that are held together purely by gravitational forces (that is, not counting the cohesive forces within the material). The rings are probably of icy particles in regions where denser, rocky satellites of small size can survive. S 26 and S 27 are only about 100 km in diameter. They certainly could not survive considerably closer to Saturn, where the tidal force is stronger. However, smaller rocky bodies that could cling together with nongravitational cohesive forces could exist even in the inner ring system. (After all, pebbles and even large boulders can exist on the earth's surface, well within its Roche limit, although mountains with heights of many tens of kilometers could not.) As described above, a large number of such moonlets might account for some of the apparently bizarre structure of Saturn's rings.

(d) Satellites

The Jovian planets sport five giant satellites—the four Galilean moons of Jupiter and Titan of Saturn. Titan is the only satellite in the solar system known to possess an appreciable permanent atmosphere. Until Voyager 1, it was also rated as the largest satellite in the solar system, but we now know that its large apparent radius (2900 km) is due to light scattered from the top of its dense atmosphere. Voyager measurements showed its solid surface to have a radius of 2560 km, compared with 2635 km for Ganymede—Jupiter's largest satellite.

Voyager scientists had hoped to be able to view the surface of Titan, but its atmosphere proved to be too dense. The total pressure at the surface of that moon is about 1.5 times sea-level pressure on the earth, but because of Titan's lower surface

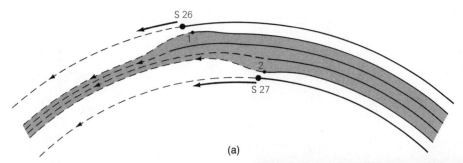

(a)

Figure 18.11 *(a)* Small satellites S 26 and S 27 straddle the F ring of Saturn. Ring particles coming close to these satellites, which would otherwise diffuse outward from the ring, are perturbed back toward the middle of the ring, maintaining its structure. *(b)* Voyager 1 photograph of the complex F ring. (NASA/JPL)

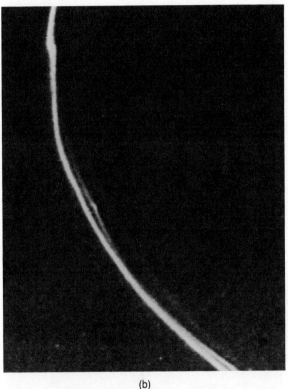

(b)

gravity, the actual mass of atmosphere over each square meter of its surface exceeds that over each square meter of the earth by about 9 times. Gerard Kuiper had first detected methane in the atmosphere of Titan in the 1940s by means of spectrographic studies from earth. Voyager confirmed the methane but showed that it makes up at most only 2 percent of the total atmosphere of the satellite, the bulk of the rest being nitrogen. There are also traces of other gases. At the surface temperature of Titan (measured at 97 K) we expect that methane may exist there in all three states: solid, liquid, and gas.

Saturn's other satellites fall into two classes: seven with sizes intermediate between the Galilean satellites and the many small moons of Jupiter—these have radii from 150 to 765 km—and the rest with radii of 100 km or less. Not much is known about the many small ones in the latter category. Those in the first group all have low mean densities—from 1.0 to 1.4 times that of water—which suggests that they are mostly ice. Indeed, some ice has been identified spectroscopically.

Saturn's second largest moon is Rhea, with a diameter of 1530 km, and an orbit of radius 527,000 km, just under half the size of Titan's. Rhea is very heavily cratered, as are nearly all of Saturn's satellites whose surfaces have been observed.

More remote than Titan from Saturn is Iapetus, a most remarkable object because of the very large difference in reflecting power between its two hemispheres—a ratio of 10 to 1! It is the contrast in brightness between the different sides of Iapetus that inspired Arthur Clarke to select it as an extraterrestrial base in his novel *2001*. It is thought that the bright side of Iapetus is freshly deposited ice, perhaps from a recent flooding and refreezing after an impact with another body. The dark side, which reflects about as much light as asphalt, may be a deposit of low-reflecting material on the surface, or an older part of the surface that is heavily eroded and contaminated.

Dione, the fourth largest satellite, is about 60 percent ice and displays large white wispy regions that may be fractures. Tethys, similar to Dione in size but closer to Saturn (orbit radius of 295,000 km), has huge trough-like valleys, larger than any such features seen on any other satellite in the solar system. A great canyon system extends at least two thirds of the way around that moon. It is thought that it may have resulted from cracking of the surface as the largely-water satellite froze and ex-

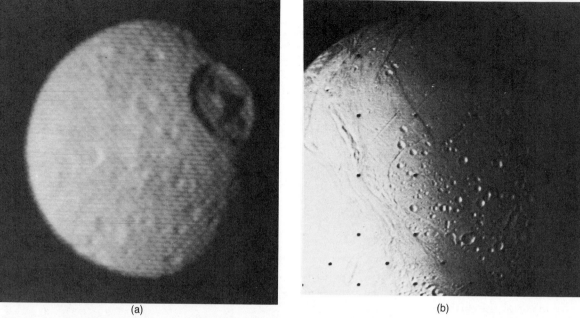

(a) (b)

Figure 18.12 Voyager views of some of Saturn's satellites: (a) Mimas, photographed by Voyager 1 from 425,000 km. (b) Enceladus, seen by Voyager 2 from 112,000 km. (c) Voyager 1 view of Mimas from 129,000 km. (d) Image of Enceladus from Voyager 2 at a range of 119,000 km, enhanced to show the contrast between the smooth and cratered regions. (e) Voyager 2 view of Tethys from 594,000 km, showing the huge canyon system on the right. (f) Dione, photographed by Voyager 1 from 240,000 km. (g) Iapetus, seen by Voyager 1 from 3.2 million kilometers, showing the strong contrast between its bright and dark regions. (h) Phoebe, Saturn's outermost moon, photographed by Voyager 2 on September 4, 1981, from 2.2 million kilometers, as the spacecraft left the Saturnian system to begin its journey to Uranus. (NASA/JPL)

(c) (d)

(e)

(f)

(g)

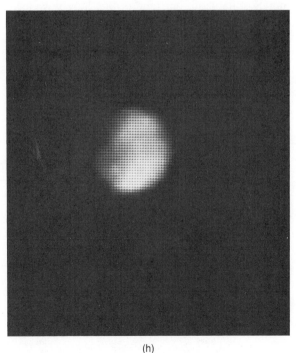

(h)

panded—much like a bottle of soda pop placed in the freezer. Tethys also has a crater about 400 km across—the largest crater known in the solar system.

Enceladus is the most reflective moon of Saturn and has large areas completely devoid of craters. Terrain of several different ages is apparent on Enceladus. Evidently, recent flows of water, melted

perhaps by an impact, submerged older features and then refroze. The innermost of Saturn's large satellites, Mimas, is not only very heavily cratered, like most of the others, but has one most remarkable looking "eye"-like crater about 130 km across. It is estimated that any impact that could produce a much larger crater on Mimas would likely shatter the little moon to bits.

Some data on Saturn's satellites and rings are summarized in Table 18.2.

(e) Many New Worlds

Before the space program, our earth and moon were the only celestial objects that we knew rather intimately. All other worlds were seen as points of light, or as indistinct fuzzy balls through our telescopes, or revealed only the outer layers of their atmospheres. By the late 1970s we had added Mercury and Mars to the list of objects whose surfaces we had mapped, bringing the total to four, if we do not count the radar mapping of Venus. Now the Voyagers have revealed the surfaces of 11 good-sized satellites of Jupiter and Saturn, and in detail far clearer than we could ever see the surface of Mars with telescopes from earth. We have explored at least 15 worlds so far (not counting the very little ones). They display a very wide range of properties, but none except the earth has conditions even remotely suitable for our kind of life. Surely we have gained some insight and perspective which alone has made the exploration of space worthwhile. These are exciting times!

18.3 URANUS

Uranus is a planet that must surely have been seen by the ancients and yet was unknown to them. It was discovered on March 13, 1781, by the German-English astronomer William Herschel, who was making a routine telescopic survey of the sky in the constellation of Gemini. Herschel noted that through his telescope the planet did not appear as a stellar point but seemed to present a small disk. He believed it to be a comet and followed its motion for some weeks. Several months later, a preliminary solution for its orbit was computed and it was found to be a nearly circular one, lying beyond that of Saturn; the object was unquestionably a new planet.

Uranus can be seen by the unaided eye on a dark clear night, but is near enough to the limit of visibility so that it is indistinguishable from a very faint star. It is so inconspicuous that its motion escaped notice until after its telescopic discovery. However, it turned out that Uranus had been plotted as a star on charts of the sky on at least 20 previous occasions since the year 1690. These earlier observations were later of use in the determination of how perturbations were altering the planet's orbit.

Herschel proposed to name the newly discovered planet *Georgium Sidus*, in honor of George III, England's reigning king. Others suggested the name Herschel, after the discoverer; the name finally adopted, in keeping with the tradition of naming planets for gods of Greek mythology, was Uranus, father of the Titans, and grandfather of Jupiter.

Table 18.2 **Known Satellites of Saturn**

SATELLITE OR OUTER RING	DIAMETER (km)	MEAN DENSITY	ORBIT RADIUS (km)
A Ring (outer radius)			136,200
1980 S 28*	300 (?)	?	137,300
1980 S 27*	220 ± 100	?	139,400
F ring			140,600
1980 S 26*	200 ± 100	?	141,700
G Ring			170,000
1980 S 3*	90 × 40	?	151,422
1980 S 1*	100 × 90	?	151,472
Mimas	390 ± 10	1.1 ± 0.1	188,224
Enceladus	500 ± 20	1.1 ± 0.6	240,192
1980 S 13*	?	?	289,600
Thethys	1050 ± 20	1.0 ± 0.1	296,563
1980 S 6*	80 ± 40	?	378,600
Dione	1120 ± 20	1.4 ± 0.1	379,074
Rhea	1530 ± 20	1.3 ± 0.3	527,828
Titan	5120	1.9	1,221,432
Hyperion	290 ± 40	?	1,502,275
Iapetus	1440 ± 40	1.1 ± 0.1	3,559,400
Phoebe	140 ± 80	?	10,583,200

*These are temporary designations. The satellites will be given permanent names by the International Astronomical Union.

(a) Bode's Law

The discovery of Uranus brought Herschel great fame. It also brought delight to the German astronomer Johann Bode (pronounced Bō'-da), because it fitted beautifully into a sequence of numbers announced in 1772 by Daniel Titius, which describe the approximate distances of the planets from the sun. Bode had been so impressed with Titius' progression that he published it in his own introductory astronomy text, and the sequence became known as "Bode's law." The sequence is obtained by writing down the numbers: 0, 3, 6, 12, . . . , each succeeding number in the sequence (after the first) being double the preceding one. If 4 is now added to each number and the sum divided by 10, the resulting numbers are the approximate radii of the orbits of the planets in astronomical units, as can be seen in Table 18.3. The rule breaks down completely for Neptune and Pluto, but these planets were, of course, unknown in Bode's time.

The fact that Uranus fit so well into the scheme suggested to Bode that the progression was a law of nature, which led him to expect an unknown planet in an orbit of semimajor axis 2.8 AUs. Most of the asteroids have orbits near 2.8 AUs, so we shall have more to say about Bode's law in the following chapter; it also played a role in the discovery of Neptune (Section 18.4).

(b) Orbit and Satellites

The orbit of Uranus lies more nearly in the plane of the ecliptic than that of any of the other planets, its inclination being only 46'. Its eccentricity is 0.051, and it varies by a little more than 125 million km from its median distance of 2871 million km from the sun. Being so far from the sun, its orbital speed is low (6.8 km/s), and its period is long (84 years).

Uranus has five known satellites; the last to be discovered, and the faintest, was found by Kuiper in 1948. Herschel himself found the two brightest moons of the planet. None of the satellites is probably much over 1000 km in diameter. Their distances range from 123,000 to 586,000 km from the center of Uranus. They are thought to be icy bodies.

(c) Gross Characteristics

The mass of Uranus is found from the motions of its satellites and from its perturbative effects upon the motion of Saturn; it is $14\frac{1}{2}$ times the mass of the earth. It presents a small angular diameter, just under $4''$, because of its great distance from the sun. Its linear diameter, therefore, is difficult to measure accurately, but it is about 51,000 km. With these values for its mass and size, its mean density is calculated to be $1\frac{1}{3}$ times that of water, its surface gravity about the same as on the earth, and its velocity of escape about 21 km/s. Micrometric measures and also studies of the perturbations on its satellites both show Uranus to be somewhat oblate. The accepted value for the difference in equatorial and polar diameters is 1 part in 30, but this figure is quite uncertain.

Uranus has a high albedo (0.50), so probably it is surrounded by a cloud layer that reflects sunlight. The spectrum of the planet reveals a strong concentration of methane, and also a series of infrared absorption lines attributed to molecular hydrogen. The identification of hydrogen was achieved in 1952. Uranus is the first planet in which hydrogen was identified spectrographically. Ammonia does not appear in the spectrum, probably because it is frozen out of the atmosphere; the temperatures to be expected on Uranus are below 90 K.

Table 18.3 **Bode's Law**

TITIUS' PROGRESSION	PLANET	ACTUAL MEAN DISTANCE (ASTRONOMICAL UNITS)
(0 + 4)/10 = 0.4	Mercury	0.387
(3 + 4)/10 = 0.7	Venus	0.732
(6 + 4)/10 = 1.0	Earth	1.000
(12 + 4)/10 = 1.6	Mars	1.524
(24 + 4)/10 = 2.8		
(48 + 4)/10 = 5.2	Jupiter	5.203
(96 + 4)/10 = 10.0	Saturn	9.539
(192 + 4)/10 = 19.6	Uranus	19.19
(384 + 4)/10 = 38.8	Neptune	30.1
(768 + 4)/10 = 77.2	Pluto	39.5

Figure 18.13 Uranus and three of its satellites, photographed with the 3.05-m telescope. *(Lick Observatory)*

The internal structure of Uranus is probably similar to that of Jupiter and Saturn. Its mean density is high enough, however, to show that it must contain a far smaller proportion of hydrogen and helium than do the giant planets, and a larger proportion of ice.

(d) Appearance and Rotation

Uranus appears as a greenish disk when seen through the telescope. The green color is probably due to its atmospheric methane. A few observers report faint markings, but these are too indefinite to indicate the rotation. The rotation period, therefore, must be obtained from the Doppler shift in the spectrum of light from different parts of its disk. The most recent value, very uncertain, is between 10 to 25 hours.

A unique feature about Uranus is that its axis of rotation lies almost in the plane of its orbit; during some parts of its revolution, it is so oriented that we look almost directly at one or the other of its poles. The actual inclination of its equatorial plane to that of its orbit is 82°. Its direction of rotation is the same as that of the revolution of its satellites, and both are in the *reverse* direction from the rotation of all the other planets except Venus. Its direction of orbital revolution, however, is normal, that is, from west to east.

(e) Uranus' Ring System

Until 1977, only one planet, Saturn, was thought to have rings. In that year, however, a small ring system about Uranus was discovered. The Voyager discovery of a ring about Jupiter brings to a total of three the number of ringed planets.

As is the case for Saturn, the particles in the rings of Uranus revolve about the planet in its equatorial plane. They reflect too little sunlight to be seen directly; the first five were discovered when Uranus passed in front of a star on March 10, 1977. The rings are not solid and did not occult the star completely, but the star did dim somewhat as each ring portion passed by. Observations of subsequent occultations have confirmed the existence of the rings and have indicated that there are more than five. The present number identified is nine. They are all very narrow, reminiscent of the F ring of Saturn; they may also be "herded" by close-lying satellites.

The outer diameter of Saturn's bright rings (the A ring) is 275,000 km, 2.31 times that planet's diameter. The rings of Uranus have a maximum diameter of 102,000 km, a little more than twice that of Uranus itself.

We can look forward to far more information about Uranus and its rings when Voyager 2 rendezvous with the planet in 1986.

Figure 18.14 Neptune and its larger satellite Triton, photographed with the 3-m telescope. *(Lick Observatory)*

18.4 NEPTUNE

Whereas the discovery of Uranus was quite unexpected, Neptune was found as the result of mathematical prediction. The discoveries of the two planets could hardly have been made under more different circumstances, yet in other respects Uranus and Neptune are very much alike.

(a) Discovery

By 1790, an orbit had been calculated for Uranus (first by Delambre) based on observations of its motion in the decade following its discovery. By early in the 19th century, it was realized that all was not well with Uranus! Even after allowing for the perturbative effects of the known planets, it was found that it was not possible to account for the motion of Uranus over the entire period for which it had been observed—including the observations back to 1690. The discrepancies between the predicted positions of the planet, based on gravitational theory, and its observed positions grew worse and worse, until by 1844 they amounted to a full 2 minutes of arc. Two Uranuses that far apart in the sky would not be seen

as separate objects by the average eye; yet the difference was totally unacceptable in gravitational theory.

Several possible explanations were suggested: Perhaps Uranus was knocked off course by collision with a comet, or perhaps it is in mutual revolution about an invisible satellite. But by 1840 all such notions were shown not to be possible, and the remaining alternatives were either that the law of gravitation breaks down over such large distances, or that there is an unknown planet providing gravitational perturbations on Uranus in addition to those of the known planets.

Most knowledgeable astronomers favored the latter hypothesis, but to have enough data to calculate where such a planet would have to be was believed to require observations of Uranus over the next few hundreds of years. Still, the problem was intriguing, and it appealed to two young mathematicians, both highly talented, both relatively unknown, and each without any knowledge of the other. They were Urbain Jean Joseph Leverrier (1811–1877) in France and John Couch Adams (1819–1892) in England. Both Adams and Leverrier saw that the peculiar way in which Uranus had been first ahead of and then behind its predicted position was an important clue to finding where a hypothetical perturbing planet would have to be. They both solved the problem and obtained very nearly identical results. They both had been misled into believing that the unknown planet would be as far from the sun as is suggested by the next planet in the series in Bode's law—39 AU. The perturbing planet was actually closer to the sun by about 21 percent, but fortunately this error almost cancels out in the solution for the direction of the planet in the sky, so both Adams and Leverrier had predicted correctly where to look for the new object.

Unfortunately, the actual discovery of Neptune was not so straightforward. Adams finished his calculations first and had attempted to deliver them in person to the Astronomer Royal, Sir George Biddell Airy, at the Royal Greenwich Observatory, in late September 1845. Airy, at the time, was on the Continent and had not received word of Adam's call. Adams tried again on October 21, but on that morning Airy was out of his office. Adams left a brief summary of his results, and a message that he would call later that same day. Regrettably the message about the later call was never delivered to Airy, and when Adams came again, Airy was at dinner, and his overprotective butler would not allow him to be disturbed. Adams, feeling rebuffed, left.

When Airy got around to reading Adams' summary, he was skeptical because he was not familiar with the detailed theory himself, and because Adams was still young and unproven. Thus when Airy replied to Adams, on November 5, he attempted to put the young man to the test by asking him about what seemd to Adams to be an almost irrelevant and trivial problem concerning the earlier observations of Uranus. Probably annoyed, Adams did not reply (until more than a year later, by which time the discovery of the new planet was history).

Meanwhile, on November 10, Leverrier had completed the first part of his own calculations and had published them in the appropriate French journal. Airy saw Leverrier's paper and was most impressed with its clarity and style. During the winter and spring of 1846, Leverrier completed his analysis, and his published solution reached Airy in late June 1846. Airy noted that Leverrier's prediction for the position of the new planet agreed to within 1° with Adams', and he immediately sent Leverrier the same question about the orbit of Uranus that he had sent Adams. Airy received a prompt and authoritative reply from Leverrier on July 1 (the mails must have been more efficient in those days!). That convinced Airy that the new planet must exist.

Airy sent two letters to James Challis, Director of the Cambridge Observatory, pointing out that the Observatory's Northumberland telescope, with its 11.75-inch lens, was the most ideal telescope for the search for the suspected planet. (That instrument is one of the oldest working telescopes in the world and is still used on clear nights by Cambridge undergraduates.) Airy's second letter, dated July 13, expresses the urgency with which Airy finally regarded the search for the new planet. "In my opinion," he wrote, "the importance of this inquiry exceeds that of any current work, which is of such a nature as not to be totally lost by delay." Challis accordingly began a systematic search of the area of the sky indicated by Airy, based on the predictions of Adams and Leverrier. The Cambridge astronomers did not possess up-to-date star charts against which to compare suspected planets, so their procedure was to survey systematically the designated part of the sky on dates separated by several days and then to compare the records of the positions of the observed stars to see if one, by virture of its motion, had revealed itself as a planet. Challis began trial observations on July 29.

Leverrier completed his final calculations and presented his paper to the French Academy on August 31. In that paper he gave detailed information on where to look for the new planet. Now, Adams had experienced difficulty convincing the established authorities (Airy, at least) of his competence and was not taken seriously at first. Leverrier did not have that problem; his work was warmly received. On the other hand, Leverrier had no success whatsoever convincing the French astronomers that they should bother to look for the new planet. In desperation, Leverrier wrote to a young assistant at the Berlin Observatory, Johann Gottfried Galle, from whom he had had some correspondence a year earlier. Galle was much excited about the prospect of discovering a new planet, and after much effort finally received permission from the Observatory Director, Johann Franz Encke, to try a search the very night he received Leverrier's letter—September 23. Galle agreed to allow a young student astronomer, Heinrich Louis d'Arrest, to join him.

The 9-inch Fraunhofer refractor, pride of the Observatory, was turned to the predicted position: right ascension 22^h46^m, declination $-13°24'$. At first they found nothing, but Galle and d'Arrest located new star charts of that region of the sky, and while Galle looked through the telescope and described the stars he saw, d'Arrest compared Galle's comments with the stars on the chart. After so checking out a few stars, Galle described an eighth-magnitude star at right ascension $22^h53^m25\overset{s}{.}84$, and d'Arrest excitedly pointed out that that star was not on the map; it was, at last, the eighth planet!

When word reached Cambridge of the Berlin discovery of the new planet, Challis checked through his old records and found that the planet had actually been observed, but not recognized, six weeks earlier at Cambridge. It appeared in the observing records of August 12; a marginal comment, which Challis added later, points out, "This is the planet. It does not appear in this place on page 3." On page 3 of the observing notebook we find the observations of the same part of the sky made 12 days earlier, on July 30th.

Today the honor for the discovery of Neptune is justly shared by those two brilliant and tenacious mathematicians, Adams and Leverrier, who successfully predicted its existence. That prediction ranks high in the history of science, for it was an extraordinary triumph of Newtonian gravitational theory. But by the time it was announced the discovery was not so much of a surprise as a fulfillment of expectations. On September 10, 1846, 13

Saturn and its ring system, photographed by Voyager 2 on August 11, 1981, at a distance of 13.9 million km. *(NASA/JPL)*

Saturn's northern mid-latitude region photographed by Voyager 2 on August 16, 1981, at a range of 9.3 million km. Note the strangely curled cloud region attached by a ribbonlike cloud to the white band of clouds. *(NASA/JPL)*

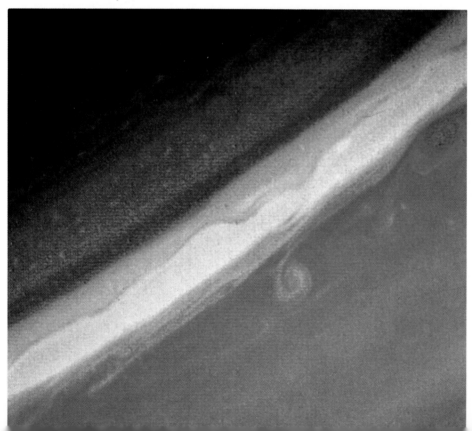

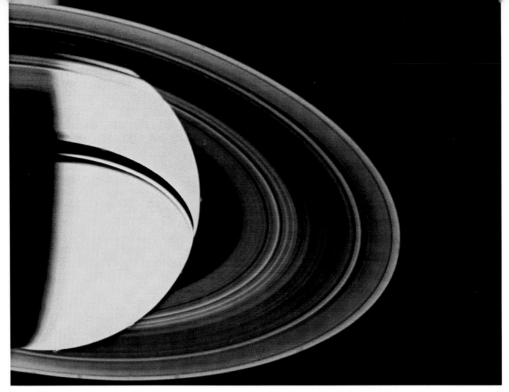

Voyager 2 photograph of Saturn's ring plane taken from a distance of 3.4 million km beyond the planet. *(NASA/JPL)*

Computer-enhanced photograph of Saturn's rings as photographed by Voyager 2. The exaggerated color shows the reflection and transmission of different wavelengths of light by particles of different sizes in various parts of the ring system. *(NASA/JPL)*

days before Galle's observation of the planet to be named Neptune, Sir John Herschel, son of the discoverer of Uranus, remarked in his valedictory address as president of the British Association, "We see it as Columbus saw America from the shores of Spain. Its movements have been felt, trembling along the far-reaching line of our analysis with a certainty hardly inferior to ocular demonstration."

(b) Pre-discovery Observations

We have seen that Uranus had been observed a number of times before its discovery but was mistaken for a star. A search of old records showed that there were also two pre-discovery observations of Neptune—on May 8 and 10, 1795, by Joseph Lelande. These observations were used in 1847 by the American astronomer Sears Cook Walker to calculate an improved orbit for Neptune. It was then for the first time that it was realized that Neptune is only about 30 AUs from the sun and that Bode's law does not work for all of the planets. Table 18.4 compares some of the orbital elements calculated by Adams and Leverrier and the modern values.

It is now believed that Neptune was also observed a full 234 years before its "official" discovery by no less a person than Galileo. Astronomer Charles Kowal of Caltech and historian Stillman Drake of Toronto had investigated close conjunctions of planets that occurred in historical times and found that there was a close conjunction of Neptune and Jupiter in January 1613, about the time Galileo was observing Jupiter and its satellites. A check of Galileo's notebook showed that he had indeed seen an eighth-magnitude star near Jupiter on several occasions and had even noticed that it seemed to move. He evidently was not sure enough of the motion to realize the significance of what he had observed. If Galileo's plot of Neptune was as accurate as those of Jupiter's satellites, we must conclude that Neptune's position then differed by a full minute of arc from the prediction of the best orbital elements available for it today. It is interesting that the much more recent observations by Lelande also show a discrepancy, in that case of 7 arc seconds.

We shall return to this tantalizing subject in Section 18.6.

(c) Orbit, Satellites, and Rotation

Neptune is almost exactly 30 AU from the sun. Its orbit is very nearly circular, having an eccentricity of only 0.010; its distance from the sun varies by only 39 million km on each side of its median value of 4497 million km. Traveling with an orbital speed of only 5.4 km/s, it requires 165 years to complete one revolution around the sun. It will not have completed its first revolution since its discovery until 2011.

Its axial rotation period, determined from the infrared brightness variations, is about 18 hours. Its equatorial plane is inclined about 29° to that of its orbit. Its oblateness is small, the equatorial diameter exceeding the polar diameter by only about 1 part in 40.

Neptune has two satellites. *Triton*, the larger, may have a greater size than does our moon. Its distance from the center of Neptune is only about 353,000 km, and its motion is backward, that is, east to west. Neptune's other satellite, Nereid, discovered in 1949, is smaller, and has the most eccentric orbit of any satellite in the solar system; its distance from Neptune varies from 20 to 10 million km.

(d) Other Properties

Neptune's mass is 17.2 times that of the earth. Since it subtends an angle of only about 2″, its diameter is difficult to measure, but a recent occultation observation indicates that it is about 50,450 km, with an uncertainty of about 200 km. Its mean density is about $1\frac{1}{2}$ times that of water. The internal structure of Neptune is thought to be similar to that of Uranus: considerable water, probably a rocky core, and surrounding the water liquid hydrogen and helium, but in much lower abundance than in the sun.

Neptune appears telescopically as a small greenish disk. There are no conspicuous markings. Its high albedo of 0.5 to 0.6, however, suggests that light is reflected from a gaseous atmosphere. Both methane and hydrogen have been detected spectrographically, as in Uranus. Helium is probably also present but cannot be observed in the spectrum. Ammonia appears to be absent in the gaseous state, as we would expect, since the temperature of

Table 18.4 **Orbital Elements for Neptune**

	ADAMS	LEVERRIER	MODERN
Semimajor axis (AU)	37.25	36.15	30.134
Eccentricity	0.12062	0.10761	0.005
Period (years)	227.3	217.387	164.74
Mass (Sun = 1.0)	1/6666	1/9300	1/19,350

Figure 18.15 Two photographs of Pluto, showing its motion among the stars in a 24-hour period; photographed with the 200-inch telescope. *(Caltech/Palomar Observatory)*

Neptune must be near 60 K. Neptune, like Jupiter and Saturn, appears to have an internal heat source.

It will be interesting indeed to see what Voyager 2 has to show us of Neptune if it is still operative in 1989.

18.5 PLUTO

After the perturbative effects of Neptune upon Uranus had been taken into account, the discrepancies between the observed and predicted positions of Uranus were reduced to only $1/60$ of what they had been. Today those remaining discrepancies are realized to be smaller than the errors of the original observations and thus are almost certainly not real. Yet several investigators made attempts to account for those remaining deviations by the gravitational influence of a ninth planet, beyond the orbit of Neptune. Among them were Gaillot, W. H. Pickering, and Percival Lowell. It was Lowell's solution to the problem that led to the discovery of Pluto.

The orbital elements, calculated by Lowell, Pickering, and Gaillot, along with the actual elements of Pluto, are shown in Table 18.5.

Table 18.5 **Orbital Elements of Pluto**

	GAILLOT 1909	LOWELL 1915	PICKERING 1928	ACTUAL
Semimajor axis (AU)	66	43.0	30	39.5
Period (years)	536	282	165	249
Eccentricity	(0)	0.20	0.20	0.25
Perihelion date	—	1991	1974	1990
Magnitude (see Chap. 24)	—	12–13	12.2	15
Mass (earth = 1)	24	6.6	0.8	0.002

(a) Discovery

By the beginning of the 20th century, when Lowell made his calculations, Neptune had moved such a short distance in its orbit that a knowledge of the perturbations upon it was not yet available. Therefore Lowell, as did the others, based his calculations entirely on the minute remaining irregularities of the motion of Uranus. His computations indicated two places where a perturbing planet could be, the more likely of the two being in the constellation of *Gemini*. Lowell predicted a mass for the planet intermediate between that of the earth and that of Neptune (his calculations gave about 6.6 earth masses), and also an angular diameter in excess of 1″. Lowell searched for the unknown planet at his Arizona Observatory from 1906 until his death in 1916, without success. Subsequently, Lowell's brother donated to the observatory a 33-cm photographic telescope that could record a 12° by 14° area of the sky on a single photograph. The new camera went into operation in 1929, and the search was continued for the ninth planet.

Unfortunately, Gemini lies near the Milky Way, and some 300,000 star images were recorded on each exposure. It was an immense task to compare all the star images on each of two or more photographs of the same field in the hope of finding one image that changed position with respect to the rest, revealing itself as the new planet. The job was facilitated by the invention of the *blink microscope*, a device in which are placed two different photographs of the same region of the sky. The operator's vision is automatically shifted back and forth between corresponding parts of the two photographs. If the star patterns are the same on the two plates, the observer sees a constant, although flickering picture. However, if one object has moved slightly in the interval between the times the two plates were taken, the image of that object appears to jump back and forth as the view is transferred from one photograph to the other. In this way, moving objects can quickly be picked out from among the many thousands of star images. The blink microscope is also used as a means of locating stars that vary in brightness.

In February 1930, Clyde Tombaugh, comparing photographs made on January 23 and 29 of that year, found an object whose motion appeared to be about right for a planet far beyond the orbit of Neptune. It was within 6° of the position Lowell predicted for the unknown planet; subsequent investigation of the object showed its orbit to have elements very similar to those Lowell had calculated. Announcement of the discovery was made on March 13, 1930. The new planet was named for Pluto, the god of the underworld. (Appropriately, the first two letters of Pluto are the initials of Percival Lowell.)

(b) Orbit of Pluto and its Satellite

Pluto's orbit has the highest inclination to the ecliptic (17°9′) of any planet and also the largest eccentricity (0.248). Its median distance from the sun is 39.53 AU, or 5913 million km, but its aphelion distance is over 7000 million km and its perihelion distance under 4500 million km. Part of its orbit is closer to the sun than the orbit of Neptune. There is no danger of collision between the two planets, however; because of its high inclination, Pluto's orbit clears that of Neptune by 385 million km. Pluto is not, at present, the most distant planet, for in winter 1979 it moved in closer than Neptune, and it will remain within Neptune's orbit until the spring of 1999. Pluto completes its orbital revolution in a period of 248.6 years, poking along at a mean velocity of only 4.7 km/s.

In 1978 James W. Christy of the United States Naval Observatory noticed a peculiarity on a routine photograph of Pluto taken with the Observatory's 155-cm astrometric telescope at its Flagstaff Station. The image of Pluto appeared elongated, but those of the stars on the same photograph did not (Figure 18.16). Christy and R. S. Harrington, also at the Naval Observatory, then checked other photographs on file and found that some of them showed the same effect, but that most did not. It occurred to them that Pluto may have a satellite that is almost but not quite resolved as a separate object from Pluto on photographs taken when the

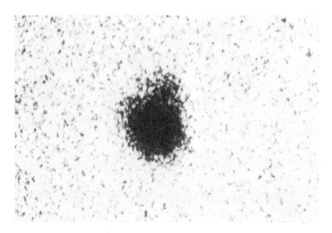

Figure 18.16 Highly enlarged image of Pluto on photograph made at the U. S. Naval Observatory, Flagstaff Station. The "bump" on the upper right is Pluto's satellite, Charon. *(Courtesy, U. S. Naval Observatory)*

two have their maximum separation as seen from earth. Christy followed up on this hunch and found that the image was indeed that of a satellite revolving about Pluto in a period of 6.387 days. The proposed name for the satellite is Charon, for the mythological boatman on the River Styx, who transports passengers to Pluto's realm.

Charon's orbit, so far as is known, is roughly circular with a radius of about 20,000 km, and is inclined at 65° to the plane of Pluto's orbit. The direction of Charon's revolution is retrograde—from east to west.

(c) Nature of Pluto and Charon

Pluto's mass, long a subject of controversy, is now known from the distance and period of its satellite, Charon. Pluto's mass is 0.0023 times the mass of the earth, which makes Pluto by far the least massive planet in the solar system. Its size has also long been a mystery, but recent observations are providing some information about its dimensions. In 1965, Pluto *failed* to occult a star, which showed that it cannot be more than 5500 km across. Then in 1976 spectral features were observed that show Pluto to have frozen methane on its surface. Frozen methane has a high reflecting power, and to account for Pluto's brightness its diameter need be only about 3000 km or so. Finally, in 1979, speckle interferometry (Section 11.4b) with the 200-inch Hale telescope on Palomar was attempted on Pluto, with the result that Pluto's diameter was found to be in the range 3000 to 3500 km. With the possible dimensions for Pluto, we find that its mean density is probably not more than that of water; we cannot rule out (as yet) that Pluto is the least dense planet in the solar system.

We have mentioned that Pluto has frozen methane on its surface. It may also have an extremely thin atmosphere of methane, if observations made at the University of Arizona in 1980 are correct; they show evidence of a sparse distribution of gaseous methane above the solid surface. The gaseous phase is unexpected for a place as cold as Pluto is expected to be—some 40 K. In any case, the gas has a very low concentration and may be only temporary, especially now that Pluto is somewhat nearer the sun than usual.

Charon's size may have been measured, at least indirectly. On April 6, 1980, astronomers in South Africa measured an occultation of a star that, according to the United States Naval Observatory, was only one arc second northwest of Pluto at the time, just where Charon was at the time. The star disappeared for about 50 seconds. If it passed behind Charon, it gives a lower limit to the diameter of that moon of 1200 km. From the relative brightness of Charon and Pluto, we estimate that the moon does not have a diameter more than about half that of Pluto, so it all fits.

Although Pluto is thought to have highly reflecting methane ice on its surface, the ice cannot comprise a uniform covering, because Pluto varies in brightness with a period of 6.39 days. This variation was first measured by R. H. Hardie and M. Walker and is interpreted to be caused by a rotation of the planet displaying alternately light and dark sides to our view. But this period of rotation is exactly the orbital revolution period of Charon; Charon is far too small to reflect enough light to us to cause the variations by itself, so we must conclude that Pluto rotates with the same period that it mutually revolves about its satellite. Because of this synchronous rotation, an observer on Pluto who could see Charon in the sky would see it always in the same part of the sky, never either rising or setting.

The picture that emerges is that both Pluto and its satellite are small bodies—about 3000 km in diameter for Pluto and something like half or less of that for Charon. Charon probably has a mass less than a tenth that of Pluto, and both objects more closely resemble the intermediate satellites of Saturn than they do the other planets. There is some speculation that Pluto may not even have been a primordial planet but is an escaped satellite of Neptune. Still, Pluto is a planet at present, and it has a satellite of its own.

(d) The Problem of Pluto's Discovery

Pluto's discovery was based on alleged residuals in the positions of Uranus after account was taken of the perturbations that could have been produced on Uranus' motion by Neptune and other known planets. These residuals, only about 4″ to 5″, were in observations of the positions of Uranus made in the early 18th century before it had been recognized as a planet. They were, in fact, smaller than the errors expected in observations made that long ago, and have now been generally discredited by the experts as having no significance. In view of this, one may well wonder whether there can have been any real validity in the calculations of the predicted orbit of Pluto.

Even if the remaining residuals in the positions of Uranus *were* real, to produce them with perturbations by Pluto, Pluto would need a mass substantially greater than that of the earth (note that Lowell predicted a mass of 6.6 earth masses). On the other hand, Pluto's mass is now known to be too small to have produced perturbations in the amount required, and hence the prediction of Pluto's position, although mathematically correct, was based on invalid data; Pluto's discovery was accidental!

The fact that Pluto *was* discovered, with so nearly the orbital elements that Lowell predicted, is an interesting coincidence. Be that as it may, it was Lowell's faith and enthusiasm that led to our knowledge of Pluto, and he has justly earned the honor for its discovery.

(e) The Solar System from Pluto

From Pluto, the solar system must appear a bleak and empty place. The sun would appear as a bright star, although it would still provide Pluto with 250 times as much light as we on earth receive from our full moon. The earth as seen from Pluto would scarcely ever be at more than 1° angular separation from the sun and would be slightly too faint to be seen with the unaided eye. Even Jupiter could only be 7° from the sun and would appear only as a medium-bright star.

18.6 ARE THERE UNKNOWN PLANETS?

The possibility of undiscovered planets exists. However, a careful search by Tombaugh after his discovery of Pluto failed to reveal any other trans-Neptunian planet. He should have picked up any object as large as Neptune within a distance of 270 AU, if it were not too far from the ecliptic.

(a) The Orbit of Mercury: Vulcan

The major axis of the orbit of a planet is called the *line of apsides*. One of the effects of perturbations by other planets is to produce a slow rotation of the line of apsides in the orbital plane, thereby gradually shifting the point of perihelion of a planet. It has long been known that the line of apsides of the orbit of Mercury is changing orientation by the amount of 574″ per century. However, when Newtonian gravitational theory is applied, and due account is taken of perturbations by known bodies, it is found that Mercury's advance of perihelion should amount to only 531″, a discrepancy of 43″.

This problem was first pointed out by Leverrier, co-discoverer of Neptune. Leverrier had based his work on the times of transits of Mercury across the sun's disk, which provided a very accurate way of determining where Mercury was in its orbit. With the phenomenal success of the discovery of Neptune in mind, it seemed that the strange motion of Mercury could be accounted for by an intermercurial planet, perturbing Mercury. The hypothetical body was given the name *Vulcan* (the Roman god of fire). Vulcan was even reportedly observed once, during a total solar eclipse. However, the observation has never been confirmed, and it is now generally discredited.

We shall see in Chapter 35 that Einstein's general theory of relativity predicts that 43″ discrepancy with Newtonian theory; in fact, Einstein suggested the advance in the perihelion of Mercury as a test for relativity. The strange behavior of the orbit of Mercury is thus completely explained by general relativity, and the "need" for Vulcan has vanished.

(b) But What About "Planet X"?

To Lowell, Pluto was "Planet X"—the object needed to remove the discrepancies in the motion of Uranus. We have seen that Pluto cannot be Lowell's Planet X; its mass is far too small. On the other hand, the observations on which its existence was based were probably not significant anyway.

Now, however, half a century after Pluto's discovery and nearly a century and a half after the discovery of Neptune, we can reassess the motions of all of the planets. This has been carefully done by astronomers at the United States Naval Observatory as part of their regular work on planetary orbits. Over the past decade it has been clear that there are minor problems with the motions of the outer planets, from Jupiter to Neptune, but these could possibly be explained if the mass of Pluto were great enough. Now, however, we have a reliable mass for Pluto from the motion of its newly discovered satellite, and we realize that Pluto is far too insignificant to account for the discrepancies, which range from a fraction of an arc second for Jupiter to several seconds for Neptune, and far more if we include the pre-discovery observations of Lelande and Galileo.

These discrepancies can be explained, according to Naval Observatory scientists, if there is a single undiscovered planet with the right characteristics, beyond the orbits of Neptune and Pluto. Thomas Van Flandern, leading investigator in the Naval Observatory team, described the situation at the January 1981 meeting of the American Astronomical Society at Albuquerque, New Mexico. The planet should have a mass between two and five times that of the earth and should lie at a distance of from 50 to 100 astronomical units from the sun. The residuals between theory and observation also suggest that it should have a relatively highly inclined orbit, as does Pluto. Van Flandern hopes that future studies, and especially the pre-discovery observations of Neptune, may help tie down the region of the sky in which to search for the planet. The search area could still be quite large (perhaps 10 degrees or more on a side), and the planet could be quite faint. Yet he feels there is a good chance that we might be able to locate it.

Clyde Tombaugh, discoverer of Pluto and then nearly 75 years old, was present at that session. He reminded the listeners of the search for unknown planets he had made, and what the limits on its visibility would have to be. Yet all agreed, if the object were at a great enough distance from the ecliptic, and not too bright, it could have escaped our notice. What an excitement it would be if the prediction of the Naval Observatory astronomers is correct, and especially if Tombaugh could live to see the discovery of Lowell's *real* Planet X!

EXERCISES

1. Which satellite would have the greatest period, one 1 million km from the center of Jupiter or one 1 million km from the center of earth? Why?

2. The satellite Io of Jupiter is 421,600 km from the center of the planet and has a nearly perfectly circular orbit and a period of $1^d18^h28^m$. It requires 3.50 minutes from the time it starts to enter Jupiter's shadow until it is completely eclipsed. What is its diameter? Show your reasoning.
Answer: About 3640 km

3. How often are Saturn's rings turned edge on to us? Assume, for this exercise, that Saturn's orbit is in the plane of the ecliptic. What is the relevance of this assumption?

4. Saturn's rings extend from a distance of 64,000 to 137,000 km from the center of the planet. What is the approximate variation factor for the periods of time required for various parts of the rings to revolve about the planet?
Answer: About 3:1

5. If Saturn's mass is 95 times the earth's, how does it happen that its surface gravity is so nearly the same as the earth's?

6. Describe the seasons of Uranus.

7. The inclination of the equator plane of Uranus to its orbit is often quoted as 98° rather than 82°. Can you suggest why?

8. Calculate the surface gravities of Uranus and Neptune and compare them to that of the earth.

9. Determine the number of Neptune days in that planet's sidereal year. (Assume a rotation period of 18 hours.)
Answer: 80,714 Neptune days

10. Calculate the perihelion and aphelion distances of Pluto if its orbit has a semimajor axis of 40 AU and an eccentricity of 0.25.

19

THE ASTEROIDS

Of the many thousands of minor planets, nearly all appear telescopically like stars—that is, as small points of light. The term *asteroid* (meaning "starlike") is therefore often applied to these tiny worlds. The term *minor planet*, preferred by the International Astronomical Union, is used by most specialists in orbital mechanics, and I used it in previous editions of this text. Today, however, both *asteroid* and *minor planet* are in common use, even by specialists in the field, with *asteroid* probably being the more common.

Most of the minor planets have orbits between those of Mars and Jupiter. From the time of Kepler it was recognized that this region of the solar system constituted a considerable gap in the spacing of the planetary orbits. We have already seen (Section 18.3a) that the sequence of numbers known as Bode's law predicts a planet at 2.8 AU from the sun. The fact that the newly discovered Uranus fit the law so well generated great faith that the sequence was a law of nature, and by the end of the 18th century it precipitated an organized search for the "missing planet."

19.1 DISCOVERY OF THE ASTEROIDS

The view that there was another planet was not held by everyone. Georg Wilhelm Hegel (with typical philosophical insight) argued that there could be only seven planets because, among other things, there are only seven openings in the human head. Nevertheless, in 1800 a methodical search for the missing planet was organized by Baron Francis Xavier von Zach. The plan was to divide the zodiac—that band around the sky centered on the ecliptic, and through which the known planets' or-

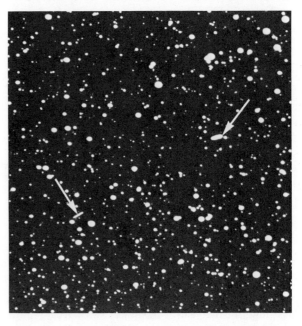

Figure 19.1 Time exposure showing trails left by two minor planets (marked by arrows). *(Yerkes Observatory)*

351

bits lie—into 24 sections and to assign each section to a different European astronomer to search. One of these was the Sicilian astronomer Giuseppe Piazzi.

(a) Discovery of Ceres

Piazzi, however, had not yet been informed of his role in the search and had not received the charts of the region of the zodiac he was to survey. He was, therefore, working independently on an entirely different project when, on January 1, 1801—the first night of the 19th century (1800 is the last year of the 18th century)—he discovered a new object not on his star charts. The next night he observed it again and thought that it had moved; on the third night he was sure of it. Piazzi, like Herschel in his discovery of Uranus, thought he had found a new comet. He observed the new object regularly until February 11, when his work was interrupted by a severe illness. On January 24, however, Piazzi had written to report his new "comet" to others, including Bode in Berlin. Bode received his letter on March 20. The news created a great deal of enthusiasm, and von Zach even published a report that the missing planet between Mars and Jupiter had been discovered! But by the time the astronomers in Northern Europe were aware of the discovery, the new object was too close to the sun to be observed, and it would be September before it should again be visible.

But by summer of 1801, Bode realized that the new planet was lost. Piazzi had observed it for only six weeks, too short a time to calculate its orbit by techniques then available. Fortunately, the discovery was saved by a 23-year-old genius from Brunswick, Germany: Karl Friedrich Gauss. The extraordinary Gauss was intrigued by an account of the discovery he had read in the September 1801 issue of *Monatliche Correspondenz*. He put his other work aside and devoted September and October to devising a new method of orbit calculation that was free of any assumptions about the orbit and that would work with observations made over only a short arc of the total orbit. He finished his calculations in November and sent the results to von Zach. Von Zach caught a glimpse of a possible object on December 7, but bad weather prevented him from confirming the observation until December 31, the last night of the year of the object's discovery; there it was, in the constellation of Virgo, almost precisely where Gauss had predicted it to be. At

Piazzi's request, the object was named *Ceres*, for the goddess of agriculture and protecting goddess of Sicily. The fame earned by Gauss for his efforts in the problem eventually led, in 1807, to his appointment as director of the Göttingen Observatory, where he remained the rest of his life.

(b) Subsequent Discoveries

Ceres was widely assumed to be the missing planet predicted by Bode's law. It came as a complete surprise, therefore, when in March 1802 Heinrich Olbers discovered a second moving starlike object— the minor planet to be named *Pallas*. It was a natural speculation that if there were room for *two* minor planets, there could be room for others as well, and a search for such objects began in earnest. The discovery of *Juno* followed in 1804 and of *Vesta* in 1807. It was 1845 before Karl Hencke discovered the fifth minor planet after 15 years of search. Subsequently, new ones were found with increasing frequency, until by 1890 more than 300 were known.

In 1891 Max Wolf, of Heidelberg, introduced the technique of astronomical photography as a means of searching for asteroids. The angular motion of a minor planet is large enough (especially if it is near opposition) so that during a long time exposure its image will form a trail on the emulsion. The object appears on the photograph, therefore, as a short dash rather than a starlike point image. *Brucia*, the 323rd minor planet to be discovered, was the first to be found photographically.

Today, discoveries of asteroids are usually accidental; they most often occur when the tiny objects leave their trails on photographs that are taken for other purposes. Literally thousands of minor planet trails appear on the photographs taken for systematic surveys of the sky, such as those with the 1.2-m Schmidt telescopes at Palomar and at Siding Spring, Australia. The majority of these trails are of objects that have never been catalogued. Most of them have been ignored, because at least three separate observations of an asteroid are required, preferably separated by intervals of several days or weeks, before its orbital elements can be determined.

(c) Total Number of Asteroids

It would be a formidable task to discover, determine orbits for, and catalogue all the minor planets bright enough to be observed with modern telescopes. Nevertheless, the total number of such objects can be estimated by systematically sampling selected regions of the sky.

Several investigators have estimated the number of observable minor planets by photographing selected regions of the zodiac, within which most minor planets are found. The total number of asteroids bright enough to leave trails on photographs taken with an efficient instrument such as the Palomar Schmidt telescope is estimated to be about 100,000.

(d) Naming the Asteroids

By modern custom, after a newly found asteroid has had its orbit calculated, and has been observed again after another circuit of the sun since its first discovery, it is given a name and a number. The number is a running index that indicates the order of discovery among the minor planets. The discoverer is customarily given the honor of supplying the name. The full designation of the asteroid contains both the number and the name, thus: 1 Ceres, 2 Pallas, 433 Eros, 1566 Icarus, and so on.

Originally, the names were chosen from gods in Greek and Roman mythology. These, however, were soon used up. After exhausting the names of heroines from Wagnerian operas, discoverers chose names of wives, friends, flowers, cities, colleges, pets, and even favorite desserts. One is named for a computer (NORC). The thousandth minor planet to be discovered was named Piazzia, and number 1001 is Gaussia. Other asteroids bear the names Washingtonia, Hooveria, and Rockefellia.

19.2 ORBITS OF ASTEROIDS

A standard reference in which the orbital elements and ephemerides of asteroids are published is the Soviet *Minor Planet Ephemerides*. The 1982 edition lists 2297 minor planets. The number of those that are numbered and named may well reach 4000 by the end of the 20th century.

The asteroids all revolve about the sun in the same direction as the principal planets (from west to east), and most of them have orbits that lie near the plane of the earth's orbit; the mean inclination of their orbits to the plane of the ecliptic is 9°.5. A few, however, have orbits inclined more than 25°;

the orbit of 2102 Tantalus is the most inclined (64°) to the ecliptic.

The main asteroid belt contains minor planets with orbits of semimajor axes in the range 2.2 to 3.3 AU, with corresponding periods of orbital revolution about the sun of from 3.3 to 6 years. The mean value of the eccentricities of the main belt asteroid orbits is 0.15, not much greater than the average for the orbits of the planets.

Some asteroids have orbits rather far outside the main asteroid belt. A few with semimajor axes around 1.9 to 2.0 AU are called the *Hungarias* (for the prototype 434 Hungaria), and a few with larger orbits, near 4 AU, are called the *Hildas* (for 153 Hilda). The Hungarias tend to be relatively small asteroids with orbits of relatively large inclination to the ecliptic (typically, 22°).

There are asteroids with even more extreme orbits, some of which cross the orbit of the earth, and a few that cross the orbit of Jupiter. We shall return to these objects later.

(a) Kirkwood's Gaps

An interesting characteristic in the distribution of asteroid orbits is the existence of several clear areas or gaps (Figure 19.2). These were interpreted in 1866 by Daniel Kirkwood as a resonance phenomenon due to the perturbative effects of Jupiter. The situation is quite similar to that of Cassini's (and other) divisions in Saturn's rings, which were originally ascribed to resonances with Saturn's innermost massive satellites (Section 18.2c). Kirkwood's gaps occur at distances from the sun where asteroids would have periods commensurable with that of Jupiter. For example, a minor planet at about $5/8$ of Jupiter's distance from the sun would have a period exactly half that of Jupiter, and every two times around the sun it would find itself relatively near Jupiter. The repeated attractions toward Jupiter, always in the same direction, would eventually perturb the orbit of such an object in a severe way. Similar Kirkwood gaps occur where asteroids would have periods of one third, two fifths, three fifths, and so on, that of Jupiter.

It is not yet fully understood, however, exactly how the perturbations of asteroids in Kirkwood's gaps affect those objects. Calculations indicate that the orbital eccentricities should increase and produce orbits that are not very consonant with smooth circular orbits about the sun. Viscous forces (which could have been important in the early solar

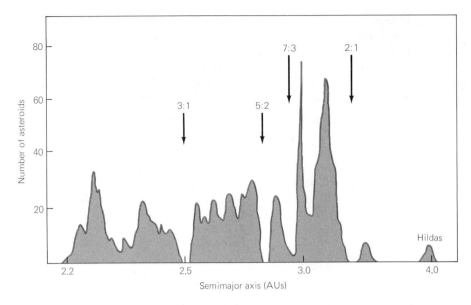

Figure 19.2 The distribution of the number of asteroids of various distances from the sun in astronomical units. Some of the resonances are indicated at gaps (Kirkwood's gaps) where the period of an asteroid would be a simple fraction of the period of Jupiter; for example, an asteroid at about 2.5 AU from the sun would have a period one-third that of Jupiter's. [Adapted from data by R. Greenberg and H. Scholl in *Asteroids*, T. Gehrels (ed.) University of Arizona Press, Tucson, 1979].

system) could result in orbits very different from circular ones in that part of the solar system. So could collisions with other asteroids (which we believe to be an important process in the evolution of asteroids; see below). Some calculations show that perturbations by Jupiter of asteroids near or in the Kirkwood gaps can result in ejecting asteroids to the part of the solar system occupied by the earth.

(b) Families

In 1917 the Japanese astronomer K. Hirayama found that a number of the asteroids fall into "families" or groups of similar orbital characteristics. He hypothesized that each family may have resulted from an explosion of a larger body or from the collision of two bodies. Slight differences in the initial velocities given the fragments of the explosion or collision account for the small spread in orbital characteristics now observed for the different asteroids in a given family.

In 1950 Dirk Brouwer extended Hirayama's investigation to the study of 1537 asteroids, and he found evidence for 29 families. He accounted for the differences among the orbital elements of members of the same family by the effect of perturbations of Jupiter and Saturn. (In other words, by working backwards, he could "reunite" the families.) Today, about half of the minor planets of well-determined orbits are identified as belonging to about 100 different families. In fact, three of the largest families contain 10 percent of the asteroids of known orbits.

Collisions among a relatively few larger objects would be extremely rare, but once they occurred subsequent collisions, because of the larger number of fragments, would be more likely. Presumably, the fragmentation of asteroids has been going on for a long time, and is continuing.

(c) The Trojans

Lagrange's solution to a particular case of three bodies (Section 6.1b) shows that there should be two points in the orbit of Jupiter at or near which an asteroid can remain almost indefinitely. These are the two points which, with Jupiter and the sun, make equilateral triangles (Figure 19.3). It seemed possible, therefore, that at either or both of the two points there could be located minor planets circling

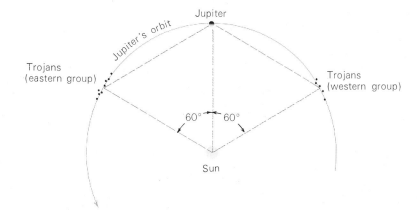

Figure 19.3 Locations of the Trojans.

the sun in the same orbit as Jupiter, and with the same period. From 1906 to 1908, four such asteroids were found. The number has increased to 22 that are numbered and listed in the *Minor Planet Ephemerides* in 1981. These asteroids are called the *Trojans*; they are named for the Homeric heroes. Those that are preceding Jupiter (that is, ahead of it in its orbit) are named for the Greek heroes (with the exception of the Trojan spy, 624 Hektor), and those following Jupiter are named for the Trojan warriors (with the Greek spy, 617 Patroclus among them). To a first approximation, the Trojan asteroids circle the sun with Jupiter's period of 12 years, only one sixth of a cycle ahead of or behind the Jovian planet. Their detailed motion, however, is very complicated; they slowly oscillate around the points of stability found by Lagrange, some of their oscillations taking as long as 140 years.

A recent search for Trojan asteroids has revealed that there are actually many of them—far more than have been catalogued. An extrapolation of the numbers actually found to take account of the small area surveyed shows that there must be close to 700 Trojan asteroids in the region preceding Jupiter (the Greek camp) that are at least 15 km in diameter and about 200 in the region following Jupiter (the Trojan camp).

(d) The Distant Asteroids

There are a few asteroids whose orbits carry them even farther than Jupiter from the sun. Until recently the most distant known was 944 Hildago, with a semimajor axis of 5.85 AU and a very large eccentricity of 0.66, which carries it, at aphelion, just beyond the mean distance of Saturn. However, in 1977 Charles Kowal, at Caltech, discovered a more remote object, now known as 2060 Chiron.

Chiron is, to this date, the most remote small object known that is circling the sun, except comets. Indeed, as we shall see, it is quite possible that Chiron is itself an old comet nucleus. It does not display the characteristics of comets (next chapter), however, so it is numbered and named as an asteroid. Its semimajor axis is 13.7 AU, and its orbit has an eccentricity and inclination of 0.38 and 6°9, respectively. At perihelion, it is 15 percent closer to the sun than is Saturn, but at aphelion it goes out nearly to the orbit of Uranus.

(e) Earth-Crossing Asteroids

There is a class of asteroids whose orbits are most remarkable, in that they either come close to or

Figure 19.4 The most remote known asteroid, Chiron, photographed 19 October 1977 with the Palomar Schmidt telescope. The bright streak in the center (marked with arrow) is the trail left by Chiron during the 75-minute time exposure. The faint, irregular streak below and to the right of Chiron is an edge-on galaxy. At the upper right is a faint trail of a "normal" asteroid. *(Courtesy Charles Kowal, Caltech/Palomar Observatory)*

cross that of the earth. Some of these are the nearest approaching natural celestial objects, save the moon and meteorites. They are divided into three groups: the *Atens* (for 2062 Aten) have orbits that cross the orbit of the earth, but lie wholly within the orbit of Mars. Four Atens are known at this writing. The *Apollos* (for 1862 Apollo) are the most famous and include 1566 Icarus, discovered at Palomar by Baade in 1949, and 1620 Geographos, named for the National Geographic Society, which financed the Palomar Sky Survey on which the little world was discovered in 1951. The Apollos are objects that cross both the orbits of the earth and of Mars; 23 are catalogued. The third group, 20 in number, are the *Amors*, for 1221 Amor, which cross the orbit of Mars but do not, at present, quite come as close as the earth's orbit. All these objects, however, are subject to rather severe perturbations by the terrestrial planets, so that while the Amors may not be earth-crossing at the present time, they may well have been in the past and may be again.

In all, it is estimated that there are about 1300 earth-crossing asteroids larger than one kilometer in diameter belonging to these three classes, if we could survey the entire sky to search for them. We think that they have been deflected to the inner part of the solar system by collisions with other asteroids in the main belt and by subsequent perturbations by major planets. Some earth-crossing asteroids have been observed at their near-earth passes but never recovered. The most famous of these is Hermes, which must have passed very close to earth in 1937. Icarus missed our planet by only 6.4 million km on June 14, 1968, and Geographos by only 10 million km in 1969. Under favorable circumstances, Geographos can come as close as Icarus did in 1968. Can any of these worlds collide with the earth? We shall return to this question in Section 19.5.

19.3 PHYSICAL NATURE OF THE ASTEROIDS

When the asteroid Vesta is at opposition while it is also at perihelion, it is faintly visible to the unaided eye. With this single exception, the minor planets are all telescopic objects. At its brightest, Ceres, the largest asteroid, is at magnitude 7.6, about four times too faint to see without optical aid. Most minor planets require time-exposure photographs with large telescopes to record. It is easy to understand why it is not easy to learn much about the true nature of the minor planets.

(a) Sizes of Asteroids

There are several methods of determining the sizes of minor planets. A few are near and large enough to present measurable disks in the telescope; in those cases we can, at least in principle, measure their true (linear) diameters from the measurements of their angular diameters, as explained in Section 9.2b. Unfortunately, however, even the largest-appearing minor planets have angular sizes of no more than 0.7 arc seconds—an angle too hard to measure with precision because of the turbulence of the earth's atmosphere. The observations, therefore, are difficult, and the results are not reliable. Some improvement is possible with the technique of speckle interferometry (Section 11.4b), and it has been tried on a few asteroids.

Perhaps the most accurate technique for measuring the size of a minor planet is timing how long it takes to pass in front of a star. Several of the larger minor planets occult stars from time to time, and recently there has been considerable effort to coordinate observations over a large geographical area to observe such occultations. From different places on earth, different parts, or chords, of an asteroid will appear to pass in front of the same remote star because the various observers see the asteroid in slightly different directions compared with the direction to the star. At least in principle, the combination of many observations of an occultation should make possible the solution for a pretty accurate value for the size of the relatively nearby asteroid. At this writing, at least 9 asteroid diameters have been measured by this technique, and for two, Pallas and Juno, values so derived are accurate to within a few percent. The occultation of a star by Juno on December 11, 1979 was widely observed; Juno was found to have an approximately elliptical cross section, of major and minor axes 299 and 246 km, respectively.

The method that works best for most asteroids is to compare their brightnesses in visible light (which is reflected sunlight) with the light they emit in the infrared, which energy, of course, they have previously absorbed from the sun. For a particular asteroid, we know its distance from the sun and therefore how much sunlight falls on each unit area of its surface. If it is assumed to be spherical, and of radius R, the total sunlight it intercepts is equal

Table 19.1   Diameters of the Largest Asteroids

ASTEROID	DIAMETER (km)	SEMIMAJOR AXIS OF ORBIT (AU)
1 Ceres	1025	2.767
2 Pallas	583	2.772
4 Vesta	555	2.362
10 Hygeia	443	3.136
704 Interamnia	338	3.059
511 Davida	335	3.183
65 Cybele	311	3.434
52 Europa	291	3.098
451 Patientia	281	3.064
31 Euphrosyne	270	3.145
3 Juno	267	2.671
15 Eunomia	261	2.643
324 Bamberga	256	2.683
107 Camilla	252	3.488
87 Sylvia	251?	3.482
45 Eugenia	250	2.719

to its cross-sectional area, πR^2 times that incident flux per unit area. Of that intercepted light, the asteroid reflects part, a fraction A (called the *albedo*), and hence absorbs the rest, the fraction $1 - A$. The asteroid reemits the energy it absorbs at wavelengths appropriate to its temperature, and hence in the infrared (Section 10.2). If we measure the infrared radiation coming from the minor planet, we are recording how much energy it must have absorbed from the sunlight falling upon it. If we compare this measure with that of the light reflected from the object, we find what its albedo—reflecting power—is. Then we can easily calculate what its size must be to account for the amount of light it reflects to us, that is, its observed brightness.

Such measurements have been made on hundreds of asteroids. We find that their albedos are rather low, ranging from about 3 percent to about 25 percent. These worlds are pretty poor reflectors. Table 19.1 gives the diameters of the largest known asteroids, derived from the latter technique in most cases, but sometimes from others (for example, the diameter of Pallas comes from occultation observations). Table 19.2 gives a summary of the size distributions of the known asteroids.

Table 19.2   Distribution of Asteroid Sizes

DIAMETER	NUMBER
More than 200 km	30
More than 158 km	58
More than 100 km	195
More than 63 km	463
More than 32 km	1100

Table 19.3   Masses and Densities for the Largest Asteroids

ASTEROID	MASS Grams	MASS Earth = 1	DENSITY (g/cm³)
Ceres	1.2×10^{24}	.00020	2.3
Pallas	2.2×10^{23}	.000036	2.6
Vesta	2.6×10^{23}	.000044	3.3

These data have been adapted from a review by B. Zellner, published in *Asteroids*, an authoritative compendium on these objects edited by T. Gehrels (U. of Arizona Press, 1979).

(b) Masses and Densities of Asteroids

Only for three of the largest asteroids—Ceres, Pallas, and Vesta—have the masses been estimated from the perturbations they produce on other asteroids. For these three objects, the derived masses, probably accurate to within 10 or 20 percent, and their densities (found from the sizes given in Table 19.1) are those listed in Table 19.3 These are adapted from data by J. Schubart and D. L. Matson in *Asteroids*.

The total mass of all the minor planets is estimated at about 0.0005 times that of the earth, or 3 $\times$ 10^{24} g. Of this, about 0.0002 times the earth's mass is tied up in Ceres itself.

(c) Rotation and Shapes of Asteroids

In 1901 it was observed that the asteroid Eros varies in brightness by about a factor of five in a period of 5^h16^m. These changes could be explained easily by the hypothesis that Eros has an elongated shape, like that of a brick, and that it rotates in a period of 5^h16^m about an axis roughly perpendicular to its long dimension. It is probably about 20 to 30 km long and about 8 km wide and thick.

Light variations attributable to rotation have now been observed for many asteroids—nearly 200 by 1979. The rotation periods range from 2.5 to 85 hours; typical periods are about 8 hours for the larger asteroids. The natures of the light variations also give some hint about the shape of the asteroid, as in the case of Eros, just mentioned.

Any variation in reflectivity or irregularity in shape causes the greatest variation in the amount of light an asteroid reflects to us if its axis of rotation is precisely perpendicular to our line of sight. Obviously, if its axis is directed pole-on to the earth, we should see no variation at all, or very little, depending on the direction of the solar illumination. Consequently, when we observe the light variations of an asteroid over a period of many months, during which time the orientation of its rotational axis with respect to the earth changes, the amplitude of its variations in light will change as well. Analysis of such data gives information about the orientations of rotational axes of asteroids. It appears that their polar axes are oriented at random in space.

Some further hint about the shapes and natures of small asteroids can be inferred from the appearance of two other small bodies of asteroidal mass and size—Phobos and Deimos, the satellites of Mars. These little moons, of maximum diameters 27 and 15 km, respectively, could conceivably even be onetime asteroids that were captured by Mars. Both are dark, are quite irregular, are heavily cratered, and spin with periods comparable to those of asteroids (7^h35^m for Phobos and 30^h17^m for Deimos). Observations of the Martian moons with resolution like that with which we observe asteroids would suggest that they are very similar, but high-resolution Viking photographs reveal that their surfaces are quite different in detail. The Phobos/Deimos experience shows us what tiny asteroids could look like, and suggests that the surfaces of minor planets could be far more interesting than would seem from existing data.

(d) Types of Asteroids

Something about the probable surface material on asteroids can be learned from the spectra of their reflected light, which have now been observed for several hundred minor planets. The majority are classed, according to their spectra, into three groups: C (the most common type), S and M. A few other classes have been used as well, but these, along with the "unclassifiables," comprise less than 10 percent of the total.

The C (carbonaceous) types are very dark, with albedos of only 3 to 4 percent. The low reflectivity is expected of a surface covered with black dust. Most C-type asteroids are thought to be carbonaceous chondrites. The S (silicaceous) types are brighter and redder and have spectra suggesting pyroxene and/or olivine mixed with nickel and iron. These are thought to be composed of comparable amounts of silicates and metals. The M (metallic) types are probably composed largely of iron and nickel. This interpretation of asteroid spectra is neither precise, 100 percent consistent, nor thoroughly established, but in a general way represents the best-informed guess. It is also suggestive in that these chemical structures correspond to those of the most common meteorites, which are believed to be of asteroidal origin (Chapter 21).

(e) Satellites of Asteroids

Occasionally asteroids occult (pass in front of) distant stars. The times of such occultations are calculated and announced in advance by the United States Naval Observatory. An active group of amateur astronomers cooperates with professionals by observing such events. Thus on June 7, 1978, J. H. McMahon observed an occultation of the star SAO 120774 by the minor planet 532 Herculina. The star disappeared, as expected, for 20.6 s while the 220-km diameter Herculina passed in front of it. But in addition, McMahon observed the star to dim six other times within 2 minutes of the main occultation, for periods that ranged from 0.5 to 4 s.

The longest secondary dimming was reported to be confirmed by photoelectric records obtained by astronomers at the Lowell Observatory in Arizona. It is interpreted to be caused by a 50-km diameter satellite at a distance of about 1000 km from 532 Herculina.

Calculations by R. P. Binzel and T. C. van Flandern show that satellites can have stable orbits up to a distance of 15,000 km from the centers of the larger asteroids, and those astronomers suggest that satellites of minor planets might be quite common. They have therefore checked through old records for indications of secondary dimmings associated with other occultations that are similar to the dimmings observed by McMahon, and have compiled a list of seven additional asteroids with a dozen probable satellites among them and ten other suggested satellites. It is possible that many minor planets have one or more satellites in revolution about them.

Whether satellites of asteroids are common, however, or even whether there is a single well-documented case, is still controversial. The best example is 532 Herculina, and it was low in the sky when the occultation was observed. Many asteroid specialists are withholding judgment on the question of satellites, pending further verification.

 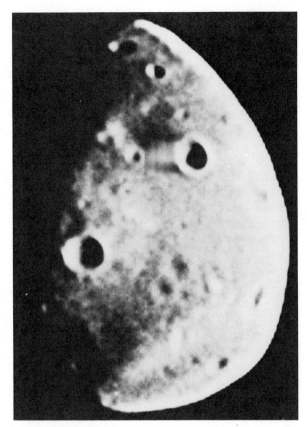

(a) (b)

Figure 19.5 *The Martian satellites Phobos (a) and Deimos (b). (NASA/JPL)*

19.4 ORIGIN AND EVOLUTION OF ASTEROIDS

At one time it was thought that the asteroids were the debris of a former planet that had exploded. There are, however, a number of problems with this theory, not the least of which is the mechanism that was responsible for the alleged explosion. Moreover, as we have seen, the estimated combined mass of all minor planets is far less than that of any of the planets—even tiny Pluto. The prevailing view today is that the minor planets are objects that never accreted to a full-sized planet, perhaps because there was too little material in that part of the solar nebula, or perhaps because of tidal disruption by Jupiter.

On the other hand, there are lots of reasons to expect the asteroid population to be undergoing evolution. Some of the bodies are perturbed out of the main belt by Jupiter, especially, it is thought, those that approach Kirkwood's gaps. The detailed orbital mechanics are not yet fully understood, but we do know that over geologic time scales earth-crossing asteroids must collide with terrestrial planets, not to mention the influx into the inner solar system of the very tiny particles that land on earth to become meteorites. If a steady state is maintained, it is estimated that about 15 Apollo-type asteroids must be brought into the inner solar system every million years or so.

We have already seen that many or most minor planets fall into families of similar orbits; presumably they are the fragments of collisions between asteroids. The collision process must be going on constantly, making little asteroids out of big ones. Quite possibly early in the history of the solar system there were relatively few rather large asteroids from which collisions produced the present huge population of them. In addition, all of the things that can happen to alter orbits of minor planets— collisions and gravitational perturbations—must be operative and must gradually be diffusing those orbits into greater ranges of size, eccentricities, and inclinations.

Finally, it is very probable that a few asteroids are very different kinds of objects from the vast majority. In the following chapter we discuss comets and shall see that they contain small, icy nuclei. In a comet ordinarily observed we see a huge glowing gas cloud (coma), which originates from the evaporation of the outer layer of its nucleus. But planetary perturbations, especially by Jupiter, alter cometary orbits drastically, and some comets can end up in orbits of only moderate eccentricity—not unlike those of asteroids. If such comets are far enough from the sun, they may never warm up enough to evaporate and may remain as ice masses masquerading as asteroids, and we would not know the difference. It is generally thought that the remote minor planets Hildago and Chiron may be such inactive nuclei of comets.

19.5 COLLISIONS BETWEEN ASTEROIDS AND THE EARTH

Iridium has an abundance in the solar system that is significantly higher than that of such familiar elements as gold and silver; it is comparable in abundance to iodine. In particular, it is found in meteorites—presumably of asteroidal origin (Chapter 21). Yet it is fairly uncommon in the earth's crust because, it is thought, in the differentiation process in the earth (wherein certain heavy elements, especially iron, sank to the earth's center) iridium separated from material in the outer mantle and crust.

However, a University of California, Berkeley, team of scientists, headed by Nobel physicist Luis Alvarez, has found that in various sites throughout the earth, sedimentary layers 65 million years old are some 30 times richer in iridium than the layers below and above them, or in the crust in general. The Berkeley group suggests, as explanation, that about 65 million years ago the earth was struck by a large earth-crossing asteroid. The resulting explosion would send the pulverized debris of the asteroid (along with a good deal of the earth's crust!) high into the stratosphere, where it could circulate over the entire surface of the earth and gradually filter to the ground, depositing matter relatively rich in iridium (present in the asteroid) on the surface, to be buried as a permanent record of the catastrophic event. The asteroid would probably have to have hit in the ocean, for the expected crater—somewhat more than 100 km in diameter—has not been found.

This scenario could well be correct, so far as we know. In fact, it is one of the favored theories

suggested to account for the excess iridium. But would such an impact have significance for life on earth? Perhaps it would. It is well known that there is a gap in the fossil record beginning about 65 million years ago, which suggests that nearly 60 percent of the plant and animal species perished rather suddenly at that time, in particular the great dinosaurs. It is known as the Cretaceous–Tertiary extinction, after the geological periods before and afterward. Alvarez and his team point out that the impact of so large an asteroid would place enough dust in the stratosphere to cut the illumination of sunlight filtering to the ground to less than that of full moonlight for a year or more. Under these circumstances, most photosynthesis would cease, and the world's food supply would drop enough that many species would perish.

The theory that a collision of an asteroid with the earth caused the extinction of the great animals of that time is by no means generally accepted. There are even other ways of pouring a large amount of dust into the atmosphere—a great volcanic eruption, for example. Nor is it known just how quickly the dinosaurs perished. Yet it is a reasonable hypothesis, and it raises the question of future asteroidal impacts on the earth.

We have enough data on the earth-crossing asteroids to make meaningful estimates. An excellent summary by Caltech scientists E. M. Shoemaker, J. G. Williams, E. F. Helin, and R. F. Wolfe appears in Gehrels' comprehensive volume, *Asteroids*, referred to previously. These authors estimate that the present rate of collisions with asteroids a kilometer or more in diameter should be about 3.5 per million years. Such objects produce craters on the earth 10 km or more across. Collisions with larger asteroids, of the sort suggested by the Alvarez team to account for the iridium-rich sediment, would be much rarer, perhaps one every 100 million years or so.

The destruction and size of the crater produced depend on the size and density of the asteroid and on its velocity at impact. Here I take rather typical values for density (3 g/cm^3) and velocity (12.5 km/s) to compute the energy released in the impact explosion and the size of the crater produced (Table 19.4). The energies are in units of megaton nuclear bombs.

It is clear from the table that the impact of a good-sized minor planet on the earth could be disastrous if the hit were in a populated area; it could probably wipe out a country the size of Switzer-

Table 19.4   Typical Asteroid Impact Energies and Crater Sizes

DIAMETER OF ASTEROID (km)	APPROXIMATE ENERGY OF IMPACT (Megaton Bombs)	APPROXIMATE DIAMETER OF CRATER (PRODUCED ON EARTH) (km)
1	2.9×10^4	10
2	2.3×10^5	20
3	7.9×10^5	30
5	3.7×10^5	50
10	2.9×10^7	90
15	9.9×10^7	125
20	2.3×10^8	160

land, to say nothing of the possible problems with the dust thrown into the atmosphere. Some inves-tigators believe that the dust produced by the impact of a 10-km asteroid could cause climatic changes great enough that most of us would die. T. Gehrels, at the University of Arizona, and E. Shoemaker, at Caltech, are preparing a "Spacewatch Camera," with which they hope to study asteroids that pass near the earth and obtain good enough orbital data for them to learn well in advance of potential collisions with our planet. Given enough time, those dangerous orbits could be changed.

On the other hand, we do not need asteroids to wipe us out. Indeed, in all probability most of us would survive the impact of a typical small asteroid. Far more likely to destroy us is the world's stockpile of nuclear weapons. It is hard to see how many life forms could survive an all-out nuclear war, let alone our modern civilization.

EXERCISES

1. Minor planets can be discovered by the trails they leave on astronomical photographs. Fainter objects could be recorded, however, if their images were points rather than trails. Explain how you might plan a photographic search for faint minor planets.

2. What would be the period of a minor planet whose distance ranges from 2.5 to 3.5 AU from the sun? *Answer:* About 5.2 years

3. How far from the sun would an asteroid be if it had a period one-third as long as that of Jupiter? Would such an object long remain in a stable orbit? Why? *Answer:* About 2.5 AU

4. Two asteroids of identical albedo are observed. One is 64 times as bright as the other. Both are at the same distance from the sun and from the earth. The brighter has a measured diameter of 240 km. What is the diameter of the other?

5. What would be your answer to Exercise 4 if the two minor planets were equally far from the sun, but if the fainter were twice as far from the earth as the brighter?

6. If you weigh 100 kg on the earth, what would you weigh on the surface of an asteroid that has a mass of about 1/10,000 that of the earth and a diameter 1/20 that of the earth? *Answer:* 4 kg

7. Assume that a minor planet rotates on its axis from west to east and that it is irregular in shape, so that its brightness varies. Assume, further, that its period of rotation is precisely constant. Will the *observed* light fluctuations be at a constant rate?

Can they ever be at a greater or lesser rate than the sidereal period of rotation of the object? Explain.

8. At times Eros fluctuates in brightness by a factor of about five. At other times its light fluctuates by a much smaller amount. Can you offer an explanation for this phenomenon?

9. The *semiminor axis*, b, of an ellipse is related to the major axis, a, and the eccentricity, e, by

$$b = a(1 - e^2)^{1/2}.$$

a) What is the eccentricity of an ellipse in which b is exactly half of a?
b) What is the semiminor axis of the orbit of Chiron?
   Answer: a) $e = 0.866$
   b) 12.67 AU

*10. If the minor planet Ceres has a diameter of 1000 km and a mass of 1/5000 that of the earth, verify that the escape velocity from its surface is about 0.4 km/s.

11. The total energy emitted by an asteroid in the infrared is four times as great as the amount of sunlight it reflects. What is its albedo?

12. The formula for kinetic energy is $mv^2/2$. If the mass, m, is in grams and the speed, v, in cm/s, the energy is in ergs. The energy from the detonation of one ton of TNT is 4×10^{16} ergs. How massive an asteroid, impacting the earth at 12 km/s, hits with the energy of a one-megaton bomb? If its density is 2 g/cm³, and if it is spherical in shape, what is its diameter? *Answer:* 5.6×10^{10} g = 5.6×10^4 tons; 3.8×10^3 cm = 38 m

Edmund Halley (1656–1742), English astronomer, was a close friend of Isaac Newton and oversaw the publication of Newton's *Principia*. Halley was active in many areas of astronomy, but he is best known for his pioneering study of comets and for the realization that the comets of 1531, 1607, and 1682 were the same one. He predicted the return of that comet in 1758. This most famous of all comets bears Halley's name. (*Yerkes Observatory*)

COMETS: NATURE'S SPACE PROBES

Comets have been observed from the earliest times; accounts of spectacular comets are found in the histories of virtually all ancient civilizations. Yet, until comparatively recently, comets were not generally regarded as celestial objects.

A typical comet that is bright enough to be conspicuous to the unaided eye has the appearance of a rather faint, diffuse spot of light, somewhat smaller than the full moon and many times less brilliant. There may be a very faint nebulous tail, extending for a length of several degrees away from the main body of the comet. Like the moon and plants, comets slowly shift their positions in the sky from night to night, remaining visible for periods that range from a few days to a few months. Unlike the planets, however, most comets appear at unpredictable times. In medieval Europe, comets were usually regarded as poisonous vapors in the earth's atmosphere and as bad omens. More fear and superstition have been attached to comets than to any other astronomical objects. Even in 1973 there were dire predictions of disaster when Comet Kohoutek passed near the sun, and numerous books about comets written by cranks and charlatans were on sale in bookstores.

20.1 EARLY INVESTIGATIONS

Perhaps the first scientific investigation of a comet, based on careful observation, was Tycho Brahe's

diligent study of the brilliant comet that appeared in 1577 (Section 3.2a). Had the comet been inside the earth's atmosphere, as was then generally supposed, changes in its apparent direction would easily have been detectable to an observer who changed his position by several kilometers. Brahe, not being able to detect any such changes, realized that the comet was a celestial object. Moreover, his failure to detect any diurnal parallax of the comet led him to the conclusion that it was at least three times as distant as the moon and that it probably revolved about the sun.

Kepler described in detail the comet of 1607 (later known as Halley's comet). He held the view that this and other comets are celestial bodies that travel in straight lines through the solar system. Two comprehensive treatises were published in 1654 and 1668 by John Hevel. These works contained systematic references to all known comets and proved highly valuable to later investigators.

(a) Newton and Comets

When Newton applied his law of gravitation to the motions of the planets, he wondered whether comets might similarly be gravitationally accelerated by the sun. If so, their orbits should be conic sections. If comets, like planets, had nearly circular orbits, they should be visible at regular and frequent intervals. On the other hand, if a comet moved in an elongated elliptical orbit of large size, it would nec-

Figure 20.1 Halley's comet in 1066, as depicted on the Bayeux tapestry. (*Yerkes Observatory*)

essarily be visible during only that relatively brief period of time during which it passed near the sun (perihelion); over most of its orbit the comet would be so far from the sun as to be invisible. Furthermore, the periods of comets moving in such orbits would be very great. A comet that had been seen for a short time, and then was seen again many tens or hundreds of years later when it next appeared near perihelion, would naturally be mistaken for a new object. One end of an ellipse, of very great length and of eccentricity near unity, is nearly indistinguishable from a parabola; thus the motion of a comet moving on a parabolic orbit would closely resemble that of a comet whose orbit is a very long ellipse. Newton concluded that comets are gravitationally attracted to the sun and move about it in orbits that are either very long ellipses or parabolas.

(b) Halley's Comet

Edmond Halley greatly extended Newton's studies of the motions of comets. In 1705 he published calculations relating to 24 cometary orbits. In particular, he noted that the elements of the orbits of the bright comets of 1531, 1607, and 1682 were so similar that the three could well be the same comet, returning to perihelion at average intervals of 76 years. If so, he predicted that the object should return about 1758.

Alexis Clairaut calculated the perturbations that this comet should experience in passing near Jupiter and Saturn and predicted that it should first make its appearance late in 1758 and should pass perihelion within about 30 days of April 13, 1759. The comet was first sighted by an amateur astronomer, George Palitzsch, on Christmas night, 1758. It passed perihelion on March 12, 1759, 31 days earlier than Clairaut's calculations predicted. The comet has been named *Halley's comet,* in honor of the man who first recognized it to be a permanent member of the solar system. Subsequent investigation has shown that Halley's comet has been observed and recorded on every passage near the sun at intervals from 74 to 79 years since 239 B.C. The period varies somewhat because of perturbations upon its orbit produced by the Jovian planets. Halley's comet last appeared in 1910, and its next perihelion passage is February 9, 1986. The comet passes fairly close to the earth twice, first on November 27, 1985 (before perihelion) at a distance of 0.62 AU, and again on April 11, 1986 (after perihelion) at a distance of 0.42 AU. For northern hemisphere observers, the comet is more favorable for viewing before perihelion (around November 1985), for it will then be fairly high in the sky and above the horizon much of the night. After perihelion, United States observers will find the comet low in the sky in the predawn hours.

It is impossible to predict accurately how bright the comet will be. The best guess, though, is that it will prove to be a disappointment to most lay people, who have heard about Halley's comet since childhood. It may be faintly visible to the naked eye on moonless nights away from metropolitan

Figure 20.2 Halley's comet. *(Yerkes Observatory)*

areas, but those who live in cities may even have difficulty sighting it with binoculars. It is not impossible, but is unlikely, that it will experience a flareup while it is close to us, making the comet a spectacular object.

The Europeans plan to intercept Halley's comet with a space probe. If successful, it will be very exciting, for it will be our first opportunity (and a rare one at that) to explore at close range one of the oldest, most numerous, and most mysterious objects in the solar system. The United States also had plans for a probe to Halley's comet, but at least at this writing those plans were scrubbed, because the United States cannot afford such a venture.

20.2 DISCOVERY

Observational records exist for about a thousand comets. Today, new comets are discovered at an average rate between five and ten per year. Some of

these are discovered accidentally on astronomical photographs taken for other purposes. Many are discovered by amateur astronomers.

Most of the new comets found each year never become conspicuous and are visible only on photographs made with large telescopes. Every few years, however, a comet may appear that is bright enough to be seen easily with the unaided eye. In 1957 there were two conspicuous naked-eye comets (Comet Arend-Roland and Comet Mrkos). Other recent bright comets were Comet Bennett in 1970, Comet Kohoutek in late 1973 and early 1974, and Comet West in 1976. About two or three times each century, there appear spectacular comets that reach naked-eye visibility even in daylight. The first "daylight comet" to appear since the brilliant comet of 1910 (which preceded the more famous Halley's comet by a few months), was Comet Ikeya-Seki in 1965.

(a) Naming Comets

A newly found comet is named for its discoverer or discoverers (there are often more than one). It is also given a temporary designation consisting of the year of its discovery followed by a lower-case letter indicating its order of discovery in that year. For example, Comet 1975a is the first comet discovered in 1975, 1975b is the second, and so on. Later, when the orbits of all recently observed comets have been calculated, each of the comets is given a permanent designation consisting of the year in which it passed nearest the sun, followed by a Roman numeral designating its order among the comets that passed perihelion during that particular year. For example, Comet 1975 I would be the first comet to pass perihelion in 1975; 1975 II, the second, and so on.

(b) Comet Kohoutek

One of the best-studied comets was discovered by the Czech astronomer Lubos Kohoutek in 1973. The story of its discovery is rather interesting. In 1971, Kohoutek, working with the Schmidt telescope at the Hamburg Observatory, had tried to find a remnant of Comet Biela (Section 20.3c). He found no trace of Biela's comet, but on three photographs he discovered 50 new minor planets. Some of the new minor planets were in a favorable part of the sky for observation in early 1973, and to study

Figure 20.3 Comet Kohoutek in a mosaic of two photographs made with the Palomar Schmidt telescope on January 15, 1974. (*Caltech/Palomar Observatory*)

them Kohoutek took a number of photographs in February and March. On a plate taken on February 27, he discovered a new comet, 1973e. He did not, however, have the opportunity to inspect all his plates immediately, and it was March 18 before he got to the ones taken on March 7. On one of these, very much to his surprise, he found a second comet—to become the famous Comet Kohoutek, 1973f.

The great importance of Comet Kohoutek is that it not only passed very close to the sun (about 0.14 AU) on December 28, but also that it was discovered when it was still 5.2 AU from the sun, more than nine months before its perihelion passage, which enabled astronomers to plan detailed observing programs. It was studied from ground-based observatories all over the world, from an airplane, from rockets, from NASA's Skylab, and even with instruments on the Venus-Mercury probe, Mariner 10. Moreover, it was the first comet for which radio observations were obtained. Comet Kohoutek was barely visible to the unaided eye in late December and early January 1974, although it did not become as bright as some early predictions had led many to expect. Its visual faintness, however, in no way detracted from its astronomical importance.

20.3 ORBITS OF COMETS

The orbit of a comet, like that of a planet, can be determined from three or more fairly well spaced observations of its position in the sky among the stars. Most comet orbits are indistinguishable from parabolas. Some are definitely elliptical; a very few comets have been observed whose orbits appear to be slightly hyperbolic.

(a) Membership in the Solar System

Because the orbits of most comets appear to be parabolic, and a few even hyperbolic, the question arises whether all comets are members of the solar system or whether some might be accidental intruders from interstellar space. The evidence is conclusive, however, that comets have always been members of the solar system.

If comets were intruders from interstellar space, their orbits should nearly all be markedly hyperbolic. An elliptical orbit is possible only for a body permanently revolving about the sun, not for one coming from outside the solar system. Even a parabolic orbit would be possible only for a body that, before being attracted toward the sun, was moving through space in almost exactly the same direction and at the same speed as the sun, that is, which was *at rest with respect to the sun*. Such motion would be extremely improbable for an interstellar visitor.

Those few comets that do appear to have moved on slightly hyperbolic orbits are believed to have approached the sun on nearly parabolic orbits that were perturbed when the comets passed near Jupiter or another planet. E. Strömgren investigated the motions of 20 such "hyperbolic" comets and has found that in every case the gravitational influences of the planets had changed the orbit from a nearly parabolic one to a hyperbola. Although a comet, through planetary perturbations, may thus escape from the solar system, there is no case where a comet is known to have *approached* the sun on a significantly hyperbolic orbit.

Moreover, the orbits of comets, unlike those of planets, are oriented at random in space. As many comets appear to approach the sun from one direction as from another. If, however, comets were interstellar objects, there should be a preponderance of them approaching from the direction of Hercules, toward which the sun is moving with a speed of about 20 km/s (see Chapter 23).

It is generally accepted, therefore, that comets are members of the solar system, and that they approach the sun on elliptical orbits, most of which are extremely long and have eccentricities near unity. A comet can be observed from the earth only when it traverses that end of its orbit where it is near perihelion. If the orbit has an eccentricity greater than about 0.99, that part of it which can be observed is often indistinguishable from a parabola. Thus "parabolic" orbits observed for comets are really the ends of very long elliptical orbits that bring comets from the farthest reaches of the solar system, tens of thousands of astronomical units away. Such comets have periods that range into millions of years.

Calculations show that observed comets with such highly elongated orbits must be approaching the sun for the first time. To have an orbit so nearly parabolic, a comet must, at each point in its orbit, travel almost exactly with its velocity of escape from the solar system. If, however, it comes within 5 or 10 AU from the sun, the perturbative effects of Jupiter are sufficient to change its speed enough that it subsequently speeds up slightly and escapes the solar system on a hyperbolic orbit, or slows down and switches to a much smaller orbit.

Indeed, some comets—about 100—have orbits of low enough eccentricities to determine their semimajor axes from the relatively small portions of the orbits that are actually observed. These comets, whose periods can be well determined, are called *periodic comets.*

(b) Periodic Comets

Only those comets whose orbits are well determined and which differ rather substantially from parabolas are technically classed as periodic. The comets of longest definitely established periods are Pons-Brooks (71 years), Halley (76 years), and Rigollet (151 years). The comet of the shortest known period is Comet Encke (3.3 years).

There are about 45 known comets whose orbits are inclined at less than 45° to the ecliptic, that travel from west to east (like the planets), and whose aphelion distances are near Jupiter's mean distance from the sun. They comprise *Jupiter's family of comets.* They have periods that range from five to ten years. They are believed to be comets that first approached the sun on very long orbits (like those of most comets) but were changed into relatively small orbits as a result of perturbations that occurred when the comets passed close to Jupiter.

(c) Disintegration of Short-Period Comets

Comets are such flimsy objects that they often suffer severe damage as the result of tidal forces produced by the sun when they pass perihelion. Some have been observed to split into two separate comets. Comet Biela, for example, with a period of seven years, was discovered in 1772. During its approach to the sun in 1846, it was observed to break into two separate comets. Both comets returned on about the same orbit in 1852, but neither has been seen since. However, spectacular meteor showers (see Chapter 21) were observed on November 27, 1872, and on November 27, 1885, on which occasions the earth passed through the orbit of Comet Biela and encountered swarms of particles traveling in the path of the disintegrated comet. The orbit of the swarm has subsequently been altered by Jupiter; consequently, spectacular displays of Bielid meteors have not been observed in the present century.

Another comet that broke up during perihelion passage was Comet West in 1976. Its nucleus was normal when last seen before the comet disappeared in the glare of the sun, but when it again became visible in the hours before sunrise, its nucleus was multiple. The several bright spots could easily be seen, even with binoculars.

We shall see that the gases in comets result from solar heat that evaporates their nuclei. It is estimated that most comets would completely evaporate in this manner after a few hundred or a few thousand perihelion passages. Thus, because of their high mortality, periodic comets cannot have been periodic for long but must originally have been comets whose orbits did not bring them near the sun. Within the recent past (perhaps the last few thousand years) they must have approached the sun for the first time and have had their orbits altered to their present relatively small size by perturbations produced by the planets, especially Jupiter. Those occasional comets that are highly spectacular, and hence cannot have suffered appreciable disintegration, usually have nearly parabolic orbits and are on their first approach to the sun.

(d) Comets of Nearly Circular Orbits

There are two known comets whose orbits are nearly circular, like those of the planets, and which are under more or less continuous observation. They are Comet Schwassmann-Wachmann, which travels in a 16-year period in an orbit of eccentricity 0.14, lying between the orbits of Jupiter and Saturn, and Comet Oterma, which travels with an eight-year period in an orbit, lying between those of Mars and Jupiter. Comet Schwassmann-Wachmann is particularly interesting. It is normally quite faint, but occasionally it displays sudden unexplained bursts of brightness in which it may brighten by more than 100 times. These intermittent flareups may possibly be associated with violent emissions of charged particles from the sun. We have already mentioned (Section 19.4) that some objects classed as asteroids, such as Hildago and Chiron, may be inactive cometary nuclei.

20.4 PHYSICAL NATURE OF COMETS

No two comets are alike, but they all have one characteristic in common—the *coma*, which appears as a round, diffuse, nebulous glow. As a comet approaches the sun, the coma usually grows in size and brightness. Typically it reaches a maximum size as large as or larger than the planet Jupiter, but often it shrinks somewhat as it continues to brighten near perihelion. Often, but not always, the small bright *nucleus* is visible in the middle of the coma. Together, the coma and nucleus constitute the *head* of the comet. Many comets, as they approach the sun, develop *tails* of luminous material that extend for millions of kilometers away from the head.

(a) The Nucleus

Far from the sun, where a typical comet spends practically all of its time, it is very cold and all of its material is frozen into the solid state in the nucleus. The generally favored model of a comet, proposed by Fred Whipple in 1950, describes it as a "dirty iceberg." Less than half of the nucleus is dusty, stony, or metallic material; most of it is ice and substances that vaporize at terrestrial temperatures. If a comet approaches within a few astronomical units of the sun the surface of the nucleus is warmed and begins to evaporate. The evaporated molecules, carrying many of the small solid particles along with them, then begin to produce a coma of gas and dust. Far from the sun, the nucleus can be seen from earth only by the sunlight it reflects. When a coma develops, however, the dust reflects still more sunlight, and the gas in the coma, absorbing ultraviolet solar radiation, begins to fluoresce. If the comet comes close enough to the sun (less than 5 AU), the light that the gas atoms reemit (by fluorescence) is usually more intense than the sunlight the solid matter reflects.

We can calculate the rate at which gas is evaporating from the nucleus into the coma from the amount of light the gas absorbs and reemits. This in turn tells us how much solar energy the nucleus must be absorbing to warm it up. Since we know how much solar energy reaches the comet, we can then calculate the albedo, or reflecting power, of the nucleus (the fraction of sunlight absorbed is one minus the albedo). We find the albedos of typical cometary nuclei are somewhat greater than 0.5. Finally, from a knowledge of its albedo and from its apparent brightness when it is far away and shines only by reflecting sunlight, we can calculate the size of the nucleus. The nuclei of comets (those dirty icebergs) turn out to be only 1 to 10 km in diameter. Cometary nuclei are expected to have densities about like that of water. Comet masses, therefore, are only from 10^{-10} to 10^{-12} that of the earth; a value near 10^{-11} earth mass is probably typical for bright comets.

(b) The Coma

The gas atoms in the coma have speeds of up to about 1 km/s. At these speeds they easily escape the nucleus, which has only a feeble gravitational field. The coma thus expands to an enormous size as the atoms disperse into space, typically to a diameter of 100,000 km or more. The head of a comet is, on the average, a high vacuum by laboratory standards. In 1910, stars were observed at full brilliance through some 70,000 km of the head of Halley's comet; yet starlight is appreciably dimmed by only a few kilometers of the earth's atmosphere.

When it is far from the sun, the spectrum of the light from a comet is simply that of sunlight.

The spectrum changes, however, when the comet comes within about 3 AU and its light is dominated by fluorescence. Bright emission lines (actually, molecular bands—see Section 10.6f) of the molecules and radicals carbon (C_2), cyanogen (CN), and hydroxyl (OH) are generally present, and often NH, NH_2, and other molecules as well. If a comet comes very close to the sun, as did Ikeya-Seki in 1965, the temperature near the outer surface of the nucleus can reach 4500 K, and bright emission lines produced by atoms of various metals may appear, including sodium, iron, silicon, and magnesium.

As the molecules in the coma of a comet absorb ultraviolet radiation they are dissociated, so the radicals whose spectra we commonly observe are not the original constituents of the nucleus. The parent molecules of the most abundant radicals are presumed to be water (H_2O), methane (CH_4) and ammonia (NH_3). In fact, for some years it has been thought that water (rather, *ice*) may be a major part of comet nuclei. Ludwig Biermann pointed out that if so, a large amount of hydrogen should be released from comets, not only from the dissociation of the parent molecules (especially H_2O) into radicals, but eventually from the further dissociation of OH as well. Hydrogen, however, emits most of its radiation in the ultraviolet Lyman lines (Section 10.6a) which cannot penetrate the earth's atmosphere. Thus hydrogen was not observed in comets until 1970, when it was detected around Comet Bennett with the Orbiting Astronomical Observatory (OAO). Since then, hydrogen has been observed around several other comets from outside the atmosphere. Generally the hydrogen corona is at least ten times the size of the coma on normal photographs; Mariner 10 instruments detected radiation from hydrogen from a region around Comet Kohoutek at least 40 million km across. Space observations of Comet Kohoutek also showed that water is dissociating into hydrogen and the OH radical within the inner 15,000 km around the nucleus, and that OH is further dissociating into hydrogen and oxygen beyond 45,000 km from the nucleus.

Kohoutek was the first comet to be observed at radio wavelengths. In the microwave region radio spectral lines of the molecules hydrogen cyanide (HCN) and methyl cyanide (CH_3CN) were observed. This observation is particularly significant, because these are molecules also observed in interstellar space (Chapter 28), but not in the solar system. These data support the hypothesis that comets were not formed when the planets formed from the solar nebula, but rather from fragments of the interstellar cloud before it contracted down to the planetary disk. The distribution of cometary orbits is also consistent with the idea of their formation in the very outskirts of the solar system. We shall return to this subject in Section 20.5.

Every time a comet passes by the sun and forms a coma, it loses part of its material. Calculations based on the brightness of emission lines in the coma and on the radiation from the surrounding hydrogen indicate that something like 10^{28} to 10^{29} molecules boil off the surface of the nucleus each second. At this rate it works out that a typical comet loses the outer few meters of its nucleus each perihelion passage. After at most some 1000 perihelions a periodic comet uses itself up. This is why, as explained in Section 20.3, periodic comets are relative newcomers to the inner solar system.

How much light is emitted by the coma depends on how many atoms it contains and of what kind (which in turn depends on how rapidly they evaporate from the nucleus) and how much ultraviolet solar radiation it absorbs. As comets approach the sun and heat up, not only do the nuclei evaporate faster but more solar radiation is absorbed and reemitted by the gas; consequently comets sometimes brighten many times more rapidly than if they were merely reflecting sunlight. Comets vary among each other, however, in size, in content of dust, and probably in chemical composition as well. There are so many variables that it is not possible to know in advance exactly how a comet will brighten, and predictions are risky. Many people, for example, were disappointed that Comet Kohoutek brightened more slowly near the sun than a typical comet of its size, and did not become the spectacle it was expected to be.

(c) VARIATION OF BRIGHTNESS OF A COMET

The apparent brightness B of a comet is proportional to the amount of light L leaving the comet and inversely proportional to the square of the distance Δ of the comet from the earth (because of the inverse-square law of light propagation—see Section 10.1b); that is,

$$B \propto \frac{L}{\Delta^2}.$$

When a comet is far from the sun, it shines only by reflected sunlight. In this case, L is inversely proportional to the square of the distance R of the comet from the sun, and

$$B \propto \frac{1}{R^2 \Delta^2}.$$

On the other hand, when a comet is within a few astronomical units of the sun, it shines partly by fluorescence, and brightens up more rapidly than it would if it merely reflected sunlight. The amount of light L that leaves a comet depends on the kinds of gases that compose it and on the effectiveness of the solar radiation in exciting those gases to glow; L depends, therefore, in a rather complicated manner, on R. Empirically, it is found that

$$B \propto \frac{1}{R^n \Delta^2},$$

where the exponent n is usually near 4, but may be lower than 3 for some comets and higher than 6 for others, and may even vary with time for a single comet.

(d) Tails of Comets

Many comets develop tails as they approach the sun. Some, however, especially certain short-period comets, have either lost the ability to form a tail or never possessed it. In the 16th century, the astronomers Fracastor and Apian described many comets, and remarked that their tails extend generally away from the sun. This observation was later confirmed by Kepler. Newton attempted to account for tails of comets by a repulsive force of sunlight driving particles away from the comets' heads, an idea close to our modern view.

Actually, material forming tails of comets is subject to several forces. (1) There is a radial ejection of material from the nucleus of the comet due to the thermal activity by which the coma evaporates from the nucleus. The evaporation, and thus this ejection force, is strongest on the side of the nucleus turned toward the sun, and actually results in a slight jet effect on the nucleus, opposing the gravitational attraction of the sun, analogous to the firing of a retrorocket on a space vehicle making a soft landing. The jet effect produces observable perturbations on the motions of comets; it is difficult to take into account in predicting the return of a periodic comet (like Halley's) however, partly because of the unpredictability of the precise evaporation rate, and partly because of added complications due to any rotation of the nucleus. (2) The material is subject to a repulsive force directed radially away from the sun, due in part to solar radiation pressure and in part to the solar wind. (3) The sun's gravitational attraction continues to act on each particle, which in the absence of the other forces would revolve about the sun in accordance with Kepler's laws.

A steady flow of cometary material is ejected outward by the first force, away from the head of a comet; it is picked up by the solar repulsive force and driven radially away from the sun. Thereafter, it describes a new orbit, generally hyperbolic, about the sun. The description of the orbit followed by the tail material depends on the relative magnitudes of the forces acting on it, but in any case is different from that of the parent comet; thus this material is lost permanently to the comet and, usually, to the solar system.

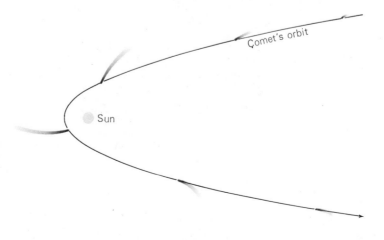

Figure 20.4 Shape of a typical comet tail as the comet passes perihelion.

Tails generally grow in size as comets near the sun; some comet tails have reached lengths of more than 150 million km. Once the tail material has left the vicinity of the comet's head, the only forces acting upon it are radial forces toward and away from the sun (gravity and the force of repulsion). As the material recedes farther from the sun, its *angular velocity* about the sun decreases, in accord with Kepler's laws, and the material lags behind the comet. In general, therefore, comet tails lie in the plane of the comet's orbit, pointing more or less away from the sun, but curving somewhat backward, away from the direction of the comet's motion (Figure 20.4). The acceleration produced by the force of repulsion from the sun varies according to the size of the particles it acts upon and according to whether the particles are gas molecules or solid grains. Material of different types, therefore, is accelerated away from the sun in different amounts. If the repulsive force exceeds the gravitational force on a certain kind of particle by 20 or 30 times, particles of that type are driven outward from the sun so rapidly that they form a tail that is nearly straight. On the other hand, if the repulsive force is only 2 or 3 times the gravitational force, the particles move outward more slowly; the difference between the orbital motions of the comet and tail material becomes apparent before the latter has receded very far from the comet's head, and the tail appears noticeably curved.

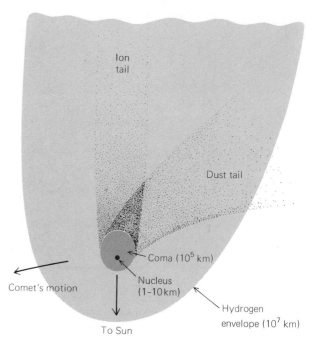

Figure 20.5 Parts of a typical comet.

Most comets have two tails that are very different from each other. One is usually nearly straight, and its spectrum shows it to be composed of ionized molecules; lines of CO^+, N_2^+, and H_2O^+ have been observed. The other tail (sometimes very broad or even multiple) is generally quite curved, and its spectrum is that of reflected sunlight, show-

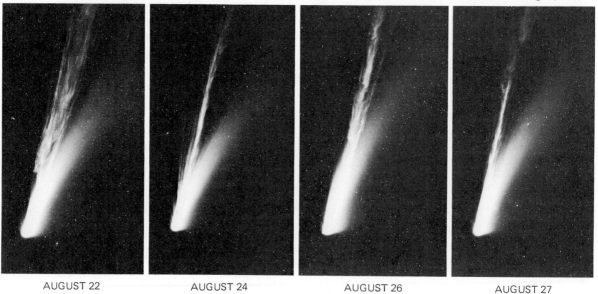

| AUGUST 22 | AUGUST 24 | AUGUST 26 | AUGUST 27 |

1957

Figure 20.6 Comet Mrkos, photographed with the Palomar Schmidt telescope. *(Caltech/Palomar Observatory)*

ing it to be composed mostly of dust (see Figure 20.5).

The nearly straight ion tails sometimes have recognizable features or knots that can be seen, from one day to the next, moving away from the comet (see the photographs of Comet Mrkos in Figure 20.6). Measures of such structures show that the streaming ions have speeds that increase from about 1 km/s near the head of the comet to more than 100 km/s toward the end of the tail. These increases in speed represent accelerations of tens or even hundreds of times that produced by the sun's gravitation at the comet's distance.

Ludwig Biermann provided the explanation for these ion tails. They are caused by the *solar wind*— ionized atoms (electrons and atomic nuclei) continually being ejected from the sun and flowing out through the solar system (Chapter 31). These solar ions carry electrical charges that interact with the ions in the coma or tail and selectively accelerate the coma ions away from the sun. The solar magnetic field lines, transported outward by the solar wind, are directed radially away from the sun at typical cometary distances, and they guide the flow of the charged particles into the straight tails. From time to time, sudden solar outbursts (*flares*) result in a greater than average flow of particles from the sun; these variations in the solar wind intensity, and variations in the ejection of material from the comet's nucleus, probably account for the irregular features in its ion tail.

The dust tails are produced by solar radiation pressure. Each photon of electromagnetic radiation has associated with it a small amount of momentum, which is numerically equal to the energy of the photon (Planck's constant times the frequency—see Section 10.3) divided by the velocity of light. If a photon strikes a particle, or is absorbed by it, the momentum of the photon is transferred to the particle, thereby accelerating the latter. On particles of moderate size, the acceleration is entirely negligible in comparison to that produced by the gravitational forces acting on them. On the very small dust particles in comets, however, the force of solar radiation can be greater than the sun's gravitational attraction. Radiation pressure is most effective on particles whose sizes are near that of the average wavelength of sunlight. If they have a density near unity (like ice crystals), the force of radiation on such particles can exceed that of gravity by about five times. Particles several times larger than the wavelength of the incident

light have masses great enough so that the force of gravitational attraction between them and the sun exceeds the outward force of radiation pressure. The effect of radiation pressure on smaller particles depends on their chemical composition; some types of smaller particles are repulsed. Many kinds of molecules also absorb radiation and are driven away from the sun.

Figure 20.7 Comet Arend-Roland photographed with the 124-cm Schmidt telescope on April 26, 1957, with the exposure of 15 minutes. Notice the prominent "antitail." (*Caltech/Palomar Observatory*)

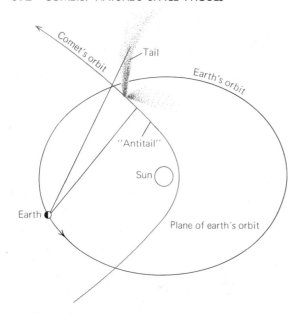

Figure 20.8 The relative positions of the earth, sun, and Comet Arend-Roland, giving rise to the seeming "antitail."

The astronauts in the Sky Lab reported a similar phenomenon for Comet Kohoutek in January 1974.

20.5 ORIGIN OF COMETS

The most widely accepted hypothesis for the origin of comets is that of the Dutch astronomer Jan H.

(Radiation pressure, incidentally, produces measurable perturbations on the motion of a balloon satellite revolving around the earth—for example, the Echo satellite, launched in the early 1960s. There is even a possibility of using radiation pressure for space travel. It has been shown that if large lightweight sails can be attached to space vehicles, "solar sailing" is a feasible mode of locomotion through interplanetary space. It would not, of course, be a sufficient means of propulsion in the strong gravitational fields near the surfaces of planets.)

The dust tails of comets are often broad or multiple because the particles of different sizes are accelerated by different amounts; radiation pressure generally imparts smaller speeds to larger particles of greater mass in the tail, so that they trail farther behind the comet. Sometimes the dust tails are distinctly fan-shaped, the trailing edge of the fan consisting of the largest particles and curving back the most. It occasionally happens that the orientation of the sun, earth, and a comet is such that we see an ion tail pointing away from the sun in the sky, while the dust tail, although actually trailing back from the comet but away from the sun, appears to us to point toward the sun. It is, of course, only a projection effect. If we view the fan of dust almost edge on, it can, at such a time, appear as a sharp spike or jet. Comet Arend-Roland had such an apparent sunward spike in April 1957, soon after the comet passed perihelion (Figures 20.7 and 20.8).

Figure 20.9 Comet Seki-Lines photographed from Frazier Mountain, California, August 9, 1962. (*Alan McClure, Los Angeles*)

Oort. He proposes that there is a huge cloud of comets revolving about the sun with orbits whose aphelia are 50,000 AU or more from the sun. Their orbits cannot have originally brought them close to the sun, or, as we have already explained, Jupiter would have transformed them to short-period comets or have ejected them from the solar system.

On the other hand, the outer fringe of the cloud of comets cannot extend beyond about 60 percent of the distance of the nearest star (Alpha Centauri) because beyond that point their orbits would be unstable. Even so, perturbations produced by those stars that pass, every few thousand years, within a few light-years of the sun must alter many of the primeval cometary orbits. Some comets are lost to the solar system as the result of such stellar perturbations, but others, retarded in their motion by the star's attraction, fall in somewhat closer to the sun. After a number of successive perturba-tions caused by other passing stars, a comet's peri-helion can reduce to a few astronomical units. Thus, Oort accounts for the steady influx of comets into the central part of the solar system.

Those comets that are so deflected toward the sun move subsequently on orbits of very high ec-centricity, until, by perihelion passage, further perturbations by the planets either convert them into "periodic" comets, or deflect them into inter-stellar space on hyperbolic orbits. In the former case, they are eventually decomposed by evapora-tion or by tidal forces; in the latter, they are per-manently lost to the solar system.

Although the supply of comets in the "cloud" must be gradually diminishing, the remaining num-ber of comets may still be very large. Calculations show that there must be at least 100 thousand mil-lion. Their combined mass is probably comparable to that of the earth.

Figure 20.10 Photograph of Comet Ko-houtek made with a 10-cm astrographic telescope on January 16, 1974, by Fa-ther Ronald Royer, amateur astronomer, Azusa, California.

20.6 COLLISIONS OF THE EARTH AND COMETS

The chance of a collision between the earth and a comet is remote in our lifetimes, but over the age of the solar system it may well have occurred many times. The coma of a comet is so tenuous that the earth would pass through it with probably no more effect than a spectacular meteor shower caused by the small particles in the coma. The only damage likely to result would be from direct impact of the nucleus, which has about the mass of a modest minor planet. A city or a small nation could be devastated by such a blow, for the energy of the impact could be that of millions of 100-megaton hydrogen bombs. Most likely, the comet would strike an ocean, and loss of life might be minimal although great tsunamis (tidal waves) could also be devastating; there would be no question, though, that we would know it had happened. It is probable that the 1908 Tunguska event in Siberia (Chapter 21) was actually a collision with a very small comet. It has also been suggested that a collision of a comet with the earth may have been responsible for the Cretaceous–Tertiary extinction, a theory similar to that of an asteroid collision proposed by Alvarez (Chapter 19).

The earth very probably did pass through the tail of Halley's comet in 1910. There was no detectable evidence of the encounter, although it is reported that at least one enterprising person made his fortune selling "comet pills."

Figure 20.11 An unusual mosaic of three photographs of Comet Kohoutek, taken 14 January 1974, showing both a curved dust tail and an exceptionally long ion tail. (*Joint Observatory for Cometary Research, NASA–Goddard Space Flight Center and New Mexico Institute of Mining and Technology*)

EXERCISES

1. Comet 1955b is also known as 1954 V. Why is it denoted by two different years?

2. Find the period of a comet which at perihelion just grazes the sun, and whose aphelion distance from the sun is: (a) 200 AU; (b) 2000 AU; (c) 20,000 AU; (d) 200,000 AU.

3. On the assumption that a periodic comet can survive 1000 perihelion passages, find the lifetime of the first comet in Exercise 2.

*4. Suppose that the brightness of a comet is given by the law

$$B \propto \frac{1}{R^4 \Delta^2}.$$

When first discovered it is exactly at opposition and at a distance of 4 AU from the earth. How much brighter will it appear when it is 1 AU both from the sun and from the earth? Compare this ratio with that by which a body would brighten if it merely reflected sunlight.

5. With the aid of a diagram, explain how you could derive the true length of the tail of a comet from the observed angular length. Assume that the tail points directly away from the sun and that the comet is 1 AU from both the earth and sun.

6. In terms of Kepler's laws, explain why the dust tail of a comet should always curve back slightly away from the direction of motion of the comet.

7. The force due to radiation pressure on a particle is proportional to the amount of radiation that the particle intercepts. Show that the ratio of the force of radiation pressure to the solar gravitational force on a spherical particle is independent of the distance of the particle from the sun.

8. If Oort's hypothesis for the origin of comets is correct, why do comets that come close to the sun for the first time always have aphelion points in the region of the cloud from which they originate?

9. Why does the material in the tail of a comet follow a hyperbolic orbit about the sun?

10. A comet is originally in a nearly circular orbit of radius 10,000 AU. A succession of passing stars slows the comet, causing it to fall toward the sun. Passing near Jupiter, the comet is perturbed into a hyperbolic orbit. Discuss the evolution of the orbital eccentricity of the comet.

11. Suppose you plan to observe a certain comet when it is at its brightest. About what time of day would you plan these observations? Why?

12. Explain why typical periodic comets tend to have small inclinations to the ecliptic.

13. Although the mass of a typical comet is comparable to that of a small asteroid, a comet can strike the earth with far more energy than can an asteroid of the same mass. Why? (*Hint:* Think about the orbits.)

21

METEOROIDS, METEORITES, AND METEORS

Although the layman often confuses comets and meteors, these two phenomena could hardly be more different. Comets can be seen when they are many millions of miles away from the earth, and may be visible in the sky for weeks, or even months, slowly shifting their positions from day to day. They rise and set with the stars, and during a single night appear motionless to the casual glance. Meteors, on the other hand, appear only when small solid particles enter the earth's atmosphere from interplanetary space. Since they move at speeds of many kilometers per second, they vaporize as a result of the high friction they encounter in the air. The light caused by the luminous vapors formed in such an encounter appears like a star moving rapidly across the sky, fading out within a few seconds. Meteors are commonly called "shooting stars."

On rare occasions, an exceptionally large particle may survive its flight through the earth's atmosphere and land on the ground. Such an occurrence is called a *meteorite fall,* and if the particle is later recovered, it is known as a fallen *meteorite.* Frequently, the term "meteorite" is reserved for the fallen particle, and the particle, when still in space, is called a *meteor.* In this book, however, we shall use the terminology recommended by the International Astronomical Union in 1961. The par-

ticle, when it is in space, is called a *meteoroid*; the luminous phenomenon caused when the particle vaporizes in the earth's atmosphere is a *meteor*; and if it survives and lands on the ground, the particle is a *meteorite.*

21.1 PHENOMENON OF A METEOR

On a typical dark moonless night an alert observer can see half a dozen or more meteors per hour. To be visible, a meteor must be within 150 to 200 km of the observer; over the entire earth, the total number of meteors bright enough to be visible must total about 25 million per day. Faint meteors are far more numerous than bright ones; the number of meteors that are potentially visible with binoculars or through telescopes, and that range down to a hundred times fainter than naked-eye visibility, must number near 4×10^{11} per day.

More meteors can generally be seen in the hours after midnight than in the hours before. Moreover, there are certain times when meteors can be seen with much greater than average frequency—up to 60 or more per hour. These unusual meteor displays are called *showers.* Meteor showers occur when the earth encounters swarms of meteor-

Figure 21.1 Photograph of the trail of a bright meteor that happened to cross the field of view of the telescope photographing the Andromeda galaxy. *(F. Klepesta)*

oids moving together in space. A distinction is made between shower meteors and nonshower, or *sporadic*, meteors.

(a) Fireballs

Occasionally an exceptionally bright meteor is reported by many observers. These bright meteors are called *fireballs*. It is estimated that tens or hundreds of thousands of fireballs appear every day over the entire earth, most of them over oceans or uninhabited regions. Some fireballs are visible in broad daylight; many are as bright as the full moon, and a few are as much as 100 times as bright. Sometimes they break up in midair with explosions that are audible from the ground. Such exploding fireballs are often called *bolides*.

Fireballs occasionally leave luminous trails or *trains* behind them, which may persist for periods ranging from one second to half an hour. The velocities of upper atmospheric winds are sometimes revealed by the twisting and distortion of meteor trains.

Bright fireballs have been recorded by very ancient peoples. There are references to them, for example, in the Book of Joshua and in records kept by the Chinese centuries before Christ.

(b) Observations of Meteors

The paths of bright fireballs through the atmosphere can occasionally be traced from an analysis of lay sightings. The untrained observer, to be sure, does not often provide reliable accounts of what he saw. Fireballs usually seem many times closer than they actually are; also, their angular altitudes and speeds are frequently grossly exaggerated. Nevertheless, the paths of spectacular fireballs have been derived by comparing reports from many observers scattered over areas of hundreds of kilometers.

Far better data on the paths of meteors come from the special photographic patrols and radar observations. In the former, arrays of specially designed, high-speed, wide-angle *meteor cameras* are placed many kilometers apart, and are directed to the same region of the sky. When a meteor chances to pass through the common field of view of two of these cameras, it is recorded by both cameras. At each camera, the exposure is interrupted by a rotating propeller-type shutter, so that the trail of the meteor on the film consists of a series of dashes rather than a continuous streak. When trails of the same meteor are identified on photographs obtained simultaneously by the two cameras, the meteor's elevation and direction of motion through the atmosphere can be computed by triangulation. Since the rotation rate of the propeller shutters is known, the spacing of the dashes that comprise the meteor's trailed image indicates the speed with which the image moved across the emulsion and hence the velocity of the meteor. During the 1960s, the Smithsonian Astrophysical Observatory operated the Prairie Network, a system of 16 such meteor cameras in the midwestern United States, which covered an area of more than a million square kilometers.

Meteors are detected by radio and radar as well as visually and photographically. In the early 1940s it was found that meteors caused brief interruptions in high-frequency broadcasting reception; these interruptions took the form of "whistles," which usually fell quickly in pitch. Since then, the rate at which meteors occur has been determined by counting the occurrences of such whistles.

Meteoroids themselves are too small to reflect radar waves back to the ground. These waves are reflected, instead, by the ionized gases that are formed when meteoroids vaporize in the air. The radio or radar waves travel, of course, with the speed of light; consequently, the time required for

Figure 21.2 Photograph *(left)* obtained with a Baker Super-Schmidt camera *(right).* Trails of three meteors are visible; they are interrupted by a timing device in the telescope. *(Harvard Observatory)*

them to travel from the ground to the meteor and back again indicates the distance or *range* of the meteor from the radar station. The motion of the meteoroid toward or away from the radar station can be detected from the rate at which the range changes (*range rate*), which in turn gives the speed of the meteoroid in the line of sight. Simultaneous observations from two radar stations give the height, speed, and direction of the meteoroid.

It is found that meteoroids produce meteors at an average height of 95 km. The highest meteors form at heights of 130 km. Nearly all meteoroids completely disintegrate, and their luminous paths end, by the time they reach altitudes of 80 km. A few meteoroids doubtless skim out of the atmosphere, returning to space before they are completely burned up.

The vast majority of meteoroids have speeds between 12 and 72 km/s. The faster ones tend to be brighter and are more likely to leave trains, because of their greater energy upon entry into the earth's atmosphere. On the other hand, the faster-moving particles "burn up" at higher altitudes and are less likely to survive their flights and reach the ground.

Spectra of some meteors have been obtained by photographing them through cameras equipped with objective prisms (Section 11.2f). It is found that more than 99 percent of the light of meteors comes from emission lines, which indicates that meteoric light is produced by glowing gases rather than by incandescent solid particles. The spectral lines present show that the luminous vapors are at high temperatures—in the thousands of kelvins. Emission lines due to calcium and iron are usually present. Lines of the elements manganese, silicon, aluminum, magnesium, sodium, nitrogen, hydrogen, and ionized iron are also found.

21.2 ORBITS OF SPORADIC METEOROIDS

A knowledge of the speeds and directions of meteors in the earth's atmosphere enables us to calculate the orbits of the meteoroids in space before they encountered the earth. To determine the *heliocentric* orbit of a meteoroid (that is, its motion with respect to the sun), it is necessary to correct for (1) that component of its velocity relative to the earth that is produced by the earth's orbital motion of about 30 km/s, (2) the acceleration of the meteoroid toward the earth that is produced by the earth's gravitational attraction, and (3) that component of its relative velocity that arises because of the earth's rotation. The latter two contributions to the meteoroid's velocity can range from 0.8 to 12 km/s.

(a) Orbits of Fireballs

Meteoroids that produce fireballs are relatively rare. Nevertheless, more than 100 were photographed by

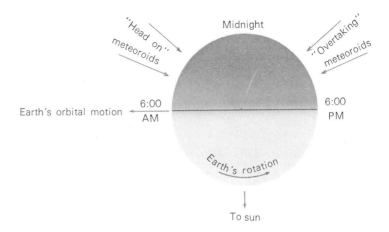

Figure 21.3 Meteors seen before midnight have overtaken the earth. Meteors seen after midnight are produced mostly by particles approaching the earth from the east.

the Prairie Network. The orbits of a few of those meteoroids producing fireballs are found to lie close to the plane of the ecliptic and have moderate eccentricities, and the revolution of the particles is *direct*, that is, from west to east, as is the revolution of the planets. These meteoroids have orbits similar to those unusual minor planets (the Apollo asteroids—section 19.2e) whose orbits bring them within 1 AU of the sun.

These meteoroids usually collide with the earth after catching up with it from the "rear," for they are moving in the same direction about the sun as the earth, but at greater speeds. These fall between noon and midnight. Some, however, have smaller heliocentric speeds than the earth's, and we overtake them; thus they fall during the morning hours. Their speeds with respect to the earth, in any case, are relatively small, and they survive their passage through the earth's atmosphere down to comparatively low altitudes before becoming hot enough to vaporize.

(b) Orbits of Typical Sporadic Meteoroids

For the vast majority of meteoroids, including even most fireballs whose orbits have been determined by photographic or radar methods, there is no preference for the ecliptic plane. Like the comets, they approach the earth from all directions. Also, like the comets, most of them are found to be moving at speeds near the velocity of escape from the solar system; that is, they travel on near-parabolic or highly eccentric elliptical heliocentric orbits.

At the earth's distance from the sun, the parabolic velocity, or velocity of escape from the sun, is about 42 km/s (see Section 5.4). The velocity with *respect to the earth* of a particle moving on a

parabolic orbit is the vector sum of its velocity of 42 km/s and the earth's orbital velocity of 30 km/s. If a meteoroid overtakes the earth from the west, it approaches the earth at a speed of about 42 *minus* 30, or 12 km/s; if it meets the earth from the east, it approaches us at about 42 *plus* 30, or 72 km/s. Most sporadic meteors have geocentric speeds in the range 12 to 72 km/s because of their near-parabolic orbits.

The reason meteors are more frequently seen, and tend to appear brighter, in the hours after midnight than in the hours before is illustrated in Figure 21.3. In the morning hours we are on the side of the earth which is turned in the direction of its orbital motion; consequently, the meteors we see then are due either to particles that run into us from the east at high relative speeds, or that we overtake. In the evening hours we are on the side of the earth which is turned away from the direction of its orbital motion; consequently, the meteors we see then arise from particles that overtake us at low relative speeds. Because of the greater frequency of head-on collisions, and because the brightness of a meteor depends strongly on the speed of its meteoroid, we are more likely to see meteors during the morning hours.

(c) Membership in the Solar System

The majority of meteoroids, if not all of them, are believed to have originated from within the solar system. The argument is exactly the same as that for the solar-system membership of comets (Section 20.3a). If meteoroids were chance intruders from interstellar space, they would have hyperbolic orbits. Only a very few meteoroid orbits have been observed to be slightly hyperbolic, and these orbits may have been altered from parabolic ones as a result of perturbations produced by other planets. Like the orbits of comets, the apparent parabolic orbits that we observe for meteoroids must actually be the ends of very long ellipses. At most, less than 1 percent of all observed meteoroids can be interstellar visitors, and probably far fewer.

21.3 METEOR SHOWERS

Several times each year meteors can be seen with much greater than average frequency. These *meteor showers* occur when the earth encounters swarms of particles moving together through space. Many such swarms of particles that the earth intercepts at regular intervals are known. Showers produced by them are predictable. On rare occasions the earth unexpectedly encounters swarms of particles that produce spectacular meteor displays.

(a) Shower Radiants

Unlike sporadic meteors, which seem to come from any direction, meteors belonging to a shower all seem to radiate or diverge away from a single point on the celestial sphere; that point is called the *radiant* of the shower. Recurrent showers are named for the constellation within which the radiant lies or for a bright star near the radiant.

The seeming divergence of shower meteors from a common point is easily explained. The meteoroids producing a meteor shower are members of a swarm; they are all traveling together in closely spaced parallel orbits about the sun. When the earth passes through such a swarm, it is struck by many meteoroids, all approaching it from the same direction. As we, on the ground, look toward the direction from which the particles are coming, they all seem to diverge from it. Similarly, if we look along railway tracks, those tracks, although parallel to each other, seem to diverge away from a point in the distance (Figure 21.4).

Actually, a meteor radiant is not a perfectly sharp point. The meteors seem to radiate away from a small region of the celestial sphere. In some showers, that region is very small—as little as 3' in diameter. In other showers, the radiant may be as large as 1° across. The size of the radiant depends on how closely packed the swarm of particles is.

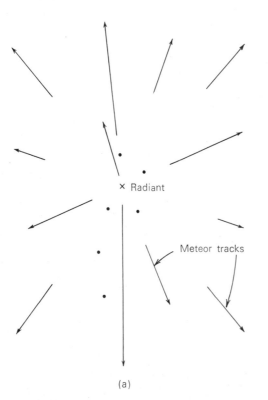

× Radiant

Meteor tracks

(a)

(b)

Figure 21.4 (a) A meteor shower radiant and (b) the apparent divergence of parallel lines.

The denser, or more closely packed the swarm, the more spectacular the meteor shower it produces. The spectacle also depends on how closely the earth approaches the densest part of the swarm.

Many meteors classed as sporadic may actually belong to showers that are so inconspicuous that they go unnoticed. Scarcely a night passes that a patient observer cannot see three or more meteors moving in the sky in directions away from a common point. Many of these may be members of minor showers.

(b) Association of Showers and Comets

The direction to the radiant of a meteor shower indicates the direction in which the swarm of particles that produces it is moving through space with respect to the earth. The velocity of the swarm is found, as explained in Section 21.1, from the velocity of the meteors. These are enough data to specify completely the orbit of the swarm.

On about August 11 of each year, the earth passes through a swarm of particles that approach from the direction of Perseus. In 1866 it was observed that the particles producing this *Perseid shower* travel in an orbit that is almost identical to that of Comet 1862 III. It was then realized that those meteoroids encountered each August are debris from the comet that has spread out along the comet's orbit.

Subsequently, it has been found that the elements of the orbits of many other meteoroid swarms are similar to those of the orbits of known comets. Not all meteor showers have yet been identified with individual comets, but it is presumed that all showers have had a cometary origin. These swarms of debris, provided by the gradual disintegration of comets, give further evidence of the flimsy nature of comets.

Twice each year the earth passes through a swarm of particles moving in the orbit of Halley's comet. Debris from that famous comet gives us the *Eta Aquarids* in May and the *Orionids* in October.

Sometimes, if a meteoroid swarm is old enough, its particles are strewn more or less uniformly along their entire orbit, as shown in Figure 21.5(a); an example is the swarm of particles that produces the Perseid shower. Consequently, the earth meets about the same number of particles each time it passes through the orbit of the swarm, and the Perseid shower is of about equal intensity each year. More often, a swarm is bunched as in Figure 21.5(b), and we experience a spectacular display only when the earth completes a rendezvous with the meteoroids as they reach that same point in their revolution about the sun. A good example is provided by the Leonid meteors in the last century, when the earth met the debris of Comet 1866 I in 1833 and again in 1866 (after an interval of 33 years—the period of the comet). Those were among the most spectacular showers ever recorded.

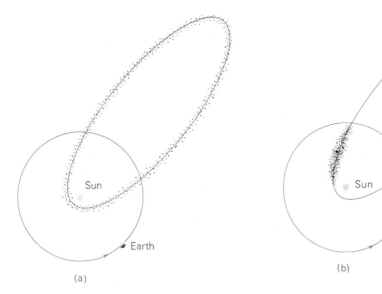

(a)

(b)

Figure 21.5 Meteoroids in a swarm may be strewn more or less uniformly along their orbit (a), or bunched up (b).

Figure 21.6 Photograph showing trails of many meteors during the shower of October 1946. Note the divergence of the trails from a radiant (off the field to upper right). Other streaks are star trails.

As many as 200,000 meteors could be seen from one place within a span of a few hours. The last good Leonid shower was on November 17, 1966, when in some southwestern states up to 140 meteors could be observed per second. Even in such dense swarms the individual particles of the swarm are separated by distances of 30 km or more; in most meteoroid swarms, the particles are more than 100 km apart.

The best meteor shower that can be depended on at present is the Perseid shower, which appears for about three days near August 11 each year. In the absence of bright moonlight, meteors can be seen with a frequency of about one per minute during a typical Perseid shower. It is estimated that the total combined mass of the particles in the Perseid swarm is near 5×10^8 tons; this gives at least a lower limit for the original mass of Comet 1862 III. The orbit of this comet (and hence of the Perseids) was nearly perpendicular to the ecliptic plane; thus the meteoroids are little perturbed by the planets, accounting for the reliability of the Perseid shower.

One spectacular shower of recent decades was the *Draconid* shower that reached maximum display on October 9, 1946. On that date the earth reached the point that Comet Giacobini-Zinner had passed 15 days earlier. Debris from the comet produced meteors that could be counted from points in the southwestern United States at a rate of two per second, even though the moon was full at the time.

The characteristics of some of the more famous meteor showers are summarized in Table 21.1. Other spectacular meteor showers can occur, however, at almost any time, just as some bright comets appear unexpectedly.

(c) Daytime Showers

We have seen that radar waves are reflected by the column of ionized gases left in the wake of a meteor. Now, some wavelengths of radio waves, those that lie in the range 4 to 5 m, are reflected efficiently only if they strike the ionized gas column nearly broadside (waves shorter than about 4 m usually are not reflected, except by the very densest ion clouds). Thus, meteors that belong to a shower can be detected by radar operating at a wavelength of 4 to 5 m only if the radar beam is directed at nearly right angles to the direction of the shower

Table 21.1   Characteristics of Some Meteor Showers

SHOWER	DATE OF MAXIMUM DISPLAY	VELOCITY (km/s)	ASSOCIATED COMET	PERIOD OF COMET (yr)
Quadrantid	Jan 3	43	—	7.0
Lyrid	Apr 21	48	1861 I	415.0
Eta Aquarid	May 4	59	Halley	76.0
Delta Aquarid	Jul 30	43	—	3.6
Perseid	Aug 11	61	1862 III	105.0
Draconid	Oct 9	24	Giacobini-Zinner	6.6
Orionid	Oct 20	66	Halley	76.0
Taurid	Oct 31	30	Encke	3.3
Andromedid	Nov 14	16	Biela	6.6
Leonid	Nov 16	72	1866 I	33.0
Geminid	Dec 13	37	—	1.6

radiant (the direction from which the meteoroids are coming). If the sky is scanned by radar, the maximum amount of energy is reflected back from meteors along a great circle in the sky that is 90° from the shower radiant. Radar observations of shower meteors, therefore, can locate the radiant even when the shower occurs in daylight when the meteors cannot be observed visually. Four daylight meteor showers have been found, thus far, which, if they were visible at night, would be at least as spectacular as those listed in Table 21.1.

21.4 FORMATION OF A METEOR

As a meteoroid plunges into the earth's atmosphere at a speed of many kilometers per second, it undergoes many collisions with air molecules. The impinging molecules penetrate the meteoroid and chip off pieces of it. The surface of the meteoroid heats up, and the dislodged particles vaporize. The whole process is called *ablation*.

The gases from the meteoroid and the air immediately around it heat sufficiently so that the atoms comprising the gases emit light in the form of bright emission lines. The *meteor* is the region of glowing gas; it may be anywhere from a few to several hundred meters in diameter. A meteoroid in space, before entering the atmosphere, must be fairly cold; by the methods outlined in Section 14.2c we find its temperature should be about 280 K. On the other hand, the emission lines observed in the light of meteors show that the luminous gases involved must have temperatures of thousands of Kelvins.

Air resistance and the ablation process rob the meteoroid of energy and slow it down. The rate at which it decelerates depends on its original speed, its mass, and the density of air. Thus, as a by-product of the study of meteors, we learn something about the density of the upper atmosphere.

(a) Masses of Meteoroids

Minimum masses of meteoroids are estimated from the amount of light they produce in their meteors. From the brightness of the photographic image of a meteor trail, the amount of light energy received from the meteor is measured. Then, from the height and distance of the meteor, the amount of luminous energy it radiated during its brief existence can be calculated. The original source of this energy is the kinetic energy (the energy of motion) of the meteoroid as it plunged into the atmosphere.

The kinetic energy of a moving particle is one-half the product of its mass and the square of its speed, that is,

$$KE = \frac{1}{2}mv^2.$$

We have already discussed how the velocities of meteoroids are determined (Section 21.1b). Thus, if we assume that the kinetic energy of the meteoroid was completely converted into luminous energy of the meteor, we can use the above formula to compute the mass of the particle before it entered the earth's atmosphere. Because some of the particle's kinetic energy is dissipated by other means than conversion to visible light, its actual mass must be even greater.

The mass of a meteoroid producing a typical bright meteor is about $\frac{1}{4}$ gm. A small fraction, however, have masses greater than a few grams. Many visible meteors are produced by particles with masses of only a few milligrams. Most meteoroids, therefore, are much smaller than pebbles.

Most meteors are observed to decelerate rapidly, which indicates that their meteoroids must be of low density, like that of ice or less. Many may, in fact, be porous.

The rate at which the earth accumulates meteoritic material is not accurately known. However, the total mass of meteor-producing meteoroids that enters the atmosphere each day is estimated to be from 10 to 100 tons.

21.5 FALLEN METEORITES

Occasionally, a meteoroid survives its flight through the atmosphere and lands on the ground; this happens with extreme rarity in any one locality, but over the entire earth probably about 500 meteorites fall each year. Fallen meteorites have been recovered throughout history; it was nearly 1800, however, before their association with "shooting stars" began to be appreciated. The general acceptance that indeed "stones fall from the sky" occurred after the French physicist B. J. Biot described the circumstances of a fall in the Orne village of l'Aigle on April 26, 1803, in which many witnesses observed the explosion of a bolide, after which many meteoritic stones were found, reportedly still warm, on the ground.

Today, fallen meteorites are found in two ways. First, sometimes bright fireballs are observed to penetrate the atmosphere to very low altitudes. A search of the area beneath the point where the fireball was observed to burn out may reveal one or more remnants of the meteoroid. *Observed falls,* in other words, may lead to the recovery of fallen meteorites.

Secondly, unusual-looking "rocks" are occasionally discovered that turn out to be meteoritic. These are termed "finds." Now that the public has become "meteorite conscious," many suspected meteorites are sent to experts each year. The late F. C. Leonard, a specialist in the field, referred to these objects as "meteorites" and "meteorwrongs." Outside of Antarctica (where hundreds are found each year), genuine meteorites are turned up at an average rate of 25 per year. Of these, from five to ten are recovered from observed falls.

(a) Types and Compositions of Meteorites

Meteorites can be classified in three general groups:

1. *Irons*—alloys of metals. From 85 to 95 percent of their mass is iron; the rest is mostly nickel.
2. *Stony irons*—relatively rare meteorites that are about half iron and half "stony" silicates.
3. *Stones*—composed mostly of silicates and other stony materials. Stones contain about 10 to 15 percent iron and nickel in metallic flakes.

Most of the stone meteorites are called *chondrites,* because they contain small rounded granules called *chondrules.* The stones that do not contain chondrules are called *achondrites.* Stone meteorites have abundances of nonvolatile elements in approximately the relative abundances of those elements found in the sun. The irons and stony irons, on the other hand, have anomalously high iron abundance and are thought to be fragments of the inner parts of larger parent bodies in which the iron and heavier metals *differentiated,* that is, sank to the interiors, much as we believe happened in the earth and other terrestrial planets.

Stones are the most common kind of fallen meteorites. Ninety-three percent of meteorites obtained from observed falls are stones. However, after erosive action, they resemble terrestrial rocks, and generally only experts can ascertain their meteoritic nature, in part from the iron pieces embedded in them. Consequently, about 58 percent of the total number of finds outside of Antarctica are irons. (Irons are seldom found, however, in those regions of the world where for centuries man has made use of iron for tools.) When irons are cut, polished, and etched, about 80 percent show a characteristic crystalline structure (*Widmanstätten figures*) that makes their identification as meteorites certain.

So far as is known, all recovered meteorites are remnants of *sporadic* meteors. There has never been

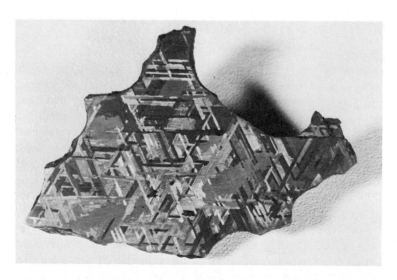

Figure 21.7 An iron meteorite cut and polished to show the Widmanstätten figures. (*Griffith Observatory*)

an observed fall from a *shower* meteor. The spectra of shower meteors, however, indicate that they contain some of the same elements as stones.

The chemical composition of many meteorites has been determined by laboratory analysis. It is found that meteorites contain the same common chemical elements that we find in the crust of the earth. When irons, which are not typical, are eliminated, the following abundances (by weight) of elements are derived: iron, 22 to 27 percent; oxygen, 30 percent; silicon, 15 percent; magnesium, 12 percent; sulfur and nickel, 2 percent each; calcium, sodium, and aluminum, 1 percent; and small amounts of carbon, lead, chlorine, potassium, titanium, chromium, manganese, cobalt, copper, and other elements.

Meteorites often show evidence of ablation that occurred during their atmospheric flights. A skin-deep layer on the surface of a meteorite may show evidence that the surface has been molten. Meteoroids pass through the air too quickly, however, for their interiors to heat up. By the time a meteoroid is slowed, by air friction, to 3 km/s, the molten surface of the iron hardens; when its speed has slowed to 1 km/s, it stops glowing altogether. Popular opinion to the contrary, meteorites are not "red hot" when they strike the ground; meteorites that have been picked up just after they have fallen are generally cool enough to handle. The surfaces of the stones may also become molten during their atmospheric passage and turn glassy. Both stones and irons are sometimes ablated to conical shapes.

Tektites are somewhat rounded glassy bodies that have been found in Indonesia, Australia, and elsewhere. A few investigators still suspect them to be of extraterrestrial origin, but most favor the hypothesis that they are formed of terrestrial material that is melted and thrown out by the explosion produced when a large meteoroid strikes the earth. Tektites show evidence of having been molten at high pressure, and having undergone rapid cooling. Some contain small spheres of iron and nickel, and schreibersite, a mineral known otherwise only in meteorites.

(b) Ages of Meteorites

Natural radioactivity provides a means of determining the ages of meteorites just as it provides a means of determining the ages of rocks in the earth's crust (Section 15.4b). The derived age, of course, ap-

plies only to the solid particle in its present mineral state. The ages of meteorites are found to be about 4.5×10^9 years—presumably the same as that of the earth and the rest of the solar system.

Meteoroids in space are continually subject to bombardment by cosmic rays. These high-energy particles break up certain of the nuclei of the atoms of the meteoroids they strike, forming, for example, an isotope of argon, ^{38}Ar. Thus the content of ^{38}Ar in a meteorite provides a measure of how long the particle was exposed to cosmic rays in space. These *exposure ages* of stones are generally 1 to 5 $\times 10^6$ years, and not over 5×10^7 years, whereas those of irons are generally in the range 200 to 1000 million years. This must be the time since the meteorite examined was released from its parent body.

(c) Largest Recovered Meteorites

The largest meteorite ever found on the earth is *Hoba West*, near Grootfontein, Namibia (formerly South-West Africa). It has a volume of about 7 cubic meters and an estimated mass of about 60 tons. The largest meteorite on display in a museum has a mass of 31 tons; it was recovered by Peary from Greenland in 1897 and is now at the American Museum of Natural History in New York. The largest fallen meteorite found in a single piece in the United States was discovered in a forest near Willamette, Oregon, in 1902; it has a mass of 13 tons. (The discoverer spent about three months hauling the meteorite to his own property, where he put it on display for an admission price. Among those in-

Figure 21.8 Tektites from Thailand. The largest is about 8 cm long. *(Ronald Oriti, Griffith Observatory)*

terested in the new exhibit were the attorneys of the Oregon Iron and Steel Company, owner of the land on which the meteorite was found. After litigation, it was decided that the company was the rightful owner of the object.)

21.6 METEORITE FALLS

A total of 1800 meteorite falls are listed in the *Hey Catalogue of Meteorites*; of these, 785 are falls and 1015 are finds, but the Antarctic recoveries are changing these statistics rapidly.

(a) Recent Spectacular Falls

A very spectacular fall in recent times took place in Tunguska, Siberia. On June 30, 1908, a brilliant fireball occurred that was seen in broad daylight. The impact produced shock waves that were registered on seismographs in distant Europe, and also set up an air-compression wave that was detected around the world! Trees were seared of their branches and were felled from the impact over an area more than 30 km in radius. About 1500 reindeer were killed, and a man standing on the porch of his home 80 km away was knocked down. No pieces of the original object have been recovered, and no crater is found in the area; evidently the impacting body exploded in the air. Its original mass before it entered the earth's atmosphere is estimated to have been about 10^5 tons. Because the mass completely disintegrated, it is believed that it was the nucleus of a small comet, but the object would have been only about 60 m or less in diameter—many times smaller than the nuclei of comets that become visible in the sky or on photographs.

A second spectacular Siberian fall occurred on February 12, 1947, near Vladivostok. The approaching fireball was described as "bright as the sun." The impact produced 106 craters and impact holes ranging in size up to 28 m across. Trees were felled radially around each of the large craters. The entire region covers nearly 5 square kilometers. More than 23 tons of iron meteorite fragments have been recovered from the area.

There is no authenticated case of the killing of a human being by a meteorite fall; however, there have been some close calls. On September 29, 1938, a woman in an Illinois town heard a crash in the back yard. Later, it was found that a meteorite had pierced the roof of a garage 15 m away. A car was in the garage at the time, and the meteorite was found buried in the cushion of the car seat. In 1954 an Alabama woman was struck and injured by a ricocheting meteorite—the only modern case of injury.

(b) Meteorite Craters

It is estimated that in each century several meteorites strike the earth with enough force to produce craters more than 10 m across. To produce a crater of this diameter requires the fall of a meteorite with a mass of 2 tons or more; such a body would be a half-meter in diameter and could be photographed with telescopes when it was as far away as the moon. It would differ from a minor planet only in size. A large meteorite produces a crater larger than itself, for when it strikes the ground, its kinetic energy is dissipated with explosive violence.

The most famous crater definitely known to be meteoritic is the Barringer Meteorite Crater, near Winslow, Arizona. The crater is 1300 m across and 180 m deep, and its rim rises 45 m above the level of the surrounding ground. In the area over 25 tons of iron meteorite fragments have been found, some buried near the crater and many more scattered at distances up to 7 km. Sizable chunks of iron can still be found shallowly buried in the vicinity of the crater. Drillings have been made into the earth beneath the crater floor in search of a main body of the meteorite, but none has been recovered, in keeping with the view that the meteorite blew up completely into small pieces when it exploded, forming the crater. The age of the crater is estimated at about 25,000 years.

Dozens of even larger craters are believed to be meteoritic. After many tens of thousands of years, erosion makes them hard to recognize, but fossil craters many millions of years old are known, some of them tens of kilometers in diameter.

21.7 MICROMETEORITES

Many tiny meteorites strike the earth each day. Those that are only a few microns in diameter (1 micron is 10^{-4} cm) are slowed in the air before they have a chance to heat up; they eventually settle to the ground. These particles, too small to make meteors, are called *micrometeorites*.

There are two kinds of evidence for the existence of micrometeorites. (1) Many small particles,

Figure 21.9 The Barringer Meteorite Crater in Arizona. *(American Meteorite Museum)*

rich in iron and also containing silicon and magnesium, can be collected from the ground, from roofs, from rainwater, and even from the ocean floors. Although the fact has not yet been proved, these particles may be meteoritic; that is, they may be micrometeorites. (2) Impacts of small particles have been detected by space vehicles that are equipped with microphones or other types of micrometeorite detectors. These particles must be meteoritic, for they are encountered in space.

In the vicinity of the earth, micrometeorites may number from 10 to a few hundred per km^3, but this estimate is very uncertain. The annual accretion of micrometeorites by the earth is estimated at 10^5 tons, or more.

21.8 THE INTERPLANETARY MATERIAL

We see, then, that the space between the planets contains a vast number of micrometeorites. These particles comprise a distribution of *interplanetary dust*. Micrometeorites are a few microns or more in diameter. Particles less than 1 micron in size are "blown" out of the solar system by radiation pressure from the sun, just as small particles are blown out of comets in the form of comet tails (Section 20.4d). Particles the size of micrometeorites are not blown away but revolve about the sun as tiny planets. Their orbits are gradually diminishing in size, however, and they consequently spiral inward toward the sun, because of a perturbative effect known as the *Poynting-Robertson effect*.

(a) THE POYNTING-ROBERTSON EFFECT

For particles the size of micrometeorites, the sun's attractive gravitational force exceeds the repulsive force of the sun's radiation pressure. Like the planets, therefore, the micrometeorites revolve about the sun in Keplerian orbits. If the force of radiation pressure acting on such a particle were *exactly radial* (away from the sun), its only effect would be to reduce slightly the sun's attraction, resulting in a somewhat increased period of orbital revolution. However, the direction of the force of radiation pressure is not exactly radial but rather has a small component in a direction opposite to that of the particle's motion.

Suppose one runs through the rain. Even if the rain is falling vertically, it appears to strike the runner's face obliquely, because *relative to him* the raindrops have a horizontal component of motion—namely, a velocity equal and opposite to his velocity with respect to the ground. Similarly, to objects revolving about the sun, photons do not appear to come from an exactly radial direction, away from the sun, but have a slight component of velocity in a direction opposite to that of the objects' own motions. Thus, radiation pressure produces a slight "backward" force upon them. The effect of radiation pressure is negligible on particles of large mass (like the planets), but small particles, such as micrometeorites, are appreciably perturbed. The "backward" force acts like a "drag" on the orbital motion of such a particle, first making its orbit more and more circular and then gradually causing the orbit to diminish in size until the particle ultimately spirals in to the sun. A particle 1 mm in diameter that originates in the region of the minor planet belt (between the orbits of Mars and Jupiter) spirals into the sun in only 10 million years. A particle even as large as 10 cm in diameter at the earth's distance from the sun will spiral into the sun in about 10^8 years. The fact that we find small particles around the earth is evidence that they are either newly formed or have newly arrived in our part of the solar system.

Attention was first directed to the effect by J. H. Poynting in 1903; it was confirmed by a rigorous application of relativistic theory by H. P. Robertson in 1937. It is thus called the *Poynting-Robertson effect*.

Figure 21.10 Zodiacal light behind a telescope dome. The time exposure also shows star trails. (*Yerkes Observatory*)

(b) Further Evidence of Interplanetary Material

In addition to micrometeorites, or interplanetary dust, space vehicles have found evidence for interplanetary gas. Large numbers of charged atoms (ions) have been encountered. It is now well established that these ions have been expelled from the sun, and this outflow of gas is known as the *solar wind*. In the neighborhood of the earth, the ion density is from one to ten particles per cubic centimeter. Further evidence of the solar wind is provided by its effect on comets in producing the straight tails.

There is also additional evidence for the interplanetary dust (micrometeoritic material): the *zodiacal light* and the *gegenschein*.

(c) Zodiacal Light

The zodiacal light is a faint glow of light along the zodiac (or ecliptic). It is brightest along those parts of the ecliptic nearest the sun and is best seen in the west in the few hours after sunset or in the east before sunrise. Under the most favorable circumstances, the zodiacal light rivals the Milky Way in

brilliance. It is sometimes called the "false dawn" because of its visibility in the morning hours before twilight actually begins.

The zodiacal light has the same spectrum as the sun, which shows it to be reflected sunlight. Gas molecules and atoms cannot be numerous enough in space to scatter enough sunlight to contribute appreciably to the zodiacal light. Not only are molecules inefficient scatterers of light, but what light they do scatter (that is, reflect helter-skelter) is mostly blue and violet light of short wavelength—the blue daylight sky comes from the scattering of sunlight by air molecules. The present interpretation of the zodiacal light, then, is that the interplanetary dust is concentrated most heavily in the plane of the ecliptic. The dust reflects enough sunlight to produce the faint glow along the zodiac.

The total mass of the material responsible for the zodiacal light is estimated at 2.5×10^{19} g. Because of the rate at which these particles are spiralling into the sun due to the Poynting-Robertson effect, there must be a continual replacement, presumably by collisional fragmentation of objects

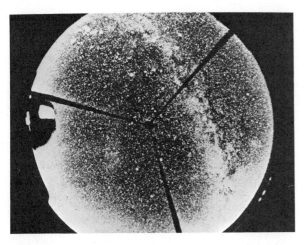

Figure 21.11 The gegenschein (patch of light just to the left and below the center) and the Milky Way, photographed by Osterbrock and Sharpless with the wide-angle Greenstein-Henyey camera. (*Yerkes Observatory*)

in the asteroid belt, in the amount of about 10 tons per second.

(d) The Gegenschein

The *gegenschein*, which means "counterglow," is a faint glow of light, centered on the ecliptic, which is exactly opposite the sun in the sky. Its angular size is from 8° to 10° by 5° to 7°. It is much more difficult to see than the zodiacal light, but it can be measured photoelectrically, and it has also been photographed. Like the zodiacal light, the gegenschein appears to be reflected sunlight.

Several explanations have been advanced for the gegenschein. The one that is now thought to be correct is that the glow is simply sunlight reflected from those solid particles opposite the sun in the sky. Such particles are viewed at the "full" phase, like the full moon, and therefore reflect more light to us than particles in other directions that are viewed at gibbous or crescent phases. It is believed that the gegenschein is, in fact, the portion of the zodiacal light where the particles reflecting sunlight are at the most favorable illumination angle.

21.9 THE ORIGIN OF METEORITIC MATERIAL

In size, the lesser bodies of the solar system range smoothly from minor planets through crater-pro-

ducing meteorites, fireball-producing meteoroids, ordinary sporadic meteoroids, and on down to micrometeorites. The numbers of particles of various sizes fit the distribution we would expect from some breakup process like rock crushing. Collisions between a few larger bodies would be extremely rare and unlikely; however, once a collision occurred and formed a larger number of smaller bodies, subsequent collisions between them would be more likely, accelerating the fragmentation.

Three recovered falls are from fireballs whose orbits were determined, and those orbits are suggestive of near-approaching asteroids, as are the orbits of some other bright fireballs. Moreover, the compositions of fallen meteorites suggest asteroidal origins. The objects could not, however, have been small meteoroids in orbit about the sun from the time the solar system was formed, for the Poynting-Robertson effect will cause rocks even 10 cm across to spiral from the asteroid belt into the sun in a thousand million years, and for smaller objects, even faster. Furthermore, meteorites are observed to have short exposure ages, indicating that they can survive in space for periods of only 20×10^6 years or less. We conclude that these meteoroids must have been formed relatively recently and that they are fragments of former asteroids.

On the other hand, whereas most or all recovered meteorites, probably micrometeorites and other matter responsible for the zodiacal light, and a small fraction of the meteoroids producing meteors, are thought to be from the asteroid belt, we recall that the overwhelming majority of meteors, including even most fireballs, are produced by particles with orbits like comets. Shower meteors, in fact, are known to be associated with cometary debris. The solid pieces from comets are expected to be of low density and very fragile, and they also tend to enter the earth's atmosphere at high speeds. Thus few, if any, could survive to the ground. We conclude that most meteor-producing meteoroids, but not those that fall to the earth's surface, have cometary origins.

Although the rocky chondritic asteroids may now be stored in orbits between Mars and Jupiter, some may have originated in the region between Mercury and Mars. An interesting meteorite research project being carried out by J. Wasson at UCLA and his associates is to search for evidence that allows particular meteorite classes to be linked to one or the other of the inner planets.

EXERCISES

1. Two meteor cameras are located 190 km apart on an east-west line. A meteor is recorded by both cameras. The westernmost camera shows its trail to begin 45° above the east point on the horizon, while the easternmost camera shows its trail to begin 45° above the west point on the horizon. At what altitude did the meteor trail begin?

2. Suppose meteoroids all moved in nearly circular orbits. What then would be the range of velocities we should observe for them, relative to the earth?

3. Comets that have been associated with meteor showers are all periodic comets. Why do you suppose showers have not been identified with comets having near-parabolic orbits?

4. Why must the perihelion of the orbit of a meteoroid swarm never be greater than 1 AU from the sun for us to see a shower produced by that swarm, and why are the perihelia of such swarms usually closer to the sun than 1 AU?

5. Why is it that a particular shower may be seen only at certain hours of the night, for example, early evening, or the hours before sunrise?

6. Suppose a meteor shower were detected in December, and that the radiant of the shower is in the constellation of Scorpio. How would the shower have been discovered?

7. From the data in Table 21.1 find the semimajor axis of the orbit of the Draconid swarm.

8. Show by a diagram how the earth encounters a swarm left by Halley's comet twice each year.

9. Two meteoroids of the same mass enter the earth's atmosphere at the same instant. Both produce observable meteors. One is moving 70 km/s, and the other is moving 35 km/s. Which meteor gives out the more light? Give an estimate of how many times as much light it produces as the other.

10. Two meteoroids enter the atmosphere side by side, moving at the same speed. An observer notes that one of the meteors produced is three times as bright as the other. Estimate the relative masses of the two objects.

11. Suppose a micrometeorite has a cubical shape and measures 4 microns on a side. If its density is five times that of water, what is its mass in grams? How many such particles would be required to make up a mass equal to that of the earth?

12. The sun's outer corona, the F corona, like the zodiacal light, appears to shine by the scattering of sunlight by particles. It is much brighter, since it is nearer the sun, than the zodiacal light. Why is it, then, that the F corona is generally observed only at times of total solar eclipses, whereas the zodiacal light is visible on many clear evenings?

13. From the data in the chapter calculate the approximate density of the largest meteorite ever found.
Answer: About 8.6 g/cm³.

14. Meteoroids can be of either cometary or minor planet origin. Which types encounter the earth with the greatest speed? Which types have a better chance of reaching the ground?

22

SURVEYING THE HEAVENS

Nearly all measurements of astronomical distances, as well as of distances on the earth, depend directly or indirectly on the principle of triangulation. Some of the concepts of triangulation were known to the Egyptians; the art was developed by the Greeks.

22.1 TRIANGULATION

Six qualities describe the dimensions of a triangle: the lengths of the three sides and the values of the three angles. It is a well-known theorem in elementary geometry that any three of these quantities in succession around the perimeter of the triangle (for example, two sides and an included angle or two angles and an included side) determine the triangle uniquely.

As an example, suppose that in the triangle ABC (in Figure 22.1) the side AB and the angles A and B are all known. The triangle can then be constructed without ambiguity, for the side AB can be laid out and the lines AE and BD can be drawn at angles A and B, respectively, to the line AB. The two lines intersect at C, which completes the construction.

(a) Application to Surveying

Suppose, in the triangle in Figure 22.1, that the point C represents an inaccessible object—say, a remote mountain peak, or an island in a large river, or the moon. The distance to C can be determined by setting up two observation stations at A and B, separated by the known distance AB. This known distance (AB) is called a *base line*. At station A the angle A is observed between the directions to B and C. At station B, the angle B is observed between the directions to A and C. Enough information is now available to construct a scale drawing of the triangle ABC. The base line (AB) is first laid

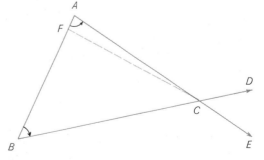

Figure 22.1 Construction of a triangle with two angles and an included side given.

391

out on the drawing at some convenient scale. Lines
AE and BD are then constructed at angles A and
B to the base line, and their intersection is at C.
The distance of C from any point on the base line
(for example, A, or B, or F) can now be measured in the drawing; since the scale of the drawing
is known, the real distances to C are determined.

The preceding two paragraphs describe a *geometrical* method for solving a triangle. In practice
the triangle can be solved more simply and more
accurately by numerical calculation. The solution of
triangles by calculation rather than by geometrical
construction is the subject of *trigonometry*.

(b) The "Skinny" Triangle

In astronomy we frequently have to measure distances that are very large in proportion to the length
of the available base line, and the triangle to be
solved is thus long and "skinny." Suppose (Figure
22.2) that it is desired to measure the distance to
the moon or a minor planet, located at O. Two observation stations, A and B, are set up on the
earth. The base line AB is known, since the size of
the earth and the geographical locations of A and B
are known. From A, O appears in direction AS,
in line with a distant point, such as some very remote star. Because of the great distance of that star,
observers at A and B would look along parallel
lines to see it; thus from B the same star is in direction BS'. Observer B sees O, on the other
hand, in direction BT, at an angle p away from
direction BS'. This angle, p, is the same as the
angle at O between the lines AO and BO (because
the line BO intercepts the parallel lines AS and
BS'). Note that p is the difference in the directions
of O as seen from A and B; this is the apparent
displacement (or parallax) of an object seen from
two different points (Section 2.2e).

Now the problem is to find the distance to O,
which lies somewhere along the line AS. This can
be done in two ways. One way is to determine the
angles OAB and OBA and then determine the distance to O by solving the triangle OAB as described in Section 22.1a. However, because the
sides AO and BO are so nearly parallel, the two
angles would have to be measured with very great
precision. It would be very difficult to obtain adequate precision, because at each station (A and B)
it would be necessary to compare the direction of O
with that of the other terrestrial observation point.

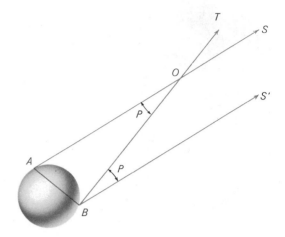

Figure 22.2 "Skinny" triangle.

An alternative procedure for finding the distance
AO is to measure the parallax p and then calculate
how distant O must be in order for the base line
AB to subtend that angle. Such a calculation is illustrated in Section 22.1c.

Because observers at A and B both see O projected against remote stars on the celestial sphere,
each can measure $O's$ direction among those stars
with great accuracy; the difference between those
directions, the parallax, can be found, therefore,
with considerable precision. In astronomical practice the greatest emphasis is thus placed on measuring the parallax, rather than the angles at the end
of the base line. Even the parallax, however, can
be determined only as accurately as the direction to
O can be measured with respect to the background
of stars on the celestial sphere. Under the most favorable circumstances parallaxes can be measured
to within a few thousandths of a second of arc. The
more distant the object, the smaller its parallax and
the greater the uncertainty of its value in comparison to its size, that is, the greater its *percentage error*, and the greater the uncertainty in the distance
to the object that is derived from its parallax. It is
advantageous, therefore, that the base line be made
as long as possible, so that the parallax will be as
large as possible in comparison to the error of its
measurement.

(c) AN EXAMPLE OF THE SOLUTION OF THE "SKINNY" TRIANGLE

Suppose it is found that the displacement in direction
of an object (at O, in Figure 22.3) viewed from op-

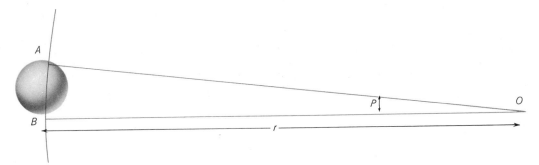

Figure 22.3 Solution of the "skinny" triangle.

posite sides of the earth is the angle p; p, then, is the angle at O subtended by the diameter of the earth. Imagine a circle, centered on O, that passes through points A and B on opposite ends of a diameter of the earth. If the distance of O is very large compared to the size of the earth, then the length of the chord AB is very nearly the same as the distance along the arc of the circle from A to B. This arc is in the same ratio to the circumference of the entire circle as the angle p is to 360°. Since the circumference of a circle of radius r is $2\pi r$, we have

$$\frac{AB}{2\pi r} = \frac{p}{360°}.$$

By solving for r, the distance to O, we find

$$r = \frac{360°}{2\pi} \frac{AB}{p}.$$

If p is measured in seconds of arc, rather than in degrees, it must be divided by 3600 (the number of seconds in 1°) before its value is inserted in the above equation. After such arithmetic, the formula for r becomes

$$r = 206,265 \frac{AB}{p(\text{in seconds})}.$$

As an example, suppose p is 18 seconds of arc (about what would be observed for the sun). Since AB, the earth's diameter, is 12,756 km,

$$r = 206,265 \frac{12,756}{18} = 1.46 \times 10^8 \text{ km}.$$

In Exercise 2 the moon's distance is to be calculated by the same procedure.

22.2 RELATIVE DISTANCES IN THE SOLAR SYSTEM

The planets and other bodies that revolve about the sun in the solar system are so far away that at any time each one is seen in almost the same direction by all terrestrial observers. The parallaxes of these bodies, are therefore, very small and are difficult to measure accurately; a larger base line than the diameter of the earth is required. As the earth moves about the sun, however, it carries us across a base line that can be as large as the diameter of the earth's orbit—about 300 million km. Thus, we can determine accurate distances to the other members of the solar system by observing them at times when the earth is in two different places in its orbit. Kepler arranged that the different observations of the same body were made at intervals of its sidereal period when it was at the same place in its orbit (Section 3.3). This is not necessary, however, for we can take into account the motion of the object during the interval between the sightings of it. It is convenient, of course, not to have to restrict our observations of a body to intervals of its sidereal period. There are various mathematical techniques (which will not be gone into here) for unscrambling the effects of the combined motions of the body and the earth.

Measures described above—measures within the solar system—are not obtained directly in kilometers however; they are found in terms of the *astronomical unit*, the semimajor axis of the earth's orbit. The foregoing procedure for surveying the distances to the planets provides us with an accurate map of the solar system, but to find the scale of the map—that is, to evaluate the astronomical unit—we must find the distance to some object that revolves about the sun, both in astronomical units and in kilometers (or some other terrestrial units).

22.3 DETERMINATION OF THE LENGTH OF THE ASTRONOMICAL UNIT

The earliest known attempt to find the length of the astronomical unit was made by Aristarchus in the third century B.C. (Section 2.3a). His value of 200 earth diameters for the distance to the sun (the earth's size was measured by Aristarchus' contemporary, Eratosthenes—Section 2.3b) was later corrected to 600 earth diameters by Hipparchus (Section 2.3c). The latter value survived until the 17th century, when Kepler estimated that it was too small by a factor of at least three because of the absence of a diurnal parallax of Mars (Section 3.3).

Today the astronomical unit is known from radar measures with high precision. Before describing the modern procedure for evaluating it, however, we shall review some of the historical approaches, which in themselves are very instructive.

(a) Older Methods of Determining the Astronomical Unit

Before radar some of the fairly accurate methods of measuring the AU were the following:

1. *Direct triangulation of the sun.* It is difficult to observe the sun's direction in the sky with respect to the stars. Occasionally, however, Mercury or Venus can be seen in transit across the sun's disk. The exact times of the start and finish of such a transit are different for observers at different places on the earth, and the differences depend on the distances of both the sun and the planet. Since the relative distances are already known, transit observations can, in principle, be used to triangulate both the sun and the planet. This method was suggested by the English astronomer Edmund Halley, but the observational problems are so great that it never yielded a satisfactory result.

2. *Direct triangulation of planets.* Because the distances to the planets are known directly in AU, there is no need to attempt to measure the sun's parallax directly; better, we can triangulate the distance to Mars, which at its closest is only $1/3$ AU away and is favorably situated for observations during an entire night. (We cannot use the moon, however, for although its distance is found directly with considerable accuracy in terrestrial units, say, kilometers, it cannot be surveyed in AU from different sides of the sun, for as the earth moves in its orbit, the moon follows along with us.)

3. *Triangulation of asteroids.* Earth-crossing asteroids (Section 19.2e) come even closer than Mars. In 1932 the British astronomer Harold Spencer Jones coordinated a worldwide effort to triangulate the distance to the asteroid Eros, which passed only 22 million km from earth and had a parallax of about $2'$.

4. *Gravitational perturbations of asteroids.* As an asteroid (for example, Eros) comes near the earth, its orbit is changed by the earth's gravitational effect on it. We can calculate the shape and size of its orbit, in AU, by surveying from various parts of our own orbit, both before and after the encounter, thereby obtaining an accurate measure of how the earth's gravitational perturbation on the asteroid changed its orbit. The earth's gravitational influence that produced this change, however, depends on how far in kilometers from the center of the earth the asteroid must have passed. Since we know how far the asteroid passed in astronomical units, the number of kilometers per AU is determined. Because the gravitational pull of the earth drops off with the *square* of the distance from earth, the actual error in the derived distance is proportional to only the *square root* of the errors in the observations of the orbital changes, so this is a more sensitive determination than is that of direct triangulation. Astronomer Eugene Rabe used this technique to find the distance to Eros when it passed in 1932.

5. *Doppler shifts.* Stars near the ecliptic show periodic changes in radial velocity during the year because of the earth alternately approaching and receding from them. These changes, measured from the changing Doppler shifts of the stars' spectral lines (Section 10.5b), tell us how fast the earth is moving in its orbit. This speed, in

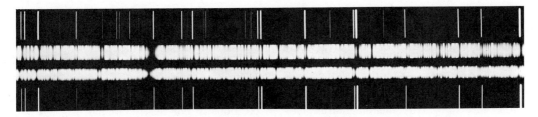

Figure 22.4 Two spectrograms of Arcturus, taken six months apart. On July 1, 1939 *(top)*, the measured radial velocity was +18 km/s; on January 19, 1940, it was −32 km/s. The difference of 50 km/s is due to the orbital motion of the earth. *(Caltech/Palomar Observatory)*

km/s, multiplied by the number of seconds in a year, gives us the circumference of the earth's orbit and hence the length of the astronomical unit.

6. *Aberration of starlight.* The 20″.5 displacement of stars in the direction of the earth's motion (Section 7.3) indicates the speed of the earth in terms of that of light. In 1862 the French physicist Jean Foucault used his measurement of the speed of light to find the earth's speed and hence the distance to the sun.

(b) Modern Determination of the Astronomical Unit

We have already seen (Section 9.2a) how measurement of the round-trip travel time of the radio waves beamed to the moon and reflected back gives the moon's distance accurately. Radar observations of Venus, the nearest approaching major planet, began in 1958. Today radar observations are routinely carried out on all but the most remote planets. The radar determination of the length of the astronomical unit requires a knowledge of the speed of light, c. The Jet Propulsion Laboratory finds 1 AU = (499.004784s) $\times$ c. For c = 299,792.458 km/s, the value of the astronomical unit is 149,597,870.7 km, with an uncertainty of about 1 km. The use of radar now replaces direct triangulation, but its accuracy is limited by our knowledge of the precise sizes of the earth and the observed planets, whose surfaces reflect the radio waves.

The *solar parallax* is defined as the angle subtended by the radius of the earth at a distance of 1 AU. The modern radar determination of the AU gives a value for the solar parallax of 8″.7940; before 1961 the best value was 8″.798, obtained by Rabe from the perturbations of Eros when it passed the earth in 1932.

22.4 SURVEYING DISTANCE TO STARS

In principle one should be able to detect the parallax of a star—a change in its apparent direction—as the earth moves from one side of its orbit to the other—over a baseline of 2 AU. The stars are so remote, however, that none has a parallax perceptible to the unaided eye. This fact alone, by the way, shows the stars to be self-luminous. At their great distances, to reflect enough sunlight back to us to account for their apparent brightnesses they would have to be extremely large and show observable disks, which they do not.

In the century and a half following the publication of Newton's *Principia*, by which time the motion of the earth was generally conceded, a number of investigators had tried to detect stellar parallaxes telescopically. One of them, the German-English astronomer William Herschel, catalogued many instances of two stars appearing close together in the sky. Herschel thought that careful observation of the brighter, presumably nearer star with respect to the fainter, presumably more remote one might reveal the parallax of the former. What the observations showed, however, was that the stars in a pair are in mutual revolution about each other; Herschel had discovered not the parallax he was searching for but rather that many stars are members of binary star systems. As we shall see in Chapter 26, Herschel's discovery was even more important than parallax, for by analyzing binary stars we learn a great deal about stellar masses and sizes; in addition, Newton's laws were shown to apply to these stars as well as to the planets.

Another serendipitous discovery was made during an attempt to detect stellar parallax by James Bradley in 1729. Bradley mounted a vertical telescope in his chimney and measured the positions of stars as the rotation of the earth carried them past an illuminated reticle or scale in the field of view of his telescope. He had hoped that some relatively nearby stars, because of parallax, would pass the reticle in different places at different times of the year. What he found instead is that *every* star shifts its direction periodically during the year by 20″5 in either direction. This phenomenon is *aberration of starlight*, already described.

Parallaxes were eventually detected, but even the nearest star shows a total annual displacement of only about 1″5. The first observation of the parallax of a star is usually credited to the German astronomer Friedrich Bessel, who measured the parallax of 61 Cygni in 1838 and obtained a value only about 6 percent too large. Later in the same year Thomas Henderson, at the Cape of Good Hope, and Friedrich Struve, in Russia, reported measures of the parallaxes of the stars Alpha Centauri and Vega, respectively. At last quantitative measurements had extended beyond the solar system, opening the new era of stellar astronomy.

(a) The Parallactic Ellipse

As the earth moves about its orbit, the place from which we observe the stars continually changes. Consequently, the positions of the comparatively near stars, projected against the more remote ones, are also continually changing. If a star is in the di-

rection of the ecliptic, it seems merely to shift back and forth in a straight line as the earth passes from one side of the sun to the other. A star that is at the pole of the ecliptic (90° from the ecliptic) seems to move about in a small circle against the background of more distant stars, as we view it from different positions in our nearly circular orbit. A star whose direction is intermediate between the ecliptic and the ecliptic pole seems to shift its position along a small elliptical path during the year. The eccentricity of the ellipse ranges from that of the earth's orbit (nearly a circle) for a star at the ecliptic pole to unity (a straight line) for a star on the ecliptic (see Figure 22.5). This small ellipse is called the *parallactic ellipse*.

(b) Stellar Parallax and Stellar Distances

The angular semimajor axis of the parallactic ellipse is called the *stellar parallax* of the star. Since the major axis of the ellipse is the maximum apparent angular deflection of the star as viewed from opposite ends of a diameter of the earth's orbit, the stellar parallax, the semimajor axis of this ellipse, is the angle, at the star's distance, subtended by 1 AU perpendicular to the line of sight.

Until now we have implied that the star whose parallax is observed is motionless with respect to

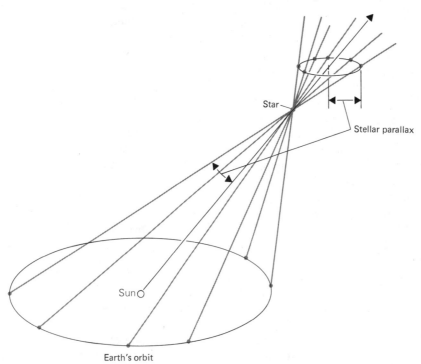

Star

Stellar parallax

Sun○

Earth's orbit

Figure 22.5 The parallactic ellipse.

the sun. Actually, the stars are all moving at many kilometers per second (Chapter 23). The effects of the relative motion of the star and the sun can be separated from the effects of the earth's motion (the star's parallax) by observing the star's direction at the same time of the year several years apart. Any observed change in its direction then must be due to its own motion relative to the sun. The change indicates the corrections that must be applied to observed total changes in the direction of the star in order to obtain its true stellar parallax.

(c) Units of Stellar Distance

If a line of length D subtends an angle of p seconds of arc as seen from a distant object, the distance r of that object is given by the formula derived in Section 22.1c:

$$r = 206{,}265\frac{D}{p}.$$

Since the parallax of a star is the angle, in seconds, subtended by 1 AU at the star's distance, the distance of a star, in astronomical units, is $206265/p$. The length 206,265 AU is defined as a *parsec* (abbreviated pc). One parsec is, therefore, the distance of a star that would have a *parallax* of one *second*. The distance of any star, in parsecs, is thus given by

$$r = \frac{1}{p},$$

where p is the parallax of the star. A star with a parallax of $1/2''$, for example, has a distance of 2 pc; one with a parallax of $1/10''$ has a distance of 10 pc. Parallaxes of stars are usually measured in seconds, so the parsec is a convenient unit of distance—a star's distance in parsecs is simply the reciprocal of its parallax. One parsec is 3.086×10^{13} km. Since 1 pc is 206,265 AU, the sun's distance is $1/206{,}265$ pc.

Another unit of stellar distance is the *light year*, which is the distance traversed by light in one year at the rate of 299,792.5 km/s. One light year (abbreviated LY) is 9.46×10^{12} km; 1 pc contains 3.26 LY.

(d) The Nearest Stars

The nearest stellar neighbors to the sun are three stars that make up a multiple system. To the naked eye the system appears as a single bright star, Alpha Centauri, which is only 30° from the south celestial pole and hence is not visible from the mainland United States. Alpha Centauri itself is a double star—two stars revolving about each other that are too close together to be separated by the naked eye. Nearby is the third member of the system, a faint star known as *Proxima Centauri*. Proxima is slightly closer to us than the other two stars of the system. All three have a parallax of about $0\rlap{.}''76$, and a distance of $1/0.76$, or about 1.3 pc (4.3 LY). The nearest star, except for the sun, visible to the naked eye from most parts of the United States is the brightest-appearing of all the stars, *Sirius*. Sirius has a distance of 2.6 pc, or about 8 LY. It is interesting to note that light reaches us from the sun in eight minutes and from Sirius in eight years.

Parallaxes have been measured for thousands of stars. Only for a fraction of them, however, are the parallaxes large enough (about $0\rlap{.}''05$ or more) to be measured with a precision of 10 percent or better. The 1969 edition of Gliese's catalogue of nearby stars lists 1049 within 20 pc, but the total number of such stars, including those not yet discovered, may be near 4000. Of those stars within about 20 pc, most are invisible to the unaided eye and actually are intrinsically less luminous than the sun. Most of the stars visible to the unaided eye, on the other hand, have distances of hundreds or even thousands of parsecs and are visible not because they are relatively close, but because they are intrinsically very luminous. The nearer stars are described more fully in Chapter 27.

22.5 OTHER METHODS OF MEASURING STELLAR DISTANCES

The vast majority of all known stars are too distant for their parallaxes to be measured, and we must resort to other methods to determine their distances. Most of these methods are either statistical or indirect; they are discussed in later chapters. For completeness, however, some of the more important procedures, other than parallax measurement, for determining stellar distances are listed:

1. *Stellar motions (Chapter 23)*. All stars are in motion, but only for comparatively nearby ones are the angular motions perceptible. Statistically, therefore, the stars that have

large apparent motions are the nearer ones; we can estimate the average distance to stars in a large sample from the average angular motions of those stars.

2. *Moving clusters (Chapter 23).* In a few cases the direction of motion through space of a cluster or swarm of stars can be determined from the apparent convergence or divergence of the directions of motions of the individual stars in that cluster. In such a case, an analysis of the apparent motions and radial velocities of the member stars gives the distance to the cluster.

3. *Inverse-square law of light (Chapter 24).* The apparent brightness of a star depends on both its intrinsic luminosity and its distance (through the inverse-square law of light). Very often it is possible to infer the intrinsic brightness of a star from its spectrum (Chapter 27) or because it is a recognizable type of variable star (Chapter 24). Then its distance can be calculated from a knowledge of its observed brightness.

4. *Interstellar lines (Chapter 28).* The space between the stars throughout much of space contains a sparse distribution of gas. Sometimes this interstellar gas leaves absorption lines superposed upon the spectrum of a star whose light must shine through the gas to reach us. The amount of the star's light absorbed by these interstellar lines indicates the total mass of the gas that must lie in the light path from the star. If we can estimate the approximate density of the gas in space (as we often can) we can tell what the total path length of the star's light must be, and hence the distance of the star.

5. *Galactic rotation (Chapter 29).* The sun and its neighboring stars are part of a vast system of stars—our Galaxy. The Galaxy is rotating; the stars that compose it all revolve about its center much as the planets revolve about the sun. The speeds with which distant stars in the Galaxy approach us or recede from us as a result of this galactic rotation depends on the directions and distances of these stars. Observations of their radial velocities, therefore, can lead to an estimate of their distances.

EXERCISES

1. At a distance of 3500 m from the base of a vertical cliff, the top of the cliff is observed to have an angular altitude of 45°. What is its height?

2. The moon's position among the stars was observed at moonrise. Six hours and 12 minutes later the moon happened to be exactly at the zenith. Its position among the stars by then had shifted to the east by 148′. On the other hand, because of the moon's orbital motion, it is known to move eastward on the celestial sphere at the rate of 33′ per hour. Find the distance to the moon. Compare your result with the value given in Section 9.2.

3. At a particular opposition of Mars, the planet was simultaneously observed from two points on the equator—one where it was rising and one where it was setting. Its direction among the stars from the two observing stations differed by 41″. What was the distance to Mars in astronomical units? *Answer:* 0.43 AU

4. Why would observations such as those described in Exercise 3 be very difficult to make?

5. When the minor planet Geographos is only 10 million km away, how large an angular displacement in direction can be observed for it from different places on earth?

6. Describe how you might organize a program to determine the length of the astronomical unit from observations of Geographos. Explain what kinds of observations would be needed, how many would be desired, and what use you would make of them.

7. A star lying on the ecliptic and exactly 90° to the west of the sun is observed to have a radial velocity of exactly 48 km/s toward the earth. Six months later the star's radial velocity is 12 km/s away from us. What is the star's radial velocity with respect to the sun? What is the earth's orbital velocity? Show how this gives the value of the astronomical unit.

8. Assume the earth's orbit to be circular, and use the orbital velocity found above to evaluate the astronomical unit in kilometers.

9. Suppose that when Mars is exactly 56 million km away, radar waves are beamed toward it. Exactly 4½ minutes later a return "beep" is picked up that the radar operator thinks is a radar echo reflected from Mars. Do you agree with him? Why?

10. Show that a light year contains 9.46×10^{12} km.

11. Show that 1 pc equals 3.086×10^{13} km and that it *also* equals 3.26 LY. Show your reasoning.

12. Make up a table relating the following units of astronomical distance: kilometer, earth radius, astronomical unit, light year, parsec.

13. Another convenient unit of angular measure is the *radian,* defined such that 2π radians = 360°. Calculate the number of degrees in a radian and the number of seconds of arc in a radian.

14. Show that if parallax is measured in radians (see Exercise 13), then distance = baseline/parallax in radians.

23

MOTIONS OF STARS

The ancients distinguished between the "wandering stars" (planets) and the "fixed stars," which seemed to maintain permanent patterns with one another in the sky. The stars are, indeed, so nearly fixed on the celestial sphere that the apparent groupings they seem to form—the constellations—look today much as they did when they were first named, more than 2000 years ago. Yet the stars are moving with respect to the sun, most of them with speeds of many kilometers per second. Their motions are not apparent to the unaided eye in the course of a single human lifetime, but if an ancient observer who knew the sky well—Hipparchus, for example—could return to life today, he would find that several of the stars had noticeably changed their positions relative to the others. After some 50,000 years or so, terrestrial observers will find the handle of the Big Dipper unmistakably "bent" more than it is now. Changes in the positions of the nearer stars can be measured with telescopes after an interval of only a few years.

23.1 ELEMENTS OF STELLAR MOTION

(a) Proper Motion

The *proper motion* of a star is the rate at which its direction in the sky changes, usually expressed in seconds of arc per year. It is almost always an angle

that is too small to measure with much precision in a single year; in an interval of 20 to 50 years, on the other hand, many stars change their directions by easily detectable amounts. The modern procedure for determining proper motions is to compare the positions of the star images on two different photographs of the same region of the sky taken at least a decade apart. Most of the star images on

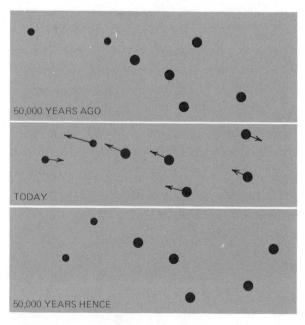

Figure 23.1 Appearance of the Big Dipper over 100,000 years.

Figure 23.2 Two photographs of Barnard's star, showing its motion over a period of 22 years. *(Yerkes Observatory)*

such photographs do not appear to have changed their positions measurably; these are, statistically, the more distant stars that are relatively "fixed," even over the time interval separating the two photographs. With respect to these "background" stars, the motions of a few comparatively nearby stars can be observed. Modern practice is to measure proper motions of stars with respect to remote galaxies, which can show no measurable proper motions themselves (Chapter 37).

The star of largest known proper motion is *Barnard's star,* which changes its direction by 10″.25 each year. This large proper motion is partially due to the star's relatively high velocity with respect to the sun but is mostly the result of its relative proximity. Barnard's star is the nearest known star beyond the triple system containing Alpha Centauri; its distance is only 1.8 pc, but it emits less than $1/2500$ as much light as the sun and is about 25 times too faint to see with the unaided eye. There are a few hundred stars with proper motions as great as 1″.0. The mean proper motion for all naked-eye stars is less than 0″.1; nevertheless, the proper motions of most stars are larger than their stellar parallaxes.

The complete specification of the proper motion of a star includes not only its angular rate of motion, but also its *direction* of motion in the sky.

(b) Radial Velocity

The *radial* velocity (or line-of-sight velocity) of a star is the speed with which it approaches or recedes from the sun. This can be determined from the Doppler shift of the lines in its spectrum (Section 10.5b). Unlike the proper motion, which is observable only for the comparatively nearby stars, the radial velocity can be measured for any star that is bright enough for its spectrum to be photographed. The radial velocity of a star, of course, is only that component of its actual velocity that is

projected along the line of sight, that is, that carries the star toward or away from the sun. Radial velocity is usually expressed in kilometers per second and is counted as *positive* if the star is moving *away* from the sun, and *negative* if the star is moving *toward* the sun. Since motion of either the star or the observer (or both) produces a Doppler shift in the spectral lines, a knowledge of the radial velocity alone does not enable us to decide whether it is the star or the sun that is "doing the moving" (indeed, as we saw in Chapter 13, it does not even make sense to ask which is moving). What we really measure, therefore, is the speed with which the distance between the star and sun is increasing or decreasing, that is, the star's radial velocity *with respect to the sun.*

Actually, the radial velocity of a star with respect to the sun is not obtained directly from the measured shift of its spectral lines, because we must observe the star from the earth, whose rotational and orbital motions contribute to the Doppler shift. Since the direction of the star is known, however, as are the speed and direction in which the moving earth carries the telescope at the time of observation, it is merely a problem in geometry (albeit a slightly complicated one) to correct the observed radial velocity to the value that would have been found if the star had been observed from the sun.

(c) Tangential Velocity

Radial velocity is a motion of a star along the line of sight, while proper motion is produced by the star's motion *across,* or at right angles to, the line of sight. Whereas the radial velocity is known in kilometers per second, the proper motion of a star does not, by itself, give the star's actual *speed* at right angles to the line of sight. The latter is called the *tangential* or *transverse* velocity. To find the tangential velocity of a star, we must know both its

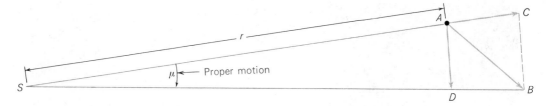

Figure 23.3 Relationship of proper motion, radial velocity (AC), and tangential velocity (AD).

proper motion and *distance*. A star with a proper motion of 1″.0, for example, might have a relatively low tangential velocity and be nearby, or a high tangential velocity and be far away.

The relation between tangential velocity and proper motion is illustrated in Figure 23.3. As seen from the sun S, a star A, at distance r, is in direction SA. During one year it moves from A to B and then appears in direction SD, at an angle μ (the proper motion) from SA. The star's radial motion is AC—it has moved a distance AC farther away during the year; its *tangential* motion is AD—it has moved that distance across the line of sight. The motions shown in the figure are grossly exaggerated; actual stars do not move enough to change their distances by an appreciable percentage, even in 100 years (see Exercise 3).

The tangential motion AD can be approximated very accurately by a small arc of a circle of radius r centered on the sun. The arc AD is the same fraction of the circumference of the circle, $2\pi r$, as the proper motion is of 360°. The proper motion is expressed in seconds of arc per year, so we have

$$\frac{\mu}{1,296,000} = \frac{AD}{2\pi r}$$

(there are 1,296,000″ in 360°). If we solve for AD in km/yr, and note that AD is the product of the star's tangential velocity in km per second and the number of seconds in a year (3.16×10^7), we obtain

$$T = \frac{\mu r}{6.52 \times 10^{12}} \text{ km/s,}$$

where r must be in kilometers. If r is to be expressed in parsecs, we must multiply the right-hand side of the above equation by 3.086×10^{13},

the number of kilometers per parsec. The formula for the tangential velocity then becomes

$$T = 4.74 \, \mu r = 4.74 \frac{\mu}{p} \text{ km/s,}$$

where p is the stellar parallax, which is merely the reciprocal of the distance in parsecs (Section 22.4c). The above equations show how the tangential velocity of a star is related to its proper motion and distance (or parallax).

(d) Space Velocity

The *space velocity* of a star is its total velocity in kilometers per second with respect to the sun. The radial velocity is the distance the star moves toward or away from the sun in one second; the tangential velocity is the distance it moves at right angles to the line of sight in one second. The space velocity, therefore, the total distance the star moves in one second, is simply the hypotenuse of the right triangle whose sides are the radial and tangential velocities (Figure 23.4). It is found immediately by the theorem of Pythagoras,

$$V^2 = V_r^2 + T^2$$

where V and V_r are the space and radial velocities, respectively.

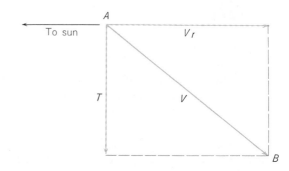

Figure 23.4 Space velocity; the star moves from A to B in one second.

23.2 SOLAR MOTION AND PECULIAR VELOCITIES OF STARS

It might be expected that the sun, a typical star, is in motion, just as the other stars are. We turn now to the motion of the sun, and how it influences the apparent motions of the stars.

(a) The Local Standard of Rest

As we know, the sun is a member of our Galaxy, a system of a hundred thousand million stars. The Galaxy is flat, like a pancake, and is rotating. The sun, partaking of this general rotation of the Galaxy, moves with a speed of about 250 km/s to complete its orbit about the galactic center in a period of about 200 million years. At first thought it might seem that the galactic center is the natural reference point with which to refer the stellar motions. However, our observations of the proper motions and radial velocities of the stars that surround the sun in space, all in our own so-called local "neighborhood" of the Galaxy, do not give us directly the motions of these stars about the galactic center. The reason is that the stars' orbits around the galactic center and their orbital velocities are both nearly the same as those of the sun. The motions we observe are merely small residual or *differential* motions of these stars with respect to the sun. These small residual motions arise because our neighboring stars' orbits about the galactic center are not absolutely identical to our own. We are overtaking and passing some stars, while others are passing us; the slightly different eccentricities and inclinations of our respective orbits bring us closer to some stars and carry us farther from others. We can study these residual motions without knowing anything about the actual motions of stars around the center of the Galaxy. Our situation is analogous to that of our driving an automobile on a busy highway. All the cars around us are going the same direction and at roughly the same speed, but some are changing lanes and others are passing each other. More or less like the highway traffic, the residual motions of the stars around us seem to be helter-skelter.

Astronomers have defined a reference system (that is, a coordinate system) within which the motions of the stars in the solar neighborhood—within a hundred parsecs or so—average out to zero. In other words, those stars in our neighborhood appear, on the average, to be at rest in this system; it is thus called the *local standard of rest*. The local standard of rest is not really at "rest," of course, but shares the average motion of the sun and its neighboring stars around the center of the Galaxy. We would not realize, however, that we are part of a huge rotating galaxy if we observed only stars near us. In fact, for many purposes it is useful to forget about the rotation of the Galaxy as a whole and to consider only the motions of stars, and of the sun, with respect to the local standard of rest, which we can pretend, for these purposes, to be really at rest.

(b) The Solar Motion and Solar Apex

We deduce the motion of the sun with respect to the local standard of rest by analyzing the proper motions and radial velocities of the stars around us. The easiest way to understand how the sun's motion is found is to consider the effect it has on the apparent motions of the other stars.

First, consider the radial velocities of stars with respect to the sun. Note that we have defined the local standard of rest as being stationary with respect to the average of the motions of the stars in the solar neighborhood. Therefore, if we could correct the observed space motions of the stars to those values they would have if the sun were not moving in the local standard of rest, they would then average out to zero. Now it is clear that in a direction that is at right angles to the direction in which the sun is actually moving, the solar motion cannot affect the observed radial velocities of the stars. In those directions, indeed, we find as many stars approaching us as receding from us—their radial velocities *do* average to zero. On the other hand, if we look in the direction *toward* which the sun is moving, we find that most of the stars are approaching us, because, of course, we are moving forward to meet them. The only stars in that direction that have radial velocities of recession are those that are moving in the same direction we are going, but at a faster rate, so that they are pulling away from us. The observed radial velocities of all the stars in the direction toward which the sun is moving do not average to zero, but to −20 km/s, showing that we are moving toward them at about 20 km/s. Similarly, stars in the opposite direction have an average radial velocity of about +20 km/s, because we are pulling away from them at that speed.

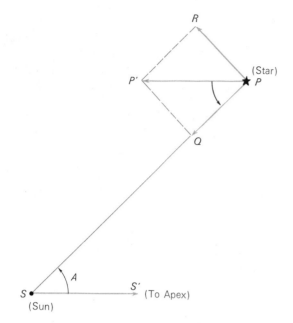

Figure 23.5 How the solar motion affects a star's space motion.

Now consider the proper motions of stars. Part of a star's proper motion, in general, is due to its own motion and part is due to the sun's motion. However, the sun's motion can contribute nothing to a star's proper motion if the star happens to lie in the direction toward which the sun is moving. Therefore, if we look at stars that lie in a path along the direction of the solar motion, as many of their proper motions should be in one direction as any other; the average of the motions of many stars in those directions, therefore, should be zero. On the other hand, the maximum effect on the proper motions of stars should occur in directions that are at right angles to the direction of the solar motion. If the stars were at rest, they would all show a backward drift due to our forward motion. As it is, the stars have motions of their own, but only those moving in the same direction we are and at a faster rate appear to have "forward" proper motions—the rest, by far the majority, *do* appear to drift backward.

William Herschel was the first to attempt to detect the direction of the solar motion from the proper motions of stars. In 1783 he analyzed the proper motions of 14 stars and deduced that the sun was moving in a direction toward the constellation Hercules—a nearly correct result.

Modern analysis of the proper motions and radial velocities of the stars around the sun has shown that the sun is moving approximately toward the

direction now occupied by the bright star Vega in the constellation of Lyra. The value found for the sun's speed depends somewhat on what stars are observed to determine it. Analysis of most of the stars in the standard catalogues gives the *standard solar motion*, which is 19.5 km/s (4.14 AU/yr). However, the sun's motion relative to its nearest neighbors, called the *basic solar motion*, is a little less—about 15.4 km/s. The direction in the sky toward which the sun is moving is called the *apex* of solar motion,[*] and the opposite direction, away from which the sun is moving, is called the *antapex*.

(c) Peculiar Velocities of Stars

The velocity of a star with respect to the local standard of rest is called its *peculiar velocity*. The *space velocity* of a star, its motion with respect to the sun, is made up of both the star's peculiar velocity and a component due to solar motion. Since the solar motion is known, the peculiar velocity can be calculated for a star of known space velocity.

Consider, for example, the special case of a star that is not moving with respect to the local standard of rest and which therefore has zero peculiar velocity. The entire space motion of that star, then, is merely a reflection of the solar motion; its space velocity is a vector of magnitude exactly equal to the solar velocity but directed toward the *solar antapex*. In Figure 23.5, the vector SS' and PP' represent the solar velocity and that star's space velocity, respectively; A is the angle at the sun between the directions to the star and to the solar apex. Vector PQ, the projection of PP' onto the line of sight from the sun to the star, is the star's radial velocity. Vector PR, the component of PP' that is perpendicular to the line of sight, is the star's tangential velocity. Since this hypothetical star has no peculiar velocity, PQ and PR are due entirely to the solar motion and are, therefore, the corrections that must be applied to the observed radial and tangential velocities of *any* star in that same direction to obtain, respectively, its *peculiar radial velocity* and *peculiar tangential velocity*. The angle A can always be observed; consequently, the desired corrections to the radial and tangential velocities can always be found geometrically or calcu-

---

[*]The equatorial coordinates of the *solar apex* are (1950) $\alpha = 18^h 4^m \pm 7^m$, $\delta = +30° \pm 1°$.

lated trigonometrically.† The correction to the star's observed proper motion that is required to obtain its *peculiar proper motion* can be found from the relation between proper motion and tangential velocity given by the formula in Section 23.1c. The peculiar proper motion can be found only for stars of known distance.

23.3 DISTANCES FROM STELLAR MOTIONS

The proper motions of stars can be expected to be largest, statistically, for the nearest stars. If, for example, a star is only a few parsecs away, its proper motion will almost certainly be observable after a few years (but see Exercise 12). The proper motion of a very distant star, on the other hand, may only be detectable after a long time, and then only if the star has a very great space velocity. Searches for nearby stars, therefore, are usually conducted by searching for stars of large proper motion. Conversely, remote stars can be identified by their lack of observable proper motions; they serve as standards against which we can measure the parallaxes of the nearby stars to determine their distances. The proper motion of an individual star does not in itself indicate its distance uniquely. However, proper motions and distances of stars are inversely correlated, and investigations of such motions do give some information about stellar distances.

(a) SECULAR PARALLAXES

Trigonometric parallaxes of stars are based on triangulation with a base line of only 2 AU, the diameter of the earth's orbit. The solar motion, however, amounts to 4.1 AU/yr. If we wait long enough between observations, the motion of the sun will carry us over as long a base line as we desire. For example, after 20 years we move 82 AU; with such a base line we should be able to detect parallaxes of stars 41 times as distant as we can with a base line of 2 AU. By this method, accurate distances to many stars could be determined if they did not have peculiar velocities of their own. Un-

fortunately, without knowing a star's distance in advance, we have no way of knowing how much of its proper motion is due to the solar motion and how much to its own peculiar velocity. On the other hand, the peculiar velocity of a large number of stars, spaced well about the sky, should average out to zero, for peculiar velocities are measured with respect to the local standard of rest, which is *defined* to have zero average velocity with respect to stars in the sun's neighborhood. Consequently, the *average* of the space velocities for a large sample of stars must be due entirely to the solar motion. The proper motion corresponding to that average space velocity depends upon the average of the distances to those stars. This motion is an average angular drift of the stars in the sample toward the solar antapex; it results from the fact that the sun carries us over a long base line. Since we know how far the sun carries us in any interval of time, we can calculate the average of the distances for the stars in question (more accurately, we calculate the *mean parallax*). The procedure is called the method of *secular parallaxes*.

At first thought, it might not seem particularly useful to know only the average of the distances to a large number of stars, because some of those stars will be much more distant than others and we have no way of knowing which are which. Suppose, however, that we have chosen the sample in such a way that we have reason to believe that all the stars in it have more or less equal luminosities (as determined, say, from their spectra). In that case, the brighter-appearing stars would be the nearer and the fainter-appearing stars the more distant. The inverse-square law of light, in other words, allows us to calculate the relative distances of stars of similar luminosities. A knowledge of their mean distance, found from the method of secular parallaxes, then enables us to find both their actual distances and luminosities. (The relation between the apparent brightness, luminosity, and distance of a star will be further explained in Chapter 24.) The method of secular parallaxes has proved very useful for determining the luminosities of relatively rare stars of certain types which are not represented by examples near enough for direct parallax measurement.

(b) STATISTICAL PARALLAXES

The method of secular parallaxes is based on the detection of the average backward drift, due to solar motion, in a large sample of stars. Now, because the sun moves toward the apex, that part of a star's motion due to the solar motion is a drift toward the antapex. However, the solar motion cannot affect the component of a star's proper motion that is in a direction *perpendicular* to a line between the apex and antapex. It is convenient, therefore, to regard the proper

---

†The corrections to the observed radial and tangential velocities are $V_0 \cos A$ and $V_0 \sin A$, respectively, where V_0 is the solar velocity (about 19.5 km/s).

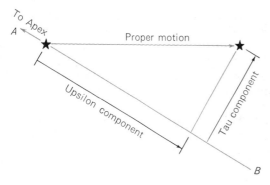

Figure 23.6 Components of proper motion; the figure is in the plane of the sky, perpendicular to the line of sight. The line *AB* is part of a great circle running through the star and the solar apex and antapex.

motion of a star as having two components: one—the *upsilon component*—along a great circle on the celestial sphere from the star through the antapex, that is, parallel to the direction to the solar apex, and another—the *tau component*—perpendicular to the direction of the apex (Fig. 23.6). The upsilon component of the proper motion of a star can be augmented or diminished by the solar motion, but the tau component can involve only the star's peculiar motion—in fact, only that part of it that contributes to the star's tangential velocity perpendicular to the direction of the solar apex. If we had some way of knowing the value of this perpendicular component of the star's tangential velocity, $T_\perp$, we could find its parallax from the tau component, τ, of its proper motion, by applying the formula of Section 23.1c:

$$T_\perp = 4.74 \frac{\tau}{p} \text{ km/s.}$$

There is no way, in general, of knowing $T_\perp$ for an individual star. The stars in a large sample, however, should be moving at random in the local standard of rest—as many are going in any one direction as in any other. We should expect, therefore, that in such a sample the average value of $T_\perp$ should be the same as the average value of the *peculiar radial velocities,* and so $T_\perp$ can be replaced by the latter (without regard to sign) in the above equation. If the quanity τ is then replaced by the average value of the many individual values of τ for the stars in the sample, the average value of the parallaxes of those stars can be computed. This procedure is called the method of *statistical parallaxes.*

Like secular parallaxes, statistical parallaxes do not indicate distances of individual stars but only the average distances for large numbers of stars. The method is very useful, however, if applied to a sample of stars that can all be expected to have the same intrinsic brightness. It gives greater accuracy than does the method of secular parallaxes, when applied to stars whose peculiar velocities are statistically larger than the sun's velocity.

(c) MOVING CLUSTERS

The stars that belong to a group or cluster, moving more or less together through space, all have about the same space velocity. If such a cluster is close enough to the sun, however, the proper motions of its member stars may not all be exactly parallel. If, for example, the cluster is approaching us, its stars appear to radiate away from a distant point in the direction from which the cluster is coming. Conversely, if the cluster recedes from us, the stars appear to converge toward the direction in which the cluster is moving. In either case, the proper motions of the member stars

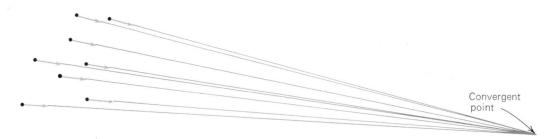

Figure 23.7 Proper motions of the stars in a moving cluster. Note that here we see the cluster as seen from the earth. We see the projections of the stellar motions on the plane of the sky. Although the stars are moving parallel to one another in space, their proper motions appear to converge toward a point infinitely far away. The direction from the earth to that convergent point is also parallel to the space motions of the cluster stars.

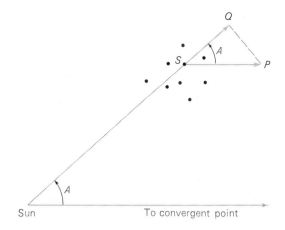

Figure 23.8 *Space velocities of the stars in a moving cluster.*

any star in the cluster, such as star S. Its space velocity, radial velocity, and tangential velocity are SP, SQ, and QP, respectively. Angle SQP is a right angle, and angle A and the star's radial velocity are both observed. Thus the tangential velocity QP can be calculated by solving the triangle.‡ Then the parallax (or distance) of the star S, and thus that of the cluster, can be found from the observed proper motion of the star and from the formula relating tangential velocity and proper motion:

$$QP = 4.74\mu r = 4.74 \frac{\mu}{p} \text{ km/s.}$$

There are several groups and clusters that are near enough for us to observe the convergence or divergence of the proper motions of their member stars. The three best-known examples are the *Hyades* in Taurus, the *Ursa Major* group (which contains most of the naked-eye stars in the Big Dipper), and the *Scorpio-Centaurus* group. These three clusters contain, in all, more than 500 stars. This is a sizable fraction of the number of stars whose parallaxes can be directly observed to an accuracy of 10 percent or better. Thus, this method of using moving clusters to obtain stellar distances is very important and fruitful. It has contributed greatly to our knowledge of fundamental data about stars.

---

‡$QP = SQ \tan A.$

indicate the direction in which the cluster is traveling. The situation for a hypothetical cluster is illustrated in Figure 23.7, which shows the apparent convergence of the proper motions of its stars toward a distant "convergent point," and in Figure 23.8, which shows the actual space velocities of the stars. In these diagrams the cluster is viewed from two different directions.

In studying a moving cluster for which the direction of convergence or divergence can be observed, we have the great advantage of knowing the angle A (Fig. 23.8) between that direction of motion and the line of sight from the sun to the cluster. Now consider

---

EXERCISES

1. In 50 years a star is seen to change its direction by $1'40''$. What is its proper motion?

2. What factors must be considered in converting the observed radial velocity of a star to the value that would be observed from the sun (a) at the ecliptic pole? (b) at the celestial pole?

3. Suppose a star at a distance of 10 pc has a radial velocity of 150 km/s. By what percentage does its distance change in 100 years?

4. Find the tangential velocities of the following stars:
   a) Proper motion = $1''.5$; distance = 20 pc.
   b) Proper motion = $0''.01$; distance = 1000 pc.
   c) Proper motion = $0''.01$; distance = 20 pc.
   d) Proper motion = $0''.01$; *parallax* = $0''.001$.

5. The wavelength of a particular spectral line is normally 6563 Å; in a certain star, the line appears at 6565 Å. How fast is that star moving in our line of sight? Is its motion one of approach or recession?

6. On June 22 a star is observed that is exactly in the direction of the autumnal equinox; its radial velocity appears to be $+36$ km/s. On December 22, the same star appears to have a radial velocity of -24 km/s. What is the radial velocity of the star as seen from the sun?

7. Show by a diagram how two stars can have the same radial velocity and proper motion but different space motions.

8. Suppose a star has a parallax of $0''.001$ but zero peculiar velocity. How long must we wait to see the star apparently change its direction by $4''.1$?

9. Do we need to correct the observed proper motion of a star for the earth's motion around the sun to

obtain the star's proper motion as seen from the sun? If so, explain how we might do so.

10. If a star moves 150 km/s, how many astronomical units does it move per year?

11. A star has a proper motion of 3″00, a parallax of 0″474, and a radial velocity of 40 km/s. What is its space motion?
Answer: 50 km/s

12. Under what conditions would the proper motion of a nearby star be zero: (a) if it were in the direction of the solar apex? (b) if it were *not* at the solar apex?

13. What is the space velocity of a star with a distance of 10 pc, a radial velocity of −8 km/s, and a proper motion of 0″5?

14. A certain star is 90° from the solar apex. The observed radial velocity is +12 km/s, and the observed tangential velocity is 24.5 km/s in the direction of the solar antapex. What is the peculiar velocity of this star?

*15. Review the conditions that a moving cluster must satisfy in order that we can find its distance by the method described in Section 23.3c.

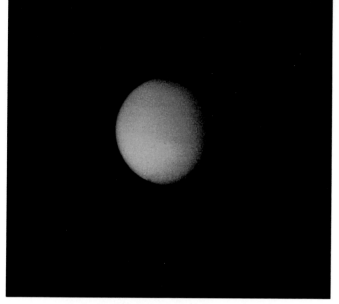

Titan, Saturn's largest satellite, photographed by Voyager 1 on November 9, 1980 from 4.5 million km. *(NASA/JPL)*

The night side of Titan, as seen from Voyager 2 at a range of 907,000 km on August 25, 1981. Sunlight can be seen filtering through the satellite's atmosphere. *(NASA/JPL)*

Voyager 2 view of Saturn's satellite Hyperion from 500,000 km on August 24, 1981. Note the irregular shape of Hyperion, which has been compared to a hamburger patty. The largest diameter of Hyperion is about 360 km. *(NASA/JPL)*

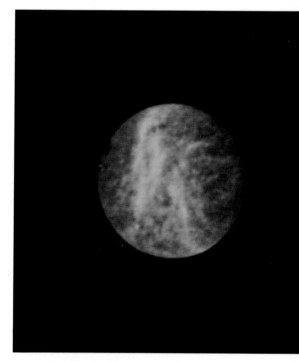

Saturn's satellite Rhea, photographed by Voyager 1 on November 11, 1980 from 1.7 million km. *(NASA/JPL)*

Voyager 2 photographs of Enceladus, taken on August 25, 1981. Note the large smooth regions free of craters. *(NASA/JPL)*

The innermost of Saturn's large satellites, Mimas, photographed by Voyager 1 on November 12, 1980, from 129,000 km. The smallest craters visible are 2 km across. *(NASA/JPL)*

Voyager 2 photograph of Iapetus, outermost of Saturn's large satellites. The bright regions have a reflectance ten times as great as that of the dark ones. *(NASA/JPL)*

Henrietta Swan Leavitt (1868–1921) joined the staff of the Harvard College Observatory in 1892. While studying variable stars in the Clouds of Magellan, she discovered the period-luminosity relation for Cepheid variable stars, which later made it possible for Edwin Hubble to demonstrate that our Galaxy is but one among billions in the universe. *(Harvard College Observatory)*

24

THE LIGHT FROM STARS

The most casual glance at the sky shows that stars differ from one another in apparent brightness. They differ not only because of actual differences in their output of luminous energy but also because they are at widely varying distances. Measures of the amounts of light energy, or *luminous flux*, received from stars are among the most important and fundamental observational data of astronomy; they are used in estimating both the distances and the actual output of energy of stars.

24.1 STELLAR MAGNITUDES

In the second century B.C., Hipparchus compiled a catalogue of about a thousand stars (Section 2.3). He classified these stars into six categories of brightness, which are now called *magnitudes*. The brightest-appearing stars were placed by him in the *first magnitude;* the faintest naked-eye stars were of the *sixth magnitude*. The other stars were assigned intermediate magnitudes. This system of stellar magnitudes, which began in ancient Greece, has survived to the present time, with the improvement that today magnitudes are based on precise measurements of apparent or total luminosity rather than arbitrary and uncertain eye estimates of star brightness.

(a) Measuring Starlight— Photometry

That branch of observational astronomy which deals with the measurement of the intensity of starlight is called *photometry*. In the latter part of the 18th century, William Herschel devised a simple and direct, although only approximate, method of stellar photometry. The principle of his method depends on the fact that the light-gathering power of a telescope is proportional to the area of its aperture (lens or mirror). Thus, if the same star is viewed through two telescopes, identical in construction but different in size, more light energy will be received through the larger telescope than through the smaller one, and the ratio of these two energies will be in the same proportion as the areas of the apertures of the two telescopes. Conversely, if two stars appear of the same brightness when each is viewed with a separate telescope, then the earth must really receive *less* light from the star that is viewed with the larger of the telescopes. Actually, it is not necessary to use more than one telescope. The aperture size of a telescope can be varied at will by the use of circular diaphragms or an iris diaphragm, like the one on a camera. As an example, consider two stars that appear of different brightness in the field of view of a telescope. Suppose,

Figure 24.1 Effect of telescope aperture and exposure time. (*Top left*) 36 cm, 1 min. Stars to 12th magnitude. (*Top right*) 152 cm, 1 min. Stars to 15th magnitude. (*Lower left*) 152 cm, 27 min. Stars to 18th magnitude. (*Lower right*) 152 cm, 4 hr. Stars to 20th magnitude. (*Mount Wilson and Las Campañas Observatories*)

now, that the dimmer of the stars becomes just barely visible when the telescope is closed down to 20 cm diameter, while the brighter one becomes barely visible after the aperture is stopped down to 10 cm. The two stars, when made just barely visible, appear equally bright. However, since only one fourth as much light enters the telescope with the 10 cm diaphragm as with the 20 cm one, we must, in fact, actually be receiving four times as much light from the brighter star as from the fainter. By such reasoning, Herschel estimated the relative amounts of luminous flux received from different stars. He determined, for example, that the average first-magnitude star delivers to the earth somewhat more than 100 times as much light as a

star that is just barely visible on a dark night, that is, a star of approximately sixth magnitude.

A widely used method of comparing the amounts of light received from stars is *photographic photometry*, in which the degrees of blackness and the sizes of star images on photographic negatives are measured. The most common of several techniques in use is to pass a collimated light beam (one whose rays are parallel) through the negative, with the star image to be measured centered in the beam. The blacker, larger image of a brighter star attenuates the beam more than the smaller, less black image of a fainter star; the attenuation of the beam, measured with a photoelectric cell, thus serves as a measure of the star's light.

Actually, even large telescopes do not resolve the stars into disks. In principle, stars make point-like images; the photographic images of bright stars, however, are larger than those of faint ones because of the *turbidity* of the photographic emulsions—the fact that during the time exposure the light from a bright star image is scattered about within the emulsion, exposing a larger area than that occupied by the true image. Bright star images, in other words, are very much overexposed in long time exposures. Atmospheric turbulence (seeing) also enlarges star images. To a lesser extent optical imperfections (aberrations) and diffraction account for some image enlargement. Finally, images of the faintest stars, those which are just barely recorded photographically, would not be geometrical points even if these other effects were absent because of the finite resolution of the photographic emulsions used. On the most common types of photographic plates and films used in telescopic photography the smallest images attainable, in good seeing, are usually about 0.02 mm in diameter.

The modern method of stellar photometry, *photoelectric photometry*, has come into widespread use since World War II. The light from a star, coming to a focus at the focal plane of the telescope, is allowed to pass through a small hole in a metal plate and thence onto the photosensitive surface of a photomultiplier (Section 11.2g). The electric current generated in the photomultiplier is amplified and recorded and provides an accurate measure of the light passing through the hole. Because even the darkest night sky is not completely dark, however, not all the light striking the photomultiplier is due to the star, but some is provided by the light of the night sky, which is, of course, also gathered by the telescope and passed through the hole. Consequently, the hole is next moved to one side of the star image, so that only light from the sky passes through and is recorded; the difference between the first and second readings is a measure of the star's light entering the telescope.

(b) The Magnitude Scale

In 1856, after early methods of stellar photometry had been developed, Norman R. Pogson proposed the quantitative scale of stellar magnitudes that is now generally adopted. He noted, as did Herschel, that we receive about 100 times as much light from a star of the first magnitude as from one of the

sixth, and that, therefore, a difference of five magnitudes correspoonds to a ratio in luminous flux of 100:1. It is a widely accepted assumption in the physiology of sense perception that what appear to be equal intervals of brightness are really equal *ratios* of luminous energy. Pogson suggested, therefore, that the ratio of light flux corresponding to a step of one magnitude be the fifth root of 100, which is about 2.512. Thus, a fifth-magnitude star gives us 2.512 times as much light as one of sixth magnitude, and a fourth-magnitude star, 2.512 times as much light as a fifth, or 2.512×2.512 times as much as a sixth-magnitude star. From stars of third, second, and first magnitude, we receive 2.512^3, 2.512^4, and 2.512^5 ($= 100$) times as much light as from a sixth-magnitude star. By assigning a magnitude of 1.0 to the bright stars Aldebaran and Altair, Pogson's new scale gave magnitudes that agreed roughly with those in current use at the time.

Table 24.1 gives the approximate ratios of light flux corresponding to several selected magnitude differences. Note that a given ratio of light flux, whether between bright or faint stars, always corresponds to the same magnitude interval. Further, note that the numerically *smaller* magnitudes are associated with the *brighter* stars; a numerically *large* magnitude, therefore, refers to a faint star.[*]

Table 24.1 **Magnitude Differences and Light Ratios**

DIFFERENCE IN MAGNITUDE	RATIO OF LIGHT
0.0	1:1
0.5	1.6:1
0.75	2:1
1.0	2.5:1
1.5	4:1
2.0	6.3:1
2.5	10:1
3.0	16:1
3.5	25:1
4.0	40:1
4.5	63:1
5.0	100:1
6.0	251:1
10.0	10,000:1
15.0	1,000,000:1
20.0	100,000,000:1
25.0	10,000,000,000:1

---

[*]If m_1 and m_2 are the magnitudes corresponding to stars from which we receive light flux in the amounts l_1 and l_2, the difference between m_1 and m_2 is defined by

$$m_1 - m_2 = 2.5 \, log \frac{l_2}{l_1}.$$

With optical aid, stars can be seen that are beyond the reach of the naked eye. The *limiting magnitude* of a telescope is the magnitude of the faintest stars that can be seen with that telescope under ideal conditions. A 15-cm telescope, for example, has a limiting magnitude of about 13. The *photographic limiting magnitude* of a telescope is the magnitude of the faintest stars that can be photographed with it. The photographic limiting magnitude of the 5-m telescope on Palomar Mountain is about 24.

The so-called first-magnitude stars are not all of the same apparent brightness. The brightest-appearing star, Sirius, sends us about ten times as much light as the average star of first magnitude, and so has a magnitude of $1.0 - 2.5$ (see Table 24.1), or of about -1.5. Several of the planets appear even brighter; Venus, at its brightest, is of magnitude -4. The sun has a magnitude of -26.5. Some magnitude data are given in Table 24.2. It is of interest to note that the brightness of the sun and Sirius differ by 25 magnitudes—a factor of 10,000 million (10^{10}) in light energy, or flux—and that we also receive 10^{10} times as much light from Sirius as from the faintest stars that can be photographed with the 5-m telescope. The entire range of light flux represented in Table 24.2 covers a ratio of about 10^{20} to 1.

(c) Magnitude Standards

Magnitudes, as explained above, are defined in terms of ratios of the light received from stars (or other celestial objects); thus magnitude differences between objects indicate the *relative* amounts of luminous flux received from them. To set the scale unambiguously, however, it is necessary to pick out some standard or standards with respect to

which all other stars are to be compared. Originally, it was planned to define the brightest stars as having first magnitude. The brightest stars, on the other hand, are not always conveniently disposed for observation. At one time, by agreement among astronomers, it was decided to define as standards several stars in the vicinity of the north celestial pole, which is always visible from northern observatories. This group of stars was called the *North Polar Sequence*. The magnitudes of these stars were intended to serve as a permanent basis of the magnitude system. The actual magnitude values were assigned so that the brightest-appearing stars in the sky, in accordance with previous convention, would still be reasonably near the first magnitude. At the Mount Wilson Observatory, 139 secondary standard regions, known as *selected areas,* were set up over most of the sky by comparing stars in those areas with the North Polar Sequence. Standard regions were also set up by the Harvard College Observatory, which included regions in the southern sky observed by the Southern Stations of the Observatory in South America and South Africa.

Since the development of photoelectric photometry, however, it became clear that the North Polar Sequence and the selected areas, which had been based largely upon photographic photometry, were inadequate for calibrating the precise measures that are obtained today. Consequently, the old standards have been superseded by accurate photoelectric measures of a large number of new standard stars distributed over the sky.

24.2 THE "REAL" BRIGHTNESSES OF STARS

Even if all stars were identical, and if interstellar space were entirely free of absorbing matter, stars would not all appear to have the same brightness, because they are at different distances from us, and the light that we receive from a star is inversely proportional to the square of its distance (Section 10.1b). The apparent brightnesses of stars therefore do not provide a basis for comparing the amounts of light that they actually emit into space. To make such a comparison we would first have to calculate how much light we would receive from each star if all stars were at the same distance from us. Fortunately, if we know a star's distance we can make such a calculation.

Table 24.2 **Some Magnitude Data**

OBJECT	MAGNITUDE
Sun	-26.5
Full moon	-12.5
Venus (at brightest)	-4
Jupiter, Mars (at brightest)	-2
Sirius	-1.5
Aldebaran, Altair	1.0
Naked-eye limit	6.5
Binocular limit	10
15-cm telescope limit	13
5-m (visual) limit	20
5-m photographic limit	24

(a) Absolute Magnitudes

The sun gives us thousands of millions of times as much light as any of the other stars, but on the other hand, it is hundreds of thousands of times as close to us as any other star is. To compare the intrinsic luminous outputs of the sun and of other stars, we first have to determine what magnitude the sun would have if it were at a specified distance, typical of the distances of the other stars. Suppose we choose 10 pc as a more or less representative distance of the nearer stars. Since 1 parsec is about 200,000 AU, the sun would be 2,000,000 times as distant as it is now if it were removed to a distance of 10 pc; consequently, it would deliver to us $(^1/_{2,000,000})^2$ or $1/(4 \times 10^{12})$ of the light it now does. A factor of 4×10^{12} corresponds to about $31^1/_2$ magnitudes (which can be verified by raising 2.512 to the 31.5 power). The sun, therefore, if removed to a distance of 10 pc, would appear fainter by some $31^1/_2$ magnitudes than its present magnitude of -26.5; the sun would then appear as a fifth-magnitude star.

Similarly, we can use the inverse-square law of light to calculate how luminous all other stars of known distance would appear if they were 10 pc away. Suppose, for example, that a tenth-magnitude star has a distance of 100 pc. If it were only 10 pc away, it would be only one-tenth as far away, and hence 100 times as bright—a difference of five magnitudes. At 10 pc, therefore, it too would appear as a fifth-magnitude star.

We now define the *absolute magnitude* of a star as the magnitude that star would have if it were at the standard distance of 10 pc (about 32.6 LY). The absolute magnitude of the sun is about $+5$. Most stars have absolute magnitudes that lie in the range 0 to $+15$. The extreme range of absolute magnitudes observed for normal stars is -10 to $+19$, a range of a factor of more than 10^{11} in intrinsic light output. The absolute magnitudes of stars are measures of how bright they really are; they provide a basis for comparing the actual amount of light emitted by stars.

(b) Distance Modulus

The absolute magnitude of a star is the magnitude that the star *would* have if it were at a distance of 10 pc, and therefore is a measure of the actual rate of emission of visible light energy by the star, which, of course, is independent of the star's actual distance. On the other hand, the magnitude of a star (sometimes called the *apparent magnitude* to avoid confusion with *absolute magnitude*) is a measure of how bright the star *appears* to be, and this depends on both the star's actual rate of light output and its distance. The *difference* between the star's apparent magnitude, symbolized m, and absolute magnitude, symbolized M, can be calculated from the inverse-square law of light and from a knowledge of how much greater or less than 10 pc the star's distance actually is. The difference $m - M$ therefore depends only on the distance of the star and is called its *distance modulus*. The distance modulus of a star, then, is a measure of its distance, from which the actual distance, in parsecs, can be calculated.† For example, suppose the distance modulus of a star is 10 magnitudes. Ten magnitudes (See Table 21.1) corresponds to a ratio of 10,000:1 in light. Thus, we actually receive from the star $^1/_{10,000}$ of the light that we would receive if it were 10 pc away; it must, therefore, be 100 times as distant as 10 pc, or at a distance of 1000 pc. In succeeding chapters we shall see that the absolute magnitude of a star can often be inferred from its spectrum or from some other observable characteristic. Since the apparent magnitude of a star can be observed, a knowledge of its absolute magnitude is equivalent to a knowledge of its distance.

Most of us make use of this same principle, subconsciously, in everyday life. Every experienced motorist has an intuitive notion of the actual brightness of a stop light. If, while driving down the highway at night, he sees a stop light, he judges its distance from its apparent faintness. In other words, the difference between the light's *apparent* and *real* brightness indicates its distance. The com-

---

†Let $l(r)$ be the observed light of a star at its actual distance, r, and $l(10)$ the amount of light we would receive from it if it were a distance of 10 pc. From the definition of magnitudes, we have

$$m - M = 2.5 \log \frac{l(10)}{l(r)},$$

and from the inverse-square law of light,

$$\frac{l(10)}{l(r)} = \left(\frac{r}{10}\right)^2.$$

Combining the above equations, we obtain

$$m - M = 5 \log \frac{r}{10}.$$

The quantity, $5 \log (r/10)$, then, is the distance modulus.

putation of a star's distance from its distance modulus is analogous.

The calculation of a stellar distance by the inverse-square law, as described above, is actually valid only if the space between the star and us is completely transparent. We shall see (Chapter 28) that the light from many stars is attenuated by absorbing dust in interstellar space. Fortunately, however, we can usually tell when cosmic dust is dimming the light from a star, and can correct for it.

24.3 STARS THAT VARY IN LIGHT

Most stars shine with constant light. A minority, however, are variable in magnitude. The standard international index of stars that vary in light is the Soviet *General Catalogue of Variable Stars*. The 1968 edition of this catalogue (the most recent at the time of writing) lists 20,448 known variable stars in our Galaxy, but supplements of this catalogue increase the number yearly.

(a) Designation

Variable stars are designated in order of time of discovery in the constellation in which they occur. If a star that is discovered to vary in light already has a proper name or a Greek-letter designation, it retains that name; examples are Polaris, Betelgeuse (α Orionis), Algol, and δ Cephei. Otherwise, the first star to be recognized as variable in a constellation is designated by the capital letter R, followed by the possessive of the Latin name of the constellation (for example, R Coronae Borealis). Subsequently discovered variables in the same constellation are designated with the letters S, T, . . . , Z, RR, RS, . . . , RZ, SS, ST, . . . , SZ, and so on, until ZZ is reached. Then the letters AA, AB, . . . , AZ, BB, BC, . . . , BZ, and so on, are used up to QZ (except that the letter J is omitted). This designation takes care of the first 334 variable stars in any one constellation. Thereafter, the letter V followed by a number is used, beginning with V 335. Examples are V 335 Herculis and V 969 Ophiuchi.

(b) The Light Curve

A variable star is studied by analyzing its spectrum and by measuring the variation of its light with lapse of time. Some stars show light variations that are apparent to the unaided eye. Generally, however, the apparent brightness of a variable star is determined by telescopic observation. The three techniques most commonly employed are the following:

1. The magnitude of the variable is estimated by visual observation through the telescope, by comparing its brightness with the brightnesses of neighboring stars of known magnitudes.
2. The magnitude of the variable star is measured by comparing its image with the images of comparison stars on a telescopic photograph (photographic photometry).
3. The magnitude of the variable is determined by photoelectric photometry.

A graph that shows how the magnitude of a variable star changes with time is called a *light curve* of that star. An example is given in Figure 24.2. The *maximum* is the point on the light curve where the maximum amount of light is received from the star; the *minimum* is the point where the least

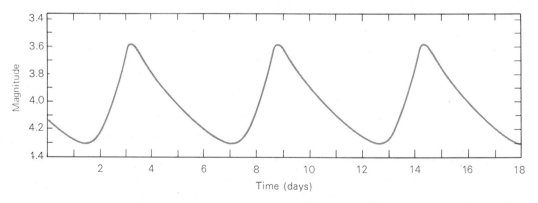

Figure 24.2 Light curve of a typical cepheid variable.

amount of light is received. If the light variations of a variable star repeat themselves periodically, the interval between successive maxima is called the *period* of the star. The *median light* of a variable star is the amount of light it emits when it is halfway between its maximum and minimum brightness. The *amplitude* is the difference in light (usually expressed in magnitudes) between the maximum and the minimum. The amplitudes of variable stars range from less than 0.1 to several magnitudes.

(c)   Types of Variable Stars

The *General Catalogue of Variable Stars* lists three types of variable stars: (1) pulsating variables, (2) eruptive variables, and (3) eclipsing variables. *Pulsating variables* are stars that periodically expand and contract, pulsating in size as well as in light. *Eruptive variables* (including novae and supernovae—see Chapter 34) are stars that show sudden, usually unpredictable, outbursts of light, or, in some cases, diminutions of light. *Eclipsing variables* (or eclipsing binaries) are binary stars whose orbits of mutual revolution lie nearly edge-on to our line of sight and which periodically eclipse each other. Eclipsing variables are not, of course, true variable stars; they will be discussed in Chapter 26. The catalogued numbers of different kinds of variable stars (in 1968) are summarized in Table 24.3.

(d)   Long-Period Variables

The largest group of pulsating stars consists of *Mira*-type stars; these are named for their prototype, Mira, in the constellation of Cetus. Other large groups of pulsating stars are the *RR Lyrae* variables, the *semiregular* variables, and the *irregular* variables.

The Mira or *red variables* are giant stars that pulsate in very long or somewhat irregular periods of months or years. Because they are not highly predictable, an important service is provided by amateur astronomers who keep track of the magnitudes

of these stars. It would require far too much of the time of professional astronomers to maintain constant vigil on all of them. The *American Association of Variable Star Observers* has a well-planned program of careful surveillance of these long-period and irregular variables and has been gathering valuable data on them for years.

We shall return to the physical properties of variable stars and to the causes of their pulsations later (Chapter 34). Some characteristic and easily recognized variable stars, however, have well determined absolute magnitudes at median light; thus when they are identified in remote clusters or systems of stars, they provide a means of finding the distances to those systems. We describe here the more important of these objects.

(e)   Cepheid Variables

Although relatively rare, the cepheid variables are very important in astronomy. They are large yellow stars named for the prototype and first known star of the group, δ Cephei. The variability of δ Cephei was discovered in 1784 by the young English astronomer John Goodricke just two years before his death at the age of 21. The magnitude of δ Cephei varies between 3.6 and 4.3 in a period of 5.4 days. The star rises rather rapidly to maximum light and then falls more slowly to minimum light (see Figure 24.2).

More than 700 cepheid variables are known in our galaxy. Most cepheids have periods in the range 3 to 50 days and absolute magnitudes (at median light) from -1.5 to -5. The amplitudes of cepheids range from 0.1 to 2 magnitudes. Polaris, the *North Star*, is a small-amplitude cepheid variable that varies between magnitudes 2.5 and 2.6 in a period of just under four days.

(f)   The Period-Luminosity Relation

The importance of cepheid variables lies in the fact that a relation exists between their periods of pulsation (or light variation) and their median luminosities, that is, their absolute magnitudes at median light. The relation was discovered in 1912 by Henrietta Leavitt, an astronomer of the Harvard College Observatory. Some hundreds of cepheid variables had been discovered in the Large and Small Magellanic Clouds, two great stellar systems that are actually neighboring galaxies (although they

Table 24.3   **Numbers of Variable Stars**

TYPE	NUMBER
Pulsating	13,782
Eruptive	1,618
Eclipsing	4,062
Unclassified or unstudied	986
All kinds	20,448

were not known to be galaxies in 1912—see Chapter 37).

Many photographs of the Magellanic Clouds had been taken at the Southern Station of the Harvard College Observatory in South America. (These stellar systems are too far south in the sky to be observed from the United States.) On some of the photographs, Miss Leavitt identified 25 cepheids in the smaller cloud and plotted light curves for them. She found that the periods of the stars were related to their relative brightness; the brighter-appearing ones always had the longer periods of light variation. Since the 25 stars are all in the same stellar system, they are all at about the same distance— the distance to the Small Magellanic Cloud—and their relative apparent brightnesses, therefore, indicate their actual relative luminosities. Miss Leavitt, in other words, had found that the median luminosities, or absolute magnitudes, of the cepheid variables were correlated with their periods. Subsequent investigation showed that this relation exists between all the cepheids in the Large and Small Magellanic Clouds.

Now, the cepheid variables known in our Galaxy are intrinsically very luminous stars. If those cepheids in the Magellanic Clouds are the same kind of object, they must also be highly luminous. They appear very faint, however, of 15th or 16th apparent magnitude, which indicates that these stellar systems must be very remote. In 1912 the actual distances to the Clouds of Magellan were not known. While it was known then that the periods of cepheids are correlated with their absolute magnitudes, it was not known what those absolute magnitudes are; only *differences* between absolute magnitudes of cepheids of different periods could be determined. For example, if one cepheid had a period of three days and another of 30 days, it was known that the one of longer period was about two magnitudes, or six times, brighter than the star of shorter period. Unfortunately, however, the actual absolute magnitude of neither star was known, beyond the fact that they were both very luminous; hence, the true distances to the stars remained inaccessible.

If the distances to only one or a few cepheid variables could be determined by independent means, however, the absolute magnitudes of those few stars could be found, and if it is assumed that the period-luminosity relation applies to cepheids in the Galaxy, as well as in the clouds, the scale of the relation would be set. Thereafter, the distance could be determined to any system or cluster of stars in which a cepheid variable had been identified. Harlow Shapley was one of the astronomers who recognized the importance of cepheids as distance indicators, and he pioneered the work of determining distances to some of them in our galaxy. His work was extended by others, and by 1939 it appeared that the distances to these stars were well determined. Unfortunately, however, the determination of distances to cepheid variables is a very difficult observational problem. In the first place, there is not a single cepheid near enough so that its trigonometric parallax can be measured (Section 22.4b); therefore, statistical methods involving the proper motions and radial velocities of these stars must be employed (Section 23.3). In the second place, most of the cepheid variables in our Galaxy lie close to the plane of the galactic system, where clouds of interstellar dust heavily obscure their light (see Chapter 28). Corrections must be applied, therefore to the measured apparent magnitudes of these cepheids. Nevertheless, today we use cepheid variables as fundamental indicators of distance to other galaxies (Chapter 37). We shall return to the absolute magnitudes of cepheid variables in Chapter 30.

(g) Cepheids in Globular Clusters

In a very few globular star clusters, and also outside of clusters in the galactic halo (Chapters 29 and 30), there are variable stars with periods in the range 10 to 30 days. The light curves of these stars are similar to those of the cepheid variables except that they fall from maximum light to minimum more slowly, and the stars are somewhat bluer in color. These objects are sometimes called *W Virginis* stars, after the prototype W Virginis, or sometimes *type II cepheids*, to distinguish them from ordinary or "classical" cepheids, which are called *type I cepheids*. Although only a few dozen type II cepheids are known, a period-luminosity relation also seems to hold for them, and it resembles that for type I cepheids, in that stars of longer period are the more luminous. Because of the similar characteristics of type I and type II cepheids, they were originally thought to be the same kind of star, and only since about 1950 has it been realized that type I and type II cepheids are actually different.

(h) RR Lyrae Stars

Next to the long-period variables, the most common variable stars are the *RR Lyrae* stars, named for RR Lyrae, best known member of the group. Nearly 4500 of these variables are known in our Galaxy. Almost all of them are found in the nucleus or the halo of our Galaxy (Chapter 29) or in globular clusters. In fact, nearly all globular clusters contain at least a few RR Lyrae variables, and some contain hundreds; these stars, therefore, are sometimes called *cluster-type* variables.

The periods of RR Lyrae stars are less than one day; most periods fall in the range 0.3 to 0.7 day. Their amplitudes never exceed two magnitudes, and most RR Lyrae stars have amplitudes less than one magnitude. Several subclasses of RR Lyrae stars are recognized, but the differences between these subclasses are small and need not be considered here.

It is observed that the RR Lyrae stars occurring in any particular globular cluster all have about the same median apparent magnitude. Since they are all at approximately the same distance, it follows that they must also have nearly the same absolute magnitude. Because the RR Lyrae stars in different clusters are all similar to each other in observable characteristics, it is reasonable to assume that *all* RR Lyrae stars have about the same absolute magnitude. If we could learn what that absolute magnitude is, we could immediately calculate the distances to all globular clusters that contain these stars.

Unfortunately, as is true for cepheids, not a single RR Lyrae star is near enough to measure its parallax by direct triangulation. Like the cepheids, distances to RR Lyrae stars have also had to be determined by statistical means, that is, by analyzing their proper motions and radial velocities. The early distance investigations, carried out from 1917 to about 1940, seemed to indicate an absolute magnitude of about 0 for these important stars—they are, therefore, about 100 times as luminous as the sun, whose absolute magnitude is about +5. More recent work shows that different RR Lyrae stars may differ from each other slightly in median absolute magnitude, and that the average value may be a little fainter than was once thought—somewhere between 0 and +1. These recent revisions do not vitiate the important results concerning the nature of our Galaxy derived from the study of RR Lyrae stars (Chapter 29).

Figure 24.3 shows the form of the period-luminosity relation for the three kinds of variable stars discussed in the last four subsections.

24.4 COLORS OF STARS

Until now, this discussion has ignored the fact that stars have different colors and that all colors do not produce an equal response in the human eye. The apparent brightness of a star can depend to some extent, therefore, upon its color.

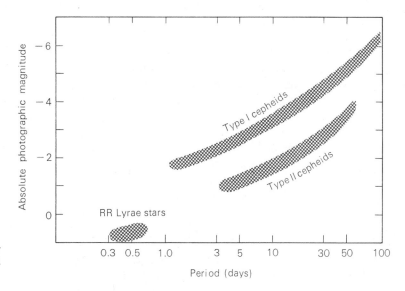

Figure 24.3 Period-luminosity relation for type I cepheids, type II cepheids, and RR Lyrae stars.

(a) Color Response of the Eye and Other Detecting Devices

Every device for detecting light has a particular color or spectral sensitivity. The human eye, for example, is most sensitive to green and yellow light; it has a lower sensitivity to the shorter wavelengths of blue and violet light and to longer wavelengths of orange and red light. It does not respond at all to ultraviolet or to infrared radiation. The eye, in fact, responds roughly to the same kind of light that the sun emits most intensely; this coincidence is probably not accidental—the eye may have evolved to respond to the kind of light most available on earth.

Another detecting device is the photographic plate (or film). The early photographic emulsions, before the development of yellow- and red-sensitive and panchromatic emulsions, were sensitive only to violet and blue light and did not respond to light of wavelengths longer than about 5000 Å (in the blue-green). The basic photographic emulsion is still sensitive to violet and blue; dyes must be added to the basic emulsion to make it sensitive to longer wavelengths.

Suppose, now, that the total amount of light energy entering a telescope from each of two stars is exactly the same if light of all wavelengths is considered, but that one star emits most of its light in the blue spectral region and the other in the yellow spectral region. If these stars are observed visually (that is, by looking at them through the telescope) the yellow one will appear brighter, that is, will have a numerically smaller magnitude, because the eye is less sensitive to most of the light emitted by the blue star. If the stars are photographed on a blue-sensitive photographic plate, however, the blue star will produce the more conspicuous image; measures of the photographic images will show the blue star appearing brighter and having the smaller magnitude. Consequently, when a magnitude system is defined, it is necessary also to specify how the magnitudes are to be measured, that is, what detecting device is to be used.

Magnitudes, whether apparent or absolute, that are based on stellar brightness as they are observed with the human eye are called *visual magnitudes* or *absolute visual magnitudes* and are symbolized m_v and M_v, respectively. In the past, magnitudes based on measures of star images on standard violet- and blue-sensitive photographic emulsions were called *photographic magnitudes* and *absolute photographic magnitudes* and were symbolized m_{pg} and M_{pg}. When green- and yellow-sensitive photographic emulsions were perfected, it was found that they could be used in conjunction with a suitable color filter to approximate closely the spectral response of the human eye; stellar photographs so obtained provide *photovisual magnitudes* (m_{pv} and M_{pv}). Today photographic plates or photomultipliers can be used in conjunction with many kinds of color filters to produce a great variety of different magnitude systems, for example, red magnitudes, ultraviolet magnitudes, infrared magnitudes, and so on. A certain few spectral bands have become more or less standard, however, and are now widely used to define magnitudes.

(b) Color Indices

Since about 1960 the most commonly used magnitudes have been U (ultraviolet), B (blue), and V (visual). The U and B magnitudes are obtained from measures of the flux from stars through certain standardized ultraviolet and blue filters with a common type of photomultiplier. The visual magnitude (V) is measured through a filter that approximates the response of the human eye. The *difference* between any two of these magnitudes, say, between blue and visual magnitudes $(B - V)$, is called a *color index*. Thus the standard U, B, V-system provides two independent color indices, $U - B$, and $B - V$.

Since the inverse-square law of light applies equally to all wavelengths, the color index of a star would not change if the star's distance were changed. The color index of a star, therefore, can be defined in terms of either its apparent or absolute magnitudes; for example,

$$B - V = M_B - M_V.$$

A very blue star measures brighter through a blue filter than through a yellow one; its blue magnitude, therefore, is algebraically *less* than its visual magnitude, and its $B - V$ color index is *negative*. A yellow or red star, on the other hand, has a brighter (smaller) visual magnitude, and a *positive* color index. Color indices, therefore, provide measures of the *colors* of stars. Colors, in turn, indicate the temperatures of stars. Ultraviolet, blue, and visual magnitudes are adjusted to be equal to each other, so that they give a color index of zero to a star with a temperature of about 10,000 K (a spec-

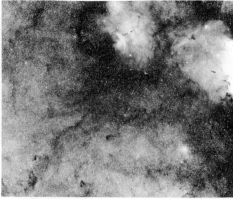

Figure 24.4 Two photographs of the same region of the Milky Way taken with the 124-cm Schmidt telescope: *(left)* on a photographic emulsion sensitive to blue light, *(right)* through a red filter and on a photographic emulsion sensitive to red light. Note the difference with which the stars and nebulae show up in the different colors. *(National Geographic Society–Palomar Observatory Sky Survey; reproduced by permission from Caltech/Palomar Observatory)*

tral type A star—see Chapter 25). The $B - V$ color indices of stars range from -0.4 for the bluest to more than $+2.0$ for the reddest.

If a star field is photographed on emulsions and through filters chosen to isolate the blue and visual spectral regions, the blue and visual magnitudes and color indices so obtained comprise a powerful tool for determining the temperatures of all the stars appearing on the photographs. Since the star images on these two black and white photographs can be accurately measured, the color-index method is a more precise means of obtaining colors and temperatures of stars than the method of color photography would be.

24.5 BOLOMETRIC MAGNITUDES AND LUMINOSITIES

All the magnitude systems so far discussed refer only to certain spectral regions. A magnitude system based on *all* the electromagnetic energy reaching the earth from the stars would seem to be more fundamental. Magnitudes so based are called *bolometric magnitudes*, m_{bol}, and the bolometric magnitudes that stars would have at a distance of 10 pc are *absolute bolometric magnitudes*, M_{bol}. Unfortunately, bolometric magnitudes are difficult to observe directly because some wavelengths of electromagnetic energy do not penetrate the earth's atmosphere. It is true that most of the radiation from stars such as the sun does reach the earth's

surface, and for such stars bolometric magnitudes can be determined approximately. A large part of the energy from stars that are substantially hotter or cooler than the sun, however, lies in the far ultraviolet or in the infrared, which is blocked by the earth's atmosphere and cannot be observed from the ground; for those stars, bolometric magnitudes can only be estimated or calculated from theoretical considerations unless the stars are observed from rockets or satellites.

The difference between the visual and bolometric magnitudes of a star is called its *bolometric correction, BC:*‡

$$V - m_{bol} = M_V - M_{bol} = BC$$

A bolometric correction, then, is a sort of color index, except that for stars of most temperatures it is not obtained from direct observation but from theoretical calculation. Bolometric magnitudes are scaled so that bolometric corrections are nearly zero for stars like the sun. They are positive quantities for stars that are much hotter or cooler than the sun because these stars would appear brighter (would have numerically smaller bolometric than photovisual magnitudes) if all of their radiant energy could be observed. For the hottest stars, calculated bolometric corrections reach as much as eight magnitudes.

‡Sometimes bolometric corrections are defined as $m_{bol} - V$, in which case they are always negative, rather than positive as in the convention used in this book.

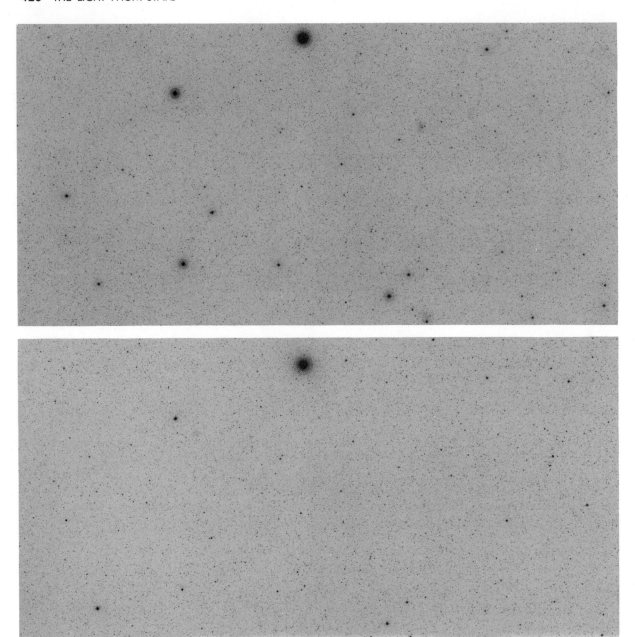

Figure 24.5 Two photographs of the same region of the sky (in Coma Berenices). The upper is in blue light, and the lower in red; note how the colors of the stars are apparent from comparison of the two. These are negative prints, the stars showing as black dots on a white sky. (*National Geographic Society–Palomar Observatory Sky Survey; reproduced by permission from Caltech/Palomar Observatory*)

(a) Stellar Luminosities

The absolute bolometric magnitude of a star is a measure of the rate of its entire output of radiant energy. The rate at which a star pours radiant energy into space, usually expressed in ergs per second, is called its *luminosity*. Absolute bolometric

magnitudes of stars bear the same relation to their luminosities as their visual magnitudes do to the amount of light we receive from them.§ The steps

---

§If two stars of absolute bolometric magnitudes, $M_{bol}(1)$ and $M_{bol}(2)$, have luminosities L_1 and L_2,

$$M_{bol}(1) - M_{bol}(2) = 2.5 \log \frac{L_2}{L_1}.$$

required to determine the luminosity of a star are the following;

1. Its apparent visual magnitude is observed.
2. The bolometric correction is found from rocket or satellite data or is calculated from theory and from a knowledge of the star's temperature. The latter can be estimated, say, from the spectrum or the color of the star. The bolometric correction is applied to the star's photovisual magnitude to obtain its bolometric magnitude.
3. The absolute bolometric magnitude is calculated from the star's apparent bolometric magnitude and its distance.
4. The luminosity is found by comparing the star's absolute bolometric magnitude to that of some other star (say, the sun) whose luminosity is already known.

(b) The Luminosity of the Sun

The first step in determining the luminosity of the sun is to measure the rate at which its radiation falls upon the earth. There are various devices for measuring this quantity. In one method, the sun's radiation is allowed to fall upon the blackened surface of some metal (for example, platinum or silver), which absorbs this radiant energy and rises in temperature. The rate of the temperature rise can be measured by the increased resistance that the metal offers to an electric current. A simpler device measures the rate at which the temperature of water rises as the sun's radiation is absorbed by it. The measures obtained of the sun's radiation must, in any case, be corrected for the attenuation of the sunlight passing through the earth's atmosphere. (Measures of stellar magnitudes must similarly be corrected.)

It is found that a surface of 1 cm^2 just outside the atmosphere and oriented perpendicular to the direction of the sun receives from the sun 1.37×10^6 ergs/s. This value is the solar constant (Section 14.2c).

The total energy that leaves the sun during an interval of one second diverges outward, away from the sun, in all directions. Since one astronomical unit is 1.49×10^{13} cm, the area of the spherical surface over which the solar radiation has spread by the time it reaches the earth's distance from the sun (about eight minutes later) is 2.8×10^{27} cm^2. The solar constant of 1.37×10^6 ergs/s/cm^2 is the energy that crosses just one of those square centimeters. The total energy that leaves the sun in one second—its luminosity—is thus $1.37 \times 10^6 \times 2.8 \times 10^{27} = 3.8 \times 10^{33}$ ergs/s. The corresponding absolute bolometric magnitude of the sun is $+4.6$.

A dramatic illustration of the magnitude of that amount of energy is obtained by imagining a bridge of ice 3 km wide and 1.5 km thick and extending over the 150 million-km span from the earth to the sun. If all the sun's radiation could be directed along the bridge, it would be enough to melt the entire column of ice in one second.

EXERCISES

1. Suppose that star A is just barely visible through a 15-cm telescope and that star B is just barely visible through a 60-cm telescope. Which star gives us more light and by what factor? What is the approximate difference in magnitude of the stars?

2. Show that if all stars were of the same intrinsic brightness, the "depth" in space that could be penetrated by a telescope would be proportional to its aperture.

3. What magnitude would be assigned to an object from which we receive
   a) 100 times as much light as from Venus when it is at its brightest?
   b) $^1/_{10}$ as much light as a star at the naked-eye limit?
   c) 100,000 times as much light as from stars at the photographic limit of the 5-m telescope?

4. Find the absolute magnitude of each of the following stars of apparent magnitude m and distance r:
   a) $m = 7$; $r = 10$ pc
   b) $m = 20$; $r = 100$ pc
   c) $m = 0$; $r = 100$ pc
   d) $m = 3^1/_2$; $r = 5$ pc
   e) $m = 17^1/_2$; $r = 20$ pc
   f) $m = -5$; $r = ^1/_{10}$ pc (there is no such star as this)

5. What are the distances of stars of the following apparent magnitudes m and absolute magnitudes M?
   a) $m = 10$; $M = 5$
   b) $m = 5$; $M = 10$
   c) $m = 13.5$; $M = 15$
   d) $m = 20$; $M = 10$
   e) $m = -26.5$; $M = 5$

6. Suppose obscuring matter between a star and an observer dims the star so much that only 10 percent of the star's light he would otherwise receive reaches him. If the observer were unaware of the obscuration, what would be the error in the distance modulus he would assign to the star?
Answer: His assigned modulus would be 2.5 magnitudes too great.

7. If a star has a color index of $B - V = 2.5$, how many times brighter in visual light does it appear than in blue light?

8. What would be the designations of the fourth, 40th, and 400th discovered variable stars in the constellation Ophiuchi? (Ignore proper names and Greek letter designations.)
Answer: U Ophiuchi, VV Ophiuchi, V 400 Ophiuchi

9. Suppose a type I cepheid variable is observed in a remote stellar system. The cepheid has a period of 50 days and an apparent magnitude of +20. What is the distance to the stellar system? (See Figure 24.3.)

10. Suppose a type I cepheid and a type II cepheid are both observed to have the same apparent photographic magnitude and a period of 10 days. Consult Figure 24.3 and Table 24.1 and estimate how many times further away the type I cepheid is.

11. Draw a diagram showing how two stars of equal bolometric magnitude, one of which is blue and the other red, would appear on photographs sensitive to blue and to yellow light.

12. What is the distance of a star for which $B = 12.4$, $M_v = 6.8$, and $B - V = 0.6$?

13. If a star has a bolometric correction of six magnitudes, what fraction of its radiant energy is observed in its visual magnitude?

14. The apparent visual magnitude of a star is 10.4. Its bolometric correction is 0.8. Its parallax is 0″.001. What is its luminosity?
Answer: 3.8×10^{35} ergs/s

15. By what steps do you suppose the absolute bolometric magnitude of the sun is found?

Annie Jump Cannon (1863–1941), who during her long career at Harvard College Observatory personally classified more than 500,000 stellar spectra and arranged them in the spectral sequence: O, B, A, F, G, K, M. She received many honors for her work, which laid the groundwork for modern stellar spectroscopy, so fundamental to our understanding of the stars. (*Harvard College Observatory*)

25

ANALYZING STARLIGHT

About 1665 Newton showed (perhaps not for the first time) that white sunlight is really a composite of all colors of the rainbow (Section 10.4c), and that the various colors, or wavelengths, of light could be separated by passing the light through a glass prism. William Wollaston (Section 10.5a) first observed dark lines in the solar spectrum, and Joseph Fraunhofer catalogued about 600 such dark lines. As early as 1823, Fraunhofer observed that stars, like the sun, also have spectra that are characterized by dark lines crossing a continuous band of color. Sir William Huggins, in 1864, first identified some of the lines in stellar spectra with those of known terrestrial elements.

25.1 CLASSIFICATION OF STELLAR SPECTRA

When the spectra of different stars were observed, it was found that they differed greatly among themselves. In 1863 the Jesuit astronomer Angelo Secchi classified stars into four groups according to the general arrangement of the dark lines in their spectra. Secchi's scheme subsequently was modified and augmented, until today we recognize seven such principal *spectral classes*.

(a) The Spectral Sequence

As we have seen (Sections 10.5 and 10.6), each dark line in a stellar spectrum is due to the presence of a particular chemical element in the atmosphere of the star observed. It might seem, therefore, that stellar spectra differ from each other because of differences in the chemical makeup of the stars. Actually, the differences in stellar spectra are due mostly to the widely differing temperatures in the outer layers of the various stars. Hydrogen, for example, is by far the most abundant element in all stars, except probably in those at an advanced stage of evolution (Chapter 34). In the atmospheres of the very hottest stars, however, hydrogen is completely ionized, and can, therefore, produce no absorption lines. In the atmospheres of the coolest stars hydrogen is neutral and can produce absorption lines, but in these stars practically all of the hydrogen atoms are in the lowest energy state (unexcited), and can absorb only those photons that can lift them from that first energy level to higher ones; the photons so absorbed produce the *Lyman series* of absorption lines (Section 10.6a), which lies in the unobservable ultraviolet part of the spectrum. In a stellar atmosphere with a temperature of about 10,000 K, many hydrogen atoms are not ionized; nevertheless, an appreciable number of them are excited to the second energy level, from which they can absorb additional photons and rise to still higher levels of excitation. These photons correspond to the wavelengths of the *Balmer series*, which is in the part of the spectrum that is readily observable. Absorption lines due to hydrogen, therefore, are strongest in the spectra of stars whose atmospheres have temperatures near 10,000 K, and they are less conspicuous in the spectra of both hotter and cooler stars, even though hydrogen is, roughly, equally abundant in all the stars. Similarly, every other chemical element, in each of its possible stages of ionization, has a characteristic temperature at which it is most effective in producing absorption lines in the observable part of the spectrum.

Once we have ascertained how the temperature of a star can determine the physical states of the gases in its outer layers, and thus their ability to produce absorption lines, we need only to observe what patterns of absorption lines are present in the spectrum of a star to learn its temperature. We can therefore arrange the seven classes of stellar spectra in a continuous sequence in order of decreasing temperature. In the hottest stars (temperatures over 25,000 K) only lines of ionized helium and highly ionized atoms of other elements are conspicuous. Hydrogen lines are strongest in stars with atmos-

pheric temperatures of about 10,000 K. Ionized metals provide the most conspicuous lines in stars with temperatures from 6000 to 8000 K. Lines of neutral metals are the strongest in somewhat cooler stars. In the coolest stars (below 4000 K), bands of some molecules are very strong. The most important among the molecular bands are those due to titanium oxide, a tenacious chemical compound which can exist at the temperatures of the cooler stars. The sequence of spectral types is summarized in Table 25.1 and Figure 25.1. (Hot stars—types O, B, A—are sometimes referred to as having *early* spectral types, and cool stars—G, K, M—as having *late* spectral types. This jargon derived from old ideas of stellar evolution that are now discarded, but the terminology is still in wide use.)

The spectral classes of stars listed in Table 25.1 can be subdivided into tenths; thus a star of spectral class A5 is midway in the range of A-type stars, that is, halfway between stars of type A0 and F0. The sun is of spectral class G2—two tenths of the way from class G0 to K0.

The spectral sequence, ranging smoothly from O to M with decreasing temperature, was established through the classification of hundreds of thousands of stellar spectra in the years 1918 to 1924 by Harvard astronomer Annie Cannon. The famous American astronomer Henry Norris Russell

Table 25.1 **Spectral Sequence**

SPECTRAL CLASS	COLOR	APPROXIMATE TEMPERATURE (K)	PRINCIPAL FEATURES	STELLAR EXAMPLES
O	Blue	>25,000	Relatively few absorption lines in observable spectrum. Lines of ionized helium, doubly ionized nitrogen, triply ionized silicon, and other lines of highly ionized atoms. Hydrogen lines appear only weakly.	10 Lacertae
B	Blue	11,000–25,000	Lines of neutral helium, singly and doubly ionized silicon, singly ionized oxygen and magnesium. Hydrogen lines more pronounced than in O-type stars.	Rigel Spica
A	Blue	7,500–11,000	Strong lines of hydrogen. Also lines of singly ionized magnesium, silicon, iron, titanium, calcium, and others. Lines of some neutral metals show weakly.	Sirius Vega
F	Blue to white	6,000–7,500	Hydrogen lines are weaker than in A-type stars but are still conspicuous. Lines of singly ionized calcium, iron, and chromium, and also lines of neutral iron and chromium are present, as are lines of other neutral metals.	Canopus Procyon
G	White to yellow	5,000–6,000	Lines of ionized calcium are the most conspicuous spectral features. Many lines of ionized and neutral metals are present. Hydrogen lines are weaker even than in F-type stars. Bands of CH, the hydrocarbon radical, are strong.	Sun Capella
K	Orange to red	3,500–5,000	Lines of neutral metals predominate. The CH bands are still present.	Arcturus Aldebaran
M	Red	<3,500	Strong lines of neutral metals and molecular bands of titanium oxide dominate.	Betelgeuse Antares

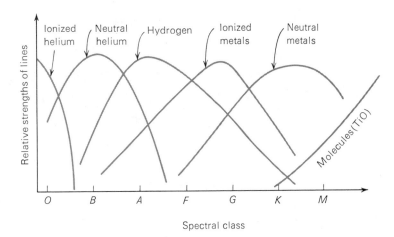

Figure 25.1 Relative intensities of different absorption lines in stars at various places in the spectral sequence.

proposed a scheme by which every student can remember the order of classes in the spectral sequence: The class letters are the first letters in the words, "Oh, Be A Fine Girl, Kiss Me!"

(b) THE ROLE OF PRESSURE

Although temperature is the most important factor that determines the characteristics of a stellar spectrum, other factors are present. Two stars whose atmospheres have the same temperatures but different pressures, for example, will have somewhat different spectra. The degree to which atoms of a particular kind are ionized in a gas depends on the rate at which those at-

oms can recapture electrons, and this, in turn, depends on how closely the atoms, ions, and electrons in the gas are packed together. If the density of the gas is high, the particles are close together; it is easier for ions to capture electrons, and the fraction of atoms that are ionized at any instant of time is then lower than if the gas density is low. At any given temperature, the density of a gas is proportional to its pressure (*Boyle's law*); consequently, in a star whose atmospheric gases are at a high pressure, a smaller fraction of each kind of atom will be ionized than in a star of the same temperature whose atmospheric gases are at a low pressure.

Atmospheric pressure, either on a planet or a star, results from the *weight* of the atmosphere, that is, from the gravitational attraction exerted on the atmosphere. The weight of a unit mass upon the surface of

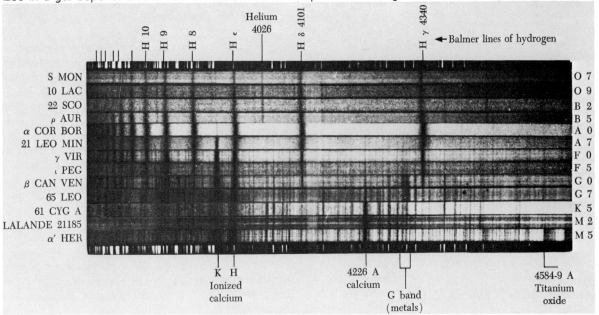

Figure 25.2 Spectra of several stars of representative spectral classes. (*UCLA Observatory*)

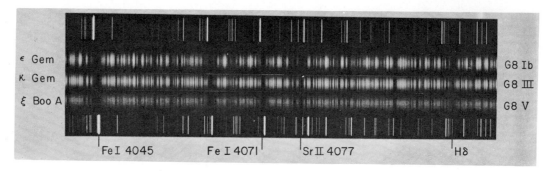

ε Gem — G8 Ib
κ Gem — G8 III
ξ Boo A — G8 V

FeI 4045   Fe I 4071   Sr II 4077   Hδ

Figure 25.3 Spectra of three G8 stars; from top to bottom, a supergiant, a giant, and a main-sequence star. Note the subtle differences in spectra due to the lower surface gravities of the larger stars. *(Lick Observatory)*

a spherical body is called the *surface* gravity of that body; it is proportional to the mass of the body and is inversely proportional to the square of its radius (Section 4.3c). We shall see in Chapter 26 that stars differ enormously from one another in radius, but only moderately in mass. Two stars that differ very greatly in radii may thus have masses that are only slightly different, and consequently these stars must have very different surface gravities (see Exercise 4). The photosphere, where most of the spectral lines are formed, may extend in depth through a different total mass of material in a large star than in a small one, but the difference is generally small compared to the difference in surface gravity. Consequently, the weight of the atmosphere of the larger star, and hence its pressure, is less than in the smaller star. Even if the two stars are at the same temperature, therefore, the gases in the larger one must be at a higher degree of ionization than those in the smaller one. The spectrum of a larger star thus resembles that of a smaller star of greater temperature.

The resemblance is not, however, perfect. The lower atmospheric pressure of the larger star partially compensates for its lower temperature, but the compensation is not exactly the same for all elements; there are subtle differences in the fractions of the atoms of the various elements that are ionized in the two stars. Spectroscopic evidence for these differences was first noted in 1913 by Adams and Kohlschütter at the Mount Wilson Observatory. A trained spectroscopist can tell from its spectrum whether a star is a giant with a tenuous atmosphere, or a smaller, more compact star of somewhat higher temperature. The importance of the subtle spectroscopic differences between large and small stars will become clear when we consider the method of spectroscopic parallaxes in Chapter 27.

(c) SPECIAL SPECTRAL CLASSES

The majority of all known stars fit into the spectral sequence outlined in Section 25.1a. Some relatively rare stars, however, require special classification. Among the most important of these groups are the following:

The Wolf-Rayet stars. The Wolf-Rayet stars are O-type stars that have broad emission lines in their spectra. These emission lines are presumed to originate from material that has been ejected from the star at a high velocity. The ejected gas absorbs light from the star and reemits it as emission lines; the different Doppler shifts of the light coming from different parts of the expanding envelope of gas smear each emission line into a broad "band." Some of the more prominent broad emission lines, or bands, are due to carbon in some stars and to nitrogen in others. Wolf-Rayet stars are objects of unusual composition, in which the original hydrogen is thought to have been depleted by nuclear reactions (Chapter 33).

Early emission-line stars. The Of, Be, and Ae stars' spectra display bright emission lines of hydrogen. These lines, like those of the Wolf-Rayet stars, are presumed to come from extended gaseous envelopes surrounding the stars.

Peculiar A stars. The peculiar A stars are stars of spectral type A that show abnormally strong lines of certain ionized metals. The intensities of these absorption lines sometimes vary periodically with time. These effects may be due to magnetic fields in the stars, or possibly to unequal distribution of the chemical elements over the stars' surfaces.

R stars. R stars are stars with spectral characteristics of K stars, except that molecular bands of C_2 and CN (carbon and cyanogen, respectively) are present.

N stars. The N stars are like M stars, except that bands of C_2, CN, and CH are strong, rather than those of titanium oxide. R and N stars are also called *carbon stars* and are often grouped together as a carbon sequence.

S stars. The S stars are like M stars except that molecular bands of zirconium oxide and lanthanum oxide are present in addition to, or instead of, bands of titanium oxide.

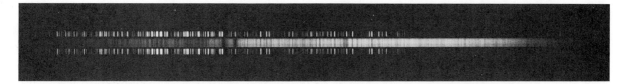

Figure 25.4 A typical stellar spectrogram. The bright streak along the middle, crossed by dark lines, is the spectrum of the star. The bright lines on either side form the comparison (emission) spectrum of iron.

25.2 SPECTRUM ANALYSIS AND THE STUDY OF THE STELLAR ATMOSPHERES

Perhaps half of the observing time of many large telescopes is assigned to the study of stellar spectra. Analyses of stellar spectra yield an enormous amount of information about the stars. Unfortunately, spectrum analysis is a very complicated subject; in the following sections the approach is only briefly outlined. Sections 22.2a, b, and c can be skipped over by those not interested in the underlying physics. Some of the more important data that can be gleaned from the detailed study of stellar spectra are summarized at the end of this section.

(a) HOW STELLAR SPECTRA ARE OBSERVED

The design of a spectroscope or spectrograph and how spectra are observed are described in Section 11.2e.

If the spectrum is observed with a modern detecting device, the intensity at various wavelengths is re-corded electronically and can be displayed by a computer or plotted out. If a stellar spectrum has been photographed, the blackness of the image of the spectrum at various wavelengths indicates the intensity of the starlight at those wavelengths. The intensity of the light drops down, for example, where there is an absorption line. These variations in intensity along a stellar spectrum, and through its various absorption lines, can be exhibited conveniently if the spectrogram (the photograph of the spectrum) is scanned with a *microphotometer* (or *microdensitometer*). In this device a narrow beam of light is passed through the spectrogram. The intensity of the transmitted beam is a measure of the blackness of the image; it can be recorded as various parts of the spectrogram are passed along the light beam so that the beam scans along different wavelengths. Figure 25.5 illustrates the appearance of a recording of a portion of a spectrogram made with a microphotometer. The same sort of plot results from a computer display of a spectrum recorded electronically.

We note that the spectral lines are not perfectly sharp but have a finite width (the reasons will be explained below). The *profile* of a spectral line is the exact shape the line would have on the recording of a microphotometer scan that was calibrated in such a way as to indicate directly the intensity of the light in the spectrum at various wavelengths. Because the pro-

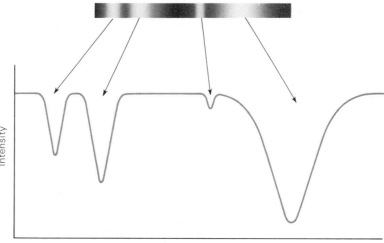

Figure 25.5 Appearance of the recording of a microphotometer scan of a portion of the negative of a spectrography.

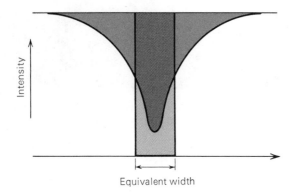

Figure 25.6 Equivalent width of a line. The regions of two different shadings have the same area.

files of different lines differ in shape, it is convenient to define some measurable quantity that can be used to calculate the total amount of light energy that is abstracted (or subtracted) from the spectrum by a line. The most widely used measure is the *equivalent width*, the width of a hypothetical line with rectangular profile of zero intensity along its entire width. The equivalent width represents the same subtraction of light from the stellar spectrum as is removed by the actual line (Figure 25.6).

(b) ANALYSIS OF THE CONTINUOUS SPECTRUM

The continuous spectrum of a star consists of radiation of a wide range of wavelengths that escapes from a relatively shallow layer of gases in the outer part of the star, called the photosphere. One of the problems of astrophysics is to understand the atomic processes that give rise to this emission of light.

In any given region in a stellar photosphere the gas atoms must reach equilibrium and emit radiation at the same rate that they absorb it. The problem of understanding the emission of light in stellar photospheres is thus equivalent to that of understanding how the photospheric gases absorb light, that is, what their source of opacity is.

Calculations show that only hydrogen is sufficiently abundant in the atmospheres of stars of most spectrial classes to contribute appreciably to the opacity of their outer layers. We have seen (Section 10.6e) that hydrogen atoms can absorb light over a range of wavelengths in the process of becoming ionized. The ionization of hydrogen atoms by the absorption of photons, then, is a possible source of opacity in a stellar photosphere. On the other hand, in stars as cool as the sun, or cooler, most atoms of hydrogen are in their lowest energy states. They can be ionized only if they absorb the invisible radiation of wavelengths less than 912 Å (which possesses energy greater than the ionization energy of hydrogen). Opacity at visible wavelengths can result only from the ionization of hydrogen atoms that are already excited to their third or higher energy levels, so that less energetic photons, of longer wavelengths, can be absorbed. Those excited hydrogen atoms are so rare in the solar photosphere that if they were the only source of opacity, the photosphere should be far more transparent than it actually is; we should be able to see into far hotter and deeper layers of the sun than we do.

The problem was solved by the realization that in stars of temperatures near that of the sun and lower, an occasional hydrogen atom can capture and temporarily hold a passing electron. Such a hydrogen atom, possessing *two* electrons, is called a *negative hydrogen ion*. The negative hydrogen ion can absorb visible light and dissociate into an ordinary hydrogen atom and a free electron again. Only about one hydrogen atom in 100 million in the solar photosphere has an extra electron at any given time. This number is enough, however, to account for the opacity of the sun's atmospheric gases in the visible spectrum. Negative hydrogen ions re-form at the same rate that they dissociate, and in doing so emit the visible continuous spectrum that we observe.

We can probe the outer layers of the sun by observing to different depths of its photosphere. This is possible because of the phenomenon of *limb darkening*. The *limb* of the sun (or of any celestial body) is its apparent edge as it is seen in the sky. The limb of the sun appears darker than the center of the disk, because when we look at the center of the disk we see into deeper, hotter layers of the sun's photosphere than when we look at the limb. The explanation for this phenomenon is seen in Figure 25.7. Point A is relatively deep in the photosphere and is beneath the exact center of the sun's disk. Suppose photons that are emitted at point A travel, on the average, a distance d_a before being absorbed. The only photons that have an even chance of escaping from the photosphere are those that are moving radially outward, a direction that happens to be toward the earth. We see, then, sunlight emerging from points as deep as point A only at the center of the sun's disk. Suppose that at point B, photons travel, on the average, a distance d_b before being absorbed. To get into the line of sight to the earth they must travel outward obliquely, rather than radially, and hence can reach the earth only if point B is closer to the surface of the sun than point A. To reach the earth from point C, photons would have to traverse an extremely long path through the solar photosphere unless point C were practically at the outermost photospheric level. The depths to which we see into the sun at various distances from the center of its disk are indicated (although not to scale) by the dashed line.

The gases of the sun are not equally opaque at all wavelengths. Light from the limb of the sun of *any* wavelength reaches the earth only if it originates from

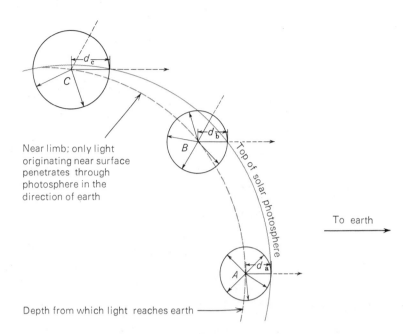

Figure 25.7 Illustration of limb darkening.

the *highest photospheric layers;* on the other hand, we find that at short wavelengths (blue and violet) the distance we can see into the photosphere at the center of the disk is greater than at long wavelengths (at least over the visible spectrum). The detailed manner in which the limb darkening varies with wavelength enables us to calculate from observations how the opacity of the photospheric gases varies with wavelength. It is found that the propensity of the atoms in the sun's outer layers to absorb light at various wavelengths is just what we would expect if negative hydrogen ions were the main source of opacity. The fact that negative hydrogen ions in the sun account for its continuous spectrum, therefore, is quite well established.

Once we know the source of opacity in a stellar photosphere, we can calculate how the temperature, pressure, and density of its gases increase with depth. One clue is provided by the condition that the pressure of the gases in a photosphere must be just great enough to balance the weight of the overlying layers of gases (the condition of *hydrostatic equilibrium*—see Chapter 31). The laws of gases (also see Chapter 31) tell us how density, pressure, and temperature are related. Finally, the known opacity of the gases tells us how they impede the flow of radiation through them. It is the opacity of the gases, blocking in the radiation, that maintains the increase of temperature inward through the stellar photosphere. Thus, a knowledge of the opacity enables us to calculate the temperature at each depth in the photosphere. When he puts the clues together, the astrophysicist can calculate what is called a *model photosphere* or *model atmosphere* for a star. In practice, the calculations are difficult and the results obtained for the match of pressure, temperature, and density through a photosphere are some-

what uncertain. Model photospheres do indicate, however, at least roughly, the physical nature of the outer layers of a star.

(c) ANALYSIS OF THE LINE SPECTRUM

The analysis of the absorption-line spectrum (sometimes called the *Fraunhofer spectrum*) of a star is even more difficult than that of the continuous spectrum, but it yields a great deal of additional information about the star. The object of the analysis is to account in detail for the appearance of each spectral line and to extract such information as is possible concerning the physical and chemical characteristics of the star.

Before we can account for the shape, or profile, of a spectral line, we must consider those processes that *broaden* it, that is, that keep it from being an infinitely sharp line of one precise wavelength only. The principal sources of *line broadening* are the following.

Natural broadening. We saw in Section 10.6 that atoms have discrete energy levels. The levels are not *perfectly* sharp, however. The nominal energy levels of an atom, and the wavelengths of the photons it can absorb, are actually only the most probable energies (or wavelengths). There is a small range of energy about each nominal energy that the atom can have.★

★According to the Heisenberg uncertainty principle, the mean range in energy of an energy level, ΔE, and the average time an atom spends in that level, Δt, are related by the formula

$$\Delta E \Delta t = h,$$

where h is Planck's constant (6.6×10^{-27} erg · s). A typical value of Δt is 10^{-8} second; the corresponding value of ΔE is 10^{-18} erg.

There is, therefore, a small range of wavelengths that can be absorbed or emitted. The resulting (usually) slight range of wavelengths over which an atom can absorb radiation in the vicinity of a line is called the *natural width* of the line.

Doppler broadening. Photospheric atoms are in rapid motion because of the high temperatures of the outer layers of stars. Atoms, moving at various speeds with respect to the earth, "see" the wavelengths of the photons they encounter as different from the lengths we would measure on earth, because of the Doppler shift. At any instant of time, some atoms approach us and others recede from us. Some, therefore, absorb photons of wavelengths that we measure to be a little shorter than the nominal value for an absorption line, and others absorb photons that we measure to have slightly longer wavelengths. The result is that the line is smeared out into a finite width. This *Doppler broadening* is the main source of broadening of most "weak" lines, that is, lines in which only a little energy is subtracted from the spectrum.

Collisional broadening. The energy levels of an atom can be *perturbed* by the presence of other atoms and ions that pass near it or that collide with it. These perturbations of the energy levels shift them to slightly different energies, so that atoms then absorb at wavelengths slightly different from their usual ones. As in Doppler broadening, the effect is to smear out each absorption line. *Collisional broadening* is the most important source of line broadening for the majority of "strong" lines.

Zeeman effect. Other effects can also perturb energy levels; in particular, magnetic fields cause energy levels to split into two or more separate levels, which are shifted from the nominal energies by amounts that depend on the strength of the field. Although this *Zeeman effect* splits each line into two or more lines (see Figure 2.1), the several components are not usually resolved in stellar spectra, so the effect merely appears to broaden a line.

Were it not for the broadening, very little energy could be removed from the continuous spectrum by line absorption. Since, however, atoms can absorb radiation over a small range of wavelengths near each line, they can subtract enough energy from the continuous spectrum so that the lines can be observed easily and analyzed in considerable detail.

The strength of an absorption line depends not only on the total abundance of the relevant atomic species, but also on the fraction of those atoms that are in the right state of ionization and excitation to produce the line. From a model photosphere for a star, we have the temperatures and pressures as functions of depth in its atmosphere. For each depth in the photosphere, therefore, we can calculate what fraction of the atoms are in the relevant ionization and excitation state (Sections 10.6c and 10.6e), and also how that particular absorption line is broadened at each depth. Since an absorption line arises from atoms at all depths throughout the photosphere† we must consider how the contributions from atoms at all layers combine to predict the observed profile of the line. We can then predict how the profile depends on the total number of atoms of the element in question. It is found that for a relatively low abundance, the equivalent width of a line is approximately proportional to the total number of atoms of the element present. For a great enough number of atoms, the line "saturates" and, with the addition of a still larger number of atoms, its strength increases only moderately. If the abundance of the element is very great, however, broad "wings" appear on the line as the result of collisional broadening. Figure 25.8 shows how the profiles of successively stronger lines appear.

(d) What is Learned from Stellar Spectra

Having described some of the procedures by which stellar spectra are analyzed, we now summarize some of the more important kinds of data that can be obtained from spectrum analysis.

Temperature

The kind of temperature measured by an ordinary thermometer is called *kinetic temperature;* it is a measure of the kinetic energies of the atoms or molecules. Since we cannot place a thermometer in the photosphere of a star, we cannot measure its kinetic temperature directly but must infer its temperature from the quality of the radiation that we receive from the star. Because that radiation emerges from a variety of depths in the stellar photosphere, it cannot be expected to indicate a unique temperature that corresponds to any one layer in the star. The temperature that we derive is a representative temperature, corresponding to some representative depth in the photosphere. What that depth is depends on the manner in which the temperature is measured. If the star's continuous spectrum, or a portion of it, is fitted to the spectral energy distribution of a black body (perfect radiator), for ex-

---

†At one time, stellar photospheres were regarded as discrete *surfaces* in the atmospheres of stars, and absorption lines were described as being produced in a *reversing layer* floating immediately above the photosphere. The concept of a reversing layer, in the literal sense, has now generally been discarded, although the concept is still useful for the mathematical representation of some spectral lines.

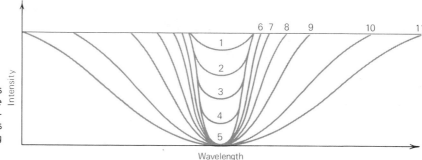

Figure 25.8 Profiles of spectral lines of different equivalent widths. The profiles of larger numbers correspond to larger numbers of atoms that can produce the line being present in the stellar photosphere.

ample, the temperature of that black body is called the star's *color temperature*. A temperature that is estimated from the extent to which lines of ionized atoms appear in the spectrum is called an *ionization temperature*. A temperature that is estimated from the extent to which lines of certain atoms in certain states of excitation are present is called an *excitation temperature*. The temperature of a perfect radiator that would emit the same total amount of radiant energy per unit area as the star does is the *effective temperature* of the star; the effective temperature cannot be determined directly from the spectrum alone, but is found, rather, from the luminosity and radius of a star if they are known (see Chapter 26). If a star were a perfect radiator and were in complete thermodynamic equilibrium, all these temperatures would agree. As it is, the different kinds of temperatures usually do agree with each other fairly well—a circumstance that results from the fact that the observable radiation from most stars emerges from a relatively small range of depth and temperature in their photospheres.

Pressure

We have already discussed how the pressure in a stellar photosphere can affect its spectrum and how stars of high photospheric pressure and high temperature can be distinguished from those of low photospheric pressure and a somewhat lower temperature. It is generally possible, therefore, to

make a rough estimate of the pressures of the gases in a stellar photosphere.

Chemical Composition

Dark lines of a majority of the known chemical elements have now been identified in the spectra of the sun and stars. Of course, the lines of *all* elements are not observable in the spectrum of a star, nor are lines of any one element visible in the spectra of *all* stars. As we have seen, because of variations among the stars in temperature and pressure of the photosphere, only certain of the prevailing kinds of atoms are able to produce absorption lines in any one star. The absence of the lines of a particular element, therefore, does not necessarily imply that that element is not present. Only if the physical conditions in the photosphere of a particular star are such that lines of an element *should* be visible, were the element present in reasonable abundance in that star, can we conclude that the absence of observable spectral lines implies low abundance of the element. On the other hand, spectral lines of an element, in the neutral state or in one of its ionized states, certainly imply the presence of that element in the star.

Relative Abundances of the Elements

Once due allowance has been made for the prevailing conditions of temperature and pressure in a

Figure 25.9 A portion of the solar spectrum and a comparison spectrum of iron photographed with the same spectrograph. Notice that many of the dark lines in the solar spectrum are matched by the bright lines in the comparison spectrum, showing that iron is present in the sun. *(Lick Observatory)*

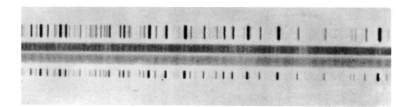

Figure 25.10 Rotational broadening of spectral lines: *(top)* spectrum of WZ Ophiuchi, which does not rotate rapidly and whose lines are relatively sharp; *(bottom)* spectrum of UV Leonis, a rapidly rotating star whose lines are very broadened and have a washed-out appearance. These spectrograms are negative prints.

star's photosphere, analyses of the strengths of absorption lines in its spectrum can yield information regarding the relative abundances of the various chemical elements whose lines appear. In practice, abundances of elements in stars are determined by comparison of observed line strengths with theoretical curves of growth. It is found that the relative abundances of the different chemical elements in the sun and in most stars (as well as in most other regions of space that have been investigated) are approximately the same. Hydrogen seems to comprise from 60 to 80 percent of the mass of most stars. Hydrogen and helium together comprise from 96 to 99 percent of the mass; in some stars they comprise more than 99.9 percent. Among the 4 percent or less of "heavy elements," neon, oxygen, nitrogen, carbon, magnesium, argon, silicon, sulfur, iron, and chlorine are among the most abundant. Generally, but not invariably, the elements of lower atomic weight are more abundant than those of higher atomic weight.

Radial Velocity

The radial or "line-of-sight" velocity of a star can be determined from the Doppler shift of the lines in its spectrum (Section 10.5b). Huggins made the first radial-velocity determination of a star in 1868. He observed the Doppler shift in one of the hydrogen lines in the spectrum of Sirius and found that the star was receding from the solar system.

Rotation

If a star is rotating, unless its axis of rotation happens to be directed exactly toward the sun, one of its limbs approaches us and the other recedes from us, relative to the star as a whole. For the sun or a planet, we can observe the light from one limb or the other and measure directly the Doppler shifts that arise from the rotation. A star, however, appears as a point of light, and we are obliged to analyze the light from its entire disk at once. Nevertheless, if the star is rotating, part of the light from

it, including the spectral lines, is shifted to shorter wavelengths and part is shifted to longer wavelengths. Each spectral line of the star is a composite of spectral lines originating from different parts of the star's disk, all of which are moving at different speeds with respect to us. The effect produced by a rapidly rotating star is that all its spectral lines are broadened so that their profiles have a characteristic "dish" shape (see Exercise 8). Fortunately, this dish shape is highly characteristic and usually can be distinguished from broadening produced by other sources. The amount of rotational broadening of the spectral lines, if observable, can be measured, and a lower limit to the rate of rotation of the star can be calculated (see Exercise 9).

Turbulence

If large masses of a star's photospheric layers have vertical turbulent motion, this motion, like rotation, causes a broadening of the spectral lines. Anomalously high equivalent widths of the lines in the spectra of some stars have been attributed to such turbulence in their photospheres.

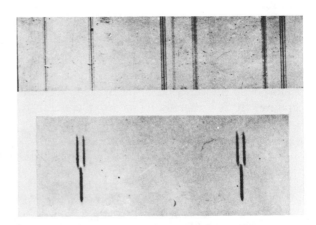

Figure 25.11 The Zeeman effect. *(Above)* the splitting of several iron lines in a magnetic field; *(below)* two yellow mercury lines in the presence *(upper half)* and in the absence *(lower half)* of a magnetic field. *(Yerkes Observatory)*

Magnetic Fields

The Zeeman effect splits or broadens the spectral lines produced by gases in a strong magnetic field. The actual splitting of the spectral lines, for example, can be observed in light coming from the magnetic fields associated with spots on the sun (Chapter 31). However, in the spectra of those stars that possess appreciable magnetic fields, the lines are usually only broadened, because the separate components of a line are not resolved. Zeeman broadening can still be differentiated, however, because the light that is not absorbed in the components of the split lines is polarized. The amount by which a spectral line is broadened by a magnetic field can be detected, therefore, by photographing the star's spectrum through a polarizing filter. A few stars have been found to have very strong magnetic fields. In some cases, these field strengths are variable. One possible explanation is that the field on such a star is localized, and rotation of the star alternately carries the face with the magnetic field into and out of our view.

Shells and Ejected Gases

In the spectra of some stars, absorption lines are observed that do not appear to originate in the photospheres. Sometimes such lines can be associated with shells or rings of material ejected by the star. If enough material is ejected, and if the star has radiation of high enough energy to excite or ionize the gas, the latter produces emission lines instead of absorption lines, superposed on the stellar spectrum. Examples are Wolf-Rayet stars and early type emission-line stars. Stars that are ejecting matter are further discussed in Chapter 34.

EXERCISES

1. What are the probable approximate spectral classes for stars whose wavelengths of maximum light have the following values (see Section 10.2c)?
   a) 2.9×10^{-5} cm   d) 12.0×10^{-5} cm
   b) 0.5×10^{-5} cm   e) 15×10^{-5} cm
   c) 6.0×10^{-5} cm

2. What are the probable approximate spectral classes of stars described as follows?
   a) Balmer lines of hydrogen are very strong; some lines of ionized metals are present.
   b) Strongest lines are those of ionized helium.
   c) Lines of ionized calcium are the strongest in the spectrum; hydrogen lines show with only moderate strength; lines of neutral and ionized metals are present.
   d) Strongest features are lines of C_2, CN, and CH; titanium oxide is absent.

3. The spectrum of a star shows lines of ionized helium and also molecular bands of titanium oxide. What is strange about this spectrum? Can you suggest an explanation?

4. Most stars have masses in the range 0.1 to 10 times the sun's mass. Stellar radii range, however, from 0.01 to 1000 times that of the sun. What is the ratio of the surface gravities of two stars, each of one solar mass, but one with a radius of 0.01 times the sun's radius, and one with a radius of 1000 times the sun's radius?

*5. Sketch how the microphotometer tracing of a spectrogram might appear if the spectrum showed *both* absorption and emission lines.

*6. At what part of the solar disk would the temperature of the visible gases correspond to the "boundary temperature" of the sun? How could you define a "boundary"?

*7. Explain why extremely little light could be subtracted from a continuous spectrum if there were no line broadening.

8. Explain (with a diagram) how stellar rotation broadens spectral lines.

9. Why is it that only a lower limit to the rate of stellar rotation can be determined from rotational broadening, rather than the actual rotation rate?

*10. Consider a small star and a large star of the same mass and spectral type. Would you expect the large or the small star to have sharper (narrower) lines in its spectrum? Why?

*11. Outline the steps by which you would analyze the spectrum of a star to determine the relative abundances of the chemical elements in its outer layers.

12. Draw a diagram showing a shell of gas that has been ejected from a star. Indicate on the diagram the direction of the earth. Also show which parts of the shell give rise to emission lines superimposed on the star's spectrum, and which part or parts can produce absorption lines in the star's spectrum. (*Hint:* See Chapter 34.)

Edward Charles Pickering (1846–1919), American astronomer, was a pioneer in the study of stellar spectra, and especially in the study of the spectra of binary stars. He was the first to demonstrate the existence of spectroscopic binaries, and he also carried out fundamental work in the investigation of eclipsing binary stars—so important to our knowledge of the fundamental properties of stars. (*Yerkes Observatory*)

DOUBLE STARS: WEIGHING AND MEASURING THEM

Roughly half the stars around the sun are found in pairs (binary stars) or in systems of three or more, ranging up to clusters of thousands—each star moving under the combined gravitational influence of the other stars in the same system. The circumstance is fortunate, because analyses of these systems provide us with our best means of learning stellar masses and sizes. Although by direct methods we have accurate data for relatively few stars in such systems, they make it possible for us to calibrate indirect methods that we can use to estimate the masses and radii of many others. Because binary stars, the simplest systems, have lent themselves best to analysis, this chapter is especially concerned with them.

26.1 DETERMINATION OF THE SUN'S MASS

The masses of stars must be inferred from their gravitational influences on other objects. We can measure the force exerted by the sun, for example, on the earth and other planets, and the mass of the sun can be determined more reliably than that of any other star.

The most direct way to calculate the mass of the sun is from the acceleration of the earth. For the sake of illustration, let us assume that the orbit of the earth is circular. Then the acceleration required to keep the earth in its orbit is the centripetal acceleration (Section 4.2):

$$a = \frac{v^2}{R},$$

where v is the earth's orbital speed and R is the radius of its orbit. This acceleration must be provided by the gravitational attraction of the sun on a unit mass at the earth's distance; that is,

$$a = \frac{v^2}{R} = \frac{GM_s}{R^2},$$

where G is the universal gravitational constant and M_s is the desired mass of the sun. Solving the above equation for the sun's mass, we obtain

$$M_s = \frac{v^2 R}{G}.$$

Both v and R are known from observation and G is determined from laboratory measurements (Sec-

434

tion 4.3e). In metric units, $v = 3 \times 10^6$ cm/s, $R = 1.49 \times 10^{13}$ cm, and $G = 6.67 \times 10^{-8}$. Substituting these values into the above formula, we find for the mass of the sun,

$$M_s = 2 \times 10^{33} \text{ g} = 2 \times 10^{27} \text{ tons.}$$

Because the earth's orbit is nearly circular, the value thus found for the mass of the sun is very nearly the correct one. Of course, the mass of the sun can be found more accurately by using the instantaneously correct acceleration of the earth and its exact distance from the sun at a given moment in its elliptical orbit.

A simple way to calculate the mass of the sun relative to that of the earth is to apply Kepler's third law, as it was refined by Newton (Section 5.3), to two systems, the sun-earth system in which the sun is the accelerating mass, and the earth-moon system in which the earth is the accelerating mass. Since the earth's mass is negligible compared with the sun's, Kepler's third law, when applied to the earth-sun system, becomes

$$M_s P_{es}^2 = K a_{es}^3,$$

where P_{es} is the period of revolution of the earth about the sun (sidereal year), a_{es} is the semimajor axis of the relative orbit of the earth and sun (the length of the astronomical unit), and K is a known constant of proportionality whose value depends on the units used to measure mass, time, and length. The same formula may be applied to another pair of mutually revolving bodies, the earth-moon system:

$$(M_e + M_m) P_{em}^2 = \frac{82.3}{81.3} M_e P_{em}^2 = K a_{em}^3,$$

where M_e is the earth's mass, which is 81.3 times the mass of the moon (M_m), P_{em} is the period of revolution of the moon about the earth (the sidereal month), and a_{em} is the semimajor axis of the relative orbit of the earth-moon system. If the first of the above equations is divided by the second, the result is

$$\frac{M_s}{M_e} = \frac{82.3}{81.3} \left(\frac{P_{em}}{P_{es}} \right)^2 \left(\frac{a_{es}}{a_{em}} \right)^3 .$$

The sidereal month is about 27.32 days, or about $^1/_{13.4}$ sidereal years, and the astronomical unit is about 389 times the moon's distance from the earth. If these figures are inserted in the above equation,

$$\frac{M_s}{M_e} = 333,000.$$

That is, the sun is about one third of a million times as massive as the earth. Since the earth's mass is 6×10^{27} gm (Section 4.3e), the sun's mass is 2×10^{33} gm, as found above.

26.2 BINARY STARS

We are not able to observe directly planets revolving around other stars, but we do observe many double stars and cases of three or more stars that belong to the same dynamical system.

(a) Discovery of Binary Stars

In 1650, less than half a century after Galileo turned a telescope to the sky, the Italian Jesuit astronomer Giovanni Baptista Riccioli observed that the star Mizar, in the middle of the handle of the

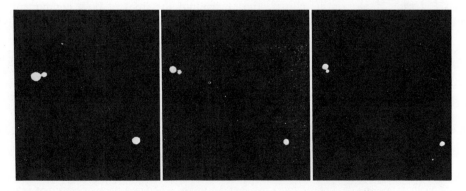

Figure 26.1 Three photographs, covering a period of about 12 years, which show the mutual revolution of the components of the double star Kruger 60. (*Yerkes Observatory*)

Big Dipper, appeared through his telescope as *two* stars. Mizar was the first *double star* to be discovered. In the century and a half that followed, many other closely separated pairs of stars were discovered telescopically.

Between 1782 and 1821 William Herschel, thinking double stars would be useful for measuring stellar parallaxes (Section 22.4), published three catalogues, listing more than 800 double stars. However, the majority of these systems are physical pairs of stars revolving about each other, and hence are not suitable for parallax determination. This had been suggested as early as 1767 by John Michell, who pointed out the extreme improbability of there being more than a very few close pairs of stars that are really only lined up in projection.

One famous double star is Castor, in Gemini. The telescope reveals Castor to be two stars. They were separated by an angle of nearly 5″ in 1804, when Herschel had noted that the fainter component of Castor had changed, slightly, its direction from the brighter component. Here, finally, was observational evidence that one star was moving about another; it was the first evidence that gravitational influences exist outside the solar system. Herschel had failed in his program to facilitate parallax determinations but had found something of far greater interest. As he put it, he was like Saul, who had gone out to seek his father's asses and had found a kingdom. His son, John Herschel, continued the search for double stars, and prepared a catalogue (published posthumously) of more than 10,000 systems of two, three, or more stars.

If the gravitational forces between stars are like those in the solar system, the orbit of one star about the other must be an *ellipse*. The first to show that such is the case was Felix Savary, who in 1827 showed that the relative orbit of the two stars in the double system ξ *Ursae Majoris* is an ellipse, the stars completing one mutual revolution in a period of 60 years.

Another class of double stars was discovered by E. C. Pickering, at Harvard, in 1889. He found that the lines in the spectrum of the brighter component of Mizar (the first double star to be discovered) are usually *double*, but that the spacing of the components of the lines varies periodically, and at times the lines even become single. He correctly deduced that the brighter component of Mizar (Mizar A) itself is really *two* stars that revolve about each other in a period of 104 days. When one star is approaching us, relative to the center of mass of the two, the other star is receding from us; the radial velocities of the two stars, and therefore the Doppler shifts of their spectral lines, are different, so that when the composite spectrum of the two stars is observed, each line appears double. When the two stars are both moving across our line of sight, however, they both have the same *radial* velocity (that of the center of mass of the pair), and hence the spectral lines of the two stars coalesce.

Stars like Mizar A, which appear as single stars when photographed or observed visually through the telescope, but which the spectroscope shows really to be double stars, are called *spectroscopic binaries*; systems that can be observed visually as double stars are called *visual binaries*. In 1908 Frost found that the fainter component of Mizar, Mizar B, is also a spectroscopic binary.

Almost immediately following Pickering's discovery of the duplicity of Mizar A, Vogel discovered that the star Algol, in Perseus, is a spectroscopic binary. The spectral lines of Algol were not observed to be double, because the fainter star of the pair gives off too little light compared with the brighter for its lines to be conspicuous in the composite spectrum. Nevertheless, the periodic shifting back and forth of the lines of the brighter star gave evidence that it was revolving about an unseen companion; the lines of both components need not be visible in order for a star to be recognized as a spectroscopic binary.

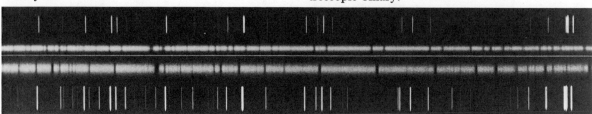

Figure 26.2 Two spectra of the spectroscopic binary κ Arietis. When the components are moving at right angles to the line of sight *(bottom)*, the lines are single. When one star is approaching us and the other receding *(top)*, the spectral lines of the two stars are separated by the Doppler effect. *(Lick Observatory)*

The proof that Algol is a double star is significant for another reason. In 1669 Montonari had noted that the star varied in brightness; in 1783 John Goodricke established the nature of the variation. Normally, Algol is a second-magnitude star, but at intervals of $2^d20^h49^m$ it fades to one third of its regular brightness; after a few hours, it brightens to normal again. Goodricke suggested that the variations might be due to large dark spots on the star, turned to our view periodically by its rotation, or that the star might be eclipsed regularly by an invisible companion. Vogel's discovery that Algol is a spectroscopic binary verified the latter hypothesis. The plane in which the stars revolve is oriented nearly edgewise to our line of sight, and each star is eclipsed once by the other during every revolution. The eclipse of the fainter star is not very noticeable because the part of it that is covered contributes little to the total light of the system; the second eclipse can, however, be observed. A binary such as Algol, in which the orbit is nearly edge on to the earth so that the stars eclipse each other, is called an *eclipsing binary*.

Binary stars are now known to be very common; they may be the rule, not the exception. In the stellar neighborhood of the sun, somewhere between one half and two thirds of all stars are members of binary or multiple star systems.

We may summarize the types of binary stars before proceeding to explain how they are used to determine stellar masses.

Optical Doubles

As stated above, an optical double consists of two stars, in nearly the same line of sight, of which one is far more distant than the other; they are not true binary stars, and are not discussed further here.

Visual Binaries

A visual binary is a gravitationally associated pair of stars; the members are either so near the sun, or so widely separated from each other (usually, both), that they can be observed visually (in a telescope) as two stars. Typical separations for the two stars in a visual binary system are hundreds of astronomical units; thus the orbital speeds of the stars are usually quite small and their orbital motion may not be apparent over a few decades of observation. Nevertheless, two closely separated stars are generally assumed to comprise a visual binary system if there is no reason to doubt that they are at the same distance from us and if they have the same proper motion and radial velocity, indicating that they are moving together through space. Over 64,000 such systems are catalogued.

Astrometric Binaries

Sometimes one member of what would otherwise be a visual binary system is too faint to be observed; its presence may be detected, however, by the "wavy" motion of its companion, revolving about the invisible center of mass of the two stars as they move through space. In 1844 Bessel discovered that the bright star Sirius displays such a motion with a period of 50 years. Sirius remained an *astrometric binary* until 1862, when Alvan G. Clark found its faint companion—a member of the class of stars known as *white dwarfs* (discussed in Chapter 27).

Spectroscopic Binaries

When the binary nature of a star is known only from the variations of its radial velocity (or of both radial velocities if the spectral lines of both stars are visible), it is said to be a spectroscopic binary. Over 700 systems have been analyzed.

Spectrum Binaries

If the orbit of what would otherwise be a spectroscopic binary is oriented nearly "face on" to us (that is, perpendicular to our line of sight), or if the masses of the member stars are so low that they

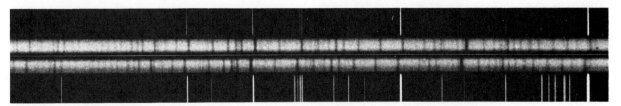

Figure 26.3 Two spectrograms of the single-line spectroscopic binary α^1 Geminorum. The upper part shows the spectrum of the brighter star when it is moving away from us at nearly its maximum orbital speed, and the lower one when it is moving toward us. (*Lick Observatory*)

have very small orbital velocities, we can see no radial-velocity variations. It may still be obvious, however, that there are two stars if the composite spectrum contains lines that are characteristic of both hot and cool stars and which would not be expected to occur in the spectrum of a single star. Such systems are called spectrum binaries.

Eclipsing Binaries

If the orbit of a binary system is oriented nearly edge on to us, so that the stars eclipse each other, it is called an eclipsing binary. More than 4000 are catalogued.

The different kinds of binaries are not mutually exclusive. An eclipsing binary, for example, may *also* be a spectroscopic binary, if it is bright enough that its spectrum can be photographed, and if its radial velocity variations have been observed. Also, a small number of relatively nearby spectroscopic binaries can also be observed as visual binaries.

(b) Mass Determinations of Visual Binary Stars

The two stars of a binary pair revolve mutually about their common center of mass (or barycenter), which in turn moves in a straight line among the neighboring stars. Each star, therefore, describes a wavy path around the course followed by the barycenter. With careful observations it is possible to determine these individual motions of the member stars in a visual binary system. It is far more convenient, however, simply to observe the motion of one star (by convention the fainter), about the other (the brighter). The observed motion shows the *apparent relative orbit*. The periods of mutual revolution for visual binaries range from a few years to thousands of years, but only for those systems with periods less than a few hundred years can the apparent relative orbits be determined with much precision; even then, a long series of observations covering a number of decades is usually necessary. The data observed are the angular separation of the stars and the *position angle*, which is the direction, reckoned from the north, around toward the east, of the fainter star from the brighter one. These data can be measured on a photograph if the separation of the stars is not too small; in any case, they can be measured directly at the telescope. A typical apparent relative orbit, assembled from many observations of separation and position angle, is shown in Figure 26.4.

The true orbit of the binary system does not, in general, happen to lie exactly in the plane of the sky (that is, perpendicular to the line of sight). Consequently, the apparent relative orbit is merely the *projection* of the *true relative orbit* into the plane of the sky. Now it is easy to show that when an ellipse in one plane is projected onto another plane (that is, is viewed obliquely), the projected curve is also an ellipse, although one of different eccentricity. Moreover, the foci of the original ellipse do *not* project into the foci of the projected ellipse. Therefore, the brighter star, although it is located at one focus of the *true* relative orbit, is not at a focus of the *apparent* relative orbit. This circumstance makes it possible to determine the inclination of the true orbit to the plane of the sky. The problem is simply one of finding the angle at which the ellipse of the true relative orbit must be projected in order to account for the amount of displacement of the brighter star from the focus of the apparent relative orbit (Figure 26.5). There are several techniques for solving this geometry problem, so that the shape and angular size of the true orbit can be found. The period of mutual revolution, of course, is observed directly if the system has completed one revolution during the interval it has been under observation; otherwise, the period must be calculated from the rate at which the fainter star moves in the relative orbit.

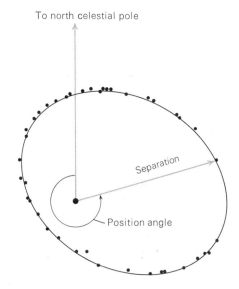

Figure 26.4 Separation and position angle in a visual-binary star.

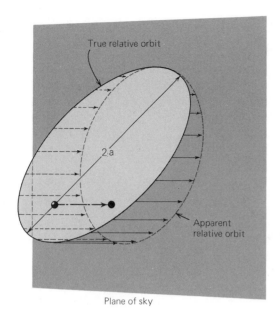

Plane of sky

Figure 26.5 Relation between a true and an apparent relative orbit, showing the apparent displacement of the focus.

If the semimajor axis of the true relative orbit (that is, the semimajor axis the orbit would appear to have if it were seen face on) has an angular length, in seconds of arc, of a'', and if the system is at a distance of r pc, the semimajor axis, in astronomical units, is $r \times a''$ (see Exercise 7). The sum of the masses of the two stars, in solar units, is given by Kepler's third law:

$$m_1 + m_2 = \frac{(r \times a'')^3}{P^2}.$$

To find what share of the total mass belongs to each star, it is necessary to investigate the individual motions of the stars with respect to the center of mass of the system. The distance of each star from the barycenter is inversely proportional to its own mass.

As an example, consider the visual binary Sirius A and Sirius B. The semimajor axis of the true relative orbit is about $7\frac{1}{2}''$ and the distance from the sun to Sirius is 2.67 pc. The period of the binary system is 50 years. The sum of the masses of the stars, then, is

$$m_1 + m_2 = \frac{(2.67 \times 7.5)^3}{(50)^2} = 3.2 \text{ solar masses.}$$

Sirius B, the fainter component, is about twice as far from the barycenter as Sirius A, and so has only about half the mass of Sirius A. The masses of Sirius A and Sirius B, therefore, are about 2 and 1 solar masses, respectively.

(c) The Possible Discovery of Planets

The barycenter of the sun-Jupiter system is about 0.001 of the distance from the center of the sun to the center of Jupiter, which puts it outside the surface of the sun. Thus every 12 years the sun revolves about a tiny orbit about 0.005 AU in radius. An observer on a hypothetical planet about even a nearby star would probably not be able to detect such a motion of the sun with equipment comparable to that available to our astronomers. Yet if the sun were a somewhat less massive star, and if Jupiter were a little more massive, the sun's motion might just be barely detectable.

Many nearby stars have been carefully observed to see if any display an orbital motion revealing a companion of planetary mass. At this writing, there has been no confirmed detection of such a planet. We suspect, however, that many planetary systems exist (Chapter 36), and there is slight hope that with its superior resolution the space telescope may indirectly observe planets about other stars.

(d) Spectroscopic Binaries

If the two stars of a binary system have a small linear separation, that is, if their relative orbit is small, there is little chance that they will be resolved as a visual binary pair. On the other hand, they have a shorter period, and their orbital velocities are relatively high, compared with the stars of a visual binary system; unless the plane of orbital revolution is almost face-on to our line of sight, there is a good chance that we will be able to observe radial velocity variations of the stars due to their orbital motions. In other words, they comprise a spectroscopic binary system.

Most spectroscopic binaries have periods in the range from a few days to a few months; the mean separations of their member stars are usually less than 1 AU. If the two stars of a spectroscopic binary are not too different in luminosity, the spectrum of the system displays the lines of both stars, each set of lines oscillating in the period of mutual revolution. More often, lines of only one star are observed. A graph showing the radial velocity of a member of a binary star system plotted against time

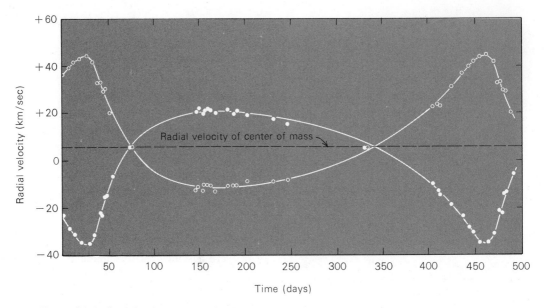

Figure 26.6 Radial velocity curves for the spectroscopic binary system φ Cygni. *(Adapted from Rach and Herbig)*

is called a *radial velocity curve*, or simply, a *velocity curve.*

In general, the orbit of each star about the barycenter is an ellipse, and the radial velocity curve is skewed (Figure 26.6). Both stars of the system, of course, have the same period, for they stay on opposite sides of the barycenter as they revolve about it. Also, the orbits of both stars have the same eccentricity (see Exercise 10) but differ in size by a ratio that is in inverse proportion to their masses (since the masses of the stars are in inverse proportion to their distances from the barycenter—Section 5.1). Consequently, if the spectral lines of both stars are visible, so that a radial velocity curve can be plotted for each, those velocity curves are mirror images of each other, differing only in scale (or in *amplitude*); this is true for the binary system whose velocity curves are shown in Figure 26.6. From the shape of each radial velocity curve, it is possible to determine the eccentricity of the orbits of the stars, as well as the orientation of the major axis of the orbit in the orbital plane.

Here, however, we shall only consider, as an example of the analysis of a spectroscopic binary, the simple case in which the orbits of the stars are circular, and in which the spectral lines of both stars are observed. Figure 26.7 shows the two stars in their orbits, the Doppler shifts of a hypothetical spectral line, and the two radial velocity curves. Because the orbits of the stars in this example are circular, the velocity curves are symmetrical; they

have shapes that are known, technically, as *sine curves.* A complete cycle of variation of radial velocity—the period of the system—is 17.5 days. At position 1, star *A* has its maximum possible component of velocity toward the sun, and star *B* has its maximum possible component of velocity away from the sun. The conditions are reversed at position 3. At positions 2 and 4, both stars are moving across our line of sight and neither has a component of velocity in our line of sight due to orbital motion; both have the same radial velocity as that of the center of mass of the system, 40 km/s. The radial velocity of star *B* ranges from +115 to −35 km/s, a range of 150 km/s. The maximum difference between the radial velocities of star *B* and that of the barycenter, then, is 75 km/s. The corresponding value for star *A* is only 25 km/s. Because both stars have the same period and since star *A* moves only one third as fast as star *B*, with respect to the barycenter, star *A* must have only one third as far to go to get around its orbit; its orbit must be one third the size of that of star *B*, and its mass, therefore, must be three times as great.

Stars *A* and *B* are moving in opposite directions with respect to their center of mass; thus the maximum radial velocity (as observed from the solar system) of star *B* with respect to star *A* must be the sum of the radial velocities of the two stars with respect to the barycenter, or 100 km/s. If the orbital plane of the system were in our line of sight, this would be the relative orbital velocity of one star

Comet Ikeya-Seki, photographed by J. B. Irwin in Chile, October 1965.

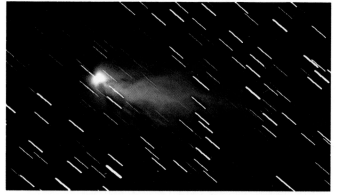

Comet Humason (1961a), photographed with the Palomar Schmidt telescope. *(California Institute of Technology/Palomar Observatory)*

The trail of a meteor in the Perseid shower, photographed by Ronald Oriti. *(Courtesy Ronald Oriti)*

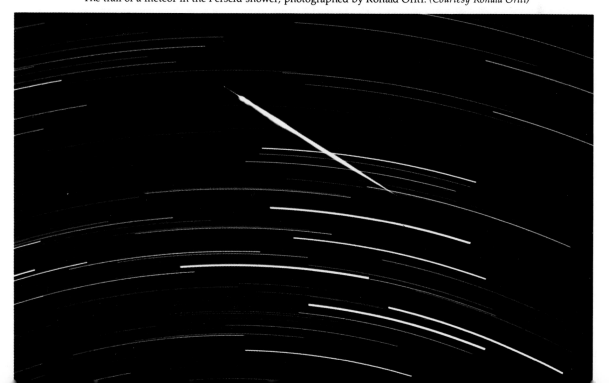

Slice of the Kamkas iron meteorite, which has been polished and then etched with a dilute nitric acid solution to show the criss-cross Widmanstätten figures. *(Photograph by Ivan Dryer)*

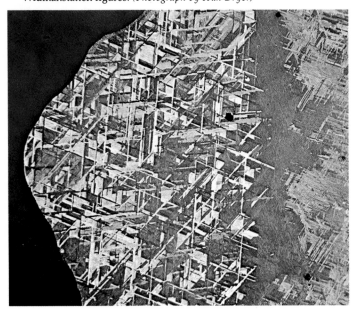

Stony-iron meteorite, Glorieta Mountains, New Mexico. The specimen has been polished and etched to show the metallic structure. *(Photograph by Ivan Dryer)*

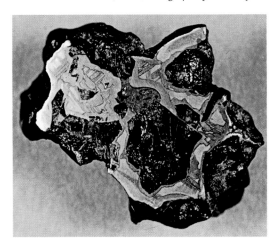

Polished slice of the Albin, Wyoming, stony meteorite. This type of meteorite consists of nickel-iron metal with inclusions of the green mineral olivine. *(Photograph by Ivan Dryer. All meteorite specimens on this page are from the collection of Ronald Oriti and are reproduced with his kind permission.)*

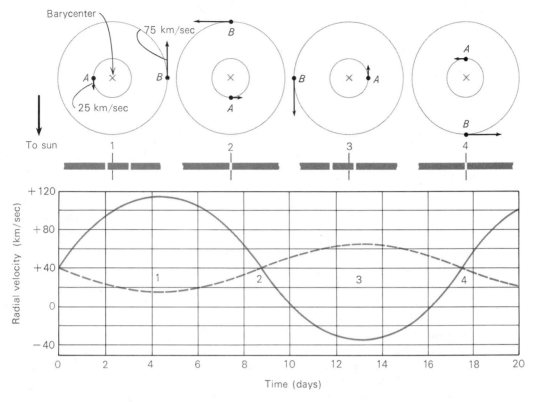

Figure 26.7 Hypothetical spectroscopic binary system with circular orbits.

with respect to the other. As it is, however, the orbital plane is tilted at some unknown angle to our line of sight, and the 100 km/s is only the maximum *radial component* of the relative velocity; the actual relative velocity will be greater by an unknown factor. Now, the distance around the relative orbit—its circumference—is the relative orbital velocity multiplied by the period—the time one star takes to get around the other. The distance between the stars, *a*, is the circumference of the orbit divided by 2π; that is,

$$a = \frac{V \times P}{2\pi},$$

where *V* is the relative velocity. If we substitute the observed lower limit to the relative velocity—in our example, 100 km/s—for *V* in the above equation, we will obtain a lower limit to the separation of the stars. If this lower limit to *a* is applied to Kepler's third law, we obtain a lower limit to the sum of the masses of the stars:

$$m_1 + m_2 = \frac{a^3}{P^2}.$$

The calculation for the numerical example given in Figure 26.7 may be illustrated as follows: A velocity of 100 km/s is equivalent to about 20 AU/yr, and a period of 17.5 days is about 0.048 yr; thus a lower limit to *a* is given by

$$a = \frac{20 \times 0.048}{2\pi} = 0.153 \text{ AU}.$$

A lower limit to the sum of the masses, then, is

$$m_1 + m_2 = \frac{0.153^3}{0.048^2} = 1.6 \text{ solar masses.}$$

Since star *A* is three times as massive as star *B*, it has 75 percent of the total mass; lower limits to the individual masses of the star are therefore

$$m_A \geq 1.2 \text{ solar masses}$$

and

$$m_B \geq 0.4 \text{ solar mass,}$$

where the symbol "$\geq$" means "greater than or equal to."

The analysis is more difficult in the general case of elliptical, rather than circular, orbits, but it can nevertheless be carried out. The results are similar except that a is then the semimajor axis of the relative orbit instead of the constant separation between the stars that exists if the orbits are circular. In either case, individual masses of the stars are not found, only lower limits to their masses.* If the spectral lines of only one star are visible, we do not even find lower limits to the masses of the individual stars but only a relation between their masses, known as the *mass function*.†

The mass function does, however, indicate a lower limit to the mass of the star whose spectral lines are *not* observed in terms of the one whose spectral lines *are* seen. If the spectrum of the visible star gives an indication of its mass (as is often the case) we have at once an estimate of the lower limit to the mass of the fainter companion. We shall see (Chapter 35) that this analysis can be very important for finding candidates for black holes.

The analysis of spectroscopic binaries, we have seen, yields only lower limits to stellar masses. To find the masses of individual binary systems we must have some way of determining the angle at which the orbital plane of the system is inclined to our line of sight (or to the plane of the sky); we can find this angle only when the orbit is almost edge on, so that the spectroscopic binary is also an eclipsing binary (or in a few cases, when it is also a visual binary).

(e) Eclipsing Binaries

To simplify our discussion of eclipsing binaries, we shall assume that the two stars in such a system revolve about each other in circular orbits. Actually the assumption is not so bad. Most stars in binary systems that are seen to eclipse each other are relatively close to each other. Their mutual proximity enhances the effects of their perturbations on each

---

*What is actually found for each star is $m \sin^3 i$, where i is the inclination angle of the plane of orbital revolution to the plane of the sky.

†The *mass function* is

$$\frac{m_2^3 \sin^3 i}{(m_1 + m_2)^2}$$

where m_2 is the mass of the star whose spectrum is not observed.

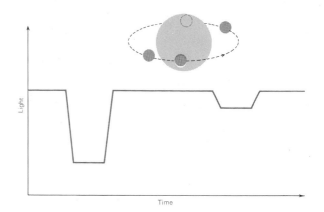

Figure 26.8 *Schematic light curve of a hypothetical eclipsing binary star with alternating total and annular eclipses.*

other's motions, and these perturbations tend to produce circular, or nearly circular, orbits.

During the period of revolution of an eclipsing binary, there are two times when the light from the system diminishes—once when the smaller star passes behind the larger one and is eclipsed, and once when the smaller star passes in front of the larger one and eclipses part of it. If the smaller star goes completely behind the larger one, that eclipse is *total* and the other eclipse half a period later is *annular* (see Figure 26.8). If the smaller star is never completely hidden behind the larger star, both eclipses are partial (Figure 26.9). Each interval during an eclipse when the light from the system is farthest below normal is called a *minimum*. Both minima are not, in general, equally low in light. The same area of each star is covered during the time it is eclipsed; for example, if the eclipses are total or annular, an area equal to the total cross-sectional area of the smaller star is eclipsed in each minimum. The relative amount of light drop at

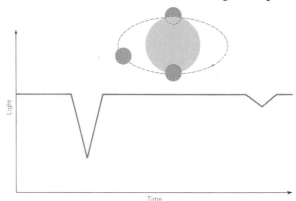

Figure 26.9 *Schematic light curve of a hypothetical eclipsing binary star with partial eclipses.*

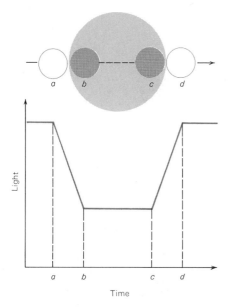

Figure 26.10 Contacts in the schematic light curve of a hypothetical eclipsing binary with central eclipses.

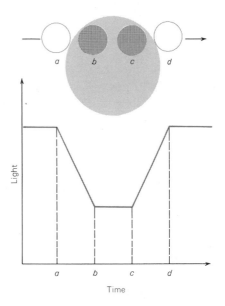

Figure 26.11 Contacts in the schematic light curve of a hypothetical eclipsing binary with noncentral eclipses.

each minimum, however, depends on the relative surface brightnesses of the two stars, and hence on their temperatures. *Primary minimum* occurs when the hotter star is eclipsed (whether it is a total, an annular, or a partial eclipse), and *secondary minimum* occurs when the cooler star is eclipsed. A graph of the light from an eclipsing binary system, plotted against time through a complete period, is called a *light curve*.

The most important data that are obtained from the analysis of the light curve of an eclipsing binary system are the sizes of the stars, relative to their separation, and the inclination of the orbit to our line of sight. To illustrate how the sizes of the stars are related to the light curve, we may consider a hypothetical eclipsing binary in which the stars are very different in size, in which the orbit is exactly edge on, so that the eclipses are *central* (Figure 26.10), and in which the orbit is large enough that we can assume the stars to move in approximately straight lines during the eclipses. When the small star is at point *a (first contact)*, and is just beginning to pass behind the large star, the light curve begins to drop. At point *b (second contact)*, the small star has gone entirely behind the large one and the total phase of the eclipse begins. At *c (third contact)* it begins to emerge, and when the small star has reached *d (last contact)* the eclipse is over. During the time interval between first and second contact (or between third and last contacts), the

small star has moved a distance equal to its own diameter. That time interval is in the same ratio to the period of the system as the diameter of the small star is to the circumference of the relative orbit. During the time interval from first to third contacts (or from second to last contacts), the small star has moved a distance equal to the diameter of the large star; that time interval is to the period as the diameter of the large star is to the circumference of the relative orbit. We see, therefore, that the light curve alone gives the sizes of the stars in terms of the size of their orbit. If the lines of both stars are visible in the composite spectrum of the binary, both radial velocity curves can be found. Then the size of the relative orbit can be found, and we can determine the actual (linear) radii of the stars. In other words, the velocity of the small star with respect to the large one is known, and, when multiplied by the time intervals from first to second contacts and from first to third contacts, gives, respectively, the diameters of the small and large stars.

In actuality the orbits are not, generally, exactly edge on, and the eclipses are not central. To see how the inclination affects the light curve, consider the system shown in Figure 26.11. Here the smaller star does not pass behind the larger one in a direction perpendicular to its limb, but instead passes obliquely and thus takes longer to be eclipsed. Moreover, it does not pass behind a full

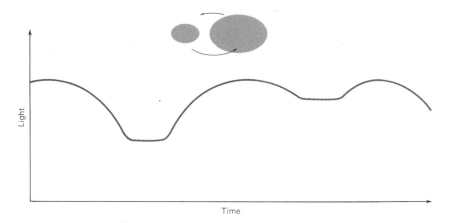

Figure 26.12 Effect of tidal distortion on the light curve in a binary system.

diameter of the disk of the large star but along a shorter chord, so that the entire eclipse, especially the total phase, is of shorter duration than when the eclipses are central. The effects of the sizes of the stars and of the inclination of the orbit are interdependent. However, it is a relatively simple geometry problem, at least in principle, to separate these effects, and from the depths of the minima and the exact instants of the various contacts to calculate both the inclination of the orbit and the sizes of the stars relative to their separation.

The foregoing discussion applies only to eclipses that are total and annular. If they are partial, the analysis is far more difficult, although it can still be accomplished.

There are various other complications which have been ignored here. For example, stars, like the sun, exhibit limb darkening (Section 25.2b), which affects the rate of the drop of light during eclipse. Also, frequently, the two stars in an eclipsing binary are so close together that they suffer severe tidal distortion and have shapes more like footballs than spheres. The light from such systems is not constant, even outside of eclipse, but is greatest when the stars' longest dimensions are turned "broadside" to us, and is less just before and just after a minimum (Figure 26.12). Quite a few such close binaries are found to lose or exchange matter as a result of their interactions. There are many other complications as well, which the specialist must concern himself with.

To summarize: From the analysis of the light curve of an eclipsing binary we can find the inclination of the orbit and the sizes of the stars relative to their separation. If, in addition, we can measure the Doppler shifts of the spectral lines of both stars during their period of revolution, we can obtain their velocity curves. The analysis of the velocity

curves, as described in the preceding section, leads to a determination of lower limits to the masses of the stars. The knowledge of the inclination of the orbit now allows us to convert these minimum values for the masses to actual masses for the individual stars. We can also convert the lower limit to the separation of the stars (or the semimajor axis of their relative orbit) to the actual value when the inclination is known; since the sizes of the stars relative to this separation are found from the light curve, we find their actual diameters or radii. Finally, from the relative depths of the primary and secondary minima, we can calculate the relative surface brightnesses of the stars and hence their effective temperatures (the surface brightness of a star is proportional to the fourth power of its effective temperature—Sections 10.2d and 25.2d). Note that we do not need to know the distance to an eclipsing-spectroscopic binary to determine its mass, as we do in the case of a visual binary. Among the thousands of known eclipsing binaries, however, only a few dozen are so favorably disposed for observation that all the necessary data can be obtained. A recent (1980) review by UCLA astronomer D. M. Popper lists just under a hundred individual stars in eclipsing systems and 28 additional ones in visual binary systems, for which the derived masses are reliable to 15 to 20 percent.

(f) Mass Exchange in Close Binaries

We saw in Section 6.1b that in Lagrange's solution to the three-body problem, in which two stars revolve about each other in circular orbits, there are equilibrium points where a small body can remain stationary with respect to the two stars. Should the small body be shoved closer to one of the revolving

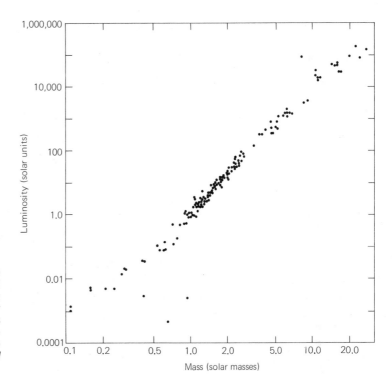

Figure 26.13 Mass-luminosity relation. The plotted points show the masses (abscissas) and luminosities (ordinates) of stars for which both of these quantities are known to an accuracy of 15 to 20 percent. The open circle represents the sun. The three points lying below the sequence of points are all white dwarf stars (see Chapter 27). (*Adapted from data compiled by D. M. Popper*)

stars, however, it is drawn to it by gravitational attraction. We shall also see that as stars age the generation of nuclear energy in their interiors causes them to distend their outer layers greatly, so that those stars become giants (Chapter 33). If such a star is a member of a close binary system, the atoms in its expanding outer layers may reach and pass through one of the points (the *inner Lagrangian point*) where they do not belong to either star. Subsequently the matter from the expanding star can flow to the other star.

Mass exchange is believed to occur between many stars in close binary systems. This exchange of mass can have profound effects on the evolution of the stars in a system. D. Morton at Princeton was the first to study the effects of mass exchange on binary star evolution. His work was followed independently by M. Plavec in Czechoslovakia (now at UCLA), by B. Paczynski in Poland, and by R. Kippenhahn in West Germany, and today it is an important part of the study of stellar evolution. Not only may matter stream from one star to another in a close binary system, possibly accounting for such phenomena as novae and supernovae, but it can also form a large circumstellar disk or ring of material around the binary system, and even be ejected from the system altogether. It also may be involved in the creation of neutron stars and black holes (Chapters 34 and 35).

26.3 THE MASS-LUMINOSITY RELATION

Studies of binary stars have provided knowledge of the masses of over a hundred individual stars. When the masses and luminosities of those stars for which both of these quantities are determined are compared, it is found that, in general, the more massive stars are also the more luminous. This relation, known as the *mass-luminosity relation*, is shown graphically in Figure 26.13, plotted from data compiled in Popper's review (referred to above). The mass-luminosity relation provides a useful means of estimating the masses of stars of known luminosity that do not happen to be members of visual or eclipsing binary systems. We shall return to the physical explanation of the relation in Chapter 33.

It should be noted how very much greater the range of stellar luminosities is than the range of stellar masses. Luminosities of stars are roughly proportional to their masses raised to the 3.5 to 4.0 power. Most stars have masses between one tenth and 50 times that of the sun; according to the mass-luminosity relation, however, the corresponding luminosities of stars at either end of the range are respectively less than 0.01 and about 10^6 solar luminosities. The intrinsically faintest known star has a luminosity near 10^{-6} that of the sun; its mass is probably greater than 0.01 of the sun's. If two stars differ in mass by a factor of 2, their luminosities would then be expected to differ by a factor of 10 to 15.

26.4 DIAMETERS OF STARS

The eclipsing binary systems, as we have seen, provide one means of determining the diameters of stars, but of course only of those stars that happen to be members of eclipsing systems for which the necessary analysis can be carried out. It would be convenient if the angular sizes could be measured directly for many stars of known distances; then their linear diameters could be calculated just as they are for the moon or planets. The sun is unfortunately the only star whose angular size can be resolved visually and whose diameter can be calculated simply. There are many other stars, however, whose angular sizes are only slightly beyond the limit of resolution of the largest telescopes and which can be measured with special techniques.

(a) MEASURING ANGULAR DIAMETERS OF STARS

The earliest procedure for measuring angular diameters of stars was with the stellar interferometer, invented by the American physicist A. E. Michelson. The instrument consists of a long beam, which supports two movable mirrors whose separation can be varied. Each gathers light from a star and reflects it, via secondary mirrors, to a telescope, which forms an image of the light from the two movable mirrors. If the star were a true point source, the image would be a diffraction pattern (Section 10.1a) of a point source as seen by the two separate mirrors. But a star is not, in reality, a point, but a disk, consisting of many parts, and the way the diffraction patterns of light from the different parts of the star combine depends on the angular size of the star and on the spacing of the mirrors on the beam. In its application, the spacing of the mirrors in the interferometer is varied, and the changes in the composite diffraction pattern are observed. The way the pattern changes permits the determination of the angular size of the stellar image under observation. The instrument works only for stars near and/or large enough that they would present a resolved disk as seen through an ideal telescope of aperture equal to the separation of the movable mirrors. There are

not many such stars, but in 1920 Michelson and F. G. Pease, using a 5-meter beam mounted on the 100-inch telescope on Mount Wilson, were able to measure the angular diameters of seven giant stars. The one of largest angular diameter is Mira (0."056).

An electronic analog of the stellar interferometer has been applied more recently to the measurement of star diameters. Separate optical telescopes, placed up to hundreds of meters apart and equipped with photomultiplier tubes, are used to observe the same star simultaneously. The electric currents generated in the two tubes are brought together to a single amplifier. The electrical impulses carry information that can be made to interfere in a way somewhat analogous to the production of fringes in the optical interferometer. The radio astronomers Brown and Twiss first used the device to measure the angular diameter of the star Sirius.

The Brown-Twiss type of interferometer is known as an *intensity interferometer*. The largest model, assembled in Australia, uses two telescopes that can be separated up to 188 m. Each uses a mosaic mirror 6.7 m in diameter, which in turn consists of 251 separate small mirrors, all mounted to reflect the light from a star to one point. These special-purpose mirrors are not of high optical quality and could not be used for ordinary telescopic observations. They are adequate, however, to focus starlight on the light-sensitive surface of a photomultiplier. This large intensity interferometer can achieve a resolution down to 5×10^{-4} seconds of arc and has been used to measure the angular diameters of nearly a hundred stars; the smallest have an angular diameter of only 1.6×10^{-4} seconds of arc.

Other techniques for measuring the angular sizes of stars involve speckle interferometry (Section 11.4b) and high-speed photoelectric observations of the dimming of light from stars being occulted by the moon or by planets.

In all, about a hundred angular stellar diameters have been measured by one or more of the above methods. There are some difficulties with each technique (for example, in connection with limb darkening—see Section 25.2b). Still, these data give us some confidence in the correctness of the less direct determinations of stellar diameters (see below) from radiation laws.

(b) Finding Stellar Radii from Radiation Laws

For most stars we must use an indirect method by which we can calculate their radii from theory. The theory involved is the Stefan-Boltzmann law (Section 10.2d); we calculate the radius of a spherical perfect radiator that has the same luminosity and effective temperature that a star does.

The luminosity of a star can be obtained by the procedure discussed in Chapter 24, and the temperature of a star can be obtained in various ways, as from its color or its spectrum (Section 25.2d). Now, the energy emitted per unit area of a star (given by Stefan's law), multiplied by its entire surface area, gives the star's total output of radiant energy, that is, its luminosity. Since the surface area of a sphere of radius R is $4\pi R^2$, the luminosity of a star is given by

$$L = 4\pi R^2 \times \sigma T^4.$$

The above equation can be solved for the radius of the star.

Note that the temperature appearing in the above formula is raised to the fourth power; if it is in error, therefore, the computed value of the star's radius can be substantially incorrect. In particular, because stars are *not* perfect radiators, values of stellar temperatures as determined by different methods do not all agree precisely. There is no assurance that the temperature estimated from the star's spectrum, is the most appropriate value to use in the computation of stellar radii from radiation laws. In the relatively few cases (described above) in which both the radius and luminosity of a star can be determined independently, the radius and luminosity of the star can be used with the above relation to derive the star's temperature. A temperature so determined, the *effective temperature* (T_{eff}), is the temperature of a black body that emits the same amount of energy per unit area as the star does, and thus is the appropriate kind of temperature to use in the calculation of stellar radii. Those effective temperatures which have been determined for stars enable us to see what small corrections must be applied to temperatures derived by other means (for example, from colors) to make the determination of stellar radii more reliable. Different kinds of stellar temperatures were discussed more fully in Section 25.2d.

We shall illustrate the use of Stefan's law for the computation of stellar radii with two examples. Consider, first, a star whose red color indicates that it has a temperature of about 3000 K, roughly half the temperature of the sun. Each square centimeter of the star, therefore, emits only $^1/_{16}$ as much light as the sun (for the light emitted is proportional to the fourth power of the temperature). Suppose,

however, that the star is, nevertheless, 400 times as luminous as the sun. It must be many times larger than the sun to emit more light despite its much lower surface brightness. We can find its radius, in terms of the sun's, by noting that $L \propto R^2 T^4$, and thus (since the constants of proportionality cancel in each of the ratios),

$$\frac{R_*}{R_\odot} = \sqrt{\frac{L_*}{L_\odot}}\left(\frac{T_\odot}{T_*}\right)^2 = \sqrt{400} \times 4 = 80.$$

(The subscripts $_*$ and $_\odot$ refer to the star and sun, respectively.) This star has 80 times the sun's radius; if the sun were placed at its center the star's surface would reach past the orbit of Mercury.

Next, consider a star whose blue color indicates a temperature of about 12,000 K—twice the sun's temperature. Suppose, however, that this star has a luminosity of only $^1/_{100}$ that of the sun. Now we find, for the star's radius,

$$\frac{R_*}{R_\odot} = \sqrt{\frac{L_*}{L_\odot}}\left(\frac{T_\odot}{T_*}\right)^2 = \sqrt{\frac{1}{100}}\left(\frac{1}{2}\right)^2 = \frac{1}{40}.$$

The star has only $^1/_{40}$ the sun's radius—less than three times the radius of the earth. These two examples are by no means extreme cases, but are more or less typical of *very red giants* and *white dwarfs*, respectively (Chapter 27).

(c) Summary of Stellar Diameters

The few score good geometrical determinations of stellar radii come from (1) direct measure of the sun's angular diameter, (2) measures of the angular diameters of seven nearby giant stars with the stellar interferometer, (3) measures of the angular sizes of nearly 100 other stars with an electronic adaptation of the stellar interferometer, (4) a few stars whose diameters have been measured with speckle interferometry or from observations of occultations of stars by the moon or planets, and (5) analyses of the light curves and radial velocity curves of eclipsing binary systems. All other determinations of stellar radii make use of the radiation laws; the validity of this indirect method is verified by noting that it gives the correct radii for those stars whose sizes can also be measured by geometrical means.

The temperatures obtained from observations of the colors or spectra of stars are not found to be exactly the same as the effective temperatures which must be used with Stefan's law to compute stellar radii. The necessary corrections to color or spectral temperatures that are required to obtain effective temperatures are known for those stars whose radii are determined geometrically. Unfortunately, many classes of stars do not contain objects whose sizes have been measured by direct or geometrical methods, and effective temperatures for them must be predicted from theoretical studies of stellar atmospheres. Although radii computed for those stars are somewhat uncertain, we can use Stefan's law to find the sizes of most stars with a precision of 10 to 20 percent.

EXERCISES

1. Many eclipsing binaries can be observed which do not appear in catalogues of spectroscopic binaries. Can you suggest an explanation?

2. A few stars are both visual binaries *and* spectroscopic binaries (their radial velocity variations can be detected). Why do you suppose such stars are rare? Can you suggest a way of determining the distance to such a system? (*Hint:* Consider the method of determining the parallax to a moving cluster, Section 23.3c.)

3. Describe the apparent relative orbit of a visual binary whose true orbital plane is edge on to the line of sight. Describe the apparent motions of the individual stars of the system among the background stars in the sky.

4. What, approximately, would be the periods of revolution of binary star systems in which each star had the same mass as the sun, and in which the semimajor axes of the relative orbits had these values? (a) 1 AU; (b) 2 AU; (c) 6 AU; (d) 20 AU; (e) 60 AU; (f) 100 AU.

*5. In each of the binary systems in Exercise 4, at what distance would the two stars appear to have an angular separation of 1″? (Assume circular orbits.)

6. Why do most visual binaries have relatively long periods and most spectroscopic binaries relatively short periods? Under what circumstances could a binary with a relatively long period (over a year) be observed as a spectroscopic binary?

7. Show that the semimajor axis of the true relative orbit of a visual binary system, in astronomical units, is equal to its angular value, in seconds of arc, times the distance of the system, in parsecs.

8. The true relative orbit of ξ Ursae Majoris has a semimajor axis of 2″.5, and the parallax of the system is 0″.127. The period is 60 years. What is the sum of the masses of the two stars in units of the solar mass?
Answer: 2.1 solar masses

9. In a particular visual-spectroscopic binary the maximum value of the radial velocity of one star with respect to the other is 60 km/s, the inclination of the orbital plane to the plane of the sky 30°, and the period is 22 days. If the stars in the system have circular orbits, what is the sum of their masses?
Answer: 3.8 solar masses

10. Explain why the two stars in a binary system must have orbits of identical shape (eccentricity), differing only in size.

*11. The observed component of a hypothetical astrometric binary system is found to move in an elliptical orbit of semimajor axis 1 AU about the barycenter of the system in a period of 30 years. If the visible star is assumed to have a mass equal to that of the sun, what is the mass of its unseen companion?
Answer: About 0.11 solar masses

12. A hypothetical spectroscopic-eclipsing binary star is observed. The period of the system is three years. The maximum radial velocities, with respect to the center of mass of the system, are as follows: Star *A*, $^4/_3\pi$ AU/ year; Star *B*, $^2/_3\pi$ AU/year.
(a) What is the ratio of the masses of the stars?
(b) Find the mass of each star (in solar units). Assume that the eclipses are central.

13. In an eclipsing binary in which the eclipses are exactly central, and in which a small star revolves about a considerably larger one, the interval from first to second contacts is one hour and from first to third contacts is four hours. The entire period is three days. The centers of the stars are separated by 11,460,000 km. What are the diameters of the stars?
Answer: 1,000,000 and 4,000,000 km

14. Although the periods of known eclipsing binaries range from 4^h39^m to 27 years, the average of their periods is less than the average period of all known spectroscopic binaries. Can you suggest an explanation?

15. How many times as massive as the sun would you expect a star to be that is 1000 times as luminous? What if it were 10,000 times as luminous? (Assume that the mass-luminosity relation holds for these stars.)

*16. The diameters of the stars measured by Michelson and Pease are given in the following table:

STAR	ANGULAR DIAMETER	DISTANCE (*pc*)	LINEAR DIAMETER (*In Terms of Sun's*)
Betelgeuse	0″034*	150	500
(α Orionis)	0.042		750
Aldebaran			
(α Tauri)	0.020	21	45
Arcturus			
(α Bootis)	0.020	11	23
Antares			
(α Scorpii)	0.040	150	640
Scheat			
(β Pegasi)	0.021	50	110
Ras Algethi			
(α Herculis)	0.030	150	500
Mira			
(o Ceti)	0.056	70	420

*Variable in size.

Which of the stars are larger than the orbit of the earth? Are any larger than the orbit of Mars? Of Jupiter?

17. Show how the measured angular diameters and observed energy fluxes of stars can be used to measure their effective temperatures. (*Hint:* Use the inverse-square law of light, and recall Stefan's law.)

18. What is the radius of a star (in terms of the sun's radius) with the following characteristics:
   (a) Twice the sun's temperature and four times its luminosity?
   (b) Eighty-one times the sun's luminosity and three times its temperature?

19. Assume the wavelength of maximum light of the sun to be exactly 5000 Å, its temperature exactly 6000 K, and its absolute bolometric magnitude exactly 5.0. Another star has its wavelength of maximum light at 10,000 Å. Its apparent visual magnitude is 15.5, its bolometric correction is 0.5, and its parallax is 0″01. What is its radius in terms of the sun's?
   Answer: $R/R_\odot = 0.4$

Ejnar Hertzsprung (1873–1967), Danish astronomer, spent most of his productive career in Göttingen (Germany) and Leiden (the Netherlands). He specialized in the study of binary stars and star clusters and the properties of stars, and he was the first to recognize the distinction between giant and dwarf stars. *(Nordisk Pressefoto A/S and the American Institute of Physics)*

Henry Norris Russell (1877–1957), American astronomer, was professor at Princeton University and director of the observatory. His many interests included the study of stellar evolution. Hertzsprung and Russell independently discovered the main sequence of stars, best illustrated on the famous diagram now known as the *Hertzsprung-Russell diagram.* *(Princeton University Archives)*

27

THE STELLAR POPULATION: A CELESTIAL CENSUS

Having described the methods by which we are able to obtain basic information about individual stars, we now summarize the data that have been gathered about the stellar population.

27.1 THE NEAREST AND THE BRIGHTEST STARS

Let us consider first our most conspicuous stellar neighbors, the brightest-appearing stars in the sky. Appendix 14 lists some of the properties of the brightest 20 stars. Many of these are double or triple star systems; data are given, in such cases, for each component.

(a) The Brightest Stars

The most striking thing about the brightest-appearing stars is that they are bright not because they are

nearby, but because they are actually of high intrinsic luminosity. Of the brightest 20 stars listed in Appendix 14, only six are within 10 pc of the sun. The absolute magnitude of a star (Section 24.2a) is the apparent magnitude that it would have if it were at a distance of 10 pc. Since the brightest 20 stars are of apparent magnitude 1.5 or brighter, the 14 of them that are more distant than 10 pc must have absolute magnitudes *less* (that is, brighter) than 1.5. Even among the approximately 3000 stars with apparent magnitudes less than 6.0, only about 60 are within 10 pc. Most naked-eye stars are tens or even hundreds of parsecs away and are many times as luminous as the sun. Indeed, among the 6000 stars visible to the naked eye, at most 50 are intrinsically fainter than the sun. Figure 27.1 is a histogram showing the distribution among various absolute visual magnitudes of the 30 brightest-appearing stars (the absolute visual magnitude of the sun is +4.8).

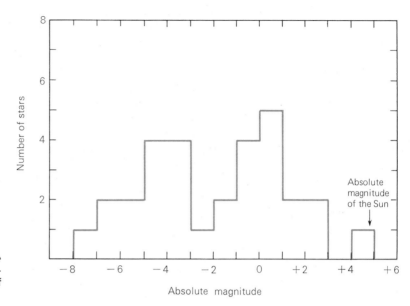

Figure 27.1 Distribution among absolute magnitudes of the 30 brightest-appearing stars. The units are the numbers of stars per unit of absolute magnitude.

From Appendix 14, or Figure 27.1 we might gain the impression that the sun is far below average among stars in luminosity. This is not so. Most stars, as we shall see in the next subsection, are much less luminous than the sun is. They are too faint, in fact, to be conspicuous unless they are nearby. Stars of high luminosity are rare—so rare that the chance of finding one within a small volume of space, say, within 10 pc of the sun, is very slight. Why, then, are the most common, intrinsically faint stars not among the most common naked-eye stars, while rare, highly luminous stars are?

The question is best answered with the help of some numerical examples. The sun, whose absolute visual magnitude is +4.8, would appear as a very faint star to the naked eye if it were 10 pc away. Stars much less luminous than the sun would not be visible at all at that distance. Stars with absolute magnitudes in the range +10 to +15 are very common, but a star of absolute magnitude +10 would have to be within 1.6 pc to be visible to the naked eye. Only Alpha Centauri is closer than this. The intrinsically faintest star observed has an absolute magnitude of about +19. For this star to be visible to the naked eye, it would have to be within 0.025 pc, or 5200 AU. The star could not be photographed even with the 5-m telescope if it were more distant than 100 pc. It is clear, then, that the vast majority of nearby stars, those less luminous than the sun, do not send enough light across interstellar distances to be seen without optical aid.

In contrast, consider the highly luminous stars. Stars with absolute magnitudes of 0 have luminosities of about 100 times that of the sun. They are far less common than stars less luminous than the sun, but they are visible to the naked eye even out to a distance of 160 pc. A star with an absolute magnitude of −5 (10,000 times the sun's luminosity) can be seen without a telescope to a distance of 1600 pc (if there is no dimming of light by interstellar dust—see Chapter 28). Such stars are very rare, and we would not expect to find one within a distance of only 10 pc; the volume of space included within a distance of 1600 pc, however, is about 4 million times that included within a distance of only 10 pc. Hence many stars of high luminosity are visible to the unaided eye.

(b) The Nearest Stars

Evidently, the brightest naked-eye stars do not provide a representative sample of the stellar population. Let us turn then to the nearest known stars. Appendix 13 lists the 44 known stars within 5 pc (some are double or multiple systems) from data provided by the U. S. Naval Observatory. (Additional nearby stars are discovered from time to time; the total number of stars within 5 pc may be double the number listed. Moreover, the measurements of distances, luminosities, and so on for nearby stars are being continually refined, but the table does indicate the general characteristics of the sun's nearest stellar neighbors.)

First the table shows that only three of the 43 stars (other than the sun) are among the brightest 20 stars: Sirius, Alpha Centauri, and Procyon. This fact is further confirmation that the nearest stars are not the brightest-appearing stars. The nearby stars also tend to have large proper motions, as would be expected (Section 23.3). In fact, the large proper motions of many of these stars led to the discovery that they are located nearby. Another interesting observation is that 13 of the 44 stars are really binary or multiple star systems; the table thus contains, actually, 59 rather than 44 stars. Twenty-eight of these 59 stars, or nearly half, are members of systems containing more than one star. Moreover, two or three other stars in the list are suspected of having astrometric companions.

The most important datum concerning the nearest stars is that most of them are intrinsically faint. Only ten of the nearest stars are individually visible to the unaided eye. Only three are as intrinsically luminous as the sun; 43 have absolute magnitudes fainter than +10. If the stars in our immediate stellar neighborhood are representative of the stellar population in general, we must conclude that the most numerous stars are those of low luminosity; in a random sample, only about one star in 60 is as intrinsically luminous as the sun.

An estimated lower limit can be established for the mean density of stars in the solar neighborhood. There are at least 59 stars within 5 pc (counting the members of binary and multiple star systems and the sun). We would expect eight times as many stars within 10 pc (for that distance includes a volume of space eight times as large), and about 64 times as many stars within 20 pc. W. Luyten estimates there to be more than 500 stars within 10 pc and some of the fainter stars within that distance must have escaped discovery. Within 20 pc there may be 4000 or more stars. A sphere of radius 5 pc has a volume of $\frac{4\pi(5)^3}{3}$, or about 524 pc^3. Since this volume of space contains at least 59 stars, the density of stars in space is at least one star for every 9 pc^3; the actual stellar density, of course, can be greater than this figure if there are undiscovered stars within 5 pc. The mean separation between stars is the cube root of 9, or about 2.1 pc. If the matter contained in stars could be spread out evenly over space, and if a typical star has a mass 0.4 times that of the sun, the mean density of matter in the solar neighborhood would be about 3×10^{-24} g/cm^3.

The nearest stars comprise a much more nearly representative sample of the stellar population than do the brightest stars. We are still not sure, however, that we have identified all of the faintest stars in the solar neighborhood. Moreover, there do not happen to be any stars of high luminosity in this "tiny" volume of space. Yet we can identify all the luminous stars, with a reasonable degree of completeness, out to a much greater distance. If we make allowance for the different volumes of space that we must survey to catalogue large samples of stars of different intrinsic luminosities, we can gain some indication of their relative abundances. For example, within 10 pc there are about 12 known stars brighter than absolute magnitude +4, while within 5 pc there are about 57 known stars fainter than absolute magnitude +4. We would expect, however, some 8×57, or 456 stars fainter than absolute magnitude +4 within 10 pc; therefore, the ratio of stars with absolute magnitude greater (fainter) than +4 to the number of more luminous stars is about 456 to 12, or 38:1. This calculation is only an example and may not indicate the precise ratio that exists in the stellar population.

(c) The Luminosity Function

Once the numbers of stars of various intrinsic luminosities have been found, the relative numbers of stars in successive intervals of absolute magnitude within any given volume of space can be established. This relationship is called the *luminosity function*. Figure 27.2 shows the luminosity function for stars in the solar neighborhood, as it has been determined by W. J. Luyten. Compare Figure 27.2 with Figure 27.1.

The sun, we see, is more luminous than the vast majority of stars. Most of the stellar mass is contributed by stars that are fainter than the sun. On the other hand, the relatively few stars of higher luminosity than the sun compensate for their small numbers by their high rate of energy output. It takes only ten stars of absolute magnitude 0 to outshine 1000 stars fainter than the sun, and only one star of absolute magnitude −5 to outshine 10,000 stars fainter than the sun. Most of the starlight from our part of space, it turns out, comes from the relatively few stars that are more luminous than the sun.

27.2 THE HERTZSPRUNG-RUSSELL DIAGRAM

In 1911 the Danish astronomer E. Hertzsprung compared the colors and luminosities of stars within several clusters by plotting their magnitudes against

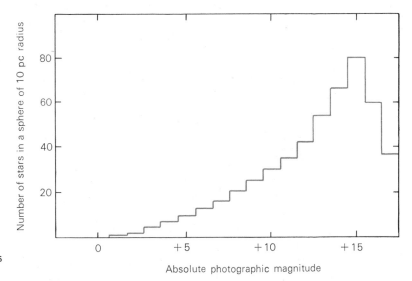

Figure 27.2 Luminosity function of stars in the solar neighborhood.

their colors. In 1913 the American astronomer Henry Norris Russell undertook a similar investigation of stars in the solar neighborhood by plotting the absolute magnitudes of stars of known distance against their spectral classes. These investigations by Hertzsprung and by Russell led to an extremely important discovery concerning the relation between the luminosities and surface temperatures of stars. The discovery is exhibited graphically on a diagram named in honor of the two astronomers—the *Hertzsprung–Russell* or *H–R diagram*.

(a) Features of the H–R Diagram

Two easily derived characteristics of stars of known distances are their absolute magnitudes (or luminosities) and their surface temperatures. The absolute magnitudes can be found from the known distances and the observed apparent magnitudes. The surface temperature of a star is indicated either by its color or its spectral class. Before the development of yellow- and red-sensitive photographic emulsions and, of course, photoelectric techniques, spectral classes of stars were usually used to indicate their temperatures. Now that stellar colors can be measured with precision, the color index is more often employed, even though the use of spectral classes is still of great value.

If the absolute magnitudes of stars are plotted against their spectral classes, an H–R diagram like that of Figure 27.3 is obtained. The most significant feature of the H–R diagram is that the stars

are not distributed over it at random, exhibiting all combinations of absolute magnitude and temperature, but rather cluster into certain parts of the diagram. The majority of stars are aligned along a narrow sequence running from the upper left (hot, highly luminous) part of the diagram to the lower right (cool, less luminous) part. This band of points is called the *main sequence*. A substantial number of stars, however, lie above the main sequence on the H–R diagram, in the upper right (cool, high luminosity) region. These are called *giants*. At the top part of the diagram are stars of even higher luminosity, called *supergiants*. Finally, there are stars in the lower left (hot, low luminosity) corner known as *white dwarfs*. To say that a star lies "on" or "off" the main sequence does not refer to its position in space, but only to the point that represents its luminosity and temperature on the H–R diagram.

An H–R diagram, such as Figure 27.3, that is plotted for stars of known distance does not show the relative proportions of various kinds of stars, because only the nearest of the intrinsically faint stars can be observed. To be truly representative of the stellar population, an H–R diagram should be plotted for all stars within a certain distance (see Exercise 11). Unfortunately, our knowledge is reasonably complete only for stars within a few parsecs of the sun, among which there are no giants or supergiants. It is estimated that about 90 percent of the stars in our part of space are main-sequence stars and about 10 percent are white dwarfs. Less than 1 percent are giants or supergiants. We shall return to the theoretical interpretation of the distribution of stars on the H–R diagram in Chapter 33.

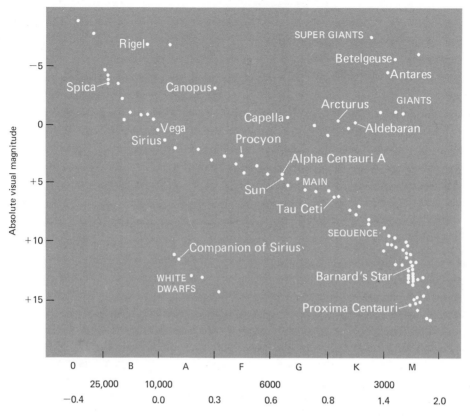

Figure 27.3 Hertzsprung-Russell diagram for a selected sample of stars.

(b) Method of Spectroscopic Parallaxes

One of the most important applications of the H–R diagram is in the determination of stellar distances. Suppose, for example, that a star is known to be a spectral class G2 star on the main sequence. Its absolute magnitude could then be read off the H–R diagram at once; it would be about +5. From this absolute magnitude and the star's apparent magnitude, its distance can be calculated (Section 24.2b).

In general, however, the spectral class alone is not enough to fix, unambiguously, the absolute magnitude of a star. The G2 star described in the last paragraph could have been, for example, a main-sequence star of absolute magnitude +5, a giant of absolute magnitude 0, or a supergiant of still higher luminosity. We recall, however (Section 25.1b), that pressure differences in the atmospheres of stars of different sizes result in slightly different degrees of ionization for a given temperature. It will be seen in the next subsection that giant stars are larger than main-sequence stars of the same spectral class and that supergiants are larger still. In 1913

Adams and Kohlschütter, at the Mount Wilson Observatory, first observed the slight differences in the degrees to which different elements are ionized in stars of the same spectral class but different luminosities (and therefore different sizes). It is thus possible to classify a star by its spectrum, not only according to its temperature (spectral class) but also according to whether it is a main-sequence star, a giant, or a supergiant.

The most widely used system of classifying stars according to their luminosities is that of W. W. Morgan and his associates at the Yerkes Observatory. In favorable cases it has been found possible to divide stars of a given spectral class into as many as six categories, called *luminosity classes.*

These luminosity classes are:
- Ia Brightest supergiants
- Ib Less luminous supergiants
- II Bright giants
- III Giants
- IV Subgiants (intermediate between giants and main-sequence stars)
- V Main-sequence stars

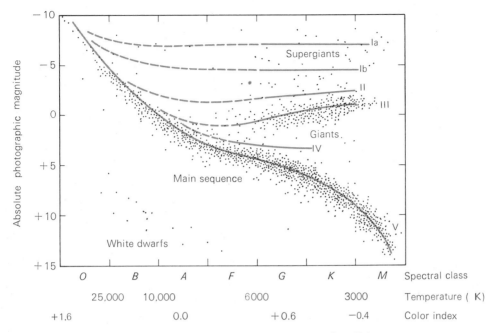

Figure 27.4 Luminosity classes on the Hertzsprung-Russell diagram.

A small number of stars that may lie below the normal main sequence are called *subdwarfs* (Sd). The white dwarfs are much fainter. Main-sequence stars (luminosity class V) are often termed "dwarfs" to distinguish them from giants. The term "dwarf" is even applied to main-sequence stars of high luminosity, which may have diameters several times as great as the sun's. The term "dwarf," when applied to a main-sequence star, should not be confused with its use as applied to a white dwarf. The full specification of a star, including its luminosity class, would be, for example, for a spectral-class F3 main-sequence star, F3 V. For a spectral class M2 giant, the specification would be M2 III. Figure 27.4 illustrates the approximate mean positions of stars of various luminosity classes on the H–R diagram. The dashed portions of the lines represent those spectral classes (for a given luminosity class) for which there are very few or no stars.

With both its spectral class and luminosity class known, a star's position on the H–R diagram is uniquely determined. Its absolute magnitude, therefore, is also known, and its distance can be calculated. Distances determined this way, from the spectral and luminosity classes, are said to be obtained from the *method of spectroscopic parallaxes*.

There are some stars that do not fit into the standard classification scheme (see, especially, Chapter 33). The method of spectroscopic parallaxes does not work for them.

(c) Extremes of Stellar Luminosities, Radii, and Densities

The most massive stars have absolute magnitudes of −6 to −8. A few stars are known that have absolute bolometric magnitudes of −10; they are a million times as luminous as the sun. These super-luminous stars, most of which are at the upper left on the H–R diagram, are very hot spectral-type O and B stars, and are very blue. These are the stars that would be the most conspicuous at very great distances in space.

Consider now the stars at the upper right corner of the H–R diagram. A red, cool supergiant has a surface temperature of 3000 K and an absolute bolometric magnitude of −5. This star has 10,000 times the sun's luminosity but only half its surface temperature. Since each unit area of the star emits only 1/16 as much light as a unit area of the sun (Section 10.2d), its total surface area must be greater than the sun's by 160,000 times. Its radius, therefore, is 400 times the sun's radius. If the sun could be placed in the center of such a star, the star's surface would lie beyond the orbit of Mars. Even more extreme are infrared supergiants.

Red giant stars have extremely low mean densities. The volume of the star described in the last paragraph would be 64 million times the volume of the sun. The masses of such giant stars, however, are probably at most only 50 solar masses, and very likely much less. Plaskett's star, a spectral-type O star with a mass of at least 50 solar masses, is one of the most massive stars known. If we assume that a supergiant star with 64 million times the sun's volume has only 10 times its mass, we find that it has just over 1 ten millionth the sun's mean density, or only about 2 ten millionths the density of water; the outer parts of such a star would constitute an excellent laboratory vacuum.

In contrast, the very common red, cool stars of low luminosity at the lower end of the main sequence are much smaller and more compact than the sun. As an example, consider such a red dwarf, the star Ross 614B, which has a surface temperature of 2700 K and an absolute bolometric magnitude of about $+13$ ($\frac{1}{2000}$ of the sun's luminosity). Each unit area of this star emits only $\frac{1}{20}$ as much light as a unit area of the sun, but to have only $\frac{1}{2000}$ the sun's luminosity, the star must have only about $\frac{1}{100}$ the sun's surface area, or $\frac{1}{10}$ its radius. A star with such a low luminosity also has a low mass (Ross 614B has a mass about $\frac{1}{12}$ that of the sun), but still would have a mean density about 80 times that of the sun. Its density must be higher, in fact, than that of any known solid found on the surface of the earth.

The faint red main-sequence stars are not the stars of the most extreme densities, however. The white dwarfs, at the lower left corner of the H–R diagram, have the highest densities of the normal stars known to be common.

(d) The White Dwarfs

The first white dwarf to be discovered is the companion to Sirius, the brightest appearing star in the sky. From its wavy proper motion, Sirius was known to have a companion since 1844 (Section 26.1a); it was first seen telescopically in 1862. Sirius is the brightest star in the constellation Canis Major, Orion's big dog. It is interesting that Procyon, the brightest star in Orion's other dog, Canis Minor, also has a white dwarf companion. A third nearby star with a white dwarf companion is 40 Eridani.

Figure 27.5 The Milky Way in Sagittarius. (*Yerkes Observatory*)

The companion of Sirius has a mass of about 94 percent that of the sun. From its temperature and luminosity, however, we find its diameter to be only about 2 percent of the sun's, or about twice that of the earth. The white dwarf has a mean density more than 100 thousand times that of the sun, and a sixth of a million times that of water. Some white dwarfs have much higher mean densities, and many have central densities in excess of 10^7 times that of water. A teaspoonful of such material would weigh nearly 50 tons.

The British astrophysicist Sir Arthur Eddington described the first known white dwarf this way: "The message of the companion of Sirius, when decoded, ran: I am composed of material three thousand times denser than anything you've come across. A ton of my material would be a little nugget you could put in a matchbox. What reply could one make to something like that? Well, the reply most of us made in 1914 was, 'Shut up; don't talk nonsense.'"

We shall describe the theory of white dwarfs in Chapter 34, where we shall also learn that there are stars millions of times denser yet!

27.3 THE DISTRIBUTION OF THE STARS IN SPACE

In the immediate neighborhood of the sun, the stars seem to be distributed more or less at random (except for their tendency to form small clusters). The larger the volume of space we survey, the more stars we find, and if allowance is made for the fact that the faintest stars become invisible at larger distances, it is found that the number of stars we can count is roughly proportional to the cube of the distance to which we look. Eventually, however, the stars do thin out more rapidly in some directions than in others. The way they thin out is a clue to the nature of the stellar system to which the sun belongs. The idea that the sun is a part of a large system of stars was suggested as early as 1750 by Thomas Wright in his *Theory of the Universe*. Immanuel Kant, the great German philosopher, suggested the same hypothesis five years later. It was the German-English astronomer William Herschel, however, who first demonstrated the nature of the stellar system.

(a) Herschel's Star Gauging

Herschel sampled the distribution of stars about the sky by a procedure he called *star gauging*. He observed that in some directions he could count more stars through his telescope than in other directions. In 1784 and 1785 he presented two papers giving the results of gauges or counts of stars that he was able to observe in 683 selected regions scattered over the sky. While in some of these fields he could see only a single star, in others he was able to count nearly 600. Herschel reasoned that in those directions in which he saw the greatest numbers of faint stars, the stars extended the farthest, and in other directions they thinned out at relatively short distances. As a result of his star gauging, Herschel arrived at the conclusion (only partially correct, as we shall see) that the sun is inside a great sidereal system, and that the system is disk-shaped, roughly like a grindstone, with the sun near the center.

(b) The Phenomenon of the Milky Way

All of us who have looked at the sky on a moonless night away from the glare of city lights are aware of the Milky Way, a faint, luminous band of light

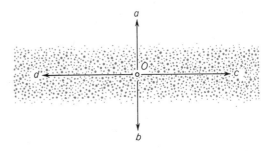

Figure 27.6 We see the Milky Way when we look from the earth at *O* edge on through our Galaxy (directions *c* and *d*).

that completely encircles the sky. Galileo turned his telescope on the Milky Way and saw that it really consists of a myriad of faint stars. Herschel's grindstone explains why the Milky Way appears as a band all the way around the sky.

It must be recalled that we view our sidereal system from the inside. Figure 27.6 shows an idealized portion of the "grindstone," viewed edge on. The sun's position is at *O*. If we look from *O* toward either face of the wheel, that is, in directions *a* or *b*, we see only those stars that lie between us and the nearest boundary of the stellar system. In these directions in the sky, therefore, we see only scattered stars. On the other hand, if we look edge on through the wheel, say in directions *c* or *d*, we encounter so many stars along our line of sight that we get the illusion of a continuous band of light. Since the greatest dimensions of the grindstone extend in all directions along its flat plane, the band of light extends completely around the sky. This band of light is the Milky Way; it is simply the light from the many distant stars that appear lined up in projection when we look edge on through our own flattened stellar system. Figure 27.7 is a copy of one of Herschel's diagrams, showing a cross section of the stellar system, as he derived its shape.

(c) The Galaxy

We call our stellar system the *Galaxy*, or sometimes the *Milky Way Galaxy*, or even just *Milky Way*. The Galaxy is far more complicated than Herschel's image of the grindstone. It is a vast, wheel-shaped system of some 10^{11} stars, with a diameter that probably exceeds 30,000 pc (100,000 LY). The flattened shape of the Galaxy is a consequence of its rotation. The sun, about two thirds of the way from the center out to the rim of the wheel, moves at a speed of about 250 km/s to complete its orbital revolution about the galactic center in some

Figure 27.7 A copy of a diagram by Herschel, showing a cross section of the Milky Way system of stars. The large circle shows the location of the sun.

200 million years. (Unlike a wheel, the Galaxy does not rotate as a solid body—see Chapter 29).

At the hub of the Galaxy is a huge nuclear bulge of stars, where the stars are somewhat closer together than they are in the solar neighborhood (although still light-months or light-years apart). Extending outward from the nucleus, and winding through the disk of the Galaxy, like the spirals of light in a gigantic pinwheel, are the *spiral arms*. The spiral arms consist of vast clouds of gas and cosmic dust—the interstellar medium. Associated with these gas and dust clouds are many young stars, a few of which are very hot and luminous. In the interstellar gas and dust clouds of the spiral arms star formation is taking place.

In addition to individual stars and clouds of interstellar matter, the Galaxy contains many *star clusters*—groups of stars, presumably having a common origin and age. The most common star clusters, numbering in the thousands, are the *open* or *galactic clusters*. Typically, an open cluster consists of a few hundred stars, loosely held together by their mutual gravitation, and moving together through space. The open clusters are located in the main disk of the Galaxy and are usually in or near spiral arms. Besides the open clusters, there are over a hundred *globular clusters*—beautiful, spherically symmetrical clusters, each containing hundreds of thousands of member stars. Most of the globular clusters are scattered in a roughly

spherical distribution about the main wheel of the Galaxy, grouped around it rather like bees around a flower. They form a more or less spherical *halo* surrounding the main body of the Galaxy.

Our Galaxy is not alone in space. Today we know that far beyond its borders there are countless millions of other galaxies, extending as far as we can see in all directions. Our Galaxy, interstellar matter, star clusters, and other galaxies are the subjects of some of the later chapters.

Figure 27.8 Omega Centaurus globular star cluster. (*Kitt Peak National Observatory*)

EXERCISES

1. (a) At what distance would a star of absolute magnitude $+15$ appear as a fifth-magnitude star? (b) At what distances would a star of absolute magnitude -10 and one of absolute magnitude $+15$ appear brighter than the fifth apparent magnitude?

2. If the brightest-appearing star, Sirius, were three times its present distance, would it still make the list of "Twenty Brightest Stars" (Appendix 14)? What about the second brightest star, Canopus?

3. Would any of the stars within five parsecs (Appendix 13) be naked-eye stars at a distance of 100 parsecs? If so, which ones?

4. Describe an everyday situation that is analogous to the fact that most naked-eye stars are of far more than average stellar luminosity.

5. If stars of all kinds were uniformly distributed through space, what would the approximate luminosity function have to be in order for intrinsically faint stars to be the most common among naked-eye stars?

6. Given the luminosity function, how would stars have to be distributed in space in order for the intrinsically faint ones to be most common among naked-eye stars?

★7. Verify that the mean density of stellar matter in the solar neighborhood is about 3×10^{-24} g/cm³.

8. Suppose that within 10 pc there were 11 stars of absolute magnitude $+5$, and that within 30 pc there were 11 stars of absolute magnitude 0. Estimate the true ratio of the number of stars of absolute magnitude $+5$ to the number of absolute magnitude 0 in a given volume of space.

9. From the data in Appendix 13 (the nearest stars), plot the luminosity function for stars nearer than 5 pc. Compare your plot with Figure 27.2.

10. Why are most faint-appearing stars blue?

11. Plot a Hertzsprung-Russell diagram for the stars within 5 pc of the sun. Use the data of Appendix 13. How does this H–R diagram differ from the one in Figure 27.3? Explain the reasons for these differences.

12. Find the distances to the following stars (see Figure 27.4):
    (a) $m = +10$, spectral designation A0 Ib;
    (b) $m = +5$, spectral designation K5 III;
    (c) $m = 0$, spectral designation G2 V.

13. Consider the following data on five stars:

STAR	m	SPECTRUM
1	15	G2 V
2	20	M3 Ia
3	10	M3 V
4	15	B9 V
5	15	M5 V

(a) Which is hottest? (b) coolest? (c) most luminous? (d) least luminous? (e) nearest? (f) most distant? In each case, give your reasoning.

14. Suppose you had data on the apparent magnitudes and colors of several hundred stars in a cluster. Explain how you could use these data to determine the distance to the cluster.

15. For normal stars what is the approximate range (in order of magnitude) of (a) effective temperature; (b) mass; (c) radius; (d) luminosity?

16. Suppose you weigh 70 kg on the earth. How much would you weigh on the surface of a white-dwarf star, the same size as the earth, but having a mass of 300,000 times the earth's (nearly the mass of the sun)?

17. Why do you suppose that most visual binaries are stars of low luminosity?

18. Sometimes our Galaxy is called, simply, the "Milky Way." Why is this poor terminology? Where, exactly, is the Milky Way? Does the question make sense? Why?

19. Suppose the Milky Way were a band of light extending only halfway around the sky (that is, in a semicircle). What then would you conclude about the sun's location in the Galaxy?

Bengt Georg Daniel Strömgren (1908–   ), Danish astronomer, spent much of his productive career in the United States, especially as Professor and Director of the Yerkes Observatory (University of Chicago). He has received many honors for his fundamental work in the study of the structure and evolution of stars, and especially for his pioneering investigation of the physics of the interstellar medium. (*John B. Irwin*)

28

BETWEEN THE STARS: GAS AND DUST IN SPACE

By earthly standards the space between the stars is empty, for in no laboratory on earth can so complete a vacuum be produced. Yet throughout large regions of space this "emptiness" consists of vast clouds of gas and tiny solid particles. Most of this interstellar material is found between the stars in the spiral arms of our own and other galaxies. The density of the interstellar matter in the arms of our galaxy, for example, in the neighborhood of the sun, is estimated to be from three to 20 times that of the interarm regions. The gas and dust is not distributed uniformly, however, but has a patchy, irregular distribution, being denser in some areas than in others, hence forming "clouds."

Sometimes, these tenuous clouds are visible, or partially so, in the form of *nebulae* (Latin for "clouds"). More often, they are invisible, and their presence must be deduced. In the spiral arms, on the average, there is about one atom of gas per cubic centimeter in interstellar space, and from a few hundred to a few thousand tiny particles or "grains," each less than a thousandth of a millimeter in diameter, per cubic kilometer. In some of the denser clouds, the densities of gas and dust may exceed the average by as much as a thousand times, but even this is more nearly a vacuum than any at-

tainable on earth. In air, for contrast, the number of molecules per cubic centimeter at sea level is of the order 10^{19}.

It is interesting to inventory the energy content of space in the solar neighborhood. The most important, by far, is the kinetic energy associated with the revolution of stars, gas, and dust about the center of the Galaxy. Otherwise the six major sources of energy, all of comparable orders of magnitude, are the heat contained in the interstellar medium, cosmic rays, starlight, interstellar magnetic fields, the random motions of gas clouds, and the cosmic background radiation left over, we think, from the big bang (Chapter 39).

28.1 COSMIC "DUST"

The tiny solid grains in interstellar space, commonly called interstellar dust, are manifested in the following ways: (1) dark nebulae, (2) general obscuration, (3) reddening of starlight, (4) reflection of starlight, (5) polarization of starlight, (6) infrared radiation around certain stars, and (7) interstellar molecules, which we theorize must have formed on dust grains.

(a) Dark Nebulae

Relatively dense clouds of the solid grains produce the *dark nebulae*, the opaque-appearing clouds that are conspicuous on any photograph of the Milky Way. Even in the densest clouds, the particles are very sparse, but the clouds extend over such vast regions (measured in parsecs) that they absorb or scatter a considerable portion of the starlight passing through them. Such concentrations of dust often have the appearance of dark curtains, greatly dimming or completely obscuring the light of stars behind them.

The "dark rift," running lengthwise down a long part of the Milky Way and appearing to split it in two, is an excellent example of a collection of such obscuring clouds. The obstruction of light from the stars located behind it is so great that a century ago astronomers thought that it was a sort of "tunnel" through which they could see beyond the Milky Way, into extragalactic space. Today, we know that the dark rift is not such a tunnel; the Galaxy extends far beyond such observable dark nebulae (Chapter 29).

The conspicuous obscuring dust clouds are relatively close to us, within 1000 pc. The more distant opaque clouds are difficult to discover because the large number of stars lying between them and us reduces the contrast produced by their own obscuration of starlight. The contrast produced by remote absorbing clouds is further reduced by the general obscuration caused by the foreground dust, which pervades the entire spiral-arm region of the Galaxy.

In addition to the large dark clouds, many very small dark patches can be seen on Milky Way photographs, silhouetted against bright backgrounds of star fields or glowing gas clouds. Many of these patches, called *globules*, are round or oval and have angular diameters of down to a few seconds of arc. The sharp contrast of their dark boundaries shows that they cannot be more distant than about 1000 pc, which allows us to calculate upper limits for their linear diameters. They probably range from a few thousand to a hundred thousand astronomical units across. The high opacity of the globules (they dim background objects by five magnitudes or more) implies that they must be very dense compared with the usual interstellar material. It has been suggested that the globules may be condensations of matter that may ultimately form into stars.

Figure 28.1 A portion of the Milky Way in Cygnus, showing the "dark rift." *(Caltech/Palomar Observatories)*

(b) The General Obscuration

Although the distribution of the interstellar dust is spotty, and dense clouds produce conspicuous dark nebulae, some of the dust is thinly scattered more or less evenly throughout the spiral arms of the Galaxy. As a result, some absorption of starlight occurs even in regions where dark clouds are not apparent. Unfortunately, the presence of such sparse absorbing matter is not obvious, and it has been the cause of considerable difficulty in the determination of stellar distances.

It has been shown (Section 24.2b) that the distance to a star can be calculated from comparison of its apparent and absolute magnitudes. If light from a star has had to pass through interstellar dust to reach us, however, it is dimmed, much as a traffic light is dimmed by fog. We therefore underestimate the true apparent brightness of the star—that is, we assign to it too large (that is, faint) an apparent magnitude, and the distance we calculate for the star, corresponding to its known (or assumed) absolute magnitude, is too large. Analogously, a motorist may overestimate the distance to a stoplight that he views through fog.

Astronomers once tried to calculate the extent of the Galaxy by counting the numbers of visible stars in various directions and calculating the distances required to account for their observed apparent magnitudes. Because of the general obscuration, however, they overestimated the distances of the stars, and the error increased with increasing actual distances (and hence increasing obscuration). In other words, interstellar absorption of light produces an *apparent* thinning out of stars with distance. The early investigators arrived at the erroneous conclusion that the Galaxy was centered on the sun, and thinned out to its "edge" at a distance of only about 10,000 LY. In actual fact, we do not even see (in visible light) as far as the Galaxy's brilliant central nucleus. Were it not for the obscuring dust in space, we would be able to read at night by the light of the Milky Way.

The presence of the general interstellar obscuration can be demonstrated in several simple ways. We now have independent knowledge that the Galaxy does not thin out a few thousand parsecs from the sun (Chapter 29). The numbers of stars counted in any given region of the sky would be expected to increase, therefore, as the limiting brightness to which stars are counted is decreased (see Exercise 2). As we count to fainter and fainter magnitudes in any direction along the Milky Way, we count more and more stars, but their numbers do not increase as rapidly as we would expect for a more or less uniform stellar distribution because of the dimming of the stars by the dust. It is possible, in fact, to estimate the average absorbing power of the interstellar material by making such star counts in directions at different angles to the plane of the Galaxy.

Interstellar obscuration also shows up in the apparent distribution of external galaxies. Because of the disklike shape of our Galaxy, we would expect to encounter more absorption in the direction of the Milky Way (in the plane of the system) than at right angles to it (out either face of the disk). Throughout a region near the Milky Way, the absorption is so heavy that practically no external galaxies show through. Hubble called this region the "zone of avoidance." However, more and more galaxies can be observed as one turns away from the Milky Way, the greatest number at roughly 90° from the plane of the Galaxy (see Chapter 38). Light from this direction is probably dimmed by less than 30 percent. However, the varying numbers of galaxies in various directions also indicate that there is a spotty, irregular distribution of the absorbing material.

(c) Interstellar Reddening

It is a fortunate circumstance for observational astronomy that the interstellar obscuration is *selective;* that is, light of short wavelengths is obscured more readily than that of long wavelengths. Seventy years ago, astronomers were puzzled by the existence of stars whose spectra indicate that they are intrinsically hot and blue, of spectral-type B, although they actually appear as red as cool stars of spectral-type G. We know today that the light from these stars has been reddened by the interstellar absorbing material; most of their violet, blue, and green light has been obscured, leaving a greater percentage of their orange and red light, of longer wavelengths, which penetrates through the obscuring dust. This *reddening* of starlight by interstellar dust not only shows that the stars are dimmed, but also provides a means of estimating the amount of obscuration they have suffered.

The manner in which the absorption depends on wavelength can be evaluated by comparing, at various wavelengths, the relative brightnesses of two appropriate stars. Stars are chosen whose spectra show them to be approximately identical. One, however, is dimmed and reddened by interstellar dust, while the other is not, being in a direction in the sky that is relatively free of interstellar obscuration. As an illustration, suppose the nearer of the two stars, in the absence of obscuration, would be brighter than the other by half a magnitude at all wavelengths. If this nearer star were dimmed by dust, then, at a wavelength of 10,000 Å it might be, say, only 0.2 magnitude brighter than the unobscured star; at 5000 Å it would appear about half a magnitude fainter, and at 3300 Å it would be about one magnitude fainter than the more distant, unobscured star. From a study of several pairs of such obscured and unobscured stars, the absorbing power of the interstellar material at various wavelengths has been determined. Over the visible spectral region, it turns out that the absorption, expressed in magnitudes, is roughly inversely proportional to wavelength. The absorbing properties of interstellar dust are more complicated, however, outside the visible spectrum. Infrared observations show absorption features (bands) characteristic of silicates at 9.7 μ (1 μ = 1 micron = 10^4 Å); a few clouds show bands at 3.1 μ due to water ice. Ultraviolet observations from satellites show an absorption feature at 2200 Å characteristic of small particles of graphite. We shall return shortly to the significance of these observations.

We can estimate the total amount by which a star is dimmed from the amount that it is reddened. The reddening of the light from a star increases its apparent color index (the redder the star, the greater the color index—see Section 24.3b). The difference between the *observed* color index and the color index that the star *would have* in the absence of obscuration and reddening is called the *color excess*. The $B - V$ color excess, for example, is the amount by which the difference between the blue and visual magnitudes of a star is increased by reddening. In most directions in the Galaxy, the total absorption, in visual magnitudes, is found empirically to be about three times the $B - V$ color excess.

The calculation of the distance to a star that is dimmed by obscuration may be illustrated with a numerical example.★ Suppose a spectral-type G2 star is observed to have an apparent blue magnitude of 14.4 and an apparent visual magnitude of 12.8. Its *observed* color index, therefore, is $+1.6$. Now it is known that this type of star has an absolute visual magnitude of $+4.8$, an absolute blue magnitude of $+5.4$, and therefore, an intrinsic color index of $+0.6$. Its color excess is $1.6 - 0.6 = 1.0$, and its obscuration, in visual magnitudes, is about three times its color excess, or about 3.0 magnitudes. In the absence of obscuration, therefore, we estimate that its apparent visual magnitude would be $12.8 - 3.0 = 9.8$, just five magnitudes fainter than its absolute magnitude (its distance modulus is $+5$—see Section 24.2b). In the absence of interstellar dimming, in other words, the star would appear five magnitudes, or 100 times, fainter than it would appear at a distance of 10 pc; we estimate the true distance of the star, therefore, at 100 pc.

(d) Reflection Nebulae

Until now, the term "absorption" has been used loosely. The tiny interstellar grains actually absorb

---

★The apparent distance modulus of a star dimmed by interstellar obscuration is given by

$$V - M_v = 5 \log \frac{r}{10} + 3CE,$$

where V and M_v are its apparent and absolute visual magnitudes, r its distance in parsecs, and CE its blue minus visual color excess.

Figure 28.2 Reflection nebula about the star Merope in the Pleiades. (*Caltech/Palomar Observatories*)

some of the starlight they intercept, but at least half of it they merely scatter—that is, they redirect it helterskelter in all directions. Since the starlight that is scattered, as well as that which is truly absorbed, does not come directly to us, the effect is the same as if the loss were all due to actual absorption. The whole process is more correctly termed *interstellar extinction*. The scattered or reflected light comes to us from the direction of the scattering dust, thus betraying its presence. Consequently, even the darkest dark nebulae are not completely dark but are illuminated by a faint glow of scattered starlight that can actually be measured. It is estimated that about one third of the light of the Milky Way is diffused starlight, scattered by interstellar dust.

The light scattered by a particularly dense cloud of dust around a luminous star may be bright enough to be seen or photographed telescopically.

Such a cloud of dust, illuminated by starlight, is called a *reflection nebula*. One of the best known examples is the nebulosity around each of the brightest stars in the Pleiades cluster.

Blue light is scattered more than red by the dust. A reflection nebula, therefore, usually appears bluer than its illuminating star. A reflection nebula could be red only if the star which is the source of its light were very red. However, it takes a bright star to illuminate a dust cloud sufficiently for it to be visible to us, and in the regions of the Galaxy where dust clouds are found, the brightest stars are usually blue main-sequence stars. Most of the reflection nebulae that are conspicuous, therefore, are very blue since they are illuminated by blue stars. Sometimes, of course, an intrinsically blue star illuminating a reflection nebula may appear much redder than it actually is because of interstellar absorption and reddening.

From the measured brightness and size of a reflection nebula we can calculate how much light it actually reflects. We find that this is usually a large fraction of the starlight intercepted by the dust. Thus, much of the obscured starlight is scattered (or reflected) by the dust and is not truly absorbed. The interstellar dust, in other words, has a rather high reflecting power or *albedo*. The fraction of the light that is truly absorbed by the dust eventually is reradiated in longer wavelengths.

(e) Interstellar Polarization

Molecules of gas scatter light and polarize it at the same time. The light of the blue daylight sky, for example, is highly polarized. The dust particles of interstellar space also polarize light, but not so much. The light from reflection nebulae is observed to be polarized typically by 10 to 20 percent, and occasionally as much as 50 percent, which means that the brightness of a reflection nebula as observed through a polaroid filter varies by that percentage as the filter is rotated through various angles. The light of stars dimmed by interstellar dust is also slightly polarized. Evidently, the dust grains preferentially scatter light in a particular plane of vibration.

The interstellar polarization can be understood if the particles are elongated in shape and at least partially aligned with each other. The mechanism by which the elongated particles become aligned, while not understood in detail, is undoubtedly associated with the presence of interstellar magnetic fields (see also Chapter 29).

(f) Circumstellar Dust

Infrared surveys of the sky reveal a large number of stars that appear very much brighter at infrared wavelengths than in visible light. It is now thought that at least many of these objects are red giant stars surrounded by circumstellar dust clouds. Light from such a star is greatly dimmed by the dust, but the absorbed radiation heats up the dust itself, which then reradiates the energy at infrared wavelengths. This hypothesis is strengthened by the fact that infrared spectra of some of these stars show emission of the 9.7 μ band due to silicates. To emit as strongly as it does, the dust must absorb considerable energy from the star, which means it must be quite near it.

(g) Nature of the Interstellar Grains

The preceding paragraphs have described various phenomena associated with the interstellar dust. These phenomena reveal something of the nature of the solid grains themselves.

First, it is found that the absorption of light is accomplished by *solid particles*, and not by interstellar gas. Atomic or molecular gas is almost transparent. Consider the earth's atmosphere; despite its incredibly high density compared with interstellar gas, it is so transparent as to be practically invisible. The absorbing power of the interstellar medium exceeds that of an equal mass of gas by more than 100,000 times. The quantity of gas that would be required to produce the observed absorption in space would have to be many thousands of times the amount that can possibly exist. The gravitational attraction of so great a mass of gas would produce effects upon the motions of stars that would be easily detected; such effects, however, are not observed.

Moreover, molecules or atoms (of a gas) scatter light quite differently from the interstellar material. Scattering by gas molecules, as happens in the earth's atmosphere, is called *Rayleigh scattering*. Rayleigh scattering discriminates very strongly among colors, blue light being scattered very efficiently, as we see from the brilliant blue of the daytime sky; the scattering efficiency is inversely proportional to the fourth power of the wavelength.

Interstellar particles, too, scatter selectively, but not so strongly as the earth's atmosphere. Interstellar extinction, as we have seen, is approximately inversely proportional to the first power of the wavelength.

Whereas gas can contribute only negligibly to absorption of light, we know from our everyday experience that tiny particles can be very efficient absorbers. Water vapor in the gaseous state in the air is quite invisible. When some of that vapor condenses into tiny water droplets, however, the resulting cloud is opaque. Dust storms, smoke, and smog furnish other familiar examples of the opacity of solid particles.

Calculations confirm that widely scattered solid particles in interstellar space must be responsible for the observed dimming of starlight. We have already seen that there is direct spectral evidence that they contain silicates and other substances. The complicated physical theory that deals with the scattering of light by solid particles indicates that a variety of sizes and shapes of particles probably exists. At least some must be elongated to produce the observed polarization. Those most responsible for the extinction of light must have diameters of the order of 10^3 Å.

28.2 INTERSTELLAR GAS

Although interstellar gas is estimated to be a hundred times as abundant by mass, on the average, as the dust, because of its high transparency it is not visible by reflected starlight, nor does it contribute to the general interstellar absorption. It does manifest itself, however, in several other ways. Because of the process of *fluorescence*, clouds of gas near hot stars often shine brightly. The gas also produces narrow absorption lines superposed upon the spectra of stars whose light passes through the gas. Moreover, the gas is responsible for the emission of radio waves over a broad range of wavelengths and for many emission and absorption lines at radio wavelengths, including the important line at 21 cm, and those of dozens of molecules in space. Finally, a small fraction of the gas is probably very hot (10^6 K) and emits low-energy X rays.

(a) General Gas in Space

Interstellar gas is distributed generally throughout the regions of the spiral arms of the Galaxy, and to

Figure 28.3 Orion nebula. *(Lick Observatory)*

a lesser extent in other regions. The very hot gas (that emits X rays) is found throughout the disk of the Galaxy and even on either side of the disk, sometimes as far as 10^3 pc from the galactic plane. Hydrogen makes up about three quarters of the gas, and hydrogen and helium together compose from 96 to 99 percent of it by mass. Most of the gas is cold and nonluminous. Near very hot stars, however, it is ionized by the ultraviolet radiation from those stars. Since hydrogen is the main constituent of the gas, we often characterize a region of interstellar space according to whether its hydrogen is neutral—an "H I region"—or ionized—an "H II region."

The gas in the H II regions glows by the process of fluorescence. The light emitted from these regions of ionized gas consists largely of emission lines, so they are also called *emission nebulae*. Those emission nebulae in which the gas happens to be much denser than average (it occasionally reaches densities of 10^3 or 10^4 atoms per cubic centimeter—still an extremely high vacuum on earth) are especially conspicuous. The best-known example is the Orion nebula (Figure 28.3), which is barely visible to the unaided eye, but easily seen with binoculars,

Figure 28.4 The Lagoon nebula in Sagittarius, photographed in red light with the 5-m telescope. *(Caltech/Palomar Observatories)*

in the middle of the sword of the hunter. Other famous emission nebulae are the North America nebula in Cygnus and the Lagoon nebula in Sagittarius (Figure 28.4).

(b) Fluorescence in H II Regions

All ultraviolet radiation of wavelength 912 Å or less can be absorbed by neutral hydrogen, and in the process the hydrogen is ionized (Section 10.6a). An appreciable fraction of the energy emitted by the hottest stars lies at wavelengths shorter than 912 Å. If such a star is embedded in a cloud of interstellar gas, the ultraviolet radiation from that star ionizes the hydrogen in the gas, converting it into positive hydrogen ions (protons) and free electrons. These detached protons and electrons are then a part of the gas, each of them acting like an individual molecule, that is, a free particle. Protons in the gas are continually colliding with electrons and capturing them, becoming neutral hydrogen again. As the electrons cascade down through the various energy levels of the hydrogen atoms on their way to the ground states, they emit light in the form of emission lines. Lines belonging to all the series of hydrogen (Section 10.6a) are emitted—the Lyman series, Balmer series, Paschen series, and so on—but the lines of the Balmer series are most easily observed from the surface of the earth because of the opacity of our atmosphere to most wavelengths.

Part of the invisible ultraviolet light from the star is thus transformed into visible light in the Balmer emission lines of hydrogen. After an atom has captured an electron and emitted light, it loses that electron again almost immediately by the subsequent absorption of another ultraviolet photon from the star. Thus, although neutral hydrogen absorbs and emits light in H II regions, almost all the hydrogen, at any given time, is in the ionized state.

The interstellar gas, of course, contains other elements besides hydrogen. Many of them are also ionized in the vicinity of hot stars and are capturing electrons and emitting light, just as the hydrogen does. Of these, only helium is abundant enough to contribute an appreciable amount of light to an emission nebula by the process of electron capture that we have described.

On the other hand, certain blue and violet emission lines of doubly ionized oxygen and nitrogen are present in strengths not expected from the abundances of these elements. The explanation (supplied by the American physicist I.S. Bowen) is that the oxygen ions are raised to an excited level by photons in one of the ultraviolet helium lines that, by coincidence, happen to have just the right energy. By further coincidence, one of the resulting ultraviolet emission lines of oxygen is just right in energy to excite the nitrogen ions. There are also additional emission lines of the *forbidden radiation*, described in Section 28.2d.

(c) Intermixture of Interstellar Gas and Dust

Gas and dust are generally intermixed in space, although the proportions are not everywhere exactly the same. The presence of dust is apparent on many photographs of emission nebulae. Clouds of dark material can be seen silhouetted on the Orion nebula, actually hiding a large part of the H II region from our view. Foreground dust clouds produce the lagoon in the Lagoon nebula, and the Atlantic Ocean and Atlantic coastline in the North America nebula. Although the dust is most conspicuous when it is in front of an emission nebula and is silhouetted against it, the dust is also intermixed with the gas. Spectra of H II regions often reveal the faint continuous spectrum (with absorption lines) of the central star, whose light is reflected to us by the dust associated with the gas. In other words, emission nebulae are generally superimposed upon *reflection nebulae*.

Both the emission component (due to the gas) and the reflection component (due to the dust) are brighter the farther the central star is up the main sequence. The brightness of an emission nebula, however, is far more sensitive to the kind of central, exciting star than is that of a reflection nebula. Stars cooler than about 25,000 K have so little ultraviolet radiation of wavelengths shorter than 912 Å (that is, which can ionize hydrogen) that the reflection nebulae around such stars outshine the emission nebulae. The dust around a star with a surface temperature of only 10,000 K scatters more then 5000 times as much visible light as is emitted by the gas around the same star, although even the reflection component would probably not be conspicuous unless the star were a supergiant. Stars hotter than 25,000 K emit enough ultraviolet energy so that the emission nebulae produced around them generally outshine the reflection nebulae. The dust around a star of 50,000 K reflects less than one tenth the amount of light that is emitted by the gas that is expected to be present.

(d) FORBIDDEN RADIATION

It was explained in Section 10.6c that an atom or ion can be excited in either of two ways, by absorbing radiation or by collision with another particle. Atoms of singly ionized nitrogen and singly and doubly ionized oxygen all contain energy levels that correspond to low energies above their ground states. The ions are easily excited to these "low-energy" levels by collisions with free electrons in the H II regions (most of these electrons have been freed from hydrogen atoms by ionization). Ordinarily, observed emission lines originate from atoms or ions that have remained excited for only a very brief period—on the order of a hundred millionth to a ten millionth of a second—before becoming de-excited by the emission of radiation. The levels to which oxygen and nitrogen ions are excited by collision, however, are said to be *metastable* levels, because the ions will normally remain in them for periods of hours before radiating energy and dropping to their ground states. Consequently, emission lines corresponding to such atomic transitions are exceedingly weak compared with other lines and are not usually observed in the laboratory; they are known as *forbidden lines*.

In the interstellar gas, however, the cards are stacked in favor of forbidden radiation. At the temperatures of H II regions many of the free electrons have just the right kinetic energy to excite oxygen and nitrogen ions to their metastable levels. Although transitions from these levels are slow to occur, so many ions are excited to them at any time that many such "forbidden" transitions occur in an H II region. Now the gas is transparent to visible light, so the photons emitted through the entire depth of an H II region contribute to visible emission lines; indeed, the forbidden radiation often comprises half or more of the observable light from H II regions. The most important forbidden lines in the spectra of emission nebula are two green lines (5007 and 4959 Å) due to doubly ionized oxygen. Other important forbidden lines are two ultraviolet lines (near 3727 Å) due to singly ionized oxygen, two red lines (6584 and 6548 Å) due to singly ionized nitrogen and two ultraviolet lines (3867 and 3968 Å) due to doubly ionized neon.

When the green forbidden oxygen lines were first observed in the spectra of emission nebulae, their origin was a mystery. For a time, they were ascribed to an unknown element, *nebulium*, named for its apparent prevalence in gaseous nebulae. The correct explanation of the "nebulium lines" was provided by I. S. Bowen in 1927.

(e) SIZE AND BRIGHTNESS OF AN H II REGION

If a cloud of gas surrounding a hot star is not very extensive, some of the ultraviolet radiation emitted by the star that is capable of ionizing hydrogen may leak out through the gas, and the apparent boundary of the emission nebula will be the actual edge of the gas cloud. Such a nebula is said to be *optically thin*, and the H II region is said to be *gas bounded*. Usually, however, an H II region is *optically thick*, which means that all of the star's ultraviolet radiation (of wavelengths less than 912 Å) is absorbed within the gas, and the H II region is said to be *radiation bounded*. In this case, the boundary of the H II region is merely the limiting distance through the gas to which the star's ultraviolet radiation penetrates.

If the interstellar gas were distributed with absolute uniformity, and if it and the stars were all at rest, every emission nebula would be a spherical H II region exactly centered on a hot star. Because the distribution of the gas is patchy and irregular, and because the ionizing stars are often moving through it, actual H II regions are only approximately spherical, with irregular boundaries corresponding to the irregularities in the gas density; where the gas is denser, the ionizing radiation is consumed, and the H II region ends closer to the star. Further irregularities may result when two or more stars are responsible for the radiation, and their H II regions overlap. The theory of H I and H II regions was worked out in detail by the astronomer B. Strömgren. The more or less spherical emission regions are therefore sometimes called *Strömgren spheres*.

The linear size of an H II region depends on two things: (1) the "ultraviolet" luminosity of the central star, that is, how much energy it emits per second in wavelengths less than 912 Å; and (2) the density of the gas. The higher the density, the more hydrogen per unit volume there is to ionize, and the shorter is the distance through the gas that the ultraviolet energy can penetrate before it is completely absorbed. If the gas density is very low, and if the star is very hot and luminous, the H II region can be very large; if the density is one atom per cubic centimeter, a main-sequence spectral-type O6 star can ionize a region more than 100 pc in diameter. Main-sequence stars of types B0 and A0 would produce in the same gas H II regions having diameters of 40 and 1 pc, respectively.

The amount of light emitted by a unit volume of gas in an H II region depends on the rate at which the ionized atoms capture electrons, which depends in turn on the gas density; the higher the density, the greater the rate of recombinations of ions and electrons, and the more light that is emitted. Since visible light travels freely through interstellar gas, the observed surface brightness of a glowing H II region is proportional both to the amount of light emitted per unit volume within the region, and to the length of the column along the observer's line of sight intercepted by the region. The surface brightness of an H II region, therefore, depends on both the density of the gas within the region and its size.

(f) TEMPERATURES OF H I AND H II REGIONS

When we speak of the temperature of a gas, we usually mean its *kinetic temperature*, which is a measure of the energy with which typical particles in the gas are moving. To understand the temperature of the interstellar medium, we must first consider the processes that heat the gas and then those that cool it. Finally, we calculate the equilibrium temperature that exists when these two processes—the heating and the cooling of the gas—exactly balance each other.

In H II regions the heating is principally by ionization of hydrogen. The neutral hydrogen atom, in becoming ionized, can absorb *any* photon of wavelength shorter than 912 Å. The energy absorbed by the atom, in excess of that required for its ionization, is converted to kinetic energy of the freed electron. By collisions, the electrons gradually share their energy with the other particles of an H II region. Excess energy absorbed in the process of ionization of hydrogen, therefore, is slowly converted into heat in the gas.

Electrons can lose energy by collisions with ions of oxygen and nitrogen, because these ions can be excited by such collisions and can eventually radiate the energy away in the form of "forbidden" emission lines. This mechanism, in other words, takes energy *from* the gas, and therefore *cools* it. Calculations show that the heating (by ionization) and cooling (by the emission of forbidden lines) should balance each other at an equilibrium temperature in the range of 7000 to 20,000 K. If the temperature should drop much below 10,000 K, the rate of collisional excitations of oxygen and nitrogen would decrease (because the electrons would be moving more slowly) and the gas would heat up. If the temperature rose much above 20,000 K, the collisional excitations would become so numerous that the gas would cool rapidly. The heating and cooling mechanisms, therefore, act like a thermostat keeping the gas in the H II region at a relatively even temperature.

We may measure the temperature of H II regions by comparing the intensities of two or more lines originating from different collisionally excited levels of the same kind of atom or ion. Measures of intensities of the forbidden lines of doubly ionized oxygen (the "nebulium" lines) are especially useful for this purpose. Temperatures of H II regions can also be estimated from the intensities of their radio emission at different wavelengths. Most such measurements suggest that H II regions have temperatures between 8000 and 10,000 K.

The temperature of H I regions is also dictated by a balance of heating and cooling mechanisms, but the processes are more complicated, and they vary in effectiveness from place to place, depending, among other things, on the shielding of the matter from starlight by dust obscuration. Heating can occur by ionization of atoms of such heavier elements as carbon and silicon and by cosmic rays. Cooling mechanisms include collisional excitation of carbon and other atoms, and collisions between atoms and solid grains. Measures mostly derived from radio observations indicate that some cold clouds have temperatures below 20 K, while other regions are at temperatures of more than 125 K.

In any case we can conclude that H II regions are hot, with temperatures on the order of 10,000 K, and that H I regions are cold, with temperatures on the order of 100 K, or less.

The very hot gas is believed to be heated by shock waves produced by various mechanisms—for example, by the expanding shells of supernovae pushing through it (Chapter 34), by stellar winds (Chapter 31), by collisions between gas clouds, perhaps by density waves (Chapter 29), and by the expansion of H II regions, described in the following paragraphs. The violent events that heat that gas may also be responsible for shoving some of it far from the main plane of the Galaxy.

(g) EXPANSION OF H II REGIONS

It will be seen in Chapter 33 that the very hot, luminous stars—the type that are generally responsible for producing H II regions—are relatively short-lived. H II regions, therefore, are temporary phenomena; most existing emission nebulae have been formed within the past few million years. Consider, now, what happens when a very hot, luminous star is first formed, and rather suddenly ionizes the gas surrounding it in space, thereby producing an emission nebula. The ionized gas, although initially of the same density as the unionized gas in the adjacent H I region, is about 100 times hotter than the cool gas. Since the pressure in a gas is proportional to the product of its density and temperature (see Chapter 32), the H II region has a pressure that is also about 100 times greater than that of the H I region. Consequently, the H II region expands, decreasing its own density and pushing outward, away from the central star, against the cold gas in the H I region. At the same time, the radiation pressure exerted by the light from the star upon the interstellar dust particles pushes them outward, away from the star, much as sunlight pushes out the dust tail of a comet.

The expanding front of an emission nebula—the interface between the H I and H II regions—may move at first very rapidly. As a result, there is often a buildup in the density of the gas and dust at the boundary of the nebulae. The bright edges visible in many emission nebulae constitute observational evidence for the higher densities of the gas at the expanding front.

Eventually, the increasing density of the material in the H I region surrounding the hot, expanding emission nebula offers sufficient resistance to slow down the expansion. Sometimes points of higher than average density in the H I region break through the expanding front of hot gas and appear as "intrusions" into the emission nebula. Such intrusions generally look like dark cones with bright edges pointing toward the star

at the center of the nebula; the emission nebula Messier 16 in Serpens is an excellent example. The dark intrusions are sometimes referred to as "comet-tail" or "elephant-trunk" structures. They are not, of course, actually moving inward; the hot gas expands *outward*, around them. Dark globules are often found near, or as dissociated extensions of, such elephant-trunk structures (Figure 28.5).

The fact that stars exist which we have reason to believe are very young (Chapter 33) implies that stars must be continually forming. The only material for them to form from is the interstellar gas and dust. Regions within the interstellar medium, therefore, must be the "birthplaces" of new stars. Places where the densities of the interstellar gas are unusually high may be future sites of star formation.

(h)   Interstellar Absorption Lines

The cold interstellar gas—that in the H I regions—is not visible by reflected or emitted light, nor does it appreciably dim the light of stars shining through it. Yet it often reveals its presence by leaving dark absorption lines superposed upon the spectra of stars that lie beyond it. There are several ways of knowing that the interstellar lines seen in the spectrum of a star do not originate in the star itself. Since the interstellar gas is cold, most of its atoms are neutral and are practically all in the state of lowest energy; the lines they produce, therefore, are generally not the same as the ones that are produced by the atoms of the hot gases in stellar photospheres. Moreover, the lines of the cold interstellar gases are very sharp, while those formed in stellar photospheres show the characteristic broadening associated with the spectra of hot gases at relatively high pressure (Section 25.2c). Finally, an interstellar gas cloud does not, in general, move with the same radial velocity as the star whose spectrum is observed, and the Doppler shifts of the interstellar lines are thus different from those of the stellar lines.

Interstellar lines have been found of most of those elements for which observable lines would be expected. The most conspicuous optical interstellar lines are produced by sodium and calcium. Lines are also observed of some other common elements as well as bands of CN, CH, and CH^+. Satellite ultraviolet observations have detected lines of carbon, hydrogen, oxygen, nitrogen, and other elements; of molecular hydrogen; and of CO (carbon monoxide). The strengths of interstellar lines lead to estimates of the relative abundances of the ele-

Figure 28.5 Messier 16, nebula in Serpens. Photographed in red light with the 5-m telescope. *(Caltech/Palomar Observatories)*

ments that produce them. For some elements such estimates do not differ markedly from their relative abundances in the sun and other stellar photospheres. For others the relative abundance is noticeably lower, especially for elements that most easily condense into solids at relatively high temperatures (notably aluminum, calcium, and titanium).

Sometimes the strength of an interstellar line seen in the spectrum of a star provides an indication of the distance to that star. A very strong interstellar line, for example, indicates that the starlight has traversed a considerable amount of interstellar gas. Because the density of gas in space is very low, the starlight would have to travel a long distance to encounter that much material; that is, the star would have to be very far away. Sometimes the interstellar lines are double or even multiple, which indicates that the starlight has traversed two or more gas clouds, moving with respect to each other, whose different radial velocities produce different Doppler shifts.

(i) Molecular Hydrogen in Space

Until recently, it was thought that in H I regions hydrogen was in the atomic form. In 1970, however, Carruthers observed a band of the hydrogen molecule, H_2 in an ultraviolet stellar spectrum obtained from a rocket. Subsequently, satellite obser-

vations have shown the H_2 bands in the spectra of many stars. It has now been demonstrated that about half of the interstellar hydrogen is in the molecular form.

Ultraviolet radiation from stars easily dissociates the hydrogen molecule into two hydrogen atoms. Thus the H_2 is found mostly in interstellar clouds that are shielded from ultraviolet photons by interstellar dust. We shall see that the dust particles are probably necessary for the formation of molecular hydrogen as well as for its protection.

(j) Radio Emission from Interstellar Gas

We receive from the Milky Way a large amount of radio energy in the frequency range 10 to 300 MHz (or wavelength range 1 to 30 m). This emission from the Milky Way was, in fact, the first radio radiation of astronomical origin to be observed (Section 12.12). Similar radiation at radio frequencies is received from nearby galaxies. The radio energy from the Milky Way does not originate in stars. The sun, to be sure, is an apparently strong source of radio waves, but the sun is very close to us. If its radio emission is typical of that of the other stars, all the stars in the Galaxy would emit less than 10^{-9} of the radio energy actually observed. The radio energy originates in the interstellar gas.

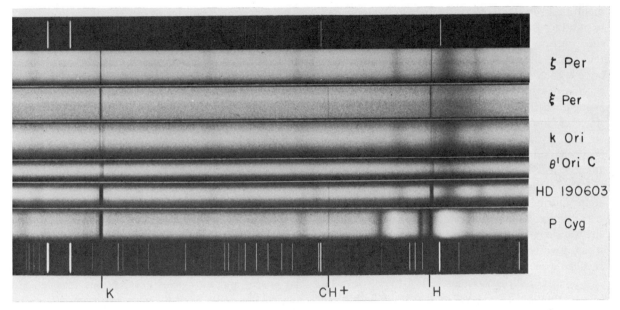

Figure 28.6 Interstellar H and K lines of ionized calcium and a line of the CH-radical showing in the spectra of several stars. *(Lick Observatory)*

Many strong sources of radio energy have been identified with individual gaseous nebulae. One example is the Crab nebula (Chapter 34); another is a group of faint gaseous filaments in Cassiopeia (known as the Cassiopeia A source). The optical spectra of the different filaments in the Cassiopeia A source show Doppler shifts that indicate that some blobs are moving with respect to one another with speeds in excess of 4000 km/s. Both the Crab nebula and Cassiopeia A are the remnants of supernova outbursts (Chapter 34)—gas shells blown out by the explosions of dying stars. There are many other supernova remnants in the Galaxy; some are visible both optically and in radio wavelengths, but the majority are observed only by radio, their optical light being obscured by intervening interstellar dust. Other discrete radio sources are associated with the more conspicuous nebulae (like the Orion nebula). Superimposed on them all is the general background of radio radiation from those regions where interstellar gas prevails.

(Thousands of discrete radio sources have been catalogued that are in directions in the sky away from the Milky Way. Most of these are extragalactic sources, and will be discussed in more detail in Chapters 37 and 38.)

One cause of galactic radio emission is free-free transitions (bremsstrahlung—Section 10.6e) in H II regions. An electron, passing near an ionized hydrogen atom (proton), can change from its original hyperbolic orbit to another hyperbola, emitting or absorbing in the process a photon of a definite wavelength. The many photons of different wavelengths arising from such free-free transitions in the interstellar gas comprise continuous radiation, a good part of which lies in the radio part of the spectrum. This is called *thermal radiation*—the normal emission from hot gas. On the other hand, some galactic radio sources are due to synchrotron radiation (Chapter 29), which results when electrons spiral around magnetic lines of force with speeds near that of light. In particular, the radio sources associated with old supernovae (such as the Crab nebula) are produced by the synchrotron mechanism.

(k) Radio Lines: the 21-cm Line of Hydrogen

A number of emission and absorption lines at radio wavelengths can now be observed. The first radio line to be detected was the 21-cm line of neutral hydrogen.

Some of the most important radio observations carried out in the interstellar medium are of the spectral line of hydrogen at the radio wavelength of 21.11 cm. A hydrogen atom possesses a tiny amount of angular momentum by virtue of the axial spin of its electron and the electron's orbital motion about the nucleus (proton). In addition, the proton has an axial spin of its own, and the angular momentum associated with this spin may either add to or subtract from that of the electron, depending on the direction of spin of the nucleus, with respect to that of the electron. If the spins of the two particles oppose each other, the atom as a whole has a very slightly lower energy than if the two spins are aligned. Ordinarily, an atom of hydrogen is in the state of lower energy. If the requisite minute amount of energy is imparted to the atom, however, the spins of the proton and electron can be aligned, leaving the atom in a slightly *excited state*. If it loses that same amount of energy again, the atom returns to its ground state. The amount of energy involved is that associated with a photon of 21-cm wavelength. An atomic transition of this type, which involves the spin of the nucleus of an atom, is called a *hyperfine transition*.

Neutral hydrogen atoms in H I regions can be excited to this 21-cm level by collisions with electrons and other atoms. Such collisions are extremely rare in the sparse gases of interstellar space; an individual atom may wait many years before such an encounter reverses its nuclear spin, that is, aligns it with the electron spin. Nevertheless, over many millions of years a good fraction of the hydrogen atoms are so excited. An excited atom can then lose its excess energy either by a subsequent collision or by radiating a photon of 21-cm wavelength. It happens, however, that hydrogen atoms are extremely loath to do the latter; an excited atom will wait, on the average, about 10 million years before emitting a photon and returning to its state of lowest energy (it is a highly forbidden transition). On the other hand, there is a definite chance that the atom will radiate before a second collision carries away its energy of excitation. In 1944 the Dutch astronomer H. C. van de Hulst predicted that enough atoms of interstellar hydrogen would be radiating photons of 21-cm wavelength to make this radio emission line observable. Equipment sensitive enough to detect the line was not available until 1951. Since that time, "21-cm astronomy" has been a very active field of astronomical research.

Observations at 21 cm show that the neutral hydrogen in the Galaxy is confined to an extremely flat layer, most of it in a sheet less than 100 pc thick, extending throughout the plane of the Milky Way. The strength of the line indicates that neutral interstellar hydrogen makes up from 1 to 2 percent of the mass of the Galaxy. The measurement of the Doppler shifts of the 21-cm radiation from various directions also helps us detect the spiral structure of our galactic system (Chapter 29).

Since the discovery of the 21-cm line, other radio spectral lines have been observed. Among them is a multiplet of four closely spaced lines due to the OH radical—a bond of one oxygen and one hydrogen atom. Also, very high-level hydrogen emission lines are observed. The various series of hydrogen lines are described in Section 10.6a. Transitions from energy level 110 to 109 of the hydrogen atom, for example, give rise to a radio line of 6-cm wavelength; those from level 157 to 156 produce a line at 17 cm.

(l) Molecules in Interstellar Space

Years ago optical observations revealed the molecules CN and CH and the radical CH^+ in interstellar space, and ultraviolet observations revealed H_2 and CO as well. Now we have also seen that radio lines of OH are observed. In 1968, radio lines of interstellar polyatomic molecules were seen for the first time. Since then, observations of many absorption and emission lines at radio wavelengths have led to the exciting new field of interstellar chemistry.

The radio radiation from OH near 18 cm seemed at first to be anomalous, in that it is enormously stronger than would be expected from the possible abundance of the OH radical in space. The explanation is that the intensity of this radiation is greatly amplified by maser action. Large numbers of the molecules are excited by stellar radiation to a metastable energy level from which they undergo spontaneous transitions to their lowest (ground) energy states only very slowly; thus, a large concentration of the molecules in that metastable level is built up. However, radiation from the occasional molecules that do jump to their ground states induces or *stimulates* other excited molecules to make the same transitions by emitting photons of the same wavelength and direction as the ones stimulating them. This process of stimulated emission repeats until the radiation is amplified into a powerful beam. Meanwhile, molecules in the ground state are continually reexcited or "pumped" back to the metastable level by the radiation from surrounding stars. Water molecules produce another such *interstellar maser*. (Masers are described more fully in Chapter 10.)

By 1981 about 50 kinds of molecules and radicals had been identified in space. Common atoms of hydrogen, oxygen, carbon, nitrogen, and sulfur make up molecules of water, carbon monoxide,

Horsehead nebulae in Orion, NGC 2024, photographed with the 48-inch Schmidt telescope. *(California Institute of Technology/Palomar Observatory)*

Orion nebula, NGC 1976, Messier 42. Photographed with the 200-inch telescope. *(California Institute of Technology/Palomar Observatory)*

The Rosette nebula in Monoceros, NGC 2237, photographed with the Palomar Schmidt telescope. *(California Institute of Technology/Palomar Observatory)*

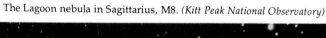

The Lagoon nebula in Sagittarius, M8. *(Kitt Peak National Observatory)*

North America nebula in Cygnus, NGC 7000. *(California Institute of Technology/Palomar Observatory)*.

ammonia, hydrogen sulfide, and such common organic molecules as formaldehyde, hydrogen cyanide, methyl cyanide, and simple alcohols.

Like the hydrogen molecule (H_2), most of these more complex molecules are dissociated by short-wave stellar radiation. Hence they are found in relatively dense, dark interstellar clouds where dust shields the region from ultraviolet starlight. Such clouds are estimated to have a thousand to a million times the mass of the sun and to be several light years across. The molecules in the clouds are excited to various states of vibration and rotation by collisions among themselves; they subsequently transform to lower energy states with the emission of the infrared and radio radiation by which we observe them. In the process, energy is removed from the dark clouds, which causes them to cool, contract, and become denser. Particularly cold clouds, rich in interstellar molecules, are found behind the Orion nebula, near the galactic center, and elsewhere, and are believed to be prime sites of star formation (Chapter 33).

Relatively heavy molecules, such as HC_9N are found in some cold clouds, and ethyl alcohol, C_2H_5OH, is quite plentiful—up to one molecule for every cubic meter in space. The largest of the cold clouds have enough ethyl alcohol to make 10^{28} fifths of 100-proof liquor. Wives or husbands of future interstellar astronauts, however, need not fear that their spouses will become interstellar alcoholics; even if a spaceship were equipped with a giant funnel 1 km across and could scoop through such a cloud at the speed of light, it would take about a thousand years to gather up enough alcohol for one standard martini!

The cold interstellar clouds also contain cyanoacetylene and acetaldehyde, generally regarded as starting points for the formation of amino acids necessary for living organisms. The presence of these organic molecules does not, of course, imply the existence of life in space. On the other hand, as we learn more about the processes by which they are produced, we gain an increased understanding of similar processes which must have preceded the beginnings of life on the primitive earth some thousands of millions of years ago.

28.3 ORIGIN OF THE INTERSTELLAR MATTER

In part, the interstellar material must represent leftover matter that did not form completely into stars when the galaxy condensed from a protogalactic cloud some 10^{10} years ago (perhaps much as the solar system formed from the solar nebula). A large part of the interstellar material, however, has certainly been ejected into space from stars, especially those near their final evolutionary states (Chapter 34). Nuclear transformations within stars, as we shall see, have altered the chemical composition of the interstellar medium—in particular they have enriched its content of the heavier elements. New stars forming from gas and dust clouds of space—as we think the sun and planets formed 5×10^9 years ago—are thus *second generation* stars (or stars of the third or a higher generation). According to this picture the atoms that compose our bodies have been produced by nuclear reactions in stellar interiors!

Interstellar gas densities, however, appear to be too low to permit an appreciable amount of the gas to have condensed into dust particles. Rather, many investigators think that the dust particles condense from matter ejected from the outer layers of extreme red giants. From some of these stars matter is observed to be streaming away into space (Chapter 34). Thus it may be that as this material flows out and cools, those elements that most easily condense, such as aluminum, calcium, and titanium, begin to collect in solid grains. Condensation theory predicts that other substances, including graphite, magnesium, silicates, silicon carbide, and metallic iron also become locked up in the grains.

After the dust grains become part of the interstellar cloud they interact with the gases in space. According to a model by G. Field, atoms such as those of hydrogen, oxygen, carbon, and nitrogen strike the grains. The last three tend to stick and unite with the impinging hydrogen atoms, building up a mantle of ice and other molecules. Some of the hydrogen forms H_2 molecules, which are easily dislodged from the grains and escape into space again as molecular hydrogen. Ultraviolet photons from stars striking the grains break up some of the molecules there. These molecular fragments may unite with other molecules or fragments to form some of the more complex molecules observed in space. Further chemical changes can occur through interactions of cosmic rays with the grains or with the molecules in space.

Despite its low density, the interstellar medium is evolving rapidly on the time scale of the age of the Galaxy, and it is evidently undergoing active and varied chemical reactions.

EXERCISES

1. Identify several dark nebulae on photographs of the Milky Way in this book. Give the figure numbers of the photographs, and specify where the dark nebulae are to be found on them.

2. Suppose all stars had the same absolute magnitude, and that they were distributed uniformly through space. Show that the number of stars from which we receive an amount of light b is inversely proportional to b raised to the $3/2$ power. Now suppose that stars of many absolute magnitudes exist, but that the numbers of stars of various absolute magnitudes exist in the same relative proportions everywhere in space. Is the proportionality between the number of stars appearing brighter than b and $b^{-3/2}$ changed? Explain.

*3. Suppose stars are counted to increasingly fainter limiting brightnesses. Make a sketch showing how the numbers of stars counted will increase as b is decreased, both in the absence and in the presence of general interstellar obscuration.

4. A spectral-type A star normally has a color index of 0.0. The blue and visual magnitudes of an A star are $B = 11.6$ and $V = 10.8$, respectively.
   a) What is the color excess of the star?
   b) What is the total absorption of its light in visual magnitudes?
   c) What is the total absorption of its light in blue magnitudes?

5. The sun is observed from a distant star to have an apparent blue magnitude of 14.4 and an apparent visual magnitude of 12.8. How far away, approximately, is the star? Assume that the sun's color index is $B - V = 0.6$, and that its absolute visual magnitude is $+4.8$.

6. Suppose a bright reflection nebula appears yellow. What kind of star probably is producing it?

7. The red color of the sun, when seen close to the horizon, and the blue color of the daytime sky provide analogies to the reddening of starlight and the blue color of reflection nebulae. Discuss this analogy more fully, and also explain how it breaks down.

*8. The amount of light emitted in the hydrogen emission lines of an emission nebula depends on the amount of ionized hydrogen. This in turn depends on the amount of ultraviolet radiation, capable of ionizing hydrogen, that is emitted by the central star. The star's brightness in the visible part of the spectrum can be observed and compared with the amount of light in the hydrogen lines of the nebula. Does this suggest a way by which the temperature of the star might be estimated? Explain. (See Chapters 10 and 25.)

*9. Explain why "nebulium" lines are not observed in the solar spectrum. Might they be possible in the spectrum of the very rarefied upper atmosphere of the earth?

*10. Describe in detail the appearance of the spectrum of an emission nebula such as the Orion nebula. Would you expect any continuous spectrum to be present? Explain.

*11. Explain in detail why emission nebulae are brighter in regions where the gas is denser, and why their brightness depends primarily upon the gas density.

*12. Suppose you examined the spectrum of some nebulosity surrounding a main-sequence spectral type O star and found that it contained no emission lines, only the continuous spectrum of the star. What conclusions could you draw about the nature of the interstellar material around that star?

*13. If an H II region at a temperature 10,000 K is in pressure equilibrium with an adjacent H I region at a temperature of 100 K, what can you say about the relative densities of the two regions?

29

THE GALAXY

In 1610 Galileo described his telescopic observations of the Milky Way, which showed it to be composed of a multitude of individual stars. In 1750 Thomas Wright published a speculative explanation, which turned out to be substantially correct—that the sun is located within a disk-shaped system of stars, and that the Milky Way is the light from the surrounding stars that lie more or less in the *plane* of the disk. The disk shape of the stellar system to which the sun belongs—the *Galaxy*—was demonstrated quantitatively in 1785 by Herschel's "star gauging." We have already described the results of Herschel's investigations (Section 27.3a).

It was the second decade of the 20th century before astronomers had deduced, approximately, the true size of the Galaxy and the fact that the sun is located far from the center of the disk-shaped system. In 1925 the existence of other, similar, stellar systems was proved; the "pinwheel" shape of many of these other galaxies suggested that our own system might also contain spiral arms winding outward from a massive central nucleus. Only within the last few decades, however, has substantial progress been made in identifying the spiral arms and in mapping them. This chapter deals with 20th century investigations of the structure of our Galaxy.

29.1 SIZE OF THE GALAXY, AND OUR POSITION IN IT

Until early in the 20th century, the Galaxy was generally believed to be centered approximately at the sun and to extend only a few thousand light years from it. The shift from the "heliocentric" to the "galactocentric" view of our system, as well as the first knowledge of its true size, came about largely through the efforts of Harlow Shapley and his investigation of the distribution of the globular clusters. Globular clusters (Section 27.3c) are great symmetrical star clusters, containing tens of thousands to hundreds of thousands of stars each. Because of their brilliance, and the fact that they are not confined to the central plane of the Galaxy where they would otherwise be largely obscured by interstellar dust, they can be observed (with telescopes) to very large distances.

(a) Distribution of the Globular Clusters

Most globular clusters contain at least a few RR Lyrae variable stars (or cluster-type variables—Section 24.3h), whose absolute magnitudes are known

475

to lie between 0 and +1. The distance to an RR Lyrae star in a globular cluster, and hence to the cluster itself, can therefore be calculated from its observed apparent magnitude (Section 24.2b). Shapley was able to determine the distances to the closer globular clusters that contained RR Lyrae stars that could be observed individually. Distance estimates to globular clusters that do not contain RR Lyrae stars, or that are too far away to permit resolution of the variables, were obtained indirectly. Shapley measured angular diameters of globular clusters of known distance, thus obtaining their true diameters. Assuming a statistical average for the true diameter of the clusters, he was then able to obtain distance estimates for the remote ones from their observed angular diameters.

From their directions and derived distances, Shapley, in 1917, mapped out the three-dimensional distribution in space of the 93 globular clusters then known. He found that the clusters formed a roughly spheroidal system with the highest concentration of clusters at the center. That center was not at the sun, however, but at a point in the middle of the Milky Way in the direction of Sagittarius, and at a distance of some 25,000 to 30,000 LY (Figure 29.1). Shapley then made the bold—and correct—assumption that the system of globular clusters represented the "bony frame" of the entire Galaxy; that not only is the distribution of clusters centered upon the center of the Galaxy but, moreover, that the extent of the galactic system is indicated by the cluster distribution. Today the assumption has been verified, not only by the spiral structure deduced from the 21-cm radiation from neutral hydrogen but also by the observed distributions of globular clusters in other spiral galaxies. A schematic of the Galaxy is shown in Figure 29.2. Although the sun lies far from the galactic center, the main disk of the Galaxy probably extends a nearly equal distance beyond the sun and comprises a gigantic system 100,000 LY across. The exact size of the disk is not known, however, nor is the exact distance of the sun from its center. Today, the center of the galactic nucleus is estimated to be from 8000 to 10,000 pc from the sun (or from 25,000 to 30,000 LY).

(b) The Galactic Halo

The main body of the Galaxy is confined to a relatively flat disk, while the globular clusters define a more or less spheroidal system superimposed upon the disk. A sparse "haze" of individual stars—not members of clusters but far outnumbering the cluster stars—also exists in the region outlined by the globular clusters. This haze of stars and clusters forms the galactic halo, a region whose volume exceeds that of the main disk of the Galaxy by many times. The presence of stars in the halo was first suspected when RR Lyrae stars not belonging to clusters were found lying in such directions and at such distances as to place them far from the galactic

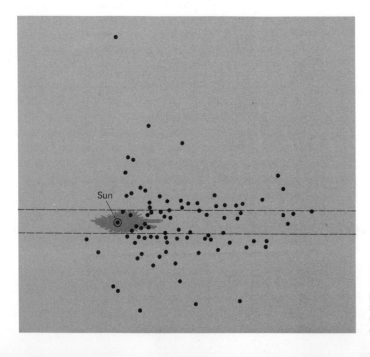

Figure 29.1 Copy of a diagram by Shapley showing the distribution of globular clusters in a plane perpendicular to the Milky Way, and containing the sun and the center of the Galaxy. Herschel's diagram (Figure 27.7) is shown centered on the sun, and approximately to scale.

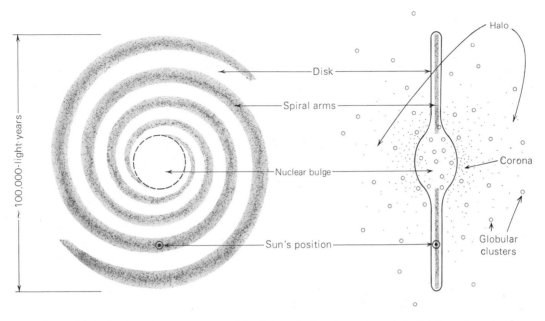

Figure 29.2 Schematic representation of the Galaxy. In the planar view on the left (seen from the south side of the galactic plane), the sun's revolution is counterclockwise.

plane. These variables, of course, represent only an extremely minute fraction of all halo stars, but they serve as tracers, indicating the distribution of stars in the halo, not only because their distances can be determined easily, but because they have fairly high luminosities and can be seen to relatively large distances. Some of the galactic radio emission also originates in the halo, showing that there is gas there as well as stars. X-ray radiation from the gas indicates that it is very hot—of the order of 10^6 K—too hot to produce the emission lines seen in ordinary H II regions. The gas, probably heated by supernova shells and/or stellar winds (Section 28.2f), extends only into the inner part of the halo and defines a region now usually referred to as the *galactic corona*.

The spatial density of stars and clusters in the halo of the Galaxy increases toward the Milky Way plane, particularly toward the galactic nucleus. When we look to either side of the Milky Way (that is, slightly above or below its plane) in the directions of Scorpius, Ophiuchus, and Sagittarius, our line of sight skims near the nuclear bulge in the middle of the disk of the Galaxy. In those directions we find the greatest numbers of globular clusters and stars in the halo. The largest number of RR Lyrae stars seen in these directions have apparent magnitudes near 15, which means that they must be at distances of from 8000 to 10,000 pc.

Individual RR Lyrae stars have been found as far away as 30,000 to 50,000 LY on either side of

the galactic plane, which shows that the halo must have an overall thickness of at least 100,000 LY. A few globular clusters discovered on the Palomar Sky Survey appear to have distances from the sun of more than $\frac{1}{4}$ million light years. We are not sure whether they are true members of the Galaxy or are intergalactic objects. In any event, the possibility remains that the halo extends to very great distances in some directions. The halo is either a spherical or spheroidal system at least 100,000 LY thick, and in the direction of the galactic plane may have a diameter of two or three times this figure, extending far beyond the "rim" of the main disk of the Galaxy. Halos of some other galaxies have been traced to similar distances. We shall see in Section 29.4 that modern data on the rotation of our Galaxy give good reason to think that the halo contains a large fraction of the Galaxy's mass.

29.2 REVOLUTION OF THE SUN IN THE GALAXY

Like a gigantic "solar system," the entire Galaxy is rotating. The sun, partaking of the galactic rotation, moves with a speed of 220 to 250 km/s in a nearly circular path about the nucleus. The method has been described (Chapter 23) by which we determine the motions of stars with respect to the sun (*space motions*), or with respect to the local standard of rest (*peculiar velocities*). Proper motions, however, and thus space motions and peculiar velocities, can be detected only for those stars that are relatively near the sun and that occupy a volume of space which is very small compared with the size of the Galaxy. These are stars moving along with us about the center of the Galaxy; their space motions, most of which are less than a few tens of kilometers per second, result from the slight differences between the inclinations, eccentricities, and sizes of their galactic orbits and the sun's. Only by observing far more distant objects can we determine our own true motion in the Galaxy.

(a) The Sun's Galactic Orbit

The motion of the sun in the Galaxy is deduced from the apparent motions of objects surrounding us that do not share in the general galactic rotation. The globular clusters are the most convenient such objects. These clusters are moving, to be sure, but the fact that they are found in a spheroidal distribution, rather than being confined to the flat plane of the Galaxy, is evidence that the system of globular clusters as a whole is not rotating as rapidly as the disk of the Galaxy. By analyzing the radial velocities of the globular clusters in various directions, we can determine the motion of the sun with respect to them very much in the same way that we determine the solar motion with respect to the local standard of rest (Section 23.2b). In one direction, the globular clusters, on the average, seem to be approaching us, while in the opposite direction they seem to recede from us. This procedure yields only a lower limit for the speed of the sun's revolution about the galactic center if there is any net rotation of the system of globular clusters.

The motion of the sun in the Galaxy can also be deduced from the radial velocities of the nearby external galaxies. The Large Magellanic Cloud, for example, has a radial velocity of recession of about 270 km/s while the Andromeda galaxy has a velocity of approach of about 300 km/s (see Chapter 39). Studies of the radial velocities of other nearby galaxies show that if they are moving at random their individual velocities with respect to each other and to the center of the Galaxy are relatively small, and that most of the high velocities observed for them are actually due to the orbital motion of the sun about the center of our own galaxy.* We conclude,

therefore, that the sun is moving in the general direction away from the Large Magellanic Cloud and toward the Andromeda galaxy.

When the data from various sources are combined, they indicate that the sun is moving in the direction of the constellation Cygnus, with a speed that, although somewhat uncertain, probably lies in the range 220 to 250 km/s. This direction lies in the Milky Way and is about 90° from the direction of the galactic center, which shows that the sun's orbit is nearly circular and lies in the main plane of the Galaxy. As viewed from the north side of the galactic plane, the orbital motion of the sun is clockwise. The period of the sun's revolution about the nucleus, the *galactic year*, can be found by dividing the circumference of the sun's orbit by its speed; it comes out roughly 200 million (2×10^8) of our terrestrial years. We can observe, therefore, only a "snapshot" of the Galaxy in rotation; we do not actually see stars traverse appreciable portions of their orbits.

(b) High- and Low-Velocity Stars

As already mentioned, the majority of the stars near the sun move nearly parallel to the sun's path about the galactic nucleus, and their speeds with respect to the sun are generally less than 40 or 50 km/s. These are said to be *low-velocity stars*. The radial velocities of nearby gas clouds, as indicated by the Doppler displacements of the interstellar absorption lines and the bright lines of emission nebulae (Section 28.2), are also low and show that they, like the sun, move in roughly circular orbits about the galactic nucleus. In other words, the interstellar material in our part of the Galaxy also belongs to the class of low-velocity objects.

Some stars, on the other hand, have speeds relative to the sun in excess of 80 km/s and are called *high-velocity stars*. They move along orbits of rather high eccentricity that cross the sun's orbit in the plane of the Galaxy at rather large angles (Figure 29.3). Nearby stars moving on such orbits are passing through the solar neighborhood and are only temporarily near us. Stars in the galactic halo and globular clusters also have orbits very different from the sun's and are high-velocity objects.

The term "high velocity" or "low velocity" refers to the speed of an object *with respect to the sun* and has nothing to do with its motion in the Galaxy. Most high-velocity stars lag behind the sun in its motion about the galactic center and hence are ac-

---

*This statement applies only to the very nearest external galaxies—those that belong to the Local Group; see Chapters 37 and 38.

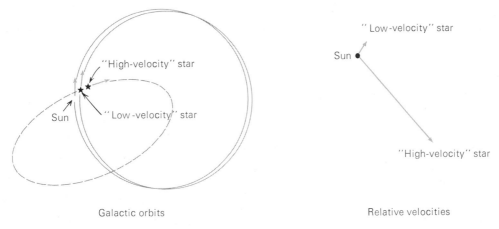

Figure 29.3 *Galactic orbits of high- and low-velocity stars (left), and their relative velocities.*

tually revolving about the Galaxy with speeds *less* than those of the low-velocity stars near the sun.

The component of velocity of an individual star in a direction perpendicular to the plane of the Galaxy (sometimes called the z component of its velocity) cannot, in general, be determined from its radial velocity alone. However, the average of the z-velocity components for stars of a given class or group can be found from a statistical analysis of their radial velocities. It is learned that low-velocity stars usually have lower velocity components perpendicular to the galactic plane than do high-velocity stars. Consequently, high-velocity stars tend to be less strongly concentrated to the plane of the Galaxy than low-velocity stars. Stars in the halo are extreme examples and usually have very high z-velocity components. It is, of course, the high velocities of coronal objects perpendicular to the galactic plane that accounts for their spheroidal distribution in the Galaxy. Globular clusters, in particular, are believed to revolve about the nucleus of the Galaxy in orbits of high eccentricity and inclination to the galactic plane, perhaps rather like the comets revolving about the sun in the solar system. A globular cluster must pass through the plane of the Galaxy twice during each revolution. The large distances between stars, both within the cluster and within the Galaxy itself, make stellar collisions during the cluster's penetration of the galactic disk exceedingly improbable.

The different kinds of motion of stars in the Galaxy may be related to the times and places of formation of those stars; the high-velocity objects are believed to be among the older members of the Galaxy (see Chapter 33).

29.3 SPIRAL STRUCTURE OF THE GALAXY

Observations have been made of many other spiral galaxies that contain interstellar matter and in which individual stars can be resolved (Chapter 37). In these systems both the interstellar matter and the most luminous resolved stars are generally concentrated in the spiral arms. The association of the brightest stars and the gas and dust is not mysterious; highly luminous stars are relatively young objects and have recently formed from clouds of interstellar gas and dust (Chapter 33). In mapping out the spiral structure of our own galaxy, therefore, we can use the gaseous nebulae and the very luminous main-sequence O and B stars as "tracers" to identify spiral arms.

(a) The Spiral Structure from Optical Observations

Because of our position in the disk of the Galaxy, and because we are surrounded by interstellar dust, it is difficult for us to identify even the nearby spiral arms optically. Nevertheless, in 1951 short pieces of three nearby arms were detected by W. W. Morgan and his students at the Yerkes Observatory from the observed directions and distances of a large number of O and B stars and from the distribution of emission nebulae. The fragmentary data on the spiral structure of the Galaxy, as deduced from such optical observations, are summarized in Figure 29.4.

The sun appears to be near the inner edge of an arm called the *Orion arm*, which contains such

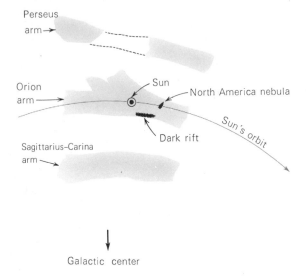

Figure 29.4 Spiral structure of the Galaxy as deduced from optical observations.

conspicuous features as the North America nebula, the Coalsack (near the Southern Cross), the Cygnus Rift (great dark nebula in the summer Milky Way), and the Orion nebula. More distant, and therefore less conspicuous, emission nebulae can be identified in the *Sagittarius-Carina* and *Perseus arms*, located, respectively, about 2000 pc inside and outside the sun's position with respect to the galactic nucleus.

(b) DIFFERENTIAL GALACTIC ROTATION

At the sun's distance from its center, the Galaxy does not rotate as a solid wheel. Stars in larger orbits do not keep abreast of those in smaller ones, but trail behind. This effect produces a shearing motion in the plane of the Galaxy, called *differential galactic rotation*. We can detect the differential galactic rotation from observations of proper motions and radial velocites of stars around us.

As an illustration, let us consider the effect of differential rotation on the radial velocities of stars in different directions in the plane of the Galaxy. Let us assume that all stars have circular orbits (only partly true). If we look in the direction of the galactic center or in the opposite direction, the stars are moving across our line of sight and show no radial velocity. Neither do stars directly in front of us or behind us in our galactic orbit, for they are moving the same as we are. Stars in front of us, but at an angle of only 45° from the direction to the galactic center, on the other hand, are pulling away from us, for we, in our larger orbit, are not moving fast enough to keep up with them, so they show positive radial velocities. We, in turn, are pulling ahead of stars in the opposite direction (behind us, but 135° from the center), so they too have positive radial velocities. We are gaining, however, on stars ahead of us in larger orbits (say, 135° from the center), so they show negative velocities, as do stars opposite them that are gaining on us.

If we prepare a plot of radial velocities of stars of a given distance but in different directions in the plane of the Galaxy, we obtain a curve like that of Figure 29.5. The abscissas are galactic longitude, the angle measured eastward in the plane of the Galaxy from the direction to its center. Such a curve is called a *double sine* curve, because it goes through two cycles in 360°. The amplitude of the curve (the height of its waves) depends on the distances to the stars observed.

(c) THE ROTATION CURVE OF THE GALAXY

A graph of the orbital speeds of stars in the Galaxy against orbital radius is called the *rotation curve*. The

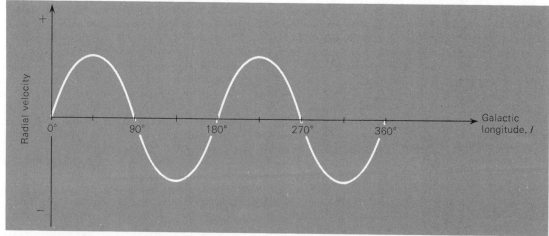

Figure 29.5 Plot of the observed radial velocities of stars in the plane of the galactic equator at a certain distance from the sun against their galactic longitudes.

inner part of the curve has been known from the 1950s; it was best obtained from observations of the Doppler shifts of the 21-cm radiation from neutral hydrogen in the Galaxy (Section 28.2k), because in most directions absorption by interstellar dust limits our range of optical and ultraviolet observations.

Suppose we point a radio telescope at galactic longitude *l* in the plane of the Galaxy (Figure 29.6). If *l* is between 0° and 90° (as in the figure), all gas clouds along the line of sight that are within the sun's orbit must be moving *away* from the sun, and the 21-cm line from each will be shifted to a slightly longer wavelength. We actually receive 21-cm radiation simultaneously from many clouds at various distances, moving away from us at various speeds. That radiation, therefore, is received as a *band* covering a small range of wavelength, each part of the band being 21-cm radiation from a different cloud which is along our line of sight, and is therefore shifted by a different amount. The long-wavelength edge of the band corresponds to radiation coming from the interstellar hydrogen moving away from us with the greatest radial velocity. It can be shown (if all galactic orbits are assumed to be circular) that at a given galactic longitude, the gas with the greatest radial velocity will be at position *A* (in Figure 29.6), where our line of sight passes closest to the galactic center. The distance, *R*, of *A* from the galactic center is a simple function of the radius of the sun's orbit, R_0, and the galactic longitude, *l*.† Since our line of sight is tangent to the orbit of a cloud at *A*, the maximum observed radial velocity, after correction for the sun's motion, is the actual orbital speed of a cloud moving on a circular orbit of radius *R*. By directing the radio telescope at various other galactic longitudes, we find the circular orbital speeds at other distances from the galactic center.

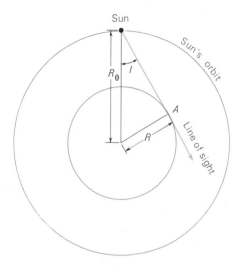

Figure 29.6 Determination of the orbital velocities of objects with smaller orbits than that of the sun.

Beyond the sun's orbit it is more difficult to detect the rotation curve because the distances to the emitting interstellar clouds are less well determined. Nevertheless, observations since 1980 have provided pretty good evidence for the rotation curve out to nearly 20,000 pc from the galactic center.

In its inner parts, the Galaxy rotates like a solid body (or wheel), but by the distance of the sun's orbit the curve flattens out. Beyond the sun's orbit, the orbital speeds gradually rise again, reaching about 300 km/s at 20,000 pc (Figure 29.7). Still further out, radial velocities of distant globular clusters indicate that the rotational velocity of the galaxy remains near 300 km/s to much greater distances—possibly to as far as 60,000 pc. We shall see (Chapter 37) that many other galaxies have similar flat rotation curves to very great distances from their centers.

---

†$R = R_0 \sin l$.

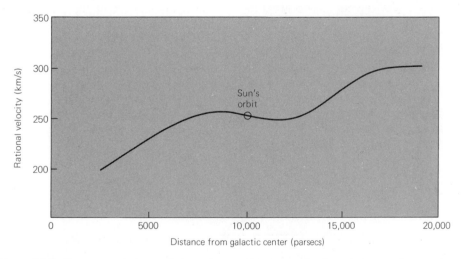

Figure 29.7 Rotation curve for our Galaxy, showing the rotational velocity at various distances from the galactic center.

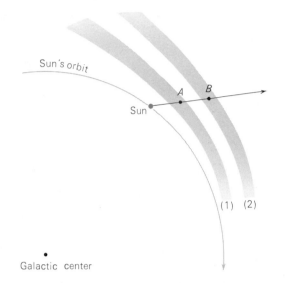

Figure 29.8 Observation of spiral arms at a wavelength of 21 cm.

(d) SPIRAL STRUCTURE AS FOUND FROM 21-CM OBSERVATIONS

Once the rotation of the Galaxy at various distances from its center is known, we can use 21-cm observations to map out the spiral structure of our stellar system. The problem is actually very complicated because the neutral hydrogen clouds are known not to revolve about the galactic center in perfectly circular orbits. In principle, however, the procedure is straightforward. Suppose a radio telescope is pointed in the direction shown in Figure 29.8; our line of sight (in this example) passes through two spiral arms, arm (1) at point A and arm (2) at point B. Most of the interstellar gas in that line of sight is concentrated at those points. Now energy gathered by a radio telescope is generally passed into a radio receiver that can be swept in frequency, or wavelength, much as a home radio receiver can be changed in frequency to tune in different stations. As the receiver is tuned through wavelengths near 21 cm, the signal strength increases whenever radiation is received from neutral hydrogen. In Figure 29.8, the strongest radiation, coming from points A and B, is received at wavelengths λ_1 and λ_2, respectively; λ_1 and λ_2 are both slightly less than 21 cm because the sun is overtaking A and B in their galactic revolution, giving them *negative* radial velocities. A plot of observed signal strength versus wavelength would look something like Figure 29.9. The difference between λ_1 (or λ_2) and the nominal wavelength of the 21-cm line (actually 21.11 cm) gives the radial velocity of point A (or B); these data, plus a knowledge of how the Galaxy rotates, enable the distances to A and B to be determined and thus locate them in the galactic plane. Similar observations made at other galactic longitudes reveal the distances of spiral arms in other directions. Thus, a map of the spiral structure of the Galaxy is built up. Such is shown in Figure 29.10.

In practice, uncertainties in the outer part of the rotation curve and particularly the noncircular motions of the gas clouds make the determinations of the locations of spiral arms very uncertain, but at least observations of this sort have shown that the spiral structure exists. Moreover, 21-cm maps of some other galaxies show that the neutral hydrogen outlines their spiral structures very clearly.

(e) Formation and Permanence of Spiral Structure

It is not surprising that the interstellar material is concentrated into spiral arms. No matter what the original distribution of the material might be, the

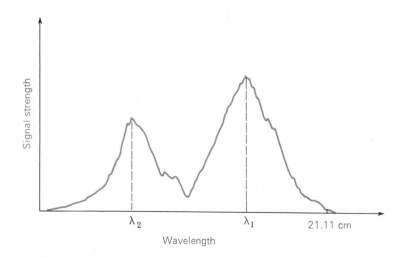

Figure 29.9 Strength of received radio signals versus wavelength for the observation illustrated in Figure 29.8.

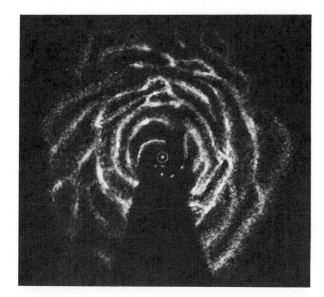

Figure 29.10 A drawing of the spiral structure of the Galaxy, deduced from 21-cm observations made at Leiden, The Netherlands, and at Sydney, Australia. The large and small circles show the location of the galactic center and the sun, respectively. (Courtesy of Gart Westerhout)

differential rotation of the Galaxy would be expected to form it into spirals. Figure 29.12 shows the development of spiral arms from two irregular blobs of interstellar matter, as the portions of the blobs closest to the galactic center move fastest, while those farther away trail behind.

It is harder to understand, however, why the arms are not wound tighter than they are. At the sun's distance from the center of the Galaxy, the Galaxy rotates once in about 2×10^8 years. Its total age, however, is believed to be about 10^{10} years, in which case a point near the sun has made at least 50 revolutions. With so many turns, we would ex-

pect the spiral arms to be wound very tightly and to lie very much closer together than they do.

A step toward an understanding of how spiral structure can be maintained is provided by a theory developed by C. C. Lin at the Massachusetts Institute of Technology and his associates Frank Shu and Chi Yuan. Lin and his coworkers have calculated the manner in which stars and gas clouds would move if they had circular paths about the galactic center and were influenced both by the gravitational fields produced by the Galaxy as a whole and by the matter forming the spiral arms themselves. They find that objects should slow down slightly in the regions of the spiral arms, and linger there longer than elsewhere in their orbits, thus building up a wave of higher than average density where the spiral arms are. This *density wave* model predicts that an equilibrium state is reached in which there are two trailing arms, inclined about 6° to the circular orbits of the stars and gas clouds, spaced about 3000 pc apart—in rough agreement with observations. The entire spiral pattern rotates more slowly than the actual material in the Galaxy, so that the stars, gas, and dust pass slowly through the spiral arms.

As gas and dust clouds approach the inner boundaries of an arm and encounter the higher density of slower moving matter, they collide with it, producing a shock boundary. It is here that the theory predicts star formation is most likely to occur, with protostars (like the solar nebula—Section 14.3) condensing in higher density regions. In some other galaxies we do see the highly luminous, and hence certainly young, stars along with the densest dust clouds near the inner boundaries of spiral arms, as well as radio radiation from the shocks.

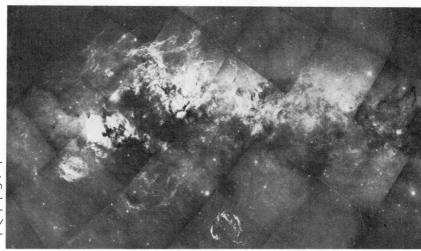

Figure 29.11 Mosaic of photographs of the Milky Way in Cygnus, photographed with the 124-cm Schmidt telescope. (National Geographic Society—Palomar Observatory Sky Survey, reproduced by permission of Caltech/Palomar Observatory)

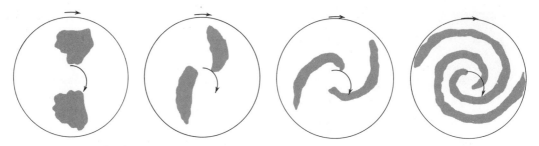

Figure 29.12 Hypothetical formation of two spiral arms from irregular clouds of interstellar material.

(f) Different Stellar Populations in the Galaxy

Striking correlations are found between the characteristics of stars and other objects and their locations in the Galaxy. Some classes of objects, for example, are found only in regions of interstellar matter, that is, in the spiral arms of the Galaxy. Examples are bright supergiants, main-sequence stars of high luminosity (spectral classes O and B), Wolf-Rayet stars, type I cepheid variables, and young open star clusters.

The distributions of some other classes of objects show no correlation with the location of spiral arms. These objects are found throughout the disk of the Galaxy, with greatest concentration toward the nucleus. They also extend into the sparse galactic halo. Examples are planetary nebulae, novae, type II cepheids, RR Lyrae variables, and Mira variables of periods less than 250 days. The globular clusters, which also belong to this group of objects, are found almost entirely in the halo and nuclear bulge of the Galaxy.

Main-sequence stars of spectral types F through M exist in all parts of the Galaxy, as do red giants and, probably, white dwarfs (observations of white dwarfs, of course, are limited to our immediate neighborhood because of their low luminosities).

There are also differences in the chemical compositions of stars in different parts of the Galaxy. Nearly all stars appear to be composed mostly of hydrogen and helium, but the residual abundance of the heavier elements seems to spread over a large range for different stars. In the sun and in other stars associated with the interstellar matter of the spiral arms, the heavy elements (elements heavier than hydrogen and helium) account for about 1 to 4 percent of the total stellar mass. Stars in the galactic halo and in globular clusters, however, have much lower abundances of the heavy elements—often less than one tenth, or even one hundredth that of the sun. There is also some evidence that many stars in the disk of the Galaxy whose galactic orbits show that they are not associated with interstellar matter may have lower heavy-element abundances than the sun.

The stars associated with the spiral arms are sometimes said to belong to *population I*, while those found elsewhere in the Galaxy are said to belong to *population II*. The terms "population I" and "population II" were first applied to different classes of stars by W. Baade, late astronomer at the Mount Wilson and Palomar Observatories. During World War II, Baade was impressed by the similarity of the stars in the nuclear bulge of the Andromeda galaxy to those in the globular clusters in the halo of our own Galaxy; he concluded that the stars situated in spiral arms must, collectively, display different properties from those located elsewhere in the Galaxy.

Today we can interpret the phenomenon of different stellar populations in the light of stellar evolution. As has been mentioned, and as will be discussed further (Chapter 33), only in the interstellar matter of spiral arms is star formation expected to take place. Thus, population I comprises stars of many different ages, including some that were recently formed or are still forming. Population II, on the other hand, consists entirely of old stars, formed, probably, early in the history of the Galaxy.

It is clear today that two stellar populations are insufficient to account completely for the distribution of all the different kinds of stars in the Galaxy. Modern investigators now generally define several different stellar populations, ranging from "extreme population I" (spiral-arm objects), through "disk population" objects, to "extreme population II" (halo objects).

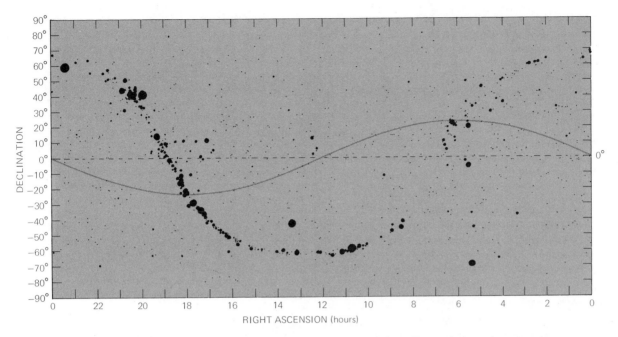

Figure 29.13 The distribution of discrete radio sources in the Galaxy. The symbols on this map of the sky show the positions of galactic radio sources emitting continuous radiation near 21 cm (not the 21-cm line). The larger symbols denote stronger sources. The dashed line across the middle is the celestial equator and the solid wavy line is the ecliptic. Note how most of the radio sources are concentrated along the plane of the Milky Way, which, although really a great circle in the sky, appears on this kind of map as a large open U. (*Courtesy G. Verschuur*)

29.4 THE MASS OF THE GALAXY

We can make an estimate of the mass of the inner part of the Galaxy (lying inside the sun's orbit) with an application of Kepler's third law (as modified by Newton). Assume the sun's orbit to be circular and the Galaxy to be roughly spherical so we can treat it as though its mass internal to the sun were concentrated to a point at the galactic center. If the sun is 10,000 pc from the center, its orbit has a radius of 2×10^9 AU (there are 2×10^5 AU in one parsec). Since its period is 2×10^8 yr, we have (Section 5.3)

$$
\text{Mass (Galaxy)} = \frac{(2 \times 10^9)^3}{(2 \times 10^8)^2}
$$
$$
= 2 \times 10^{11} \text{ solar masses.}
$$

More sophisticated calculations based on complicated models give a similar result.

It must be emphasized that this is only the mass contained in the volume inside the sun's orbit. In the light of recent observations yielding the Galaxy's rotation curve outside the sun's orbit, we now realize that the total mass of the Galaxy is actually very much greater, because otherwise its rotation curve would fall off with distance, as do the planetary speeds in the solar system. To account for the roughly flat rotation curve beyond the sun, the Galaxy's mass contained within a given radial distance from its center must be approximately proportional to that distance. At present, the total mass is thought to be from 1 to 2×10^{12} solar masses, but this value is very uncertain.

29.5 MAGNETISM IN THE GALAXY

There is ample evidence that magnetism is an important phenomenon in the Galaxy. In the sun, we shall see (Chapter 31) that there are very strong magnetic fields associated with the active regions around sunspots, and a general solar field of about the strength of the earth's. We shall also see (Chapter 34) that some stars have extremely strong magnetic fields, and that objects like pulsars must have incredibly strong fields. But in addition to magnetism associated with particular objects in the Galaxy, there are general fields whose force lines usually run along the spiral arms, and possibly a weak field throughout the halo as well. These general interstellar magnetic fields of the Galaxy are of low intensity, but cover such vast regions of space that they contain a great deal of total energy. We consider now some of the clues by which we can detect indirectly the general magnetism of the Galaxy.

(a) Synchrotron Radiation

We have already described some of the mechanisms by which atoms can radiate electromagnetic energy (Chapter 10). In addition, charged particles, such as electrons and ions, radiate electromagnetic energy if they are accelerated by a magnetic field. When a charged particle enters a magnetic field, the field compels it to move in a circular or spiral path around the lines of force; the particle is thus accelerated and radiates energy. If the speed of the particle is nearly the speed of light, the particle is said to be *relativistic*. In this case, the energy it radiates is called *synchrotron radiation*, because particles so radiate when they are accelerated to relativistic speeds in a laboratory synchrotron. Any solid or liquid body or gas that is not at absolute zero temperature radiates electromagnetic energy (Section 10.2). To distinguish it from this normal *thermal* radiation from gases or bodies, synchrotron radiation is sometimes called *nonthermal* radiation.

Both the intensity and frequency of synchrotron radiation are greater, the greater the energy of the particle and the stronger the magnetic field. This nonthermal radiation has properties that make it easy to recognize: it has a distinctive distribution of intensity with wavelength, it is highly polarized, and the energy radiated by a particle is primarily in the direction of the particle's instantaneous motion (see Figure 29.14). Both atomic nuclei (positive ions) and electrons radiate when accelerated, but the nuclei are thousands of times as massive as electrons, and consequently for the same energy have much lower speeds. Thus the nuclei do not generally move fast enough to emit significant synchrotron radiation.

We find many astronomical examples in which relativistic electrons are spiraling through magnetic fields and are emitting synchrotron radiation, although we do not yet, in all cases, know the origin of these energetic electrons, nor the mechanisms that give them their great speeds. We have already described the radiation belt around the planet Jupiter (Section 18.1d) and the lesser one about the

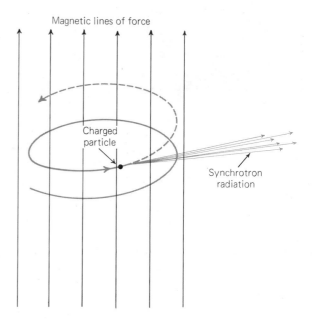

Magnetic lines of force

Charged particle

Synchrotron radiation

Figure 29.14 The emission of synchrotron radiation by a charged particle moving at nearly the speed of light in a magnetic field.

earth (Section 15.3), in which relativistic electrons in the magnetic fields of those planets emit synchrotron radiation at radio wavelengths. Such energy is emitted far more strongly from electrons in the magnetic fields associated with remnants of supernovae (Chapter 34). One such supernova remnant is the *Crab* nebula, in which many of the electrons have such high energy that much of the synchrotron radiation is in visible, ultraviolet, and even X-ray, as well as in radio wavelengths.

The Galaxy, in fact, abounds in sources of nonthermal radiation. Some are in its very nucleus (Section 29.6). In certain other galaxies (Chapter 37) we find synchrotron radiation emanating from their inner halos, and often from invisible regions far outside the parts of those galaxies that we can observe optically. Evidently those galaxies have very extended magnetic fields that sometimes reach far beyond their visible images.

The nonthermal radiation shows us that there are interstellar magnetic fields in our Galaxy. When polarized radiation passes through ionized gas in a magnetic field, the plane of the polarization rotates as the radiation moves forward—an effect called *Faraday rotation*. The amount of Faraday rotation, however, depends on the frequency. The highly polarized synchrotron radiation from nonthermal sources shines through the partially ionized gases of interstellar space to reach the earth. We find that different wavelengths of this radiation have undergone different amounts of Faraday rotation, showing that even the nearly empty interstellar space has magnetic fields. We can estimate what fraction of the interstellar gas is ionized and measure the amount of Faraday rotation produced at different wavelengths. With these data we can calculate the intensity of the magnetic field in space. Because of their inherent uncertainties the calculations are not very precise, but they indicate an average field in the spiral arms of the Galaxy of from 10^{-6} to 10^{-5} gauss; by comparison, the field at the surface of the earth is a little less than 1 gauss.

(b) Polarization of Starlight

We have seen (Section 28.1e) that interstellar dust polarizes starlight shining through it, which is evidence that some of the dust particles are elongated

and are aligned in space. An interstellar magnetic field is the only mechanism suggested to account for this alignment of dust grains. Moreover, we find that the directions of the polarization for all of the stars we observe in the general direction toward or away from the galactic center are roughly parallel to each other. This is what we would expect if the grains were aligned by magnetic fields whose lines of force run along the spiral arms, for in those directions we are looking more or less broadside at spiral arms. On the other hand, the polarization of light from stars that lie in directions nearly at right angles to the galactic center, and hence along our own spiral arm, has random orientation. Again this is what we expect, for here we look along the lines of force of the interstellar field so that there is no preferential alignment of particles across our line of sight. Theory suggests that the magnetic field strength required to produce enough alignment of dust grains to account for the observed polarization

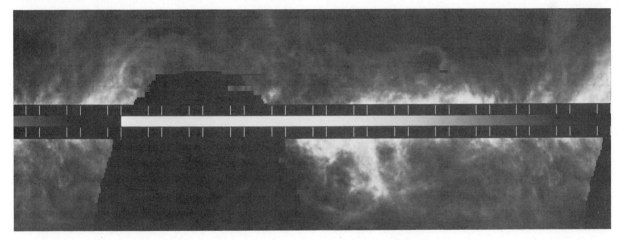

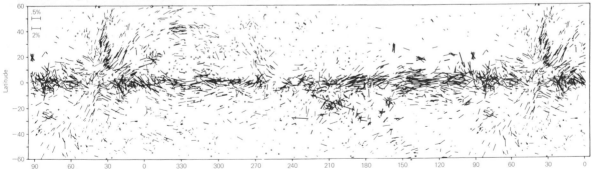

Figure 29.15 Two charts plotted in galactic coordinates, in which the plane of the Galaxy (Milky Way) runs horizontally across the middle. *(above)* The lightness indicates the relative strength of 21-cm radiation, which is emitted by neutral hydrogen in the spiral arms. *(below)* Mathewson and Ford's map showing the relative strength and direction of starlight polarization. Note how the polarization directions match many of the filaments of neutral hydrogen, showing the alignment of polarization with the spiral arms. *(Courtesy C. Heiles, University of California, Berkeley, and E. B. Jenkins, Princeton University)*

effects is of the same order as that calculated from the Faraday rotation of synchrotron radiation.

(c) Containment of Cosmic Rays

Cosmic rays approach the earth almost equally from all directions in space. Most of them cannot originate from the sun or another part of the solar system, for then they would come from certain directions, and, for example, show strong differences in influx between day and night. Some may come from beyond the Galaxy, but it is hard to believe that very many do, for we cannot imagine what sources could fill all of extragalactic space with cosmic rays to the observed density. Most investigators, therefore, consider it probable that most cosmic rays originate in our own Galaxy. One likely source is believed to be supernovae (Chapter 34).

On the other hand, the atomic nuclei that compose most cosmic rays travel at very nearly the speed of light. If they travel in straight lines and originate in the Galaxy they would escape it in at most a few tens of thousands of years. Now, the suggested sources of cosmic rays are all very unusual objects (such as supernovae), and there cannot be enough of them situated uniformly around us to provide a continuous, nearly isotropic flux of cosmic rays all traveling in straight lines. We conclude that cosmic rays must be produced by scattered sources over the entire Galaxy, and that they are trapped or stored in it for very long periods of time, moving about in interstellar space in all possible directions.

Because cosmic rays are charged particles (atomic nuclei) they can be captured into spiraling orbits around magnetic lines of force, and this, in fact, is the only way we know of that they can be kept in the Galaxy. Calculations show that interstellar fields of strength like that predicted from Faraday rotation and polarization can trap cosmic rays of energy up to 10^{17} or 10^{18} eV in the Galaxy for hundreds of millions of years. Cosmic rays of very much higher energy—say 10^{20} eV—would escape the Galaxy in a relatively short time, but ones of such very high energy are observed more rarely at the earth, and could originate from relatively few nearby sources, such as supernovae that occurred within the past few thousand years. Some very high-energy cosmic rays, alternatively, could have an extragalactic origin. In any case, we can understand the general cosmic ray phenomena only if the Galaxy has extensive magnetic fields.

We recall from the last chapter the interesting coincidence that in interstellar space the energy associated with the heat of the interstellar gas, with starlight, with cosmic rays, and with galactic magnetism are all the same to an order of magnitude. Thus the magnetic fields in the Galaxy comprise a major force and play an important role in galactic astronomy.

29.6 THE NUCLEUS OF THE GALAXY

Near the center of the Galaxy is a large concentration of stars, generally called the *nuclear bulge*. At its center, lying behind the constellation Sagittarius, is the *nucleus* of the Galaxy. We cannot see the nucleus in visible light or in the ultraviolet, because those wavelengths are absorbed by the intervening interstellar dust. High-energy X rays and gamma rays, however, force their way through the interstellar medium and are recorded by instruments on rockets and satellites. Also the infrared and radio radiation, whose wavelengths are long compared with the sizes of the interstellar grains, flow around them and reach us from the center of the Galaxy. The very bright radio source in that region is known as *Sagittarius A*.

(a) Infrared Radiation and Star Densities

At about 2 microns (20,000 Å) infrared radiation from the galactic center comes from an extended region 2° or so in diameter, and the distribution of its intensity about the center of the Galaxy is just about the same as that of visible light from the central bulge of the neighboring Andromeda galaxy (Chapter 37). It is assumed, therefore, that this radiation, amounting to some 10^{41} erg/s, is emitted directly from stars in the nuclear bulge of our Galaxy. From it, we can derive the approximate density of

Table 29.1   **Distribution of Stars near the Galactic Center**

DISTANCE FROM CENTER (*pc*))	NUMBER OF STARS PER CUBIC PC	TOTAL MASS WITHIN THIS DISTANCE (*Solar Masses*)
0.1	3×10^7	3×10^5
1.0	4×10^5	4×10^6
10.0	7×10^3	7×10^7
20.0	2×10^3	2×10^8

stars at various distances from the galactic center; the results are given in Table 29.1

We see that near the center of the Galaxy the stars are still light-months or more apart, but within the inner parsec the average separation of stars is only a few thousand astronomical units— only one or a few light-weeks. At the very center of the galactic nucleus, the stars are only a few hundred astronomical units apart. Yet if the earth were a planet revolving about a sun in almost any part of the galactic nucleus we would still find the neighboring stars far enough away to appear as points of light. In most parts of the nuclear bulge the night sky would appear almost as dark as it does from our present location in the Galaxy. On the other hand, in the inner few parsecs there would probably be enough nearby bright stars to illuminate our night sky as much as moderate moonlight does, and at the very center there would probably be some stars near enough to give us a sky brighter

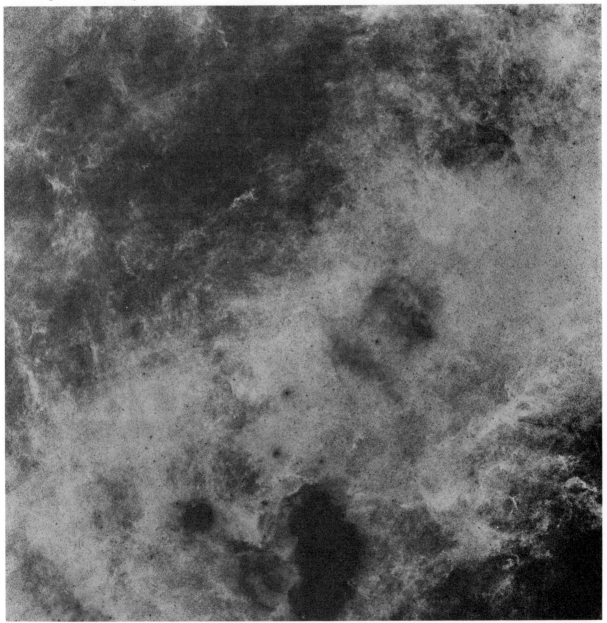

Figure 29.16 A region of the Milky Way near the galactic center. Negative print, photographed in red light with the 124-cm Schmidt telescope. The white regions are absorbing dust clouds. *(Copyright, National Geographic Society—Palomar Observatory Sky Survey, reproduced by permission of Caltech/Palomar Observatory)*

than at full moon. It is an interesting commentary, however, that if the typical city dweller on earth could be transferred to a planet around a star at the galactic center, he would probably be unaware of any difference in the brightness of the night sky.

The strongest emission of infrared radiation (some 10^{42} erg/s or more) comes at wavelengths longer than 100 microns, and it seems to be spread over a large area, as are the stars. This long-wave infrared energy is thought to be emitted by a sparse distribution of interstellar dust simply reradiating the light it absorbs from stars. At the intermediate wavelengths of 5 to 20 microns, on the other hand, the infrared radiation seems to come from several discrete sources. Each of them probably emits as much as 10^6 times the energy of the sun. They are believed to be dust clouds surrounding stars—either dust condensing from matter ejected from those stars (Section 28.1f), or dust in nebulae from which new stars are forming.

(b) Radio Radiation from the Nucleus

The Sagittarius radio source contains synchrotron radiation, which means that relativistic electrons and magnetic fields are present at the galactic center. Many of the most complex interstellar molecules observed at radio wavelengths (Section 28.21) are found in this region as well. The molecules may well have an origin associated with the circumstellar dust clouds inferred from the 5 to 20 micron radiation. The region emitting radiation from molecules seems to be expanding from the center of the nucleus. Thus we have evidence for expanding H II regions, magnetic fields, interactions between solid grains and interstellar gas, and processes giving rise to relativistic electrons, all in the small region of the galactic center.

We also receive 21-cm radiation from neutral hydrogen in the region around the galactic center. The Doppler shifts of this radiation show that the neutral hydrogen is flowing away from the center at a speed of about 50 km/s. Therefore, the nucleus of our Galaxy is emitting radiation from gamma rays to radio waves, both thermal and nonthermal, and is ejecting gas as well. Some stars there are evidently ejecting matter, while others may be forming from interstellar matter.

(c) Nuclear Activity

The emission of many kinds of radiation as well as the ejection of matter from the nucleus of our Galaxy leads us to compare it with the nuclei of other galaxies. There exists a class of galaxies called *Seyfert galaxies* (Chapter 37), which are spiral stellar systems like our own, except that they have nuclei that emit very strongly in visible light, infrared radiation, and sometimes radio waves. Seyfert galaxy nuclei evidently have regions of very hot gas that may be associated with violent activity; the energy radiated from these nuclei is typically 10^{44} to 10^{45} erg/s. The well known Seyfert galaxy NGC 1068 has a nucleus from which gas is also flowing at a rate of about 600 km/s.

Other galaxies have nuclei of still more explosive properties. The irregular galaxy M82 may have suffered an explosion in prehistoric times in its nucleus, from which gas is still streaming out (Chapter 37), and the galaxy M87 has a jet of material evidently ejected from its nucleus. Most spectacular, however, are the quasars (also described in Chapter 37), which emit energy at the rate of 10^{47} to 10^{48} erg/s. Some investigators suspect that quasars are extreme examples of explosive events at the centers of remote galaxies whose main parts are too faint to observe because of their great distance.

There seems, in other words, to be a continuum of nuclear phenomena, ranging from the mighty quasars, through the Seyfert galaxies, down to the nuclei of apparently "normal" galaxies like our own. It is not ruled out that a violent nuclear event might occur, from time to time—or possibly at least once—in *any* galaxy. In other words, it is conceivable that our own Galaxy is in a quiescent period, before or after which it too could display the more violent characteristics of Seyfert galaxies or even quasars.

Until recently the evolution of a galaxy was thought of largely in terms of the evolution of its stellar content. Now we know that this is not all of the picture. The existence of events such as those described above shows that the life history of a galaxy may be influenced, and perhaps even largely determined, by large-scale collective phenomena affecting the structure of the galaxy as a whole rather than just the evolution of its individual components.

EXERCISES

1. Sketch the distribution of globular clusters about the Galaxy, and show the sun's position. Show how they would appear on a Mercator map of the sky, with the central line of the Milky Way chosen as the "equator."

2. The globular clusters probably have highly eccentric orbits, and either oscillate through the plane of the Galaxy or revolve about its nucleus. Suppose the latter is the case; where would the clusters spend most of their time? (Think of Kepler's second law.) At any given time, would you expect most globular clusters to be moving at high or low speeds with respect to the center of the Galaxy? Why?

3. The period of the sun's revolution about the center of the Galaxy was calculated from its measured speed and distance from the center of the Galaxy. How would the period be changed if the sun's distance from the galactic center were 20 percent greater than the figure assumed?

4. If the galactic halo has an overall radius of 100,000 LY, what is the volume occupied by the Galaxy in cubic parsecs? If the mass of the Galaxy is 1×10^{12} suns, and if it were uniformly distributed throughout this volume, what would be the mean density?
   Answer: 1.5×10^{13} pc³; 6.6×10^{-2} solar masses/pc³

5. What would the mass of the Galaxy interior to the sun be if the sun's distance from the center were 10,000 pc but its period of revolution about the nucleus were only 100 million years?

6. Suppose we correctly knew the sun's distance from the center of the Galaxy but had derived a value for the speed of the sun in its orbit that is too high by 10 percent. How much would our calculated mass of the Galaxy interior to the sun be in error, and in which direction would the error be?

7. Suppose the mean mass of a star in the Galaxy were only ⅓ solar mass. Using the value for the mass of the Galaxy found in the text, find how many stars the system contains. What did you assume about the total mass of interstellar matter in finding your answer?

★8. If its orbital speed is 300 km/s, what is the period of a gas cloud moving in a circular orbit of radius 20,000 pc?

★9. Suppose 21-cm observations are made in the galactic plane in a direction 45° from that of the galactic center. What is the distance from the galactic center to the gas cloud with the maximum observed radial velocity?
   Answer: About 7000 pc

★10. Why are we not able to map out the spiral structure of the Galaxy in directions $l = 0°$ and 180° from 21-cm observations? Why do you suppose we *are* able to map out its spiral structure in directions $l = 90°$ and 270°?

11. Describe the details of an experiment in which you witness the formation of "spiral arms" of cream in a cup of coffee that you have stirred vigorously before putting in the cream.

12. Distinguish clearly between the orbital motion of the sun, toward galactic longitude 90°, and the *solar motion*, toward the *solar apex*, which was described in Chapter 23.

13. Use the data of Table 29.1 to estimate the mean separation between stars at (a) 0.1 pc from the galactic center, and (b) 10 pc from the galactic center.
   Answer: (a) 650 AU; (b) 10,000 AU

Harlow Shapley (1885–1972) began his career as a newspaper reporter. In his twenties he enrolled in the University of Missouri, where he searched through the catalogue for a suitable major and found Astronomy in the "A's." After earning his bachelor's degree he went on to Princeton for his Ph.D. Subsequently, at Mount Wilson, his study of globular clusters revealed the true extent of our Galaxy. (*Yerkes Observatory*)

STAR CLUSTERS

The study of star clusters is very important, because the stars in a single cluster are all at about the same distance from the earth (the distance to the cluster); consequently, their luminosities, colors, and so on can be intercompared easily and accurately. Moreover, the stars of a cluster probably have had a common origin, being formed at about the same time from the same prestellar material. Investigations of clusters of different ages are a great aid to the study of stellar evolution.

A number of clusters bear proper names. Some of these are names of mythological characters (the Pleiades); other clusters bear the names of the constellations in which they appear (the Double Cluster in Perseus). Most of the conspicuous clusters are listed in the early catalogues of star clusters and nebulae.

30.1 DESCRIPTIONS OF STAR CLUSTERS

Clusters that contain a great many stars are said to be *rich;* those that contain comparatively few are said to be *poor.* Rich clusters are likely to be conspicuous, and their identification as genuine stellar systems is certain. Poor clusters, on the other hand, are much more difficult to pick out against the background of the general star field. Sometimes a real cluster may not be identifiable as such against

the background. Other groups of stars that appear to be real systems may actually be stars at different distances seen close together in projection. Most of the clusters that are catalogued, however, contain a high enough density of stars to stand out against the background so that there is virtually no chance of their being accidental superpositions of stars at different distances. Even so, it is often difficult or impossible to say with certainty whether a given individual star is a member of the cluster or not. In general, therefore, a few of the stars studied as cluster members are actually stars in the foreground or background, that is, stars which belong to the field.

(a) Globular Clusters

About a hundred globular clusters are known, most of them in the halo and nucleus of our Galaxy. All are very far from the sun, and some are found at distances of 60,000 LY or more from the galactic plane. A few, nevertheless, are bright enough to be seen with the naked eye; they appear as faint, fuzzy stars. One of the most famous naked-eye globular clusters is M13, in the constellation of Hercules, which passes nearly overhead on a summer evening at most places in the United States. Through a good pair of binoculars the more conspicuous globular clusters resemble tiny moth balls. A small telescope reveals their brightest stars, while a large telescope

Figure 30.1 The globular cluster M3. *(Lick Observatory)*

shows them to be beautiful, globe-shaped systems of stars. Visual observation, however, even through the largest telescope, does not reveal the multitude of fainter stars in globular clusters that can be recorded on telescopic photographs of long exposure.

A good photograph of a typical globular cluster shows it to be a nearly circularly symmetrical system of stars, with the highest concentration of stars near its own center (a few globular clusters, such as Omega Centauri, appear slightly flattened). Most of the stars in the central regions of the cluster are not resolved as individual points of light but appear as a nebulous glow. Two photographs of a globular cluster made on emulsions sensitive to two different colors of light, say red and blue, show that the brightest stars are red. These stars are two or three magnitudes brighter than the RR Lyrae variable stars that are almost always found in globular clusters (also called cluster variables—see Section 24.3h); since RR Lyrae stars average about absolute

Figure 30.2 The Pleiades. North is to the left. *(Lick Observatory)*

magnitude 0 to +1, the brighter red stars must be red giants. Other kinds of variables sometimes found in globular clusters include type II Cepheids and RV Tauri stars. One cluster (NGC 7078) contains a planetary nebula (see Chapter 34).

Distances to globular clusters are sometimes calculated from the apparent magnitudes of the RR Lyrae stars they contain (Section 29.1a). From their angular sizes (typically a few minutes of arc) their actual linear diameters are found to be from 20 to 100 pc or more. In one of the nearer globular clusters more than 30,000 stars have been counted, but if those stars too faint to be observed are considered, most clusters must contain hundreds of thousands of member stars. The combined light from all these stars gives a typical globular cluster an absolute magnitude in the range -5 to -10.

The average star density in a globular cluster is about 0.4 star per cubic parsec. In the dense center of the cluster the star density may be as high as 100 or even 1000 per cubic parsec. There is plenty of space between the stars, however, even in the center of a cluster. The "solid" photographic appearance of the central regions of a globular cluster results from the finite resolution of the telescope, seeing effects of the earth's atmosphere, and the scattering of light in the photographic emulsion. A bullet fired on a straight line through a point near the center of a cluster would have far less than one chance in 10^{11} of striking a star. If the earth revolved not about the sun, but about a star in the

densest part of a globular cluster, the nearest neighboring stars, light-months away, would appear as points of light. Thousands of stars, however, would be scattered uniformly over the sky. The Milky Way would be hard, if not impossible, to see, and even on the darkest of nights the brightness of the sky would be comparable to faint moonlight.

The motions of globular clusters were described in Chapter 29. They are high-velocity objects that do not partake of the general galactic rotation. They are believed to revolve about the nucleus of the Galaxy on orbits of high eccentricity and high inclination to the galactic plane (rather like the orbits of comets in the solar system). Obeying Kepler's second law, a cluster spends most of its time far from the nucleus; a typical cluster probably has a period of revolution of the order of 10^8 years.

Because most globular clusters lie outside the plane of the Milky Way, probably a good fraction of them have been discovered, although a few dozen, hidden by the obscuring dust clouds, may remain undiscovered in the disk and nucleus of the Galaxy.

(b) Open Clusters

In contrast to the rich, partially unresolved globular clusters, *open clusters* appear comparatively loose and "open" (hence their name). They contain far

Figure 30.3 The open star cluster M67 (NGC 2682) in Cancer. Photographed with the 5-m telescope. *(Caltech/Palomar Observatory)*

fewer stars than globular clusters and show little or no strong concentration of stars toward their own centers. Although open clusters are usually more or less round in appearance, they lack the high degree of spherical symmetry that characterizes a globular cluster; some open clusters actually appear irregular. On photographs made with long-focus telescopes, the stars in these clusters are usually fully resolved, even in the central regions.

Open clusters, as mentioned previously, are found in the disk of the Galaxy, often associated with interstellar matter. Because of their locations, they are sometimes called *galactic clusters* rather than open clusters (this term should not be confused, however, with *clusters of galaxies*—Chapters 37 to 39). They are low-velocity objects and belong to stellar population I; they are presumed to originate in or near spiral arms. Over 1000 open clusters had been catalogued as of 1982, but many more are identifiable on good search photographs such as those of the Palomar Sky Survey. Yet only the nearest open clusters can be observed, because of interstellar obscuration in the Milky Way plane. We conclude, therefore, that we see only a small fraction of the open clusters that actually exist in the Galaxy; possibly tens or even hundreds of thousands of them escape detection.

Several open clusters are visible to the unaided eye. Most famous among them is the *Pleiades,* which appears as a tiny group of six stars (some people see more than six) arranged like a tiny dipper in the constellation of *Taurus;* a good pair of binoculars shows dozens of stars in the cluster, and a telescope reveals hundreds. (The Pleiades is *not* the Little Dipper; the latter is part of the constellation of *Ursa Minor,* which also contains the North Star.) The *Hyades* is another famous open cluster in Taurus. To the naked eye, the cluster appears as a V-shaped group of faint stars, marking the face of the bull. Telescopes show that the Hyades actually contains more than 200 stars. The naked-eye appearance of the *Praesepe,* in Cancer, is that of a barely distinguishable patch of light; this group is often called the "Beehive" cluster, because its many stars, when viewed through a telescope, appear like a swarm of bees.

Typical open clusters contain several dozen to several hundred member stars, although a few, such as M67, contain more than a thousand. Compared with globular clusters, open clusters are small, usually having diameters of less than 10 pc. Bright supergiant stars of high luminosity in some

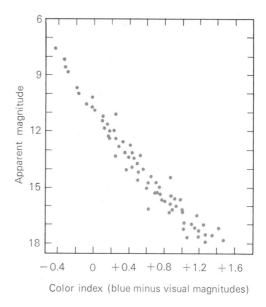

Figure 30.4 Color-magnitude diagram for a hypothetical open star cluster.

open clusters, however, may cause them to outshine the far richer globular clusters. The RR Lyrae stars are never found in open clusters, but other kinds of variable stars, such as type I Cepheids, are sometimes present.

(c) Associations

For more than 50 years it has been known that the most luminous main-sequence stars—those of spectral types O and B—are not distributed at random in the sky but tend to be grouped into what are now called *associations,* lying along the spiral arms of our Galaxy. In the decade following World War II, interest was greatly revived in these groups of hot stars, especially by the Soviet astronomer V. A. Ambartsumian, who called attention to many of them and pointed out that they must be very young groups of stars. Because the stars of an association lie in the galactic plane and are spread over tens of parsecs, each revolves about the galactic center with a slightly different orbital speed; Ambartsumian showed that the different orbital speeds of the different members of an association would completely disrupt the group after a few million years. All associations, therefore, must be very young objects on the astronomical time scale.

We distinguish between two kinds of associations: those containing O and B stars are called *O-associations;* the others contain T Tauri stars and are called *T-associations.* T Tauri, or RW Aurigae stars, are red eruptive variables of intermediate to low luminosity that show rapid and irregular increases in brightness (Chapter 33). They are always associated with interstellar matter and are believed to be very young stars that have just formed from it. They are often present in extremely young star clusters. Sometimes, as in Orion, T-associations and O-associations coexist.

An O-association appears as a group of several (say, 5 to 50) O stars and B stars, and sometimes Wolf-Rayet stars, scattered over a region of space some 30 to 200 pc in diameter. Because these stars are rare, it would be very unlikely for so many of them to exist by chance in so relatively small a volume of space. It is assumed, therefore, that the stars in an association are either physically associated or at least have had a common origin. Stars of other spectral types may also belong to associations, but these more common stars are not conspicuous against the general star field and do not attract attention to themselves as belonging to any particular group.

Often a small open cluster is found near the center of an association. It is presumed, in such cases, that the stars in the association are the outlying members of the cluster, and that they probably had a common origin with the cluster stars.

Some associations are expanding—their member stars moving radially outward from the association center. An example of such an association is one in Perseus, known as the *Zeta Persei association*. More than a dozen of its member stars are bright enough so that their positions have been measured accurately for more than a century. Most of the stars seem to be moving outward, from the center of the group. If we extrapolate their motions backward, we find an age for the expansion of about $1\frac{1}{2}$ million years. Another famous association is in *Orion*, centered on a group of stars in the middle of the Orion Nebula. Rather far from the main association are three luminous O and B stars that seem to be moving away from it at speeds of from 70 to 130 km/s. It is tempting to postulate that these stars originated with those of the association, but there is as yet no satisfactory hypothesis to explain their extremely high speeds. Analysis of the statistics of associations and O and B stars, however, have led some investigators to conclude that all such stars have originated in associations.

About 70 associations are now catalogued. Like ordinary open clusters, however, they lie in regions occupied by interstellar matter, and many others must be obscured. There are probably several thousand undiscovered associations in our galaxy.

(d) Summary of Clusters

The foregoing descriptions of star clusters are summarized in Table 30.1. The numbers of globular clusters, open clusters, and associations are taken from the second edition of the *Catalogue of Star Clusters and Associations*, published by the Hungarian Academy of Sciences. The *Catalogue* includes all objects reported up to 1970 and is the most comprehensive star cluster catalogue available. The sizes, absolute magnitudes, and numbers of stars listed for each type of cluster are approximate only and are intended as representative values.

30.2 DYNAMICS OF STAR CLUSTERS

The problem of two bodies moving under the influence of their mutual gravitational attraction was solved by Newton. We have seen (Chapter 5) that each member of such a two-body system moves on

Table 30.1 Characteristics of Star Clusters

	GLOBULAR CLUSTERS	OPEN CLUSTERS	ASSOCIATIONS
Number known in Galaxy	125	1055	70
Location in Galaxy	Halo and nuclear bulge	Disk (and spiral arms)	Spiral arms
Diameter (pc)	20 to 100	<10	30 to 200
Mass (solar masses)	10^4 to 10^5	10^2 to 10^3	10^2 to 10^3?
Number of stars	10^4 to 10^5	50 to 10^3	10 to 100?
Color of brightest stars	Red	Red or blue	Bue
Integrated absolute visual magnitude of cluster	-5 to -10	0 to -10	-6 to -11
Density of stars (solar masses per cubic parsec)	0.5 to 1000	0.1 to 10	<0.01
Examples	Hercules Cluster (M13)	Hyades, Pleiades, h and χ Persei, Praesepe	Zeta Persei, Orion

a path (a conic section) that can be described by a simple algebraic equation. A star cluster, on the other hand, is a system of many bodies. Each star in a cluster moves under the influence of gravitational forces exerted upon it simultaneously by all the other stars. Since all the stars in a cluster are moving, the gravitational forces they exert on each other are constantly changing, and it is an enormously complicated problem to predict the future path of any one.

In principle, the motion of an individual star in a cluster could be computed in detail. In fact, the individual motions in hypothetical clusters of up to hundreds of stars have been calculated with electronic computers to study theoretically the dynamical evolution of such systems. Similar calculations must be performed in space science applications, for example, to compute the motion of a space vehicle moving in the combined gravitational fields of the earth, moon, and sun. The particular wanderings of an individual star in a cluster, however, are not of much interest. Far more significant is the way all the cluster members move on the average, for their average motion depends upon certain fundamental characteristics of the cluster—its mass, size, and structure. This section will describe, briefly, the statistical methods that enable us to learn some of the properties of clusters from the average motions of their member stars.

(a) THE VIRIAL THEOREM AND MASSES OF CLUSTERS

If a stone is raised high above the ground and released, the force of the earth's gravity upon it accelerates it downward; as it falls, it picks up more and more speed, or energy of motion. The energy associated with the motion of an object is its *kinetic energy*. The potential ability of gravity to accelerate a body and give it kinetic energy is called *gravitational potential energy*. The greater the height from which it is dropped, the greater is its potential energy before it is released.

Similarly, a star in a cluster feels itself attracted generally toward the center of the cluster, and so has gravitational potential energy. The actual amount of its potential energy depends upon the strength of the total resultant gravitational force acting on it and its distance from the cluster center. As a star moves about in a cluster, it sometimes decreases its distance from the center, and thus speeds up, converting some of its potential energy to kinetic energy. At other times the star

increases its distance from the cluster center, pulling against gravity, and therefore slows down; it then converts kinetic to potential energy. The actual path of the star may be very complicated, because occasionally it may pass near another star, and its direction of motion is then deflected by the gravitational attraction between the two. Such a deflection is called a *gravitational encounter*. An "encounter" in this case is not a real collision; because of the distances separating the stars in a cluster, actual stellar collisions are extremely improbable.

There is associated with a star cluster a certain total kinetic energy, which is defined as the sum of the individual kinetic energies of its member stars. (The kinetic energy of a star of mass m and speed v is $\frac{1}{2} mv^2$.) By convention, the potential energy of a star in a cluster at a given instant is defined as the work that must be done on the star (that is, the energy that must be given it) to remove it from its location in the cluster at that instant to a point infinitely far away, pulling, in the process, against the gravitational attraction between the star and cluster. The potential energy of the entire cluster is the energy required to separate all the stars infinitely far apart. Actually, as defined, the potential energy of a star cluster is the potential energy its stars would have given up if they had all fallen together under their mutual gravitational attraction, from a configuration in which they were extremely widely separated to their present configuration in the cluster. Clearly, more potential energy would have been released if the cluster were smaller and more concentrated than it is, and less if it were more spread out.

After a cluster has existed for a long enough time (usually an interval of hundreds or thousands of millions of years) there will have been enough encounters between stars to divide the total energy of the cluster (potential plus kinetic) approximately evenly among the stars. The energy will never be distributed exactly equally; there will always be a few stars that by virtue of recent encounters have more energy than average, and others that have less energy than their share. On the average, however, each star will have its share of the total energy of the cluster; no one *kind* of star would be expected to have more energy than any other kind. The cluster is then said to be in a state of *statistical equilibrium*.

Under the above conditions of statistical equilibrium, the total potential energy of a cluster (as defined above) is twice its kinetic energy. This statement is called the *virial theorem*. An important application of the virial theorem in astronomy is in the estimation of masses of systems of stars presumed to be in statistical equilibrium. The underlying physical idea can be summarized briefly as follows: If we assume that a cluster is in equilibrium, its members are held together and their motions are determined by their mutual gravitation. The virial theorem tells us how much mass the cluster must have (and hence how much gravitation) in order for its stars to move with their observed speeds, and in such paths that the cluster has its observed shape.

(b) STABILITY OF STAR CLUSTERS

A condition for the stability of an isolated cluster is that its total potential energy be greater than its kinetic energy. Otherwise, the average stellar speed exceeds the escape velocity from the cluster, which dissipates into space. If the potential energy of the cluster exceeds its kinetic energy, it is gravitationally bound—that is, its member stars cannot all escape. (See Exercise 11.)

Clusters, however, are not completely isolated but move in various orbits in the Galaxy. Thus, an added condition for the stability or permanence of a cluster is that it be bound together with gravitational forces that are stronger than the disrupting tidal forces of the Galaxy, or other nearby stars, upon it. The more compact a cluster, the greater is its own gravitational binding force compared to the disrupting forces, and the better chance it has to survive to old age.

Globular clusters are highly compact systems and are, consequently, very stable. Most globular clusters can probably maintain their identity almost indefinitely. Even these clusters lose some stars, however—especially those of relatively small mass. A few stars in a cluster are always moving substantially faster than average. Every now and then one of them, through an encounter, will be given enough speed to escape the cluster. Some of the stars in the galactic halo must be stars that have, in the past, escaped in this way from globular clusters.

When a star escapes it carries off energy, leaving less energy than before for the stars remaining in the cluster. The result is that, over time, the cluster develops a tightly bound core surrounded by a rarefied halo of stars. In the dense core, stars occasionally collide and some of the debris eventually coalesces. It is predicted that this dynamical evolution will lead to the development of a massive black hole at the cluster center (Chapter 35). Meanwhile, a few stars in the outer parts of the cluster continue to escape.

The escape rate and dynamical evolution for the rich globular clusters, however, is so slow that the clusters can survive for many thousands of millions of years. The situation is analogous to the evaporation of molecules from a more or less permanent planetary atmosphere.

Matters are very different for most open clusters. Those that we have discovered are relatively close to the sun; at our distance from the galactic center, the tidal force of the Galaxy will shear a cluster apart in short order if the star density within it is much less than about one star per cubic parsec. The Pleiades, for example, is probably just stable in its central regions,

while its outer parts are probably dissipating. The Hyades is on the verge of instability. Typical open clusters have maximum lifetimes as clusters of only a few hundreds of millions of years; a handful of the richer, denser ones, like M67, can be expected to survive for thousands of millions of years. Most open clusters, in other words, are relatively young stellar groups, and only a small fraction are very old.

The case for stellar associations, which have very low star densities, is even more extreme. These loose groups cannot possibly be permanent stellar systems. If they are actually expanding, as one or two appear to be, they would be highly unstable even in the absence of disruptive galactic tidal forces.

30.3 DETERMINATION OF DISTANCES TO CLUSTERS

We described how distances to globular clusters are found in Chapter 29. In Section 23.3c we saw how distances to three moving open clusters—the Hyades, the Ursa Major Group, and the Scorpio-Centaurus Group—have been found from the apparent convergence or divergence of their member stars as they move across the sky.

If such techniques are not available, the most useful method of determining the distance to a star cluster is from a plot of the apparent magnitudes versus the color indices of its member stars, as shown in Figure 30.4. The time required to photograph spectra of many individual cluster stars is sometimes prohibitive, but their colors can be observed very quickly. Recall that the color index of a star depends on the star's temperature and is thus related directly to its spectral class. Therefore, the plot of magnitude versus color index (called a *color-magnitude diagram*) is, with one exception, like a Hertzsprung–Russell diagram for the cluster (Section 27.2). The exception is that a normal H–R diagram is a plot of the *absolute* magnitudes (rather than apparent magnitudes) of stars against their spectral classes (or color indices). All the stars in a cluster, however, are at the same distance, and the difference between the apparent and absolute magnitudes is the same for every star—the *distance modulus* of the cluster (Section 24.2b). Most of the stars in a cluster generally lie along a *main sequence* similar to that defined by stars in the neighborhood of the sun. We find a main sequence, therefore, in the color-magnitude diagram of a cluster (see Figure 30.4), with the bluer, brighter-appearing stars

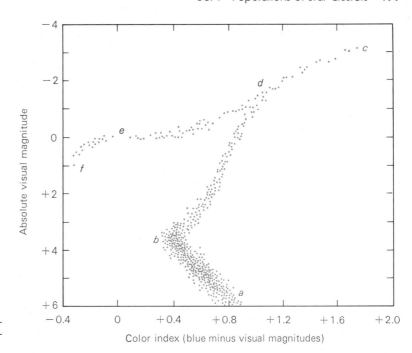

Figure 30.5 Hertzsprung–Russell diagram for a hypothetical globular star cluster.

(which are really the more luminous main-sequence stars in the cluster) farther up on the diagram than the redder, fainter-appearing stars. From the ordinary H–R diagram, we know what absolute magnitudes correspond to various color indices along the main sequence. The difference, at any given color index, between the apparent magnitude of the cluster stars and the absolute magnitude of known main-sequence stars of the same color is the distance modulus of the cluster.

As a numerical example, consider the hypothetical cluster whose color-magnitude diagram is shown in Figure 30.4. A main-sequence cluster star of color index +0.6 is seen to have an apparent magnitude of +15. But this is a star like the sun, whose absolute magnitude is +5. At a distance of 10 pc, therefore, this star would appear ten magnitudes brighter than it does at the actual distance of the cluster. Since 10 magnitudes corresponds to a factor of 10,000 in light, the cluster must be 100 times as distant as 10 pc or must be 1000 pc away.

In actual practice the apparent magnitudes and color indices must first be corrected for effects of interstellar absorption and reddening (Chapter 28).

30.4 POPULATIONS OF STAR CLUSTERS

The Hertzsprung–Russell diagrams, or color-magnitude diagrams, of star clusters are extremely use-

ful in the study of stellar evolution. Here we shall describe some of the properties of the color-magnitude diagrams for different kinds of clusters and some other properties of the stellar populations of clusters. The interpretation in terms of stellar evolution is discussed in Chapter 33.

(a) Color-Magnitude Diagrams of Globular Clusters

Globular clusters nearly all have very similar appearing color-magnitude diagrams. Figure 30.5 shows the appearance of the color-magnitude diagram for a typical globular cluster of known distance, for which the apparent magnitudes have been converted to absolute magnitudes. The region from a to b is the main sequence. Presumably, the main sequence would extend farther down than a if the cluster were near enough for us to observe its fainter stars. Above point b, however, the main sequence seems to terminate; in most globular clusters, this point occurs at about absolute magnitude, $M_v = +3.5$. From b to c there extends a sequence of stars that are yellow and red giants; the brightest and reddest of them (at c) at $M_v = -3$ are brighter than typical red giants in the solar neighborhood. A third sequence of stars extends from d to f; it is called the *horizontal branch* of the H–R diagram for a globular cluster. There is a gap in the horizontal branch at $M_v = 0$ (point e), where no star of constant light output is found. The

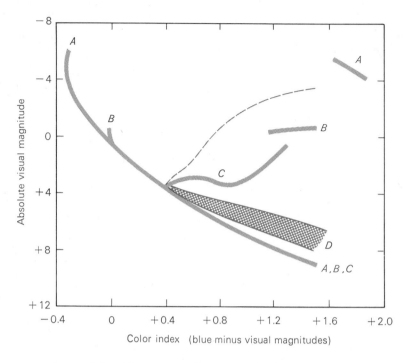

Figure 30.6 Composite Hertzsprung–Russell diagram for four hypothetical open clusters, A, B, C, and D. The dashed line shows the location of the giants in globular clusters.

stars observed in this gap are the RR Lyrae variables. The color-magnitude or H–R diagram of a globular cluster is more or less similar to that of any system of stars which belongs to stellar population II.

(b) Color-Magnitude Diagrams of Open Clusters

Whereas globular clusters nearly all have very similar H–R diagrams, those of open clusters differ widely from one another. Figure 30.6 shows, schematically, superposed H–R diagrams for four hypothetical open clusters, which will serve as examples.

In cluster A the main sequence extends to high-luminosity stars, and the highly luminous cool stars in the cluster are red supergiants. The double cluster h and χ Persei has an H–R diagram similar to that of cluster A. The main sequence of cluster B (representative of the open clusters M11 and M41) extends to less luminous stars, and the red giants of that cluster are less luminous than those in A as well. Cluster C (characteristic of the open cluster M67) has a main sequence that terminates at about the same absolute magnitude as do those of globular clusters. Cluster C has a sequence of yellow and red giants that resembles, but is less luminous than, that of the globular clusters (dashed line). The red giants in cluster C have about the

same luminosity as typical red giants near the sun. At least one open cluster of the "C type" (M67) has a suggestion of a horizontal branch.

Note that in the three clusters discussed so far, the upper ends of the main sequences terminate at different absolute magnitudes and that the red giants (if present) usually lie off to the right of the top of the main sequence in each H–R diagram. In a few clusters (such as the Pleiades), no red giants are observed, but such clusters are not generally very rich. We note that there is a gap between the top of the main sequence and the red giants in the H–R diagrams of clusters A and B. This gap, called the *Hertzsprung gap*, is broadest in the color-magnitude diagrams of clusters whose main sequences extend to high luminosities, and narrows for clusters whose main sequences terminate at successively lower luminosities. The gap finally disappears in the color-magnitude diagrams of clusters whose main sequences extend only as far as that of cluster C. We shall see (Chapter 34) that at this gap stars are in stages of evolution in which they are unstable and either pulsate or evolve very rapidly.

The least luminous stars observed in clusters A, B, and C all lie on the main sequence. In cluster D, however, the brighter stars are on the main sequence, while the faintest observed ones lie off to the *right*. The open cluster NGC 2264 has such an H–R diagram. We shall see (Chapter 33) that the faintest stars in this cluster (and others like it) are

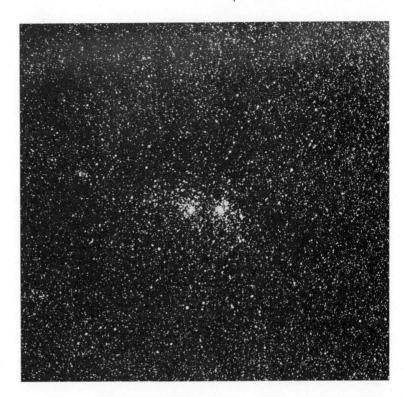

Figure 30.7 The double open star cluster h and χ Persei. (Yerkes Observatory)

believed to be still in the process of contraction from interstellar matter.

(c) Differences in Chemical Composition of Stars in Different Clusters

Hydrogen and helium, the most abundant elements in stars in the solar neighborhood, are also the most abundant constituents of the stars in all kinds of clusters. The exact abundances of the elements heavier than helium, however, vary from cluster to cluster. In the sun, and most of its neighboring stars, the combined abundance (by mass) of the heavy elements seems to be between 1 and 4 percent. The strengths of the lines of heavy elements in the spectra of stars in most open clusters show that they, too, have 1 to 4 percent of their matter in the form of heavy elements.

Globular clusters, however, are a different story. Spectra of their brightest stars often show extremely weak lines of the heavy elements. The heavy-element abundance of stars in typical globular clusters is found to range from only 0.1 to 0.01 percent, or even less. Other stars of population II also have spectra that often indicate low abundances of heavy elements, although the difference between them and the sun is not usually as extreme as for some of the globular cluster stars. Apparently, differences in chemical composition are related to differences in stellar population. The probable explanation of these phenomena is discussed in Chapter 34.

EXERCISES

*1. The RR Lyrae stars in a particular globular cluster appear at apparent magnitude +15. How distant is the cluster? (Ignore interstellar absorption and assume that the RR Lyrae stars have absolute magnitudes of 0.)

2. Where in the Galaxy do you suppose undiscovered globular clusters may exist?

3. Suppose globular clusters have orbits about the galactic center with very high eccentricities—near

unity. When a particular globular cluster is at its farthest from the center of the Galaxy, its distance from the center is 10^4 pc. What is its period of galactic revolution? (*Hint:* 1 pc = 2 × 10^5 AU. Assume that the mass of the Galaxy is 10^{12} solar masses.)

4. Table 30.1 indicates that stellar associations can emit even more light than a globular cluster. How is this possible if the associations have so few stars?

5. What color would a globular cluster appear? Why?

6. From the data of Table 30.1, estimate the average mass of the stars in each of the three different cluster types.

7. What is the density in solar masses per cubic parsec of the following clusters: (a) a globular cluster 50 pc in diameter containing 10^5 stars? (b) a stellar association of 100 solar masses and 20 pc in radius? *Answer:* (a) 1.5 solar masses/pc^3, (b) 0.003 solar masses/pc^3

8. Why do you suppose it is sometimes said that the problem of dealing with the motions of more than two bodies interacting gravitationally has no solution? Is the statement true? Explain.

*9. At what time of year is the earth's potential energy greatest? At what time of year is its kinetic energy greatest?

*10. Could the virial theorem be used to compute the mass of a flock of birds flying together in formation? Why?

*11. Compare the question of the stability of a star cluster to the question of whether a body moving in the solar system has a closed elliptical, or an open hyperbolic, orbit. Is the orbital potential energy of the earth less or greater than its kinetic energy? Why?

12. A main-sequence star of color index 0 has an absolute magnitude of about +1. In the color-magnitude diagram of a certain cluster, it is noted that stars of 0 color index have apparent magnitudes of about +6. How distant is the cluster? (Ignore interstellar absorption.)

13. It is often possible to observe fainter main-sequence stars in open clusters than in globular clusters. Why do you suppose this is the case?

Sir Arthur Stanley Eddington (1882–1944), British mathematician and astrophysicist, organized two expeditions to observe the total solar eclipse of 1919 in order to test a prediction of Einstein's general theory of relativity. Eddington is best known among astronomers for his development of the theoretical methods of investigating the internal structure of the sun and stars. (*Yerkes Observatory*)

31

THE SUN: AN ORDINARY STAR

The sun is not only the dominant object in the solar system, but its relative stability is vital to all life on earth. Here we examine the sun's gross properties and also the very chaotic and changeable conditions at its surface. Some solar data, obtained by techniques described in previous chapters, are summarized in Table 31.1

31.1 OUTER LAYERS OF THE SUN

The only parts of the sun that can be observed directly are its outer layers, collectively known as the sun's *atmosphere*. There are three general regions, each having substantially different properties: the *photosphere*, the *chromosphere*, and the *corona*.

(a) The Solar Photosphere

What we see when we look at the sun is the solar photosphere. As stated in Section 25.2b, the photosphere is not a discrete surface but covers the range of depths from which the solar radiation escapes. Most of the absorption of visible light in the solar photosphere, as we have seen, is done by negative hydrogen ions. As one looks toward the limb of the sun, his line of sight enters the photosphere

at a grazing angle, and the depth below the outer surface of the photosphere to which he can see is even less than at the center of the sun's disk. The light from the limb of the sun, therefore, comes from higher and cooler regions of the photosphere. Analysis of this *limb darkening* can be used to determine the variation of temperature with depth in the photosphere, as described in Section 25.2. From the data obtained from observations of limb darkening, using our knowledge of the physics of gases and the way in which atoms absorb and emit light, we can calculate a *model solar photosphere*. Such a model solar photosphere is given in Table 31.2.

It is evident from Table 31.2 that within a depth of about 260 km the pressure and density increase by a factor of 10, while the temperature climbs from 4500 to 6800 K. At a typical point in the photosphere, the pressure is only a few hundredths of sea-level pressure on the earth, and the density is about one ten-thousandth of the earth's atmospheric density at sea level.

(b) Chemical Composition

More than 60 of the elements known on the earth have now been identified in the solar spectrum. Those that have not been identified in the sun either do not produce lines in the observable spec-

Table 31.1 **Solar Data**

DATUM	HOW FOUND	TEXT REFERENCE (Section)	VALUE
Mean distance	Radar reflection from planets	22.3b	1 AU
			149,597,892 km
Maximum distance from earth			1.521×10^8 km
Minimum distance from earth			1.471×10^8 km
Mass	Acceleration of earth	26.1	333,400 earth masses
			1.99×10^{33} g
Mean angular diameter	Direct measure	9.2b	$31'59''.3$
Diameter of photosphere	Angular size and distance	9.2b	109.3 times earth diameter
			1.39×10^{11} cm
Mean density	Mass/volume	4.1g	1.41 g/cm^3
Gravitational acceleration at photosphere (surface gravity)	$\dfrac{GM}{R^2}$	4.3c	27.9 times earth surface gravity
			27,300 cm/s^2
Solar constant	Measure with instrument such as bolometer	24.5b	1.96 cal/min/cm^2
			1.368×10^6 ergs/s/cm^2
Luminosity	Solar constant times area of spherical surface 1 AU in radius	24.5b	3.8×10^{33} ergs/s
Spectral class	Spectrum	25.1a	G2V
Visual magnitude			
Apparent	Received visible flux	24.1b	-26.7
Absolute	Apparent magnitude and distance	24.2a	$+4.8$
Rotation period at equator	Sunspots, and Doppler shift in limb spectra	31.2	24^d16^h
Inclination of equator to ecliptic	Motions of sunspots	31.2	$7°10'.5$

trum or are so rare on the earth that they cannot be expected to produce lines of observable strength on the sun unless, proportionately, they are far more abundant there. Most of the elements found in the sun are in the atomic form, but more than 18 types of molecules have been identified. Most of the molecular spectra are observed only in the light from the cooler regions of the sun, such as the sunspots.

The relative abundances of the chemical elements in the sun are similar to the relative abundances found for other stars. About three quarters of the sun (by weight) is hydrogen, and about 98 percent is hydrogen and helium. The remaining few percent is made up of the other chemical elements, in approximately the amounts that are described in Section 25.2 (See also Appendix 19.)

(c) The Chromosphere

There is a change in the physical state of the gases just above the photosphere. The photosphere ends at about the place where the density of negative hydrogen ions has dropped to a value too low to result in appreciable opacity. Gases extend far beyond the photosphere, but they are transparent to most radiation. The region of the sun's atmosphere that lies immediately above the photosphere is the chromosphere.

Until this century the chromosphere was best observed when the photosphere was occulted by the moon during a total solar eclipse. In the 17th century several observers described what appeared to them as a narrow red "streak" or "fringe" around

Table 31.2 **Model Solar Photosphere**

DEPTH BELOW SURFACE (km)	PERCENT OF LIGHT THAT EMERGES FROM THAT DEPTH	TEMPERATURE (K)	PRESSURE (Earth Atmospheres)	DENSITY (g/cm^3)
0	100	4500	1.0×10^{-2}	2.8×10^{-8}
50	95	4800	1.7×10^{-2}	4.2×10^{-8}
100	91	5000	2.6×10^{-2}	6.2×10^{-8}
140	82	5300	3.8×10^{-2}	8.7×10^{-8}
170	67	5600	5.4×10^{-2}	11.5×10^{-8}
225	37	6200	8.3×10^{-2}	16.0×10^{-8}
260	13	6800	11.2×10^{-2}	20×10^{-8}

The globular cluster M13. *(U.S. Naval Observatory)*

The open cluster M16 and associated nebulosity in Serpens. *(California Institute of Technology/Palomar Observatory)*

The Pleiades and associated nebulosity in Taurus, photographed with the Palomar Schmidt telescope. *(California Institute of Technology/Palomar Observatory)*

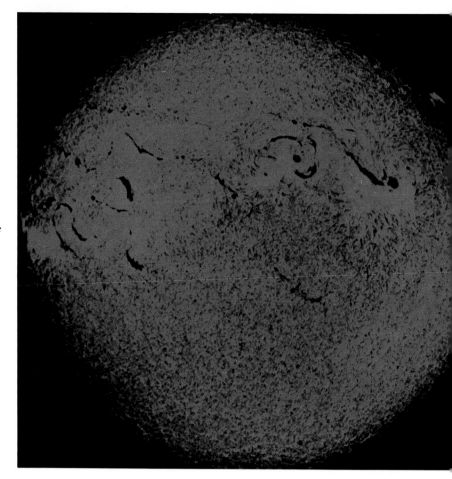

The sun in the red light of the first Balmer line of hydrogen. *(Carl Zeiss)*

Raymond Davis, Jr's, ''neutrino telescope.'' *(Brookhaven National Laboratory)*

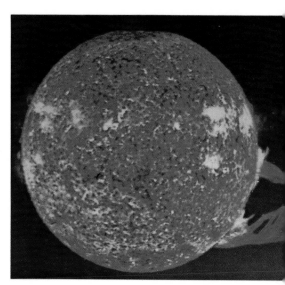

Ultraviolet photograph of the entire sun, showing several flares and a large prominence, taken by Skylab. *(NASA)*

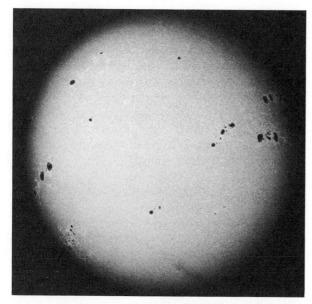

Figure 31.1 The sun, photographed under excellent conditions, showing a large number of sunspots, on September 15, 1957. *(Mount Wilson and Las Campañas Observatories)*

one limb of the moon during a brief instant after the sun's photosphere had been covered. Not until the careful observations of the solar eclipses of 1842, 1851, and 1860, however, was much attention paid to the chromosphere. Some of these observations were made by photographic methods and established beyond doubt the existence of the chromosphere. In 1868 the spectrum of the chromosphere was first observed; it was found to be made up of bright lines, which showed that the chromosphere consists of hot gases that are emitting light in the form of emission lines. These bright lines are difficult to observe against the bright light of the

photosphere but appear in the spectrum of the light from the extreme limb of the sun just after the moon has eclipsed the photosphere. They disappear within a few seconds, when the moon has covered the chromosphere as well. Because of the brief instant during which the chromospheric spectrum can be photographed during an eclipse, its spectrum, when so observed, is called the *flash spectrum*. The element *helium* (from *helios*, the Greek word for "sun") was discovered in the chromospheric spectrum before its discovery on earth in 1895.

Today it is possible to photograph both the chromosphere and its spectrum outside of eclipse. One instrument used for this purpose is the *coronagraph*, a telescope in which a black disk at the focal plane occults the photosphere, producing an artificial eclipse. The chromosphere can also be photographed in the light of its strongest emission lines through monochromatic filters. In addition, the chromosphere is regularly observed in ultraviolet, X-ray, infrared, and radio wavelengths.

The chromosphere is about 2000 to 3000 km thick, but in its upper region it breaks up into a forest of jets (called *spicules*), so the position of the upper boundary of the chromosphere is somewhat arbitrary. Its reddish color (whence comes its name) arises from one of the strongest emission lines in the visible part of its spectrum, the bright red line due to hydrogen (the Hα line—first line in the Balmer series—Section 10.62). The density of the chromospheric gases decreases upward above the photosphere, but spectrographic studies show that the temperature *increases* through the chromosphere, from 4500 K at the photosphere to 100,000 K or so at the upper chromospheric levels. The processes that heat the corona (see below) evidently

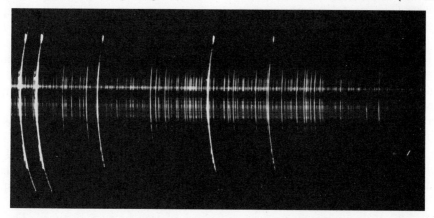

Figure 31.2 Flash spectrum of the eclipse of 2 February 1968. The bright arcs are monochromatic images of the solar limb. The irregularities are prominences. *(Courtesy Sacramento Peak Observatory, Air Force Cambridge Research Laboratories)*

also heat the chromosphere. Far ultraviolet observations of the chromosphere, such as those made from the Skylab, have been especially helpful in revealing information about its structure.

(d) The Corona

The chromosphere merges into the outermost part of the sun's atmosphere, the corona. Like the chromosphere, the corona was first observed only during total eclipses, but unlike the chromosphere, the corona has been known for many centuries; it is referred to by Plutarch and was discussed in some detail by Kepler. Many of the early investigators regarded the corona as an optical illusion, but photography confirmed its existence in the 19th century. Its spectrum was first observed in 1869 by the American astronomers Harkness and Young. The corona extends millions of miles above the photosphere and gradually thins to a sparse wind of ions and electrons flowing outward through the entire solar system (the *solar wind;* Section 31.5). The corona emits half as much light as the full moon; its invisibility, under ordinary circumstances, is due to the overpowering brilliance of the photosphere. Like the chromosphere, the corona can now be photographed, with the coronagraph and other instruments, under other than eclipse conditions.

The corona is also observed at radio wavelengths. In Great Britain in 1942, unexpected noise was picked up on radar receivers. It was subsequently learned that the source of this noise was the sun. Since World War II, radio observations of the sun have been made regularly at many radio astronomical observatories. Shortwave radio energy (near 1 cm wavelength) can escape the sun from the lower chromosphere. The corona is more and more opaque, however, to longer and longer radio wavelengths. Those of 15 m escape the sun only if they originate high in the corona. Thus, by observing the sun at different radio wavelengths, we observe to different depths in the corona and chromosphere and can determine the heights in the solar atmosphere at which various disturbances giving rise to radio emission occur.

If the sun is typical among stars as a radio source, we would expect radio emission from other stars, because of their distances, to be too feeble to detect. The various cosmic radio sources known are not ordinary stars; we shall discuss them in later chapters.

Not only do radio waves originate in the corona, but the corona produces scintillations in distant radio sources when they are observed through its outer part. The phenomenon is somewhat analogous to the scintillation of the light from stars caused by the earth's atmosphere. Scintillations of remote radio sources observed 90° away from the sun in the sky show that the corona actually reaches out beyond the earth. Coronal atoms are also regularly detected in the vicinity of the earth by space vehicles (Section 31.4).

The visible light from the corona is about 10^{-6} that of the photosphere of the sun. It has three components. The spectrum of the first is simply that of reflected sunlight which shows the same dark Fraunhofer lines as the solar photospheric spectrum; this part is called the F (Fraunhofer) corona. It extends far from the sun and gradually fades into interplanetary space. The F corona is believed to be caused by dust particles—presumably an inner extension of the zodiacal light. Superimposed on the F corona is the K corona, whose light dominates in the corona's inner, brighter region, less than two solar radii from the center of the sun. The K corona is photospheric light reflected by free electrons. However, all of the absorption lines are washed out, leaving the K corona with a pure continuous spectrum (the "K" designation comes from the German word *Kontinuum*). The washing out of the spectral lines is due to the very high random speeds of the electons, redistributing the reflected light, thereby filling in the absorption lines by many different Doppler shifts. The inferred high speeds of the electrons provide evidence that the corona is extremely hot—in the millions of Kelvins.

The third spectral component of the corona consists of bright emission lines superimposed on the light of the K and F coronas. In 1942 the Swedish physicist B. Edlen identified these lines as forbidden lines of calcium, iron, and nickel (see Section 28.2d for a discussion of forbidden radiation). There are also many ordinary permitted emission lines, but most of these lie in the far ultraviolet, where they have been mapped by the Orbiting Solar Observatories and by Skylab.

Analysis of the coronal spectral lines confirms the K-corona inference that the corona is very hot—millions of Kelvins. The atoms are all very highly ionized—for example, there are ultraviolet lines of iron ionized 16 times. The density, however, is very low. At the base of the corona there are about 10^9 atoms/cm^3, compared with 10^{16}/cm^3 in the upper photosphere and 10^{19}/cm^3 at sea level in the earth's atmosphere. Thus, despite the high temperature of the corona (a measure of how fast the particles are moving), its density is so low that the actual heat (energy content per cubic centimeter) is very low; in that near-vacuum it would take a long time for the hot coronal gases to warm up a cup of coffee (of course, the radiant energy from the nearby photosphere would do the job in a hurry).

Observations from Skylab revealed that there are sometimes large regions of the corona that are relatively cool and quiet. These *coronal holes* are places of extremely low density and are usually (but not always) found in the polar regions of the sun. They cause the empty spaces that can be seen on some of the eclipse photographs of the solar corona (see Figures 9.14 and 9.15).

31.2 SOLAR ROTATION

Galileo first demonstrated that the sun rotates on its axis by recording the apparent motions of the sunspots as the turning sun carried them across its disk. He found that the rotation period of the sun is a little less than one month. In 1859 Richard Carrington found that in its equatorial regions the sun rotates in about 25 days, but that at a latitude of 30° the period is about 2½ days longer.

The sun's rotation rate can be determined also from the difference in the Doppler shifts of the light coming from the receding and approaching

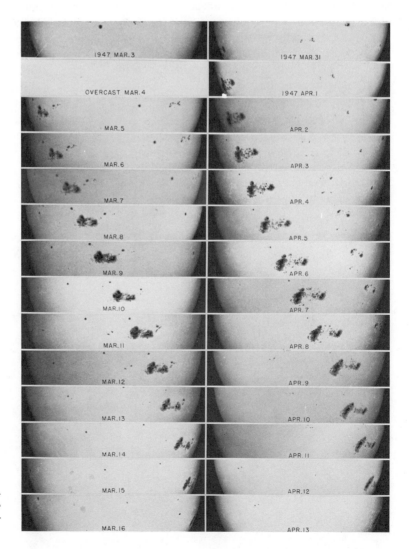

Figure 31.3 Series of photographs showing the motions of sunspots, indicating the solar rotation. (*Mt. Wilson and Las Campañas Observatories*)

limbs (Section 25.2d). From 1887 to 1889, Duner, at Uppsala, determined the solar rotation by spectroscopic methods at latitudes beyond those at which spots are found, and confirmed that the sun rotates most rapidly at low latitudes. Very sensitive Doppler shift measurements over the interval 1973–1977 showed the period to average 25.8 days at the equator, 28.0 days at latitude 40°, and 36.4 days at latitude 80°, in the direction west to east (like the orbital motions of the planets). There are, however, variations in the rotation rate over time at any given latitude, corresponding to speeds of up to a few meters per second. Those variations appear to be correlated with the solar cycle (next section). The rotation may be quite different beneath the surface, as well. Of course the sun, being a fluid gas, need not rotate as a solid body.

The apparent motions of sunspots are not usually straight lines across the sun's disk but rather slight arcs, because the axis of rotation of the sun is not exactly perpendicular to the plane of the ecliptic. The angle of inclination of the solar equator to the ecliptic is about 7°.

31.3 PHENOMENA OF THE SOLAR ATMOSPHERE

In its gross characteristics the sun is quite stable, but the detailed features of its atmosphere are constantly changing.

(a) Photospheric Granulation and Supergranulation

Direct telescopic observation and photography show that the photosphere is not perfectly smooth but has a mottled appearance resembling rice grains—this structure of the photosphere is now generally called *granulation*. Typically, granules are 700 to 1000 km in diameter; the smallest observed are about 300 km across. They appear as bright spots surrounded by narrow darker regions.

The motions of the granules can be studied by the Doppler shifts in the spectra of gases just above them. It is found that the granules themselves are columns of hotter gases arising from below the photosphere. As the rising gas reaches the photosphere it spreads out and sinks down again. The darker intergranular regions are the cooler gases sinking back. The centers of the granules are hotter than the intergranular regions by 50 to 100 K. The vertical motions of gases in the granules have speeds of about 2 or 3 km/s. Individual granules persist for about 8 minutes. The granules, then, are the tops of convection currents of gases rising through the photosphere.

The granules themselves form part of a structure of still larger scale called *supergranulation*. Supergranules are cells that average about 30,000 km in diameter, within which there is a flow of gases from center to edge. The structure that outlines the

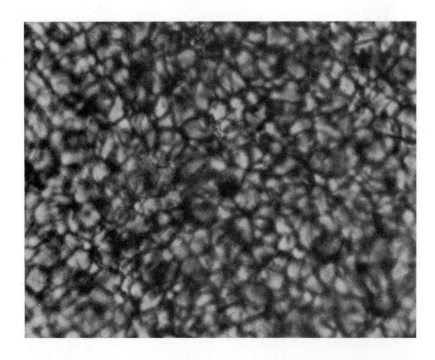

Figure 31.4 A highly magnified photograph of solar granulation. (Sacramento Peak Observatory. Association of Universities for Research in Astronomy, Inc.)

supergranules persists throughout the upper photosphere and chromosphere. Moreover, magnetic fields are concentrated at supergranule boundaries. Most solar astronomers now regard supergranules as being fundamental regions of the photosphere.

In addition to the vertical currents of gases in the granules and the center-toward-edge flow in the supergranules, in each region of the sun (typically up to tens of thousands of kilometers across) the gases rhythmically pulse up and down with speeds of about $\frac{1}{3}$ km/s, taking about 5 minutes for a complete cycle—a phenomenon known as the *five-minute oscillation*.

(b) Sunspots

The most conspicuous of the photospheric features are the *sunspots*. Occasionally, spots on the sun are large enough to be visible to the naked eye, and such spots have been observed for many centuries. Galileo first showed that sunspots are actually on the surface of the sun. In 1774 the Scot Alexander Wilson suggested that spots were "holes" through which we could see past the photosphere into a cooler interior of the sun. William Herschel held a similar view; he imagined that the sun had a cool,

probably inhabited, interior that was surrounded by two cloud layers.

Actually the spots are regions of the photosphere where the gases are up to 1500 K cooler than those of the surrounding photosphere. Sunspots are nevertheless hotter than the surfaces of many stars. If they could be removed from the sun, they would be seen to shine brightly; they appear dark only by contrast with the hotter, brighter surrounding photosphere.

Individual sunspots have lifetimes that range from a few hours to a few months. They are first seen as small dark "pores" somewhat over 1500 km in diameter. Most of them disappear within a day, but a few persist for a week or occasionally much longer. If a spot lasts and develops, it is usually seen to consist of two parts: an inner darker core, the *umbra*, and a surrounding less dark region, the *penumbra*. Many spots become much larger than the earth, and a few have reached diameters of 50,000 km. Frequently spots occur in groups of from two to twenty or more. If a group contains many spots, it is likely to include two large ones, one approximately east of the other, and many smaller spots clustered around the two principal ones. The principal spot to the east is most often the largest one of

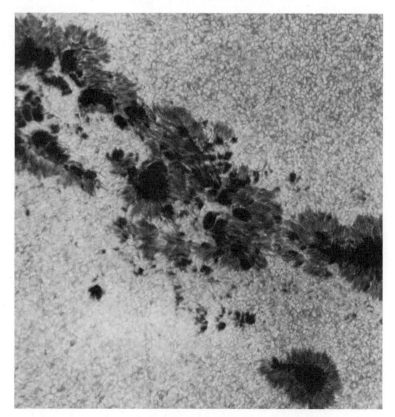

Figure 31.5 An excellent photograph of sunspots and solar granulation in the surrounding photosphere. *(Sacramento Peak Observatory. Association of Universities for Research in Astronomy, Inc.)*

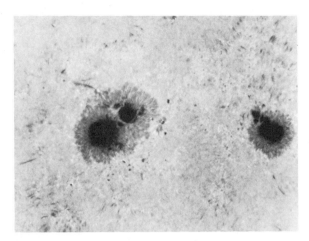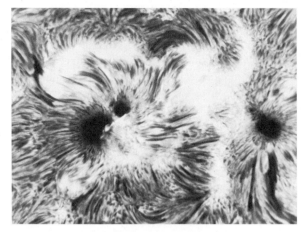

Figure 31.6 Bipolar sunspot group photographed May 21, 1972, at Big Bear Solar Observatory; (a) in white light; (b) in the red light of the hydrogen Hα line. (*California Institute of Technology*)

the group. The largest groups are very complex and may have over a hundred spots. Like storms on the earth, sunspots may move slowly on the surface of the sun, but their individual motions are slow when compared with the solar rotation, which carries them across the disk of the sun.

(c) The Sunspot Cycle

In 1851 Heinrich Schwabe, a German apothecary and amateur astronomer, published an important conclusion he had reached as a result of his observations of the sun over the previous decade. He found that the number of sunspots visible, on the average, varied with a period of about ten years. Since Schwabe's work, the *sunspot cycle* has been clearly established. Although individual spots are short-lived, the total number of spots visible on the sun at any one time is likely to be very much greater during certain periods, the periods of *sunspot maximum*, than at other times, the periods of *sunspot minimum*. Sunspot maxima have occurred at an average interval of 11.1 years, but the intervals between successive maxima have ranged from as little as eight years (from 1830 to 1838) to as long as 16 years (from 1888 to 1904). During sunspot maxima, more than 100 spots can often be seen on the sun at once. During sunspot minima, the sun sometimes has no visible spots. The last maximum (at this writing) was in 1979; the next is expected in 1991.

At the beginning of a cycle, just after a minimum, a few spots or groups of spots appear at latitudes of about 30° on the sun. As the cycle pro-

gresses, the successive spots occur at lower and lower latitudes, until, at the maximum of the cycle, their average latitude is about 15°. Near minimum, the last few spots of a cycle appear at about 8° latitude. About the same time, the next cycle begins with a few spots occurring simultaneously at higher latitudes. Sunspots almost never appear at latitudes greater than 40° or less than 5°. The locations of sunspots on the sun in both the northern and southern hemispheres are related to the sunspot cycle in the same way; however, sunspot activity in one hemisphere may dominate for long periods. We shall return later to the interpretation of the sunspot cycle.

(d) Magnetic Fields on the Sun

As stated in Section 25.2c, a spectral line is usually split up into several components in the presence of a magnetic field, the phenomenon known as the Zeeman effect. In 1908 the American astronomer George E. Hale observed the Zeeman effect in the spectrum of sunspots and found them to possess strong magnetic fields. The magnetic fields observed in sunspots range from 100 to nearly 4000 gauss. This is as great as the field of a good alnico magnet and, moreover, is spread over a region tens of thousands of kilometers across. The magnetic field is present in the region surrounding a spot as well as in the spot itself and persists even after the spot has disappeared.

Whenever sunspots are observed in pairs or in groups containing two principal spots, one of the

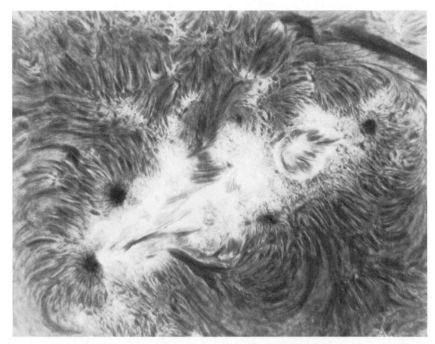

Figure 31.7 An active sunspot group photographed in the red light of the Hα line at Big Bear Solar Observatory on October 11, 1970. *(California Institute of Technology)*

spots usually has the magnetic polarity of a north-seeking magnetic pole, and the other has the opposite polarity. Moreover, during a given cycle, the leading spots of pairs (or leading principal spots of groups) in the northern hemisphere all tend to have the same polarity, while those in the southern hemisphere all tend to have the opposite polarity. During the next sunspot cycle, however, the polarity of the leading spots is reversed in each hemisphere. For example, if during one cycle the leading spots in the northern hemisphere all had the polarity of a north-seeking pole, the leading spots in the southern hemisphere would have the polarity of a south-seeking pole; during the next cycle, the leading spots in the northern hemisphere would have south-seeking polarity and those of the southern hemisphere would have north-seeking polarity. We see, therefore, that the sunspot cycle does not repeat itself as regards magnetic polarity until *two* 11-year maxima have passed. The sunspot cycle is therefore sometimes said to last 22 years, rather than 11.

The strong magnetic fields on the sun are associated with sunspots. When the spots die, their fields spread out towards the poles, contributing to a more general, but far weaker, solar magnetic field, with the polarity of the western spots of each group of the previous cycle. H. W. and H. D. Bab-cock, of the Mount Wilson Observatory and the Hale Solar Observatory, investigated magnetic fields on the sun for many years, and in the 1950s first established the existence of the general polar magnetic field with a strength of a few gauss. Because the sunspot polarity reverses with each cycle, the general magnetic field of the sun also reverses, but in a rather irregular way.

The solar magnetic field is not uniform. The motions of gas in the supergranules, for example, cause the field strength to build up to strengths of 1500 to 2000 gauss at the supergranule boundaries. Magnetism is still not a dominant force in the photospheric layers, but in the chromosphere and corona, where the gas density and pressure are enormously less, the magnetic fields are relatively strong and play an important role in influencing the motions of ionized gases. Far out in the corona magnetic lines of force manifest themselves by organizing ionized coronal gases into streamers, which are easily seen and photographed during total solar eclipses. Low in the corona, magnetic fields guide the motions of ions in solar prominences (see below). The solar magnetic field even extends into interplanetary space and is measured in the vicinity of the earth and other planets with magnetometers carried on space probes. It can even be deduced from changes in the earth's field.

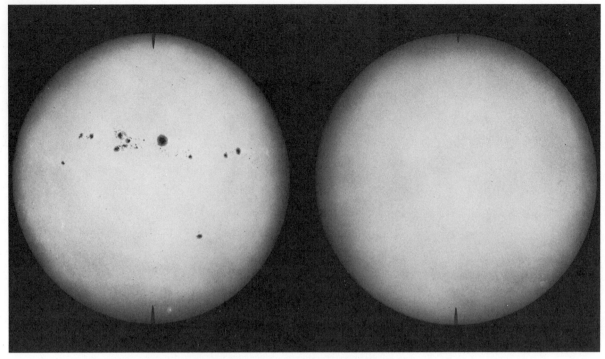

Figure 31.8 Direct photographs of the sun near the time of sunspot maximum *(left)* and near sunspot minimum *(right)*. *(Mt. Wilson and Las Campañas Observatories)*

(e) Filtergrams

In order to see regions of the sun that lie directly above the photosphere, we may observe in spectral regions to which the photospheric gases are especially opaque—at the centers of strong absorption lines such as those of hydrogen and calcium.

In 1892 Hale and, independently, Deslandres in France invented the *spectroheliograph,* with which the sun can be photographed in light comprising only a very narrow spectral region, generally that corresponding to a small part of a particular line of some element. The instrument produces a spectrum of the sun and then isolates the particular line or spectral region by masking off all other light in the spectrum with a second slit. The entire apparatus, spectrograph and slit, is then scanned across the telescopic image of the sun. The light that passes through the slit isolating a small part of the spectrum simultaneously scans across a photographic plate, leaving an image of the sun as it appears in that narrow spectral region. The photograph so obtained is called a *spectroheliogram.*

It is most common in practice to place the slit in the spectrum so that it isolates one of the absorption lines of ionized calcium in the ultraviolet (the K line of ionized calcium) or the Hα line of hydrogen in the red. These spectral lines appear dark when viewed against the rest of the solar spectrum, but they are not completely dark; what light remains in the centers of the lines is emitted from atoms of calcium or hydrogen, respectively, in the chromosphere. Thus, spectroheliograms reveal the appearance of the chromosphere in the light of calcium, or hydrogen. Today there are special filters which pass only light in narrow spectral regions, and now astronomers routinely photograph the sun through such *monochromatic* filters. These photographs are called *filtergrams.*

Spectacular motion pictures have been taken through monochromatic filters, or with spectroheliographs. Time-lapse photographs, in which frames are exposed every few seconds or every few minutes and then run through a projector at normal speed, show in a dramatic way changes that occur in the solar chromosphere.

(f) Plages (Flocculi) and Faculae

Filtergrams in the light of calcium and hydrogen show bright "clouds" in the chromosphere in the magnetic-field regions around sunspots. These

Figure 31.9 The sun in the light of the first Balmer line of hydrogen (Hα). *(Sacramento Peak Observatory. Association of Universities for Research in Astronomy, Inc.)*

bright regions (formerly called *flocculi*—"tufts of wool") are known as *plages*. Calcium and hydrogen plages are also sometimes seen in regions where there are no visible sunspots, but these regions are generally those of higher than average magnetic fields.

The plages are not, of course, concentrations of calcium or hydrogen, but are regions where calcium and hydrogen happen to be emitting more light at the observed wavelengths. These elements are partially ionized throughout most of the visible chromosphere, and some of the atoms emit light as they capture electrons and become neutral (or less ionized) or as those atoms (or ions) cascade down through the various excited energy levels. The plages, then, are regions where some of the atoms of the element observed are changing their states of ionization or excitation and are emitting more light than in the surrounding areas. Plages of hydrogen and calcium usually occur in approximately the same projected regions at the same time.

Plages sometimes emit light at many wavelengths and can be seen in the direct image of the sun. These "white-light" plages are called *faculae* ("little torches") and were first described by Galileo's contemporary Christopher Scheiner. Faculae are seen best near the limb of the sun where the photosphere is not so bright and the contrast is more favorable for their visibility.

(g) Spicules

The chromosphere also contains many small jetlike spikes of gas rising vertically through it. These features, called *spicules,* occur at the edges of supergranule cells; when viewed near the limb of the sun so many are seen in projection that they give the effect of a forest. They show up best when the chromosphere is viewed in the light of hydrogen. They consist of gas jets moving upward at about 30 km/s and rising to heights of from 5000 to 20,000 km above the photosphere. Individual spicules last only ten minutes or so. Through the spicules matter continually flows into the corona. They are now believed to be of fundamental importance in the energy and momentum balance in the solar atmosphere.

(h) Prominences

Among the more spectacular of coronal phenomena are the *prominences.* Prominences have been viewed telescopically during solar eclipses for centuries. They appear as red flamelike protuberances rising

Figure 31.10 Solar spicules, photographed in the light of Hα. (*Sacramento Peak Observatory. Association of Universities for Research in Astronomy, Inc.*)

above the limb of the sun. Prominences can now be viewed at any time on spectroheliograms and filtergrams. Motions of prominences are exhibited in motion pictures. The gross features of some, the *quiescent* prominences, may remain nearly stable for many hours, or even days, and may extend to heights of tens of thousands of km above the solar surface. Others, the more active prominences, move upward or have arches that surge slowly back and forth. The relatively rare *eruptive* prominences appear to send matter upward into the corona at speeds up to 700 km/s, and the most active *surge* prominences may move upward at speeds up to 1300 km/s. Some eruptive prominences have reached heights of over one million km above the photosphere. When seen silhouetted on the disk of the sun, prominences have the appearance of irregular dark filaments.

Superficially, prominences appear to be material ejected upward from the sun, but the motion pictures show that whereas a prominence may grow in size and rise higher and higher above the photosphere, the actual material in the prominence most often appears to move downward in graceful arcs, evidently along lines of magnetic force. Apparently, most prominences form from coronal material that

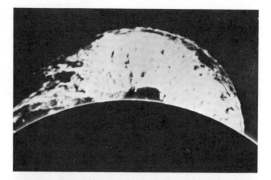

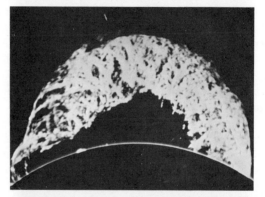

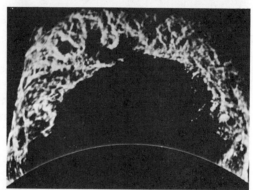

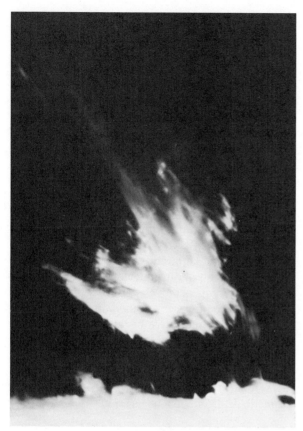

Figure 31.11 A solar prominence 160,000 km high, photographed in the light of Hα at Big Bear Solar Observatory on June 12, 1972. (*Mt. Wilson and Las Campañas Observatories*)

cools and moves downward, even though the disturbance that characterizes the prominence may move upward.

Prominences are cool and dense regions in the corona where atoms and ions are capturing electrons and emitting light. Their origin is unknown, but it is significant that they usually originate near regions of sunspot activity and lie on the boundary between regions of opposite magnetic polarity. Quiescent prominences are supported by coronal magnetic fields, and eruptive prominences evidently result from sudden changes in the magnetic fields. Prominences seem to be further symptoms of the same general disturbances that produce spots and plages, that is, local magnetic fields.

(i) Flares

Occasionally, the chromospheric emission lines (the Hα and ionized calcium lines in particular) in a small region of the sun brighten up to unusually high intensity. Such an occurrence is called a *flare*.

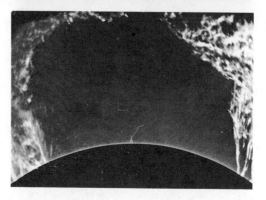

Figure 31.12 Four successive spectroheliograms of the great explosive prominence of June 4, 1946, taken with a coronagraph. The total elapsed time between the first (*top*) and the last (*bottom*) picture was one hour. (*Harvard Observatory*)

A flare is usually discovered on a filtergram that is being made in the light of one of the spectral lines that brighten. It appears as an intensely bright spot on the photograph a few thousand or tens of thousands of kilometers in diameter. Very rarely, the continuous spectrum of that part of the solar surface affected also brightens during a flare. These white-light flares are among the most intense observed. A flare usually reaches maximum intensity a few minutes after its onset. It fades out slowly, and after an interval ranging from a few minutes to a few hours, it disappears. Near sunspot maximum, small flares occur several times per day and major ones may occur every few weeks.

During major flares, an enormous amount of energy is released. The visible light emitted is, of course, very small compared to that from the entire sun, but a flare covering only a thousandth of the solar surface can actually outshine the sun in the ultraviolet. X rays and gamma rays are emitted as well and are observed from space probes and satellites. In addition, matter is thrown out at speeds of 500 to 1000 km/s.

Photographs of the sun at X-ray wavelengths show the corona both outside the disk of the sun, and silhouetted on the sun's disk in hot regions above flares. Moreover, the radio radiation from the sun shows sudden bursts at times of flares. Some of this radiation is nonthermal, showing that high-speed electrons are ejected into coronal magnetic fields, but thermal energy also is emitted from coronal regions near flares. From the intensity of the thermal X-ray radiation we calculate that those regions of the corona have been heated to temperatures near 20 million K. Initially the energy of radio bursts is received only at short wavelengths, but at successively greater intervals of time after the start of the burst, energy is received over longer and longer wavelengths, which shows that the source of the radio energy is rising higher and higher in the corona; the outward velocity of such a source may reach 1500 km/s. About 50 hours later the matter ejected reaches the earth.

Flares are most frequent in the regions of complex sunspot groups with complicated magnetic field structure. Material thrown into the corona from a flare expands and cools and, as ions and electrons recombine, it may emit light and be seen as a loop prominence over the active region of the sun. Loop prominences are the hottest of all prominences and are indicative of rather violent activity in the sun's atmosphere. As is the case with almost all solar phenomena, we do not know the energy source of flares or the mechanism that triggers them, but they are somehow connected with the magnetic fields of sunspot regions.

(j) Theory of the Solar Cycle

There are many solar phenomena that seem to be associated with a particular kind of disturbance in the solar atmosphere. The disturbances are accom-

Figure 31.13 A solar flare at its peak intensity, photographed in the light of Hα on August 7, 1972, at Big Bear Solar Observatory. (*Mt. Wilson and Las Campañas Observatories*)

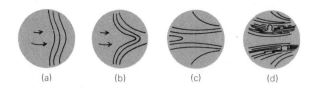

Figure 31.14 Babcock's hypothesis for the solar cycle. (a) The lower loops of the lines of force of the solar magnetic field are buried shallowly beneath the photosphere. (b) and (c) The solar differential rotation stretches the lines out, winding them around the sun. (d) Eventually, the wound-up force lines are buoyed to the surface and break through as active regions. In all figures, north is at the top.

panied by strong magnetic fields, sunspots, plages, prominences, and solar flares. All these phenomena are related to the semiregular sunspot cycle of 11 (or 22) years. Even the shape of the corona varies with the sunspot cycle. During sunspot maximum it is nearly spherical and distended, while at sunspot minimum it is contracted but extends to relatively greater distances near the plane of the solar equator.

The most successful theory yet advanced to account for the solar cycle is that of H. D. Babcock, E. Parker, and others. They suggest that early in the cycle the inner loops of the lines of force of the general solar field are buried a few hundred kilometers below the photosphere, with the outer loops arching through the corona. As the sun rotates, it carries the buried magnetic field with it. However, as we have seen, the sun does not rotate as a solid body, but faster at the equator. Thus at low latitudes the field lines are carried forward faster than at high latitudes, and after a large number of solar rotations, they are stretched out around the sun many times (Figure 31.14). Adjacent field lines, wound ever closer together, build up strong local fields; the lines may even be twisted together by turbulence in the outer solar layers. The matter in

the region of these strong fields expands, lowers in density, and rises, carrying loops of the "cables" of field lines through the surface of the photosphere to form active regions. As the differential rotation of the sun continues to wind up the lines at lower and lower latitudes, the spots occur nearer the equator. Eventually the field lines coalesce and disperse, dissipating the strong fields, and the half cycle is over. Detailed analysis shows that as the local field neutralizes, the general field of the sun reverses, so that spots in the next half cycle are of opposite polarity from the previous one.

We cannot, of course, directly observe or count spots on other stars. However, in the 1970s Olin Wilson, at the Mount Wilson Observatory, was able to estimate chromospheric activity in some other stars by observing changes in the intensity at the centers of certain strong lines in the spectra of those stars—the same lines in which we observe plages on the sun. The periodic variations of those intensities strongly suggest that those other stars go through magnetic cycles similar to the sun's.

(k) The Maunder Minimum

There is considerable evidence that solar activity ceased entirely, or nearly entirely, from 1645 to 1715. During that 70-year interval there were no records of sunspots except for two small groups in 1705; but there are ample reports of sunspots preceding 1645. Furthermore, the solar corona that was visible during the solar eclipses that occurred in those 70 years had no streamers, and reached out only 1' to 3'. This interval of quiescence in solar activity, recently brought to the attention of astronomers by John Eddy, of the High Altitude Observatory in Boulder, Colorado, was first noted by Gustav Spörer in 1887 and by E. W. Maunder in

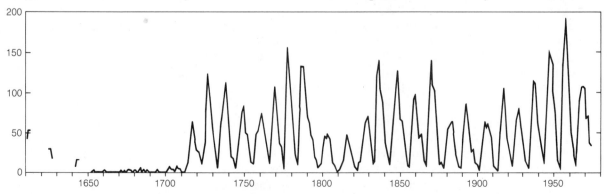

Figure 31.15 The relative numbers of sunspots, as a function of time. Note the absence of sunspots from 1645 to 1715. (*Courtesy John Eddy, National High Altitude Observatory*)

1890, and is now called the *Maunder minimum*. See Figure 31.15, showing the incidence of sunspots over the past four centuries.

31.4 IS THE SUN VARIABLE?

Many stars are known to be variable in size, temperature, and luminosity (Chapter 33). Except for the activity associated with the solar cycle, previously described, the sun is stable enough that no general variability had been detected up through the 1960s. On the other hand, solar physicists have long suspected that the sun might show subtle changes.

There is growing evidence, not yet conclusive, that the sun does change its gross characteristics. The best evidence for variations in the solar luminosity comes from observations obtained with NASA's Solar Maximum Mission Satellite, launched in February 1980. Analysis of solar luminosity measures over a period of 153 days shows almost continuous variations of a tiny fraction of a percent and several dips of up to 0.2 percent, lasting from eight to ten days each. These latter dips are almost certainly due to large sunspot groups blanketing some of the solar radiation, but it is not yet certain that the small-scale changes can be so explained.

One claim for a change in the solar radius is based on a comparison of three solar eclipses—those of 1715, 1976, and 1979. The width of the path of totality depends on the angular size of the sun, and for an observer near the edge of the path, the length of time the sun is completely hidden is very sensitive to his location with respect to that edge. If the sun were slightly larger and the path very slightly narrower, such an observer would miss seeing totality altogether. Sir Edmond Halley organized a large group to observe the eclipse of 1715, which passed through England, and the edge of the path of totality can be located from the reports of the observers. Comparison with the paths of the recent eclipses suggests that the sun may have shrunk in size in that 264-year interval by 0″.34. The uncertainties are fairly large, however, and are sensitive to the location of lunar valleys through which the light of an otherwise eclipsed sun can be glimpsed. Moreover, the precise location of a key observer was not recorded. Consequently, the result is open to question. Even if real, it would suggest that the sun oscillates a bit in size, rather than continually shrinking, for in the latter case the sun would be collapsed in a million years or so.

Yet, the observations are tantalizing and encourage very precise measurements in the years to come. We may learn that the sun, as some predict, is not the pillar of stability we have grown accustomed to expect, and could even be involved with long-term climatic changes on the earth.

31.5 THE SOLAR WIND

In 1852, one year after Schwabe announced his discovery of the solar sunspot cycle, three investigators, Edward Sabine in England, and Rudolf Wolf and Alfred Gautier in Switzerland, independently discovered that sunspot activity was correlated with magnetic storms on the earth. During geomagnetic storms, the earth's magnetic field is disturbed, and the compass needle shows fluctuations. Today we know that long-range shortwave radio interference and displays of the aurora are also correlated with geomagnetic storms and the sunspot cycle. Some investigators are of the opinion that even long-term changes in climate may be connected with solar activity.

These effects are due to the ultraviolet and X-ray radiation from the sun and also to the *solar wind*. In addition to electromagnetic radiation, the sun emits corpuscular radiation in the form of charged atomic particles, mostly protons and electrons. The solar wind is actually an extension of the solar corona in the form of a more or less continuous outflowing of ions and electrons. They have been recorded by instruments carried on artificial earth satellites and space probes; their effect on comet tails has been known for years (Section 20.4d).

The speed of the solar wind near the earth's orbit averages about 400 km/s, and its density is usually from 2 to 10 ions/cm^3. Both the speed and density of the solar wind, however, are highly variable. The density has been observed to be as low as 0.4 and as high as 80 ions/cm^3, and speeds in excess of 700 km/s have been measured.

During solar flares, high-energy particles are sometimes ejected from the sun into space. Some of these, with energies of 10^9 to 10^{10} eV, are in the low-energy cosmic ray range. Thus the sun (at times of flare activity) is an occasional source of weak cosmic rays. These, however, comprise only a tiny fraction of the total cosmic ray influx to the earth.

At the surface of the earth, we are protected from the solar wind and solar cosmic rays by the atmosphere and by the earth's magnetic field. Some of the particles may be trapped temporarily in the earth's magnetic field. Many others are deflected toward the earth's magnetic poles. Only the most energetic of the particles strike the upper atmosphere of the earth above the equatorial regions. In any case, they interact with molecules of the air and do not penetrate directly to the earth's surface.

These charged particles, along with electromagnetic radiation from the sun interacting with the upper atmosphere of the earth, do, however, disrupt the ionized layers of gas in the *ionosphere* (Section 15.2). These ionized layers reflect shortwave radio waves back to the earth, making long-range radio communication between distant stations on the earth possible. When the ionospheric layers are disturbed, they may disrupt the reflection of radio waves transmitted from the ground; this results in radio "fadeouts."

The rain of particles impinging upon the upper atmosphere and on the radiation belts of the earth is also responsible for the aurora (northern and southern lights). As the particles strike atoms and molecules in the upper atmosphere, they excite them. Radiation from the ions and atoms in the atmosphere gives rise to the auroral emission of light. The most spectacular auroras occur at altitudes of 75 to 150 km.

31.6 THE SOLAR INTERIOR; STELLAR STRUCTURE

One circumstance that greatly facilitates the computation of conditions in the interiors of the sun and stars is that stars, in most cases at least, are completely gaseous throughout. Not only are the temperatures too high to permit molecules to exist in stellar interiors, but even the atoms are almost completely ionized. Consequently, the overwhelming majority of the particles of which stars are made are free electrons and atomic nuclei, and most of the latter are simple protons. When we recall (Section 1.1) how extremely tiny these particles are compared with the sizes of neutral atoms, we can see that even in stars where the gases are compressed to enormous densities, there is mostly empty space between the electrons and atomic nuclei. For this reason the idealized gas laws, described below, hold throughout the interiors of most stars to a high degree of accuracy.

(a) The Perfect Gas Law

The particles that comprise a gas are in rapid motion, frequently colliding with each other and with the walls of the container of the gas. This constant bombardment is the *pressure* of the gas. The pressure is greater, the greater the number of particles within a given volume of the gas, for of course the combined impact of the moving particles increases with their number. The pressure is also greater the faster the molecules or atoms are moving; since their rate of motion is determined by the temperature of the gas, the pressure is greater the higher the temperature.

Most students have run across these concepts in high school, in the form of Boyle's law, which states that the pressure of a gas at constant temperature is inversely proportional to the volume to which it is constrained (that is, is proportional to its density), and Charles' law, which states that the pressure (at constant volume) is proportional to the temperature of the gas. These two ideas combine to give us the *perfect gas law* (also called the *equation of state* for a perfect gas). The perfect gas law provides a mathematical relation between the pressure, density, and temperature of a perfect, or ideal, gas (one in which intermolecular or interatomic forces can be ignored) and states that the pressure is proportional to the product of the density and temperature of the gas. The gases in most stars closely approximate an ideal gas; thus, they must obey this law. The exceptions are very massive stars, where radiation pressure (Section 20.4d) can play an important role, and collapsed stars or the collapsed cores of stars, where the matter is *degenerate*. We shall describe degeneracy in Chapter 34.

(b) Hydrostatic Equilibrium

The sun, like the majority of other stars, is *stable*, that is, neither expanding nor contracting. Such a star is said to be in a condition of *equilibrium;* all the forces within it are balanced, so that at each point within the star the temperature, pressure, density, and so on, are maintained at constant values. We shall see (Chapters 33 and 34) that even these stable stars, including the sun, are changing as they evolve, but such evolutionary changes are so gradual that to all intents and purposes the stars are still in a state of equilibrium.

The mutual gravitational attraction between the masses of various regions within a star produces tremendous forces that tend to collapse the star toward its center. Yet, since stars like the sun have remained more or less unchanged for billions of years, the gravitational force that tends to collapse a star must be exactly balanced by a pressure from within. Most of it is the pressure of the gases themselves, although in some very luminous stars the pressure of radiation also contributes appreciably.

If the internal pressure in a star were not great enough to balance the weight of its outer parts, the star would collapse somewhat, contracting and building up the pressure inside. If the pressure were greater than the weight of the overlying layers, the star would expand, thus decreasing the internal pressure. Expansion would stop, and equilibrium would be reached, when the pressure at every internal point again equaled the weight of the stellar layers above that point. An analogy is an inflated balloon, which will expand or contract until an equilibrium is reached between the excess pressure of the air inside over that of the air outside and the tension of the rubber. This condition is called *hydrostatic equilibrium*. Stable stars are all in hydrostatic equilibrium; so are the oceans of the earth, as well as the earth's atmosphere. The pressure of the air keeps the air from falling to the ground.

(c) Minimum Pressures and Temperatures in the Solar Interior

For mathematical purposes, we can regard the sun (or a star) as being composed of a large number of concentric spherical shells (like the layers in an onion). The sun is not actually stratified, of course; we speak of these shells in the same sense that we speak of levels in the ocean. Now suppose we knew how matter is distributed within the sun—that is, what fraction of its mass is included within each shell. Since the weight of a shell is the gravitational attraction between it and all the underlying layers, we could then calculate the weight of each shell. From the condition of hydrostatic equilibrium, we could next calculate how the pressure must increase downward through each shell to support its weight. At the surface of the sun, where there are no overlying layers of stellar matter, the pressure is zero.

By simply adding up the increases of pressure through the successive layers inward, we would be able to find the pressure at each point within the sun, in particular at its center. Using the pressures and densities thus determined at all points along the radius of the sun, we could then find the corresponding temperatures from the perfect gas law. In other words, if we only knew how the material within the sun is distributed, we would be able to calculate the density, pressure, and temperature at all its internal points.

It is not known in advance, of course, how the matter in a particular star is distributed. On the other hand, some ways that it is *not* distributed can be specified. Internal gravity, for example, must force the gases comprising the sun into higher and higher compression at deeper and deeper levels in its interior. The material is expected to show high central concentration; the densities of outer layers would certainly not exceed those of inner layers. To assume that the matter in a star is distributed with uniform density, therefore, would certainly be to underestimate its central compression, and the values calculated for its internal pressures and temperatures would certainly be lower than the true values. Here, then, is a method by which *lower limits* can be found for the pressures and temperatures of the solar interior.

Thus, with only the assumption of hydrostatic equilibrium and a knowledge of the perfect gas law, it is possible to learn something of the conditions inside the sun. We find that the mean pressure is at least 500 million times the sea-level pressure of the earth's atmosphere, that the central pressure is at least 1.3×10^9 times that of the earth's atmosphere, and that the mean temperature is at least 2.3 million Kelvins. Since these pressures and temperatures would exist if the sun were uniform in density, the actual values must be much higher. Under such conditions all elements are in the gaseous form, and the atoms cannot be combined into molecules. Moreover, most of the atoms are almost completely ionized—that is, stripped of their electrons (Section 10.6e). These electrons, freed from their parent atoms, become part of the gas itself, moving about as individual particles.

(d) Thermal Equilibrium

From observation we know that electromagnetic energy flows from the surfaces of the sun and stars.

According to the second law of thermodynamics, heat always tries to flow from hotter to cooler regions. Therefore, as energy filters outward toward the surface of a star, it must be flowing from inner hotter regions. The temperature cannot ordinarily decrease inward in a star, or energy would flow in and heat up those regions until they were at least as hot as the outer ones. We conclude that the highest temperature occurs at the center of a star and that temperatures drop to successively lower values toward the stellar surface.* The outward flow of energy through a star, however, robs it of its internal heat and would result in a cooling of the interior gases were that energy not replaced. There must therefore be a source of energy within each star.

If a star is in a steady state (that is, in hydrostatic equilibrium and shining with a steady luminosity), the temperature and pressure at each point within it must remain approximately constant. If the temperature were to change suddenly at some point, the pressure would similarly change, causing the star to contract suddenly, or to expand, or otherwise to deviate from hydrostatic equilibrium. Energy must be supplied, therefore, to each layer in the star at just the right rate to balance the loss of heat in that layer as it passes energy outward toward the surface. Moreover, the rate at which energy is supplied to the star as a whole must, at least on the average, exactly balance the rate at which the whole star loses energy by radiating it into space; that is, the rate of energy production in a star is equal to the luminosity. We call this balance of heat gain and heat loss for the star as a whole and at each point within it the condition of *thermal equilibrium*.

(e) Heat Transfer in a Star

There are three ways in which heat can be transported: by *conduction*, by *convection*, and by *radiation*. The rate at which heat passes through gases by conduction, however, is so low that this mode of transfer can be ignored in stellar interiors, unless the gas is degenerate (Chapter 34).

Stellar convection occurs as currents of gas flow in and out through the star. While these convection currents travel at moderate speeds and do not upset the condition of hydrostatic equilibrium, they nevertheless carry heat outward through a star very efficiently. However, convection currents cannot be maintained unless the temperatures of successively deeper layers in a star increase rapidly in relation to the rate at which the pressures increase inward. Convection does occur, nevertheless, in certain parts of many stars, and convection currents may travel completely through some of the least luminous stars.

Unless convection occurs, the only significant mode of energy transport through a star is by electromagnetic radiation, which gradually filters outward as it is passed from atom to atom. However, radiative transfer is not an efficient means of energy transport, because under the conditions that prevail in stellar interiors gases are very opaque—that is, a photon does not go far before it is absorbed by an atom (typically, in the sun, about 1 cm). The energy absorbed by atoms is always reemitted, to be sure, but most of it is reemitted in random directions. A photon that is traveling outward in a star when it is absorbed has almost as good a chance of being reradiated back toward the center of the star as toward its surface. A particular quantity of energy being passed from atom to atom, therefore, zigzags around in an almost random manner and takes a long time to work its way from the center of the star to the surface; in the sun, the time required is of the order of a million years.

The measure of the ability of a gas to absorb radiation is called its *opacity*. It should be no surprise that the gases in the sun are opaque. If they were completely transparent, we would be able to see all the way through the sun. We have discussed earlier (Section 10.5) the processes by which atoms and ions can interrupt the flow of energy—such as by becoming ionized and by bremmstrahlung (free-free transitions). In addition, individual electrons can scatter radiation helter-skelter. For a given temperature, density, and composition of a gas, all of these processes can be taken into account, and the opacity can be calculated. The computations are very complicated, so require large computers.

Once the opacity is known, we can find how each layer or shell or the sun or a star impedes the outward flow of radiation. Of course there is such a net outward flow, or the star would have no luminosity. Thus from the opacity we calculate how the temperature must increase inward through the shell to force the observed radiation out, and thereby

---

*The high temperature of the sun's corona may therefore appear to be a paradox. The actual heat energy in the sun's corona is relatively small because the corona is a highly rarefied gas. Its high temperature is believed to be maintained by shock waves or by some other process that would not exist for a gas in thermodynamic equilibrium.

learn the temperature distribution throughout the interior.

If the temperature difference across some regions of a star should be high enough to support convection, convection currents, rather than radiation, carry most of the energy. Within those regions the variation of temperature with depth is determined by the expansion of outward-moving masses of gas and the contraction of inward-moving ones. Here again, knowledge of the energy transport mechanism within a star makes possible calculation of the temperature distribution.

In this section we have described some of the ways we probe the interiors of the sun and stars. The story is not yet complete, however, for the various physical variables (such as temperature, density, and pressure) are interdependent, and so far we have not specified enough conditions to separate and solve for all the unknowns. To do so, we need to know the source of energy of the sun and stars. This is the subject of the next chapter, so we shall take a short recess from our consideration of the structure of a star while we explore what makes it shine.

EXERCISES

1. Describe the principal spectral features of the sun.

2. (a) What is the distance to the sun in parsecs? (b) What is the distance modulus of the sun? (c) How many times farther away would the sun be if it were removed to a distance of 10 pc? (d) How many times fainter would the sun appear at 10 pc?

3. How might you convince an ignorant friend that the sun is not hollow?

4. Give at least three good arguments that refute the view proposed by Herschel that the sun has a cool interior that is inhabited.

5. Suppose an eruptive prominence rises at 150 km/s. If it did not change speed, how far from the photosphere would it extend in three hours?

6. Would the material in the prominence in Exercise 5 escape the sun? Why? (See Section 5.4).

7. Suppose you were to take two photographs of the sun, one in light at a wavelength centered on a strong absorption line, and the other at a wavelength region in the continuum away from strong lines. In which photograph would you be observing deeper, hotter layers? Why?

8. From the Doppler shifts of the spectral lines in the light coming from the east and west limbs of the sun, it is found that the radial velocities of the two limbs differ by about 4 km/s. Find the approximate period of rotation of the sun.

9. If the rotation period of the sun is determined by observing the apparent motions of sunspots, must any correction be made for the orbital motion of the earth? If so, explain what the correction is and how it arises. If not, explain why the earth's orbital revolution does not affect the observations.

10. If the corona, which is outside the photosphere, has a temperature of 1,000,000 K, why do we measure a temperature 6000 K for the surface of the sun?

11. Show with a diagram why the duration of totality is extremely sensitive to the distance of the observer from the edge of the eclipse path.

12. From the data in Section 31.5, find how long it takes solar wind particles, on the average, to reach the earth from the sun.

13. Give some everyday examples of hydrostatic equilibrium. It is known that the pressure in a container of water increases with depth in the container. Is this a consequence of hydrostatic equilibrium? Explain. Compare the pressure-depth relation in water with that in the earth's atmosphere. Why is the case much simpler for water?

14. If the atmospheric pressure were the same on two different days, but if one day were much hotter than the other, what could you say about the relative density of the air on the two days?

15. If, in a vacuum chamber, the pressure is only one millionth of sea-level pressure, how does the density of the gas in the chamber compare with the average density of air at sea level?

16. Give everyday examples of convection and radiation of heat through a gas.

32

STELLAR ENERGY: THE ATOMIC NUCLEUS

The rate at which the sun emits electromagnetic radiation into space, and thus the rate at which energy must be generated within it, is about 4×10^{33} ergs/s (Section 24.5b). Moreover, the power output of the sun has been about the same throughout recorded history and, according to geological evidence, not very different since the formation of the earth thousands of millions of years ago. Our problem is now to find what sources can provide the gigantic amounts of energy required to keep the stars like the sun shining for so long.

32.1 Thermal and Gravitational Energy

Two large stores of energy in a star are its internal heat, or *thermal energy*, and its *gravitational energy*. The heat stored in a gas is simply the energy of motion (kinetic energy) of the particles that compose it. If the speeds of these particles decrease, the loss in kinetic energy is radiated away as heat and light. This is how a hot iron cools after it is withdrawn from a fire (except that the atoms in a solid vibrate within a crystalline structure, rather than moving freely, as in a gas).

Because a star is bound together by gravity, it has gravitational potential energy, as does a star cluster (Chapter 30). If the various parts of a star fall closer together, that is, if the star contracts, it converts part of its potential energy into heat, some of which can be radiated away. About the middle of the 19th century, the physicists Helmholtz and Kelvin postulated that the source of the sun's luminosity was indeed the conversion of part of its gravitational potential energy into radiant energy.

The sun cannot be infinitely old, of course, for no source of energy can last forever. Sometime in the past the sun must have formed as the solar nebula gradually fell or gravitated together giving up its potential energy. It can be shown by thermodynamics that about half the potential energy released by a contracting star goes into radiation (or luminosity) and the other half goes into heating up its interior. Thus the internal heat or thermal energy of a star is numerically equal to about half the potential energy it has given up in its contraction.

Helmholtz and Kelvin showed that because of its enormous mass, the sun need contract only extremely slowly to release enough gravitational potential energy to account for its present luminosity. In fact, over the time span of recorded history, the decrease in the sun's size resulting from its contraction would be so negligible as to escape detection. It seemed to these researchers, therefore, that the sun's gravitational and thermal energies were sufficient to keep it shining for an extremely long time, and were certainly the source of its power.

The amount of potential energy that has been released since the presolar cloud began to contract is of the order of 10^{49} ergs. This is the amount, according to the Helmholtz and Kelvin theory, that it could have converted to thermal energy and luminosity. Since the present luminosity of the sun is 4×10^{33} ergs/s, or about 10^{41} ergs/yr, its contraction can have kept it shining at its present rate for a period of the order of 100 million years. It is only within the present century that it has been learned that the earth, and hence the sun, has an age of at least several thousand million years, and therefore that the sun's gravitational energy is grossly inadequate to account for the luminosity it has generated over its lifetime.

Einstein's special theory of relativity shows that there is an equivalence between mass and energy and that one can be converted to the other. About 1928 it was suggested that the conversion of matter to energy might account for the sun's luminosity. Such conversion is possible through nuclear reactions. These reactions involve the fundamental particles of physics, the study of which is one of the most exciting frontiers in modern physics. Before proceeding, therefore, let us digress briefly for a quick review of that field.

32.2 THE ELEMENTARY PARTICLES AND FORCES

Our present physical description of nature reduces to the fundamental interactions among the basic entities of mass and energy—the fundamental particles. The interactions are manifestations of the forces of nature, of which four are known (Chapter 1): strong, weak, electromagnetic, and gravitational. The particles are the things on which the forces act, as well as particles by whose exchange the forces are transmitted.

(a) The Forces of Nature

The known forces of nature were alluded to in Chapter 1. Their properties are summarized in Table 32.1. Gravitation and electromagnetism are both long-range forces whose strengths drop with the square of the distance. Gravitation, however, acts on all particles, while the elctromagnetic force acts only between charged particles. Because bulk matter is usually electrically neutral, the most important manifestation of the electromagnetic force is in binding together the charged parts of atoms, or in binding atoms together in molecules or in solids,

Figure 32.1 Aerial view of the European Center for Nuclear Research (CERN), a major internationally operated installation, located in Geneva, for carrying out experimental probes of the atomic nucleus. *(Photo CERN)*

Table 32.1 **The Forces of Nature**

FORCE	RELATIVE STRENGTH	RANGE	INTERACTING PARTICLES	EXCHANGE PARTICLE	IMPORTANT ASTRONOM- ICAL APPLICA- TIONS
gravitation	10^{-38}	universe	all	graviton	planets, stars, galaxies and the universe
electromagnetic	1	usually effective over 10^{-8} cm	all charged	photon	atoms, molecules, and solids
weak	10^{-12}	10^{-13} cm	leptons	intermediate vector bosons (?)	big bang; stellar energy sources
strong	10^2	10^{-13} cm	hadrons	π mesons	big bang; stellar energy sources

and the interacting particles are separated only by atomic dimensions (10^{-8} cm). (We have seen, however, that magnetism can be important over at least the dimensions of the Galaxy.)

The strong force (or strong nuclear force) binds together the heavy particles in the atomic nucleus and is effective only over nuclear dimensions (10^{-13} cm). We shall see that most (all?) heavy particles are unstable, in that they decay (change) into other particles spontaneously. Often these decays occur over extremely short times (typically, 10^{-23} s), and these also involve the strong force. Other decays, however, are much slower (typically, 10^{-10} s, but sometimes very much slower), and involve the weak (or weak nuclear) force. These weak interactions also always accompany the formation of a neutrino. The weak force, like the strong, is effective only over nuclear dimensions.

It is now believed that each of the four forces is mediated through the exchange of certain particles. The exchange particle for the electromagnetic force is the familiar photon, and that for the gravitational force is the hypothetical *graviton* (not yet observed). Both photons and gravitons are massless. The exchange particle for the strong force is the neutral π meson, with a mass of 264 times that of the electron. The exchange particle for the weak force has not yet been observed but is believed to be very heavy—in excess of 150,000 times the mass of the electron.

(b) Elementary Particles

The most familiar of the other particles are the proton, neutron, and electron, which, as we have seen, are the constituent particles of ordinary atoms. But each of these can exist as a separate entity, as in cosmic rays, the solar wind, in laboratory particle accelerators, or, for that matter, in the ionized gases that make up the bulk of stars. We have learned in the 20th century, however, that these are by no means *all* the particles that exist.

First, for each kind of particle, there is a corresponding *antiparticle*. If the particle carries a charge, its anti has the opposite charge. The antielectron is the *positron*, of the same mass as the electron but positively charged. The antiproton has a negative charge. The antineutron, like the neutron, has no charge, but interacts with other matter opposite to the way the neutron does. Some particles—among them the photon and the graviton—are their *own* antiparticles; for them the particles and antiparticles are identical. Whole atoms could exist of positrons, antiprotons, and antineutrons. They constitute what is called *antimatter*. Such atoms do not exist around here because when a particle comes in contact with its antiparticle the two annihilate, turning into energy. Antimatter in our world of ordinary matter, therefore, is highly unstable (in large doses, it would be mighty dangerous!), but individual antiparticles are found in cosmic rays and can be formed in the laboratory.

In addition, hundreds of other kinds of particles are formed in the nuclear reactions produced in the laboratory when particles are smashed into each other at very high speeds. (One way of observing colliding particles and their debris is from the tracks they leave in bubble chambers—Section 10.6c.) The host of particles now known are classed into two groups: the *leptons* (light-weights) and *hadrons* (heavy-weights). Electrons are examples of leptons. Hadrons are further subdivided into *baryons*, which include the proton and neutron, and *mesons*, which are intermediate in mass and include a particle known as the π-*meson*, or *pion*, whose mass is about 270 times that of the electron. Pions can be positive, negative, or neutral; their antis have the opposite sign, except the neutral pion, which, like the photon, is its own antiparticle. Baryons make up the nuclei of atoms.

Most hadrons are extremely unstable and decay into other particles in very short times. The neutron is stable in the atomic nucleus, but when isolated it decays with a half-life of about 11 minutes into a proton, an electron, and an antineutrino. Most hadrons decay very much more rapidly than the neutron.

In contrast to the hundreds of known hadrons, only six leptons (and their antis) are known. They consist of three kinds of electrons—the familiar one, the heavier muon (which we encountered in Chapter 13), and the recently discovered, much heavier tau—and associated with each kind of electron is a neutrino. The muon and tau are unstable and decay eventually to electrons and neutrinos.

The existence of neutrinos was originally postulated in 1933 by physicist Wolgang Pauli to account for small amounts of energy that appeared to be missing in certain nuclear reactions. They were presumed to be massless and to move with the speed of light. They interact very weakly with other matter and so are very difficult to detect; most of them pass completely through a star or a planet

without being absorbed. Yet, the neutrinos associated with the electron and the muon have now been detected. The third kind of neutrino, that associated with the tau electron, has not yet been observed but is presumed to exist. Each of the three neutrinos has a corresponding antineutrino.

Recent experiments suggest that neutrinos may have a tiny mass and hence do not travel quite with the speed of light. This conclusion is not generally accepted because the experiments are very difficult, and the results are open to various interpretations. If neutrinos do turn out to have mass, however, it could have interesting consequences for cosmology (Chapters 38 and 39).

The properties of a few of the principal elementary particles are summarized in Table 32.2.

(c) Quarks

The existence of hundreds of kinds of hadrons suggests that they may not all be fundamental particles. A substantial simplification and unification was provided by an idea suggested independently in 1963 by Murray Gell-Mann and George Zweig, both at the California Institute of Technology. These physicists postulated that all hadrons are composed of various combinations of a small number of more fundamental particles called *quarks*. (The name comes from James Joyce's *Finnegan's Wake*: "Three quarks for Muster Mark . . . .")

Originally three kinds of quarks were postulated: "up," "down," and "strange," each with its corresponding antiquark. Baryons are made of three quarks each, and antibaryons of three antiquarks.

Table 32.2 **Properties of Some of the Elementary Particles**

FAMILY	PARTICLE	SYMBOL	RELATIVE MASS	CHARGE	ANTIPARTICLE SYMBOL	CHARGE
	graviton	—	0	0	same	0
	photon	γ	0	0	same	0
leptons	electron	e	1	−1	e^+ (positron)	+1
	electron neutrino	ν_e	0(?)	0	$\bar{\nu}_e$	0
	muon	μ	207	−1	μ^+	+1
	muon neutrino	ν_μ	0(?)	0	$\bar{\nu}_\mu$	0
	tau	τ	3500	−1	τ^+	+1
	tau neutrino	ν_τ	0(?)	0	$\bar{\nu}_\tau$	0
hadrons baryons	proton	p	1836	+1	$\bar{p}$	−1
	neutron + many others	n	1839	0	$\bar{n}$	0
mesons	pion	π^+	273	+1	π^-	−1
		π^0	264	0	same	0
	+ others	π^-	273	−1	π^+	+1

A meson is made of one quark and its own anti-quark in combination. Since the introduction of the theory, evidence has been found for two additional quarks: "charm" and "bottom." It is generally expected that a sixth quark, "top," will eventually be found to bring the total number equal to the number of leptons.

The different kinds of quarks are referred to as different "flavors." In addition, each quark is presumed to come in three different "colors"—three states otherwise indistinguishable from each other. (The terms "flavor" and "color" as used here, have, of course, nothing to do with the common usage of those words.) The quarks making up each baryon must represent each of the three colors. Unlike other elementary particles, quarks have fractional charge: up and charm quarks have a charge of $+\frac{2}{3}$, down and strange of $-\frac{1}{3}$, and bottom and top (if it exists) of $+\frac{1}{3}$. Quarks of the many various combinations of possible colors and flavors then can account for the hundreds of known hadrons.

Isolated quarks have never been detected in the laboratory (or at least their detection has not been confirmed), but strong evidence for their existence comes indirectly from the way the hadrons interact with other particles. It is postulated that quarks are bound together by forces that increase with separation, so that they can never be separated from their hadrons—or if they were separated with the application of very great energy, a pair of new quarks would be produced, creating two hadrons from one. The hypothetical exchange particle that binds quarks is called the *gluon*, and it would provide the underlying basis for the strong force.

(d) GUTs

At the forefront of theoretical physics is the speculation that the forces of nature are different manifestations of the same thing, and that there is a basic unity to all particles and forces; the concept is an extension of Einstein's dream of finding a unified field theory. Already there has been progress in developing such a theory—in particular in achieving a partial unification of the weak and electromagnetic forces. The theories that lead to the ultimate grand unity are called *grand unified theories (GUTs)*.

One of the predictions of the grand unified theories is that baryons are not really fundamental and can decay to leptons. The only baryon known to be stable to present experimental accuracy is the proton (to which other baryons decay). Modern exper-

iments show that if the proton *does* decay it can be only after an average time of 10^{30} years. But the lifetime for the proton predicted by GUTs is about 10^{31} years. A human would have to live well over a hundred years before a single proton in his body had an even chance of decaying—not very exciting fireworks. Yet experiments are now being prepared with which it is hoped to detect proton decay if the proton's lifetime does not exceed 10^{33} years. It will be exciting to see if the prediction of the theory is realized.

32.3 NUCLEAR ENERGY

It is hadrons (protons and neutrons) that make up an atomic nucleus. In the nuclei of most atoms, the number of neutrons is roughly equal to the number of protons, but the most common kind of atom of hydrogen contains one proton but no neutron in its nucleus. Atoms of the same chemical element (whose nuclei have the same number of protons) but with different numbers of neutrons are said to be different *isotopes* of that element. Thus *deuterium* is an isotope of hydrogen whose atoms contain nuclei of one proton and one neutron each, and *tritium* is an isotope of hydrogen whose atomic nuclei have one proton and two neutrons each.

(a) Binding Energy

Just as the gases give up gravitational potential energy when they come together to form a star, so particles release energy in uniting to form an atomic nucleus. This energy given up is called the binding energy of the nucleus.

The binding energy is greatest for atoms with a mass near that of the iron nucleus, and it is less both for the lighter and heavier atoms. Thus heavy nuclei, like those of atoms of lead and uranium, are held together with less energy per atomic mass unit than iron is, and so are the lighter nuclei, like carbon and lithium. In general, therefore, if light atomic nuclei come together to form a heavier one (up to iron), energy is released; this is called *nuclear fusion*. On the other hand, if heavy atomic nuclei can be broken up into lighter ones (down to iron), an increase in the total binding energy results, with the release of that much potential energy; this is called *nuclear fission*. Nuclear fission sometimes occurs spontaneously, as in natural radioactivity.

Mass and energy, of course, are equivalent, so the binding energy must correspond to a *mass defect*. Indeed, we find that the mass of every nucleus (other than the simple proton nucleus of hydrogen) is less than the sum of the masses of the nuclear particles that are required to build it; this slight deficiency in mass is always only a small fraction of a mass unit (the unit of mass, u, is $^1/_{12}$ the mass of an atom of carbon). The mass defect per nucleon (nuclear particle) is greatest for the nucleus of the iron atom and is less for both more massive and less massive nuclei. A *nuclear transformation* is a buildup of a heavier nucleus from lighter ones, or a breakup of a heavier nucleus into lighter ones. In any such nuclear transformation, if the mass defect increases the equivalent amount of energy is released. That energy, of course, is the difference in mass defect times the square of the speed of light. On the other hand, if, in the nuclear transformation, the mass defect *decreases*, a corresponding amount of energy must be put into the system.

(b) Nuclear Energy in Stars

It was suggested about 1928 that the energy source in stars might be fusion of light elements into heavier ones. Since hydrogen and helium account for about 98 or 99 percent of the mass of most stars, we logically look first to these elements as the probable reactants in any such fusion reaction. Helium atoms are about four times as massive as hydrogen atoms, so it would take four atoms of hydrogen to produce one of helium. The masses of hydrogen and helium atoms are 1.007825 u and 4.00268 u, respectively. Let us compute the difference in initial and final mass. (Here we include the mass of the entire atoms, not just the nuclei, because the electrons are involved as well, even though, in stellar interiors, the hydrogen and helium are completely ionized. When hydrogen is converted to helium, two positrons are created in the nuclear reactions, and these annihilate with two free electrons, adding to the energy produced.)

$$4 \times 1.007825 = \quad 4.03130 \text{ u (mass of initial}$$
$$\text{hydrogen atoms)}$$
$$\underline{-4.00268} \text{ u (mass of final}$$
$$\text{helium atom)}$$
$$0.02862 \text{ u (mass lost in the}$$
$$\text{transaction)}$$

The mass lost, 0.02862 u, is 0.71 percent of the mass of the initial hydrogen. Thus if 1 g of hydro-

gen turns into helium, 0.0071 g of material is converted into energy. The velocity of light is 3×10^{10} cm/s, so the energy released is

$$E = 0.0071 \times (3 \times 10^{10})^2$$
$$= 6.4 \times 10^{18} \text{ ergs.}$$

This 6×10^{18} ergs is enough energy to raise the 5-m telescope 150 km above the ground.

To produce the sun's luminosity of 4×10^{33} ergs/s some 600 million tons of hydrogen must be converted to helium each second, with the simultaneous conversion of about 4 million tons of matter into energy. As large as these numbers are, the store of nuclear energy in the sun is still enormous. Suppose half of the sun's mass of 2×10^{33} g is hydrogen that can ultimately be converted into helium, then the total store of nuclear energy would be 6×10^{51} ergs. Even at the sun's current rate of energy expenditure, 10^{41} ergs/year, the sun could survive for more than 10^{10} years.

There is little doubt today that the principal source of energy in stars is thermonuclear reactions. Deep in the interiors of stars, where the temperatures range up to many millions of degrees, nuclei of atoms are changing from one kind to another with an accompanying release of energy. The most important of these changes is the conversion of hydrogen to helium.

(c) Rate of Nuclear Reactions

For a nuclear reaction to occur, the nuclei of the reacting atoms must first collide with each other. As a result of this collision, a temporary "compound" nucleus is formed which either turns into a single nucleus of a different kind of atom from those originally taking part in the collision, or breaks into two or more less massive nuclei, again of different atomic types from those of the original reactants. How rapidly such events occur depends on (1) how fast the nuclei in the gas are moving (which depends on the temperature), (2) what the effective sizes of the nuclei are (which is computed from quantum theory for the simplest kinds of nuclei and measured in the nuclear physics laboratory for the more complicated ones), (3) how close together the nuclei are (which depends on the density of the gas), and (4) the probability that a particular compound nucleus that is formed will break into the relevant kinds of new nuclei (which is also determined either from theory or laboratory experiment).

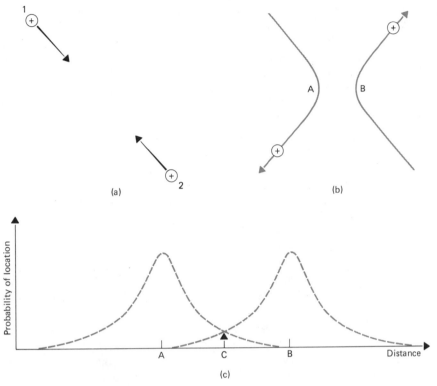

Figure 32.2 An encounter of two protons.

With a rigorous mathematical treatment of the factors discussed in the preceding paragraph, nuclear physicists combine theory and experimental data to derive formulas that predict the rate of nuclear-energy production. Such formulas enable astronomers to predict the rate of energy release in a given region of a star in terms of the chemical composition of that region, its temperature, and the gas density. Since the total rate of energy release from a star (its luminosity) is known, the formulas obtained from nuclear physics give new information about the physical conditions in the stellar interior, which, when combined with our knowledge of the conditions of hydrostatic and thermal equilibrium, provide additional clues about the structure of the star.

Under conditions that prevail at the centers of most stars, there are two chains of nuclear reactions possible by which hydrogen can be converted to helium. One is the *carbon cycle*, also called the *carbon-nitrogen-oxygen (CNO) cycle*, in which carbon, nitrogen, and oxygen nuclei are involved in collisions with hydrogen nuclei (protons), eventually ending with carbon again and a new helium nucleus. The CNO cycle is important at temperatures above 15×10^6 K. At temperatures that prevail in the sun and in less massive stars, most of the en-

ergy is produced by what is called the *proton-proton* chain. Here, by any of several possible branches, protons collide directly to build into helium nuclei. The details of the more important series of nuclear reactions that occur in stars are given in Appendix 8.

It happens that the proton-proton chain, very important in the sun, begins with a most improbable event: the collision of two protons resulting in the formation of a nucleus (a *deuteron*) of the heavy isotope of hydrogen called *deuterium*. Usually the formation of a compound nucleus of two protons simply breaks up into two protons again, rather than ejecting a positron and turning into a deuteron, and very many compound nuclei must form to produce appreciable amounts of deuterium. But even at the high temperatures of stellar interiors (up to 15×10^6 K in the sun), it is extremely hard for two positively charged nuclei to come together to undergo any kind of reaction.

The reason is that the nuclei (protons, for example) are not moving fast enough to overcome their mutual electrostatic repulsion (due to their like positive charges) and come in contact. One might not expect nuclear reactions to occur at all in stars. Note in Figure 32.2(a) that two protons on a collision course would be expected to deflect each other without touching (b). A reaction *can* occur all the same, because of the Heisenberg uncertainty principle (Section 10.6d). In Figure 32.2(c) are shown (schematically) graphs of the probability of each proton having various positions when the two are at closest approach. While the most likely positions for the protons are at A and B, there is a small but finite chance that they will actually be at the same place, for example, at C. Only rarely does probability permit such a union of two nuclei, but there are lots of them in a stellar interior, so it nevertheless happens often enough for nuclear reactions to work and supply the stars with energy.

Most of the electromagnetic radiation released in these nuclear reactions is at very short wavelengths—in the form of X rays and gamma rays. Nuclear reactions are important, however, only deep in the interior of a star. Before this released energy reaches the stellar surface, it is absorbed and reemitted by atoms a very great number of times. Photons of high energy (short wavelength) that are absorbed by atoms are often reemitted as two or more photons, each of lower energy. By the time the energy filters out to the surface of the star, therefore, it has been converted from a relatively small number of photons, each of very high energy, to a very much larger number of photons of lower energy and longer wavelength, which constitute the radiation we actually observe leaving the star.

(d) Solar Neutrinos

The proton-proton chain, which is thought to provide almost certainly most of the sun's energy, involves the emission of neutrinos. In fact, the neutrinos should comprise about 3 percent of the energy released from the sun. Their detection, however, because of their low interaction with matter, is an extremely difficult business.

Nevertheless, Raymond Davis, Jr., and his colleagues at Brookhaven National Laboratory have devised a technique by which they are detecting solar neutrinos. On rare occasions a neutrino of the energy of some of those emitted from the sun should react with the isotope chlorine-37 to transmute it to argon-37 and an electron. Davis has placed a tank containing 378,000 liters of tetrachloroethylene, C_2Cl_4 (ordinarily used as cleaning fluid), 1.5 km beneath the surface of the earth in a gold mine at Lead, South Dakota. Even though an individual neutrino is extremely unlikely to react with the chlorine in the cleaning fluid, calculations show that about one atom of argon-37 should nevertheless be produced daily. Because argon-37 is radioactive (about half of it decaying in 35 days) it is possible to isolate and detect most of those few argon-37 atoms from the more than 10^{30} atoms of chlorine in the tank.

In about 40 experimental runs during the 1970s, Davis found that the flux of neutrinos from

Figure 32.3 Raymond Davis, Jr.'s, 378,000-liter neutrino detector in a mine at Lead, South Dakota. (*Brookhaven National Laboratory*)

the sun is three to four times less than that expected. At the time of writing (late 1981), the discrepancy is unexplained. Careful analysis of the laboratory procedure has failed to suggest obvious sources of experimental error. If neutrinos should have a small mass, they may oscillate between various states, and some would arrive at Davis' tank of cleaning fluid in a state that cannot be detected by his technique. The discrepancy would then be reduced, but not eliminated. Or possibly there are as yet unexplained defects in the experiment. Otherwise present theory and/or the presumed chemical composition of the sun may need to be reconsidered. Any conclusion, however, must await a more critical analysis of the theory and observational data.

32.4 MODEL STARS

We now have enough theory to determine the internal structure of a star. We must combine the principles we have described: hydrostatic equilibrium, the perfect gas law, thermal equilibrium, energy transport, the opacity of gases, and the rate of energy generation from nuclear processes. These physical ideas are formulated into mathematical equations which are solved to determine the march of temperature, pressure, density, and other physical variables throughout the stellar interior. The set of solutions so obtained, based upon a specific set of physical assumptions, is called a theoretical model for the interior of the star in question.

(a) COMPUTATION OF A STELLAR MODEL

There are many ways to formulate mathematically the physical principles that govern the structure of a star and to solve the resulting equations to obtain a stellar model. Here we shall illustrate a particular procedure that has been used widely.

Four quantities are chosen to describe the physical conditions at any distance, r, from the star's center: the pressure, $P(r)$, the temperature, $T(r)$, the mass, $M(r)$, contained within a sphere of radius r concentric with the star's center, and the contribution to the star's total luminosity, $L(r)$, that is generated within this sphere. Once the pressure and temperature are known, the density can be calculated from the perfect gas law. The opacity and rate of energy generation at each point in the star involve no new parameters, for they,

like the density, can also be expressed as functions of the pressure, temperature, and chemical composition of the stellar material at that point.

The four quantities $P(r)$, $T(r)$, $M(r)$, and $L(r)$ are then combined into four equations that express the physical principles involved. Each equation describes how one of the quantities changes through a small radial distance within the star (say, across one of the imaginary layers or shells described in Section 31.6b). The change in pressure across such a layer is given by the condition of hydrostatic equilibrium. The mass of the shell is given by the density of the layer, which in turn is given by the pressure and temperature of the layer and the perfect gas law. The change in luminosity across the layer is given by the rate of energy generation within the shell. The change in temperature is governed by the mode of energy transport; if in that region of the star the energy is transported by radiation, the opacity determines the temperature variation, while if that region is in convection, the expansion and contraction of the gases determine the change in temperature. Since there are four equations that hold for each point within a star, their simultaneous solution determines the four desired quantities at that point.

The solution of the equations may begin at the surface of the star, where the physical conditions are known. The mass and luminosity included within the surface are, of course, the total mass and luminosity of the star. The pressure and temperature often can be considered zero to a sufficient approximation, since their actual values in the photospheric layers of a star are very small compared to those in the interior. If higher precision is desired, "surface" values of pressure and temperature can be taken from the solution of a model photosphere of the star (Section 25.2b).

The four equations are then used to calculate how the values of pressure, temperature, mass, and luminosity change over a short distance inward, beneath the surface of the star, thus yielding the values of these quantities at that new depth. Next, the equations are used to calculate the changes over the next short distance inward. So, step by step, the pressure, temperature, mass, and luminosity are found at successively deeper layers in the star, until the center is reached.

At the center, of course, the mass and luminosity should be zero, for no mass or luminosity can be contained within a point. This would be true if all the physical laws governing a star were precisely understood, and if the chemical composition were precisely known at each depth in the star. In practice, neither the physical details entering into the opacity and nuclear-energy generation formulas nor the chemical composition are known with absolute accuracy. Consequently, the solution of the four equations of structure may not lead to zero values of mass and luminosity at the center. The physical laws, therefore, are expressed as accurately as knowledge permits, and trial adjustments to the chemical composition are made until a set of solutions is found for which $M(r)$ and $L(r)$ do equal zero at the center. Then the runs of the pressure, temperature, mass, and luminosity throughout the star constitute a finished model for its interior.

Since a possible model is found only for a "correct" choice of chemical composition, something is learned of the distribution of the various chemical elements in a star as well as of the physical conditions in its interior. Hydrogen and helium are thus found to comprise (usually) more than 95 percent of the mass of a star, and more than two-thirds of that 95 percent is hydrogen. Unfortunately, however, the physical parameters in the opacity and nuclear physics theory are not yet known accurately enough to determine the chemical composition within a star with high precision from such studies.

The solution of the equations of structure to obtain a stellar model is a difficult and tedious business. Until the 1950s it often took as long as a year to compute a stellar model, and such a computation was a satisfactory topic for the dissertation of a student earning his Ph.D. degree. Now, however, high-speed electronic computers enable the calculation of a model in a few minutes or, in some cases, even in a few seconds.

(b) A Model for the Sun

The sun is the most studied of all stars, and models of its interior have been calculated for several decades. Each new model of the sun represents a refinement resulting from an improvement in our knowledge of physics or of computing methods or both. The general run of the physical parameters in the sun, however, was fairly well established even in the early approximate models. The temperature within the sun increases gradually toward its center and reaches a value of about 15 million Kelvins at the center. The density (like the pressure), on the other hand, increases very sharply near the center of the sun (indicating a high degree of central concentration of its material) and reaches a maximum value over 100 times the density of water.

As time goes on, the thermonuclear conversion of hydrogen to helium in the sun's central regions gradually changes its chemical composition. The exact temperature and density of the material at the sun's center that are required to account for its observed size and luminosity depend on how much the composition of that material has been changed—that is, on the age of the sun.

Series of models that trace the past history of the sun, therefore, have been calculated (more of this in Chapter 33). Table 32.3 and Figure 32.4 exhibit a model appropriate to the present-day sun that results from one such set of calculations. The model is based on the best physical data available in 1981, and on the assumption that the sun was originally 73 percent hydrogen and 24.5 percent helium. According to this model, the outer layers of the sun are in convection, while the inner parts transport energy by radiation. The hydrogen abundance at the very center has been reduced (by nuclear reactions) to only 38 percent, and the present age of the sun is about 4.5×10^9 years.

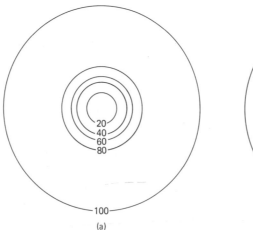

(a) (b)

Figure 32.4 (a) Distribution of mass within the sun, according to the model in Table 32.3; numbers show what percentage of the sun's mass is included within the radial zones shown. (b) Distribution of energy generation in the sun according to the same model. Successively smaller circles show the regions within which 100, 80, 60, 40, and 20 percent of the sun's energy is produced.

Table 32.3 **Model for the Structure of the Sun***

FRACTION OF RADIUS	FRACTION OF MASS	FRACTION OF LUMINOSITY	TEMPERATURE (Millions of K)	DENSITY (g/cm³)	FRACTION HYDROGEN (By Weight)
0.00	0.000	0.00	15.0	148	0.38
0.05	0.011	0.10	14.2	125	0.47
0.10	0.076	0.45	12.5	86	0.59
0.15	0.19	0.78	10.7	56	0.67
0.20	0.33	0.94	9.0	36	0.71
0.30	0.61	1.00	6.5	12	0.73
0.40	0.79	1.00	4.9	4	0.73
0.60	0.95	1.00	3.1	0.5	0.73
0.80	0.99	1.00	1.3	0.1	0.73
1.00	1.00	1.00	0.0	0.0	0.73

*Adapted from R. Ulrich.

(c) The Russell–Vogt Theorem and the Interpretation of the Main Sequence

If a star is in hydrostatic and thermal equilibrium, and if it derives all its energy from nuclear reactions, then its structure is completely and uniquely determined by its total mass and by the distribution of the various chemical elements throughout its interior. This is not to imply that we have the ability to compute perfect models for such stars; the stars themselves, nevertheless, must conform to the physical laws that govern their material and will adjust themselves to unique configurations. It should not, of course, surprise us that mass and composition, the very properties that a star is "born" with, and over which it "has no control," are just those that determine its structure. This important principle is known as the *Russell-Vogt theorem*.

Suppose a cluster of stars were to form from a cloud of interstellar material whose chemical composition was similar to the sun's. All condensations that become stars would then begin with the same chemical composition and would differ from each other only in mass. Suppose now that we were to compute a model for each of these stars for the time at which it became stable, and derived its energy from nuclear reactions, but before it had had time to alter its composition appreciably as a result of these reactions. The admixture of chemical elements would be the same, then, at all points within the star; such a composition is *homogeneous*.

The models we would calculate for these stars would indicate, among other things, their luminosities and radii. From Stefan's law we know that the luminosity of a star is proportional to the product of its surface area and the fourth power of its effective temperature (Section 26.4b). We can, therefore, calculate the temperature for each of the stars and plot it on the Hertzsprung-Russell diagram (Section 27.2). We would find that the most massive stars were the hottest and most luminous and would lie at the upper left corner of the diagram, while the least massive were coolest and least luminous and would lie at the lower right. The other stars would all lie along a line running diagonally across the diagram—the *main sequence*. The main sequence, then, is the locus of points on the H–R diagram representing stars of similar chemical composition but different mass. The observed fact that most stars in the Galaxy do lie along the main sequence is evidence that they have compositions similar to the sun's and are nearly chemically homogeneous. The observed scatter about the main sequence represents slight differences in the chemical compositions of stars. This explanation of the main sequence was first presented in the 1930s by the Danish astrophysicist B. Strömgren.

If we now plot the masses of the stars in our hypothetical cluster against their luminosities, we also find that the points lie along a line—the *mass-luminosity* relation. We have seen (Section 26.3) that most real stars do, indeed, obey a mass-luminosity relation; these stars are also of similar chemical composition. The locus of points on a plot of the masses of stars against their luminosities is simply the main sequence, plotted on a different kind of diagram. (Actually, it is possible for stars of the same mass but different chemical composition to have nearly the same luminosity—although they will differ in radius and temperature. Some stars, therefore, may obey the mass-luminosity relation even though they are *not* main-sequence stars, but this circumstance is fortuitous.)

Those stars that do not lie on the main sequence in the Hertzsprung-Russell diagram (for example, red giants and white dwarfs) must differ somehow from the majority in their chemical compositions, or else they are not stable and are not shining by nuclear energy alone. We have seen, however, that as stars age they convert hydrogen to helium, and so change their compositions, especially near their centers. Chapters 33 and 34 will describe how most non-main-sequence stars can be interpreted either as stars that are still forming from interstellar matter and are not yet deriving all their energy from nuclear sources, or as stars that, by virtue of nuclear transformations, have altered their chemical compositions and hence their entire structures.

EXERCISES

1. In what important respect (or respects) is a neutrino very different from a neutron?

2. If the nuclear force is the strongest of the known natural forces, why do you suppose it escaped discovery until the 20th century?

3. What do you suppose are the decay products of an antineutron?

4. Consider a nucleus of 11 protons. The total electrostatic repulsion between the 11 protons is expected to be about 10 times as strong as that between two protons of comparable separation. If the strong nuclear force is about 100 times as strong as the electrical force, but acts only between adjacent particles, what is the largest number of protons you would expect a nucleus to be able to have and still remain stable? What do you know about nuclei that actually contain larger numbers of protons?

5. Which of the following transformations is fusion and which is fission: the transformation of (a) helium to carbon; (b) carbon to iron; (c) uranium to lead; (d) boron to carbon; (e) oxygen to neon?

6. Verify that some 600 million tons of hydrogen are converted to helium in the sun each second.

7. Stars exist that are as much as a million times as luminous as the sun. Consider a star of mass 2×10^{35} g, and luminosity 4×10^{39} ergs/s. Assume that the star is 100 percent hydrogen, all of which can be converted to helium, and calculate how long it can shine at its present luminosity. There are about 3×10^7 seconds in a year.

8. Perform a similar computation for a typical star less massive than the sun, such as one whose mass is 1×10^{33} g and whose luminosity is 4×10^{32} ergs/s.

9. Why do you suppose so great a fraction of the sun's energy comes from its central regions? Within what fraction of the sun's radius does practically all of the sun's luminosity originate? (See Table 32.3.) Within what radius of the sun has its original hydrogen been partially used up? Discuss what relation the answers to these questions bear to each other.

10. Why do we not expect nuclear fusion to occur in the surface layers of stars?

*11. Let ΔP denote the pressure increase inward through a spherical shell of inner radius r and thickness Δr. Multiply ΔP by the area of the inner surface of the shell to find the total outward force on the shell. Now equate this outward force to the weight of the shell pulling it inward, and show that

$$\Delta P = \frac{GM(r)}{r^2} \rho \, \Delta r,$$

where $M(r)$ is the mass of the star interior to r and ρ is the density of the shell.

*12. Use the equation derived in Exercise 11 to make a very rough estimate of the pressure at the center of the sun. For this estimate suppose Δr to be the entire radius of the sun (that is, the entire sun is taken as one shell) and ΔP to be the increase in pressure from the surface to the center. For ρ use the mean density of the sun, and for $M(r)$ the entire solar mass. See Appendix 6 for needed data. Your estimate cannot be expected to be better than a few orders of magnitude, but it should give some indication of the amount of pressure.
Answer: About 3×10^{15} dynes/cm^2

*13. What is the minimum uncertainty in the momentum of a particle whose position is specified to within 1 Å (10^{-8} cm)? If the particle is a proton, what is the uncertainty in its velocity? See Appendix 6 for needed data.
Answer to Second Part: About 6×10^4 cm/s

Sir Fred Hoyle (1915–   ), the well-known British astrophysicist and cosmologist, is also well known for his science fiction, and even for an opera libretto. Hoyle was one of the pioneers in the modern study of stellar evolution, and in the 1950s his brilliant deduction about the nature of the carbon nucleus enabled us to understand where the atoms of our bodies originated (Chapter 34). *(Floyd Clark, California Institute of Technology)*

33

STAR FORMATION AND EVOLUTION

No star that is shining today can be infinitely old, for eventually it exhausts its sources of energy. Moreover, the stars of highest known luminosity (100 thousand to a million times that of the sun) can continue to exist at the rate they are now expending energy for only a few million years. Had they been formed when the sun was formed, thousands of millions of years ago, they would long since have burned themselves out. At least some stars, therefore, have formed recently (in the cosmic time scale), and there is every reason to expect that stars are still forming today.

33.1 FORMATION OF STARS

Here and there, in comparatively dense regions of interstellar matter, small condensations begin to form—atoms of gas and particles of dust slowly begin to collect under the influence of their mutual gravitation. (Local interstellar magnetic fields may also play a role.) The trick, then, to forming a star is to arrange for a relatively dense cloud of interstellar matter. Most of the gas and dust in space is clearly at too low a density to collapse into stars, or there would be no interstellar matter left. How, then, do stars form? Several different scenarios by

which protostar condensations may get started are now seriously considered.

(a) Some Mechanisms for Star Formation

1. Direct collisions of interstellar clouds can cause an increase of density that could lead to stellar condensations. One way that such collisions can occur is by gas clouds, in normal galactic rotation, encountering the density waves of spiral arms. Since, according to the density-wave theory (Section 29.3e), the pattern of spiral structure rotates more slowly than the normal galactic rotation, gas clouds should be plowing into the arms along their trailing edges. Consistent with this idea, in some other galaxies we see superluminous young stars concentrated along the trailing edges of those galaxies' arms.

2. At the interfaces between H II regions and the surrounding H I regions, we expect a buildup of density due to the expansion of the hot gas in the H II regions. We have described (Section 28.2g) how we observe "elephant trunk" intrusions of cooler gas and dust in the advancing fronts of the hot

H II regions. The young star cluster NGC 2264 appears to have been formed recently in such a region.

3. Isolated small dark clouds of mass estimated at from 20 to several hundred solar masses—globules—have already been mentioned (Section 28.2g). An excellent example of a globule is in the Coalsack, a dark region of the Milky Way in the direction of the Southern Cross. Astronomer Bart Bok has long called attention to such globules as probable clouds of gas and dust collapsing into stars or clusters of stars.

4. Supernova explosions (Chapter 34) release an enormous amount of energy both in electromagnetic radiation and in the form of a violent stellar wind. This energy, like the expanding matter in H II regions, can be an effective mechanism for compressing surrounding matter into protostars.

5. Among the most promising sites of star formation are the cold molecular clouds (Section 28.2ℓ). Part of the energy of the particles in these clouds excites states of rotation and vibration in the molecules, after which those molecules radiate that energy into space as infrared and radio waves. In this way, energy is removed from the cloud, which cools it. As it cools, the cloud must contract, until it becomes gravitationally unstable and forms protostars. A well-studied cold molecular

cloud behind the Orion nebula is believed to be such a site of star formation.

Once a stellar condensation has started, it is like the solar nebula, from which the sun and solar system formed (Section 14.3). Probably the contracting cloud has at least some rotation, if for no other reason than that it formed from material undergoing differential rotation in the Galaxy. Early on, the rotation is likely to be exceedingly slow, but to conserve angular momentum the cloud must spin faster and faster as it contracts. The angular momentum, in fact, will probably prevent the cloud from collapsing entirely to a single star. In the solar system, the nebula flattened to a disk, and planets accreted in the disk. Today the planets possess about 98 percent of the angular momentum of the entire system. The formation of planets may be commonplace, but often at least (perhaps half the time or so) the cloud must split and form two or more stars, whose orbital motions about each other contain most of the original angular momentum. It may be that formation of a planetary system or a multiple star system are the two alternatives open to a condensing cloud.

(b) Young Stars

Let us now follow the evolution of such a stellar condensation that has solved its angular momentum problem and is on its way to becoming a normal star. As its matter contracts, its density increases until eventually it becomes opaque to electromag-

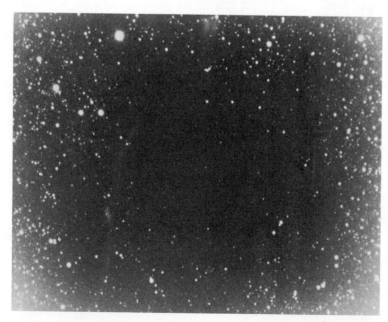

Figure 33.1 A fine globule in the southern Coalsack, photographed in the near infrared with the Cerro Tololo 4-meter telescope. The diameter is about a third of a parsec, and the mass is estimated at 20 solar masses. In the center the absorption by dust of visible light is near 20 magnitudes. (Courtesy Bart Bok, University of Arizona)

netic radiation. The very contraction, however, releases gravitational potential energy, and when the protostar becomes opaque, all of that energy cannot be radiated away, and some becomes trapped, which heats the interior and raises the internal pressure. When those pressures become high enough to support the weights of the outer material that has been falling inward, hydrostatic equilibrium is reached.

Calculations by R. B. Larson show that equilibrium is reached in the central regions first, while matter from outer parts of the nebula is still falling in, compressing, heating, and joining the condensed core. Thus, a stellar embryo forms inside, which is surrounded by a collapsing envelope that has not yet come to hydrostatic equilibrium. At least in some stars, dust should condense in that envelope, which may completely hide the star in visible light. However, the energy radiated from the hot embryo and then absorbed by the dusty envelope must be reradiated at wavelengths characteristic of the lower temperature of the envelope, that is, in the infrared. For this reason, observational searches for extremely young stars are made at infrared wavelengths.

The young star itself (whether or not it is still surrounded by infalling matter) is not yet self-sustaining with nuclear reactions but derives its energy from its gravitational contraction—by exactly the process proposed for the sun by Helmholtz and Kelvin (Section 32.1). As radiation filters out through the opaque star and is eventually radiated into space, the internal temperature and pressure would drop, upsetting hydrostatic equilibrium, unless the star contracts slightly. Half of the energy released by that contraction escapes as radiation, contributing to the star's luminosity, while the other half heats the interior, continually building up the internal pressure to support the increasing weights of the layers in the star (the weight of each shell of material is inversely proportional to the square of the radius of that shell).

In the theoretical study of stellar evolution, we compute a series of models for a star, each successive model representing a later point in time. Given one model, we can calculate how the star should change (in the case of the young stars under discussion, due to gravitational contraction), and hence what the star will be like at a slightly later time. At each step we find the luminosity and radius of the star and from these its surface temperature, and we can find where that star (or its embryo) should be represented on the Hertzsprung-Russell diagram. We thus follow the theoretical evolution of the star from its calculated track on the H-R diagram.

In its early contraction phases, a star transports its internal energy not by radiation but by leisurely convection currents (Section 31.6e). The Japanese astrophysicist C. Hayashi first showed that such stars must lie in a zone on the H-R diagram extending nearly vertically from the lower main sequence to the right extreme of the regions occupied by red giants and red supergiants (shaded region in Figure 33.2). There can be no stable star such that the point representing it on the H-R diagram lies to the right of this zone. In accord with the Hayashi theory, stars in the initial stages of their evolution contract and move (on the H-R diagram) downward in the zone along *Hayashi lines*. Representative tracks for stars or stellar embryos of several masses and of chemical composition more or less like the sun's are shown in Figure 33.2.

Except for a star of low mass, after a period of some thousands or millions of years, the convection currents cease at the center of the star, and energy must be transported by radiation in those regions. The central zone in radiative equilibrium gradually grows in size, while the convection currents extend less and less deeply beneath the stellar surface. In this stage of its evolution, the star or embryo, still slowly shrinking and deriving its energy from gravitational contraction, turns sharply on the H-R diagram and moves left, almost horizontally, toward the main sequence. Eventually, as the release of gravitational energy continues to heat up the star's interior, its central temperature becomes high enough to support nuclear reactions. Soon this new source of energy supplies heat to the interior of the star as fast as energy is radiated away. The central pressures and temperatures are thus maintained and the contraction of the star ceases; it is now on the main sequence. By this time the infall of matter is complete and the star is fully formed. The small hooks in the evolutionary tracks of the stars shown in Figure 33.2, just before they reach the main sequence, are the points (according to theory) where the onset of nuclear-energy release occurs. Larson's calculations show that stars more massive than the sun would not be visible to us during most of their pre-main sequence evolution because the light they emit is absorbed by the surrounding dust in the infalling material.

By the time stars of mass appreciably greater than the sun's have reached the main sequence,

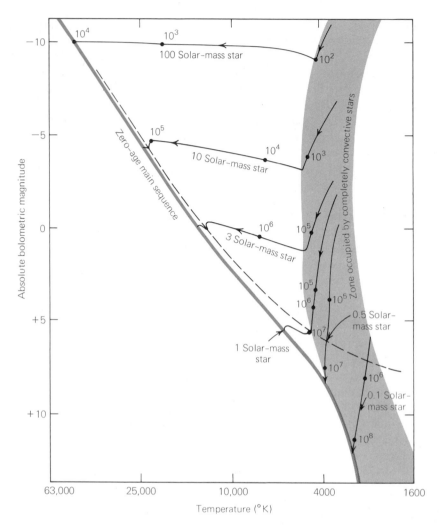

Figure 33.2 Theoretical evolutionary tracks of contracting stars or stellar embryos on the Hertzsprung-Russell diagram. According to calculations by Larson, stars or embryos lying roughly above the dashed line are still surrounded by infalling matter and would be hidden by it.

their outer convection zones have disappeared, but new cores of convection exist at their centers. Main-sequence stars of mass near that of the sun still have appreciable regions in their outer layers in convection, with their deep interiors in radiative equilibrium. Stars of rather low mass remain in complete convective equilibrium throughout and follow their Hayashi lines right down to the main sequence, where nuclear reactions finally stop their contraction. Stars of extreme low mass, on the other hand, never achieve high enough central temperature to ignite nuclear reactions. They continue to contract until (after an extremely long time) they are so dense that their matter becomes degenerate, and they reach the white dwarf state (Chapter 34). The lower end of the main sequence is considered to be that point at which stars have a mass just barely great enough to sustain nuclear reactions at a suffi-

cient rate to stop gravitational contraction; this critical mass is calculated to be near $^{1}/_{12}$ that of the sun.

At the other extreme the upper end of the main sequence terminates at the point where the mass of a star would be so high and the internal temperature so great that radiation pressure would dominate (Section 20.4d). The radiation produced from nuclear reactions would be so extreme that when absorbed by the stellar material it would impart to it a force greater than that produced by gravitation; hence, such a star could not be stable. The upper limit to stellar mass is calculated to be in the range 60 to 100 solar masses.

In general, the pre-main-sequence evolution of a star slows down with time; the numbers labeling the points on each evolution track in Figure 33.2 are the times, in years, required for the embryo stars to reach those stages of contraction. The time for the whole evolutionary process, however, is highly mass-dependent. Stars of mass much higher than the sun's reach the main sequence in a few thousand to a million years; the sun required millions of years; tens of millions of years are required for stars to evolve to the lower main sequence. For all stars, however, we should distinguish three evolutionary time scales (although the first two, as we have seen, may overlap):

1. The initial gravitational collapse from interstellar matter is relatively quick. Once the condensation is, say, 1000 AU in diameter, the time for it to reach hydrostatic equilibrium is measured in thousands of years.
2. Pre-main-sequence gravitational contraction is much more gradual; from the onset of hydrostatic equilibrium to the main sequence requires, typically, millions of years.
3. Subsequent evolution on the main sequence is very slow, for a star changes only as thermonuclear reactions alter its chemical composition. For a star of a solar mass, this gradual process requires thousands of millions of years. All evolutionary stages are relatively faster in stars of high mass and slower in those of low mass.

33.2   EVOLUTION FROM THE MAIN SEQUENCE TO GIANTS

As soon as a star has reached the main sequence, it derives its energy almost entirely from the thermonuclear conversion of hydrogen to helium. It remains on the main sequence for most of its "life." Since only 0.7 percent of the hydrogen used up is converted to energy, the star does not change its mass appreciably, but in its central regions, where the nuclear reactions occur, the chemical composition gradually changes as hydrogen is depleted and helium accumulates. This change of composition forces the star to change its structure, including its luminosity and size. Eventually, the point that represents it on the H-R diagram evolves away from the main sequence. The original main sequence, corresponding to stars of homogeneous chemical composition, is called the *zero-age main sequence*.

(a)   Evolution on the Main Sequence

As helium accumulates at the expense of hydrogen in the center of a star, calculations show that the temperature and density in that region must increase. Consequently, the rate of nuclear-energy generation increases, and the luminosity of the star slowly rises. A star, therefore, does not remain indefinitely *exactly* on the original zero-age-main sequence. In fact, the main sequence of a star cluster gradually rises in the H-R diagram as the cluster ages. The most massive and luminous stars alter their chemical composition the most quickly; thus, the main sequence rises most rapidly at the bright end, but scarcely at all at the faint end, even after billions of years. This stage of evolution does not cause the main sequence of a star cluster to deform appreciably, because a star increases its luminosity only by a moderate amount—probably less than a magnitude—before subsequent more rapid changes alter its structure enormously.

When the hydrogen has been depleted completely in the central part of a star, a core develops containing only helium, "contaminated" by whatever small percentage of heavier elements the star had to begin with. The energy source from hydrogen burning* is now used up, and with nothing more to supply heat to the helium core, it begins again to contract gravitationally. Once more the star's energy is partially supplied by potential energy released from the contracting core; the rest of its energy comes from hydrogen burning in the region immediately surrounding the core. These changes result in a substantial and rather rapid readjustment of the star's entire structure, so that the star leaves the vicinity of the main sequence altogether. About ten percent of a star's mass must be depleted of hydrogen before the star evolves away from the main sequence. The more luminous and massive a star, the sooner this happens, ending its term on the main sequence. Because the total rate of energy production in a star must be equal to its luminosity, the core hydrogen is used up first in the very luminous stars. The most massive stars spend less than 1 million years on the main sequence; a star of one solar mass remains there for about 10^{10} years, and a spectral-type M0 V star of

---

*The term "burning" is often used to describe the depletion of an element by nuclear reactions. This "nuclear burning" is not, of course, burning in the literal chemical sense.

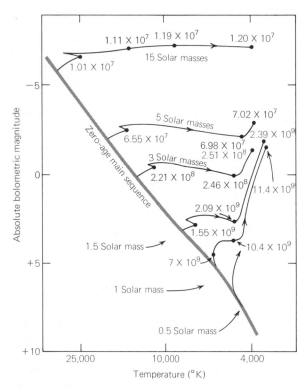

Figure 33.3 Predicted evolution of stars from the main sequence to red giants. See text for explanation. (Based on calculations by I. Iben.)

about 0.4 solar mass has a main-sequence life of some 2×10^{11} years.

(b) Evolution to Red Giants

As the central core contracts, it releases gravitational potential energy, which is absorbed in the surrounding envelope, thereby forcing the outer part of the star to distend greatly. The star as a whole, therefore, expands to enormous proportions; all but its central parts acquire a very low density. The expansion of the outer layers causes them to cool, and the star becomes red. Meanwhile, some of the potential energy released from the contracting core heats up the hydrogen surrounding it to ever higher temperatures. In these hot regions the conversion of hydrogen to helium accelerates, causing most stars actually to increase in total luminosity. After leaving the main sequence, then, stars move to the upper right portion of the H-R diagram; they become red giants.

Figure 33.3, based on theoretical calculations by Illinois astonomer Icko Iben, shows the tracks of evolution on the H-R diagram from the main se-

quence to red giants for stars of several representative masses and with chemical composition similar to that of the sun. The broad band is the zero-age-main sequence. The numbers along the tracks indicate the times, in years, required for the stars to reach those points in their evolution after leaving the main sequence.

33.3 CHECKING OUT THE THEORY

Star clusters of different ages provide an excellent way of checking theoretical calculations like those described above. Not only are the many stars in a cluster at the same distance, so that their luminosities can be directly compared, but they are essentially all of the same age. It is possible, therefore, to compare H-R diagrams plotted for the stars in a cluster to H-R diagrams prepared from the calculated properties of stars of the same composition and age, but of different mass, and hence advanced to different stages of evolution. One such set of theoretical H-R diagrams has been prepared by Rudolf Kippenhahn and his associates at Munich to enable them to follow the evolution of a hypothetical star cluster whose members have a chemical composition similar to the sun's. The Kippenhahn group originally named their cluster M007, to emphasize its fictitious identity. Because few readers today will recall James Bond's Agent 007, I have changed the name of the cluster to M2001 in the H-R diagrams displayed here—all adopted from Kippenhahn's original plots. Let us see how they match the H-R diagrams of real clusters. The dashed lines in the figures represent the location of the zero-age main sequence.

(a) Very Young Clusters

What should the H-R diagram be like for a cluster whose stars have recently condensed from an interstellar cloud? After a few million years, the most massive stars should have completed their contraction phase and be on the main sequence, while the less massive ones should be off to the right, still on their "way in." Figure 33.4 shows the H-R diagram of M2001 at an age of 3 million years.

There are real star clusters that fit the description. The first to be studied (about 1950) is NGC 2264, a cluster still associated with nebulosity (Figure 33.6). Figure 33.5 shows its H-R diagram,

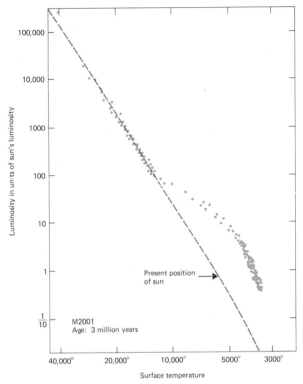

Figure 33.4 The H-R diagram of M2001 at an age of 3 million years.

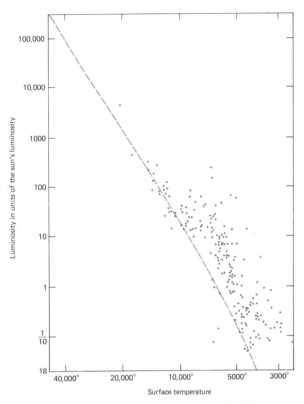

Figure 33.5 The H-R diagram of cluster NCG 2264, from data by M. Walker.

adapted from data by M. Walker at the Lick Observatory. Among the several other star clusters in such an early evolution stage is one in the middle of the Orion nebula.

Often the member stars of low luminosity in very young clusters have characteristic spectral peculiarities and often display rapid and irregular variations in light. These are called T Tauri stars (Appendix 16); when there is a considerable number of these stars in a single young cluster, it is called a T-association (Section 30.1c). T Tauri stars often emit a large fraction of their energy in the infrared, suggesting that they may still be surrounded by dust, as is expected for young stars according to Larson's analysis.

(b) Older Star Clusters

After a short time—less than a million years after reaching the main sequence—the most massive stars use up the hydrogen in their cores and evolve

Figure 33.6 Nebulosity in Monoceros, situated in the south outer region of the young cluster NGC 2264. Photographed in red light with the 5-m telescope. *(Palomar Observatory; California Institute of Technology)*

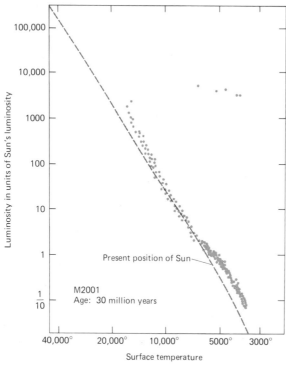

Figure 33.7 The H-R diagram of M2001 at an age of 30 million years.

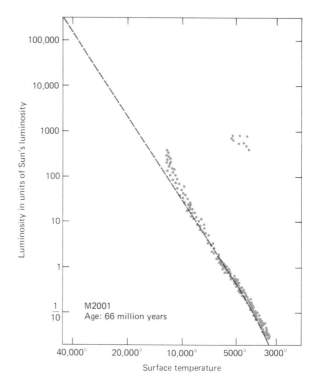

Figure 33.8 The H-R diagram of M2001 at an age of 66 million years.

off the main sequence to become red giants. As more time goes on, stars of successively lower mass leave the main sequence, making it seem to burn down, like a candle. Meanwhile, even after 30 million years, the least massive stars will not have completed their gravitational contraction to the main sequence (Figure 33.7).

Figure 33.8 shows the H-R diagram of hypothetical cluster M2001 at an age of 66 million years, and Figure 33.9 shows the H-R diagram of the real cluster, M41, adapted from data by A. N. Cox. We judge the age of M41 to be a little less than 100 million years. The Hyades, whose H-R diagram is shown in Figure 33.10, has an age of a few hundred million years.

At 4240 million years, M2001 has the H-R diagram shown in Figure 33.11. Stars only a few times as luminous as the sun have begun to leave the vicinity of the main sequence. The total main sequence lifetime for the sun is expected to be about 10^{10} years. Still older than this are the globular clusters (Section 30.1a), whose ages are calculated to be in the range 10 to 16 $\times$ 10^9 years. The H-R diagram for the globular cluster M3, adapted from data by H. C. Arp, W. A. Baum, and A. R. San-

dage, is shown in Figure 33.12. A few open clusters (for example NGC 188 and M67) approach, but do not reach, globular clusters in age.

Figure 33.13 shows a composite H-R diagram for several star clusters of different ages. On the left side is shown the absolute visual magnitude scale; on the right side is a scale that gives the approximate ages of star clusters corresponding to the points where their main sequences terminate.

As expected, most clusters have red giants. Those few that do not may just not happen to have any stars of the proper mass to be entering that stage of their evolution. In the younger clusters the red giants have magnitudes about the same as those of the brightest main-sequence stars; for stars of those masses and compositions, the tracks of evolution from the main sequence to the giant stages are approximately to the right across the diagram. In the older clusters, however, the giants are brighter than the brightest main-sequence stars and must, therefore, have increased in luminosity during their evolution from the main sequence. Note the large gap in the center of the diagram where very few cluster stars appear (the Hertzsprung gap—Section 30.4b). There are theoretical reasons

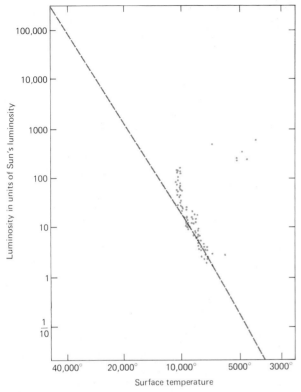

Figure 33.9   The H-R diagram of M41, from data by A. N. Cox.

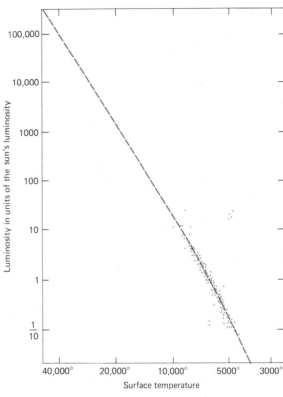

Figure 33.10   The H-R diagram of the Hyades star cluster.

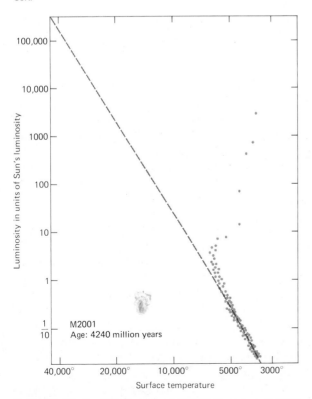

Figure 33.11   The H-R diagram of M2001 at an age of 4240 million years.

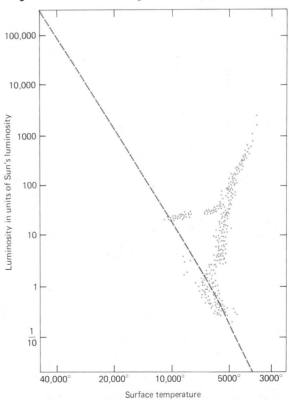

Figure 33.12   The H-R diagram of the globular cluster M3, from data by Arp, Baum, and Sandage.

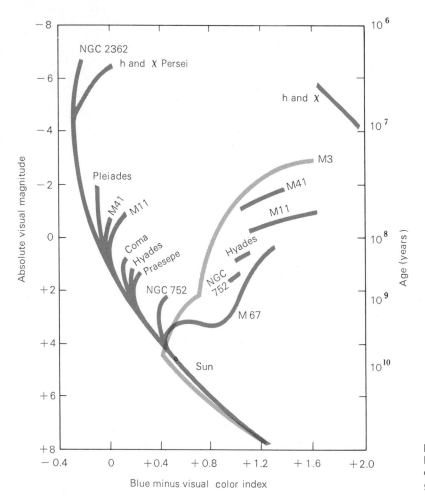

Figure 33.13 Composite Hertzsprung-Russell diagram for several star clusters of different ages. *(Adapted from a diagram by Sandage)*

to expect stars of relatively high mass to become unstable as they leave the main sequence; they would be expected, therefore, to readjust themselves to the more stable red giant configuration very rapidly. The observed absence of stars in the Hertzsprung gap does indeed suggest that stars of high mass evolve very quickly from the main sequence to the red giant domain.

The red giants in globular clusters are more luminous than are those in the oldest open clusters, such as M67. Calculations show these differences to be due to differences in chemical composition; globular-cluster stars (population II) are, on the average, lower in abundance of heavier elements than the open-cluster stars of population I (Section 30.4c).

Counts of stars of various kinds in old clusters show that the number of giants is very small compared with the number of main-sequence stars, and that all the present giants in these clusters must

have evolved from a very short segment of the original main sequence, just above its present termination point. (In Figure 33.12, to avoid crowding, points are plotted for only about one in ten of the main-sequence stars that lie below the giant branch turnoff.) In other words, all the giants in a single cluster are expected to have nearly the same mass and also to have had almost the same luminosity when they were on the main sequence. Those stars at the top of the giant branch on a cluster H-R diagram are, to be sure, further evolved than those, say, only halfway up to the top, but they started from only very slightly greater luminosities on the main sequence, and so had only a slight "head start." We can conclude, therefore, that the sequence of stars forming the giant branch in the H-R diagram of an old cluster lies very nearly along the evolutionary tracks of individual stars.

In any case, the red giant stage must be a relatively brief part of a star's life. In the youngest

Figure 33.14   *The globular cluster M13 in Hercules. (Palomar Observatory; California Institute of Technology)*

clusters these stars are red *supergiants* of high luminosity. In the older clusters, stars *increase* their luminosities as they become red giants. In this stage of evolution, therefore, a star's nuclear fuel is consumed relatively quickly, and further evolutionary changes soon follow.

The later stages of the evolution of stars are more complicated and are less thoroughly understood. It is there, however, that we find most variable stars, eruptive stars, mass ejection, enrichment of the heavy elements in the Galaxy, and—a subject at the very frontier of modern astrophysics—the final phases of a star's life, and its death. We turn to these subjects in the next two chapters.

EXERCISES

1. Where on the H-R diagram does a star *begin?*

★2. Suppose stars contracting from interstellar matter evolved exactly to the left across the H-R diagram (that is, at constant luminosity). The luminosities of main-sequence stars are approximately proportional to the cubes of their masses. Show that more massive stars would contract faster and reach the main sequence sooner than less massive stars.

3. The H-R diagram for field stars (that is, stars all around us in the sky) shows very luminous main-sequence stars and also various kinds of red giants and supergiants. Explain these features and interpret the H-R diagram for field stars.

4. In the H-R diagrams for some young clusters, stars of very low and very high luminosity are off to the right of the main sequence, while those of intermediate luminosity are on the main sequence. Can you offer an explanation? Sketch an H-R diagram for such a cluster.

5. If the sun were a member of the cluster NGC 2264, would it be on the main sequence yet? Why?

6. Roughly, how much older is the Hyades cluster than the double cluster h and χ Persei?

7. From the data in this chapter and from a knowledge of the apparent magnitudes of the brightest stars in the clusters M3 and M67, describe how you could find the relative distances of the two clusters.

8. Suppose a star cluster were at such a large distance that it appeared as an unresolved spot of light on telescopic photographs. What would you expect the color of the spot to be if it were the image of the cluster immediately after it was formed? How would the color differ after 10^{10} years? Why?

9. Explain the statement in the text that the tracks of evolution on the H-R diagram of globular cluster stars from the main sequence to the red giant branch lie very nearly along the observed sequence of stars in the diagram from the main sequence to red giant?

10. Explain how you could decide whether red giants seen in a star cluster probably had evolved away from the main sequence or were still evolving along their Hayashi lines.

11. What form of energy transport is occurring in the cores and envelopes of the following main-sequence stars: (a) O9; (b) G2; (c) M8?

12. If all the stars in a cluster have the *same* age, how can clusters be useful in studying evolutionary effects?

13. Assume the sun to be 73 percent hydrogen by mass. What mass of hydrogen will then be "burned" by the time the sun moves away from the main sequence? If the sun's luminosity were constant during this time, how long would it take for the sun to convert this amount of hydrogen into helium? (See Section 32.3b and Appendix 6.)
Answer to second part: 7.6×10^9 years

Jocelyn Bell Burnell, British astronomer, discovered the first known pulsar in 1967, while she was a graduate research student at Cambridge University.She analyzed 400 feet of chart recordings each week and found the sky to be heavily populated with compact radio sources, one of which was flaring. This first pulsar led to the realization that neutron stars exist. (*Sky and Telescope*)

34

EVOLUTION AND DEATHS OF OLD STARS

The preceding chapter followed the evolution of stars up to the point where they leave the main sequence to become red giants. Now let us explore the subsequent stages of stellar evolution. We shall find that gases in the inner parts of some giant stars, and throughout nearly the entire masses of many stars near the ends of their lives, reach enormous density and behave very differently from a perfect gas. First, therefore, we consider the nature of matter under these more extreme conditions.

34.1 DEGENERATE MATTER

The electrons in a neutral atom occupy certain allowed *states*, each of which involves a certain energy level for the atom as a whole. We saw in Section 10.6 how an electron can change states when the atom absorbs or emits energy. When the atom is ionized, the electrons are free, and we speak of them as occupying a *continuum* of states. Even free electrons, however, do not have an infinite number of states available to them.

Electrons, protons, and neutrons belong to a class of particles (there are others) that must obey a

rule of quantum mechanics, which insists that no two of them occupy the same state. What this means in plain English is that no two electrons can be in the same place at the same time doing the same thing. The rule is known as the *Pauli exclusion principle*, after the Austrian physicist Wolfgang Pauli, who enunciated it. We specify the *place* of the electron by its precise position in space, and we specify what it is doing by its momentum and the way it is spinning.

But recall from Section 10.6d that the simultaneous position and momentum of an electron (or anything else) cannot be known any more precisely than is allowed by the Heisenberg uncertainty principle. Specifically, this means that the precision of its position, or the amount by which the x-coordinate of its location is uncertain, Δx, and the precision of the x-coordinate of its momentum, Δp_x, are governed by the condition

$$\Delta x \Delta p_x \geq \frac{h}{2\pi},$$

where h is Planck's constant (6.626×10^{-27} erg · s). The state of an electron cannot be specified—indeed, cannot even exist—more precisely; the states, therefore, are fuzzy. The Pauli principle permits only one electron in each of these fuzzy states (actually two, because an electron can spin in either of two possible directions).

Imagine the free electrons in an ionized gas with a certain temperature. That temperature determines the distribution of velocities, and hence the range of momenta, for the electrons. If the temperature is high enough, the momentum range is large, and the electrons have plenty of possible momentum-position states to occupy without violating the Pauli principle. On the other hand, what if the temperature is *not* high enough? (How high is "high enough" depends on how many electrons are crowded into a given volume—that is, their density.) Then two or more electrons (of the same spin) would have to occupy the same fuzzy state, which will not happen, for the electrons will resist such crowding with overwhelming pressure. When all the available states (of position and momentum) are occupied, the electrons are said to be *degenerate*, and the gas is an *electron-degenerate gas*.

The electrons in a degenerate gas move about, as do particles in any gas, but not with freedom. A particular electron cannot change position or momentum until another electron in an adjacent state gets out of the way. It is as if all the particles were geared together. Crystalline solids are analogous; there, each electron occupies a particular state in the latticework, and all states are occupied (the lattice is also degenerate), so that one electron cannot move over until another next to it moves, producing a "hole." If electrons are removed from one side, however, others can move over, and still others behind them, producing a current through the solid. In a similar way, an electron-degenerate gas is highly conducting, and heat can flow through it with great ease.

Now what has all this to do with stars? Simply that if part or all of a star contracts and increases in density it will become electron-degenerate unless the temperature also rises enough to make available a sufficient range of momentum states. The contracting star (or the contracting core of a star) does release gravitational potential energy, half of which heats the interior, but this released heat is not enough to prevent degeneracy unless a lot of gravitational energy is released, which requires the star (or core) to be quite massive. We shall return later to just what the required mass is. When a star becomes electron-degenerate, it can contract no further.

It is noteworthy that in a typical stellar environment even a degenerate electron gas is mostly empty space. The electrons are by no means packed into contact; their densities are (typically) a million to a thousand million times lower than the density of the atomic nucleus. The nuclei themselves still move about freely among the electrons, obey the usual perfect gas law, and exert the normal pressure of particles of their masses and temperature. The pressure exerted by the degenerate electrons, however, generally swamps that of the nuclei, so it is the electron pressure that dominates and controls the structure of that region of the star.

Other particles become degenerate as well—in particular, a *degenerate neutron gas* is important in astronomy. But neutrons are nearly 2000 times as massive as electrons and at the same temperature have very much greater momentum. Neutrons, therefore, must be crowded to enormously higher densities before filling their available momentum states and becoming degenerate.

34.2 THE HELIUM FLASH

Now back to our story of the star that is becoming a red giant. We recall from the discussion in the preceding chapter that the star's core is shrinking, while its outer envelope distends. Gravitational energy released in the contracting core heats it, until by the time the star reaches the top of the red giant branch on the H-R diagram its central temperature exceeds 100 million Kelvins.

At such a high temperature, nuclear processes other than the carbon cycle and proton-proton chain are possible. The most important of these is the formation of a carbon nucleus by three helium nuclei (the *triple-alpha process*—so named because the nucleus of a helium atom is also called an *alpha particle*). Successive bombardments of a carbon nucleus by helium nuclei can build up other, still heavier nuclei. The astrophysicists G. Burbidge, E. Burbidge, W. Fowler, and F. Hoyle have found mechanisms whereby virtually all the chemical elements up to iron can be synthesized in the centers of red giant stars, in approximately the relative abundances with which they occur in nature. It now seems quite probable that a gradual buildup of the elements heavier than helium is continually going on in the hot centers of at least the more massive red giants.

The triple-alpha process is expected to begin abruptly in the central core of a red giant. As the core evolves, not only does it get very hot but also very dense, and the matter in the innermost part becomes electron-degenerate. Meanwhile, the surrounding matter soon exhausts its hydrogen and also contracts until it becomes degenerate and joins the core. With its increased mass and consequent release of additional gravitational energy, the core becomes smaller and the nondegenerate nuclei become hotter. Thus the degenerate core continues to contract and heat. As soon as the temperature becomes high enough to start the triple-alpha process going, the extra energy released is transmitted quickly through the entire degenerate core, producing a rapid heating of all the helium there. With the sudden rise in temperature, the helium burning accelerates; the phenomenon is called the *helium flash*.

The new energy released removes the degeneracy, expands the core, and reverses the growth of the outer parts of the red giant. The star then shrinks rapidly and increases in surface temperature. Calculations indicate that the point representing the star on the H-R diagram takes on a new position either to the left of its place as a red giant or somewhat below it. Usually a newly formed carbon nucleus is joined by another helium nucleus to produce a nucleus of oxygen. As soon as the helium is exhausted in the central region, however, the energy release from the triple-alpha process is shut off, and we have a situation analogous to that of a main-sequence star when its central hydrogen is

used up and the hydrogen burning ceases in its center. Now, however, we have a core of carbon and oxygen (and perhaps heavier elements) surrounded by a shell where helium is still burning; farther out in the star is another shell where hydrogen is left and is still burning. The star now moves on the H-R diagram back to the red giant domain. Calculations indicate that a star may actually move first to the left across the H-R diagram, and then back to be a red giant several times, each time as a consequence of the onset of new nuclear reactions or of nuclear energy released in new parts of the star. All these evolutionary stages occur in tens or hundreds of millions of years or less—a brief time compared with the stars' main-sequence lives. Some observational evidence supporting the theoretical calculations is the presence of the "horizontal branch" of stars on the H-R diagram of globular clusters and possibly in some open clusters (Section 30.4a).

Figure 34.1 shows post–main sequence evolutionary tracks on the H-R diagram of stars of several masses, as calculated by Emma Meyer-Hofmeister at the Max Planck Institute in Munich. At an age of about 57 million years, a cluster like M2001 (preceding chapter) would have an H-R diagram like that shown in Figure 34.2. Figure 34.3 shows the H-R diagram of the real cluster, NGC 1866; note the similarity to Figure 34.2.

34.3 PULSATING STARS

Some stars are unstable against pulsation. Most types of stellar variability are now thought to rep-

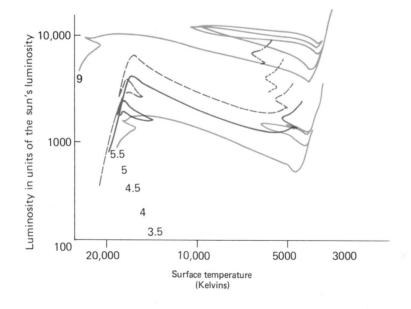

Figure 34.1 Theoretical tracks on the H-R diagram showing the evolution of stars of various masses from the main sequence to the red giant phase. *(Adapted from a diagram by E. Meyer-Hofmeister)*

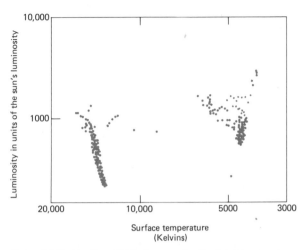

Figure 34.2 Predicted H-R diagram for the brighter stars in a cluster like M2001 at an age of about 57 million years. *(Adapted from E. Meyer-Hofmeister)*

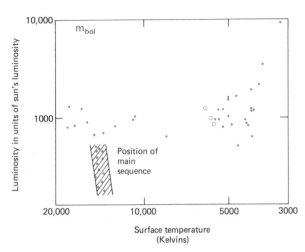

Figure 34.3 H-R diagram for the brighter stars in the star cluster NGC 1866. *(Adapted from E. Meyer-Hofmeister)*

resent stages of evolution of stars of various masses or compositions, in which they become unstable and pulsate. For example, the Hertzsprung gap (Section 30.4b) of the H-R diagram is a region in which most single stars, when their post-red-giant evolution places them there, are variable. Those stars in the gap in the horizontal branch of the H-R diagram for a globular cluster, for example, are all variable—they are RR Lyrae stars.

(a) The Pulsation of Cepheid Variables

Cepheid variables (Section 24.3e) provide a good example of pulsating variables. A hint of the physical nature of the variability of these stars comes from their spectra. At maximum light, their spectral classes correspond to higher surface temperatures than at minimum light. The light variations of a cepheid, therefore, are due in part to variations in the temperature of its radiating surface. Further spectroscopic evidence reveals that the temperature fluctuations are accompanied by actual pulsations in the sizes of these stars. The lines in the spectrum of a cepheid show Doppler shifts that vary in exactly the same period as that of the star's light fluctuations. Evidently, the changes in light are associated with a periodic rise and fall of the cepheid's radiating surface.

A graph that displays changes in the Doppler shifts of the spectral lines of a cepheid with lapse of time is called a *radial-velocity curve* (Figure 34.4). It is like the radial-velocity curve of a spectroscopic binary star, except that in a cepheid the Doppler

shifts are due to the periodic rising and falling of its radiating surface rather than to the orbital motion of the star as a whole. The mean value of the apparent radial velocity corresponds to the line-of-sight motion of the star itself; the photospheric pulsations cause variations about this mean value. When the photosphere expands, it is approaching us with respect to the rest of the star, and each spectral line is shifted to slightly shorter wavelengths than that of its mean position; when the photosphere contracts, the lines are shifted to slightly longer wavelengths. When the photosphere reaches its highest or lowest point—that is, when the star is at its largest or smallest size—the position of the spectral lines corresponds to the radial velocity of the star itself.

We can calculate the total distance through which the cepheid's photosphere rises or falls by multiplying the velocity of its rise at each point in the pulsation cycle by the time it spends at that velocity, and then adding up the products.* For δ Cephei, for example, we find by such a calculation that the photosphere pulsates up and down over a distance of somewhat under 3 million km. On the other hand, the mean diameter of δ Cephei as calculated from Stefan's law (Section 26.4b) is about 40 million km. During a pulsation cycle, therefore, the radius of the star changes by about 7 or 8 percent.

We might expect a pulsating star to be hottest when it is smallest and most compressed. δ Cephei,

---

*Or, technically, by *integrating* the velocity curve over the time of rise or fall.

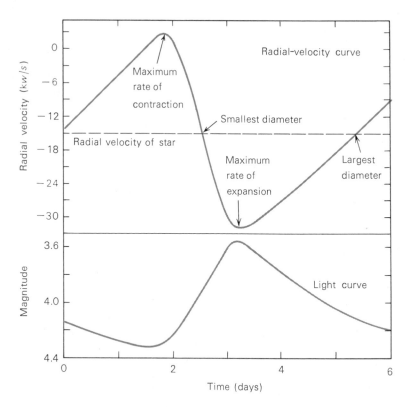

Figure 34.4 Radial-velocity curve *(top)* and light curve *(bottom)* of the star δ Cephei.

however, like other cepheid variables, is hottest and brightest at about the time when its radiating surface is rushing outward at its maximum speed. Evidently, the greatest compression of the star as a whole does not correspond to the maximum temperature at its surface; the explanation is related to the mechanism by which energy is transferred outward through the outer layers of the star.

The RR Lyrae stars are similar, in their pulsation characteristics, to cepheids, except that they are smaller, denser stars and pulsate more rapidly—in about half a day.

(b) Long-Period Variables

Most common among the pulsating stars are the *long-period variables.* These stars, all red giants or red supergiants, are sometimes called "red variables." The best known, Mira (o Ceti), is ordinarily at apparent magnitude 8 to 10, too faint to be visible to the unaided eye. About every 11 months, however, it brightens to naked-eye visibility and averages magnitude 2.5 at maximum. (Once it reached magnitude 1.2.)

The spectral type of Mira varies from M6 to M9, which corresponds to a variation in surface temperature of 2600 to 1900 K. With this moderate range of surface temperature, one would expect (from Stefan's law) a variation in luminosity of only about a factor of $3^{1}/_{2}$. Yet in the visible spectral region, Mira varies in light by about 100 times or more. One reason is that most of the light emitted by the star lies in the invisible infrared spectral region. Although the *total energy* emitted by the star at maximum light may be only a few times greater than at minimum, the higher temperature at maximum causes that energy to be emitted at shorter wavelengths (Section 10.2b). A far greater fraction of it is thus in the visible part of the spectrum than at minimum. Another reason for the enormous light variations of Mira may be that when it is at minimum, more molecules can form in its outer atmosphere, and much of its visible light may be veiled by increased absorption in molecular bands, or even by temporary condensations of solid or liquid particles.

Mira is one of the few stars whose angular diameters have been measured with the stellar interferometer (Section 26.4a). Its diameter is 420 times that of the sun. The radius of Mira changes by 20 percent during a pulsation cycle. The mean density of the star is less than a ten-millionth that of the sun, and its outer parts would comprise a high vacuum.

Other long-period variables have periods in the range 80 to 600 days, and amplitudes ranging from 2.5 to more than 7 magnitudes. The periods of these red-giant variables tend to be somewhat unstable; Mira's period, for example, *averages* 330 days, but the interval between successive maxima has varied from this period by as much as several weeks. In general, the longer the period of a red variable, the less regular is its pulsation.

(c) Irregular and Semiregular Variables

Many red variables show marked irregularity, and some exhibit no trace whatever of regularity in their light variations. The latter are designated *irregular variables*. Those stars that usually exhibit periodic variability, but are at times unpredictably disturbed, are classed as *semiregular variables*. Generally, semiregular and irregular variables have light amplitudes that are less than two magnitudes, that is, a smaller range than the range of ordinary long-period variables. To the class of semiregular variables belongs the famous star Betelgeuse, which marks Orion's right shoulder.

(d) "SPECTRUM" VARIABLES

We mention, finally, some variable stars that do not actually pulsate. For example, some peculiar main-sequence spectral-type A stars comprise a class called *spectrum variables*. They show unusually strong lines of the ions of strontium, silicon, and chromium, or of ionized rare-earth elements. These lines vary in strength, however, in periods of from 1 to 25 days. Often one set of lines (such as those of rare-earth elements) weakens, while another set (such as those of chromium) strengthens. At the same time the brightness of the star varies, but with an amplitude of only about 0.1 magnitude. Moreover, these stars have strong magnetic fields, which vary in the same period.

An example of the group, α^2 Canum Venaticorum, has a period of 5.5 days. H. W. Babcock, of the Hale Observatories, has observed a magnetic field associated with this star which varies from -4000 to $+5000$ gauss. (By comparison, the magnetic field of the earth has a strength of less than 1 gauss.) One hypothesis that explains some of these phenomena, suggested by the late A. J. Deutsch, is that the star (like the earth) has a magnetic axis that is inclined to its axis of rotation and that the stellar rotation, in a period of 5.5 days, alternately carries first the north, then the south magnetic pole into the view of earth, causing the observed variation of the magnetic field. According to Deutsch's hypothesis, the lines of certain elements vary in intensity because these elements tend to congregate in the vicinity of one or the other magnetic pole.

(e) Summary of Variable Stars

It is neither feasible nor appropriate to describe in detail the many other kinds of variable stars in a survey of astronomy such as this book. However, a tabular summary of the more important types of variables is given in Appendix 15.

(f) Cause of Pulsation

In a stable star the weights of the constituent layers, bearing toward the center of the star under the influence of gravity, are just balanced by the pressure of the hot gases within (Chapter 31). A pulsating star, on the other hand, is something like a spring: as the star contracts, its internal pressures build up until they surpass the weights of its outer layers. Eventually, these pressures start the star pulsing outward, but because of their inertia, the outward-moving layers overshoot the equilibrium point where their weights will just balance the internal pressures. As the star expands further, the weights of the overlying layers decrease, but the internal pressure decreases faster. Hence, the overlying layers are not supported adequately, and the star begins to contract. As it does so, it overshoots again, and this time it becomes too highly contracted. Once more the inner pressures cause the star to expand—and so the pulsation continues.

Stars would pulsate indefinitely if there were no dissipation of energy. Most stars do not pulsate in this way, however, because with each pulsation some of the energy is radiated away, converted to convection of stellar gases, or otherwise lost. Thus if a star were to start oscillating, its pulsations would quickly die out due to these damping forces unless it experienced some kind of driving force.

The energy needed to drive the pulsation comes, of course, from the nuclear, thermal, or gravitational energy sources that make the star shine. The mechanism that continually feeds this energy into pulsation for certain stars is a sort of valve action provided by the ionization and de-ionization of some abundant elements such as hydrogen and helium.

The pulsation of a cepheid variable, for example, is evidently produced by the ionized gases in its outer envelope cooling and becoming neutral when it is at its largest. As the gas cools it radiates its heat into space and the pressures within it drop, allowing the outer parts of the star to fall inward. As they do so, they absorb energy from the interior of the star and become ionized again. The reionization heats the gas and raises its pressure again, causing it to expand and cool; thus the process is repeated.

In most stars the neutral hydrogen and helium is either in too narrow a layer near the surface to be important or there is too little energy from the stellar interior to ionize it anyway. Thus this mechanism does not cause most stars to pulsate. Still, it is interesting to ask what the pulsation period would be for an ordinary star if it were unstable. It turns out that the period is greater for a giant star of low mean density than for a smaller compact star of higher density—just as a long piano string vibrates more slowly than a short one. It can be shown that for pulsating stars of any one type, the period of a particular star is inversely proportional to the square root of its mean density.

We can derive the relation between the period and mean density of a pulsating star from a consideration of Kepler's third law. When a star is at its maximum size and there is no longer enough pressure to support its outer layers they simply fall inward under the influence of the star's gravitation. Thus the matter at the surface of such a star is temporarily in free fall toward the star's center, as if it were in an elliptical orbit about the center of the star (but a degenerate ellipse that is a straight line). The period of revolution of such material in this hypothetical orbit is of the same order of magnitude as the period with which the star would pulsate. By Kepler's third law, this period, P, is related to the mass, M, of the star and the semimajor axis of the orbit—which is proportional to the radius, R, of the star—by the equation (Section 5.3c)

$$MP^2 \propto R^3.$$

Upon solving for the period, we find

$$P \propto R^{3/2}/M^{1/2}.$$

Now the mass of the star is the product of its volume, $4\pi R^3/3$, and its mean density, ρ; therefore,

$$P \propto R^{3/2}/(R^{3/2}\rho^{1/2}),$$

or

$$P\sqrt{\rho} = \text{constant}.$$

The surface of the star does not, of course, complete an actual orbit about its center. We have seen that in a cepheid, for example, the radius changes by less than 10 percent. On the other hand, if a point on the stellar surface *could* complete such a full orbit it would still spend most of its time in the outer part of that orbit (in accord with Kepler's second law); thus the representation is not as bad as it might seem. Even if the actual pulsation period differed by five or ten times from that derived by the above procedure, it would still be the case that the ratios of pulsation periods of different stars would be roughly inversely proportional to the square roots of their mean densities. (The proportionality would be exact if the different stars were exactly similar to each other, that is, were built on the same model.)

The shortest period with which a star could rotate is, in order of magnitude, the same as the period with which it would pulsate. This relation is easily seen; the speed at the equator of a rotating star cannot exceed the speed of a body at that point on a circular orbit, and that circular orbit has only twice the semimajor axis of the rectilinear elliptical orbit of major axis equal to the star's radius.

Detailed calculation shows that if the sun were to pulsate its period would be about one hour. Thus we can estimate the periods of pulsation of other kinds of stars by comparing their mean densities to the sun's. Table 34.1 shows the results of such a comparison for stars all of one solar mass but of

Table 34.1 **Pulsation Periods for Various Stars of One Solar Mass**

RADIUS (Solar Radii)	PERIOD	EXAMPLES
1	1 hr	Sun
1000	4 yr	Red supergiants
100	1 month	Cepheid
10	1 day	RR Lyra star
0.1	2 min	
0.01	4 s	White dwarf
10^{-5}	10^{-4} s	Neutron star

different radii. Because most stars have masses that differ from the sun's by less than a factor of 10, their pulsation periods would differ from the tabulated values for one solar mass stars of the same radii by only at most a factor of 3.

In sum, the shorter the period of pulsation of a star, or the minimum period with which it can rotate, the higher the mean density of the star. Most known variable stars are giant and supergiant stars of low mean density. This should not surprise us, for most variable stars are discovered to vary by comparing photographs of them made at different times, and most astronomical photographs are time exposures of at least half an hour. A star as dense as the sun or denser would pulsate through one or more complete cycles during a single exposure and its variation would never be detected on separate photographs. On the other hand, modern electronic observing techniques have revealed some stars to vary in light in periods of a few seconds, and in one case (the Crab pulsar—Section 34.9) in only 0.03 s.

34.4 MASS EXCHANGE IN BINARY STARS

On the line between the centers of the two stars of a binary system there is a point where the gravitational attraction between a small body and one of the stars would be equal to that between the body and the other star. If the two stars were of equal mass, for example, that point would lie halfway between them. There are, in fact, entire surfaces that partially or completely surround either or both stars; a small body can move freely along one of these surfaces under the combined effects of the gravitational forces of the two stars and the mutual revolution system (see Section 6.1b).

Now suppose that one of the stars of a binary system, in becoming a red giant, increases in size, so that its outer surface extends through one of these regions between the stars. The stellar material along such a surface "doesn't know which star it belongs to" and is no longer gravitationally bound to the original star. An exchange of material from one star to the other may result. Also, some material may flow away from both stars and form a ring encircling both, or even a common envelope. Such a gaseous envelope surrounding a binary system can reveal itself by producing absorption lines superposed upon the spectra of both stars, lines which do not show large variations in Doppler shift, since the material in the outer envelope is not moving as rapidly as the two stars are in their mutual revolution.

Of the many binary systems in which the member stars are exchanging or losing matter, a famous example is the eclipsing binary, β Lyrae. The spectrum of the system shows a set of absorption lines that appear to belong to a star of spectral type B8, and which show variable Doppler shifts that yield a radial-velocity curve with a period of just under 13 days. Also in the spectrum, however, appear both dark and bright lines that do not exhibit the Doppler shifts of the lines of the B star. One model suggested for the system consists of a B8 star and an F star in mutual revolution. The F star is too faint for its spectrum to be observed, but its spectral type is surmised from the relative amounts of light lost during the two eclipses of the system. Material is presumed to be flowing from each star and forming a ring surrounding the system; this ejected material produces the absorption and emission lines that do not show large variations in Doppler shift. As a result of the mass loss, the period of mutual revolution of the two stars is increasing by about 10 s/yr. At the present rate at which they are ejecting material, the stars would be depleted of matter in about 300,000 years. The existing state of affairs for β Lyrae, however, is almost certainly temporary; the system will doubtless change its character in a very much shorter time.

There is a large group of eclipsing binary stars, known as *W Ursae Majoris stars*, in which the two members of each system are so close together that they are nearly in contact and revolve about each other in a period of less than one day. These stars are greatly distorted by their mutual tidal forces, and material from each of them forms a gaseous envelope surrounding both stars.

34.5 WHITE DWARFS: ONE FINAL STAGE OF EVOLUTION

Sooner or later a star must exhaust its store of nuclear energy. Then it can only contract and release more of its potential energy. Eventually, the shrinking star will attain an enormous density. We observe such stars—the extremely compact white dwarf stars (Section 27.2d), whose mean densities range up to over 1 million times that of water. In a white dwarf the electrons are completely degenerate throughout the star, save for a very thin layer at the surface.

(a) Structure of White Dwarfs

The structure of a white dwarf was first studied by R. H. Fowler. White dwarfs are simpler than most stars because the pressure that supports a white dwarf in hydrostatic equilibrium is supplied almost entirely by the degenerate electrons, and therefore does not depend on the temperature, but only on the density. Now as we have seen in Section 34.1, the volume to which a star can be compressed before the electrons become degenerate (that is, its density) depends on the amount of gravitational potential energy that can be released by the collapsing star, which in turn depends on its mass. The size of a white dwarf, therefore, depends on its mass—the more massive the white dwarf, the smaller its size. A white dwarf of one solar mass must have a radius of about one percent of the sun's—about the size of the earth!

In the more massive white dwarfs some of the electrons have speeds that are an appreciable fraction of that of light, and a rigorous treatment must include the effects of special relativity (Chapter 13). The first such rigorous models of white dwarfs were constructed by the Indian-American astrophysicist S. Chandrasekhar. Chandrasekhar's analysis shows that white dwarfs of masses successively greater than the sun's are successively smaller than one percent of its radius, until a mass of 1.4 solar masses is reached, at which point the electrons do not become degenerate, and hydrostatic equilibrium cannot be achieved (Figure 34.5). Thus, 1.4 solar masses is the upper limit to the mass of a white dwarf. A more massive star must continue to collapse to a far smaller size.

Now those stars that have had time to exhaust their nuclear fuel supply and evolve to the white dwarf stage must have had original masses greater than 1.4 solar masses, for those more massive stars are the very ones that use up their energy store most rapidly. But in such stars the electrons do not become degenerate, and the stars cannot become white dwarfs.

On the other hand, white dwarf stars are plentiful, and they must have come from somewhere. Moreover, the number of white dwarfs is high enough, as nearly as can be determined, to account for most or all evolved stars of original mass greater than 1.4 solar masses. It is thought therefore, that most stars eventually do, in fact, become white dwarfs. Consequently they must lower their masses somehow, before reaching that stage, by ejecting

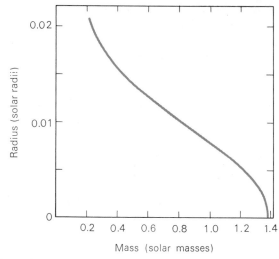

Figure 34.5 Theoretical relation between the masses and radii of white dwarf stars.

matter into space. In the next two sections we shall discuss eruptive variables and stars with extended atmospheres, both of which are examples of stars ejecting mass.

White dwarfs have hot interiors—tens of millions of Kelvins. At those temperatures and at the high densities of these stars, any remaining hydrogen would undergo violent fusion into helium, giving the stars luminosities many times higher than observed. Consequently, white dwarfs can have no hydrogen. Their most probable internal composition is a mixture of carbon and oxygen, the principal products of helium burning.

Recent theoretical studies indicate that at least some white dwarfs probably have cores in which the matter has crystallized. Some also have very strong magnetic fields—up to hundreds of millions of gauss. Moreover, a few display light variations, with periods of several minutes.

34.6 ERUPTIVE STARS

There are many types of eruptive variable stars; they range from the *flare* (or *UV Ceti*) stars, which display occasional sudden flareups in brightness, through the *novae*, to the spectacular *supernovae*. Also to the eruptive variables belong the *R Coronae Borealis variables*, stars that show sudden, unpredicted decreases in brightness. Most eruptive stars eject matter into space. Several examples are described here, and a table giving summary data for many kinds of eruptive stars is in Appendix 16.

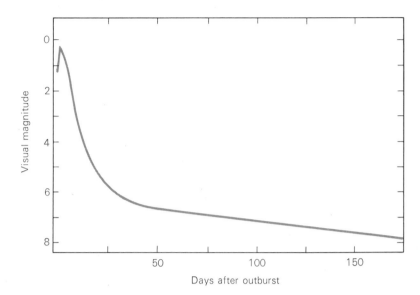

Figure 34.6 Light curve of Nova Puppis, 1942.

(a) Novae

The most famous of the eruptive variables are the *novae*. Nova literally means "new." Actually, a nova is an existing star that suddenly emits an outburst of light. In ancient times, when such an outburst brought a star's luminosity up to naked-eye visibility, it seemed like a new star. Novae remain bright for only a few days or weeks and then gradually fade. They seldom remain visible to the unaided eye for more than a few months. The Chinese, whose annals record novae from centuries before Christ, called them "guest stars." Only occasionally are novae visible to the naked eye, but, on the average, two or three are found telescopically each year. Many must escape detection; altogether there may be as many as two or three dozen nova outbursts per year in our galaxy. The light curve of a typical nova is shown in Figure 34.6.

According to the currently favored theory, novae occur in close binary star systems, in each of which one member is a white dwarf and the other member is a star transferring mass to the white dwarf by the mechanism described in Section 34.4. For a while (years to centuries) hydrogen-rich material from the outer layers of the nonwhite dwarf just piles up on the surface of the white dwarf. Gradually, however, the weight of this matter increases, and hence the temperature, until it approaches that of the degenerate interior of the dwarf. Then, explosively, hydrogen burning ignites through the CNO cycle, and like a nuclear bomb it blows off the outer layer of matter that had accumulated on the white dwarf.

After the outburst, the white dwarf settles down, but since mass is continually flowing onto it from its companion, the process eventually repeats. Some novae have long been known to be recurrent. Generally, the more intense the outburst, the longer the period of quiescence between nova flareups. According to the model, *all* novae recur on some time scale or other; the most violent *classical* novae, which reach absolute visual magnitudes of -6 to -9, may wait hundreds or thousands of years or more between outbursts.

We learn something of the physics of novae from their spectra. Just after outburst, the dark lines show sudden shifts to the violet—indicating that the star's photosphere is rising and approaching us. Shortly after the maximum light output, however, the expanding ejected shell of matter thins out enough to become transparent. It then no longer emits a continuous spectrum but still absorbs some of the star's light and reemits it in bright emission lines. These emission lines, however, originate from different parts of the shell, which are moving at different speeds along our line of sight, and consequently the lines display a wide range of Doppler shifts (Figure 34.7). Light reemitted from parts A and B of the shell, moving away from the earth, is shifted to longer wavelengths; light from parts C and D is not shifted; light from parts E and F is shifted to the violet. Only part G of the shell, directly between the earth and the star itself, produces absorption lines in the observed spectrum of the star; these are shifted farthest of all to the violet. The emitted light from the shell, therefore, does not appear *to us* in the form

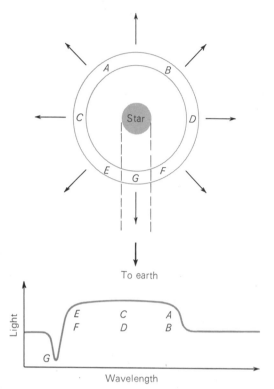

Figure 34.7 Diagram of an expanding nova shell. Below is a profile of an emission and absorption line formed by the shell.

of sharp bright lines but as broadened lines, or emission bands. At the violet end of each emission band is the sharp absorption line produced by the part of the shell directly between us and the star.

From the widths of the emission bands, or from the displacement of the absorption lines at their violet ends, we can calculate the velocity with which the shell is expanding. Velocities of ejection of up to 1000 km/s or more are found. From the total light emitted by the shell, we can calculate the amount of material it contains. A typical nova shell is found to contain from 10^{-5} to 10^{-4} solar mass. Some months or years after the outburst, the expanding envelope may become visible on telescopic photographs, as in the case of Nova Herculis (Figure 34.9).

Some post novae (or, presumably, novae between outbursts) are X-ray sources. X rays are normally emitted by a very hot gas (millions of Kelvins). Gas can be heated to such temperatures by friction if it falls together at very high speeds. The matter falling on the white dwarf star from its companion accelerates to such high speeds as it approaches the dense compact star, so the X rays are a natural consequence of mass transfer in a binary system when the recipient star is a small dense one.

(b) Flare Stars

Flare stars (or *UV Ceti stars*) are red main-sequence stars, mostly of spectral types M3 to M6. These stars show rare flareups in brightness. The total increase in light may be from 0.1 to several magnitudes. Their rise to maximum is very rapid—a few seconds—and the decline lasts for from a few minutes to an hour or so. They behave rather like miniature novae.

The light associated with a flare is blue, appropriate to a star with a temperature of 10,000 to 20,000 K. A large area at such a temperature would radiate far more energy than is observed to come from flares. Consequently, a flare must be a local-

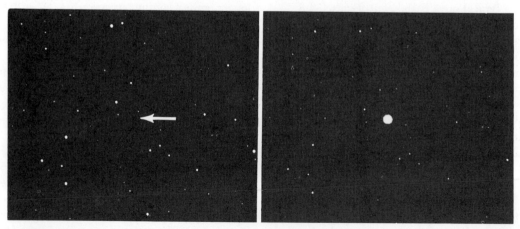

Figure 34.8 Nova Herculis, as it appeared before and after its outburst in 1934. (*Yerkes Observatory*)

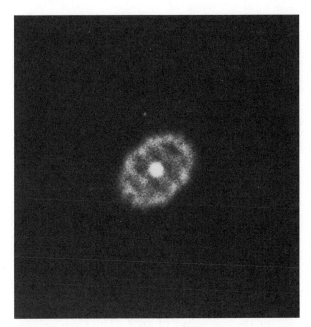

Figure 34.9 The shell of Nova Herculis photographed with the Lick Observatory's 3-m telescope in 1972. *(Courtesy H. Ford, UCLA)*

ized disturbance on a small part of a star. In total energy output, a typical flare observed on a UV Ceti-type star resembles the more spectacular solar flares (Section 31.3i).

Flares cause proportionately larger increases in the light of intrinsically faint stars than of brighter ones. Suppose a flare causes an M2 star to brighten by one magnitude—a factor of $2\frac{1}{2}$ in light. The flare itself, then, gives off (during the brief period of its existence) $1\frac{1}{2}$ times as much light per second as does its parent star. A flare of exactly the same energy would increase the light of a star two magnitudes brighter than the M2 star by only about 20 percent, and would increase the sun's light by only 1 percent, an amount too little to be noticed if the sun were as distant as the other stars. We cannot be sure, in other words, that flares do not occur on intrinsically bright, as well as intrinsically faint, stars; they are obvious only on the latter. On stars of intermediate luminosity, flares might be observed spectrographically. At least one flare was observed on a K star by means of bright emission lines that happened to be recorded while the star's spectrum was in the process of being photographed.

(c) Supernovae

Among the more spectacular of the cataclysms of nature is the *supernova*. In contrast to an ordinary nova, which increases in luminosity a paltry few thousands or at most tens of thousands of times, a supernova is a star that flares up to hundreds of millions of times its former brightness. At maximum light, supernovae reach absolute magnitude −14 to −18, or possibly even −20. The three most famous supernovae to have been observed during the last ten centuries in our Galaxy are (1) the supernova of 1054 in Taurus (described in the *Chinese Annals*), (2) Tycho's "star" of 1572 (Section 3.3) in the constellation Cassiopeia, and (3) the supernova of 1604 in Serpens, described by both Kepler and Galileo. Supernovae in other galaxies are now commonly observed. In a typical galaxy, supernovae occur at the rate of at least one every hundred years.

The 1054 supernova may also have been observed by Indians in the American Southwest. Several petroglyphs of about the right age in Arizona and New Mexico depict a bright star next to the crescent moon, in approximately the orientation the supernova and moon would have had when they rose together on July 5, 1054, the day after the outburst, and at about the time when it was most spectacular. The 1054 supernova remained visible to the unaided eye in broad daylight for several weeks and in the nighttime sky until April 1056.

The light curve of a supernova is similar to that of an ordinary nova, except for the far greater luminosity of the supernova and its greater duration of visibility. There are several different kinds of supernovae, but they all rise to maximum light extremely quickly (in a few days or less), and for a brief time one may outshine the entire galaxy in which it appears. Just after maximum, the gradual decline sets in, and the star fades in light until it disappears from telescopic visibility within a few months or years after its outburst. Bright emission lines in the spectra of supernovae indicate that they, like ordinary novae, eject material at the time of their outbursts. The velocities of ejection may be substantially greater than in ordinary novae, however, and speeds of up to 10,000 km/s have been observed. Moreover, a much larger amount of material is ejected; in fact, a large fraction of the original star may go off in the expanding envelope.

(d) The Crab Nebula

An example of the remnant of a supernova explosion is the Crab nebula in Taurus, a chaotic, expanding mass of gas, visible telescopically, and spectacular on telescopic photographs. The Crab nebula is one of the most interesting and most studied objects in the sky.

The outer parts of this gas cloud are observed to be moving away from the center at rates roughly proportional to their distances from it. V. Trimble has measured angular motions of up to 0″.222 per year. The Doppler shifts of light from the center of the nebula show the gases there to be moving toward us at speeds of up to 1450 km/s. If the nebula were to expand at the same rate in all directions, we could conclude that at the distance of the Crab, 1450 km/s produces an annual proper motion of 0″.222. In Section 23.1c we developed the formula

$$T = 4.74 \, \mu r \text{ km/s},$$

which relates tangential velocity, T, proper motion, μ, and distance, r. Upon substituting 1450 km/s for T and 0″.222 for μ, we find for the distance to the Crab nebula 1380 pc. However, the nebula is elongated in shape (Figure 34.10). Trimble advances reasons for thinking that it is actually a prolate spheroid (like a football) with its longest dimension roughly in the plane of the sky. Then the velocity of 1450 km/s would correspond more

nearly to the maximum proper motion expansion along the *minor axis* of the projected image, which is 0″.151. The Crab's distance corresponding to this assumption is about 2020 pc. When all factors are considered, a distance for the nebula of about 2000 pc is thought to be most probable.

Now if we were to assume that the nebula has always expanded at the same rate, we could derive its age by calculating how long the expansion at its edge (0″.222/yr) would take for it to reach its present maximum radius of 180″. Trimble has analyzed the motions of nearly 300 filaments, and she finds that with their present speeds they would have had to start their outward motion at about 1140 A.D. Radiation from the supernova blast should accelerate the gas, however, so that it must have started expanding at an earlier epoch. It turns out that both the location and computed time of formation of the Crab nebula are in good agreement with the occurrence of the supernova of 1054. The Crab nebula, therefore, must be the material ejected during that stellar explosion.

The Crab nebula is a strong source of radio waves, infrared radiation, X rays, and gamma rays, as well as of light. As an X-ray source, it is

Figure 34.10 The Crab nebula, photographed in red light with the 5-m telescope. (*Palomar Observatory; California Institute of Technology*)

known as Tau X-1. The radio spectrum (variation of radio energy with wavelength) has characteristics that led the Soviet astrophysicist I. S. Shklovsky, in 1953, to propose that the radiation is from the synchrotron process (Section 29.5a). The Crab nebula is the first astronomical object from which synchrotron radiation was recognized. When the nature of the radio radiation from the Crab nebula was discovered, it was suggested that some of its visible light might similarly originate from the synchrotron mechanism. If so, theory predicted that the light should be polarized. The polarization of light from the Crab nebula was observed at the 5-m telescope by W. Baade in 1956. The red filaments (see the color photograph) derive their light mostly from hydrogen ions recombining with electrons, but the white light, and the other radiation, from radio to gamma rays is synchrotron, showing the Crab nebula to possess strong magnetism, and a large source of relativistic electrons. We shall see also that at the center of the nebula is a pulsar (see Section 34.9).

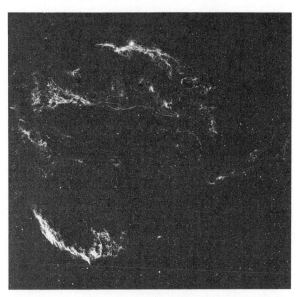

Figure 34.11 Filamentary nebula in Cygnus, photographed with the 1.2-m Schmidt telescope. The nebula is believed to be the remnant of an ancient supernova. (*Palomar Observatory; California Institute of Technology*)

(e) Other Galactic Supernovae

Radio sources are now known that correspond to positions of other supernovae, including those observed by Tycho (1572) and Kepler (1604). If the position of a temporarily appearing bright star, as given in historical records, agrees with that of an observed radio source, and if the characteristics of the radio radiation suggest synchrotron emission, the star is generally assumed to have been a supernova. The identification is further strengthened if gas filaments or nebulosity can be observed at the same position. Provisional identifications of a dozen or more supernovae in our Galaxy within the past 2000 years have so been made. Descriptions of the temporary or "guest" stars themselves appear in the annals of the Chinese (especially), the Japanese, Koreans, Arabs, and Europeans. Several of these were observed with the naked eye for more than one year, and one was seen for two years. The latter reached an estimated apparent magnitude (at brightest) of −8, many times as bright as Venus; several were easily visible in daylight. Still other radio sources have been identified with filaments of gas that are believed to be the remnants of prehistoric supernovae. An example of these is the filamentary "loop" or Veil nebula in Cygnus.

Another famous supernova remnant is the Vela X-ray source in the southern sky. Like the Crab nebula, Vela X has a pulsar at its center (Section 34.9). The distance of the remnant is about 460 parsecs, and the supernova is believed to have occurred between 11,000 and 20,000 years ago. At such a near distance the supernova would have been extremely bright in the sky; prehistoric man must have been treated to (or possibly frightened by) a star that might have been as bright as the full moon. The energy released in the supernova explosion was probably about 10^{52} ergs, and is believed to have been responsible for the formation of the Gum nebula—a region of ionized interstellar gas over 350 parsecs in radius. This nebula (named for the late astronomer, Colin S. Gum, who studied it extensively) is the largest known in our Galaxy; its nearest edge is only about 100 pc from the sun.

From the historical records of "guest stars" that most nearly fit the description of a supernova, and from the number of observed supernova remnants, we can estimate the average frequency of supernovae in our Galaxy. The estimates are uncertain, and opinions vary among specialists, but most fall in the range of one supernova every 30 to 100 years.

34.7 STARS WITH EXTENDED ATMOSPHERES

We have seen that novae, supernovae, and novalike stars eject gas shells at the times of their outbursts.

In addition, many other stars, for which no outbursts have been observed, are known to be surrounded by extended atmospheres of expanding gas shells. The extended atmosphere is usually revealed by the presence of emission lines or bands superposed on the continuous spectrum of the star, just as in nova shells (Section 34.6a). Sometimes the light absorbed and then reemitted by the gas shell is too feeble to be observed, but the shell may still reveal its presence by producing absorption lines that, because of their wavelengths or sharpness, cannot originate in the stellar photosphere. In a few cases, a gaseous envelope surrounding a star can be seen or photographed telescopically.

Many red giants have extended atmospheres, whose spectra show that the gaseous shells have been and are being ejected and are now expanding about them. The mechanism by which this material is ejected is unknown, but nevertheless it appears that some of these stars may lose a substantial fraction of their original mass over a period of a few hundred million years.

About 4000 spectral-class B stars are known whose spectra show emission lines, usually of hydrogen, and sometimes of other elements as well. These stars, known as *B emission*, or *Be*, *stars*, have evidently ejected material from their outer layers. Most of them are rapidly rotating, and the mass ejection may be related in some manner to their rapid rotation.

(a) SHELL STARS

A good example of the class of stars known as *shell stars* is the B5 star Pleione, one of the brighter members of the Pleiades cluster. In 1888, bright emission lines of hydrogen appeared in the spectrum of Pleione, but by 1905 they had disappeared. They reappeared, however, in 1938, and this time sharp dark absorption lines appeared in the centers of the emission lines. The bright lines originated from gases, apparently ejected from and surrounding the star. The dark lines were formed by the gases directly between the star and us. The significant thing about the absorption lines produced by the ejected gas is that they were sharp, while the dark lines produced in the photosphere of the star are greatly broadened by the star's rapid rotation, which produces a linear velocity of about 350 km/s at its surface. This is more than 100 times as great as the linear rotational velocity at the surface of the sun.

In such a rapidly spinning star, the equatorial regions, where the speed is greatest, are moving at nearly the circular satellite velocity (Section 5.7). A small ejection force, therefore, can cause gases in the equatorial region to move out, spiraling into ever-larger orbits. Such ejected material would produce a slowly expanding ring. It is generally supposed that Pleione and other shell stars are surrounded by *rings* of gaseous material which have been ejected from the equatorial parts of the stars. Several dozen shell stars, all more or less like Pleione, have been studied.

(b) WOLF-RAYET STARS

Nearly 200 Wolf-Rayet stars (also called *W stars*) are known. The brightest of them is the second-magnitude star, γ Velorum. Wolf-Rayet stars are named for the two astronomers who discovered the first ones in 1867.

W stars are spectral-class O stars of moderately high luminosity (absolute magnitudes of about −5) and very high surface temperatures, ranging up to estimated values of 100,000 K. In size they are more modest, however, averaging about only twice the radius of the sun.

Bright, broad emission lines appear in the spectra of Wolf-Rayet stars. These lines are usually of the elements helium, nitrogen, oxygen, silicon, or carbon. At the violet edge of each emission line is a sharp absorption line. The widths of the emission lines show that material must be ejected at speeds of from 1000 to 3000 km/s. The gas shells in Wolf-Rayet stars are similar to the envelopes ejected from novae except that they are not known to accompany a sudden outburst of light. Some investigators feel that W stars may be related to P Cygni stars (Appendix 16) differing mainly in that the former are hotter and, of course, are not known to be eruptive variables.

Many Wolf-Rayet stars are members of spectroscopic binary systems; a few are even eclipsing stars. The binary condition may be related to the unusual properties of W stars, but how is not known.

(c) Planetary Nebulae

For stars of 1.2 or so solar masses—the most massive stars among red giants in globular clusters and in the galactic corona, nucleus, and disk—one of the most important mass-ejection mechanisms may be the planetary nebula phenomenon. *Planetary nebulae* are shells of gas ejected from and expanding about certain extremely hot stars. They derive their name from the fact that a few bear a superficial telescopic resemblance to planets; actually they are thousands of times larger than the entire solar system, and have nothing whatever to do with planets.

Planetary nebulae are identified in two ways. Often they appear large enough to see or to photograph with a telescope. The most famous example is the ring nebula, in Lyra. It is typical of many planetaries in that, although actually a hollow shell of material emitting light, it appears as a ring. The explanation is that we are looking through the *thin* dimensions of the front and rear parts of the shell, while along its periphery our line of sight encounters a long path through the glowing material. Similarly, a soap bubble often appears to be a thin ring. The other way in which planetary nebulae are identified is by their spectra. Those that are very distant from us are unresolved and appear stellar, but their spectra show emission lines which indicate the existence of luminous shells of gas surrounding stars, as do spectra of nova shells. Altogether, there are about 1000 planetary nebulae catalogued. Doubtless there are many distant ones that have escaped detection, so there must be some tens of thousands in the Galaxy. Nevertheless, among the tens of billions of stars in the system, planetary nebulae must be classed as rare objects.

An appreciable amount of material is ejected in the shell of a planetary nebula. From the light emitted by the shells, we calculate that they must have masses of 10 to 20 percent that of the sun. The shells, typically, expand about their parent stars at speeds of 20 to 30 km/s.

It is an interesting problem to determine the distances to planetary nebulae. One (NGC 7078) is in a globular cluster of known distance, and one (NGC 246) belongs to a star that appears to be a member of a binary-star system. The distance to the companion star (and thus to the planetary) can be estimated from the companion's spectrum by the method of spectroscopic parallaxes (Section 27.2b). The central stars of planetary nebulae themselves, unfortunately, do not fit into the usual pattern of spectral and luminosity classes, and the method of spectroscopic parallaxes cannot be applied to them. Most planetary nebulae are too distant for their parallaxes to be measured directly; therefore, their distances are estimated by indirect means. The most widely applied technique is to assume that all planetary nebula shells have about the same mass—say, one fifth that of the sun. Then the distance to an individual nebula is estimated by calculating the distance it must have, if the mass assumed for it is correct, in order that it have its observed angular size and its observed brightness.

The linear diameters of planetary-nebula shells can be calculated from their angular diameters and distances (although the latter are known only with considerable uncertainty). A typical planetary appears to have a diameter of about $\frac{1}{2}$ LY to 1 LY. If it is assumed that the gas shell has always expanded at the speed with which it is now enlarging about its parent star, its age can be calculated. Most of the gas shells have been ejected within the past 50,000 years; an age of 20,000 years is more or less typical. After about 100,000 years, the shell is so enlarged that it is too thin and tenuous to be seen. The rarity of planetary nebulae, therefore, is due entirely to the fact that they cannot be seen for very long; they are temporary phenomena. When we take account of the relatively short time over which planetary nebulae exist, we find that they are actually very common, and that an appreciable fraction of all stars must sometime evolve through the planetary nebula phase.

The gas shells of planetary nebulae shine by the process of fluorescence. They absorb ultraviolet radiation from their central stars and reemit this energy as visible light. The physical processes are the same as in emission nebulae (Section 28.2b). During the first few tens of thousands of years after the ejection of the shell, the gas in it is dense and thick enough to prevent the star's ultraviolet radiation from penetrating all the way through it. This energy is completely absorbed, therefore, within the inner part of the shell, and only that inner part is luminous. The outer portion of the gaseous shell is dark and cold; it is transparent to visible light, however, so we see all the reradiated, longer-wavelength energy emitted from within.

The mechanism of the fluorescence process enables us to calculate, or at least to estimate, the temperatures of the central stars of planetary nebulae. All the visible light emitted by the gas shell is converted from ultraviolet energy originally emitted by the star. Knowing the details of the atomic processes of absorption and emission of light that are involved in the fluorescence phenomenon, we can calculate the rate at which ultraviolet radiation must be leaving the star to account for the visible light coming from the gas shell; it turns out to be a far greater amount of energy than the star radiates in its observable, visible spectrum. Most of the sun's radiant energy, on the other hand, is in the form of visible light. The central star of a planetary nebula, therefore, must be many times hotter than the sun for so large a fraction of its luminosity to be in the ultraviolet (see Section 10.2). Nearly all these stars are hotter than 20,000 K, and some have temperatures well in excess of 100,000 K, which makes them among the hottest known stars.

Despite their high temperatures, the central stars of planetary nebulae do not have exceedingly high luminosities—some emit little more total energy than does the sun. They must, therefore, be stars of small size; some, in fact, appear to have the dimensions of white dwarfs. Thus a planetary nebula may be the last ejection of matter by a star before it collapses to a white dwarf.

The central stars of planetary nebulae comprise a class of very hot, and usually very small, dense stars. It does not follow, however, that the stars were small and hot when the gas shells were ejected. The velocity of escape from such a small dense star is extremely high—up to thousands of kilometers per second. It is very difficult to imagine an ejection mechanism whereby a shell of gas can be shot off at such speeds, and a few thousand years later be expanding at a leisurely 20 to 30 km/s. It is more likely that the nebulae were ejected from their parent stars when the latter were, at an earlier stage in their evolution, large red giants, from which the escape velocity would be under 100 km/s.

34.8 CONSEQUENCES OF MASS EJECTION

We recall that white dwarfs cannot have masses more than 1.4 times that of the sun, but that white dwarfs are nevertheless numerous enough to account for most stars originally much more massive than the sun that, in the age of the Galaxy, can have expended their nuclear fuel and evolved to their final states. Evidently massive stars manage to qualify for white-dwarfhood by ejecting part of their matter into space. Very significantly, however, the material they shed is not the same as that from which they were formed, for the nuclear reactions by which they shine alter the chemical composition of their constituent gases.

(a) Increase of Heavy Elements in the Universe

We saw in Section 32.3 that stars convert hydrogen into helium and, moreover, that at least some stars

in some stages of their evolution are building up helium into carbon and heavier elements. Thus, inside stars, some of the lighter elements of the universe are gradually being converted into heavier ones. As these stars eject matter into the interstellar medium, that matter is richer in heavy elements than was the material from which the stars were formed. In other words, a gradual enrichment of the heavy-element abundance in interstellar matter is taking place. The heavy-element abundance in stars that are forming now is thus higher than in those that formed in the past. The fact that the oldest known stars (those in globular clusters) are the stars with the lowest known abundance of heavy elements provides evidence for this scenario.

We have reason to think (Chapter 39) that originally *all* stars in the Galaxy were formed of nearly pure hydrogen and helium, and that all the other elements were synthesized in the hot centers of stars at advanced stages of their evolution, and/or during explosive stages, such as in supernovae. Stars such as the sun, in whose outer layers heavy elements are observed spectroscopically, thus have to be of the second (or higher) "generation"; that is, they have been formed of matter that was once part of other stars. It is a grand concept that the planets (and we!) are composed of atoms that were synthesized in earlier generation stars—that we are literally made of "stardust."

(b) Evolution of White Dwarfs

A white dwarf is presumed to have exhausted its available nuclear-energy sources. It cannot contract and release gravitational potential energy because of the great pressure of the degenerate electron gas. Thus its only source of energy is the thermal energy (that is, kinetic energy) of the nondegenerate nuclei of atoms, behaving as ordinary gas particles, scattered throughout the degenerate electrons. As these nuclei slow down (cool), the electron gas conducts their thermal energy to the surface. At the boundary of the star, the very thin skinlike layer of nondegenerate gas radiates this energy into space. Only the opacity of this outer layer keeps the nuclei in the interior of the star from cooling off at once.

Gradually, however, a white dwarf does cool off, much like a hot iron when it is removed from a stove. The cooling is relatively rapid at first, but as the star's internal temperature drops, so does its cooling rate. Calculations indicate that its luminosity should drop to about 1 percent of the sun's in the first few hundred million years of its existence as a white dwarf. Since the radius of a white dwarf is constant, its surface temperature (by Stefan's law) is proportional to the fourth root of its luminosity. As it cools, its track on the H-R diagram is along a diagonal line, toward the lower right (low temperature and low luminosity).

Eventually, a white dwarf will cease to shine at all. It will then be a *black dwarf,* a cold mass of degenerate gas floating through space. (Do not confuse *black dwarf* with *black hole*—Chapter 35.) A long time may be required, however, for a star to cool off to the black dwarf stage. It may be that the Galaxy is not old enough for any star to have yet had time to become a black dwarf.

Figure 34.12 The Cambridge dipole array radio telescope used by Jocelyn Bell when she discovered the first pulsar. (*Courtesy Jocelyn Bell Burnell*)

34.9 PULSARS

In 1967, Jocelyn Bell, graduate research student at Cambridge University, was studying scintillations of radio sources with one of the Cambridge radio telescopes. (The apparatus she was using more nearly resembles an array of low wire clotheslines stretched out a few feet above the ground than a telescope; the Observatory even brings in sheep to graze the grass under the wires in summer, because there is not room for a man with a lawnmower. Yet, it *is* a radio telescope—Figure 34.12). In the course of Miss Bell's investigation she made a remarkable discovery—one that won her advisor, Antony Hewish, the Nobel Prize in physics, because his analysis of the object (and other similar ones) revealed the first evidence for *neutron stars*.

(a) Discovery of Pulsars

What Miss Bell had found, in the constellation of Vulpecula, was a source of rapid, sharp, intense, and extremely regular pulses of radio radiation, the pulses arriving exactly every 1.33728 s. For a time there was speculation that they might be signals from an intelligent civilization, and the radio astronomers half-jokingly dubbed the source, "LGM," for "little green men," and withheld announcement pending more careful study. Soon, however, three additional similar sources were discovered in widely separated directions in the sky, making it highly unlikely that they were signals. By 1981 hundreds of such sources had been discovered at radio observatories throughout the world. They are called *pulsars,* for *pulsating radio sources*.

The first pulsar to be discovered is typical. Detailed studies show each pulse to be complex and to consist of at least three much shorter pulses, each lasting only a few milliseconds, and the entire pulse sequence only a few tens of milliseconds. Although extremely regular in period, the pulses vary considerably from one to the next in intensity. The radiation in all pulses is polarized.

The pulses are emitted over a wide range of radio frequencies, presumably simultaneously, but they are received at the earth first at high frequency, with successively lower frequencies being received with successively longer delay times. The delay in the arrival of the radiation is believed to be due to the retardation of radio waves by free electrons in interstellar space. The interstellar electrons, in effect, give interstellar space an index of refraction (at radio wavelengths) greater than unity, so that the speed of radio radiation through space is reduced below its value in a perfect vacuum; the reduction in speed is greater the greater the wavelength, accounting for the observed relation between frequency and delay time of the received pulse. We can estimate the density of free electrons in space and calculate the distances to the pulsars from the relation between frequency and the delay of reception time of the pulses; typical distances for the pulsars are a few hundred parsecs. A few pulsars have 21-cm radio absorption lines (Section 28.2k), which show them to lie at least as far as certain interstellar hydrogen clouds producing those lines. Some pulsars, we find, are more than 1000 pc away. At such distances, the radio energy emitted in each pulse must be enormous for us to observe it as strongly as we do. Moreover, because of the sharpness of the pulses, that radio energy must be coming from a region at most a few hundred kilometers in diameter; otherwise, the light-travel time across the emitting region would result in lengthening the pulse time. The pulse periods of different pulsars range from $1/30$ s to about 4 s.

One pulsar is in the middle of the Crab nebula (described above). It has the shortest pulse period known (at this writing)—0.033 s, and the period is observed to be very slowly increasing, showing that pulsars evolve, pulsing gradually more slowly as they age. This pulsar is also observed to emit optical (visible) light and X-ray pulses with that same 0.033-second period. About 10 to 15 percent of the X-ray radiation from the Crab nebula comes from the pulsar. It has been identified with what appears to be a star of about 16th magnitude (Figure 34.13).

One other pulsar (in Vela—Section 34.7e) pulses in visible light. In addition to these pulsating radio sources, dozens of X-ray sources pulse in short regular periods. Of these, only the Crab pulsar is seen also in visible light and radio waves, but the other X-ray pulsars are believed to be objects similar to radio pulsars, but more energetic in their emission. At least some are members of binary star systems, and for four of these, enough information is available to calculate masses by techniques described in Chapter 26. These four X-ray pulsars have masses in the range 1.4 to 1.8 times that of the sun.

(b) Theory of Pulsars—Neutron Stars

The energy emitted by pulsars is not small; that from the Crab pulsar is considerably more than the energy emitted by the sun. Thus pulsars are like stars in their output of radiation. Yet they emit this energy in pulses of up to 30 per second, as if they were pulsating at such high frequency. But what kind of object can pulsate up to 30 times per second? Even a white dwarf would pulsate with a period of several seconds or more—hardly fast enough for the pulsars.

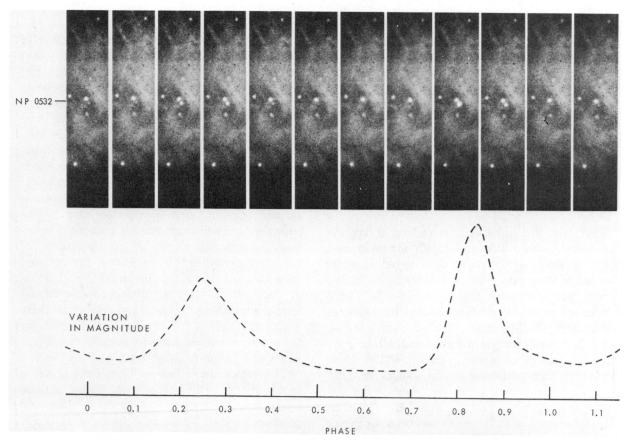

NP 0532 —

VARIATION IN MAGNITUDE

0   0.1   0.2   0.3   0.4   0.5   0.6   0.7   0.8   0.9   1.0   1.1

PHASE

Figure 34.13 A series of photographs of the central part of the Crab nebula taken by S. P. Maran at Kitt Peak National Observatory. Note the star that seems to blink on and off; it is the pulsar, which has a period of 1/30 s. *(Kitt Peak National Observatory)*

The solution to the problem suggested itself in 1932, when Chadwick discovered the neutron. Theoreticians then speculated that if the matter in a star could be subjected to such high pressure as to force the free electrons into the atomic nuclei, the star could become a body composed entirely of neutrons. A few years later, Mount Wilson astronomers Walter Baade and Fritz Zwicky suggested that supernova explosions might form neutron stars. At least some pulsars are associated with the remnants of supernovae. Could pulsars be neutron stars?

Neutrons, like electrons, obey the Pauli principle and can become degenerate if crowded into a sufficiently small volume for a given momentum range, so perhaps a star could collapse into degenerate neutrons if it somehow escaped becoming a white dwarf. The neutrons, in such a condition, cannot decay into protons and electrons, for by the time the star is that collapsed, the allowable states for electrons would be filled. The structure of a neutron star is analogous to that of a white dwarf, except that neutron stars are much smaller. A neutron star of one solar mass would have a radius of only about 10 km. Such a star would have a density of 10^{14} to 10^{15} g/cm^3—comparable to that of the atomic nucleus itself.

We note in Table 34.1 that a star of such dimensions would have a natural period of pulsation of less than one ten-thousandth of a second. Since a neutron star could rotate with any period much longer than this, we believe that pulsars must be rotating neutron stars.

There exists a mass-radius relation for neutron stars, and an upper mass limit as well; although the exact theory is not yet certain, the mass limit for a neutron star is believed to be from 2 to 3 solar masses. Since the upper mass limit for a neutron star probably exceeds that for white dwarfs, a star of mass greater than 1.4 solar masses could gravitationally contract to a neutron star, missing the white dwarf configuration.

We think, however, that neutron stars are also formed in supernova explosions. There are several mechanisms we know of by which this may occur. It seems unlikely that a star of less than 1.4 solar masses can become a supernova unless it is first a white dwarf that is a member of a close binary-star system. Suppose its mass is very close to the upper limit for the mass of a white dwarf and that the companion star is a giant that is transferring some of its matter to the dwarf. As the transferring material pours onto the white dwarf, it can raise its mass above the limiting value for a stable white dwarf; then there is nothing to stop it from collapsing to a neutron star.

Alternatively, a massive red giant star can have a small dense degenerate core whose mass may approach that of the limiting value for a white dwarf. Continued conversion of hydrogen to helium in a shell around the degenerate core can add mass to it until it exceeds the mass limit and collapses to a neutron star. In any case, gravitational energy is released from the mass collapsing to a neutron star, and that energy is probably enough to eject the outer layers of the star. Also, if the degenerate gas contains carbon, the carbon, heated by the gravitational energy released by the contracting matter, should undergo additional nuclear fusion into magnesium, sodium, and neon, which in turn undergo nuclear transformations to still heavier elements due to their absorption of released alpha particles, protons, and neutrons. The energy liberated helps eject the outer part of the star and at the same time new heavy elements are produced. Actually, the cause of supernovae is at best only poorly understood. Yet there seems to be a number of possible mechanisms that result in producing heavy elements, while ejecting a large amount of gas into space with the release of great amounts of energy.

Any magnetic field that existed in the original star is highly compressed if the core of the star collapses to a neutron star. Thus a moderate field of the order of 1 gauss in a star the size of the sun increases to the order of 10^{10} to 10^{12} gauss around the neutron star. At the very surface of the collapsed star neutrons decay into protons and electrons. Many of these charged particles should leave the stellar surface and move out in the vicinity of the magnetic poles into the circumstellar magnetic field. With such intense fields and high densities many of the particles are relativistic, especially the electrons, which emit electromagnetic energy by the synchrotron mechanism. The radiation is very directional, however, and if the magnetic field of the star happens not to be coaxial with the star's rotation (which also increases greatly during the collapse, to conserve angular momentum), the rotation carries first one and then the other magnetic pole into our view. The consequence can be that the radiation from the rotating magnetic field can be directed toward us once each time the star turns on its axis.

This, at least, is one model for the pulsed radiation from a pulsar. Meanwhile, the atomic nuclei are thought also to be accelerated by the magnetic field to relativistic speeds, producing cosmic rays. If this entire picture (or some modification of it) is correct, we can understand how supernovae produce heavy elements in the universe, nebulae like the Crab, neutron stars, pulsars, and cosmic rays.

Neutron stars may also be associated with another kind of pulsed radiation. More than two dozen X-ray sources, called *bursters,* are now known whose more or less steady emission of X radiation is suddenly enhanced by several times in a period of only a second or so. This burst of X rays then dies down over the next 10 to 20 s, and the source returns to normal. Generally the bursts recur at irregular intervals, but not more than 5 to 10 times a day. (The single exception is the so-called *rapid burster,* which for a couple of months in 1976 emitted bursts every 10 to a few hundred seconds; its activity is thought to have been produced by a different mechanism from that of the ordinary bursters.) The burst activity often continues for a few weeks and then enters a quiescent period of comparable or longer duration. The X radiation in the bursts appears to resemble that from a very hot black body (millions of Kelvins) and is thought to be produced by gas transferring to a neutron star from a companion, and either heating irregularly in an accretion disk or falling onto the surface of the degenerate neutron star and igniting in a nuclear blast—not unlike the suggested scenario by which novae are associated with white dwarfs.

(c) Evolution of Pulsars

The energy radiated by a pulsar, evidently from matter ejected from the star interacting with the stellar magnetic field, robs the star of rotational energy. Thus, theory predicts that the rotating neutron stars should gradually slow down, and that the pulsars should slowly increase their periods. Indeed, as we have already mentioned, the Crab pulsar is actually observed to be increasing the interval between its pulses; similarly, the Vela pulsar (in the center of the Gum nebula) is also found to be slowing down. According to present ideas, the Crab pulsar is rather young and of short period (we know it is only about 900 years old) while the other, older pulsars have already slowed to longer periods. Typical pulsars thousands of years old have lost too much energy to emit appreciably in the visible and X-ray wavelengths, and are observed only as radio pulsars; their periods are a second or more.

Studies by R. Lingenfelter show that supernovae within a few thousand parsecs and the pulsars resulting from at least some of them can produce a large number of cosmic rays. Lingenfelter predicts that as these cosmic rays interact with interstellar magnetic fields, they should migrate through the Galaxy and impinge on the earth with a frequency that is consistent with that of the observed cosmic ray influx.

One other possible end state for a star is that of a *black hole,* which involves one of the more bizarre predictions of general relativity theory. We postpone discussion of black holes until Chapter 35.

34.10 SS 433

One of the most remarkable objects in the Galaxy—possibly the first of many such objects to be discovered—is SS 433.

SS 433 is in a catalogue of objects showing H-alpha (hydrogen) emission lines compiled by C. G. Stephenson and N. Sanduleak in 1977. More recently, it attracted attention when it was found to be a source of variable radio emission and X rays. Moreover, it is near the center of a supernova remnant and may be related to a supernova outburst. But the really astonishing features of SS 433 became known when its optical spectrum was studied in a long series of observations—especially by B. Margon and his associates at UCLA.

The spectrum shows bright emission lines of hydrogen and helium arising from three different sources. One source appears to be the main object, and the Doppler shift of its lines indicates a relatively low speed, like that expected for an object in our Galaxy. The other two sources, however, have very high speeds with respect to the main body and are evidently streams of gas ejected from it in oppositely directed jets. Analysis shows the gas to be streaming out (in both directions) at 26 percent the speed of light. Moreover, the whole system is rotating in such a way that the directions of the jets describe a conical motion (suggesting precession) with a period of 164 days.

The object giving rise to the remarkable radiation is a member of a binary star system with a period of 13.1 days. A theory favored by some investigators is that the system consists of a collapsed object, probably a neutron star, in mutual revolution about a star that is transferring mass onto it. Because of the orbital revolution, that transferring matter does not fall directly on the neutron star,

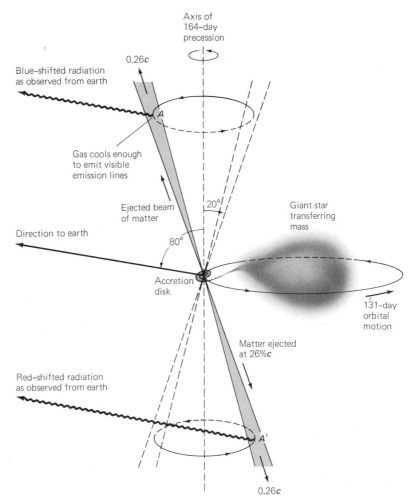

Figure 34.14 A model for SS 433, proposed by UCLA astronomers Jonathan Katz, Bruce Margon, and George Abell. A binary system consists of a giant star and a collapsed star (probably a neutron star) in orbit about each other with a period of 13.1 days. Matter from the giant transfers to the compact star and accumulates into an accretion disk, from which jets of gas are ejected at 26 percent the speed of light. The disk precesses so that the jets describe a motion along the surface of an imaginary cone of half angle 20°. SS 433 is viewed from the earth at an angle of 80° to the cone axis, so that at a typical time one jet has a component of motion toward the earth, and its radiation appears blue-shifted by the Doppler effect, whereas the other beam has a component of motion away from the earth, and its radiation appears red-shifted. The material in the jets probably does not cool enough to emit visible radiation until some distance from the disk—say, at A and A'.

but ends up in a flat rotating disk of matter surrounding the neutron star in its equatorial plane. (In this *accretion disk*, some of the matter may lose angular momentum to other matter and fall into the star.) According to the model, it is from the accretion disk that matter is ejected, perpendicularly to the disk, into the streams that give rise to the observed radiation. Under tidal forces produced by the non–neutron-star companion, the accretion disk is presumed to precess with the 164-day period, accounting for the variations in radial velocity of the radiation from the jets (Figure 34.14). This model, however, is not the only one possible and is therefore not universally accepted at this writing.

The importance of SS 433 is the study of the mechanisms for the mass ejection at so high a speed and the periodic change in direction. In some respects, SS 433 has properties that on a small scale remind us of those of radio galaxies and quasars (Chapter 37). Perhaps our eventual understanding of SS 433 will bring us the insight needed to fathom

the powerhouses of those far grander objects associated with remote galaxies.

34.11 SOLAR EVOLUTION

Figure 34.15 summarizes our current ideas on the evolution of a star of about 1.2 solar masses on the H-R diagram. In its early stages, the star contracts and moves to the left, reaching the main sequence with a size only slightly greater than that of the sun. In its subsequent evolution to the red giant stage, it grows to a radius of tens of millions of kilometers. The further evolution is uncertain. Perhaps the star goes through stages of variability, or emits material as a planetary nebula. Its final size as a white dwarf is only about that of the earth.

The sun is a typical star, and much of the theory of stellar evolution that we have discussed applies to it. From theoretical calculations we can now form a fairly clear picture of the approximate

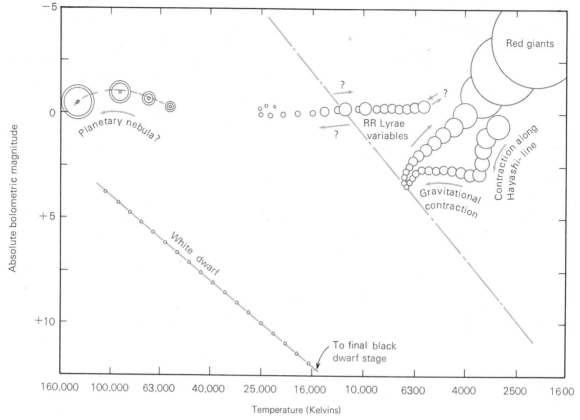

Figure 34.15 Summary of the evolutionary track on the H-R diagram for a star of 1.2 solar masses.

past history of the sun, and we can make at least educated guesses about its future.

(a) Early Solar Evolution

Several calculations of the sun's early evolution, analogous to those described in Section 33.1b, have been made. (See Section 14.3 for a discussion of the formation of the solar system.) The exact track of evolution of the sun on the H-R diagram depends on its assumed initial chemical composition and on rather uncertain data involving the opacity of the outer layers of the young sun. The time required for it to contract to the main sequence is probably a few tens of millions of years.

Since it reached the main sequence, the sun has increased somewhat in luminosity, probably by about 30 to 50 percent. During that interval of from 4 to 5 thousand million years, it has depleted much of the hydrogen at its very center, but a pure helium core has not yet had time to form. It is not certain how much more time the sun has before starting to evolve to the red giant stage, but a good guess is that it has lived out about half of its main-sequence life. We can probably look forward to at least another 5 thousand million years before the sun's structure undergoes large changes.

(b) The Future of the Solar System

All available evidence leads us to expect that sometime in the future the sun will leave the main sequence and evolve to a red giant. According to R. Ulrich, that time will occur in about 5 thousand million years, when the sun's photosphere will reach nearly to the orbit of Mars. The earth, then well inside the sun, and exposed to temperatures of thousands of degrees, will gradually vaporize. The gases in the greatly distended outer layers of the sun will, of course, be very tenuous, but they should still offer enough resistence to the partially vaporized earth to slow it in its orbital motion. Ulrich believes that the earth will spiral inward toward the very hot interior of the sun, reaching its final end

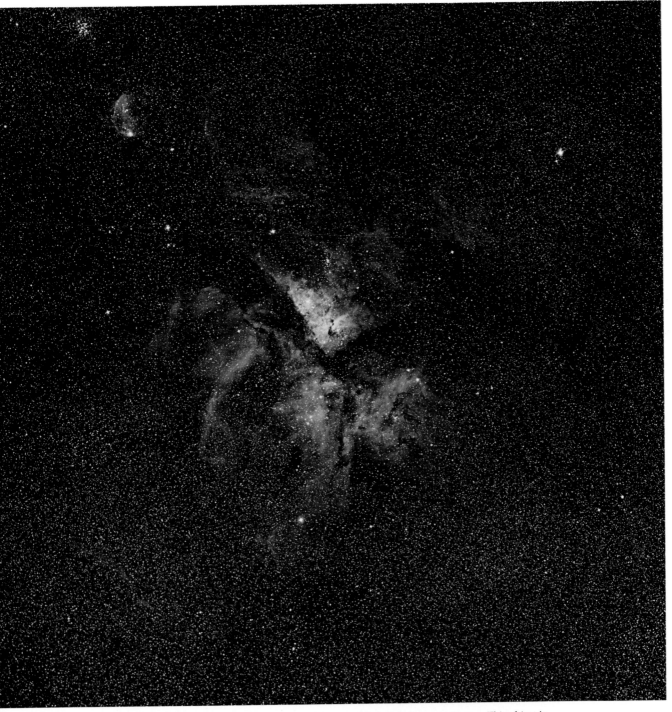

Eta Carinae, a great mass of gas, intermixed dust, and imbedded stars in Carina. This object is known to have changed greatly in apparent brightness over the last century and is suspected by some astronomers to be a region of recent star formation. Photographed at Cerro Tololo, Chile. *(Kitt Peak National Observatory)*

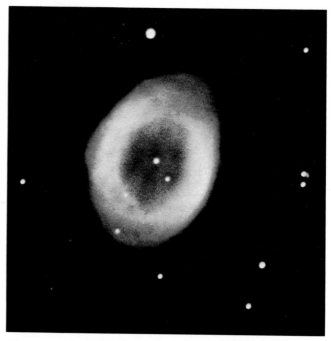

The Dumbbell nebula, a planetary nebula in Vulpecula, M27. *(California Institute of Technology/Palomar Observatory)*

The Ring nebula in Lyra, M57, photographed with the 5-m telescope. *(California Institute of Technology/Palomar Observatory)*

The Crab nebula, photographed with the 5-m telescope. *(California Institute of Technology/Palomar Observatory)*

about ten thousand years after being swallowed by the sun.

If, in that remote time, man could leave the doomed earth but remain in the solar system, he would find that most of the naked-eye stars in our 20th century sky would long since have exhausted their nuclear fuel and evolved to white dwarfs (or perhaps some other end). Main-sequence stars less massive than the sun would still be shining, but only those few of them passing temporarily through the solar neighborhood would be near enough to see with the unaided eye. It is doubtful, however, if by then *all* star formation in the Galaxy would have ceased. Luminous young stars might be shining in remote clouds of interstellar matter; a Milky Way might still stretch around the sky.

EXERCISES

1. Where on the H-R diagram does a star *end?*

2. Suppose a star spends 10×10^9 years on the main sequence and burns up 10 percent of its hydrogen. Then it quickly becomes a red giant with a luminosity 100 times as great as that it had while on the main sequence and remains a red giant until it burns up the rest of its hydrogen. How long a time would it be a red giant? Ignore helium burning and other nuclear reactions, and assume that the star brightens from main sequence to red giant almost instantaneously.

3. Verify that for the temperature range mentioned in the chapter, the luminosity of Mira can be expected to vary by only about a factor of $3\frac{1}{2}$.

4. Sketch the spectral energy distribution of Mira, both at maximum and at minimum, and show how a greater fraction of its total energy is in the visible spectral region when it is at maximum than when it is at minimum.

5. Compare the mean densities of two type I cepheids that have periods of four days and of 40 days. Which is brighter and by how much?

6. Since supernovae occur so rarely in any one galaxy, how might a search for them be conducted? (It may help to glance at Chapter 37.)

* 7. Explain why a rotating ring surrounding a shell star produces a sharp absorption line at the *center* of each emission line rather than at the violet edge, as in the case of an expanding shell of gas.

* 8. Calculate the diameter of a Wolf-Rayet star, in terms of the sun's diameter, if the star has 10,000 times the sun's luminosity and seven times its temperature.

9. Compare and contrast nova shells, supernova shells, and planetary nebulae.

10. Suppose the luminous shell of a planetary nebula is easily resolved with a telescope. Now suppose the spectrum of the nebula is photographed, with the slit of the spectrograph extending completely along one diameter of the shell. Sketch the appearance of a typical emission line in the spectrogram, and explain your sketch. It may help to look over the description of the construction of a spectrograph in Chapter 11.

11. Suppose the central star of a planetary nebula is 16 times as luminous as the sun, and 20 times as hot (about 110,000 K). Find its radius, in terms of the sun's. Does this star have the dimensions of a white dwarf?

12. The gas shell of a particular planetary nebula is expanding at the rate of 20 km/s. Its diameter is 1 LY. Find its age. For this calculation, assume that there are 3×10^7 s/yr, and 10^{13} km/LY.

*13. The angular radius of an expanding spherical shell of gas about a star is observed to increase at $0\rlap{.}''19$/year. The radial velocity of expansion of the shell, with respect to the star, is observed to be 1200 km/s. What is the distance to the object? (See Section 23.1c).
Answer: About 1330 pc

14. Assume that a pulsar is 100 pc away. Suppose that no star brighter than apparent magnitude 23 shows up in that position of the sky. What is the brighter limit to the absolute magnitude that a star associated with the pulsar could have?

15. Show that the surface temperature of a particular gradually cooling white dwarf is proportional to the fourth root of its luminosity. (See Chapter 10.)

16. By the time the sun becomes a white dwarf, the constellations familiar to us now will not be seen, even if their stars never change in luminosity. Why?

17. Calculate, very roughly, the density of a hypothetical neutron star of one solar mass.

18. Even when the jets of gas in SS 433 are moving *across* our line of sight, the wavelengths of the spectral lines are longer (as observed from earth) than they would be if the jets were motionless, or if the source were in the laboratory. Can you suggest an explanation? (*Hint:* See Chapter 13. How fast is the material moving? What happens to its time, as we observe it? What does this have to do with the wavelengths [or frequencies] of the light it emits?)

Albert Einstein (1879–1955) received the Nobel Prize in 1921, not for his theory of relativity but for the photoelectric effect. At that time his ideas on relativity were still at the frontier. Einstein believed that such seemingly diverse areas of physics as mechanics and electromagnetic phenomena—and even gravitation—were guided in the same way by underlying principles.

GENERAL RELATIVITY: CURVED SPACETIME AND BLACK HOLES

In his special theory of relativity (Chapter 13), Einstein showed that different observers in uniform relative motion perceive space and time differently—that is, they disagree with each other on measurements such as length, time, momentum, and energy. The special theory also shows how those measurements by relatively moving observers compare with each other. But the rules of special relativity do not apply to the comparison of measurements by observers that are *accelerated* with respect to each other. Einstein's general theory of relativity, published in 1916, shows how we can extend the special theory to take account of observers in relative acceleration produced by gravitation. In doing so, Einstein discloses a new formulation of gravitation, because a gravitational force on a system is indistinguishable from an acceleration of that system, that is, gravitation and acceleration are equivalent.

35.1 PRINCIPLE OF EQUIVALENCE

Galileo noted that all bodies, despite their different masses, if dropped together fall to the ground at the same rate. According to Newton's law of gravi-

tation, the earth pulls on a more massive object with a greater force than it does on a less massive one. The two objects fall together, however, because according to Newton's second law, a proportionately greater force is required to impart the same acceleration to the heavier object. The mass that appears in Newton's law of gravitation, however, is *gravitational mass*, and that which appears in his formulation of the laws of motion is *inertial mass*. The fact that bodies of different mass fall together implies that inertial mass and gravitational mass are the same.

In 1909 this equality was verified to better than 5 parts in 10^9 in a famous experiment by the Hungarian physicist, Baron Roland von Eötvös. Today, similar experiments have pushed any difference to below one part in a million million (10^{12}). It is widely assumed that gravitational and inertial mass are, in fact, identical.

Einstein had the genius to turn that isolated fact* into a powerful principle of physics—the

---

*As Julian Schwinger described it; I owe several of the examples in this chapter for describing the general theory of relativity to the unique talent of Professor Schwinger for peeling away the rind and getting to the heart of the idea that describes a fundamental physical concept.

principle of equivalence. In doing so, he took the first giant step toward formulating his general theory of relativity.

Because two bodies falling together side by side approach the ground together, they are obviously *not* accelerated with respect to each other. Thus they are aware of no force acting between them. For example, suppose a brave boy and girl simultaneously jump into a bottomless chasm from opposite sides of its banks. If we ignore air friction, while they fall they accelerate downward together and feel no external force acting on them. They can throw a ball back and forth between them, aiming always in a straight line, as if there were no gravitation, and the ball, falling along with them, would move directly to its target.

It's very different on the surface of the earth. Everyone knows that a ball, once thrown, falls to the ground. Thus in order to reach its target (the catcher) the ball must be aimed upward somewhat, so that it follows a parabolic arc—falling as it moves forward—until it is caught at the other end.

Because our freely falling boy, girl, and ball are all falling together, we could enclose them in a large box falling with them. Inside that box, no one can be aware of any gravitational force; nothing falls to the ground, or anywhere else, but moves in a straight line in the most simple natural way, obeying Newton's laws. By having our box fall with the boy and girl, we have removed the force of gravitation by selecting a coordinate system that is accelerating at just the right rate to compensate for gravitation. Here is the principle of equivalence—a force of gravitation is equivalent to an acceleration of the coordinate system of the observer, and such a force can therefore be completely compensated for by an appropriate choice of an accelerated coordinate system. Einstein himself pointed out how a rapidly descending elevator seems to reduce our weight and a rapidly ascending one increases it. In a *freely falling* elevator, with no air friction, we would lose our weight altogether.

This idea is not hypothetical. In 1973 and 1974, astronauts in the Skylab lived for months in just such an environment. The Skylab was, of course, falling freely around the earth, as it continued to do until, unmanned, it finally suffered too much friction with the earth's tenuous upper atmosphere and plunged to a fiery doom in the summer of 1979. But while in free fall the astronauts lived in a seemingly magical world where there were no outside forces. One could give a wrench a shove, and it would move at constant speed across the orbiting laboratory. One could lay a pencil in midair and it would remain there, as if no force acted on it.

Mind, there *was* a force; neither the Skylab nor the astronauts were *really* weightless, for they continually fell around the earth, pulled by its gravity. But since all fell together—lab, astronauts, wrench, and pencil—*locally* all gravitational forces were absent.

Thus the Skylab provides an excellent example of the principle of equivalence—how local effects of gravitation can be removed by a suitable acceleration of the coordinate system. To the astronauts it

Figure 35.1 In Skylab everything stays put or moves uniformly because there is no apparent gravitation acting inside the laboratory. *(NASA)*

was as if they were far off in space, remote from all gravitating objects. But what if astronauts *were* in remote space, and were to activate the engines of their ship, producing acceleration. The ship would then push up against their feet, giving the impression of a gravitational tug. If one were to drop a small coin and a hammer, the floor of the ship would move up to meet both objects at the same time; to the astronauts, though, it would seem that the hammer and coin fell to the floor together. To them it would be exactly the same situation as that isolated fact made famous by Galileo—that heavy and light objects fall together. In other words, an acceleration of one's local environment produces exactly the same effect as a gravitational attraction; the two are indistinguishable—again, the principle of equivalence.

(a) Trajectories of Light and Matter

Einstein postulated that the principle of equivalence is a fundamental fact of nature. If so, however, there must be *no* way in which an astronaut, at least by experiments within his local environment, can distinguish between his weightlessness in remote interstellar space and his free fall in a gravitational field about a planet like the earth.

But how about light? If the astronauts shone a beam of light along the length of their ship and if the ship were falling in a free-fall orbit about a planet, would the ship not then surely fall away from a straight line path, which the beam must follow, causing the light to strike above its target?

Not so, according to Einstein. If the principle of equivalence is correct, there must be no way of knowing whether one is accelerated (any more than he can detect his own absolute motion) and hence the experiment must fail. Thus the light beam *must fall with the ship* if that ship is in orbit about a gravitating body. The idea that light, as well as material bodies, must be affected by gravity led Einstein to the prediction that stars seen by light from them that passes near the sun must appear displaced be-

cause of bending of their light by the gravitational field of the sun. This prediction, when formulated precisely, was, as we shall see, eventually confirmed by observation during a solar eclipse.

(b) The Gravitational Redshift

Let us consider another possible experiment in a freely falling laboratory (the Einstein elevator). Suppose we shine a light beam—say, a laser beam of a precise frequency—upward from floor to ceiling. Now the laboratory accelerates downward, gaining speed, so by the time the light beam travels up to the ceiling, that ceiling is moving downward faster than the source on the floor was when the light left it. In other words, the receiver at the ceiling is *approaching* the source (where it was when the light left it). Therefore, wouldn't we expect to find the light at the ceiling blueshifted slightly because of the Doppler effect (Section 10.5b)? But this would violate the principle of equivalence, for the blueshift would reveal our downward acceleration and show us we could not be weightless in free space. Therefore, Einstein postulated, there must be a *redshift*, due to the light moving upward against gravity, that exactly compensates the Doppler shift that would otherwise be observed. If so, that gravitational redshift should be observed in radiation climbing upward in a gravitational field—in principle at least—even at the surface of the earth. Is such an effect observed?

The earth's gravitational field is too weak to show the effect on visible light because we know of no source for which the frequency can be so sharply defined that the extremely tiny redshift would be noticeable. Yet, it has been observed, not in visible wavelengths of electromagnetic radiation, but in gamma radiation. In the 1960s, at the Jefferson Physical Laboratory at Harvard University, gamma rays were sent from a source in the basement to a detector at the top of the building. The source of gamma rays was radioactive cobalt; the radiation was confined to a very sharp frequency interval by a technique invented by the Nobel Prize winning

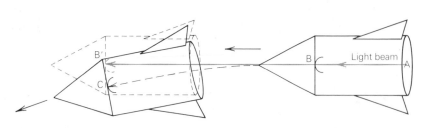

Figure 35.2 If in a spaceship moving to the left (in this figure) in its orbit about a planet, light is beamed from the rear, *A*, toward the front, *B*, we might expect the light to strike at *B'*, above the target in the ship, which has fallen out of its straight path in its orbit about the planet. Instead the light, bent by gravity, follows the curved path and strikes at *C*.

physicist Rudolf Mössbauer. If a similar detecting layer of cobalt were placed directly above the emitting layer, the gamma rays would be absorbed by the former. However, the detector was placed at the top of the building, 20 m above the source. The gamma rays, traveling upward against the earth's gravitation, suffered a gravitational redshift, and were not absorbed by the upper cobalt detector. In order to absorb them, the detector had to be moved slowly downward to produce a blueshift to compensate the earth's gravitational redshift. The actual motion of the detecting cobalt needed to make it absorb the gamma rays from the emitting cobalt in the basement was so slow that it would have required a full year to close the 20-m gap between emitter and detector. That speed produced a Doppler shift that agreed with the value needed to compensate for Einstein's predicted redshift to within 1 percent.

We can calculate the amount of gravitational redshift with elementary algebra. Let the elevator, of height h, be falling freely in, say, the earth's gravitational field, and for definiteness, near the earth's surface, where the uniform gravitational acceleration is g. Light, at speed c, takes a time $t = h/c$ to travel from the elevator floor to the ceiling. During this time, the elevator, accelerating uniformly, has increased its downward speed by

$$v = gt = gh/c.$$

The value gh/c is, then, the speed of the detector (when light reaches it) relative to the source (when light left it). The Doppler shift would then be (Section 10.5b)

$$\frac{\Delta\lambda}{\lambda} = z = \frac{v}{c} = \frac{gh}{c^2}.$$

So long as the acceleration is uniform and $v \ll c$, so that the simple formula for the Doppler shift can be used, gh/c^2 is the gravitational redshift.

In the Jefferson Laboratory experiment, $h = 20$ m $= 2 \times 10^4$ cm, so the gravitational redshift is

$$z = \frac{gh}{c^2} = \frac{980 \times 2 \times 10^4}{(3 \times 10^{10})^2} = 2.18 \times 10^{-14}.$$

The speed required to produce a Doppler shift of that amount is

$$v = 2.18 \times 10^{-14} \, c = 6.53 \times 10^{-4} \text{ cm/s}.$$

At that speed, the time for the detector to travel 2×10^4 cm is

$$t = \frac{2 \times 10^4}{6.53 \times 10^{-4}} = 3.06 \times 10^7 \text{ s} \simeq 1 \text{ year.}$$

According to the principle of equivalence, one should be able to *produce* a gravitational redshift by merely accelerating a spaceship far away from all gravitating bodies. Clearly, such a redshift would be produced: if radiation is beamed from one passenger to another in the direction of the spaceship's acceleration, the receiver, where the radiation is absorbed, is moving away from the source, where that same radiation is emitted. Thus, since the source and receiver are separating, there is a redshift (due to the Doppler effect) that is indistinguishable from that produced by a gravitational field that produces the same acceleration.

Within a freely falling spaceship (like Skylab) or in the Jefferson Laboratory the gravitational field is essentially uniform. Such is not the case, however, for the light we observe leaving a star, because that light has to pass from the strong field near the star's surface on out through the continually weakening one as it gets farther and farther from it. However, Einstein showed that we need only add up the tiny effects as the light passes through each small region within which gravity can be regarded as effectively constant to calculate the total gravitational redshift of light leaving the star. It works out that the wavelengths of light from the sun should be increased by about 2 parts in a million—a redshift too small to be distinguished from other effects.

The acceleration of gravity at a distance r from the center of a star of mass M is GM/r^2 (Section 4.3c), where G is the gravitational constant. Over a small distance, dr, the small contribution, dz, to the gravitational redshift is

$$dz = \frac{1}{c^2} \frac{GM}{r^2} \, dr.$$

Readers familiar with calculus will recognize that a simple integration—adding up of contributions—from the surface, R, of the star to a point very far away gives for the total gravitational redshift of light from the star:

$$z = \frac{1}{c^2} \frac{GM}{R}.$$

White dwarf stars, however, being very dense, have a much stronger surface gravity than the sun, and Einstein suggested that the gravitational redshift of the light from white dwarfs might be detectable. It can only be observed, however, for white dwarfs whose radial velocities are known from independent methods so that it can be separated from the Doppler shift due to the stars' motions. Fortunately, several white dwarfs are members of binary star systems, and their radial velocities can be deduced from those of their non-white dwarf companions, for which the gravitational redshift is negligible. The first reliable confirmation of the effect was made in 1954 by UCLA astronomer D. M. Popper, who measured the gravitational redshift of the white dwarf companion of the star 40 Eridani.

The precision of Popper's observations were such as to verify Einstein's predictions to within about 20 percent. Far higher accuracy has been attained recently in the near-earth environment with space-age technology. In the mid-1970s, a hydrogen maser carried by a rocket to an altitude of 10,000 km was used to detect the radiation from a similar maser on the ground. That radiation showed a gravitational redshift due to the earth's field that confirmed the relativity predictions to within a few parts in ten thousand.

(c) Limitations to the Principle of Equivalence

We have seen that the force of gravitation can be compensated in a suitably accelerated coordinate system locally—that is, over dimensions small enough that within them the acceleration produced by gravitation can be regarded as constant. Thus within a freely falling spaceship in orbit about the earth, gravity appears absent, and all bodies behave according to the rules of special relativity; everything either remains at rest or moves uniformly in a straight line—a straight line as defined by the path of a light beam. If there are two spaceships in orbit, but one, say, 100 km above the other, the principle of equivalence applies to each.

However, the motions of objects within one ship would *not* appear unaccelerated as seen from the other, for the force of the earth's gravitation varies with distance from the earth, and is appreciably different between the two ships.

Because the ships are in different gravitational fields, the gravitational redshifts within the two of them are different as well. Now all that we have said about frequency of light applies to the rate of all other physical processes as well; the rate of passing of light waves is just one of many ways to measure the passage of time. Time flows differently in different gravitational fields too. A clock in the spaceship of lower orbit (hence in a region of stronger gravity) runs more slowly than one in the other ship. Astronauts in the different ships would therefore disagree on the rate of time passage, as well as on the paths of unaccelerated bodies.

The above considerations suggest how an astronaut could tell that he was in orbit in a gravitational field—even without observing the planet beneath him. Suppose he fires a rocket probe straight ahead. For a while, the probe would continue in essentially the same orbit as the ship and would appear to hang motionless as viewed through a forward porthole. After a time, however, its motion away from the ship would carry the probe into a higher orbit where it would be accelerated downward less strongly, and it would slowly drift upward in the window, and eventually out of view. A differently accelerated reference frame now applies to the probe, just as it does to a separate orbiting spaceship.

Einstein's problem in formulating general relativity was to unify these separate descriptions of motion in different parts of a gravitational field into a connected whole, in order to find how to define a reference frame in which all objects, no matter where they are, are unaccelerated. To succeed, Einstein had to employ two ideas: spacetime and curvature.

35.2 SPACETIME

There is nothing mysterious about four-dimensional spacetime. Imagine yourself in the rear seats at an outdoor concert at the Hollywood Bowl. The sound from the orchestra in the shell, hundreds of feet away, takes a goodly fraction of a second to reach you, and the players seem to be behind the beat of the conductor. When a piece is finished, you first hear the applause from people near you, and slightly later from the front of the amphitheater. Because of the finite travel time of sound, all people do not hear the same note of music at the same time; nor do events that appear simultaneous *visually* seem to be audibly simultaneous.

Light also has a finite speed, so we never see an instantaneous snapshot of events around us (as we saw in Chapter 13). The speed of light is so great that within a single room we obtain *effectively* an instantaneous snapshot, but it is certainly not the case astronomically. We see the moon as it was just over a second ago, and the sun as it was about 8 minutes ago. At the same time, we see the stars by light that left them years ago, and the other galaxies as they were millions of years in the past. We do not observe the world about us at an instant in time, but rather we see different things about us as different *events* in spacetime.

Relatively moving observers do not even agree on the order of events. As an example, suppose you and I approach each other along an east-west direction in rapidly moving trains (Figure 35.3). When we are abreast of each other, a person on the ground pulls a switch to set off two flares, one 5 km east of us and the other 5 km west of us, and he observes the flares to go off simultaneously. I am moving toward flare B, while you are moving toward flare A. By the time the light from flare B reaches me, I am as shown in Figure 35.3 (2). You, meanwhile, are the same distance in the opposite direction from where we passed and are just receiving the light from flare A. Sometime later, I receive the light from flare A and you from flare B. Thus two happenings that appear simultaneous to one are not simultaneous to the other. Space and time are inextricably connected. We need to describe the universe not just in terms of three-dimensional space, but in terms of four-dimensional spacetime.

We can easily represent the spatial positions of objects in two dimensions on a flat sheet of paper (for example, the plan of a city). To plot three dimensions on a page, the draftsman uses projections. Architectural drawings of a home generally show three projections: floor plan and two different elevations—say, the house as seen from the east

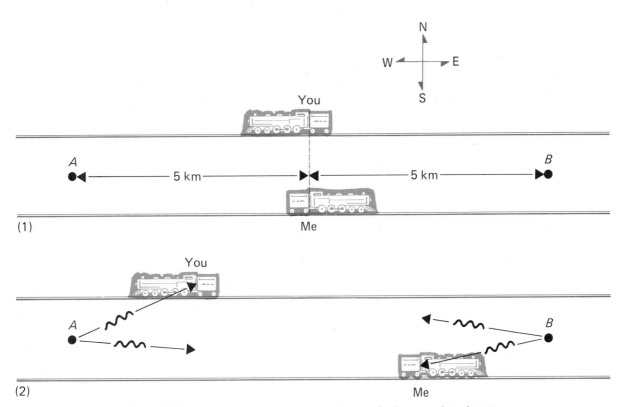

Figure 35.3 An experiment demonstrating the nonabsoluteness of simultaneity.

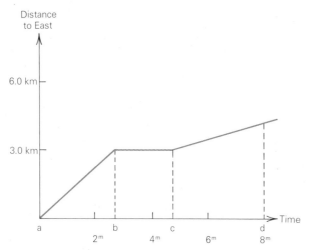

Figure 35.4 The progress of a motorist traveling east across town.

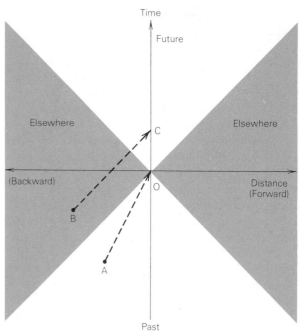

Figure 35.5 A spacetime diagram.

and from the north—to give all necessary information. By the use of perspectives (which rely on our learned experience), we can also give an impression of a three-dimensional view. There is no easy way, though, to draw a four-dimensional perspective to include time.

There is no problem, however, in showing a two-dimensional projection of four-dimensional spacetime. Figure 35.4, for example, shows the progress of a motorist driving to the east across town. How much time has elapsed since he left home is shown on the horizontal axis, and how far he has traveled eastward is shown on the vertical axis. From a to b he drove at a uniform speed; from b to c he stopped for a traffic light and made no progress, and from c to d he drove more slowly because of increased traffic.

Figure 35.5 shows a rather conventional two-dimensional representation of spacetime. Time increases upward in the figure, and one of the three spatial dimensions is shown horizontally. If we measure time in years and distance in light years, light goes one unit of distance in one unit of time, so flows along diagonal lines as shown. "Here and now" is at the origin of the diagram. At this instant we can receive information of a past event along such a line as AO; in this case the messenger was going slower than light, so he covered less distance than light would in the same time. Because nothing can go faster than light, we cannot, right now, know of something happening at point B in spacetime, for the message along BO would have to travel faster than light. We will have to wait until we are at C in the future, before a

light or radio beam can get us the word along path BC.

We can also show three dimensions of spacetime in a perspective drawing, as in Figure 35.6 Here time flows to the right, the height above the ground is upwards in the figure, and one of the two dimensions along the ground is shown obliquely as the north-south line. Suppose a batter is at point B (homeplate) and a baseball pitcher is at P, north of homeplate. The pitcher throws the ball to the batter, but the ball, traveling at a finite speed, flows along path PD in spacetime, arriving at the batter at D. Light, however, travels much faster than the ball, and arrives at the batter at C. In the gravitational field of the earth both the ball and light beam fall to the ground enroute to the plate at the same rate, but the light reaches the plate so quickly that it falls only a slight distance compared to the ball, which takes much longer. Thus in space (lower part of Figure 35.6) the path of the light is much more nearly straight than that of the ball. In spacetime, however, the path of the ball is very long compared with that of the light beam, and it is easy to see how in the uniform gravitational field (as on the baseball diamond) both ball and light fall along paths of the same shape (that is, the same *curvature*) in spacetime.

Now if that ball game were being played on a freely falling platform, that reference frame would

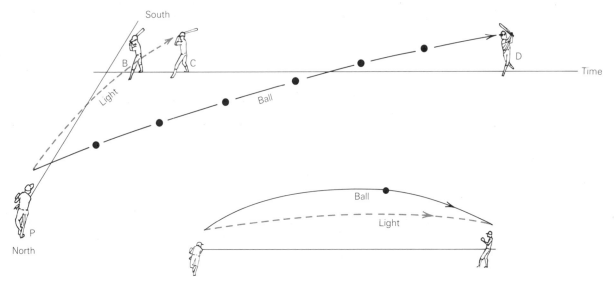

Figure 35.6 The paths of light and of a baseball in spacetime. Below are shown the corresponding paths in space alone.

be accelerating downward at the same rate the ball and light accelerate, and both would move, in that freely falling system, in straight lines, obeying the laws of special relativity. Those straight line paths of light and the ball would be the kind of straight lines Euclid was talking about in his development of plane geometry.

Actually, on the earth, where things accelerate downward, light falls like everything else. If we used light paths to define straight lines (say, by sighting along a ruler to assure that it is "straight"), the laws of Euclidean geometry would not work; if we measured the three sides of a triangle by such straight lines, we would not find its area to be one-half its base times its altitude. In the freely falling system in a uniform gravitational field, however, these difficulties are removed, for there is no force accelerating light in *that* system. In practice, of course, there is no problem on the earth, either, because light travels so fast and falls so slightly no one could measure the difference. But we *could* tell the difference on the surface of a *neutron star*—or even on a white dwarf!

35.3 CURVATURE OF SPACETIME

Because of the principle of equivalence, Euclidean geometry applies in a freely falling reference frame in a uniform gravitational field. But freely falling systems at different places on earth fall in different directions, because "down" is always toward the center of the earth. Thus we cannot describe the behavior of objects in different widely separated freely falling systems with spacetime coordinates for which Euclidean geometry holds. Let's consider a familiar analogy. A simple Mercator-type map, with lines of constant latitude running horizontally and lines of constant longitude running vertically, is fine for showing a small area of the earth—say a single city—without noticeable distortion. But such a map cannot show a large area of the curved earth without distortion; everyone knows how distorted and enlarged lands of extreme latitude (those near the poles) appear on the usual flat world maps. We cannot map the earth with Euclidean plane geometry.

Indeed, if we travel far enough in a straight line on the surface of the earth, we end back at our starting point; our path is a *great circle* (the equator is such a great circle). More generally, if we take into account the slight polar flattening of the earth, as well as effects of such irregularities as mountains, our "straight line" path is called a *geodesic*, which means "earth divider."

Einstein showed how to find spacetime coordinates within which all objects move as they would if there were no forces. In a small local region, where a gravitational field is uniform, those coordinates are Euclidean, just as a city plan can be well described with plane geometry. But to describe paths of objects over a large region, where the gravitational field varies, spacetime must be curved, just as we must use curved geometry to describe a large area of the spherical earth.

The distribution of matter determines the nature of a gravitational field, so it is the distribution of matter that determines the geometry of curved spacetime. Within this curved spacetime, everything moves in the simplest possible way as if no gravitation were there at all. In analogy with earth geometry, the paths of light and material objects in spacetime are called *geodesics*.

People usually say that the geometry of spacetime is determined by matter, but I think this description makes the subject sound unduly mysterious. What meaning can there be to "curved space" if space is the absence of matter? It is not space in itself that is curved; rather it is the system of coordinates that we can conveniently use to describe the motions of objects and light. By selecting non-Euclidean geometry—curved coordinates—we can describe the paths of light and objects as "straight" in the same sense that great circles (or geodesics) are "straight" on the curved surface of the earth.

The mathematics needed to handle the problem, on the other hand, was not available in Euclid's time. The geometry needed to describe curved spacetime was developed after the pioneering work of the great German mathematician, physicist, and astronomer, Karl Friedrich Gauss. Gauss became involved in the invention of new geometry when he was commissioned to survey the German State of Hanover by its king, George IV—also king of Great Britain. The new geometry received its full expression in the hands of Gauss' student, Bernhard Riemann.

Riemann, in applying for a university position at Göttingen, had submitted three possible topics for a lecture he would deliver. Traditionally, the judges selected one of the first two topics offered, however, so Riemann had not bothered to prepare a lecture on the third. That, however, was the very topic that Gauss had been pondering for decades. Consequently, in only a few weeks Riemann wrote out the lecture on that third topic, which was to be his masterpiece: "On the Hypotheses Which Lie at the Foundations of Geometry."

By the end of the nineteenth century the new Riemannian geometry was further facilitated with the invention of tensor calculus. By 1915, Einstein was able to use these new mathematical techniques to derive the *field equations* of general relativity, which describe the curvature of spacetime by matter, and the *geodesic* equations, which describe the unaccelerated paths of objects in spacetime.

35.4 TESTS OF GENERAL RELATIVITY

Is general relativity, then, essentially different from Newtonian gravitational theory, or merely a different but equivalent mathematical formulation? Relativity *is* different from Newtonian theory in that the signals that govern gravitational interactions are not instantaneous, but travel with the speed of light, and also, of course, in that matter and energy are equivalent, so that not only matter itself but also energy contributes to gravitation—that is, to the geometry of spacetime—and energy (light, for example) as well as mass is affected by that geometry. Naturally, where speeds are low compared with that of light, and where the gravitational field is relatively weak—both conditions of which are met throughout most of the solar system—the predictions of general relativity must agree with those of Newton's theory, which has served us so admirably in our technology and in guiding space probes to the other planets. In familiar territory, therefore, the differences between predictions of the two theories are subtle, and consequently very difficult to detect.

Einstein himself proposed three observational tests of general relativity. One is the gravitational redshift, already described, the second is the deflection of starlight that passes close to the sun, and third is a subtle effect on the motion of the planet Mercury.

(a) Deflection of Starlight

The strength of the gravitational acceleration at the surface of the sun is 28 times its value at the surface of the earth. Einstein calculated from general relativity theory that starlight just grazing the sun's surface should be deflected by an angle of $1''.75$. Stars cannot be seen or photographed near the sun in bright daylight, but with difficulty they *can* be photographed close to the sun at times of total solar eclipses. Einstein suggested an eclipse observation to test the light deflection in a paper he published during World War I. A single copy of that paper, passed through neutral Holland, reached the British astronomer Arthur S. Eddington. The next suit-

Figure 35.7 Deflection of starlight passing near the sun.

able eclipse was on May 29, 1919. The British organized two expeditions to observe it, one on the island of Principe, off the coast of West Africa, and the other in Sobral, in North Brazil. Despite some problems with the weather, both expeditions obtained successful photographs of stars near the sun. Measures of their positions were then compared with measurements on photographs of the same stars taken at other times of the year when the sun was elsewhere in the sky. The stars seen near the sun were indeed displaced and, to the accuracy of the measurements, the shifts were consistent with the predictions of relativity. It was a triumph that made Einstein a world celebrity overnight.

Eclipse observations to test the relativity effect have continued over the years, but the measures are very difficult to make and the precision of the confirmation is not high. Far higher accuracy has been obtained recently at radio wavelengths. Simultaneous observations of the same source with two radio telescopes far apart can pinpoint the direction of the source very precisely. The United States National Radio Astronomy Observatory at Greenbank, West Virginia, with radio telescopes 35 km apart, observed several remote astronomical radio sources (quasars—see Chapter 37) when the sun was nearly in front of them. The apparent directions of the quasars showed shifts similar to those of stars seen near the sun. The accuracy of these observations is high enough to confirm the Einstein prediction to within 1 percent.

Notice (Figure 35.7) how the sun acts like a lens in deflecting light passing near it. A very remote object, acting as such a *gravitational lens*, can produce distorted and multiple images of another object far beyond, but nearly in line with it. For decades astronomers have suspected that such gravitational lens effects might be found in nature, and in the late 1970s the first such example was discovered: the double quasar 0975 + 561 is believed to be two images of the same object produced when its

light passes near a foreground galaxy. It is described in Chapter 37.

In addition to sources of light or radio waves appearing displaced slightly when seen near the sun, the radiation from them is also delayed slightly in reaching the earth. We have no way of measuring that delay in light from stars or quasars, but we can detect it in the radio pulses broadcast from space probes, because we know where they are and when the signals should arrive at earth. The experiment has been performed with several planetary probes, but most precisely with the Viking landers on Mars (Chapter 17). When Mars is on the far side of the sun, signals from the Vikings must pass through a region of spacetime that is relatively strongly curved by the sun (Figure 35.8) and are observed to be delayed by about 100 microseconds, as if Mars had jumped some 30 km out of its orbit. It is just the delay expected by relativity theory to within 1 part in a thousand.

(b) Advance of Perihelion of Mercury

According to relativity, the energy and momentum associated with the motion of a body, and even its gravitational energy, all contribute to its effective mass, and hence to the force of gravitation on it. Now, Mercury has a fairly eccentric orbit, so that it is only about two thirds as far from the sun at perihelion as it is at aphelion. As required by Kepler's second law, Mercury moves fastest when nearest the sun, and all of the motional and gravitational effects add to the attraction between the two. Consequently, relativity predicts that a very tiny additional push on Mercury, over and above that predicted by Newtonian theory, should occur at each perihelion. The result of this effect is to make the *line of apsides*, which is the long dimension (major axis) of Mercury's orbit, slowly rotate in space, so that each successive perihelion occurs in

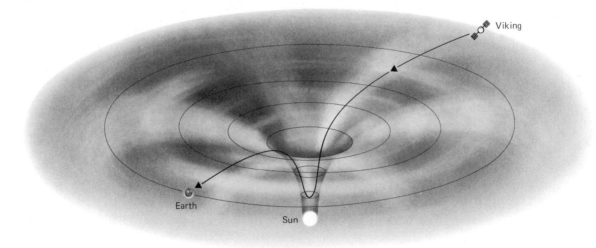

Figure 35.8   Radio signals from Viking are delayed because they have to pass near the sun, where spacetime is curved relatively strongly.

a slightly different direction as seen from the sun (Figure 35.9). The prediction of relativity is that the direction of perihelion should change by only 43″ per century; it would thus take about 30,000 years for the line of apsides to make a complete rotation.

The gravitational effects (perturbations) of the other planets on Mercury also produce an advance of its perihelion, and to a far greater extent than the relativistic prediction. According to Newtonian theory, the perihelion of Mercury should advance by 531″ per century. Even in the last century, however, it was observed that the actual advance is 574″ per century. The discrepancy was first called to attention by Leverrier, co-discoverer of Neptune, and in analogy with Neptune, it was assumed that an intramercurial planet was responsible. The hypothetical planet was even named for the god Vulcan. Vulcan, of course, never materialized, but that 43″ anomaly was entirely explained by relativity. The relativistic advance of perihelion can also be observed in the orbits of several minor planets that come close to the sun.

Additional tests for relativity theory are in the planning or experimental stages—some at the frontier of modern technology. In a satellite experiment expected to fly in the early 1980s, the behavior of a gyroscope will be carefully monitored. Relativity predicts an angular change in orientation of the axis of the gyroscope due to its motion through the earth's gravitational field of only 0″.05 per year, but it is believed that even that small change can be measured accurately enough to check the theory.

35.5   GRAVITATIONAL WAVES

Because the geometry of spacetime depends on the distribution of matter, any rearrangement of that matter must result in an alteration of spacetime—that is, it creates a disturbance. Relativity predicts that that disturbance should propagate through space with the speed of light. It is called a *gravitational wave*.

We think we should be able to detect gravitational waves, as we do other kinds of radiation. Whereas electromagnetic radiation can be detected, for example, in the way it sets charges oscillating back and forth, gravitational waves should set material objects vibrating. Detection of gravitational

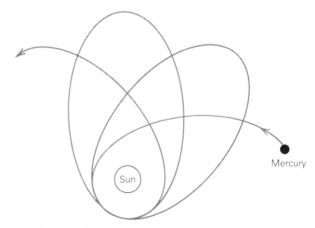

Figure 35.9   Rotation of the line of apsides of the orbit of a planet, such as Mercury, because of various perturbations.

waves, however, is extremely difficult. One reason is that gravitation is an exceedingly weak force (Chapter 1), and its radiation is correspondingly weak. But moreover, the disturbance gravitational waves should produce in an object is far more subtle than simple oscillation; rather than moving back and forth as a unit, the body suffers very slight compressions and lengthenings.

Gravitational waves produced by mass motions on the earth, or even of the earth itself, would be far too weak to detect by technology we can currently imagine. We would need a motion of very large mass, and at a speed approaching that of light. One possible source could be the sudden collapse, of a massive star, perhaps in a supernova explosion, that is, a stellar catastrophe.

The pioneering attempt to detect such gravitational waves from collapsing stars was by Joseph Weber, physicist at the University of Maryland. In the 1960s he suspended large metal cylinders equipped with very delicate sensors. Weber had hoped to detect gravitational waves by the vibrations they set up in the cylinders. To distinguish true gravitational waves from space from purely local disturbances, he placed one cylinder in Maryland and a second one at the Argonne National Laboratory outside Chicago; only signals recorded simultaneously at both stations would be of astronomical origin.

To date, Weber has not detected signals that can unequivocally be attributed to gravitational waves from space. Other laboratories, however, are developing far more sensitive detectors. At Stanford University, for example, an extremely delicate microphone is arranged to pick up the vibrations that a passing gravitational wave sets up in a metal bar. The bar is a practical detector because the two ends would be set vibrating differently, which sets up an oscillation in the bar itself. The gravitational waves passing the earth produced by the collapse of a solar mass star at the center of the Galaxy would displace the ends of the bar by only one ten-millionth the diameter of an atom, but with advanced technology, utilizing principles of superconductivity, the Stanford experiment has a good chance of detecting such gravitational waves if, indeed, they exist.

(a) The Binary Pulsar PSR1913+16

It may be that nature has already provided us with indirect evidence of gravitational radiation. In 1974, R. A. Hulse and J. H. Taylor, of the University of Massachusetts, observing with the 1000-ft radio telescope at Arecibo, Puerto Rico, discovered a remarkable pulsar, now designated PSR1913 + 16 (the numbers give the coordinates of its location in the sky). The unique thing about PSR1913 + 16 is that the period of the pulses itself shows cyclic variations over a short time interval of 7^h45^m. These period changes are due to the Doppler effect caused by the pulsar's revolution about another object. When the pulsar is approaching us in its orbit, each successive pulse has less far to travel to reach us on earth, and we receive the pulses slightly closer together than average. Conversely, when the pulsar is moving away from us, we receive the pulses slightly spread out in time. Thus we can analyze the orbital motion of the pulsar just as we do that of a spectroscopic binary star (Section 26.2d).

Such an analysis, combined with our knowledge of the expected properties of neutron stars, indicates that the pulsar is in mutual revolution in that 8-hr period about a mute companion of comparable mass that is probably either a white dwarf or another neutron star. Because the size of the orbit is only a little bigger than the diameter of the sun, it is exceedingly unlikely that the companion is a normal star, for then the pulses would be eclipsed by it during part of each revolution (contrary to observation) unless the orbit were almost face on, in which case we should not see appreciable variations in the pulse period.

Now there are two important points about this binary pulsar. First, at its orbital speed of approximately 0.1 percent that of light, it should radiate gravitational waves at a rate great enough to carry appreciable energy away from the system, causing the pulsar and its companion to spiral slowly closer together. Second, pulsars are superb clocks, remaining stable to 0.001 s over several years.

As the pulsar and its companion spiral together, their period of revolution shortens. The shortening is only about one ten-millionth of a second per orbit, but the effect accumulates like a clock that runs a little faster each day. By this writing (late 1981) the system has been under observation for nearly seven years, and the time of periastron (when the two objects are closest together) has shifted by more than a full second of time relative to the time it would now be occurring if the period had remained constant. Pulsars are such good timekeepers that this shift is easily observed.

In fact, the shift in periastron time is just what general relativity predicts it should be due to the emission of gravitational radiation by two stars, each of about 1.5 solar masses, with an orbit like that of the binary pulsar. PSR1913+16 is now regarded as providing strong evidence that gravitational waves do exist, as Einstein predicted.

35.6 BLACK HOLES

Until now, the manifestations of general relativity theory we have discussed have been so subtly different from Newtonian theory that one might well wonder if the new ideas about space and time brought about by Einstein are really important enough to concern us. Indeed, many physicists, during the first half century after Einstein introduced his revolutionary theory, regarded the subject as almost academic.

But hold! In the last chapter we encountered pulsars, which provide strong evidence for the existence of neutron stars. Neutron stars have very great density and extremely strong surface gravity. We recall that light grazing the surface of the sun is deflected by about 1.″75. Light grazing the surface of a white dwarf would be deflected by about 1′, and that grazing a typical neutron star by about 30°.

In 1796, the French mathematician Pierre Simon, Marquis de Laplace speculated about the properties of an object that had so great a gravitational field that light could not escape at all, but would be bent right around and stay with the object. Laplace's "corps obscurs" were later reconsidered by modern physicists, armed with the new rigor of general relativity theory. John Wheeler, the Princeton physicist who has become intimately associated with general relativity, has dubbed such objects "black holes."

Consider the light radiated from the surface of a neutron star. That which emerges normal (perpendicular) to the surface flows out radially from the star. That emitted at an angle of, say, 30° to the normal leaves the star at an angle somewhat greater than 30° to the normal, because of the gravitational deflection. Now imagine a more massive star that shrinks to a smaller size and higher density than a neutron star. As the surface gravity increases, the deflection of light increases too. Eventually the star reaches a size at which a horizontal beam of light enters a circular orbit. A surface of that radius is called the *photon sphere*.

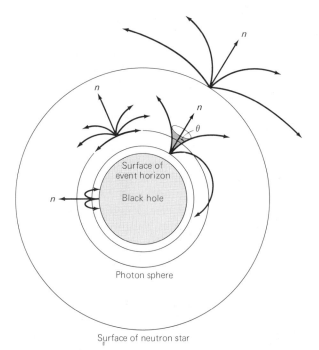

Figure 35.10 The deflection of light from a very dense star. At a radius smaller than that of the photon sphere, to escape, light must flow into a cone of half-angle θ with respect to the normal, *n*, to the surface. At the event horizon, θ = 0.

As the star shrinks to a size smaller than the photon sphere, to escape the star light must flow into a cone about the normal to the surface of half-angle θ (Figure 35.10), and light at a greater angle falls back on the star. The angle θ becomes smaller and smaller as the star collapses, until the radius of the star is two thirds that of the photon sphere, near which θ becomes zero, and no light at all can escape. At this point the velocity of escape from the star equals the speed of light. As the star contracts still more, light and everything else is trapped inside, unable to escape through that surface where the escape velocity is the speed of light. That surface is called the *event horizon*, and its radius is the *Schwarzschild radius*, named for Karl Schwarzschild, who first described the situation a few years after Einstein introduced general relativity. This surface is the boundary of the black hole. All that is inside is hidden forever from us; as the star shrinks through the event horizon it literally disappears from the universe.

The size of the Schwarzschild radius is proportional to the mass of the star. For a star of one solar mass, the black hole is about 3 km in radius; thus the entire black hole, some 6 km in diameter, is about one-third the size of a one-solar-mass neutron star.

The event horizons of larger and smaller black holes—if they exist—have greater and lesser radii, respectively. For example, if a globular cluster of 100,000 stars could collapse to a black hole, it would be 300,000 km in radius, a little less than half the radius of the sun. If the entire Galaxy could collapse to a black hole, it would be only about 10^{12} km in radius—about 0.03 pc. On the other hand, for the earth to become a black hole it would have to be compressed to a radius of only 1 cm—about the size of a golf ball. A typical minor planet, if crushed to a small enough size to be a black hole, would have the dimensions of an atomic nucleus!

It happens that the correct size of the event horizon can be calculated by pretending that Newton's laws apply. The formula for the velocity of escape (Sections 5.7a and 14.2c) is $\sqrt{2GM/R}$, where M and R are the mass and radius. By equating this to c, the speed of light, and solving for R, we find that the radius of the event horizon of a black hole of mass M is

$$R = \frac{2GM}{c^2}.$$

But should black holes exist? Stars less than about 1.4 solar masses can become white dwarfs. Those of larger mass, we think, can exist as neutron stars, but there is an upper limit to the mass of neutron stars; we think that limit is not over three solar masses. We know that a tiny fraction of all stars have still greater mass. What becomes of them when they exhaust their store of nuclear fuel? Perhaps they eject part of their mass (as a planetary nebula or in a supernova outburst) so that what is left can contract to a white dwarf or a neutron star. But what if they do not? Then we know of no other fate for such massive stars than that they become black holes. Thus we are not certain that any star *must* ever have to become a black hole, but we have good reason to expect that many massive stars, albeit a minority of *all* stars, can end up in that exotic state.

How, then, do we find a black hole, which, of course, we cannot see? We can detect it by its gravitational effects on other stars (as stars collapse into black holes they leave behind their gravitational fields), and this is most easily accomplished in a binary star system.

(a) Candidates for Black Holes

To find a black hole: (1) We must find a star whose motion (found from the Doppler shift of its spectral lines) shows it to be a member of a binary star system, and to have a companion of mass too high to be a white dwarf or a neutron star. (2) That companion star must not be visible, for a black hole, of course, gives off no light. But being invisible is not enough, for a relatively faint star might be unseen next to the light of a brilliant companion. Therefore, (3) we must have evidence that the unseen star, of mass too high to be a neutron star, is also a collapsed object—one of extremely small size—for then our theory predicts that it must be a black hole—or at least a star on the way to becoming one.

Modern space astronomy has come to the rescue in (3). One way to know we have a small object of high gravity (and possibly a black hole) is if matter falling toward or into it is accelerated to high speed. Near the event horizon of a black hole, matter is moving at near the speed of light. Internal friction can heat it to very high temperatures—up to 100 million Kelvins or more. Such hot matter emits radiation in the form of X rays. Modern orbiting X-ray telescopes—especially the Einstein telescope, HEAO 2—can and do reveal such intense sources of X-radiation.

So we want X-ray sources associated with binary stars with invisible companions of high mass. We cannot prove that such a system contains a black hole, but at present we have no other theory for what the invisible massive companion can be if, indeed, the X rays are coming from gas heated by falling toward it.

We can easily understand the origin of such infalling gas. We have already seen (Section 34.4) how stars in close binary systems can exchange mass. Suppose one star in such a double star system has evolved to a black hole and that the second star has now evolved to a red giant so large that its outer layers pass through that point of no return between the stars and some of its matter falls to the black hole. The mutual revolution of the giant star and black hole cause the material from the former to flow not directly onto the black hole, but, because of conservation of angular momentum, to spiral around it, collecting in a flat disk of matter called the *accretion disk*. In the inner part of the accretion disk the matter is revolving about the black hole so fast that its internal friction heats it up to the temperature where it emits X rays. In the course of this friction, some material in the accretion disk is given extra momentum, and escapes from the double star system, and other material loses momentum and falls into the black hole—lost forever to observation from the rest of the universe.

Another way to form an accretion disk in a binary star system is from material ejected from the companion of the black hole as a stellar wind; some of that ejected gas will flow close enough to the black hole to be captured by it into the disk. Such a case, we think, is the binary system containing the first X-ray source discovered in Cygnus—Cygnus X-1. The visible star (Figure 35.11) is a normal B-type star. The spectrographic observations, however, show it to have an unseen companion of mass near ten times that of the sun. That companion would be a black hole if it were a small, collapsed object. The X rays from it strongly suggest that it is, for we have no other explanation for the source of those X rays than gas heated by an infall toward a tiny massive object. Of course we cannot be certain that Cygnus X-1 is a black hole, but many astronomers think that it probably is.

(b) Properties of Black Holes

Much of the modern folklore about black holes is misleading. One idea is that black holes are mon-

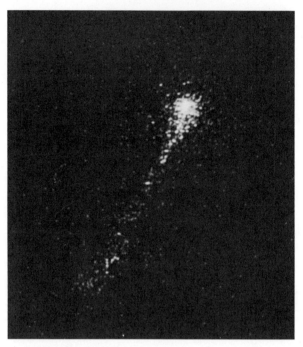

Figure 35.12 Cygnus X-1 in the light of X rays, as observed by the Einstein X-ray telescope. (*Courtesy of the Harvard–Smithsonian Center for Astrophysics*)

sters that go about sucking things up with their gravity. Actually, the gravitational attraction surrounding a black hole at a large distance is the same as that around any other star (or object) of the same mass. Even if another star, or a spaceship, were to pass one or two solar radii from a black hole, Newton's laws would give an excellent account of what would happen to it. It is only very near the surface of a black hole that its gravitation is so strong that Newton's laws break down; for a black hole of the mass of the sun, light would have to come within 4.5 km of its center to be trapped. A solar mass black hole, remember, is only 3 km in radius—a very tiny target. Even collisions between ordinary stars, hundreds of thousands of times bigger in diameter, are so rare as to be essentially nonexistent. A star would be far, far safer to us as an interloping black hole than it would have been in its former stellar dimensions.

(c) A Trip Into a Black Hole

Still, it is interesting to contemplate a trip into a black hole. Suppose that the invisible companion of the star associated with Cygnus X-1 is a black hole of ten solar masses. What would you see if I bravely fly into it in a spaceship?

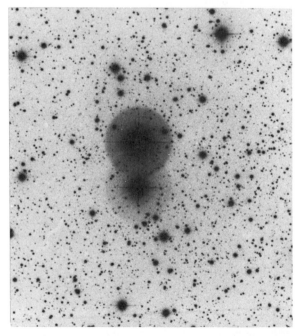

Figure 35.11 In visible light, Cygnus X-1 appears as an ordinary star. (*Palomar Observatory; California Institute of Technology*)

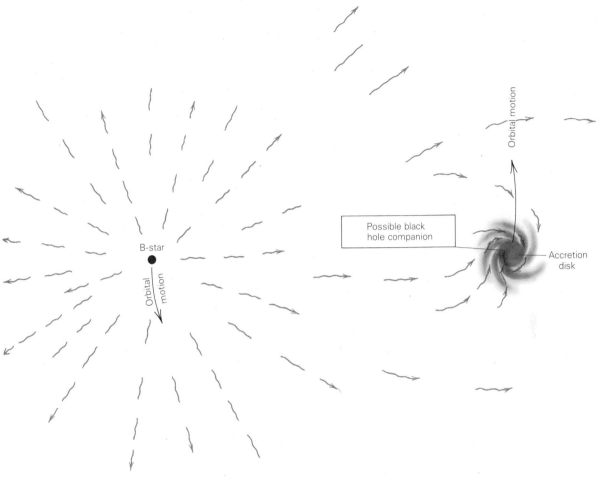

Figure 35.13 Model of Cygnus X-1.

At first I dart away from you as though I were approaching any massive star. However, when I near the event horizon of Cygnus X-1—some 30 km in radius, and presumably near the center of the accretion disk—things change. The strong gravitational field around the black hole makes my clocks run more slowly as seen by you. Signals from me reach you at greatly increased wavelengths because of the gravitational redshift. As I approach the event horizon, my time slows to a stop—as seen by you—and my signals are redshifted through radio waves to infinite wavelength; I fade from view as I seem to you to come to a stop, frozen at the event horizon. All matter falling into a black hole appears to an outside observer to stop and fade at the event horizon, frozen in place, and taking an infinite time to fall through it—including the matter of a star itself that is collapsing into a black hole. For this reason, black holes are sometimes called *frozen stars*.

This, however, is only as you, well outside the black hole, see things. To me, time goes at its normal rate and I crash right on through the event horizon, noticing nothing special as I do so, except for enormously strong tidal forces that rip me apart. At least this is true of ten-solar-mass black holes. If there were a very massive black hole, say thousands of millions of solar masses, its event horizon, although small in astronomical dimensions, would be large enough that its tidal forces are not severe, and an astronaut, in principle, could survive a trip into it.

But in no case is there an escape. Once inside, astronaut, light, and everything else is doomed to remain hidden forever from the universe outside. Moreover, current theory predicts that after entering the black hole, I race irreversibly to the center, to a point of zero volume and infinite density.

Mathematicians call such a point a *singularity*, but many physicists doubt that such singularities of infinite density can exist in the real universe and suspect that new physics will eventually emerge to help us understand what really happens to matter at extraordinary densities. In particular, mathematical solutions suggesting that black holes can collapse through *wormholes* to emerge as "white holes" in "other universes"—ideas often exploited in the popular literature—now appear to be abstractions without physical basis; recent theoretical studies, in fact, show that even general relativity theory does not predict that black holes emerge elsewhere as white holes.

On the other hand, massive stars collapsing through their own event horizons to become black holes do not have to reach densities beyond those understood by present-day physics. We can never know what goes on inside the event horizon, but we know of no reason why black holes themselves should not exist.

(d) Other Ideas About Black Holes

Black holes need not be limited to stellar masses. There has been considerable consideration of the possibilities of very large amounts of gas collecting together and collapsing into black holes in the centers of globular clusters, galaxies, or even clusters of galaxies.

A mass of gas collapsing into a black hole releases more than 100 times as much energy as can be extracted from that same mass through nuclear fusion. Thus the gravitational collapse of a million solar masses of gas into a black hole at the center of a galaxy could produce a truly prodigious amount of released gravitational potential energy. There is a great deal of speculation, nowadays, about such processes accounting for the energy of quasars and other phenomenal objects. It may well be that through massive black holes general relativity theory will be found to have profound consequences in modern astrophysics.

On the other hand, black holes may exist on the microscopic level as well. We know of no way to produce small-mass black holes today (that is, less than a few solar masses), but the brilliant British theoretical astrophysicist Stephen Hawking suggests that such black holes *could* have been produced in the big bang at the origin of the universe (Chapter 39). If so, they would involve quantum mechanics as well as relativity, and in a most amazing way.

We have seen (Chapter 32) that all fundamental particles have their antiparticles; for example, electrons and positrons, protons and antiprotons, and so on. Whenever a particle and its antiparticle come into contact, they annihilate each other, transforming completely into energy. Similarly, pure energy can be converted into pairs of particles—an electron and a positron, for example. This process is known as *pair production* and is observed regularly in the nuclear physics laboratory.

Now all this is possible because mass and energy are equivalent. But obviously mass cannot be created from nothing—we need energy to do it. Yet, according to quantum theory, it is possible for matter (or energy) to be created from nothing for an exceedingly brief period of time. This possibility comes about because of the innate uncertainty in nature, at the microscopic level, of the measures of physical quantities such as mass and energy. The principle does not violate conservation laws because any matter that comes into being almost immediately disappears again spontaneously, so on the average mass and energy (combined) are conserved.

But, Hawking points out, what if a positron and electron (say) come into existence momentarily in the vicinity of a black hole. There is a chance that one or the other will fall into the hole and hence not be able to annihilate with its antiparticle, returning the energy it "borrowed" from nature. Its antiparticle, therefore, can escape unscathed. However, many such positrons and electrons so created near black holes and escaping from them do annihilate each other, creating energy. That energy cannot come from nothing; according to Hawking's theory it must come from the black hole itself. Robbing the black hole of energy in this way robs it of mass (for they are equivalent), so the black hole must slowly evaporate through this process of pair production.

As esoteric as this idea may seem, it is generally thought to be possible by theoretical physicists. The process is only important, however, near very tiny black holes. Solar-mass black holes would evaporate in this way at an absolutely negligible rate. In fact, the only black holes that would have had time to so evaporate in the age of the universe would be those of original mass less than about 10^{15} g (like a minor planet). Smaller ones would already be gone; those of about 10^{15} g should be finishing off about now, if they were formed in the big bang in the first place. Because the evaporation rate increases as the mass of the black hole goes down, at the end one would go off explosively, emitting a final burst of gamma radiation.

Nobody knows whether such mini black holes were formed in the early universe, nor, if so, whether the evaporation process Hawking envisions would really occur. If so, we would expect to see bursts of gamma rays from exploding mini black holes from time to time. So far we have not, but the speculation remains an interesting possibility.

We have touched on many ideas in this chapter. Some, near the end, are highly speculative, but the main thrust of general relativity theory now appears to be rather firmly established, even though tests of some of its predictions are still going on. It is probably safe to say that general relativity has come of age and is an important part of modern astronomy.

And general relativity is almost singlehandedly the inspiration of one genius—Albert Einstein. For half a century it was an intellectually stimulating, but still largely academic subject. Today it is in the mainstream, and tomorrow it may well be absolutely fundamental to our understanding of the universe.

EXERCISES

1. Consider a bucket nearly full of water. A spring is attached to the middle of the inside of the bottom of the bucket, and at the other end of the spring is a cork. The cork, trying to float to the top of the water, stretches the spring somewhat. Now suppose the bucket, with its water, spring, and cork, is dropped from a high building so that it remains upright as it falls. What happens to the spring and cork? Explain your answer thoroughly.

2. A monkey hanging from a branch of a tree sees a hunter aiming a rifle directly at him. The monkey then sees a flash, telling him that the rifle has been fired. The monkey, reacting quickly, lets go of the branch and drops, so that the bullet will pass harmlessly over his head. Does this act save the monkey's life? Why?

3. Some of the Skylab astronauts exercised by running around the inside wall of their cylindrical vehicle. How could they stay against the wall while running, rather than float aimlessly inside the Skylab? What physical principles are involved?

*4. What is the radial velocity that would produce the same shift of spectral lines as the gravitational redshift of a star of one solar mass (2×10^{33} g) and of radius 4447 km? (The gravitational constant, G, is 6.67×10^{-8} dyne $\cdot$ cm^2/g^2.)
Answer: 100 km/s

5. Draw a diagram showing the progress in spacetime of an automobile traveling northward. For the first hour, in city traffic, it goes only 30 km/hr. Then, in the country, for 3 hr it goes 90 km/hr. Finally, because the driver is late, he drives the car for 1 hr at 140 km/hr.

6. Make up a new example of a geodesic in spacetime and show it on a spacetime diagram.

7. The earth moves in its nearly circular orbit of 1 AU radius at a speed of only $1/10,000$ that of light. What is the radius of the circle that most nearly matches the path of a beam of light passing the sun at the earth's distance from it?

8. As the binary pulsar loses energy through gravitational radiation, why do its members *speed up* and why does the period get *shorter?*

9. What would be the radius of a black hole with the mass of the planet Jupiter?

10. Why is the time dilation in a gravitational field equivalent to a gravitational redshift? (*Hint:* What is the definition of frequency? How is the frequency of radiation affected by a redshift?)

11. Why would we not expect X-rays from a disk of matter about an ordinary star, or even a white dwarf?

12. If the sun could suddenly collapse to a black hole, how would the period of the earth's revolution about it differ from what it is now?

13. Could any of the dark globules, small round opaque objects seen superimposed against the starlight of the Milky Way, be stars that have become black holes? Why or why not?

Frank Drake (1930–   ), the American astronomer who is a pioneer in the search for extraterrestrial intelligence. His *Project Ozma* (about 1960) was the first organized attempt to detect radio signals from extraterrestrial civilizations. That project (as expected) did not succeed, but Drake has consistently been a leader in calling attention to the possibility and importance of detecting alien intelligence.

LIFE IN THE UNIVERSE: ARE WE ALONE?

It is a legitimate question whether intelligent life, or any kind of life, exists elsewhere than the earth. At the outset we must say that we cannot answer that question with certainty. In this chapter, however, we shall try to explore the possibilities of life and of intelligent beings on other worlds.

36.1 LIFE IN THE SOLAR SYSTEM

There has been a great deal of study and a considerable advance in our understanding of the possibilities of life other than on earth in the solar system and beyond. We have learned enough about the other planets to know that we could not exist on any of them (in their natural environments), nor could the more complex forms of animal or plant life that we know on earth. Yet we cannot absolutely rule out the possibility of life forms alien to ours existing even in such seemingly hostile environments as we find on Jupiter. Although the Viking experiments (Chapter 17) found no life on Mars, laboratory experiments have shown that primitive terrestrial life forms (such as algae) can exist and multiply under the conditions that exist there. On the other hand, there is as yet no hint of

any life anywhere beyond the earth. Thus we might begin by looking at the circumstances under which life began here.

(a) Development of Life on the Earth

We have seen (Chapter 14) that the earth, along with the other inner planets, most likely formed from accretion of small solid particles in the solar nebula about 4.6×10^9 years ago. As soon as the solid earth formed, the natural decay of radioactive elements below the surface heated up the crustal rocks, causing the chemical decomposition of some of the minerals present there. In the process, water (H_2O), carbon dioxide (CO_2), and other gases were liberated from chemical compounds. These substances outgas to the surface, especially through volcanism. Water was the most plentiful compound to be released from the earth's crust; once it reached the cooler surface, it condensed to form the oceans where most of it remains today. Next most important is carbon dioxide, with about a tenth the abundance of the water. Much of it dissolved in the oceans and some recombined at the surface temperature of our planet with surface rocks to re-form carbonates, which remain on the ground today. Perhaps only one fiftieth as abundant as the carbon dioxide was nitrogen gas, but nitrogen is relatively inactive chemically, and now remains in the atmosphere as its major constituent (about 78 percent). Another even more inert gas to be released was argon-40, formed by the radioactive decay of potassium-40; argon today comprises about 1 percent of the atmosphere.

Most experts believe that in the earth's primordial atmosphere methane (CH_4), ammonia (NH_3), and, at least temporarily, hydrogen (H_2) were present, having originated either from outgassing or from chemical activity at the surface. Thus the early atmosphere of the earth may have been made up of nitrogen, carbon dioxide, argon, water, methane, ammonia, and hydrogen. The last three gases would quickly escape or decompose, but their temporary presence, bathed in solar radiation, would create a chemical environment favorable to the formation of more complex molecules. Laboratory experiments have shown that prebiological organic molecules such as amino acids and sugars will readily form under these conditions, and we can imagine how eventually the far more complex organic substances vital for living organisms (including DNA) could have been built up. Thus we begin to understand how living organisms may have developed in the primordial terrestrial atmosphere or ocean.

(b) Evolution of the Earth's Atmosphere

Of all the planets, only the earth has appreciable free oxygen in its atmosphere. We think that the earth's oxygen came about because of the development of life here. Green plant life, mainly in the oceans, flourished during the first few thousand million years of the earth's existence. This vegetation removed the carbon dioxide from the air by the process of photosynthesis, building itself with the carbon and releasing the oxygen into the atmosphere. When the vegetable organisms die, they decay (or oxidize), removing the oxygen from the air again. However, part of the dead vegetation escaped the decay process, by being preserved in the ground in the form of fossil fuels, where much of it remains today. Thus most of the oxygen is removed from the air by the decay of dead plant matter (and later by decay of animals, by combustion, and by respiration), but a little—somewhere between one part in 10^4 and one part in 10^5—of that produced by photosynthesis remains in the atmosphere, gradually building up the oxygen concentration. It is estimated that oxygen comprised only about 1 percent of the atmosphere 600 million years ago, but since then it has gradually accumulated to about 21 percent. Nitrogen makes up 78 percent, and argon the remaining 1 percent, with water vapor and carbon dioxide present in trace amounts.

If photosynthesis were suddenly to stop, it is estimated that the forces that remove oxygen from the air would take about 10^4 years to deplete our store of this precious gas. That is to say, the turnover time for oxygen in the earth's atmosphere is about ten thousand years. The present oxygen supply required some ten thousand to one hundred thousand times this long to accumulate because of the very slight margin by which production of oxygen by photosynthesis exceeds its removal by decay.

In recent years man has upset this delicate imbalance. It is estimated that now about 15 percent of oxygen removal arises from our consumption of the fossil fuels (coal and oil) dug from the earth. At the present rate it would still take thousands of years to use up our oxygen (the easily available oil and coal will be used up much sooner than this), but the resulting increase in the concentration of carbon dioxide in the atmosphere is measurable. Because this gas is very effective in blanketing the infrared radiation of the earth near the ground, a greater concentration of it could have a profound effect on the climate by increasing the world's average temperature enough to melt the polar caps and flood several countries. (This phenomenon is known as the *greenhouse effect*.)

36.2   POSSIBILITY OF INTELLIGENT LIFE IN THE GALAXY

There are probably 10^9 potentially observable galaxies in the universe, and probably many times this number of galaxies too faint to observe. Each may contain thousands of millions of stars, a large fraction of which might have planets. The possibilities of life throughout the universe, therefore, would seem to be enormously greater than in our own Galaxy. Yet because even light requires millions of years to travel the great distances between galaxies, any other societies that we have a chance of discovering are probably in our own Galaxy.

(a) The Number of Galactic Civilizations

Cornell astronomer Frank Drake has pioneered the attempt to estimate the number of potentially communicative civilizations currently extant in the Galaxy. Drake's famous equation expresses the number, N, of currently extant civilizations in the Galaxy as the product of seven factors:

$$N = n_s f_p n_p f_b f_i f_c f_l,$$

where n_s is the number of stars in the Galaxy, f_p is the fraction of those stars with planetary systems, n_p is the mean number of planets suitable for life per planetary system, f_b is the fraction of those planets suitable for life on which life has actually developed, f_i is the fraction of those planets with life on which intelligent organisms have evolved, f_c is the fraction of those intelligent species that have developed communicative civilizations, and f_l is the mean lifetime of those civilizations in terms of the age of the Galaxy. The first three factors are essentially astronomical in nature, the next two are biological, and the last two are sociological. We are able to make some educated estimates regarding the astronomical factors, we may be on shaky ground with the biological ones, and we are almost playing numbers games in trying to estimate values for the last two. Yet some interesting estimates can be made, and limits derived.

The mass of the Galaxy (Chapter 29) is believed to be from 2×10^{11} to 10^{12} solar masses. We do not know the form of much of the mass in the galactic corona, but it is a safe assumption that there are at least 4×10^{11} stars in the Galaxy. Although this figure could be several times too small, for illustration we shall adopt it for n_s.

The sun originated from a cloud of gas and dust—the solar nebula—whose rotation caused it to flatten into a disk from which the planets formed. We expect similar formation of planetary systems elsewhere to be commonplace. On the other hand, roughly half of the sun's neighboring stars are members of binary or multiple star systems. It may be that duplicity is an alternative to the formation of a planetary disk, in which case stars in double star systems might not have planets. In any case, it is unlikely that in such systems planets could have stable orbits. Therefore, we shall assume that only half of the stars in the Galaxy have planets, and adopt $f_p = 0.5$.

Not all planetary systems can contain planets suitable for the development of life. We are not sure when life first developed on the earth, but 4.5×10^9 years were required before a communicative civilization was realized. On the earth, we had time, because the total lifetime of the sun is about 10^{10} years—its main-sequence life now being about half expended. We would never have made it if the sun were a spectral-type B or A star with a lifetime of a few hundred million years or less. On the other hand, the relatively faint main-sequence M stars have luminosities of at most 0.01 that of the sun, and the majority are many times fainter still. For a planet to receive enough energy from such a star to warm it to the point that life has a good chance of developing, that planet would have to be so close to the star that tidal forces produced by the star on it would lock the planet into synchronous rotation (with one side always toward the star). Under these conditions, any atmosphere should migrate from the starlit side around to the dark side, where it would freeze out. We would, therefore, expect planets only around main-sequence stars not too unlike the sun—say, spectral classes from F2 to K5—to be likely to have conditions favorable for life. This requirement eliminates all but about 10 percent of the stars. Now in our own solar system, at least one and possibly two or three planets have suitable conditions. Other systems may have none, but if we assume that on the average a star of the right sort with a planetary system has one suitable planet, we can estimate $n_p = 0.1$.

Many biologists are of the opinion that given the right kind of planet and enough time, the development of life is inevitable. Perhaps it is so, but in the spirit of a devil's advocate, we argue that the certainty of life forming has not been demonstrated as yet. Let us consider the liberal estimate that life, given the right conditions, is certain to develop, but also the devil's advocate guess that it happens only, say, 10 percent of the time. The corresponding values of f_b are thus 1.0 and 0.1.

Similarly, given the emergence of life, there is a widespread view that with enough time and natural selection a highly intelligent species will certainly evolve. Even were it inevitable that an intelligent species evolve on every planet with life, however, how long should it take? On earth, it took 4.5×10^9 years. What if we happened to be quick about it, but that the average intelligent species takes, say, 20×10^9 years? Moreover, of the many parallel lines of evolution on earth, only one (so far) has produced a being with enough intelligence to build a technology. Certainly, one could not rule out that the probability could be as low as 10 percent. Again, we take f_i to be either 1.0 or 0.1.

Not all intelligent societies would necessarily develop a technology capable of interstellar communication. We are on the threshold of that capability and possess a natural curiosity about the rest of the universe. Insects do not appear to have any curiosity at all, however, and it is not certain if this human trait is fundamental to intelligence. Even if a society were curious, it might have good reason for wishing to have nothing to do with any other civilization. Some investigators suppose that half of all intelligent species will form communicative technological societies. As a conservative alternative, let us also assume that only one tenth of them do; that is, $f_c = 0.5$ and $f_c = 0.1$.

It is generally agreed that the final factor, f_l, is the most uncertain. Some have speculated that a technology might survive for an average of 10^9 years, so that if the age of the Galaxy is 10^{10} years, $f_l = 0.1$. The only known technology, of course, is our own, and we have only just reached the capability of interstellar communication. In Section 36.3 we shall advance arguments that our technology might well end in a few decades. If so, and if we are typical, $f_l < 10^{-8}$. In rebuttal, some contend that if a communicative society can manage to survive for 100 years, it might well maintain itself for 100 million years. A popular compromise estimate is 10^6 years for the longevity of a typical technology, which leads to $f_l = 10^{-4}$. In the spirit of the devil's advocate, we shall suppose that one can equally well defend a longevity of 10^4 years, and adopt $f_l = 10^{-6}$.

We may now take our choice of the estimates of the factors in the equation for the number of communicative civilizations, and compute N. If we substitute the more or less liberal estimates into the equation, we find

$$N = 4 \times 10^{11} \times 0.5 \times 0.1 \times 1 \times 1 \times 0.5 \times 10^{-4} = 10^6$$

This is the estimate of astronomers I. S. Shklovsky and Carl Sagan. Our devil's advocate estimates, on the other hand, lead to

$$N = 4 \times 10^{11} \times 0.5 \times 0.1 \times 0.1 \times 0.1 \times 0.1 \times 10^{-6} = 20$$

If the first estimate is correct, there is an even chance that the nearest civilization is less than 250 LY away. If the second is correct, it may lie tens of thousands of LY away. We cannot, of course, really know the number of civilizations in the Galaxy, nor the distance to the nearest one; but we do see the rationale by which estimates are made, and also see that one can get almost any answer he wants by choosing what could be defended as reasonable estimates of various factors that enter into the calculation. Indeed, it is entirely possible that we are the *only* technological society in the Galaxy.

Perhaps we are too anthropomorphic in our estimates. We have consistently thought of planets as the only places where life will develop, and that it will necessarily evolve into some kind of beings that have aspirations and interests similar to ours. For an entirely different kind of idea of what intelligent life could conceivably be like, read Fred Hoyle's science fiction novel, *The Black Cloud*.

(b) Interstellar Travel

One means of detecting other Galactic civilizations might be by interstellar travel. We have seen, however, that the nearest neighboring civilization is expected to be at least a few hundred, and probably a thousand, or even tens of thousands of light years away. Because nothing can travel faster than light, a visit to another civilization would involve at least hundreds, and more likely thousands of years.

Now to be sure, a space traveler's time slows down (with respect to ours) if he travels near enough the speed of light (Chapter 13). To make travel to other possible civilizations feasible in a human lifetime, however, the traveler's time would have to slow down at the very least by a factor of 5 (so that a 400-year round trip could be accomplished in 80 years of the crew's time). This much time dilatation requires a speed of 98 percent that of light, and the energy requirements to reach that speed are absolutely enormous.

Consider, as an example, calculations presented by S. von Hoerner: Suppose we wish to send a moderate payload of 10 tons (three to five automobiles) into interstellar space and accelerate it to $0.98c$. In this relatively small payload we must provide an environment to provide life support for the crew for several decades. We add another 10 tons for engines and propulsion systems. The total energy required, no matter how it is obtained or how fast it is expended, is about 4×10^{29} ergs—

roughly enough energy to supply the entire world's needs (at the present global expenditure rate of energy) for some 200 years. It would probably require the complete annihilation of matter, which we do not know how to accomplish at present. If we wanted the crew to reach its final speed of $0.98c$ at an acceleration equal to the earth's gravity, which would take 2.3 years, it would require the equivalent of 40 million annihilation plants of 15 million watts each, producing energy to be transmitted (with perfect efficiency) by 6×10^9 transmitting stations of 10^5 watts each, and all of this apparatus must be contained within a mass of 10 tons!

These enormous energy requirements apply only if we need to travel close to the speed of light to take advantage of the relativistic time dilation. Interstellar travel is possible if we are willing to take a long time to do it—which requires many generations for the crew. We may wonder about the morality of subjecting the crew's offspring to a life in a spaceship destined for an unknown fate generations in the future. But at least, it is possible.

On the other hand, even now we can send, and have sent, material messages into interstellar space. Both the Pioneer and Voyager spacecraft, for example, after passing Jupiter, have entered orbits on which they will eventually escape the solar system. It is unlikely that they will ever be seen or recovered by another intelligent species, but it is remotely possible. Partly for this reason, each Pioneer carries a plaque bearing line drawings of human beings, and cryptic messages describing the world from which it came, and the Voyagers contain phonograph recordings with messages from and descriptions of earth. It is doubtful that the message on the plaque will ever be decoded or the recording heard, but publicity about them has called attention to the possibility of intelligent life in the Galaxy. More important by far is the message carried by each spacecraft itself; its discovery would convey a great deal of information about the race that launched it, and the state of our technology.

A final possibility for direct communication is a visit to the earth by extraterrestrial visitors. If the nearest civilizations are hundreds of light years away, they cannot have come to see us as a result of learning about us, for even radio waves that we have inadvertently been emitting into space—our radio and television programs—have only been on their way for a few decades. They could have reached only the very nearest stars, and it is highly unlikely that anyone there has received them and dispatched space ships to look us over. If we have

been visited, it would have to have been by random selection by interstellar travellers, and it is extraordinarily unlikely that among the millions of stars in the Galaxy for every civilization, we should have been singled out for surveillance.

Yet the popular literature is full of accounts of sightings of UFOs, presumably operated by some intelligence, and even of alleged evidence for highly intelligent beings that have visited the earth and taught people to build such magnificent structures as the pyramids, Easter Island statues, and other marvels. Not only do the latter accounts fail to acknowledge the great amount of work that has been done by competent professional archeologists, but they are highly racist in their implication that earlier civilizations could not have had the talent to create great works of art.

Most scientists are highly skeptical of the extraterrestrial interpretation of reports of lights or erratically accelerating shiny objects in the sky, and of alien beings with unhuman countenances, yet with the human characteristics of two legs, two arms, a head, a mouth, nose, two eyes, and other anthropomorphic features. Hard evidence of objects from space is lacking. Scientists, more than anyone, would delight in finding concrete evidence of alien life—there is so much we have to learn from it! But we still need evidence that can be analyzed by any competent scientist qualified to judge its extraterrestrial origin. Rumors, hearsay, secondhand reports, and eyewitness accounts by lay and inexperienced observers all must be given the benefit of the doubt, but still require positive verification before being taken as final evidence for life in the universe beyond the earth.

(c) Interstellar Communication

On the other hand, if there are other Galactic civilizations, there is a very real possibility that we may be able to communicate with them by radio. We do not necessarily mean two-way communication, for the radio waves would probably require hundreds of years at the very least for their round-trip travel between each question and answer. (Interstellar communication would be between civilizations, not individuals.) On the other hand, if there are communicative civilizations in the Galaxy they may already be trying to communicate, or at least to send one-way messages to other possible civilizations, just to inform them of their existence, and probably to convey much information in addition. With even our present technology we could send such messages ourselves to other stars in the Galaxy. The likelihood of an answer within a human lifetime, however, is almost negligible. Thus, the first step is to try receiving messages.

There is a good chance that we would recognize an intelligent message—for example, a binary-coded broadcast of the number π repeated over and over. The discovery of extraterrestrial intelligence would be one of the most stupendous in the history of mankind. If there were thousands or millions of communicating societies throughout the Galaxy, we could imagine a vast system of intercommunication, whereby many civilizations are sending messages to many other places in the Galaxy, not necessarily with the hope of receiving answers, for the messages themselves may well be received centuries after the sending civilizations had ceased to exist, but to pass on information about life in the Galaxy. It is a romantic and exciting idea and many of us would like to be in on the network, if, in fact, it exists.

It is my guess that it does not exist. But we will never know if we do not investigate. So if we are willing to expend the energy to find out, despite the small chance of success, the potential rewards might well justify the effort—especially if the equipment developed for the project could be used for other kinds of astronomical research which would be certain to bring fruitful results.

In fact, with existing radio astronomy facilities, a number of limited searches for intelligently coded signals have already been made by radio astronomers in the United States and in the Soviet Union. To date, there has been no success, but no success would have been expected from such meager efforts. Just what would it take to learn if messages are being beamed in our direction from other civilizations in the Galaxy?

The problem was explored in a NASA study in 1971 by a group under B. Oliver, called *Project Cyclops*. The committee concluded that a fairly thorough search could be conducted for a cost of a few thousand million dollars. If an appreciable number of communicating civilizations exist, the Cyclops system would have a good chance of detecting one of them. The system would consist of radio telescopes in a vast array about 5 km in diameter. Even if no civilization were discovered, the system would be enormously important to radio astronomy.

36.3 LONGEVITY OF CIVILIZATION ON THE EARTH

Consider the sheet of paper on which the page you are reading is printed; it is roughly 0.1 mm thick. Imagine that you were to cut it into two sheets and stack them on top of one another. Then cut the stack of two sheets and combine the four sheets into a single stack. Next, make a third cut and stack up the eight sheets. If you were to continue the process until 100 cuts had been made, guess how thick the stack of paper would be.

Figure 36.1 Artist's conception of an aerial view of the antenna array proposed by Project Cyclops in 1971. Each of the many radio telescopes in the system would have an aperture of at least 100 m and would be designed to receive signals in the 1420 to 1660 megahertz frequency range that might originate from a distance as great as 1000 LY. *(NASA/Ames Research Center)*

Figure 36.2 An artist's rendering of a closeup of several of the antennas in the proposed Project Cyclops system for detecting radio signals. The building at the right, near the center of the array, is the proposed data collection and processing facility. At present there are no plans to build the system. (*NASA/Ames Research Center*)

The answer is 10^{10} LY—the distance to a remote quasar (Chapter 37). When we double a number at each step, it is an example of a *geometric progression*. This progression increases slowly at first, but almost suddenly it explosively rises and rapidly approaches infinity as the number of doublings increases. If a quantity (say, inflation, population, use of energy, or the price of homes) increases at I percent per year, the time for that quantity to double is approximately 70 years divided by I. Thus, if (as has been the case in recent decades) our use of energy increases by 7 percent per year, in the next ten years we will have used as much energy as has been used by all of mankind throughout history until now. Another example of a similar geometric progression is the world population, which at present is doubling about every 35 years (2 percent per year).

In New York City there are a little more than 100 square meters of land area for every person living in that city. Suppose we imagine New York to be more than 100 times as crowded as it is now, so that there is only 1 m² per person. Moreover, let us suppose that the same crowding applies over the entire land area of the earth. At the present rate of population growth, the land area of our planet would be filled to this density in about 550 years.

Perhaps one could argue about how crowded the earth can be. An uncontested ultimate limit to the number of people, however, is set by the fact that we are all made of atoms of matter, and matter is conserved; that which comprises our bodies has had to come from the earth itself. Most of the mat-

ter in our bodies is in the form of water which, we have seen, outgassed from the earth's crust. More and more of the water needed to build the bodies of people must come from the seas of the world as the population grows. At a doubling time of 35 years, the oceans would be entirely converted to people in about 1200 years (never mind about water to wash with, swim in, drink, and water our fields with, or even for fish). At the same doubling rate, in 1600 years the mass of people would be equal to the mass of the earth, and in 2300 years to that of the solar system. If we could create matter from nothing (to make people), after about 5300 years a great sphere of humanity 150 LY in radius would be expanding at its surface with the speed of light!

The above examples are not meant as predictions but to illustrate how rapidly the numbers increase with a constant doubling time. In fact, during the past thousand years the doubling time for the population of the world has actually been *decreasing* with time. An excellent representation of the world population over the past two millennia is given by a formula graphed in Figure 36.3. A literal interpretation of this formula predicts that the world population will become infinite and the space available to each of us will go to zero in 2026 A.D., on Friday, November 13—a date dubbed "Doomsday."

Of course the world population cannot become infinite; various limitations will slow its present growth rate before 2026. But the problem is that most of the growth occurs so suddenly that the population can become unmanageable without warning;

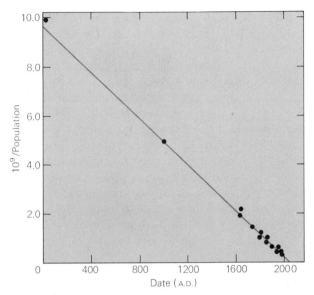

Figure 36.3 The increase of the world's population during the last 2000 years. Against the date is plotted one thousand million (10^9) divided by the population, so that the points are proportional to the reciprocal of the population, or, equivalently, to how much space per person there is on earth. The data are taken from several recent demographic studies. The straight line is a plot of an equation derived by von Förster, Mora, and Amiot, by which the population becomes infinite and the space per person drops to zero late in the year 2026 A.D.

pressure (observed, for example, in experiments with colonies of rats that are allowed to multiply without check) is in itself a threat to survival. Nuclear bombs now stockpiled in the world are equivalent to about 10 tons of TNT per person—enough energy to raise a 10,000-ton apartment house 500 m into the sky to drop on every man, woman and child now living. And the nuclear stockpile is increasing at a faster rate than the population.

In some developed countries, including the United States, the population, while not yet stable, is rising more slowly, but the worldwide growth is unabated. In most of Latin America the population doubles in 20 years. Compared with the United States, Mexico is a poor nation. Yet, to maintain their existing level of poverty, the Mexicans must double their homes, schools, hospitals, roads, factories, and in general, everything in their economy in just 20 years!

Pollution of the atmosphere and exhaustion of our fossil fuels may be altering our climate and potential for food production. The United States, if we give up eating meat, could probably feed its people for another population doubling or so, but this requires maintaining our present efficiency in farming and assumes no climatological degradation of our ability to grow food.

the problem may not be recognized until it is too late to prevent catastrophe.

Nor does expansion into space provide a solution. It has been seriously suggested that we can relieve overpopulation by emigrating to other planets. But even if we wanted to condemn our children to lives in air-tight living enclosures on the moon and Mars, without any hope of playing in the fields or hiking in the woods, we could only extend our time another 35 years, for the entire surface area of the moon and Mars combined is only about the same as that of the land area of the earth. Within about 500 years all possible planets about other stars within 150 LY could be occupied to the extent the earth is now, and to find room for our ever-increasing number of people, we would have to transport them faster than the speed of light to ever more remote planets, which, of course, cannot be done.

In short, we are in danger of using up available space (and resources) so suddenly that we will have scarcely any warning to prepare for the onslaught of aggression and suffering that almost inevitably will result. The aggression prompted by population

Figure 36.4 An ominous exhibit at the Bishop Museum, Honolulu, Hawaii. (*Bishop Museum*)

To the problems of overcrowding, the violence it breeds, and energy and food shortages, we should add the gradual increase in the mean temperature of the earth due to the heat produced by the use of energy. This factor alone could alter our climate enough to affect food production in less than a century. Another effect of the present trends of the evolution of our civilization is genetic deterioration (medicine increases the life span, but encourages survival of persons with genetically unfavorable mutations). Even if we could withstand all of these threats to our survival, there remains a possible crisis caused by the boredom and stagnation of a stable society trying to endure without substantial innovation on a completely filled planet for hundreds of centuries.

Possibly these threats to our survival can be circumvented by enlightened action. Or perhaps violent struggles for survival in a chaotic near future will result in a strong enlightened portion of society surviving to maintain a stable civilization. At present, thoughtful, rational international planning to preserve our planet as an abode for most of our race is missing. It may be overly pessimistic to predict that our technological society is doomed to an early end, but if we continue our present course, it is hard to imagine an alternative. Do other civilizations (if they exist) similarly destroy themselves, or have they learned, as we hope to learn, to preserve their longevity? Perhaps the discovery that another civilization has "made it" will one day be our salvation.

EXERCISES

1. Compare and contrast the conditions in interstellar space and at the surface of a planet as they affect formation of organic molecules.

2. Suppose water is 10^{-4} of the matter of the earth. Use the data in this chapter to estimate the mass of atmospheric argon compared to the earth's mass.

3. Why might the presence of oxygen in a planetary atmosphere suggest that life exists there?

4. Suppose we could carry on two-way radio communication with another civilization. After a question is sent, what is the time we would have to wait before expecting an answer if that civilization is:
   a) on the moon?
   b) on Jupiter (assume Jupiter to be at its nearest to the earth)?
   c) on a planet revolving around Alpha Centauri?
   d) on a planet revolving around Tau Ceti (see Appendix 13 for the distance to the star)?
   e) at the Galactic center?

5. Because of the large number of main-sequence M-type stars, the mean mass of a star in the Galaxy could be only 0.25 solar masses. What, then, would be our estimate of n_s if the mass of the Galaxy is 2×10^{11} solar masses?

6. In the case described in Exercise 5, how would our estimate of f_p and calculation of N be affected?

7. Why could a planet not have a stable orbit about a star in a close binary system?

8. How many Galactic civilizations would you expect if you use our devil's advocate estimates of the various parameters, but assume that an average civilization lasts only 100 years?

9. All considered, what would you judge to be the extreme limits for the number of civilizations in the Galaxy?

10. What wavelengths of electromagnetic radiation (other than 21-cm) might be suitable for interstellar communication, and why? What wavelengths would certainly not be suitable?

11. Suggest ways you might code a message to send into space in such a way that another civilization, if it received it, would have a good chance to recognize it as intelligently contrived.

12. Verify the statement in the text that a 0.1 mm sheet of paper cut and stacked 100 times would have a final thickness of 10^{10} LY. (See Appendix 6 for conversion factors.)

13. Take an ordinary sheet of paper, a straight edge, and a pen or pencil. Draw a line (representing a cut) that roughly bisects the paper. Now, perpendicular to your first line, bisect one of the halves. Keep this up, counting the number of times you are able to continue to bisect the smaller and smaller rectangles. Try to imagine continuing to 100 bisections.

14. A sample of bacteria in a large bottle doubles in number every minute. At 11:53 A.M. the bottle is only $1/128$ full. In the next few minutes, some bacteria scouts discover three additional identical bottles that are still empty, and reason that they now have room for four times the bacteria population that their original bottle can support. Thus they continue to reproduce at the same rate, and when their original bottle is filled, they (or their offspring) move on to the other bottles. At what time are all four bottles filled?

15. Suppose the annual inflation rate to be 12.5 percent (not far off for the United States at the time of writing). After what time has the value of currency dropped to half? To one fourth? To only one thirty-second?

Walter Baade (1893–1960), born in Westphalia, joined the staff of the Mount Wilson Observatory in 1930. He discovered the two populations of stars, expanded the distance scale and age of the universe, and was an early investigator of supernovae and radio sources. He is still recognized as one who used large telescopes to the very best advantage. (*California Institute of Technology*)

37

GALAXIES

The "analogy [of the nebulae] with the system of stars in which we find ourselves . . . is in perfect agreement with the concept that these elliptical objects are just [island] universes—in other words, Milky Ways. . . ."

So wrote Immanuel Kant (1724–1804) in 1755* concerning the faint patches of light which telescopes revealed in large numbers. Unlike the true gaseous nebulae that populate the Milky Way (Chapter 28), the nebulous-appearing luminous objects referred to by Kant are found in all directions in the sky *except* where obscuring clouds of interstellar dust intervene. Despite Kant's (and others') speculation that these patches of light are actually systems like our own Milky Way Galaxy, the weight of astronomical opinion rejected the hypothesis, and their true nature remained a subject of controversy until 1924. The realization, less than a century ago, that our Galaxy is not unique and central in the universe ranks with the acceptance of the Copernican system as one of the great advances in cosmological thought.

37.1 GALACTIC OR EXTRAGALACTIC?

The discovery and cataloguing of nebulae had reached full swing by the close of the 18th century.

A very significant contribution to our knowledge of these objects was provided by the work of William Herschel and his only son, John (1792–1871). William surveyed the northern sky by scanning it visually with the world's first large reflecting telescopes, instruments of his own design and manufacture. John took his father's telescopes to the southern hemisphere and extended the survey to the rest of the sky.

For a while, the elder Herschel himself, who had discovered thousands of "nebulae,"† regarded these objects as galaxies, like the Milky Way system; he was known to remark once that he had discovered more than 1500 "universes." He found, however, that many of the nebulae appeared as individual (although rather indistinct) "stars," surrounded by hazy glows of light. The fact that it is difficult to reconcile the appearance of such an object with that of a remote stellar system led Herschel, in 1791, to abandon the island-universe hypothesis. Still, the concept of the possibility of other galaxies never quite disappeared from astronomical thought.

---

†*Nebula* (plural *nebulae*) literally means "cloud." Faint star clusters, glowing gas clouds, dust clouds reflecting starlight, and galaxies all appear as faint, unresolved luminous patches when viewed visually with telescopes of only moderate size. Since the true natures of these various objects were not known to the early observers, all of them were called "nebulae." Today, we usually reserve the word "nebula" for the true gas or dust clouds (Chapter 28), but some astronomers still refer to galaxies as nebulae or *extragalactic nebulae*.

Universal Natural History and Theory of the Heavens.

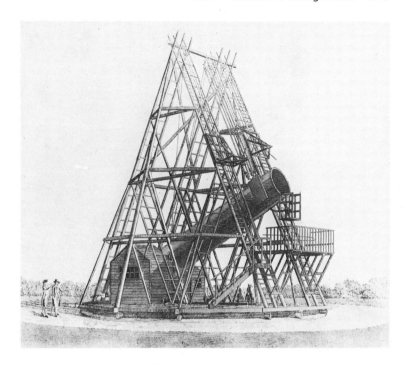

Figure 37.1   Herschel's 40-foot telescope. *(Yerkes Observatory)*

(a) Catalogues of Nebulae

One of the earliest catalogues of nebulous-appearing objects was prepared in 1781 by the French astronomer Charles Messier (1730–1817). Messier was a comet hunter, and as an aid to himself and others in his field he placed on record 103 objects that might be mistaken for comets. Because Messier's list contains some of the most conspicuous star clusters, nebulae, and galaxies in the sky, these objects are often referred to by their numbers in his catalog—for example M31, the great galaxy in Andromeda.

In the years from 1786 to 1802, William Herschel presented to the Royal Society three catalogues, containing a total of 2500 nebulae. The *General Catalogue of Nebulae*, published by John Herschel in 1864, contains 5079 objects, of which 4630 had been discovered by him and his father. The *General Catalogue* was revised and enlarged into a list of 7840 nebulae and clusters by J. L. E. Dreyer in 1888. Today most bright galaxies are known by their numbers in Dreyer's *New General Catalogue*—for example, NGC 224 = M31. Two supplements to the *New General Catalogue*, known as the first and second *Index Catalogues* (abbreviated "IC"), were published in 1895 and in 1908.

By 1908 nearly 15,000 nebulae had been catalogued and described. Some had been correctly identified as star clusters and others as gaseous nebulae (such as the Orion nebula). The nature of most of them, however, still remained unexplained. If they were nearby, with distances comparable to those of observable stars, they would have to be luminous clouds, probably of gas, possibly intermixed with stars, within our Galaxy. If, on the other hand, they were very remote, far beyond the foreground stars of the Galaxy, they could be unresolved *systems* of thousands of millions of stars, galaxies in their own right, or as Kant had described them, "island universes." The resolution of the problem required the determination of the distances to at least some of the nebulae.

(b) Arguments For and Against the Island-Universe Hypothesis

By the early 20th century the nebulae could be divided into two distinct groups: those largely irregular and amorphous, which are concentrated near the Milky Way, and those showing either elliptical or wheel-shaped symmetry, which are most numerous in parts of the sky far from the Milky Way. Those of the former type have bright-line spectra and were clearly recognized as gaseous nebulae. The few available spectra of the brighter nebulae in the latter category, on the other hand, showed absorption lines like the spectra of stars. Moreover, most of their radial velocities were found to be very

Figure 37.2 Two photographs of NGC 5457 made with the 5-m telescope. Bright nova appears on the photograph taken on February 7, 1951 (right), but not on the one taken on June 9, 1950. (This and the following photographs in the chapter are reproduced through the courtesy of the Palomar Observatory, California Institute of Technology: Figures 37.4, 37.6, 37.8, 37.9, 37.10, 37.11, 37.12, 37.13, 37.14, 37.15, 37.16, 37.17, 37.18, 37.19, 37.20, 37.22, 37.24, 37.28, and 37.29.)

high (Chapter 39), and it seemed that such rapidly moving objects would escape the Galaxy. These data supported the hypothesis that the nebulae of the second type are extragalactic.

In 1908 construction was completed of the 60-inch (1.5-m) telescope on Mount Wilson, and within a decade the 100-inch (2.5-m) telescope was also in operation. Photographs obtained with these instruments clearly resolved the brightest stars in some of the nearer "nebulae." By 1917 several novae (Section 34.6a) had been discovered in the more conspicuous nebulae. If those novae were as luminous as the 26 novae then known to have occurred in our own Galaxy, they, and the nebulae in which they appeared, would have to be at distances of about 1 million LY—far beyond the limits of our galaxy. Similar distances were found for the nebulae if the brightest resolved stars in them were assumed to have the same intrinsic brightness as the most luminous stars in the Galaxy.

On the other hand, not all astronomers agreed that real stars had actually been resolved in any of the nebulae. Even the Mount Wilson astronomers who had photographed those stars were not convinced of their true stellar nature and had described them as "nebulous stars." Moreover, two of the novae that had been observed in the nebulae were far

Figure 37.3 The Large and Small Magellanic Clouds (top and lower left). (Sky and Telescope)

The edge-on
spiral galaxy NGC
4565. (*U.S. Naval
Observatory*)

The spiral galaxy M51 in Canes Venatici. (*U.S. Naval Observatory*)

The irregular
galaxy M82 in
Ursa Major.
(*U.S. Naval
Observatory*)

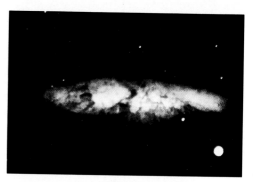

The elliptical galaxy NGC 5128 in Centaurus. (© *Association of Universities for Research in Astronomy, Inc. The Cerro Tololo Inter-American Observatory*)

The Large Magellanic Cloud, the nearest external galaxy. *(U.K. Schmit Unit, Royal Observatory Edinburgh)*

M32, a small Local Group elliptical galaxy in Andromeda. *(U.S. Naval Observatory)*

The spiral galaxy M31 in Andromeda, photographed with the Palomar Schmidt telescope. *(California Institute of Technology/Palomar Observatory)*

brighter than the rest. S Andromedae, for example, which appeared in the nebula M31 in 1885, reached magnitude 7.2. Today we recognize those two novae as *supernovae* (Section 34.6c), but supernovae were unknown in the early part of the century. If those two bright "novae" were assumed to be only as luminous as ordinary novae, the distances calculated for them turned out to be only a few thousand light-years, and the nebulae would not be extragalactic.

The most convincing evidence that seemed to prove that the nebulae were not galaxies was from measures in the early 1920s that appeared to show proper motions of brighter stars in a few of them. If those measures were valid, and the nebulae were remote enough to be extragalactic, those stars would have to have linear speeds near the speed of light or even greater. More recent and accurate measures have shown, of course, that those early results were in error, but at the time they seemed to give strong evidence for a small distance for the nebulae.

A remark should be made here concerning the Clouds of Magellan (Section 29.2a), two stellar systems that are not visible from as far north as the United States. The distances to the Clouds, determined from the cepheid variables in them, were thought at that time to be about 75,000 LY (they

are actually at least twice that distance). In the years preceding 1924, however, the diameter of our own Galaxy had been overestimated at about 300,000 LY (inadequate corrections having been made for the absorption of light from the globular clusters by interstellar dust). The Magellanic Clouds, therefore, although now regarded as neighboring external galaxies, were then considered to be outlying sections of our own Galaxy.

Two of the major protagonists in the controversy over the nature of the nebulae were Harlow Shapley, of the Mount Wilson Observatory, and H. D. Curtis, of the Lick Observatory. Their opposing views culminated on April 26, 1920, in the famous Shapley-Curtis debate before the National Academy of Sciences. Curtis supported the island-universe theory, and Shapley opposed it. Of course, the controversy was not settled by the debate; according to A. R. Sandage, "Perhaps the fairest statement that can be made is that Shapley used many of the correct arguments but came to the wrong conclusion. Curtis, whose intuition was better in this case, gave rather weak and sometimes incorrect arguments from the facts, but reached the correct conclusion."‡

---

‡A. R. Sandage, *The Hubble Atlas of Galaxies*. Carnegie Institution, Washington, D.C., 1961, p. 3.

Figure 37.4 A field of variable stars in the Andromeda galaxy, with two variables marked. Photographed with the 5-m telescope.

(c) The Resolution of the Controversy

The final resolution of the controversy was brought about by the discovery of variable stars in some of the nearer "nebulae" in 1923 and 1924. Edwin Hubble, at the Mount Wilson Observatory, analyzed the light curves (Section 24.3b) of variables he had discovered in M31, M33, and NGC 6822 and found that they were cepheids (Section 24.3e). Although cepheid variables are supergiant stars, the ones studied by Hubble appeared very faint—near magnitude 18. Those stars, therefore, and the systems in which they were found, must be very remote; the "nebulae" had been established as galaxies. Hubble's exciting results were presented to the American Astronomical Society at its 33rd meeting, which began on December 30, 1924.

37.2   THE EXTRAGALACTIC DISTANCE SCALE

One of the most important, difficult, and controversial problems in modern observational astronomy is that of the scale of distances to galaxies. Galaxies are far too remote, of course, to display parallaxes or proper motions. Hence we must resort to indirect means of surveying them. We distinguish here between the determination of the *absolute* distances to galaxies (say, in parsecs) and their *relative* distances; the former is by far the more difficult, for it involves recognizing *standard candles* or *standard yardsticks* in galaxies.

(a) The Absolute Distances to Galaxies

Some of the more important ways by which we attempt to find distances to galaxies are the following:

1. *Cepheids.* Cepheid variables gave Hubble the first clue to the remote nature of galaxies and are still our first important link to galaxian distances. That is why these relatively rare stars are so significant to us. If we can recognize a cepheid in a galaxy, we can find its absolute magnitude from its period through the period-luminosity relation (Section 24.3f). Thus, the cepheids can serve as standard candles, for comparison of their known absolute magnitudes and observed apparent magnitudes enables us to find their distances and hence the distances to the galaxies in which they occur, with the inverse-square law of light (Section 24.2b). The most luminous cepheids are at absolute magnitude -6, which makes them supergiant stars, but even so they can be detected in only about 30 of the nearest galaxies—even with the world's largest optical telescopes.

2. *Brightest stars.* The most luminous stars are brighter than cepheids and can be seen to greater distances. Thus, once calibrated in those galaxies whose distances are known from observations of cepheids, the brightest stars can be used as standard candles. Young, high-mass supergiant stars range in absolute magnitude to as bright as about -9 or even -10 and so can extend the distance scale to more than four times the distance to which cepheids can be seen, but this is still our very immediate cosmic neighborhood!

3. *Novae.* Novae can be recognized in nearby galaxies, and from their light curves we know what their approximate luminosities are. The brightest novae, however, even at maximum light, do not outshine the brightest stars.

4. *Globular clusters.* Although globular clusters range considerably in total light (because they differ in their numbers of stars), we sometimes recognize many such clusters in one galaxy. If we assume that the brightest of them is like the brightest globular cluster in our own Galaxy, that object becomes a standard candle. Distances determined this way are rather uncertain, but the method is still useful because globular clusters are often seen in galaxies without a young population of supergiant stars.

5. *Supernovae.* Because of their high luminosities, supernovae can be seen in very remote galaxies and would seem to be ideal standard candles, except that supernovae differ considerably among themselves in absolute magnitude at maximum light. Some progress is being made in calibrating different kinds of supernovae, and in the future they may well hold the key to the

extragalactic distance scale. The difficulty at present is that supernovae occur rarely in any one galaxy, and the sample that have appeared in galaxies of well-known distances is rather small.

6. *H II regions.* The total luminosities of H II regions differ greatly from one to another, and they have not proved useful as standard candles. On the other hand, some astronomers have presented evidence that the linear diameters of the largest H II regions are pretty much the same from one galaxy to the next of the same type. Thus the size of the largest H II region in a galaxy can be used as a standard yardstick, whose distance can be deduced from its angular size.

7. *21-cm line width.* A very promising technique of distance determination was developed in the late 1970s. A typical spiral galaxy contains a great deal of neutral hydrogen gas in revolution about its center. The 21-cm radiation from this hydrogen in different parts of the galaxy, moving at different speeds in our line of sight, therefore displays a range of Doppler shifts, so that the entire 21-cm line radiation from the galaxy is observed as a broad band (think of how the rotation of a star affects its spectral lines—Section 25.2d). After a simple correction for the tilt of the plane of the galaxy to our line of sight, the width of that 21-cm line gives a measure of the maximum rotational velocity in the galaxy. That rotational velocity is correlated with the galaxy's mass, and we might expect the luminosity to be correlated with the mass as well, and hence with the 21-cm line width. Indeed, it is; the correlation was first noted between the line width and the total visual light from a galaxy, but the relation is tighter if the galaxy's infrared luminosity is used instead. This method of distance determination, then, is to observe the 21-cm line width, thereby learning the galaxy's absolute infrared magnitude, then to observe the galaxy's apparent infrared magnitude, and finally to calculate the distance from the inverse-square law.

8. *Total light of galaxies.* If all galaxies were identical, they would all emit the same to-

tal amount of light, and the magnitude of a galaxy as a whole would indicate its distance. Galaxies range enormously in total luminosity; nevertheless, some types of galaxies display a relatively small range of luminosities. For those galaxies, rough distances can be estimated from their total apparent magnitudes. Of course, we can only establish what those mean luminosities are for the types of galaxies that are well represented among those whose distances can be determined from other more direct methods. Unfortunately, many galaxies, probably the majority, do not have distinguishing characteristics that enable us to estimate their absolute magnitudes; we can only tell which ones are highly luminous and which ones are not if we see a collection of them of various brightnesses, side by side in a cluster.

With so many ways of finding distances to galaxies, one might think that the distance scale is well settled. Unfortunately, it is not; the various standard candles are difficult to calibrate, and the measurements are difficult to make. There is pretty good consensus on the distances to the nearest galaxies—to within 10 or 20 percent, anyway. On the other hand, beyond about 30 million light years the distance scale is very uncertain. Experts differ in their judgment of the proper interpretation of the observations, the calibration of the standard candles and yardsticks, and on what standards are the most reliable. These differences in judgment translate to differences in the distances of remote galaxies of a factor of two.

Perhaps an uncertainty of whether a particular galaxy is two thousand million or four thousand million light years away may seem to be pretty poor precision in an exact science like astronomy, where we know the distances between the planets in the solar system to better than one part in a million. But uncertainty is common at the frontier of research. At least we are far beyond the controversy of whether or not the "nebulae" are extragalactic.

(b) Relative Extragalactic Distances

The relative extragalactic distance scale is in far better shape than the absolute scale. There are ways in which we can estimate the relative distances of two

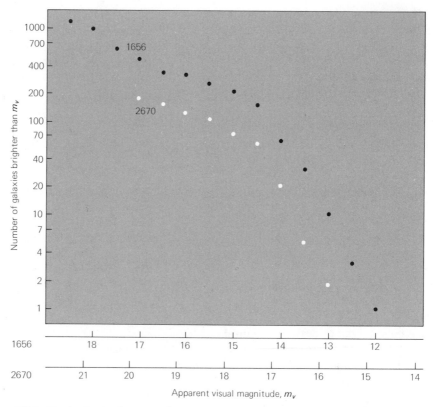

Figure 37.5 The luminosity functions of two clusters of galaxies, identified by their numbers in the Abell catalogue of rich clusters. The horizontal scale is apparent visual magnitude, with fainter magnitudes to the left. The vertical scale shows the number of galaxies in each cluster that are brighter than each corresponding magnitude. Cluster 2670 is more distant and fainter-appearing than cluster 1656, but the points have been plotted on a shifted magnitude scale to show the similarity of its luminosity function to that of the nearer cluster. The two magnitude scales are shown below; corresponding points in the luminosity function of 2670 are 2.8 magnitudes fainter than in the luminosity function of 1656. From that difference in magnitude we can easily calculate the relative distances of the two clusters.

galaxies with far more assurance than we know the actual distance to either.

One way involves clusters of galaxies. We shall see (Chapter 38) that most, if not all, galaxies are members of groups or clusters of anywhere from a few galaxies to a few thousand galaxies each. Although the galaxies in a cluster display a wide range of luminosity, if the galaxies in different clusters were formed under the same initial conditions and according to the same probabilistic laws, we would expect the brightest galaxy (or perhaps the second, or tenth brightest galaxy) in each cluster to be intrinsically similar. If, for example, the brightest galaxy in one cluster were four times as bright-appearing as the brightest galaxy in another, we

would expect the second cluster to be twice as distant as the first. Alternatively (and better, in my opinion), we can compare the distributions of the numbers of galaxies of various apparent brightnesses (or *luminosity functions*) of the two clusters. If the distributions were similar except for a shift in the apparent magnitude scale, in the sense that galaxies at some point in the distribution of one cluster were four times as bright-appearing as those at the corresponding point in the other, we could conclude that the second cluster is twice as far away as the first (see Figure 37.5).

Another way of finding relative galaxian distances is from the well-known correlation between the distances of galaxies and the Doppler shifts of the lines in their spectra to longer wavelengths. This relation, known as the *Hubble law*, is a direct consequence of the uniform expansion of the universe, and we shall return in Chapter 39 to this extremely important relation and its significance. Let it suffice for now, however, that the more remote a galaxy, the greater is the shift (to longer wavelengths) of the light coming from it, and at least for the not terribly remote galaxies, the speeds with which they are moving away from us is proportional to their distances. In visible light, a shift to longer wavelengths is a shift toward the red end of the spectrum, so the Doppler shift to longer wavelengths of a distant galaxy is usually called its *redshift*.

Knowing that a particular galaxy, moving away from us at twice the speed of another, is twice as far away does not, in itself, tell us the distance of either, but only their relative distances. The velocity-distance relation (Hubble law) must be calibrated; we must know the actual distance of a galaxy of a particular velocity, and this requires knowing the absolute extragalactic distance scale.

37.3 DETERMINATION OF GROSS PROPERTIES OF GALAXIES

The linear size of that part of a galaxy that corresponds to an observed angular size can be calculated once the distance to the galaxy is known, just as we calculate the diameter of the sun or of a planet. Also, if we know its distance, we can apply the inverse-square law of light and calculate the total luminosity, or the absolute visual magnitude, of a galaxy from the amount of light flux we receive from it or from its apparent visual magnitude. Thus a knowledge of galaxian radii and luminosities is provided by a knowledge of galaxian distances.

The determination of masses of galaxies, however, is more difficult, and, in fact, is possible (by present techniques) for only a small fraction of them. There are several techniques for measuring galaxian masses.

(a) Mass of Galaxies from Internal Motions

We determine the masses of galaxies, like those of other astronomical bodies, by measuring their grav-

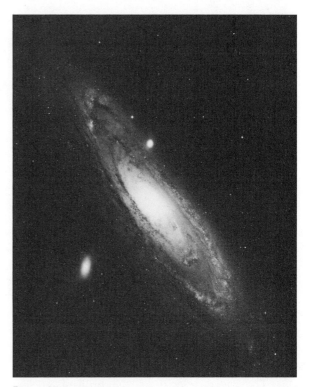

Figure 37.6 The Andromeda galaxy, M31, photographed with the 1.2-m Schmidt telescope.

itational influences on other objects or on the stars within them. We must assume, of course, that Newton's law of gravitation is valid over extragalactic distances. The internal motions within galaxies and the structures of clusters of galaxies suggest that Newton's law does indeed apply.

Internal motions in galaxies provide the most reliable methods of measuring their masses but are observable only in comparatively nearby galaxies. The most important procedure is to observe the rotation of a galaxy from the Doppler shifts of either features in the optical spectrum or the 21-cm line of neutral hydrogen, and then to compute its mass with the help of Kepler's third law.

As an illustration, we shall consider the rotation of M31, the Andromeda galaxy (Figure 37.6). M31 is probably quite similar to our own Galaxy; it has a brilliant central nucleus and conspicuous spiral arms that wind through a presumably circular disk. The galaxy is inclined at an angle of only about 15° to our line of sight, so we see it highly foreshortened. There is evidence for rapid rotation of the central part of the nuclear bulge of the galaxy, but far from the nucleus we can represent the rotation curve by the simplified version of Figure

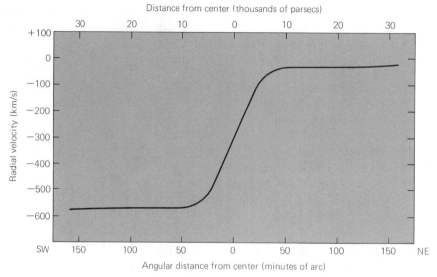

Figure 37.7 A simplified representation of the rotation curve of the Andromeda galaxy (M31).

37.7, which, while not accurately reflecting the observed details, serves better for the illustration of how the mass of the galaxy can be determined.

The radial velocity of the brilliant nucleus shows that the galaxy as a whole approaches us at nearly 300 km/s. The still more negative radial velocities of regions southwest of the nucleus indicate that that side of the galaxy is turning toward us and the northeast side away from us. We see that the maximum rotational speeds in M31 is reached at about 50′ from its center and is almost constant to 150′, at about 230 km/s with respect to the center of the galaxy. At the distance of M31 (680 thousand parsecs—680 kpc), 150′ corresponds to about 30,000 pc, or 6×10^9 AU, so the circumference of the orbit of a star at that distance would be 3.8×10^{10} AU. A speed of 230 km/s is 48 AU/yr, so the period of the star would be about 7.9×10^8 yr. If we apply Kepler's third law to the mutual revolution of such a star and M31, we find that the mass of the galaxy contained within 3×10^4 pc is about

$$\text{mass} = \frac{(6 \times 10^9)^3}{(7.9 \times 10^8)^2} = 3.5 \times 10^{11} \text{ solar masses},$$

which is the same order as the mass of our own Galaxy. The actual mass of M31 can be higher because we have not included the material in its outer halo. Lower limits to the masses of more than 50 galaxies have been found from their rotation curves by this procedure.

Many galaxies are not highly flattened and in rapid rotation. Nevertheless, the velocities of the stars in such a galaxy depend on its gravitational attraction for them, and hence on its mass. The spectrum of a galaxy is a composite of the spectra of its many stars, whose different motions produce different Doppler shifts. The lines in the composite spectrum, therefore, are broadened, and the amount by which they are broadened indicates the range of speeds with which the stars are moving with respect to the center of mass of the galaxy. Application of the virial theorem, then, enables us to calculate the mass of the galaxy, just as we calculate the mass of a star cluster (Section 30.2a).

(b) Masses of Systems of Galaxies

Like stars, galaxies are often observed in close pairs. The periods of revolution, however, are typically hundreds of millions of years, so we do not "see" the motion of one galaxy about the other, as we can observe the mutual revolution of the members of a binary star system. Nevertheless, a study of double systems tells us something about galaxian masses.

Suppose the orbit of such a binary galaxy were exactly edge on to our line of sight, and that the two galaxies were at their maximum possible projected separation, so that one would be coming straight toward us and one going straight away. Then the difference in their radial velocities would be the actual orbital speed of one galaxy about the other. From the distance to the pair, we could calculate their true separation and the circumference of the relative orbit. Division of the latter by the orbital speed would give the period of mutual revolution. Application of Kepler's third law, then, would give the sum of the masses of the two galaxies.

We have, unfortunately, no way of knowing the inclination of the orbital plane to our line of sight, nor can we know the positions of the galaxies in their orbits. The observed difference in radial velocities of the two, therefore, is less than their true relative orbital speed by an unknown amount, and the masses we calculate are less than the true masses. If we assume that the orbits are oriented at random, however, and that the galaxies in the various pairs are distributed at random in their orbits, we can calculate by how much, *on the average*, we are underestimating the masses of the galaxies. Analyses of many pairs of double galaxies, therefore, yield *average* galaxian masses. The method, first applied by T. Page, has now been used to estimate average masses for galaxies in about a hundred binary systems.

The masses of *clusters* of galaxies can be calculated by the same technique used to "weigh" star clusters. The radial velocities of many galaxies in a cluster are first measured. The average of these velocities is that of the center of mass of the cluster, and the differences between the velocities of individual galaxies and this mean value tells us how fast they are moving within the cluster. With the help of the virial theorem (Section 30.2a), we can then calculate the gravitational potential energy of the cluster, and hence its mass.

(c) Mass-to-Light Ratio

An important datum is the ratio of the mass of a galaxy, in units of the solar mass, to its light output, in units of the solar luminosity. For the sun, of course, this ratio would be unity. Galaxies, however, are not composed of stars that are all identical to the sun. The overwhelming majority of stars are less luminous than the sun, and usually these stars contribute most to the mass of a system, without accounting for very much light; thus, the mass-light ratio is generally greater than one. Only in systems, or regions within systems, where there are many young, highly luminous stars, can the ratio be less than one. Galaxies in which star formation is

still occurring tend to have mass-light ratios in the range one to ten, while in galaxies consisting mostly of an older stellar population, the ratio is larger— 10 to 30.

But the above comments refer only to the inner, more conspicuous parts of galaxies; there is growing evidence that galaxies are surrounded by massive halos, which have very high mass-light ratios. Rotation curves for spiral galaxies, for example, are usually more or less like that for M31 (Figure 37.7), which rises to a maximum velocity and then flattens. In some galaxies, a flat rotation curve (indicating constant velocity) has been traced to more than 40,000 pc from their centers. If stars and gas clouds in orbits of different size in a galaxy have the same speed, it means that the mass of the galaxy contained within those orbits is proportional to their radii (see Exercise 11). But the total light emitted from within a given distance from the center of a galaxy does not usually increase so rapidly with radius, so the ratio of mass to light must be greater in the halo. For some galaxies, ratios of total mass to light of more than 50 have been derived from rotation curves.

Statistical analyses of binary galaxies of large separation (more than 50,000 pc) also lead to large mass-light ratios, providing further evidence for dark massive halos around galaxies. Masses derived from the virial theorem for clusters of galaxies, when compared with the light from the galaxies in the clusters, sometimes yield mass-light ratios as high as 200 or 300. (Even higher values have been claimed, but are not confirmed in the best-studied clusters for which data are most complete.) We shall see (Chapter 38) that clusters often contain hot intergalactic gas, which may contribute up to half of the cluster mass, but intracluster gas cannot account entirely for the high mass-light ratios.

The outer parts of at least some galaxies, therefore, and perhaps multiple systems of galaxies as units, evidently contain matter that has not yet been identified by its light or by other means than its gravitational influence. The anomaly is sometimes called the "missing mass" problem; at present, there is no generally agreed-on solution to the problem.

37.4 TYPES OF GALAXIES

Galaxies differ a great deal among themselves, but the majority fall into two general classes: spirals and ellipticals. A minority are classed as irregular.

Figure 37.8 NGC 4565, a spiral galaxy in Coma Berenices, seen edge on. Photographed in red light with the 5-m telescope.

(a) Spiral Galaxies

Our own Galaxy and M31, which is believed to be much like it, are typical spiral galaxies. Like our Galaxy (Chapter 29), a spiral consists of a nucleus, a disk, a halo, and spiral arms. Interstellar material is usually observed in the arms of spiral galaxies. Bright emission nebulae are present, and absorption of light by dust is also often apparent, especially in those systems turned almost edge on to our line of sight (Figure 37.8). The spiral arms contain the young stars, which include luminous supergiants. These bright stars and the emission nebulae make the arms of spirals stand out like the arms of a fourth-of-July pinwheel. The individual interarm stars are usually not observable at all, save in the nearest galaxies, although their collective light may be appreciable as a uniform glow. Open star clusters can be seen in the arms of nearer spirals, and globular clusters are often visible in their halos; in M31, for example, more than 200 globular clusters have been identified. Spiral galaxies contain both young and old stars.

Some famous spirals are illustrated in these pages. Galaxies M51 and M33 (Figures 37.9 and 37.10, respectively) are seen nearly face on; NGC 4565 (Figure 37.8) is nearly edge on. Note the ab-

Figure 37.9 The Sc galaxy NGC 5194 (M51) and its irregular II companion, NGC 5195. Photographed with the 5-m telescope.

sorbing lane of interstellar dust in NGC 4565—a thin slab in the central plane of the disk—which is silhouetted against the nucleus. M81 (Figure 37.11), like M31 (Figure 37.6), is viewed obliquely.

A large minority (perhaps a third or more) of spiral galaxies display "bars" running through their nuclei; the spiral arms of such a system usually begin from the ends of the bar, rather than winding out directly from the nucleus. These are called *barred spirals.* A famous example is NGC 1300 (Figure 37.12). The bar in a barred spiral is in a sense a straight portion of spiral arm and sometimes contains interstellar matter and young stars. Studies of the rotations of some barred spirals show that their inner parts (out to the ends of the bars) are rotating approximately as solid wheels. In the absence of differential shearing rotation, the straight bar can persist, rather than winding up; the detailed structures and dynamics of barred spirals, however, are not yet thoroughly understood.

Figure 37.10 NGC 598 (M33), a spiral galaxy in Triangulum, photographed with the 5-m telescope.

Figure 37.11 NGC 3031 (M81), spiral galaxy in Ursa Major, photographed with the 5-m telescope.

In both normal and barred spirals we observe a gradual transition of morphological types. At one extreme, the nucleus is large and luminous, the arms are small and tightly coiled, and bright emission nebulae and supergiant stars are inconspicuous. At the other extreme are spirals in which the nuclei are small—almost lacking—and the arms are loosely wound, or even wide open. In these latter galaxies, there is a high degree of resolution of the arms into luminous stars, star clusters, and emission nebulae. Our Galaxy and M31 are both intermediate between these two extremes. Photographs of spiral galaxies, illustrating this transition of types, are shown in Figures 37.13 and 37.14. All spirals and barred spirals rotate in the sense that their arms trail, as does our own Galaxy (Section 29.3).

Spiral galaxies range in diameter from about 20,000 to more than 100,000 LY, and dark halos evidently extend to far greater diameters. From the limited observational data available, their masses are estimated to range from 10^9 to 10^{12} times the mass of the sun. The mass-light ratio of the inner parts of spiral galaxies is between 1 and 20. The absolute magnitudes of most spirals fall in the range -16 to -21. Our Galaxy and M31 are probably relatively large and massive, as spirals go.

Figure 37.12 NGC 1300, barred spiral galaxy in Eridanus, photographed with the 5-m telescope.

NGC 1201   Type S0
NGC 2841   Type Sb
NGC 2811   Type Sa
NGC 3031 M81   Type Sb
NGC 488   Type Sab
NGC 628 M74   Type Sc

Figure 37.13 Types of spiral galaxies.

NGC 2859   Type SB0
NGC 2523   Type SBb(r)
NGC 175   Type SBab(s)
NGC 1073   Type SBc(sr)
NGC 1300   Type SBb(s)
NGC 2525   Type SBc(s)

Figure 37.14 Types of barred spirals.

(b) Elliptical Galaxies

More than two thirds of the thousand most conspicuous galaxies in the sky are spirals. For this reason it is often said that most galaxies are spirals. Actually, however, the most numerous galaxies in any given volume of space are those of relatively low luminosity, which cannot be seen at large distances, and which, therefore, are not among the brightest-appearing galaxies. (Similarly, the most numerous stars are faint main-sequence stars, very few of which can be seen with the unaided eye— see Section 27.1). Most of these dwarf galaxies fall into the class of *elliptical* galaxies. Moreover, the rich clusters, which contain a good fraction of all galaxies, are composed mostly of ellipticals. Elliptical galaxies, therefore, are really far more numerous than spirals.

Elliptical galaxies are spherical or ellipsoidal systems that are thought to consist almost entirely of old stars; they contain no trace of spiral arms. They resemble the nucleus and halo components of spiral galaxies. Although dust and conspicuous emission nebulae are not easily observed in elliptical galaxies, some do show evidence of sparse interstellar gas in their spectra. In the larger nearby ones,

many globular clusters can be identified. The elliptical galaxies show various degrees of flattening, ranging from systems that are approximately spherical to those that approach the flatness of spirals (Figure 37.15). The distribution of light in a typical elliptical galaxy shows that while it has many stars concentrated toward its center a sparse scattering of stars extends for very great distances and merges imperceptibly into the void of intergalactic space. For this reason it is nearly impossible to define the total size of an elliptical galaxy. Similarly, it is not obvious how far the halo of a spiral galaxy extends.

The fact that elliptical galaxies are not disk-shaped shows that they are not rotating rapidly as are spirals. It is hypothesized that they are systems formed from pregalaxian material that had little angular momentum per unit mass—that is their original material had low net rotation. Consequently, as such a cloud of primeval material contracted, it did not flatten into a disk, and the density of the material was high enough that it completely (or nearly completely) condensed into stars. In a spiral, on the other hand, a considerable amount of gas (and/or dust) in the flat, rapidly rotating disk was not able to condense into stars at once. This material was presumably formed into spiral arms by the rotation

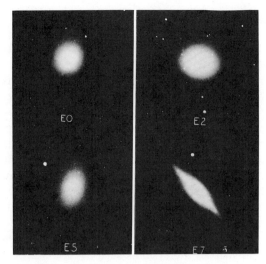

Figure 37.15 Types of elliptical galaxies.

of the galaxy, where it now still slowly condenses into stars. Elliptical galaxies probably consist entirely of old stars (at least if those galaxies are as old as our Galaxy).

Elliptical galaxies have a much greater range in size, mass, and luminosity than do the spirals. The rare giant ellipticals (for example, M87—Figure 37.16) are more luminous than any known spiral.

The brightest ellipticals in some rich clusters (for example, NGC 4886, in the Coma cluster of galaxies—see Section 38.2d) have absolute magnitudes that are brighter than -23—more than 10^{11} times the luminosity of the sun, and more than ten times the luminosity of the Andromeda galaxy. The mass data from double galaxies and from clusters suggest that the mass-light ratio for giant ellipticals is between 20 and 100. Recent studies by D. Jenner show some of these galaxies to have masses of about 10^{13} times that of the sun. While, as stated above, the diameters of these large galaxies are difficult to define, they certainly extend over at least several hundred thousand light years. The often-made statement that the largest of the galaxies are spirals like our own is incorrect.

Elliptical galaxies range all the way from the giants, just described, to dwarfs, which we think are the most common kind of galaxy. An example is the Leo II system, shown in Figure 37.17. There are so few bright stars in this galaxy that even its central regions are transparent. The total number of stars, however (most too faint to show in Figure 37.17), is probably at least several million. The ab-

Figure 37.16 NGC 4486 (M87), giant elliptical galaxy in Virgo, photographed with the 5-m telescope. Note the many visible globular clusters in the galaxy.

Figure 37.17 Leo II, a dwarf elliptical galaxy (negative print), photographed with the 5-m telescope.

Figure 37.18 NGC 6822, a nearby irregular galaxy.

solute magnitude of this typical dwarf is about −10; its luminosity is about 1 million times that of the sun. It is so near to us (about 750,000 LY) that

Figure 37.19 The Large Cloud of Magellan.

its diameter (about 5000 LY) is probably limited by the tidal force exerted on it by our Galaxy; this tidal force would pull more outlying stars away from the dwarf system.

Whether still smaller galaxies than dwarfs like Leo II exist depends on how galaxies are defined. Several globular clusters are known that are more than 200,000 LY from the nucleus of our Galaxy. It is not known whether objects like these are distributed through intergalactic space or whether they are outlying members of our Galaxy. If the former is the case, they must be galaxies in their own right. Perhaps even individual stars exist in intergalactic space.

Intermediate between the giant and dwarf elliptical galaxies are systems such as M32 and NGC 205, two near companions to M31. They can be seen in the photograph of M31 (Figure 37.6); NGC 205 is the one that is farther from M31.

(c) Irregular Galaxies

A few percent of the brightest appearing galaxies in the northern sky are classed as irregular. They show no trace of circular or rotational symmetry but have an irregular or chaotic appearance. The irregular galaxies divide into two groups. The first group, denoted *Irr I* galaxies, consists of objects showing high resolution into O and B stars and emission nebulae. The best-known examples are the Large and Small Clouds of Magellan (Figure 37.19), our nearest galaxian neighbors. We find many star clusters in these galaxies, as well as variable stars, supergiants, and gaseous nebulae; they contain both old and young stars. The Small Cloud, however, is apparently lacking in dust, although it does contain interstellar gas. The lack of conspicuous dust clouds is common in this first kind of irregular galaxy.

Galaxies of the second irregular type (*Irr II*) resemble the Irr I objects in their lack of symmetry. These objects, however, display no resolution into stars or clusters, but are completely amorphous in texture. Their spectra are continuous with absorption lines and resemble the spectra of type A5 stars, showing that the stars in these galaxies are not luminous enough to be resolved. The Irr II galaxies generally also show conspicuous dark lanes of absorbing interstellar dust. Examples are M82 (Figure 37.20) and the companion to the spiral galaxy M51 (Figure 37.9).

Figure 37.20 NGC 3034 (M82), an irregular II galaxy in Ursa Major.

(d) CLASSIFICATION OF GALAXIES

Of the several classification schemes that have been suggested for galaxies, one of the earliest and simplest, and the one most used today, was invented by Hubble during his study of galaxies in the 1920s. Hubble's scheme consists of three principal classification sequences: ellipticals, spirals, and barred spirals. The irregular galaxies (Irr I and Irr II) form a fourth class of objects in Hubble's classification.

The ellipticals are classified according to their degree of flattening or *ellipticity*. Hubble denoted the spherical galaxies by E0, and the most highly flattened by E7. The classes, E1, E2, . . ., E6, are used for galaxies of intermediate ellipticity.§ Hubble's classification of elliptical galaxies is based on the appearance of their *images*, not upon their true shapes. An E7 galaxy, for example, must really be a relatively flat elliptical galaxy seen nearly edge on, but an E0 galaxy could be one of any degree of ellipticity, seen face on. A statistical analysis of the numbers of galaxies of various apparent flattenings indicates, however, that if elliptical galaxies are *oblate* (like a pumpkin) and not *prolate* (like a football), then all degrees of real flattening are about equally represented.

Hubble classed the normal spirals as S and the barred spirals as SB. Lowercase letters a, b, and c are added to denote the extent of the nucleus and the tightness with which the spiral arms are coiled. For example, Sa and SBa galaxies are spirals and barred spirals in which the nuclei are large and the arms tightly

wound. Sc and SBc are spirals of the opposite extreme. Our Galaxy and M31 are classed as Sb.

In rich clusters, galaxies are observed which have the disk shape of spirals but no trace of spiral arms. Hubble regarded these as galaxies of type intermediate between spirals and ellipticals and classed them S0. Their possible origin is discussed in one of the next subsections.

Hubble's classification scheme for all but irregular galaxies is illustrated in Figure 37.21, in which the morphological forms are sketched and labeled, and with the three principal sequences joined at S0. The diagram is based on one by Hubble himself.

Hubble's classification scheme has been modified and expanded since his time to give a more complete description, but such refinements need not concern us here.

(e) SOME UNUSUAL CLASSES OF GALAXIES

There are also some other classes of unusual galaxies that may be briefly mentioned:

cD GALAXIES. cD galaxies are supergiant elliptical galaxies, usually E0 or E1, that are frequently found in (or near) the centers of clusters of galaxies. They are the largest galaxies known and tend to outshine the next brightest cluster galaxies by as much as a factor of 2. Often they are strong radio sources as well.

COMPACT GALAXIES. The class of compact galaxies consists of a large number of galaxies of relatively small size and high surface brightness. They are usually elliptical or irregular.

N GALAXIES. An N galaxy is a galaxy with a very bright, nearly stellar-appearing nucleus. The rest of the galaxy appears as a sort of faint, extended haze. To-

---

§Each of the numbers 0 through 7 that describe the flattening of a galaxy is defined in terms of the major and minor axes of the image of the galaxy, a and b, respectively, by $10(a - b)/a$.

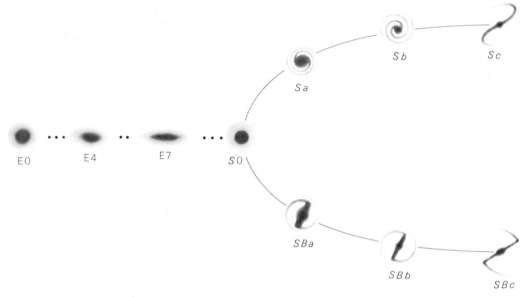

Sb

Sc

Sa

E0   *E4*   *E7*   *S0*

SBa

SBb

SBc

Figure 37.21 Hubble's classification scheme for galaxies.

day, N galaxies are regarded as belonging to a class of galaxies with active nuclei, which are described in Section 37.8.

SEYFERT GALAXIES. About a dozen galaxies of this class were first described by Seyfert, from whom the class derives its name. A Seyfert galaxy is a spiral that has a small bright region in its nucleus, whose spectrum shows broad bright emission lines arising from hot gases there. Seyfert galaxies are sometimes strong radio emitters, and may even be sources of very high-energy cosmic rays. They are also galaxies with active nuclei, described in Section 37.8.

(f) EVOLUTION OF GALAXIES

The continuity of the morphological forms of galaxies along classification sequences suggests that these different forms might represent stages of evolution for galaxies. It was speculated decades ago, for example, that elliptical galaxies gradually flatten, develop spiral arms, and become spiral galaxies. Most of the modern investigators who suggest that galaxies evolve through all types, on the other hand, would reverse this direction of evolution. They envision galaxies beginning as irregulars, evolving through various stages of spiral forms, until finally their gas and dust is completely condensed into stars; then the galaxies become ellipticals.

There is much doubt, however, that galaxies evolve from one type to another at all. The fact that different kinds of galaxies are flattened by different amounts almost certainly results from their having different amounts of angular momentum—that is, from their different rotation rates. In other words, galaxies

might always have had essentially their present forms (at least since their formation), the form of a particular galaxy depending mostly on its mass and angular momentum per unit mass.

Despite our lack of definite information concerning evolution of galaxies from one type to another, we do expect that stars within galaxies will evolve, as outlined in Chapters 33 and 34. Elliptical galaxies may always have been elliptical, but they may have had supergiant stars when they were young. Spirals may never become elliptical, but eventually their spiral arms may disappear when (and *if*) virtually all of their interstellar matter is converted into stars.

Some or all S0 galaxies may, on the other hand, have been formed from spirals that have lost their interstellar matter more dramatically. Collisions between spirals, especially in clusters, can sweep interstellar matter out of them while the widely separated stars of one galaxy can pass between those of the other almost unperturbed. Also, and probably more important, a spiral passing through intergalactic gas concentrated near the center of a cluster can have its interstellar matter effectively swept out. Finally, some interstellar matter is constantly being "blown" out of spirals by strong stellar winds of corpuscular radiation (atomic nuclei), particularly, those emitted by supernovae.

(g) Summary

The gross features of the different kinds of galaxies are summarized in Table 37.1. Many of the figures

Table 37.1 **Gross Features of Galaxies of Different Types**

	SPIRALS	ELLIPTICALS	IRREGULARS
Mass (solar masses)	10^9 to 10^{12}	10^6 to 10^{13}	10^8 to 10^{11}
Diameter (thousands of light-years)	20 to 150 or more	2 to 500 (?)	5 to 30
Luminosity (solar units)	10^8 to 10^{10}	10^6 to 10^{11}	10^7 to 2×10^9
Absolute visual magnitude	-15 to -21	-9 to -23	-13 to -18
Population content of stars	Old and young	Old	Old and young
Composite spectral type	A to K	G to K	A to F
Interstellar matter	Both gas and dust	Almost no dust; little gas	Much gas; much dust in Irr II; less and sometimes no dust in Irr I

given, especially for mass, luminosity, and diameter, are very rough and are intended only to illustrate orders of magnitude, not precise values.

37.5 GALAXIES AS RADIO SOURCES

The investigations that followed Jansky's discovery of cosmic radio waves in 1931 (Section 12.1) revealed that continuous radio energy is emitted from the disk and halo of our Galaxy. In addition, since World War II, thousands of discrete radio sources, each occupying a small region of the sky, have been discovered and catalogued. Most of these discrete sources are known to be extragalactic.

The first identification of a radio source with an extragalactic object was made in 1951. Astronomers at Palomer found that one of the strongest radio sources in the sky, discovered in 1948 and known as "Cygnus A," coincided in position with a remote cluster of galaxies in the constellation of Cygnus. In the cluster is observed what then was thought to be two galaxies in contact or in collision (Figure 37.22). A closer look shows Cygnus A to be an elliptical galaxy with a dust lane across it—perhaps not unlike NGC 5128 (Figure 37.24).

(a) Radio Galaxies

Although all galaxies emit radio waves, most are "normal"—ordinary galaxies that, like our Galaxy, emit some of their radiation at radio wavelengths. Radio energy with a continuous spectrum, for example, has been observed to emanate from the disks and halos of all nearby spirals and from the nuclei of many of them as well. Their radiation at radio wavelengths amounts, on the average, to about 10^{38} ergs/s, but ranges up to 10^{41} ergs/s, a small fraction of their energy output at visible wavelengths. In addition, radiation at the 21-cm line of neutral hydrogen (Section 28.2k) has been observed in many galaxies. The Doppler shift of this emission line indicates the same radial velocity for a galaxy as does the shift of a line in its visible spectrum.

In contrast, *radio galaxies* are galaxies which for some reason emit unusually large amounts of radio energy—often more than 10^{44} ergs/s (an amount of energy comparable to or greater than their entire optical luminosities). Some radio galaxies resemble ordinary galaxies in all their visual aspects, but others have unique properties that may be associated with their unusual radio luminosities. For example, the galaxy NGC 1068 (a Seyfert galaxy—see Section 37.8), the spectrum of whose nucleus shows emission lines, emits 100 times as much radio energy as normal spirals. Another example is M87, a giant elliptical galaxy in the Virgo cluster, which emits thousands of times as much radio energy as is typical of bright galaxies. This galaxy looks normal on photographs of full exposure. On photographs of short exposure, however, where the brighter parts

Figure 37.22 The peculiar galaxy identified with the Cygnus A radio source, photographed with the 5-m telescope.

Figure 37.23 The elliptical galaxy NGC 4486 (M87), a 3-m telescope photograph of very short exposure to show the nuclear "jet." *(Lick Observatory)*

of the galaxy are not "burned out," a luminous feature resembling a jet is seen emanating from the center of M87 (Figure 37.23). The visible light from this jet is highly polarized and is therefore assumed to be synchrotron radiation; the same is true of the radio radiation.

Some radio galaxies are very complex. An example is the Cygnus A source, which is one of the strongest *observed* radio sources despite the fact that the cluster with which it is identified has an estimated distance of about 10^9 LY. The power emitted by the Cygnus A source at radio frequencies is about 10^{45} ergs/s—many times the visible light output of the galaxy involved. Another complex and strong radio source is NGC 5128, a comparatively

nearby elliptical galaxy that appears to have a dust lane running through it.

Unless they are identified with already well-known objects (like M87), radio sources are usually known by their numbers in various catalogues of radio sources—especially the Third and Fourth Cambridge radio source catalogues; for example, 3C465 and 4C39.05, respectively.

(b) Radio Structures

In some radio galaxies, the bulk of the radio emission comes from small regions within them, while in some others—the *core-halo* sources—there are bright central sources surrounded by larger extended regions of radio emission. In about three quarters of the radio galaxies, however, the radio source is double, with most of the radiation coming from extended regions on opposite sides of the galaxy. Typically, the two emitting regions are far larger than the galaxy itself and are centered a hundred thousand parsecs away from it. The record is held by 3C 236, in which radio emitting clouds extend to two million parsecs on either side of the galaxy. Often radio observations reveal two well-delineated jets of radio radiation pointing away from the galaxy to the large extended sources (Figure 37.25). Presumably, ionized gases are shot at very great speeds into the radio "clouds" by an extremely intense source of energy in the nucleus of the galaxy. Magnetic fields must focus the ions to

Figure 37.24 The peculiar elliptical galaxy NGC 5128 in Centaurus; a strong radio source, photographed with the 5-m telescope.

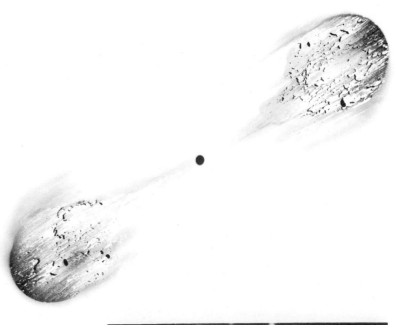

Figure 37.25 A schematic sketch of a double-source radio galaxy. The small spot in the middle represents the optical image of the galaxy, and the shaded regions on either side the relative intensity of the radio radiation. The radiation is usually most intense at the outer limits of the radio lobes. Sometimes radio "jets" are observed, as shown in the sketch.

account for the synchrotron radiation from them. It is thought that the gas eventually collides with neutral gas in the space beyond (previously emitted material?) and slows to a stop, defining the sometimes rather sharp outer edges of the emitting regions. The total magnetic energy that must be tied up in the magnetic fields to account for the observed radiation from a radio galaxy is sometimes more than 10^{60} ergs—an appreciable fraction of the total nuclear energy available in the galaxy!

Another interesting structure displayed by some radio galaxies is that of the *head-tail* source. These are probably associated only with galaxies in clusters or at least in regions of space pervaded with intergalactic gas. In head-tail radio galaxies, the two opposing radio lobes fold back along themselves, so that the radio galaxy resembles a comet (Figure 37.26). We think this is a result of the drag on intragalactic gas through which the radio galaxy is moving.

Obviously, some very energetic sources in radio galaxies are responsible for their remarkable properties. We speculate on what those power houses may be in Section 37.8.

37.6 GALAXY ENCOUNTERS AND COLLISIONS

In the existing catalogues of peculiar galaxies are many examples of strange-appearing pairs of galaxies interacting with each other. We can now un-

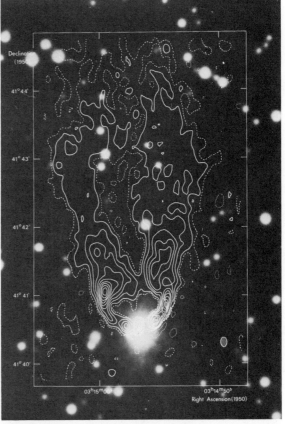

Figure 37.26 Contours showing the intensity of radio radiation at 5×10^9 Hertz of the head-tail radio source 3C83.1, in the Perseus cluster of galaxies. The map is superimposed on a print from the Palomar Sky Survey; the radio galaxy is the large image at the bottom. (*Courtesy K. J. Wellington, G. K. Miley, and H. van der Laan; photograph copyright, National Geographic Society–Palomar Observatory Sky Survey*)

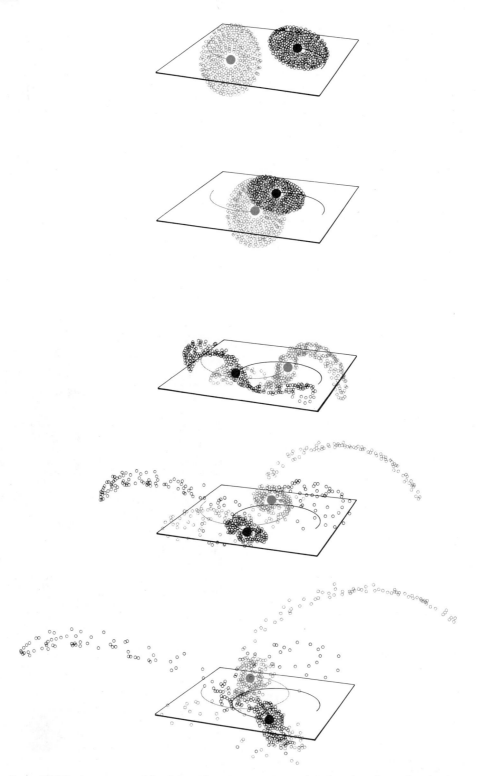

Figure 37.27 A sequence of five frames from a computer-produced motion picture that simulates the tidal distortion of two interacting galaxies. In this computer run, the initial conditions were chosen to see if the strange appearance of the pair of galaxies NGC 4038 and NGC 4039 could be accounted for in terms of tidal effects. Compare the last two frames with the photograph of the galaxies in Figure 37.28. The time interval between successive frames shown here is 200 million years. *(Courtesy Alar Toomre, MIT)*

Figure 37.28 The interacting pair of galaxies NGC 4038 and 4039.

derstand many of these in terms of gravitational tidal effects. The effects of tides between pairs of galaxies that chance to pass close to each other (especially likely in clusters) have been studied by Alar and Juri Toomre. They point out three fundamental properties of tidal interactions. (1) The tidal force is proportional to the inverse cube of the separation of the galaxies (see Section 6.4). (2) Tidal forces on an object tend to elongate it; thus there are tidal bulges on both the near and far sides of each galaxy with respect to the other. (3) The perturbed galaxies are generally rotating before the tidal encounter, and the subsequent distributions of their material must therefore reflect the conversation of their angular momenta.

At first we might expect a tidal interaction between two galaxies to pull matter out of each toward the other. Such bridges of matter may form between the galaxies, but also there are "tails" of ma-

terial that string out away from each galaxy in a direction opposite to that of the other. Because of the rotation of the galaxies, the tails and bridges can take on unusual shapes, especially when account is taken of the fact that the orbital motions of the galaxies can lie in a plane at any angle to our line of sight. The Toomre brothers have been able to calculate models of interacting galaxies that mimic the appearances of a number of strange-looking pairs actually seen in the sky (Figures 37.27 and 37.28).

(a) Galactic Cannibalism

If galaxies collide with slow enough relative speed, they may avoid the usual tidal disruption. Calculations show that some parts of slowly colliding galaxies can be ejected, while the main masses become binary (or multiple) systems with small orbits about each other. Such a newly formed binary galaxy, surrounded by a mutual envelope of stars and possibly interstellar matter, may eventually coalesce into a single large galaxy. This process is especially likely in the collisions of the most massive members of a cluster of galaxies, which tend to have the lowest relative speeds and to be concentrated toward the center. Often a rich cluster is observed to have one or more supergiant galaxies (of type cD—see Section 37.4d) near its center. Sometimes that huge galaxy has two or more nuclei. It is tempting to regard such galaxies as the result of cannibalistic encounters between the more massive slower moving members of their clusters.

37.7 QUASARS

If the sun were typical among stars as a radio emitter, we would not expect to be able to observe a single other star at radio wavelengths; the radio emission from the stars would be too feeble to detect with existing instruments. It was with considerable surprise, therefore, that in 1960 two radio sources were identified with what appeared to be stars. There seemed to be no chance that the identifications were in error, because the precise positions of the radio sources were pinned down by noting the exact instants they were occulted by the moon. By 1963 the number of such "radio stars" had increased to four. They were especially perplexing objects because their optical spectra showed emission lines that at first could not be identified with known chemical elements.

The breakthrough came in 1963 when M. Schmidt, at Caltech's Palomar Observatory, recognized the emission lines in one of the objects to be the Balmer lines of hydrogen (Section 10.6a) shifted far to the red from their normal wavelengths. If the redshift is a Doppler shift, the object must be receding from us at about 15 percent the speed of light! With this hint, the emission lines in the other objects were reexamined to see if they too might be well-known lines with large redshifts. Such proved, indeed, to be the case, but the other objects were found to be receding from us at even greater speeds. Evidently, they could not be neighboring stars; their stellar appearance must be due to the fact that they are very distant. They are called, therefore, *quasi-stellar-radio sources*, or simply *quasi-stellar sources* (abbreviated *QSS*). Later, similar objects were found, which are *not* sources of strong radio emission. Today they are all designated

by the term *quasar*.

The discovery of these peculiar objects prompted a search for others. A modern procedure is to look for stellar-appearing objects at the positions of unidentified radio sources or to examine stellar-appearing images of peculiarly blue color. By 1980 hundreds of quasars had been catalogued, and systematic surveys indicate that there must be more than 20,000 brighter than the 18th magnitude. The number of still fainter—and presumably more distant—quasars is not known, but there are certainly very many. All have spectra that show large to very large redshifts. The relative shifts of wavelength range up to $\Delta\lambda/\lambda$ of 3.53 (at this writing), and for the majority, $\Delta\lambda/\lambda$ is greater than 1.0. If we apply the exact formula for the Doppler shift (Section 10.5b), we find that $\Delta\lambda/\lambda = 3.53$ corresponds to a velocity of recession of 91 percent the speed of light.

Most investigators regard the redshifts of the quasars as indicative of their very great distances, and that they conform to the Hubble law. Quasars, however, have much higher speeds than any known galaxy and must, therefore, be even more distant.

(a) Characteristics of Quasars

The quasars are all unresolved optically—that is, they appear stellar, and most of them as very faint stars, at that. One of the brightest, 3C 273, is still several hundred times too faint to see with the unaided eye. A few quasars, though, are associated with tiny wisps or filaments of nebulous-appearing matter. Some are resolved at radio wavelengths, which indicates that the radio energy (at least for some) comes from regions outside the visible photographic images—perhaps not especially surprising, for radio radiation from galaxies, as we have seen, often originates from outside their optical images. The radio radiation is believed to be synchrotron.

Although they differ considerably from each other in luminosity, the quasars are nevertheless extremely luminous at all wavelengths. In radio energy, they are as bright as the brightest radio galaxies, and in visible light most are far more luminous than the brightest elliptical galaxies—their absolute magnitudes range down to -25 or -26. They are very blue in color—in fact, one of their recognizable characteristics is their excess amount of ultraviolet radiation, compared with normal stars and galaxies. Most surprising of all is that almost all

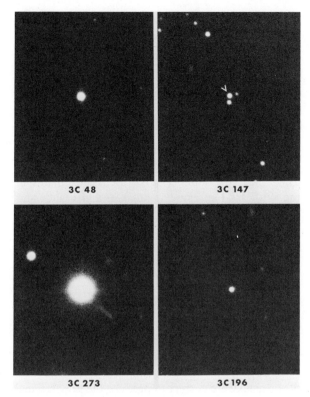

3C 48   3C 147

3C 273   3C 196

Figure 37.29 Quasi-stellar radio sources photographed with the 5-m telescope.

of them are variable, both in radio emission and visible light. Their variation is irregular, evidently at random, by a few tenths of a magnitude or so, but sometimes flareups of more than a magnitude are observed in an interval of a few weeks. Since quasars are highly luminous, a change in brightness by a magnitude (a factor of 2.5 in light) means an extremely great amount of energy is released rather suddenly. Moreover, because the fluctuations occur in such short times, the part of a quasar responsible for the light (and radio) variations must be smaller than the distance light travels in a month or so; otherwise light emitted at one time from different parts of the object would reach earth at different times (because of the range of distances light would have to travel to reach us), and we would see the increase spread over a longer time.

Following the pioneering work of A. R. Sandage at Palomar, it was found that there are many more quasars that are radio quiet than that are radio sources, although the former are harder to find, because radio emission does not call attention to them. Some astronomers think that radio-emitting quasars are temporary evolutionary stages of longer-lived quasars.

Because quasars have such high luminosities, at least if their redshifts are due to the Doppler effect, and if their distances are given by the Hubble law, some astronomers have suspected other explanations for the large shifts of the quasar spectral lines, and that the quasars are actually nearby, and not so excessively luminous. No known physical process other than the Doppler shift, however, can account for the redshifts (gravitational redshifts would produce other effects, not observed). A non-Doppler explanation for the redshift, therefore, requires new physics, waiting to be discovered! The assumption that the redshifts *do* conform to the Hubble law is called the *cosmological hypothesis* for quasars (which most astronomers consider most likely), and the observed redshifts of the quasar spectral lines, under this hypothesis, are said to be *cosmological*.

Some statistical arguments have been raised, pointing to quasars that seem to be close in direction to nearby galaxies more often than would be expected by chance, to support a noncosmological hypothesis for quasars. Authors of these studies suggest that nearby galaxies may eject quasars or may be associated with them for other reasons. The statistics, however, have been challenged by other scientists, and ideas that quasars are associated with nearby galaxies are not widely held by specialists.

(b) Superluminal Velocities in Quasars

Several quasars have been observed (about 50 by mid-1981) with very long baseline interferometry (VLBI—see Section 12.1e), and some show structure. In at least 6 cases, small discrete sources in the quasars have been found that change position from one observation to the next. The motions are generally radially outward from the center of the quasar image. The velocities have been measured for these moving sources, and if the redshifts of the quasars are cosmological, some of the velocities of the moving sources are in the range of five to ten times the speed of light. Such velocities are, of course, impossible, and for a while this was taken as evidence that the quasars must be relatively nearby.

The high measured speeds, however, are illusory. If an object is moving almost toward the observer, it can appear to have a far greater speed than it really does, because of the finite speed of light. To see how this can be, imagine a quasar

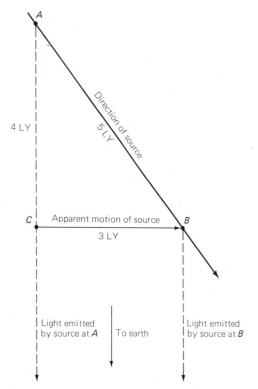

Figure 37.30 Explanation of the apparently superluminal velocities in quasars.

ejecting small sources in various directions at just under the speed of light. One such source passes through point A in Figure 37.30. The direction toward earth from the quasar is along the line $\overline{AC}$, but the source moves in a somewhat different direction, $\overline{AB}$. Let the source move 5 pc from A to B, and let A be 4 light years farther from us than B (that is, $\overline{AC}$ is 4 LY). Now the source, moving barely under the speed of light, takes barely over five years to reach B, while its radiation, emitted at A, reaches C in 4 years. That radiation continues on toward earth, being one year ahead of the radiation emitted in our direction by the source when it is at B. Perhaps billions of years later the radiation reaches our radio telescopes, that from A still one year ahead of that from B. What we observe is the source apparently moving tangentially outward from the center of the quasar, from C to B, and in one year going a distance found from the theorem of Pythagoras to be $\sqrt{(5^2 - 4^2)} = 3$ LY. The source apparently had three times the speed of light, but only because of the projection effect and because the light took four years to go from A to C while the source went from A to B in five years. If the direction of motion were more nearly toward

the earth, the effect would be even more dramatic (see Exercise 18).

(c) The Double Quasar 0957 + 561: A Gravitational Lens?

In 1979, astronomers D. Walsh, R. F. Carswell, and R. J. Weymann noticed that a pair of quasars, separated by only 6″ and known collectively as 0957 + 561 (the numbers give their coordinates in the sky), are remarkably similar in appearance and spectra. They are both at about 17th magnitude, and both have a redshift ($\Delta\lambda/\lambda$) of 1.4. The astronomers suggested that the two quasars might actually be only one, and that we are seeing two images produced by an intervening object, acting as a gravitational lens (Section 35.4a).

In 1980, a team of astronomers at Palomar found an 18th-magnitude galaxy next (in direction) to one of the quasars. In fact, the galaxy turns out to be a member of a rich cluster of galaxies, with a redshift of 0.39. The geometry and estimated mass of the galaxy are correct to produce the gravitational lens effect, and most investigators how believe that there is little question that 0957 + 561 is, indeed, a

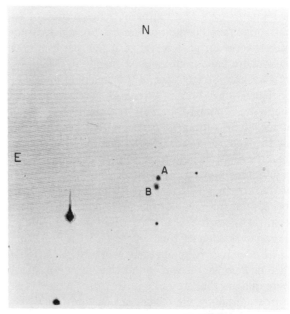

Figure 37.31 The double quasar 0957 + 561. The two components, believed to be different images of one quasar produced by a gravitational lens, are labelled A and B. (Courtesy Jerome Kristian, Mt. Wilson and Las Campañas Observatories; Carnegie Institution of Washington)

pair of images produced by a gravitational lens. A schematic of the lens is shown in Figure 37.32.

Since the discovery of 0957 + 561, at least one other candidate for a gravitational lens has been discovered. Even one case, however, is highly significant; first, it demonstrates another prediction of general relativity, but as far as quasars are concerned, we have a very remote galaxy making two images of a quasar that must be even more remote; the redshift of the latter *must* be cosmological. In Chapter 38 we discuss yet another argument for the cosmological nature of quasars.

(d) The Radiation from Quasars

The spectra of quasars have several components. First, there is a continuous spectrum, ranging all the way from radio waves to X rays (and probably gamma rays as well). The relative intensities in various regions of the spectrum vary from quasar to quasar; some are too weak in radio radiation, for example, to be observed. Data from the Einstein X-ray telescope, however, show that all or nearly all quasars are X-ray emitters.

Some quasars emit radio radiation from their central regions, but most have double radio lobes, like radio galaxies, whether or not they have central sources. Quasars, in other words, usually resemble radio galaxies in their radio properties. Of course, the radiation is synchrotron. So, evidently, is that in the visible and X-ray spectral regions. Some quasars, however, have excessive infrared radiation—more than would be expected from relativistic electrons accelerated in magnetic fields so as to produce the radiation observed in other wavelengths. The excessive infrared, when present, is thought to be caused by dust reemitting radiation that it absorbed at shorter wavelengths.

Quasars also have emission lines (with which we measure their redshifts). These must originate from ionized gas at not so high a temperature that

its atoms cannot capture electrons and emit light (Section 10.6). Moreover, these lines are all broad, suggesting that many emitting clouds have a large range of velocities in our line of sight. The best guess is that the emission is from many clouds ejected at relatively high speed in different directions from a central object.

Finally, quasars often have absorption lines. Sometimes there are two or more different sets of absorption lines displaying different redshifts. The absorption-line redshifts, however, are always less than those of the emission lines, indicating that the gas clouds producing them are between us and the source of continuous radiation and of the emission lines. These clouds are believed to be either material previously ejected by the quasar (cooled-off gas that formerly showed emission lines?) or interstellar gas in intervening unseen galaxies, through which the quasars' light must pass to reach us.

So a model of a quasar must include a source of synchrotron radiation, consisting of strong magnetic fields and hordes of relativistic particles. The source must be in a region less than a few light days or light weeks across to allow the variations in light over short intervals. Some of the particles must (in most quasars) be ejected to great distances in the surrounding space to account for the radio lobes. There must also be isolated blobs of gas containing relativistic charged particles and magnetic fields that are ejected at very high speed to produce the moving sources observed with very long baseline interferometry. Then there must be blobs of gas ejected at moderate velocities to account for the broad emission lines. There must, in some cases, be surrounding dust to absorb radiation and reemit it strongly at infrared wavelengths. And finally, there must be cool gas at larger distances between the quasar and us, perhaps formerly ejected from it, to produce the absorption lines.

What kind of marvelous animal can a quasar be? We are not sure, but there are, as we shall see

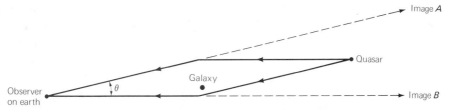

Figure 37.32 Gravitational lens associated with the double quasar 0957 + 561. Two light rays from the quasar are shown being bent in passing a foreground galaxy, and they arrive together at earth. We thus see two images of the quasar, in directions *A* and *B*. This simple schematic does not reflect the subtle complexities produced by the finite size of the galaxy, by the cluster of which it is a member, nor of relativity theory itself, but it does illustrate how we can observe multiple images of one object. The angular separation, θ, of the two images at earth is greatly exaggerated, being only 6″ in reality.

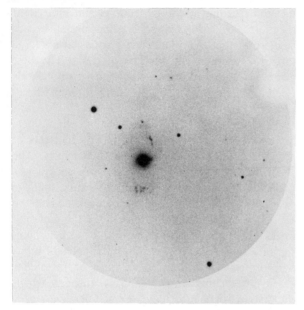

Figure 37.33 The Seyfert galaxy NGC 4151 (negative print). Although the exposure is so short that the spiral structure of the galaxy is barely visible, the brilliant nucleus is already burned out. Photographed with the Lick Observatory 3-m telescope. *(Courtesy H. Ford, UCLA.)*

in the next section, seemingly related objects to be found among more normal galaxies. We believe we have some interesting hints, and they all point to quasars being highly energetic sources in the nuclei of galaxies too faint to be seen against the glare of the quasars themselves.

37.8 ACTIVE GALACTIC NUCLEI

We have seen (Section 29.6) that our own Galaxy has a nucleus that displays many of the properties we have just described for quasars: synchrotron radiation at all wavelengths, infrared radiation, and gas clouds moving outward. In the nucleus of our Galaxy, these phenomena are on a tiny scale compared with quasars, but there are other galaxies where nuclear activity seems to be intermediate, and they may provide a link between normal galaxies and quasars.

(a) Seyfert Galaxies

In 1944 Carl Seyfert described about a dozen spiral galaxies with very unusual spectra. More than a hundred are known today. The spectrum of the light from the nucleus of a *Seyfert galaxy* has strong, broad emission lines, which indicate the presence of very hot gas in a small central region. The broad widths of the lines show that the gas is rapidly expanding, with speeds of up to thousands of km/s. Some Seyfert galaxies are also strong radio sources, and all have strong emission of infrared radiation from their nuclei. The visual luminosities of these galaxies are usually about normal for spirals, but when account is taken of the infrared energy they emit, their total luminosities are found to be 100 or so times normal. Some are known X-ray sources as well.

The unusual radiation of a Seyfert galaxy sometimes comes from several different regions of its nucleus. Some of them, for example NGC 4151, show variations in the brightness of their nuclei over periods of only a few months, which is evidence that the radiation comes from small regions, at most a few light months across. Many of the properties of Seyfert galaxies are like those of the nucleus of our own Galaxy (but on a larger scale), exhibiting discrete sources of infrared radiation, emission lines, expanding or outflowing gases, and radio emission, some of which is synchrotron radiation.

The Seyfert properties can be recognized easily only in relatively nearby galaxies. It is quite possible that 1 or 2 percent of all spiral galaxies have these active nuclei. Alternatively, it is possible that all spiral galaxies (even our own?) have these properties 1 or 2 percent of the time.

(b) Violent Activities in Galaxies

Other galaxies besides Seyfert's show evidence of explosive ejection of matter from their nuclei. M82 (Figure 37.20) has strong radio radiation from its nucleus and a complex of filaments surrounding it. The spectra of the filaments show Doppler shifts suggesting expansion of up to 1000 km/s. If they were ejected from the nucleus of M82, they would have reached their present locations in about 2×10^6 yr. (However, P. Morrison has shown that the observations can also be explained if M82 is passing through a great cloud of intergalactic dust.)

M87 is another interesting galaxy that is a strong radio source. Short exposures of it show a luminous jet directed away from its nucleus, and a faint hint of a second radial jet in the opposite direction. Both the nucleus and the brighter jet emit synchrotron radiation, indicating magnetic fields and a source of relativistic electrons.

M87 is also a strong source of X rays, implying a hot gas throughout the entire galaxy and out into its halo. The observations suggest that the halo contains an enormous amount of dark mass—perhaps more than 10^{13} solar masses. Finally, the optical spectrum of M87 shows very broad lines, indicating high velocities of the stars there, as though they were being accelerated by a very dense, massive core.

Intermediate between the quasars and such galaxies as M82 and M87 or the Seyfert galaxies is a class known as N-type galaxies. N galaxies have small nuclei that are very bright compared with the main parts of those galaxies; often they appear as stellar images superimposed on faint wispy or nebulous backgrounds. Their bright nuclei indicate that enormous amounts of energy are being emitted from those regions. N galaxies whose disks or main bodies are too faint to photograph may well be classed as compact galaxies. If we could obtain spectra of all of them, many would probably show properties of Seyfert galaxies. Most others probably would be quasars.

Objects of another class believed to be related to Seyfert galaxies and quasars are the *BL Lac* objects, named for the prototype, BL Lacertae; they are also called *Lacertids*. BL Lac is a stellar-appearing object that shows large irregular variations in luminosity, which accounts for its variable-star designation. Like other BL Lac objects, it has no spectral features, but it is a strong radio source, and its continuous radiation appears to be synchrotron. In other words, it is like a quasar in all respects, except that we have no means of determining its redshift from spectral features.

At least, we did not until 1974, when a spectrum was obtained at Palomar of the light passing through a ring-shaped aperture centered on BL Lac itself, but blocking out light from its central image. A source of faint light passing through the annular opening surrounding BL Lac, however, revealed itself to be underlying the bright object and to have a spectrum like that of a normal galaxy with a radial velocity of 21,000 km/s (or $\Delta\lambda/\lambda = 0.07$). BL Lac (and presumably other objects in its class) is evidently the brilliant nucleus of a remote galaxy.

(c) The Powerhouses of Active Galactic Nuclei

We have reviewed a number of phenomena associated with certain galaxies that hint of spectacular activity in their nuclei. These range from the quasars at one extreme to the nucleus of our own Galaxy at the other, which, although quiescent compared with quasars, has many of the same features on a small scale. But radio galaxies, Seyfert galaxies, BL Lac objects, and quasars require a source of energy with some or all of the following properties:

1. It must provide power up to 10^{47} ergs/s—equivalent to nearly 100 million million times the luminosity of the sun.
2. It must, in some cases, account for variations in the total radiated power by as much as a factor of two or more, and over time scales of years or months or, in some cases, only days.
3. In at least the objects that vary in luminosity, the powerhouse must be compact enough that light can travel across it in a time less than that of its variations.
4. It must be able to eject relativistic electrons in directed jets and in sufficient numbers to provide synchrotron radiation as intense as the total visible energy emitted by a bright galaxy.
5. It must possess powerful magnetism, and the energy in its magnetic fields must be comparable to the total nuclear energy available in all of the stars in a large galaxy.
6. At least in M87, and probably in many galaxies, it must be able to accelerate stars to high velocity in the central regions of the galaxy in which it resides.
7. In some quasars, it must be able to eject clouds of matter containing relativistic electrons every year or more often.

Perhaps no one kind of power plant accounts for all kinds of active galactic nuclei, but what they all have in common points to a small compact source of enormous energy, evidently buried in the nucleus of a galaxy. Many models have been suggested, including stellar collisions in dense galactic cores, superstars, extraordinarily powerful supernovae, and others. But most theoreticians lean to the theory that all or most active galactic nuclei derive their energy from the release of potential energy in the gravitational collapse of millions or thousands of millions of solar masses. This matter, falling together and accelerating as it does so, would reach great speeds and heat to millions of Kelvins. Ultimately, it is believed, such collapsing matter must increase in density until it has fallen through its event horizon into a black hole. New matter, falling toward the black hole, releases new energy, and probably forms an accretion disk. Perhaps the ejection of jets of matter in radio galaxies and quasars is like a large-scale version of SS 433 (Section 34.10).

Of course, these ideas are speculative. We know we have a small compact source of enormous energy, and it seems reasonable to suppose it to be associated with something like a black hole, but no one model has achieved anything close to universal acceptance.

EXERCISES

1. Why is the term "island universe" a misnomer?

2. In a hypothetical galaxy, a cepheid variable is observed, which, at median light, is at magnitude $+15$. From its period, the absolute magnitude of the cepheid is determined to be -5. What is the distance to the galaxy?
   Answer: 10^5 pc; 3.26×10^8 LY

3. In a very remote galaxy, a supernova is observed which reaches magnitude $+17$. Assume that the absolute magnitude of the supernova was -18, and calculate the distance to the galaxy.
   Answer: 10^8 pc; 3.26×10^8 LY

4. Make up a table of distance indicators for galaxies, and list the distances to which each can be used.

5. How can the visual light of an emission nebula in another galaxy exceed that from the star whose energy produced the nebula?

6. Starting with the determination of the size of the earth, outline all the steps one has to go through to obtain the distance to a remote cluster of galaxies.

7. Suppose a supernova explosion occurred in a galaxy at a distance of 10^8 pc, and that the supernova reached an absolute magnitude of -19 at its brightest. If we are only now detecting it, how long ago did the event actually occur? What is the *apparent* magnitude of the supernova at its brightest? (See Section 24.2b.)

8. Why do we use the *brightest* galaxies in a cluster as indicators of its distance, rather than average galaxies in it? (*Hint:* There are two reasons, one involving the definition of "average," the other involving the distances to typical clusters.)

9. The tenth brightest galaxy in cluster A is at apparent magnitude $+10$, while the tenth brightest galaxy in cluster B is at apparent magnitude $+15$. Which cluster is more distant, and by how many times?

10. How can we determine the inclination of M31 (the Andromeda galaxy) to our line of sight?

11. Suppose the rotation curve of a galaxy is flat beyond a certain radial distance, R, from the center of the galaxy, and suppose the galaxy to have perfect spherical symmetry. Because of the spherical symmetry, it works out that only the part of the galaxy interior to r contributes to the gravitational acceleration of a star or gas cloud at a distance r from the center.
    (a) How does that part of the galaxy's mass act, gravitationally?
    (b) If the mass of the galaxy interior to r, expressed in solar masses, is M, and if distance is expressed in AUs and time in years, what is the period of a star or gas cloud in a circular orbit at a distance r from the center?
    (c) What is the velocity of that star, or gas cloud, in AUs/year?
    (d) For r greater than R, how does M depend on r?

★12. Classify the following galaxies according to Hubble type: (a) a galaxy which is chaotic in appearance, with no symmetry and no resolved stars; (b) a galaxy with an elliptically shaped image whose major axis is twice its minor axis; (c) a galaxy with very tightly wound spiral arms and a large nucleus.

13. Where might the gas and dust (if any) in an elliptical galaxy come from?

14. If extragalactic globular clusters exist as "galaxies," as what kind of galaxies would you classify them? How about extragalactic stars?

★15. Assume that its redshift is a Doppler shift, and verify that a quasi-stellar source with a redshift of $\Delta\lambda/\lambda = 3.53$ has a radial velocity of 91 percent the speed of light (See Section 10.5b.)

16. Suppose a quasar recedes at $^{15}/_{17} (= 0.882)$ times the speed of light. What is the observed relative shift of its spectral lines $(\Delta\lambda/\lambda)$?
    Answer: $\Delta\lambda/\lambda = 3$

17. (a) Suppose a quasar has an absolute visual magnitude of -25. What is its visual luminosity in

terms of the sun's? (Assume that for the sun, $M_v = +5$.)

(b) Suppose that the quasar has a distance of 5×10^9 pc. Suppose, further, that the fact that it appears stellar implies that its angular diameter is less than 1 arc-second. What is the upper limit to its linear diameter in parsecs?

(c) If it is variable, can you suggest a way of fixing a smaller upper limit to its linear size?

Answers: (a) 10^{12}; (b) 2.5×10^4 pc

18. A quasar emits two radiating clouds in our general direction at $^{13}/_{14}$ the speed of light. They are first seen by us when they appear to have first been produced at the central powerhouse, and are subsequently observed to move, apparently radially outward in opposite directions from the center of the quasar image. Fourteen years after the quasar actually emitted the clouds, they are, in reality, 12 LY closer to us than the quasar itself.

(a) How many years, on earth, elapse between our first observation of the clouds and when we observe them at their positions when they are 12 LY nearer to us?

(b) At what speed do they appear to us to be separating?

Answer to (b): Five times the speed of light.

19. Let us imagine four classes of objects: those that are resolved in photographic images (say, whose angular sizes are more than about 1") and those that are not (smaller than 1"), and for each of these cases, those that have radial velocities less than 1000 km/s, and those that have radial velocities greater than 1000 km/s. Thus we can classify astronomical objects into four groups as follows:

	< 1"	≥ 1"	
			$V_r < 1000$ km/s
			$V_r \geq 1000$ km/s

The moon, for example, would go into the upper right-hand box. Now classify each of the following:

(a) the sun

(b) Jupiter

(c) a typical Trojan minor planet

(d) stars in our Galaxy

(e) the Orion nebula

(f) galaxies in the Coma cluster

(g) Seyfert galaxies

(h) quasars

Edwin Powell Hubble (1889–1953) left a career in law to study astronomy. In 1919 he joined the staff of the Mount Wilson Observatory, where he was the first to show that the "nebulae" are actually galaxies, like our own. He was also the first to show that the large-scale structure of the universe is homogeneous. In 1929 he gave the first evidence for the expansion of the universe. (*California Institute of Technology*)

38

THE STRUCTURE OF THE UNIVERSE

We live on a rock revolving about one star—a very insignificant star among hundreds of thousands of millions of such stars in a rather ordinary galaxy. We estimate that there may be a thousand million galaxies that could be photographed with existing telescopes if we had the time and inclination to do so. Most of those galaxies, as we have seen, fit into a sort of pattern, but there are some aberrant ones among them—if aberrant is the proper word to describe such things as Seyfert galaxies and quasars. But these details seem quite trivial in the overall organization of the universe. It is the structure of the universe as a whole that we want to look at now.

Hubble, a man of no small perspective, was interested in the same question a few years after showing that galaxies even *existed*. He was interested, for example, in how galaxies are distributed in space. Are there as many in one direction of the sky as in any other? And if we count fainter and fainter galaxies, presumably farther and farther away, do we find that their numbers increase in the way they should if galaxies are distributed uniformly in depth?

38.1 HUBBLE'S FAINT GALAXY SURVEY

Hubble had at his disposal the world's largest telescopes—the 100-inch (2.5-m) and 60-inch (1.5-m) reflectors on Mount Wilson. But although those telescopes can probe to great depths, they can do so only in small fields of view. To photograph the entire sky with the 100-inch telescope would take not just a lifetime, but thousands of years. So instead, Hubble sampled the sky in many regions, much as Herschel did with his star gauging (Section 27.3a). In the 1930s Hubble photographed 1283 sample areas or fields with the telescopes, and on each photograph he carefully counted the numbers of galaxy images to various limits of brightness, the faintest corresponding to the greatest depth of space that could then be probed.

The results of Hubble's survey are shown in Figure 38.1, which is a map of the sky shown in what are called *galactic coordinates*; the Milky Way, across the middle of the plot, defines the galactic equator, and the top and bottom of the map—the galactic poles—are 90° away from the Milky Way. The empty sectors at the lower right and left are the parts of the sky too far south to observe from Mount Wilson. Each symbol represents one of the regions of the sky surveyed by Hubble, and the size of the symbol indicates the relative number of galaxies he could observe in that area.

The first obvious thing to notice is that we do not see galaxies in the direction of the Milky Way; the obscuring clouds of dust in our Galaxy hide what lies beyond in those directions. Hubble called this part of the sky the *zone of avoidance*. Near the

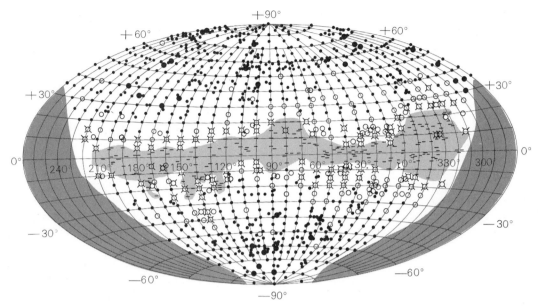

Figure 38.1 Distribution of galaxies according to Hubble's survey.

Milky Way in direction, the counts of galaxies are below average and are denoted by open circles. The farther we look from the Milky Way, the less obscuring foreground dust lies in our line of sight and the more galaxies we see. From the counts of galaxies in different directions, Hubble determined that light is dimmed by about 0.25 magnitudes (blue, or photographic magnitudes) in traversing a half-thickness of the Galaxy at the sun's position. Having derived how foreground dust dims the light from galaxies, Hubble could correct his counts to allow for the effect.

After such correction, Hubble found that on the large scale the distribution of galaxies is *isotropic*, which means that if we look at a large enough area of the sky, we find as many galaxies in one direction as in any other. From the 44,000-odd galaxies Hubble counted in his selected regions, he calculated that about 100 million were potentially photographable with the 100-inch telescope. (This means that with the larger telescopes available today, we could, given enough time, record about 1000 million.) Moreover, Hubble found that the numbers of galaxies increase with faintness about as we would expect if they were distributed uniformly in depth.

These findings of Hubble were enormously important, for they indicated, at least to the precision of his data, that the universe is isotropic and homogeneous—the same in all directions and at all distances. In other words, his results indicate that the universe is not only about the same everywhere,

but that the part we can see around us, aside from small-scale local differences, is representative of the whole. This idea of the uniformity of the universe is called the *cosmological principle* and is the starting assumption for nearly all theories of cosmology (next chapter).

38.2 CLUSTERING OF GALAXIES

The uniformity of the universe applies on the *large scale* but not on the small. One's neighborhood may be more or less typical in the large city in which he lives, but every house is not identical to every other one. The cosmological principle permits local variations; it is only over very large volumes that average conditions are pretty much the same everywhere.

One kind of small-scale variation, as we have already seen, is that galaxies tend to occur in clusters. In fact, when Hubble began his work, he was aware of several conspicuous clusters, and he made a point to avoid them in his photographic survey so that they would not bias his results. However, when he investigated the statistics of the numbers of galaxy images on the photographs, he realized that the galaxies are not distributed in space with perfect randomness, but show a clumpy distribution, as if they are clustered, even though in many cases sparse clusters at different distances, seen overlapping in projection, might pretty well wash themselves out from recognizability. The statistics, though, led Hubble to think that probably *all*, or

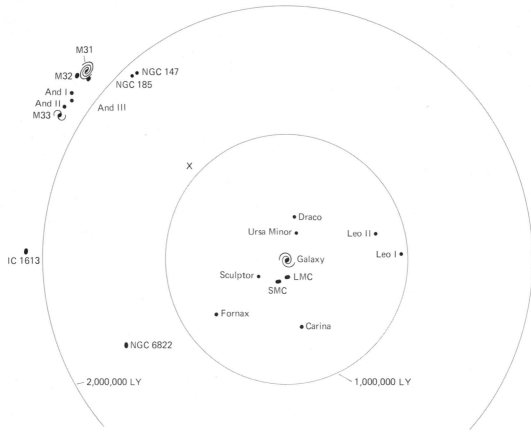

Figure 38.2 The Local Group. The "X" marks the approximate center of mass of the Group.

at least *most*, galaxies are in groups or clusters, rather than being single in space.

(a) The Local Group

One such cluster is the one to which our own Galaxy belongs—the Local Group. It is spread over about a million parsecs and contains at least 21 known members (some investigators count as members a somewhat larger number of galaxies at a somewhat greater distance). There are three large spiral galaxies (our own, the Andromeda galaxy, and M33), four irregulars, four intermediate ellipticals, and at least ten dwarf ellipticals. Dwarf ellipticals, recall, are probably the most common kind of galaxy. Appendix 17 gives the properties of the galaxies that are generally accepted to be members of the Local Group. Figure 38.2 is a plot of the Local Group; the galaxies have been projected onto an arbitrary plane centered on our Galaxy, then their distances from the center of the plot have

been increased so that they are shown at the correct relative distances from us.

The radial velocities of some of the Local Group galaxies have been measured. If we assume that they are moving at random within the Local Group, we can find the motion of the sun with respect to the center of mass of the Group. The procedure is analogous to the determination of the motion of the sun compared with its neighboring stars in our part of our own Galaxy, that is, with respect to the local standard of rest (Section 23.2a). The Andromeda galaxy, for example, approaches us at about 266 km/s, while the Large Cloud of Magellan recedes from us at about 276 km/s. Most of the apparent radial velocity of each of these objects is due to the revolution of the sun about the center of the Galaxy and the motion of the Galaxy in the Local Group. From an analysis of the radial velocities of all the other galaxies, we find that the sun's orbital velocity is about 250 km/s and that our Galaxy is moving at a speed of 100 to 150 km/s in the Local

Figure 38.3 NGC 185, an intermediate elliptical galaxy in the Local Group. (*Lick Observatory*)

Group. The average of the motions of all the galaxies in the Group indicates that its total mass is of the order 5×10^{12} solar masses.

(b) The Neighboring Groups and Clusters

Well beyond the edge of the Local Group we find other similar systems; more than 50 have been catalogued. Among them is a small group of galaxies centered on M81, and another on M51 (the galaxies shown in Figures 37.11 and 37.9, respectively). At

a distance of several tens of millions of light years, however, a cluster like the Local Group, seen in projection against the background of very many more distant galaxies, would not be noticed. At large distances, we recognize only very rich clusters (that is, clusters of many member galaxies) that stand out like "sore thumbs" against the background.

The nearest moderately rich cluster is the famous Virgo cluster, a system of thousands of members. It contains a concentration of mostly elliptical galaxies that include M87—the radio galaxy and X-ray source shown in Figures 37.16 and 37.23. Within several degrees of M87 are many spirals as well as ellipticals and, associated with the brightest galaxies, very many dwarfs, like the dwarf ellipticals in the Local Group. It is difficult to state the overall size of the Virgo cluster because it is very complex, having many concentrations of galaxies. It is probably at least several clusters, some more distant than others. All of the galaxies in the Virgo cluster (or *clusters*) are too remote for us to observe their cepheids, and distances estimated by some of the other techniques described in Section 37.2 are quite uncertain. The richest concentration, around M87, is probably between 30 and 60 million light years away (10 to 20 million parsecs). If its distance is 40 million LY, its linear diameter is about four million LY. The entire Virgo complex is much

Figure 38.4 Stephan's quintet, a small group of galaxies. (*Lick Observatory*)

Figure 38.5 The central region of the Virgo cluster of galaxies. *(Kitt Peak National Observatory)*

larger. Even the entire system, however, cannot be considered a *great* cluster—one that would be recognizable in remote parts of the universe.

(c) General Clustering

Shortly after World War II, two extensive photographic surveys were made of the sky. One was made with a 20-inch (50-cm) astrographic telescope at the Lick Observatory (in California) as part of a program to measure proper motions of stars. (A second series of photographs is now being taken with the same telescope; comparison of the first and second epoch photographs reveals the stars that have detectable proper motions.) Those photographs also reveal an enormous number of images of galaxies, and astronomer C. D. Shane, at Lick, in collaboration with statisticians Jerzy Neyman and Elizabeth Scott, at the University of California, Berkeley, made a detailed analysis of their distribution.

Shane, Neyman, and Scott found, in agreement with the earlier hunch of Hubble, that the observed distribution of galaxies in the sky, at least as revealed on the Lick photographs, is compatible with the hypothesis that all galaxies are in clusters. The hypothesis is not *proven* by the analysis, but it seems likely to be correct. Of course most clusters are unrecognizable because of confusion with the background. On the other hand, a large number of very rich clusters are easy to spot, and they can be recognized to very great distances.

The other photographic survey was the National Geographic Society-Palomar Observatory Sky Survey; it was made with the large 48-inch (1.2-m) Schmidt telescope on Palomar Mountain, and was financially supported by the National Geographic Society. The Palomar Survey reached a much greater depth in space than did the one at Lick.

While employed as observer for the Palomar survey in the 1950s, I undertook to identify on the photographs the truly great clusters of galaxies, because although rare, as clusters go, they can serve as markers of the large-scale distribution of matter in space, and moreover because we can estimate their relative distances (Section 37.2). My catalogue contains 2712 clusters in the northern three quarters of the sky that could be surveyed from Palomar; of these 2712, 1682 comprised a fairly homogeneous sample of clusters that are more or less completely identified. Among the nearest is the famous Coma cluster (Abell catalogue number 1656—Figure 38.6), which has thousands of members bright enough to show up at the cluster's distance of from 70 to 140 million parsecs. The radial velocity of the cluster is 6950 km/s, about 2 percent the speed of light. The most remote of the great clusters in my homogeneous sample are about 10

Figure 38.6 The central part of the regular Coma cluster of galaxies. *(Kitt Peak National Observatory)*

times as distant, with velocities of 20 percent that of light.

(d) Types of Clusters

Clusters of galaxies fall into two main classes: regular and irregular. The regular clusters are very rich and show marked spherical symmetry and high central concentration. They contain few if any spiral galaxies, but rather have a membership dominated by ellipticals and those galaxies that resemble spirals but without spiral arms or obvious evidence of interstellar matter (those classed as S0—Section 37.4d). A good example of a regular cluster is the Coma cluster, which has a linear diameter of at least 10 million LY and, as stated above, thousands of observable galaxies. The cluster is centered on two giant ellipticals, whose absolute visual magnitudes are between −23 and −24. The Coma cluster contains more and more members at magnitudes that are successively fainter. There is every reason to expect dwarf elliptical galaxies to be present; if so, the total number of galaxies in the cluster might be tens of thousands.

Clusters of galaxies, particularly regular clusters, are likely to be sources of X rays. The X rays from a cluster are thermal radiation from intracluster gas at a temperature of 10^7 to 10^8 K and are generally concentrated near the cluster center. The mass of gas required to produce the X-radiation is typically 10 to 20 percent, but may range up to half of the total cluster mass and so can account for a part of the missing mass in rich clusters (Section 37.3c).

There may be a significant relation between the presence of hot gas and of S0 galaxies in a regular cluster. X-ray emission lines of heavy elements such as iron have been observed at such intensity as to suggest that the heavy-element abundance in the hot gas is similar to that in the sun, rather than the matter being all or nearly all hydrogen and helium, as current theories predict for the primordial matter from which the clusters formed. This suggests that at least some of that X-ray emitting gas must have undergone nucleosynthesis in stellar interiors and that processed matter would then have been ejected into interstellar space within the cluster galaxies by such mechanisms as supernova outbursts. Finally, this material was swept from the galaxies by collisions between them and by their moving through the intracluster gas, as described in Section 37.4f. Such sweeping of interstellar matter from galaxies stops star formation in them, and the spiral arms gradually disappear, leaving the galaxies as type S0. The swept gas is hot because the galaxies collide with each other or pass through intracluster gas at speeds of up to thousands of kilometers per second.

Irregular clusters are not as well organized as the regular ones. They are generally far less rich and, in fact, range down to groups like the Local Group. The richest, though, may have as many galaxies as some regular clusters, and may be of comparable size—5 to 15 million LY across. They are not particularly symmetrical in appearance and do not have strong central concentration. They often have subclusters within them. And they contain galaxies of all types—spirals as well as ellipticals. An example is the Hercules cluster (Figure 38.7).

So up to now, we have this view of the universe: It is made up of galaxies, which range greatly in size and mass. The galaxies, in turn, are mostly (if not entirely) in groups and clusters. These systems of galaxies have diameters of from 1 to 5 million parsecs, or even a bit larger. The biggest clusters—the *great* clusters—are usually symmetrical and regular in structure and have total masses comparable to a thousand million million (10^{15}) suns.

38.3   THE LARGE-SCALE DISTRIBUTION OF MATTER: SUPERCLUSTERS

As we look out into space (on our telescopic photographs), most groups and clusters of galaxies just seem to merge together in projection into a kind of irregular background of galaxies—in all directions in the sky. But here and there very rich clusters—the great clusters—stand out and can be recognized. Figure 38.9 shows their distribution in the sky. The figure is a plot in galactic coordinates, as in Figure 38.1, except that the part of the sky too far south to survey from Palomar happens to be shown in a different part of the map. Each symbol is a great cluster of galaxies, the larger symbols being the relatively nearer clusters and the small symbols the relatively more distant ones. The most distant are from 2 to 4×10^9 LY away and occupy a volume of space at least 10^{14} times that of our own Galaxy, widely believed to be the entire universe less than a century ago.

What is found for the great clusters more or less parallels what Hubble found for galaxies in two respects. First, just as he found galaxies, on the large scale, to be distributed more or less uniformly through space, so the large-scale distribution of clusters is isotropic and homogeneous; there are as many in one direction as in any other (except, of course, for the obvious effects of absorption by dust in the Milky Way), and there is no evidence of clusters thinning out in distance. But on the *small* scale, just as Hubble found the galaxies to have a tendency to cluster, so the clusters of gal-

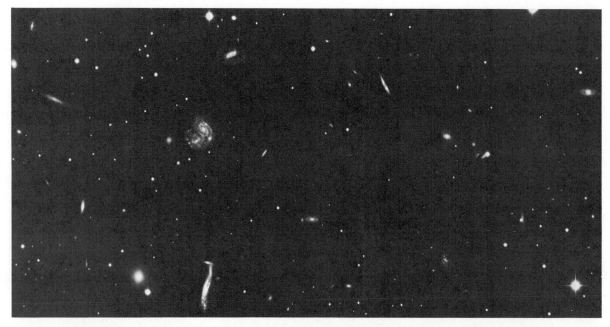

Figure 38.7   Irregular cluster of galaxies in Hercules, photographed with the 5-m telescope. *(Palomar Observatory; California Institute of Technology)*

Figure 38.8 A distant cluster of galaxies in Hydra, with a radial velocity of about 20 percent that of light. *(Palomar Observatory; California Institute of Technology)*

axies tend to clump up in what are called *superclusters*. Statistical tests show that the apparent clumping of clusters is not illusory, but that clusters of galaxies tend to be really associated with other clusters in space.

(a) Superclusters

The idea of superclusters was not new to my study. In 1937 Swedish astronomer Erik Holmberg, in in-

vestigating double and multiple galaxies, found that there seems to be a local concentration of them, and on a far larger scale than just the Local Group. Then, in the 1950s, Gerard de Vaucouleurs revived Holmberg's idea and gave the matter an intensive study, which he is continuing to this day. De Vaucouleurs has amassed an immense amount of data to demonstrate that the Local Group, many other neighboring groups, and quite a few clusters of galaxies, including the Virgo cluster, are all part of a large system he calls the *Local Supercluster*. This local concentration of galaxies and groups and clusters has been confirmed by many independent observers. The diameter of the Local Supercluster is probably 50 to 75 million parsecs, and its mass probably lies in the range 10^{15} to 10^{16} times that of the sun.

The distribution of the great clusters provided the first strong case for the existence of other superclusters. Today, evidence for superclusters comes from several sources. For example, many condensations of rich clusters can be recognized; some are evident in Figure 38.9. These are rather obvious second-order clusters or superclusters, but the cluster distribution *per se* does not suggest that *all* matter is found in superclusters, even though other evidence does, as we shall now see.

Additional information is supplied by statistical investigations of the distributions both of clusters of

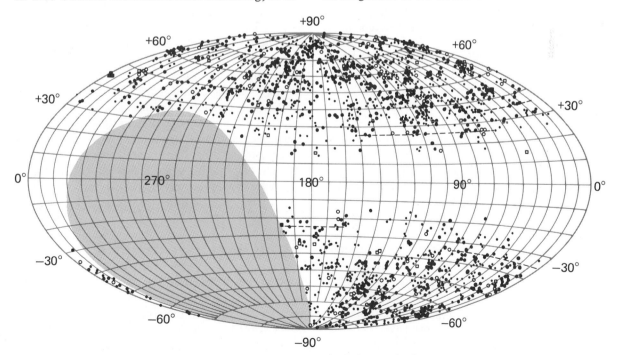

Figure 38.9 The distribution of rich clusters of galaxies.

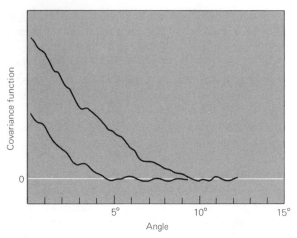

Figure 38.10 Schematic representation of the form of the covariance function as it varies with angular separation. *Upper curve:* galaxies brighter than 16th magnitude; *lower curve:* galaxies brighter than 18th magnitude.

galaxies and of individual galaxies counted in various surveys (for example, Shane's counts of galaxies on the Lick astrographic survey plates—Section 38.2c). Particularly impressive are such studies by Princeton University physicist and astronomer P. J. E. Peebles and his associates, who calculate from the observational data a statistic called the *covariance function.* The covariance function is a measure of how likely it is to find an object of one kind at a given angular separation in the sky from a second object of the same or of a different kind. Where the covariance function is zero, it means that the objects are randomly distributed and bear no relation to each other; where it is high, it means that the presence of a first object enhances the probability of finding a second one at that particular separation from the first. At angular separations that are very large compared with the angular sizes of superclusters, we would expect the covariance function to be zero, and indeed it is. However, at small angular separations, the covariance function shows that galaxies are likely to be found near other galaxies, or near clusters, or that clusters are likely to be found near other clusters.

Figure 38.10 is a schematic of the sort of thing Peebles finds. The upper curve shows the covariance function for galaxies and clusters at an intermediate distance, as would apply to those near enough to be in a catalogue of galaxies brighter than about magnitude 16. They are correlated with each other—that is, they tend to be associated—at an-

gular separations of up to more than 10°. Ten degrees is an angular size much larger than clusters of galaxies at that distance; these galaxies and clusters must be grouped together in larger systems—superclusters. More distant superclusters, of course, should subtend smaller angular sizes (at least if they are intrinsically similar to nearer superclusters). As expected, the covariance function for galaxies and clusters to magnitude 18 (lower curve in Figure 38.10) drops to zero more rapidly, suggesting superclusters at larger distance and hence smaller angular size.

Another approach is to look at examples. Intensive studies have been carried out, so far, for only about ten individual superclusters (including the Local Supercluster). In some of these studies, magnitudes and radial velocities have been measured for hundreds of individual galaxies in an effort to map their three-dimensional distribution in space. The work is very difficult, time consuming, and expensive of telescope time but is vital to our understanding of the detailed structure of the universe. One very interesting outcome from the measures of the radial velocities and magnitudes is that in a given direction there are not galaxies at *all* radial velocities; rather the velocities tend to clump around certain values, corresponding to the expected redshifts of discrete superclusters at different distances in the line of sight. This result suggests (but is not yet definitive) that *all* galaxies belong in superclusters. Typical superclusters are from 100 to 300 million light years in diameter and have estimated masses of 10^{15} to 10^{16} solar masses. Some superclusters contain a dozen or more great clusters; others contain none. The average is about two, which explains why superclustering is apparent for some but not all clusters in Figure 38.9. The radial velocity data, combined with gravitational theory, suggest that most superclusters are expanding, although perhaps not as fast as the universe as a whole; some, however, may be gravitationally bound, which means they will eventually stop expanding.

(b) Is There a Hierarchy?

If stars are in galaxies, galaxies in clusters of galaxies, and clusters in superclusters, it is natural to wonder whether the hierarchy might go on to ever higher orders of clusters of galaxies. Available evidence, though, suggests that it does not.

First, if it did, it would never be possible to define a region of space beyond which the universe is homogeneous, because larger hierarchical structures would always exist. This would violate the cosmological principle, contrary to observations, some of which have already been described.

Moreover, superclusters appear to be expanding, and the (meager) evidence to date suggests that most are not gravitationally bound. If superclusters are not bound, the next order clusters should expand more rapidly, and the order beyond them, more so yet. It is hard, in other words, to see how a hierarchical structure could be maintained. But the data are not definitive.

Further evidence is provided by the faintest radio sources. If these sources are very distant extragalactic objects, their angular distribution in the sky gives information about the large-scale uniformity of the distribution of matter in the universe. At least on scales of 1000 million parsecs, the faint sources have a remarkably random distribution in direction. Such a scale is much larger than superclusters but does rule out an indefinite hierarchy.

One of the best tests is provided by counts of galaxies of ever-increasing faintness. At bright magnitudes (nearby galaxies), the counts are quite chaotic as magnitude is increased, and differ greatly in different directions. But at faint magnitudes, characteristic of galaxies with distances of thousands of millions of light years, the counts are highly uniform in different directions in the sky. Specifically, an investigation by my former PhD student Dr. George Rainey rules out inhomogeneities on a scale as large as 500 million parsecs—that is, there cannot be "super-duper" clusters.

Finally, there is a faint background of radio radiation that we interpret as the dying glow of the big bang that started the expansion of the universe (next chapter). The uniformity of that radiation in different directions is impressive and argues for a high degree of homogeneity in the universe at large.

(c) Conclusions

Most or all of the matter of the universe appears to be clustered in systems of size up to at least 300 million light years—superclusters. But on much larger scales, the universe is very uniform, which seems to rule out a hierarchical distribution of matter in space.

The observed superclusters apparently are regions where matter exists in the universe. Typically, they are 5×10^7 to 10^8 parsecs across and contain two great clusters of galaxies. Undoubtedly, they contain many more smaller clusters and even more small groups like the Local Group. Perhaps individual galaxies exist in space between the clusters of a supercluster—that were either formed outside of clusters or escaped from clusters. It is possible, but not certain, that no galaxies exist between the superclusters. In a way, superclusters are like great metropolitan areas, containing one or two great cities, more smaller cities, many small towns and communities, and perhaps individual homes that are not parts of organized communities.

Some studies suggest that superclusters are flat, like pancakes. This model fits certain theoretical ideas about the evolution of the early universe, but the observations do not *require* that interpretation. Some observers see the distribution of matter in space as making up a vast honeycomb structure, with superclusters being the membranes and defining "holes" between them, giving the universe a cellular form. This idea probably cannot be ruled out, but if correct the honeycomb membranes are not continuous, because we see edges to at least some superclusters. In contrast, other astronomers see superclusters as isolated units in space—perhaps even gravitationally bound.

Even more fundamental is whether the universe was formed in its present fragmented state or was once a uniform mass of galaxies of stars or even of isolated stars. What came first: stars, galaxies, clusters of galaxies, or superclusters? We do not know; here we are at the frontier.

38.4 EVOLUTION IN THE UNIVERSE

As we look off into space, we look back into time, for we see remote objects as (and where) they were far in the past, when light left them to begin its long journey across space to reach our telescopes. Remote objects, therefore, are in a sense historical documents in the universe, even though we may have difficulty in interpreting their message.

(a) Distant Radio Sources

The number of radio sources increases with the feebleness of their signals. Most of these sources are believed to be extragalactic. Now, if the distribution of remote radio galaxies were random in space and time, their number should increase with the volume of space they occupy—that is, with the cube of their distance (after correction for the expansion of the universe). Then, when we apply the inverse-square law of light, we can predict how the number of sources should increase with increasing faintness (see Exercise 3). We find, however, that the actual number of sources increases far more rapidly with faintness than predicted. One could interpret this as meaning that the density of the sources in space increases with increasing distance from us, as though we were at the center of a cosmic distribution of objects. But this violates both the cosmological principle and the *Copernican cosmological principle*—namely, that we are not at a preferred location in space (for example, at the center of the universe). It is far easier to imagine that at least some of the remote radio galaxies are temporary objects, and that they represent an evolutionary phase in the history of the universe. Thus, by looking far away in space, we are looking back to a time when radio galaxies were more abundant than they are now. In this case, we need not be at any center, for any other person, anywhere else in the expanding universe, would see the same general picture. We shall return to this profound idea in the next chapter.

Quasars are a prime example. Their numbers are vastly greater at fainter magnitudes and larger redshifts than would be the case for a uniform distribution in space. After correction for the expansion of the universe, Maarten Schmidt and Richard Green, at Palomar, find that the number of quasars with redshifts greater than 2.0 exceeds the number predicted for a uniform distribution by a factor of tens of thousands! Here, incidentally, is another strong argument for the cosmological nature of quasar redshifts. If their numerical density increases rapidly with distance (whether magnitude or redshift is taken as indicative of distance is unimportant here), it suggests that we are at the center of a spherically symmetric but nonuniform distribution of quasars in space. This certainly violates our Copernican sense that we are typical, not special, in the universe, unless we assume that quasars really were more numerous in the past than they are today, so that by looking to great distances we see back into that past quasar-rich era. This assumption of great distances for the quasars, of course, implies that their redshifts are cosmological.

(b) Is There an X-Ray Background?

Before the *Einstein* orbiting X-ray Observatory, a diffuse background of X-radiation was discovered. These observations, however, were made with low angular resolution, so that it was impossible to tell whether the X rays were coming from a general distribution of hot gas in the universe at large distances or from a large number of discrete unresolved sources. If the former, it might mean that there was, in the past, a pervasive intergalactic medium of hot gas, which today, even if cooled, could contribute to the mass of the universe. If the latter, it could be due to quasars or other active galactic nuclei contributing enormously to the radiation from an earlier epoch but not necessarily to the present mass of the universe.

Recent *Einstein* observations have shown that all quasars are X-ray sources and that there are very many discrete X-ray sources. The best guess today is that quasars, too far away to be seen optically, can contribute enough X-radiation to account for the faint background and that there is no need to postulate a hot, diffuse intergalactic gas.

38.5 INVISIBLE MATTER IN THE UNIVERSE

We shall see in the next chapter that the future of the universe depends critically on the mean density of matter in space. We have described the distribution of visible galaxies and clusters of galaxies and clusters of clusters of galaxies, and of remote X-ray and radio sources. Any dark or invisible matter that might also be present would add to the mass and hence to the gravitation of the universe, but not to its light. If such invisible matter were concentrated in clusters of galaxies, it could account for their perhaps anomalously high mass-light ratios, but the total mass of such matter in the universe would nevertheless already be tallied, for it is included in the masses we derive for the clusters. On the other hand, if dark matter were to pervade the universe more or less uniformly, that part outside clusters (or superclusters) would not reveal itself by gravitational effects on other galaxies. It could, however, profoundly affect the evolution of the universe as a whole.

People have, therefore, searched for direct evidence for invisible matter, so far without success,

We cannot rule out a sparse distribution of dust throughout intercluster space, but if present, it cannot be abundant enough to add appreciably to the mean density of the universe without revealing itself by absorbing and reddening the light from galaxies. On the contrary, the colors of moderately distant galaxies, after correction for the Doppler shift of their light, are the same as those of nearby galaxies of the same type, and relative distances of moderately distant clusters of galaxies, found from the comparison of the magnitudes of their brightest member galaxies (Section 37.2b), are proportional to their redshifts. Dust in intergalactic space, therefore, cannot contribute mass to the universe in an important way.

We would expect intergalactic gas to be mostly hydrogen. But appreciable intergalactic hydrogen cannot exist in the neutral state, or we would observe a background of radiation from the 21-cm line, which we do not. We should also see the Lyman lines (Section 10.6) in absorption against the continuous spectra of remote quasars. Although many quasars show absorption lines, they are relatively sharp and discrete lines, indicating absorption in widely separated clouds. General intergalactic hydrogen, however, would have a continuous range of redshift and produce absorption lines that are very broad, and these are not observed.

Ionized hydrogen would emit bright lines and produce a luminous glow, which is not observed, unless the temperature were in the millions of Kelvins; in that case, it would emit X rays. There is a background of X-radiation, but as we have explained above, it is probably made up of discrete sources—quasars. In short, it is difficult to see how an important amount of intergalactic gas or dust could escape detection, and most astronomers doubt if it contributes appreciably to the mean density of the universe.

Another possibility would be large, nonluminous objects, perhaps intergalactic planets, collapsed stars, or black holes. We cannot prove that stars or planets do not exist or have not existed in intercluster space, but we cannot imagine how those objects could have been formed, and if they were, why many intergalactic stars of low mass (that would not burn out in the age of the universe) have never been discovered. It has been suggested that black holes may have been formed in the big bang, with (possibly) a continuous range of mass down to objects like asteroids. But the expected explosive evaporation of asteroid-mass black holes (Section 35.6d) has not been observed, and anyway the gravitation of black holes should cause them to clump up into the galaxies and clusters along with all other matter. In the latter case, they could conceivably contribute to the measured masses of clusters but would not affect the evolution of the universe as a whole.

If there is dark matter, one of the possible candidates for it might be neutrinos, if in fact they should turn out to have nonzero rest mass (Section 32.2b). If they are distributed uniformly throughout the universe, it is estimated that at present each square centimeter contains several hundred neutrinos, originally produced in the big bang (Chapter 39). If they should have no rest mass, or very small rest mass, neutrinos are expected to be distributed uniformly. A neutrino would, however, have to have a mass as much as 2×10^{-6} times that of the electron to contribute to the density of the universe enough to affect its evolution, and if they should have even that very tiny mass, neutrinos would probably clump with clusters or superclusters of galaxies. We must emphasize that (at this writing) the evidence for neutrinos having mass is only suggestive, but even if the suggestion is correct, neutrinos probably cannot be important in the overall gravitation of the universe. They could, however, play a major role in resolving the high mass-to-light ratio anomalies in galaxies and clusters.

38.6 EXTENT OF THE OBSERVABLE UNIVERSE

In all directions we see galaxies and clusters of galaxies. At greater and greater distances they appear ever fainter, and at ten thousand million light years we detect only the greatest giants among galaxies—the most luminous members of the great clusters. At still larger distances we find quasars, but at a redshift of about 3.5 (corresponding to a radial velocity of more than 90 percent of the speed of light), even quasars seem to disappear. But it is not an effect of their remoteness in space. Rather, at the distances of the farthest observed quasars, we are looking back to a time when the universe had less than 10 percent of its present age. At a much earlier time, quasars evidently did not exist. Perhaps galaxies had not even formed yet, or were only recently formed and had not had time to produce active nuclei. We have searched for protogalaxies, which might appear as faint infrared specks, but so far without success.

At great distances, in other words, and in all directions, luminous objects disappear, but it is a boundary in time, not in space. Many people think this means we must be seeing the "place" where the universe began. But there is no special place or "site" of creation. So far as we know, the universe is and always was *everywhere*. It evolves in time, to be sure, but at any given time it is homogeneous and isotropic—the cosmological principle. Observations seem to confirm the large-scale uniformity of the universe out to the distances, and back to the time, at which quasars first "turned on."

Yet, even earlier, in the first million years, the universe was young, hot, and brightly aglow, and today we see that glow, but it is redshifted to radio wavelengths. That fading glow of the big bang is today the most remote thing we observe, far off in space and back in time. This is a large part of the story of the next chapter.

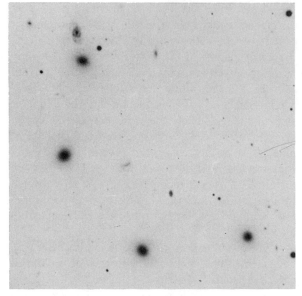

Figure 38.11 A group of galaxies in Cetus. Photographed by I. Karachentsev with the Soviet 6-m telescope (negative print). (*Courtesy I. Karachentsev, via I. Novikov*)

EXERCISES

1. Assume for the sake of illustration that the dust in our Galaxy all lies in a relatively thin flat disk, with the sun in the central plane of that disk. Under these circumstances, show by a diagram that the obscuration of distant galaxies is less and less at greater and greater directions from the Milky Way (that is, at greater and greater galactic latitudes).

2. Suppose on one survey you count galaxies to a certain limiting faintness. On a second survey you count galaxies to a limit that is four times fainter.
   (a) To how much greater distance does your second survey probe?
   (b) How much greater is the volume of space you are reaching in your second survey?
   (c) If galaxies are distributed homogeneously, how many times as many galaxies would you expect to count on your second survey?

3. Assume that all quasars have the same intrinsic luminosity and are distributed at random in space. Derive a formula for the number that should be seen that deliver to earth a light flux greater than a given flux, ℓ. Ignore the expansion of the universe and evolutionary effects.

4. If galaxies are distributed at random in space, statistical theory enables us to predict how many fields should contain a certain number of galaxies each. Hubble found that a far larger number of fields had too few galaxies, and also that a far larger number of fields had too many galaxies, than would be ex-

pected for a random distribution. Explain why this result suggests that galaxies tend to be clustered.

5. Are there subunits or subcondensations in the Local Group? If so, discuss them.

*6. From the data given in the chapter, calculate the mean number of galaxies per unit volume of space and the mean separation of galaxies in the Coma cluster of galaxies.

7. Would an indefinite hierarchy of clustering (clusters upon clusters, without limit) be consistent or inconsistent with the cosmological principle? Why?

8. From the data given in the chapter, try to make an estimate of the mean density of matter in the universe, in grams per cubic centimeter. If you cannot calculate the density unambiguously, see if you can place upper and lower limits on it. What assumptions must you make? (*Hint:* Assume [consistent with observations] that 10% of all galaxies are in great clusters.)

9. On the average there are two great clusters per supercluster. Why, then, do so many clusters, plotted on Figure 38.9, appear single?

10. Explain clearly and completely why the cosmological principle is violated if quasars are *not* temporary phenomena at an earlier epoch of the history of the universe.

Abbé Georges Lemaître (1894–1966), Belgian cosmologist, studied theology at Mechelen and mathematics and physics at the University of Leuven. It was there that he studied the expansion of the universe and postulated its explosive beginning. He predicted the Hubble law two years before its verification, and he was the first to consider seriously the physical processes by which the universe began. (*Yerkes Observatory*)

THE BIG BANG: THAT VANISHED BRILLIANCE

The Reverend Richard Bentley, chaplain to the Bishop of Worcester and contemporary of Newton, was chosen to deliver a set of eight sermons in fulfillment of a bequest of no less than £50, in proof of the Christian religion "against notorious infidels." Bentley thus wrote Newton to inquire whether a static universe with an even, finite distribution of matter would not collapse of its own gravitation without the intervention of a supreme being. Newton replied, ". . . the hypothesis of matter, being at first evenly spread through the heavens, is in my opinion inconsistent with the hypothesis of innate gravity without a supernatural power to reconcile them, and therefore it infers a deity." But when Bentley asked about an *infinite* universe, Newton replied that there might be many local collapses, which could produce the stars, but that an infinite universe as a whole could be stable.

In 1917, however, Einstein applied the new gravitational theory embodied in his general theory of relativity (Chapter 35) to the universe as a whole; he could find no solutions to the field equations that gave a homogeneous static universe, *even if it is infinite*. Einstein consequently altered the field equations with the introduction of a new term, called the *cosmological constant*, which represents a repulsion that can balance gravitational attraction over large distances and permit a static universe. There is, however, no evidence for such a repulsion in nature, and it seemed to spoil the elegance of the theory to introduce the cosmological constant *ad hoc* so that the universe would conform to preconceived notions. At the time Einstein had overlooked solutions, not needing the cosmological constant, that describe a *nonstatic* universe.

39.1 THE EXPANDING UNIVERSE

In the decade following Einstein's introduction of his static model of the universe, a number of other investigators found that the field equations also have solutions for nonstatic universes. A nonstatic universe could either contract or expand, but if it is now contracting, unless it is very young, it is hard to see why it has not already collapsed; expanding models were therefore more attractive. The most famous models for an expanding universe were those of the Dutch astronomer W. de Sitter, the Russian mathematician Aleksandr Friedmann, and the Belgian priest and cosmologist Abbe Georges Lemaître. Superficially, de Sitter's model looked static, but the coordinates he used really caused all objects to separate with time. Friedmann showed that relativity theory is compatible with an infinite, homogeneous expanding universe for all values of the cosmological constant, including zero. Today, Friedmann's solutions for zero (or no) cosmological constant are most seriously considered. Lemaître favored an expanding universe that still had a cosmological constant. All expanding solutions, however, predict that there should be a proportionality between the distances of remote objects and the speeds with which they recede from us. Lemaître, in particular, was aware of the significance of observations being made in the United States that might demonstrate the expansion of the universe.

(a) Expansion Requires a Velocity-Distance Relation

At this point, let us be quite clear why a uniformly expanding universe requires that all observers within it, no matter where they are located, must observe a proportionality between the velocities and distances of remote galaxies. Imagine a ruler made of flexible rubber, with the usual lines marked off at each centimeter. Now suppose someone with strong arms grabs each end of the ruler and slowly stretches it, so that, say, it doubles in length in one minute (Figure 39.1). Consider an intelligent ant sitting on the mark at 2 cm—intentionally *not* at either end or in the middle. He measures how fast other ants, sitting at the 4-, 7-, and 12-cm marks move away from him as the ruler stretches. The one at 4 cm, originally 2 cm away, has doubled its distance; it has moved 2 cm/min. Similarly, the ones at 7 cm and 12 cm originally 5 and 10 cm distant, have had to move away at 5 and 10 cm/min, respectively. All ants move at speeds proportional to their distance. Now repeat the analysis, but put the intelligent ant on some other mark, say on 7 or 12, and you'll find that in all cases, as long as the ruler stretches uniformly, this ant finds that every other ant moves away at a speed proportional to its distance.

For a more realistic, three-dimensional analogy, look at the raisin bread in Figure 39.2. The cook has put too much yeast in the dough, and when he sets the bread out to rise, it doubles in size during the next hour and all the raisins move farther apart. Some representative distances from one of the raisins (chosen arbitrarily, but not at the center) to several others are shown in the figure. Since each distance doubles during the hour, each raisin must move away from the one selected as origin at a speed proportional to its distance. The same is true, of course, no matter which raisin you start with. But the analogy must not be carried too far; in the bread it is the expanding dough that carries the raisins apart, but in the universe no pervading medium is presumed to separate the galaxies.

From the foregoing, it shoud be clear that if the universe is uniformly expanding, all observers everywhere, including us, must see all other objects moving away from them at speeds that are greater in proportion to their distances. It is also important to note that the best description of gravitation yet found—general relativity—in its simplest form *predicts* that the universe should be expanding (for a static universe needs that unexpected and unjustified cosmological constant). The high speeds away from us that are observed for remote galaxies are not a mystery that needs explaining; they were *predicted*. If galaxies were *not* receding from each other, then we would have to worry about that repulsive force that maintains a static universe. Now let's look at the evidence (appreciated by Lemaître and others) that the universe *is*, as expected, expanding.

(b) The Evidence: Slipher's Observations

In 1894 Percival Lowell had established his observatory in Flagstaff, Arizona, for the purpose of studying the planets and for searching for life in the universe. The Observatory's major instrument, a 24-inch telescope, was hardly designed for cosmology, yet it contributed to the 20th century's advance of our knowledge of the evolution of the uni-

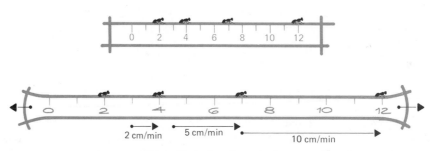

Figure 39.1   Stretching a ruler.

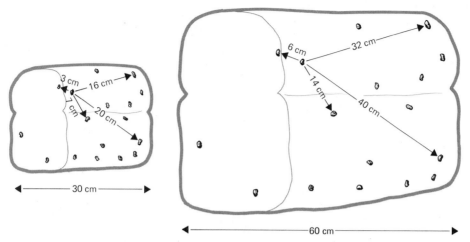

Figure 39.2 Expanding raisin bread.

verse. Lowell thought that the spiral or "white" (because their spectra did not show bright colored emission lines) nebula might be solar systems in the process of formation—like the solar nebula (Chapter 14). He therefore asked one of the Observatory's young astronomers, Vesto M. Slipher, to photograph the spectra of some of them to see if their spectral lines suggest chemical compositions of the nebulae like those expected for newly forming planets.

Slipher adapted the spectrograph of the 24-inch telescope to make it especially efficient for the study of faint sources of light. Even so, with technology available in those days, exposure times ran to from 20 to 40 hours! Slipher's first spectrum, in 1912, was of the Andromeda galaxy (M31) (not then known to be a galaxy, of course); its spectrum did not suggest new planets but did display a Doppler shift of its absorption lines that indicated a motion of that object toward the earth of about 300 km/s. We know today that most of that speed is due to the revolution of the sun about the center of our own Galaxy, carrying us roughly in the direction of M31. Nevertheless, the observed radial velocity was enormous compared with those of stars. When Slipher reported the result to Lowell, the latter instructed him to obtain spectra of additional nebulae. Over the next 20 years, Slipher photographed spectra of more than 40 nebulae (now known to be galaxies). Only a few (now recognized as members of our Local Group) are approaching us; Slipher found the overwhelming majority to be receding, and at speeds of up to 1800 km/s. When knowledge of Slipher's observations reached Europe, cosmologists like Lemaître were encouraged to believe that

they were evidence of an expanding universe. Still, it remained to be proved that the velocities of recession were proportional to the distances of the nebulae.

(c) The Hubble Law

By the end of the 1920s, Hubble had not only shown that the nebulae are galaxies but had found ways of estimating their distances. In 1929 he published a paper in the *Proceedings of the National Academy of Sciences* showing that the radial velocities of the galaxies observed by Slipher are roughly proportional to their distances, as estimated by Hubble. The correlation (Figure 39.3a), although certainly interesting, could hardly be regarded as definitive. The distances were too uncertain.

The next breakthrough came from the efforts of a remarkable man: Milton Humason. Humason began his astronomical career by driving a mule train up the trail on Mount Wilson to the observatory. In those early days, supplies had to be brought up that way; even astronomers hiked up to the mountaintop for their turns at the telescope. Humason became interested in the work of the astronomers and took a job as janitor at the observatory. After a time he became a night assistant, helping the astronomers run the telescope and take data. Eventually, he made such a mark that he became a full astronomer at the observatory.

By the late 1920s Humason was collaborating with Hubble by photographing the spectra of faint galaxies with the 100-inch telescope. Humason concentrated on galaxies in clusters whose relative distances could be estimated with some confidence

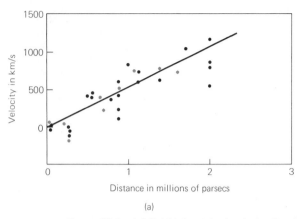
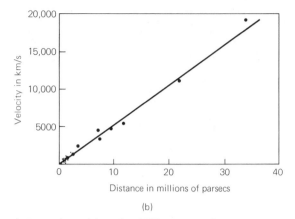

(a)　　　　　　　　　　　　　　　　　　　　(b)

Figure 39.3 (a) Hubble's original velocity-distance relation, adapted from his 1929 paper in the *Proceedings of the National Academy of Sciences*. (b) Hubble and Humason's velocity distance relation, adapted from their 1931 paper in the *Astrophysical Journal*. The small dots at the lower left are the points in the diagram in the 1929 paper (a).

(Section 37.2b). In 1931 Hubble and Humason jointly published their classic paper in the *Astrophysical Journal*, which compared distances and velocities of remote galaxies moving away from us at speeds of up to nearly 20,000 km/s. Their *law of the redshifts* (Figure 39.3b), now known as the *Hubble law*, established the expansion of the universe beyond doubt, even though the authors themselves, at that time, were cautious about so interpreting their observations.

Subsequently, more and more remote galaxies of greater and greater speed of recession have been found. The cluster of galaxies shown in Figure 39.6

Figure 39.4 Milton Humason. (*Courtesy of the Archives, California Institute of Technology*)

moves away from us with a speed of 108,000 km/s —36 percent of the speed of light. Even more remote clusters have been found; in 1981 Hyron Spinrad, at the University of California, Berkeley, reported observations of two clusters receding at about 60 percent the speed of light. The relative distances to clusters are known fairly well, and to the accuracy of the observations, remote clusters of galaxies have velocities that are proportional to their distances. The constant of proportionality, symbolized H and called the *Hubble constant*, specifies the rate of recession of galaxies or clusters of various distances. The Hubble constant is now believed to lie in the range 50 to 100 km/s per million parsecs; in other words, if H is 75 km/s per million parsecs, a cluster moves away from us at a speed of 75 km/s for every million parsecs of its distance.

Many people, concerned about such high speeds for galaxies, wonder if the large redshifts of their spectral lines might be due to some other cause than a Doppler shift. But such an explanation would require new physics, because no other mechanism is known that can produce the observed redshifts; gravitational redshifts, for example, would be accompanied by other effects not observed. But I hope the reader is convinced by now that the *easiest* way to interpret the redshifts is as Doppler shifts, indicating an expanding universe; otherwise we would have to come up with a new physical principle that could produce redshifts and also explain why the universe is *not* expanding, since existing theory predicts that it should be doing so.

The expansion of the universe does not imply that the galaxies and clusters of galaxies themselves are expanding. The raisins in our raisin bread analogy do not grow in size as the loaf expands. Similarly, their mutual gravitation holds galaxies and clusters together, and they simply separate as the universe expands, just as do the raisins in the bread. Galaxies in clusters do, of course, have individual motions of their own superimposed on the general expansion. Galaxies in pairs, for example, revolve about each other, and those in clusters move about within the clusters. In fact, a few galaxies in nearby groups and clusters move fast enough within those systems so that they are actually approaching us even though the clusters of which they are a part are moving, as units, away.

American physicist George Gamow reports that Einstein said, upon learning that his relativity theory is compatible with an expanding universe, as observed, that the introduction of the cosmological constant was "the biggest blunder of my life."

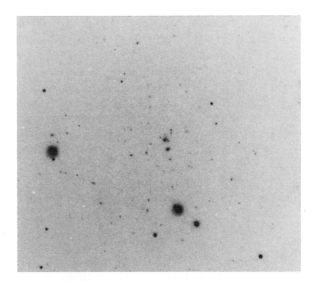

Figure 39.6 One of the remote clusters of galaxies for which there are measured redshifts (negative print). The cluster is receding from us at 36 percent the speed of light. It was discovered because its brightest member galaxy is a radio source, 3C295. Photographed with an image tube at the 3-m telescope of the Lick Observatory. *(Courtesy H. Ford, UCLA)*

39.2 COSMOLOGICAL MODELS

Let us now see what theory predicts in detail for the past and future of the universe. Common to most cosmological theories is the cosmological principle—the assumption of homogeneity of the universe—and we assume it here. We shall also adopt the best description of gravitation yet found, that of general relativity, and for simplicity, and also in common with nearly all cosmologists today, we assume that the cosmological constant is zero. Of course the universe may not be, in reality, described by so simple a model as the best theory we know today. But it is the purpose of science to find descriptions of the way nature behaves in terms of models, and cosmology is no exception. Absolute truth is not relevant and is beyond the realm of science anyway. The question is, can we find a model (theory, hypothesis) that *works?*

I think it is important to keep perspective in this respect, and not take current ideas about the origin and evolution of the universe too seriously. Those we describe in the pages to follow are consistent with the observational facts at our disposal but may turn out to be as far from reality as the epicycles and equants of antiquity. Still, it is great sport to see what our present knowledge of physical law predicts in answering the grandest questions man

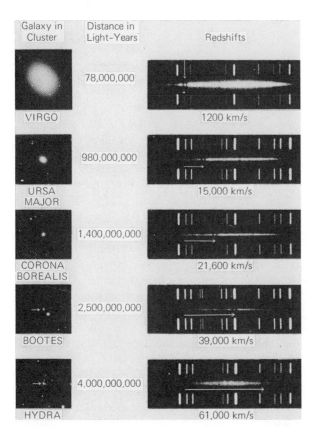

Galaxy in Cluster	Distance in Light-Years	Redshifts
VIRGO	78,000,000	1200 km/s
URSA MAJOR	980,000,000	15,000 km/s
CORONA BOREALIS	1,400,000,000	21,600 km/s
BOOTES	2,500,000,000	39,000 km/s
HYDRA	4,000,000,000	61,000 km/s

Figure 39.5 *(Left)* Photographs of individual galaxies in successively more distant clusters; *(right)* the spectra of those galaxies, showing the Doppler shift of two strong absorption lines due to ionized calcium. The distances are estimated from the faintness of galaxies in each cluster, and are provisional. *(Palomar Observatory; California Institute of Technology)*

can ask. Of course, we are extrapolating well beyond the regime in which our known physical laws have been well verified, but it is all we can do until we have new information. For the *real* truth, you will have to ask the philosopher, or a guru, or perhaps an ancient Chinese oracle. Science has no pipeline to absolute knowledge; it only can interpret observed facts in terms of rules that seem to be revealed by objective analysis. Science, therefore, has its limitations. Yet I think it is noteworthy that our admittedly tentative model of the early and present universe has proved to be consistent with no less than three solid predictions. One is that of the expansion of the universe; the other two are no less spectacular, as we shall see.

Those models of cosmology, which we shall mainly be concerned with, are the possibilities consistent with general relativity with no cosmological constant. They all presume a particular beginning of the universe at a finite time in the past, and they presume that the universe is evolving today. They are called, therefore, the *relativistic evolving models* of cosmology.

(a) Maximum Age of the Universe

As the universe expands, galaxies and clusters of galaxies separate from each other. Thus if we extrapolate backward in time, we would find them coming together, until some time in the distant past when all matter was crowded to an extreme density—a condition that marks a unique beginning of the universe, or at least of that universe we can know about. At that beginning, the universe suddenly began its expansion with a phenomenon called the *big bang*. We return in the next section to the physical conditions of the universe at and just after the big bang.

Now the total amount of matter and energy of the universe—presumed to be conserved (constant in time)—creates gravitation, whereby all objects pull on all other objects (including light). This mutual attraction must *slow* the expansion, which means that in the past the expansion must have been at a greater rate than it is today. How much greater depends on the importance of gravitation in decelerating the expansion. At the extreme, if the total mass-energy density has always been low enough that gravitation is ineffective (an essentially "empty" universe), the deceleration would be zero,

and only in that case would the universe always have been expanding at the present rate.

Clearly that extreme of an empty universe corresponds to the greatest age of the universe (since the big bang), because if the expansion were faster in the past, galaxies would have reached their present separations in a smaller time. Consequently, we can obtain an estimate of the upper limit to the age of the universe by asking how long it would take for distant galaxies, always moving away from us at their present rates, to have reached those distances. Call that maximum possible age T_o. Now the Hubble law states

$$V_r = Hr,$$

where V_r is the radial velocity of a galaxy of distance r. But velocity is just distance divided by time; that is, $V_r = r/T_o$. Hence,

$$\frac{r}{T_o} = Hr,$$

or

$$T_o = \frac{1}{H}.$$

We see, then, that the maximum age of the universe is just the reciprocal of the Hubble constant, which is believed to be no smaller than about 50 km/s per million parsecs. Since there are 3.086×10^{19} km in a million parsecs, $H = 50$ km/s per 3.086×10^{19} km, or 1.62×10^{-18} per second. The corresponding value of T_o is 6.172×10^{17} s or 1.95×10^{10} years. The actual age of the universe must be less than T_o, because the expansion has to be decelerating. Moreover, if H is greater than 50 km/s per million parsecs, as many astronomers think likely, the universal age is less yet.

Radial velocities of galaxies can be observed unambiguously from the Doppler shifts of their spectral lines, so knowledge of the Hubble constant is equivalent to knowledge of the scale of extragalactic distances. Thus, the age of the universe and the extragalactic distance scale are inextricably connected.

(b) The Scale of the Universe

By assumption (the cosmological principle), the universe is always homogeneous. Consequently, the expansion rate must be uniform (the same in all directions), so that the universe must undergo a uniform change in scale with time. It is customary to represent that scale by R, which is a function of time. The actual value of the scale is arbitrary—we could think of it as being the distance between any two representative objects or points in space, since R changes in the same way everywhere. R plays the same role as the "scale of miles" on a terrestrial map; it tells us by how much the universe has expanded (or contracted) at any time. For a static universe, R would, of course, be constant, but in an expanding universe R increases with time. The entire dynamical history of the universe is provided by the mathematical description of R with time.

The field equations of general relativity can be solved to learn how R varies with time if the amount by which the universe is decelerating is specified. The latter depends on very uncertain observational parameters: the Hubble constant (or distance scale) and the mean density of matter in space. By "mean density" we mean the mass of matter (including the equivalent mass of energy) that would be contained in each unit of volume (say, one cubic centimeter) if all of the stars, galaxies, and other objects were taken apart, atom by atom, and if all of those particles, along with the

light and other energy, were distributed throughout all space with absolute uniformity.

As the universe expands, the Hubble constant and mean density both decrease, but the current values of these observational parameters can be combined to form a single quantity, denoted q_o, and called the *deceleration parameter*.* In the evolving relativistic cosmologies under discussion, q_o must have a positive value, indicating a slowing of the expansion. We shall return in Section 39.4 to the problem of determining q_o observationally. For now, we shall treat it as a free parameter to see what are the possible solutions for R that are allowed by general relativity.

The solutions are shown in Figure 39.7. Time increases to the right and the scale, R, upward in the figure. Today, marked "present" along the time axis, R is increasing. The straight dashed line corresponds to the empty universe with no deceleration; it intercepts the time axis at a time T_o in the past. The other curves represent varying amounts of deceleration and start from the big bang at shorter times in the past.

---

*$q_o = \dfrac{8\pi G}{3} \dfrac{\rho}{H^2}$, where G is the constant of gravitation, and ρ and H are the instantaneous values of the mean density and the Hubble constant, respectively.

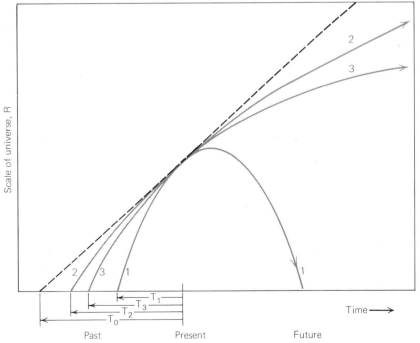

Figure 39.7 A plot of $R(t)$, the scale of the universe, against time for various cosmological models. Curve 1 represents the class of solutions for closed universes, curve 2 represents the class for open universes, and curve 3 is the critical solution for the boundary between open and closed universes. The dashed line is the case for an empty universe.

If the deceleration is above a critical value, corresponding to q_o greater than one half, R is given by curve 1; we see that in this case the universe stops expanding some time in the future and begins contracting. Eventually, the scale drops to zero, with what Princeton physicist John A. Wheeler calls the "big crunch." In this case, the universe is said to be *closed*, for it cannot expand forever. It is tempting to speculate that another big bang might follow the "crunch," giving rise to a new expansion phase, and ensuing contraction, perhaps oscillating between successive big bangs indefinitely in the past and future. Such speculation is sometimes referred to as the *oscillating* theory of the universe, but it is not really a theory, for we know of no mechanism that can produce another big bang. General relativity (and other theories) predicts, instead, that at the crunch the universe will collapse into a universal black hole. (Of course, we have no complete theory for the first big bang, either!) In any case, the oscillating theory is a speculation on a possible variation of the closed model of the universe.

Alternatively, if q_o is *less* than one half (curve 2 in Figure 39.7), gravitation is never important enough to stop the expansion, and the universe expands forever. In this case the universe is said to be *open*. At the critical value, $q_o = \frac{1}{2}$ (curve 3), the universe can just barely expand forever; it is the boundary between the families of open and closed universes. The mean density to make $q_o = \frac{1}{2}$ is referred to as the "critical density" to close the universe. The $q_o = \frac{1}{2}$ universe has an age of exactly $\frac{2}{3}T_o$—the age of the empty universe. Open universes have ages between $\frac{2}{3}T_o$ and T_o, and closed universes ages of less than T_o.

The various possibilities for the evolution of the universe are analogous to those for a rocket launched from the surface of the earth. If the launch velocity is the critical velocity of escape from the earth (about 11 km/s—Section 5.7), the rocket barely escapes the earth on a parabolic orbit; this corresponds to the universe with $q_o = \frac{1}{2}$. At lower and higher launch velocities, the rocket falls back to earth or escapes on a hyperbolic orbit—analogous to closed and open universes, respectively.

(c) Curvature of Spacetime in the Universe

We saw in Section 35.3 that the presence of matter and energy produces a curvature of spacetime, and

that in general relativity we describe objects as moving unaccelerated along curved geodesics in spacetime. The mathematics required to calculate the curvature is very complex around separated bodies such as those in the solar system. For the universe as a whole, however, the assumption of the cosmological principle vastly simplifies the problem. Locally (near stars and galaxies) the curvature is complex, but on the large scale it must be uniform and the same everywhere.

If the universe is closed, in the sense that material bodies cannot separate forever, then neither can light or other radiation. All geodesics are closed, and the curvature of spacetime is said to be positive. It does *not* mean that theoretically a light ray beamed from earth will literally come back to its source, however; first, in passing near discrete stars and galaxies it will be deflected slightly (see the discussion of gravitational lenses in Section 37.7c), and second, as the universe expands and changes scale, so does the radius of curvature of the light beam. It does mean, though, that light, as well as material particles, starting at the big bang must return to the big bang (or crunch); there is no possibility of ever communicating with an infinite universe. The universe is closed, then, in a broader sense than just that matter falls back together. The closed universe is *unbounded*, for no edge can ever be observed, but it is still *finite*.

On the other hand, if the universe is open, the curvature of spacetime is said to be negative. Geodesics are curved, but never close on themselves, like the path of a tiny being walking always straight ahead on a surface shaped like a saddle. The open universe is curved but infinite.

Spacetime in the critical universe with $q_o = \frac{1}{2}$, separating the open and closed ones, has zero curvature; in this case geodesics are the kinds of straight lines to which Euclid's geometry applies. This universe is said to be *flat* or *Euclidean*. The flat, $q_o = \frac{1}{2}$ universe is also called the *Einstein–de Sitter* universe, for it is mathematically equivalent to a particular model considered by those scientists subsequently to Einstein's early ideas about a static universe.

(d) Summary of Relativistic Evolving Models of the Universe

The possibilities described in the previous paragraphs are summarized in Table 39.1. To be spe-

Table 39.1 Parameters of Evolving Relativistic Cosmological Models for Which the Cosmological Constant is Zero and for Which $H = 50$ km/s per Million Parsecs

MODEL (Curve in Figure 39.7)	KIND OF UNIVERSE	AGE, T, SINCE BIG BANG (Units of 10^9 yr)	DECELERATION PARAMETER, q_0	SIGN OF CURVATURE, k	MEAN DENSITY, ρ (g/cm^3)
1	closed (oscillating?)	$T < 13$	$q_0 > \frac{1}{2}$	$+1$	$\rho > 5 \times 10^{-30}$
2	open	$13 < T \le 19.5$	$q_0 < \frac{1}{2}$	-1	$\rho \le 5 \times 10^{-30}$
3	flat	$T = 13$	$q_0 = \frac{1}{2}$	0	$\rho = 5 \times 10^{-30}$

cific, a particular value of the Hubble constant, H, has been assumed: 50 km/s per million parsecs. If H is actually 100 km/s per million parsecs, the ages should be one half the values given in the table and the mean densities four times as great as those given. The sign of the curvature of spacetime is conventionally indicated by the quantity k, which takes on the values -1, 0, and $+1$ for negative, zero, or positive curvature.

(e) Other Cosmologies

I emphasize that the foregoing discussion is based on particular assumptions—the cosmological principle, conservation of mass-energy, and general relativity theory with zero cosmological constant. Of course, any or all of these assumptions could be wrong, and if so, there is an infinity of other possibilities.

All evidence is compatible with the cosmological principle, but we cannot be absolutely certain that it is not violated. Among the models that violate the cosmological principle are hierarchical ones, in which there are ever higher orders of clustering. As we saw in Section 38.3b, present observations seem to rule out such a hierarchy, but not all cosmologists agree that the observations are absolutely definitive.

Another model of cosmology, which received wide attention in the 1950s, is the *steady-state* theory of Hermann Bondi, Thomas Gold, and Fred Hoyle. The Steady State is based on a generalization of the cosmological principle called the *perfect cosmological principle*, in which it is assumed that on the large scale the universe is not only the same everywhere, but for all time. In the steady state, mass-energy is *not* conserved, for as the universe expands and would otherwise thin out, new matter is continuously being created to keep the mean density the same at all times. (Hence the steady state is also called the *continuous creation* theory.) The theory also predicts at what rate mat-

ter is created to assure that there will always be the same admixture of young and old stars and galaxies. The creation would occur so gradually (presumably as individual atoms coming into being here and there) that we would not notice it. The steady-state universe is infinite and eternal and has much philosophical appeal. But we have already seen (Section 37.8) that certain objects—especially quasars—seem to suggest a secular evolution of matter in the universe, which violates the perfect cosmological principle. Moreover, we shall see below that there is direct evidence that the universe has evolved from a hot dense state, strongly supporting the idea of a big bang.

A more recent suggestion of Hoyle and the Indian astrophysicist J. V. Narlikar is that all of spacetime is divided into alternating regions in which matter has opposite properties, and within each of which all fundamental particles first increase and then decrease in mass. The boundaries between the regions are where the masses of the particles pass through the value zero. In this theory the redshifts of remote objects are not caused by expansion of the universe, but because we see them in the past when their atoms had lower mass, and hence the spectral lines occurred at lower energies. The Hoyle-Narlikar theory has not attracted wide interest, in part because it is difficult to test, but I mention it as an example of the kinds of ideas that have been suggested by serious scientists.

The models of the universe, other than those summarized in Table 39.1, that have received the most interest are those based on general relativity, but with nonzero cosmological constant. Although there is no evidence for a cosmological constant, its introduction by Einstein was nevertheless allowed mathematically in his derivation of the field equations, and general relativity could be quite correct even if the cosmological constant is *not* zero. By assigning various values to the constant, a whole new range of possibilities becomes available, and in particular, the universe can be much older than $1/H$

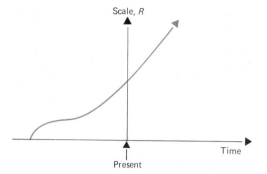

Figure 39.8 The change in scale with time for a universe with positive cosmological constant.

(Figure 39.8). Many astronomers are concerned that recent observations seem to suggest that H may be near 100 km/s per million parsecs. If so, the time since the big bang would have to be less than 10^{10} years, which is less than our current estimate for the age of the oldest globular star clusters. A way out of this difficulty is to adopt a value of the cosmological constant greater than zero. Most cosmologists, however, prefer to stick to the simplest case (zero cosmological constant) unless or until a more definitive determination of the value for the Hubble constant requires us to abandon those models.

39.3 THE BEGINNING

George Lemaître was probably the first to propose a specific model for the big bang itself. He envisioned all the matter of the universe starting in one great bulk he called the *primeval atom*. The primeval atom broke into tremendous numbers of pieces, each of them further fragmenting, and so on, until what was left were the present atoms of the universe, created in a vast nuclear fission. Lemaître thought the left-over radiation from that cosmic fireball was the cosmic rays. In a popular account of his theory he wrote, "The evolution of the world could be compared to a display of fireworks just ended—some few red wisps, ashes and smoke. Standing on a well cooled cinder we see the slow fading of the suns and we try to recall the vanished brilliance of the origin of the worlds."

We know today that the cosmic rays are not from the big bang, but probably from supernovae in our Galaxy. We also know much more about nuclear physics, and that the primeval fission model cannot be correct. Yet Lemaître's vision inspired

more modern work, and in some respects was quite prophetic.

In the 1940s the American physicist George Gamow suggested a universe with the opposite kind of beginning—nuclear fusion. He worked out the details with Ralph Alpher, and they published the results in 1948. (They added the name of physicist Hans Bethe to their paper, so that the coauthors would be Alpher, Bethe, and Gamow, a pun on the first three letters of the Greek alphabet: alpha, beta, and gamma.) Gamow's universe started with fundamental particles that built up the heavy elements by fusion in the big bang. His ideas were close to our modern view, except that the conditions in the primordial universe were not right for atoms to fuse to carbon and beyond, and only hydrogen and helium should have been formed in appreciable abundances. The heavier elements, we think, formed later in stars (Chapter 34).

(a) Standard Model of the Big Bang

The modern theory for the evolution of the early universe is called the *standard model* of the big bang. The details were worked out in a 1967 paper by Robert Wagoner, at Stanford University, and William Fowler and Fred Hoyle at Caltech. In the first extremely tiny fraction of a second, it is thought that all kinds of particles existed in equilibrium with radiation—particles and their antiparticles being produced in pairs from photons, and annihilating, reconverting to photons again. Some theoretical physicists believe that the present-day universe may provide clues about the kind of particles that could have existed in it at that very early instant, and hence information about the basic nature of matter itself. If so, studies in this area of cosmology could make the early universe a testing ground for the grand unified theories (Section 32.2d).

In any event, by the time the universe was one second old, it would have cooled to about 10^{10} K, by which time the prevalent photons were not of high enough energy to create pairs of particles. Then the matter consisted of particles such as protons, electrons, positrons, neutrons, and neutrinos. By the time the universe was 100 s old, the temperature had dropped to 10^9 K, and the particles began to combine to form some heavier nuclei. This nucleosynthesis continued for the next few minutes, during which about 25 percent of the mass of the material formed into helium. Some deuterium also formed (deuterium is an isotope of hydrogen with a nucleus containing one proton and one neutron) but only a small amount—probably less than one part in ten thousand. The actual amount of deuterium formed depends critically on the density of the fireball; if it was fairly high, most of the deuterium would have been built up into helium. Scarcely any nuclei heavier than those of helium are expected to have survived. So the composition of the fireball when nuclear building ceased is thought to have been mostly hydrogen, about 25 percent helium, and a trace of deuterium.

It is a striking success of the standard model that the predicted ratio of hydrogen to helium—three to one by mass—is just the ratio observed in stars and interstellar matter. A small enhancement of the helium must have resulted from nucleosynthesis in stars, to be sure, but by far most of the helium must be primordial—especially in the outer layers of stars. Hence the agreement between prediction and observation must be regarded as a second triumph for the big bang and relativistic evolving cosmological theories (the first being the expansion of the universe). We come to a third successful prediction in the next subsection.

For the next few hundred thousand years, the fireball was like a stellar interior—hot and opaque, with radiation being scattered from one particle to another. By about 700,000 years after the big bang, the temperature had dropped to about 3000 K and the density of atomic nuclei to about 1000 per cubic centimeter. Under these conditions, the electrons and nuclei combined to form stable atoms of hydrogen and helium. With no free electrons to scatter photons, the universe became transparent, and matter and radiation no longer interacted; subsequently each evolved in its separate way.

One thousand million years after the big bang, stars and galaxies had probably begun to form, but we are not sure of the precise mechanisms. Certainly, however, deep in the interiors of stars matter was reheated, stars began to shine, nuclear reactions were ignited, and the gradual synthesis of the heavier elements began.

Now we must emphasize that the fireball must not be thought of as a localized explosion—like an exploding superstar. There were no boundaries and no site of the explosion. It was everywhere. The fireball is still existing, in a sense. It has expanded greatly, but the original matter and radiation are still present and accounted for. The stuff of our

Figure 39.9 Arno A. Penzias (right) and Robert W. Wilson. (Bell Labs)

bodies came from material in the fireball. We were and are still in the midst of it; it is all around us.

But what happened to that radiation released when the universe became transparent at the tender age of just under a million years?

(b) The Cosmic Background Radiation

The question was first considered by Alpher and Robert Herman, both associates of Gamow. They realized that just before the universe became transparent it must have been radiating like a black body at a temperature of 3000 K. If we could have seen that freed radiation just after neutral atoms formed, it would have resembled that from a reddish star. But that was at least ten thousand million years ago, and in the meantime the scale of the universe has increased a thousandfold. The light emitted by the once hot gas in our part of the universe is now thousands of millions of light years away.

To observe that glow of the early universe, we must look out in all directions in space to such great distance—10 to 20 thousand million light years—that we are looking back in time through those 10 to 20 thousand million years. Now those remote parts of the universe, because of its expansion, should be receding from us at a speed within two parts in a million of that of light. The radiation from them would be Doppler shifted to wavelengths a thousand times those at which it was emitted.

When a black body approaches us, the Doppler shift shortens the wavelengths of its light and causes it to mimic a black body of higher temperature. When a black body recedes, it mimics a cooler black body. Alpher and Herman predicted that the glow from the fireball should now be at radio wavelengths and should resemble the radiation from a black body at a temperature of only 5 K—just a few degrees above absolute zero. But there was no way, in 1948, of observing such radiation from space, so the prediction did not attract much attention and was forgotten.

In the mid 1960s, however, the idea occurred independently to Princeton physicist Robert H. Dicke, who realized that microwave radio telescopes could then be built that might detect that dying glow of the big bang. He, P. J. E. Peebles, P. G. Roll, and D. T. Wilkinson confirmed that the theory was correct and began construction of a suitable microwave receiver on the roof of the Princeton biology building. They were not, however, the first to observe the radiation.

Unknown to them, a few miles away, in Holmdel, New Jersey, Arno Penzias and Robert Wilson, of the Bell Telephone Laboratories, were using the Laboratories' delicate microwave horn antenna to make careful measures of the absolute intensity of radio radiation coming from certain places in the Galaxy. But they were plagued with some unexpected background noise in the system that they could not get rid of. They checked everything, and eliminated the Galaxy as a source, also the sun, the sky, the ground, and even the equipment.

At one point they realized that a couple of pigeons had made their home in the antenna, and nested up near the throat of the horn where it was warmer. Penzias and Wilson could chase the birds away while they observed, but they found that the birds left, as Penzias puts it, a layer of white, sticky dielectric substance coating the inside of the horn. That substance would radiate, producing radio interference. They disassembled the horn and cleaned it, and the unwanted noise did go down somewhat, but it did not go away completely.

Finally, Penzias and Wilson decided that they had to be detecting radiation from space. Penzias mentioned it in a telephone conversation with another radio astronomer, B. Burke, who was aware of the Princeton work. Burke got Penzias and Wilson in touch with Dicke, and it was soon realized that the predicted glow from the primeval fireball had been observed. Since then the radiation has been very thoroughly checked throughout the entire radio spectrum, with observations from ground-based radio telescopes, with instruments carried aloft in balloons, and even with a receiver in a U2 reconnaissance airplane. The microwave background radiation closely matches that expected from a black body with a temperature of 2.7 K.

Penzias and Wilson received the Nobel Prize for their work in 1978. And perhaps almost equally fitting, just before his death in 1966, Lemaître learned about the discovery of his "vanished brilliance."

(c) Further Information from the Cosmic Background Radiation

The faint glow of radio radiation is now called the *cosmic background radiation* (sometimes, *CBR*). It has now been observed at many wavelengths, and all observations are compatible with the CBR being redshifted radiation emitted by a hot gas. Among other things, if correctly interpreted, the CBR shows that the universe has evolved from a hot, uniform state—as opposed, for example, to the steady-state theory.

The CBR does, however, show subtle deviations from a black body spectrum. It happens that a 2.7 K black body radiates most strongly at between two and three millimeters (the peak of the curve shown in Figure 39.10). But the millimeter wavelengths are the very ones in which the earth's atmosphere radiates very strongly at upper altitudes, which makes observations at those wavelengths very difficult. Thus the only successful observations of the spectrum of the cosmic background radiation at its peak have been made high above most of the earth's atmosphere. University of California, Berkeley scientists, P. L. Richards and D. P. Woody have made such observations with a detector carried by a balloon to an altitude of tens of thousands of meters. Their observations are the ones shown as a shaded zone in Figure 39.10. That zone shows the regions within which Richards and Woody believe the intensity of the cosmic background radiation to most probably lie.

Note that the observations of Richards and Woody lie above the smooth curve (representing the emission from a black body at 2.7 K) at the peak. A black body of 2.96 K actually fits Richards' and Woody's infrared data in this spectral region better

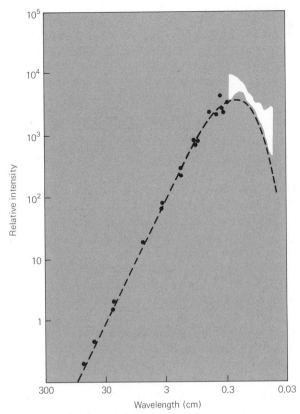

Figure 39.10 The cosmic background radiation. The dashed curve is a 2.7 K black body. The unshaded region corresponds to the infrared observations of Richards and Woody.

the radiation tells us that at an age of less than a million years the universe had to be extremely uniform in density. But at least *some* density variations had to be present to allow matter to gravitationally clump up to form stars and galaxies. The isotropy of the CBR, therefore, puts interesting constraints on theories of star and galaxy formation.

On the other hand, a large-scale anisotropy in the cosmic background radiation has now been established, in the sense that it is slightly brighter in one direction than in the opposite direction in the sky. This is because of our own motion through space. If you approach a black body, its radiation is all Doppler-shifted to shorter wavelengths and resembles that from a slightly hotter black body; if you move away from it, the radiation appears like that from a slightly cooler black body. Such an effect has been searched for and observed in the microwave background.

The measurements are very difficult because the difference in intensity is very tiny compared with the radiation of the earth's atmosphere. They must, therefore, be made at a very high altitude. Princeton physicist D. Wilkinson and his group put their apparatus in a high-flying balloon, and University of California, Berkeley, physicist George Smoot sends a similar receiver to an altitude of more than 20,000 m in one of NASA's converted U2 planes. Both have found the variation expected due to our motion, and their results are in reasonable agreement.

It might seem that we have now succeeded in doing what Michelson and Morley could not: determining the absolute speed of the earth through the universe. Does this not violate the special theory of relativity? It does not, because we still are measuring only a *relative* speed—our velocity with respect to the gas that emitted the radiation thousands of millions of years ago.

The measured velocities of the sun, as determined from the observations of the two groups, are shown in Table 39.2. The velocities are considered uncertain by about 50 km/s, and the celestial coordinates (right ascension and declination) by from 10 to 20°. The direction of the sun's motion is roughly toward the constellation Leo, but a good part of the motion is the sun's revolution about the center of the Galaxy. When we take into account the galactic rotation and the Galaxy's motion in the Local Group, we find that the Local Group as a whole is moving with a speed of about 540 km/s in the general direction of the constellation Hydra.

than one of 2.7 K, but then the observations are off at other wavelengths.

These observations are still in need of confirmation by other investigators. Should the small discrepancies turn out to be real, however, they may tell us something else about the early universe. It is possible, for example, that the deviations from a perfect black body spectrum can be due to absorption and reemission by dust formed from matter ejected from stars of a very early generation—one dubbed *population 3*. These hypothetical stars would have to have existed when the universe had expanded to only a few percent of its present scale, probably within the first thousand million years after the big bang.

At any given wavelength, the cosmic background radiation is extremely isotropic on the small scale. In directions that differ by only a few minutes of arc, any fluctuations in its intensity must be less than a few parts in 10^4; recent Soviet observations claim that the CBR is isotropic on the small scale to better than a few parts in 10^5. The uniformity of

Table 39.2 **The Speed and Direction of the Sun With Respect to the CBR**

GROUP	VELOCITY *(km/s)*	RIGHT ASCENSION	DECLINATION
Princeton	332	12.3 hours	$- 1°$
Berkeley	360	11.2 hours	$+ 19$

The motion of the Local Group at 540 km/s is with respect to the uniform expansion of the universe as a whole. It can be thought of as a peculiar motion of the Group, superimposed on the general expansion. But what can cause the Local Group to have such a high peculiar velocity? One interpretation is that it is due to the gravitational attraction of the Local Supercluster. The Supercluster is expanding, but less rapidly than the universe as a whole. Thus we find that the Virgo cluster (which may lie near the direction to the center of the Supercluster) is separating from us at about 1000 km/s, but if it were not for the gravitation of the Supercluster, we and the Virgo cluster would be flying apart at about 1400 km/s. (Some popular accounts have described the Local Group as "falling toward the Virgo cluster," but as we see, this is misleading.) If the interpretation is correct, we learn something of the dynamics of the Local Supercluster from the large-scale anisotropy of the cosmic background radiation.

(d) A Look Back to the Big Bang

Since this radiation comes from the time when the fireball first became transparent, it is at the farthest point in space to which we can presently observe. If we could see that radiation visually, it would be as if it were coming from an opaque wall, and no radiation from a more distant source could ever reach us—for that source would have to lie farther back into time where it would be behind that opaque wall.

Figure 39.11 is a schematic cross section of four-dimensional spacetime (see Chapter 35). The distance from us, in some arbitrary direction, is shown along the horizontal axis, and time increases along the vertical axis. Light paths are shown as 45° lines, as in Chapter 35. Our world line (geodesic) is upward along the time axis, and we are presently at time t_0. Times t_1, t_2, and t_3 are successively farther back into the past. The world lines of several other hypothetical observers on distant galaxies are shown and are labeled a, b, c, and so forth, in order of increasing distance. Note that they all recede from us with time as the universe expands. (Because of

the peculiarity of spacetime diagrams of an expanding universe, the diagram is not strictly correct for uniform time and distance measures along the axes, but nevertheless serves to illustrate the situation.)

When we look out into space, we look back into time along a light geodesic (diagonal lines). Note that the light we see comes from successively more distant objects successively farther into the past at times when they were actually separated less far from us in spatial coordinates. There is a limit, however, to how far back we can look. The fireball is presumed to have become transparent at time t_i, and the shaded region at the bottom of the diagram corresponds to the opaque fireball into which we cannot see. Thus the limiting distance we could observe in a particular direction is at the point where the light path to us intercepts the top of the shaded region at time t_i.

As time goes on, we can see farther and farther away, and more galaxies would come into view (if we had a large enough telescope) for, as the time from the big bang becomes greater we are looking

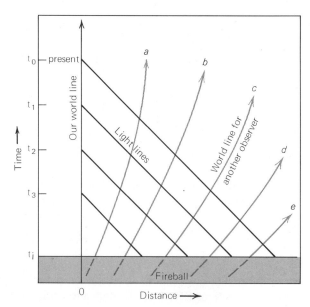

Figure 39.11 Schematic representation of world lines for us and for several other hypothetical observers on distant galaxies in the universe on a two-dimensional cross section of spacetime. The straight diagonal lines are light paths. (See text for discussion.)

farther into the past to see the fireball, and hence farther away in space. Note that at earlier times such as t_3, t_2, and t_1, we saw only relatively nearer objects, and there were fewer galaxies between us and the threshold provided by the fireball itself. Thus, not only does the universe expand with time, but the part of it accessible to observation becomes greater as well.

On earth, the microwave radiation is very feeble compared with, say, sunlight. But far off in intergalactic space, that radio background is by far and away the most intense radiation around. The observed radiation comes equally from all directions and gives no direction to a "center" of the universe; the universe, its "center" and its origin are all around us. There is a boundary to the observable universe, but it is a boundary in time, not in space.

39.4 THE FUTURE OF THE UNIVERSE

There are several observational tests by which we hope to be able to distinguish between the evolving cosmological models. As of this writing, the observations are still not critical enough to reach a definite conclusion, but it is worth describing some examples of these tests.

(a) The Mean Density

One is the determination of the mean density of matter in space. We have noted that knowledge of it is sufficient to calculate the way R changes with time from the field equations of general relativity. We can estimate the mean density from the number of galaxies and clusters we observe out to a given distance, and from a knowledge of the masses of these objects. There is considerable uncertainty in the masses of clusters of galaxies, and we do not know how much matter (if any) may exist in intergalactic space. Nevertheless, such estimates indicate a mean density less than 10^{-30} g/cm^3, and probably near 10^{-31} g/cm^3. This is below the critical density, and suggests an open universe, but the estimates are too uncertain to be sure of the conclusion.

(b) The Deuterium/Hydrogen Ratio

We also saw that the production of deuterium in the fireball is very sensitive to the density of the universe within the first few minutes after the big bang. The proportion of deuterium in interstellar space is thought to be a measure of that formed in the fireball, for in stellar interiors it is rather quickly converted to helium. It is very difficult to detect deuterium in space, but careful measures show that the ratio of deuterium to hydrogen is probably in the range 10^{-4} to 10^{-5}. From this crude estimate the density of the fireball at the critical time can be inferred, and from that knowledge it is possible to predict what present-day density would result. The calculation suggests a present density of about 10^{-31} g/cm^3, again pointing to an open universe.

(c) The Hubble Law

Another less obvious test involves the speeds of receding galaxies at great distances. The evolutionary models we have described predict the expansion to slow down due to the gravitational forces between galaxies. Because of these changes in the rate of expansion of the universe, the radial velocities of remote galaxies should deviate from a relation exactly proportional to their observed distances; that is, a graph of radial velocity versus distance should not necessarily be a straight line for very distant galaxies (see Exercise 6). The exact form of the redshift-distance relation is different as predicted by each cosmological model.

The differences between the observable quantities predicted by various models are small until very large distances are reached—that is, until we look back through an appreciable interval of time. Unfortunately, precise distances of remote objects, as well as their luminosities and other characteristics, are very difficult to determine, and critically accurate observations are required.

One procedure for obtaining relative distances of clusters is from their brightest member galaxies. Figure 39.12 is a plot of the radial velocities (on a logarithmic scale) of nearly a hundred clusters of galaxies against the apparent magnitudes of their brightest members, adapted from a diagram by A. Sandage. The magnitudes of the brightest cluster galaxies are indicators of distance (see Chapter 37); they have been corrected for interstellar absorption in our Galaxy and for certain effects of the redshift. The different lines on the plot correspond to predictions of different cosmological models. We can see from the scatter of the points that more precise ob-

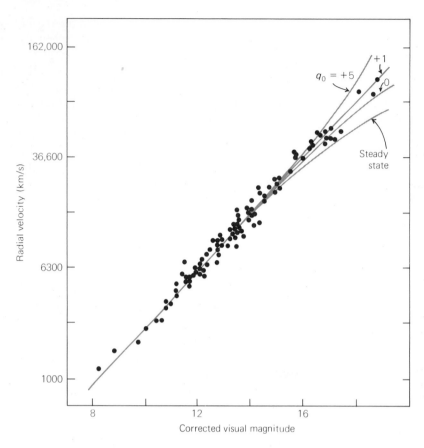

Figure 39.12 Hubble diagram for clusters of galaxies. Ordinates are radial velocities (on a logarithmic scale) and abscissas are the apparent magnitudes of the brightest cluster galaxies, corrected for interstellar extinction and certain effects of redshift. Points are observed values, and the solid curves are the predicted relations for various cosmological models. *(Adapted from a diagram by A. Sandage)*

servations are needed to apply this test. Also we do not yet know how to take account of changes in the luminosities of galaxies due to evolution of the stars in them.

In summary, observational tests of cosmological models appear to rule out the steady state, but are not yet definitive enough to choose between the various evolving models predicted by general relativity. And of course the observations may well be compatible with many other cosmological theories not yet considered.

39.5 CONCLUSION

The currently favored big bang theory results from an application of our present knowledge of the laws of physics to the universe as a whole. So far, the theory has withstood three observational tests: (1) The universe is expanding; (2) the ratio of hydrogen to helium is 3 to 1 by mass; (3) there is a cosmic background radiation. The background radiation, in particular, is strong evidence that the universe

has evolved from a hot, dense, uniform state. We can say nothing, however, about the time *preceding* the big bang.

General relativity theory, in its simplest form, allows two possibilities for the future of the universe: It may expand forever, or it may one day collapse upon itself. Present observations seem to favor the former alternative, but we are by no means certain of that conclusion.

Thus we have concluded our exploration of the universe with descriptions about how it *might* be, but no assertion of how it *is*. Science does not, nor can it, provide definitive answers to all questions. There will always be a new frontier.

The following words, written by Edwin Hubble in 1936, are still appropriate†:

Thus the explorations of space end on a note of uncertainty. And necessarily so. We are, by definition, in the very center of the observable region. We know

---

†*The Realm of the Nebulae*, Yale University Press, 1936. pp. 201–202. Quoted by permission of the publisher.

our immediate neighborhood rather intimately. With increasing distance, our knowledge fades, and fades rapidly. Eventually, we reach the dim boundary—the utmost limits of our telescopes. There, we measure shadows, and we search among ghostly errors of measurement for landmarks that are scarcely more substantial.

The search will continue. Not until the empirical resources are exhausted, need we pass on to the dreamy realms of speculation.

EXERCISES

1. A cluster of galaxies is observed to have a radial velocity of 60,000 km/s. Assuming that $H = 50$ km/s per million parsecs, find the distance to the cluster.

2. Plot the "velocity-distance relation" for the raisins in the bread analogy from the data given in Figure 39.2.

3. Repeat Exercise 2, but use some other raisin than A for a reference. Is your new plot the same as the last one?

4. Why can the redshifts in the spectra of galaxies *not* be explained by the absorption of their light by intergalactic dust?

5. Calculate the maximum possible age for the universe for the case where the Hubble constant, $H = 75$ km/s $\cdot$ 10^6 pc.

6. Assume that the radial velocities of galaxies have always been the same and are given *at this instant of time* by $V = Hr$, where $H = 50$ km/s per million parsecs. Note, however, that we do not observe the *present* distances of galaxies, but the distances they had when light left them on its journey to us. Now plot the relation between velocity and distance that would be obtained directly from *observations* (that is, corresponding to *measured* distances, not present distance). Consider several distances, out to 2×10^9 pc. Discuss the shape of the curve. How would this curve differ if the expansion rate were decreasing (say, because of gravitational attraction between galaxies)? What if it were increasing?

7. Suppose we were to count all galaxies out to a certain distance in space. If the universe were not expanding, the total number counted should be proportional to the *cube* of the limiting distance of our counts. (Why?) Taking account of the finite time required for light to reach us, describe the relation that would be *observed* between the total count and the limiting distance for (a) the steady-state theory, and (b) the evolutionary theory of the universe.

8. Draw a diagram like that in Figure 39.7 for an evolving cosmological model with $q_0 < \frac{1}{2}$ (open space). Indicate on the time axis about where the most remote quasars are, and also where the 2.7 K black-body radiation is coming from.

9. Refer to Figure 39.11. Describe how the background black-body radiation must have differed at times t_3, t_2, and t_1 from how it appears now.

10. Sketch and discuss a schematic spacetime diagram like that in Figure 39.11, showing the world lines of several galaxies for the case of a closed universe which will contract again sometime in the future.

11. The Hubble law, $V = Hr$, relates the velocity of recession and the distance to an object. But we do not actually observe velocity and distance. What are the *observable* quantities?

12. Can you see any problems in postulating a closed universe if the Hubble constant, H, were 100 km/s $\cdot$ 10^6 pc?

13. Suppose the universe will expand forever (that is, an open model). Describe what will become of the radiation from the primeval fireball.

14. Are you perplexed with the task of imagining an infinite universe? What if it is finite; then what lies beyond? Do you suppose the experts can visualize infinities and eternities any better than you can? If not, can they simply accept these concepts, because they are used to talking and thinking about them? Collect opinions of such experts if you can find them. Who are the experts on infinities and eternities?

15. Suppose $q_0 > \frac{1}{2}$, so that the universe is closed. Can there be other similar "universes" beyond the closed light paths in our own universe, with the same or even different properties? What might they be like, and how do they interrelate with our universe? Should cosmologists be concerned with studying such possibilities? Why or why not?

Appendix 1 BIBLIOGRAPHY

(Technical references are marked with an asterisk.)

Some Other General Textbooks in Astronomy

Abell, G. O., *Realm of the Universe*, 2nd ed. Philadelphia: Saunders College Publishing, 1980.

Alter, D., Cleminshaw, C. H., and Phillips, J., *Pictorial Astronomy*. New York: Crowell, 1974.

Berman, L., and Evans, J. C., *Exploring the Cosmos*. Boston: Little Brown, 1980.

Brandt, J. C., and Maran, S. P., *New Horizons in Astronomy*. San Francisco: Freeman, 1979.

Hartmann, W. K., *Astronomy: The Cosmic Journey*. Belmont, Calif.: Wadsworth, 1978.

Hoyle, F., *Astronomy and Cosmology: A Modern Course*. San Francisco: Freeman, 1975.

Jastrow, R., and Thompson, M. H., *Astronomy: Fundamentals and Frontiers*. New York: Wiley 1977.

King, I. R., *The Universe Unfolding*. San Francisco: Freeman, 1976.

Pasachoff, J. M., *Contemporary Astronomy*, 2nd ed. Philadelphia: Saunders College Publishing, 1981.

Pasachoff, J. M., and Kutner, M. L., *University Astronomy*. Philadelphia: Saunders College Publishing, 1978.

*Smith, E. v.P., and Jacobs, K. C., *Introductory Astronomy and Astrophysics*. Philadelphia: Saunders College Publishing, 1973.

Zeilik, M., *Astronomy: The Evolving Universe*. New York: Harper and Row, 1979.

Histories of Astronomy

Abell, G. O., and Singer B. (eds.), *Science and the Paranormal*. New York: Scribner's Sons, 1981. (A collection of critical essays on allegedly paranormal topics by well-known scientists and science writers.)

Berendzen, R., Hart, R., and Seeley, D., *Man Discovers the Galaxies*. New York: Neale Watson Academic Publications, 1976. (A superb history of the development of modern extragalactic astronomy.)

Culver, R. B., and Ianna, P. A., *The Gemini Syndrome: Star Wars of the Oldest Kind*. Tucson: Pachart, 1979. (An excellent account of astrology.)

Galileo, G., *Dialogue on the Great World Systems* (translated by T. Salisbury). Chicago: University of Chicago Press, 1953. (The classic work that resulted in Galileo's inquisition.)

Gardner, M., *Fads and Fallacies in the Name of Science*. New York: Dover, 1957. (A fascinating look at pseudoscience.)

Grosser, M., *The Discovery of Neptune*. New York: Dover, 1979. (A very engrossing account of the discovery of new planets in the solar system.)

Hoyle, F., *Astronomy*. New York: Doubleday, 1962. (Interesting history of astronomy.)

King, H. C., *Exploration of the Universe*. New York: New American Library, 1964. (A scholarly history; not to be confused with a famous text on astronomy of the same title.)

Koestler, A., *The Sleepwalkers*. New York: Macmillan, 1959. (A famous and very interesting history of the beginnings of modern science.)

*Lang, K. R., and Gingerich, O. (eds.), *A Source Book in Astronomy and Astrophysics, 1900–1975*. Cambridge: Harvard University Press, 1979. (A valuable collection of original papers or selections from them.)

*Newton, I., *Mathematical Principles of Natural Philosophy*. R. T. Crawford, ed. Berkeley: University of California Press, 1966. (Important classic.)

Pannekoek, A., *A History of Astronomy*. New York: Interscience, 1961. (A respected history.)

Randi, J., *Flim-Flam*. New York: Lippincott and Crowell, 1980. ("The Amazing Randi," a famous magician, exposes frauds, some of whom duped scientists as well as the public.)

Shapley, H., and Howarth, H. E., eds., *A Source Book in Astronomy*. New York: McGraw-Hill, 1929. (A collection of original papers, suitable for the general reader.)

Shapley, H. (ed.), *Source Book in Astronomy 1900–1950*. Cambridge: Harvard University Press, 1960. (An update of the previous reference.)

Standen, A., *Forget Your Sun Sign*. Baton Rouge: Legacy Publishing Co., 1977. (A delightful review of astrology.)

Struve, O., and Zebergs, V., *Astronomy of the Twentieth Century*. New York: Macmillan, 1962. (A modern history with original source material.)

Celestial Mechanics

Ahrendt, M. H., *The Mathematics of Space Exploration*. New York: Holt, Rinehart and Winston, 1965. (A relatively elementary survey of celestial mechanics.)

*Danby, J. M., *Fundamentals of Celestial Mechanics*. New York: Macmillan, 1962. (A good basic text for upper-division science students.)

*Moulton, F. R., *An Introduction to Celestial Mechanics*. New York: Macmillan, 1923. (The classic text in the subject for advanced science students.)

Ryabov, Y., *An Elementary Survey of Celestial Mechanics*. New York: Dover, 1961. (An elementary survey.)

Van de Kamp, P., *Elements of Astromechanics*. San Francisco: Freeman, 1964. (For the lower division science student.)

Telescopes

Christianson, W. N., and Hogborn, J. A., *Radio Telescopes*. London: Cambridge University Press, 1969. (Good account of the fundamentals for the serious student.)

Miczaika, G., and Sinton, W., *Tools of the Astronomer*. Cambridge, Mass.: Harvard University Press, 1961. (Authoritative book for the serious layperson, from the Harvard Series on Astronomy.)

Page, T., and Page, L. W. (eds.), *Telescopes: How to Make Them and Use Them*. New York: Macmillan, 1966. (A collection of articles by various experts on many phases of telescope making and the use of telescopes, suitable for the amateur and serious layperson.)

Earth, Sun, and Solar System

*Brandt, J. C., and Hodge, P. W., *Solar System Astrophysics*. New York: McGraw-Hill, 1964. (For the upper-division science student.)

*Gehrels, T. (ed.), *Asteroids*. Tucson: University of Arizona Press, 1979. (The proceedings of a conference, with contributions by leading experts in the field. For the serious student or professional, but many chapters are readable by more elementary students. The most comprehensive reference available on minor planets.)

*Gibson, E. G., *The Quiet Sun*. Washington, D. C.: NASA, 1973. (Excellent account of the sun under usual conditions for the science student, written by the scientist-astronaut.)

Hawkins, G. S., *Meteors, Comets, and Meteorites*. New York: McGraw-Hill, 1964. (For the interested layperson.)

*Kaula, W. M., *An Introduction to Planetary Physics*. New York: Wiley, 1968. (Text for the more advanced student.)

Menzel, D., *Our Sun*. Cambridge, Mass.: Harvard University Press, 1959. (One of the Harvard Series in Astronomy for the intelligent layperson.)

Scientific American reprints: *Continents Adrift*. San Francisco: Freeman, 1970. (Good series of semipopular articles that review the field of plate tectonics.)

Watson, F., *Between the Planets*. Cambridge, Mass.: Harvard University Press, 1956. (Account for the intelligent layperson on the small bodies of the solar system; one of the Harvard Series in Astronomy.)

Whipple, F., *Earth, Moon, and Planets*. Cambridge, Mass.: Harvard University Press, 1968. (Somewhat dated in 1981 but still a good classic account for the layperson, from the Harvard Series in Astronomy.)

Stellar Astronomy and Astrophysics

Aller, L. H., *Atoms, Stars and Nebulae* (rev. ed.). Cambridge, Mass.: Harvard University Press, 1971. (One of the best books for the interested layperson, from the Harvard Series in Astronomy.)

*Batten, A. K., *Binary and Multiple Star Systems*. Oxford: Pergammon Press, 1973. (Professional monograph, but not highly technical; suitable for the science student.)

Bok, B. J., and Bok, P. E., *The Milky Way* (4th Ed.). Cambridge, Mass.: Harvard University Press, 1973. (The classic description of our Galaxy by the world-famous astronomer and his late wife.)

Relativity and Modern Physics

Davies, P. C. W., *The Forces of Nature*. Cambridge, England: Cambridge University Press, 1979. (An excellent semipopular account of modern physics.)

Davies, P. C. W., *Space and Time in the Modern Universe*. Cambridge, England: Cambridge University Press, 1977. (A fine semipopular account of special and general relativity.)

Einstein, A., *Relativity: The Special and General Theory*. New York: Crown Publishers, 1961. (A classic popular account, by Einstein himself, of his epoch-making new physics.)

Feynman, R., *The Character of Physical Law*. Cambridge: Massachusetts Institute of Technology Press, 1965. (A series of popular lectures by one of the most articulate theoretical physicists of our time; an excellent insight into modern science.)

Gardner, M., *The Relativity Explosion*. New York: McGraw-Hill, 1966. (A very readable description of the meaning of relativity for the layman.)

Kaufmann, W. J., III., *The Cosmic Frontiers of General Relativity*. Boston: Little Brown, 1977. (A layman's account of relativity theory and of some of the possible esoteric consequences of it.)

*Misner, C. W., Thorne, K. S., and Wheeler, J. A., *Gravitation*. San Francisco: Freeman, 1973. (An advanced and impressive text on relativity that has become a classic.)

Polkinghorne, J. C., *The Particle Play*. San Francisco: Freeman, 1979. (A layman's introduction to the exciting new world of the particles of nature.)

Russell, B., *The ABC of Relativity*. New York: Mentor Book, from the New American Library, 1969. (An interpretation of relativity theory for the nonscientist by the famous philosopher.)

Sciama, D. W., *The Physical Foundations of General Relativity*. Garden City, N.Y.: Doubleday, 1969. (A description of the physical basis of Einstein's theory for the layman with only a smattering of algebra.)

Galaxies and Cosmology

*Couderc, P., *The Expansion of the Universe*. London: Faber and Faber, 1952. (An account of cosmology for those with minimal mathematics—through intermediate algebra.)

Ferris, T., *The Red Limit*. New York: William Morrow and Company, 1977. (A very perceptive account of the history of modern cosmology and the men who played a role in it, by a well-known journalist.)

Hodge, P. W., *Galaxies and Cosmology*. New York: McGraw-Hill, 1966. (A readable account of extragalactic astronomy for the nonscientist.)

Hubble, E., *The Realm of the Nebulae*. New Haven: Yale University Press, 1936; also, New York: Dover, 1958. (A classic book for the educated layman by one of the great astronomers of our time, describing his exploration of the extragalactic universe.)

*Peebles, P. J. E., *Physical Cosmology*. Princeton: Princeton University Press, 1971. (An authoritative monograph on modern cosmology for the more advanced student.)

*Peebles, P. J. E., *The Large-Scale Structure of the Universe*. Princeton: Princeton University Press, 1980. (A monograph for the more advanced student on the distribution of matter in space.)

Sandage, A. R., *The Hubble Atlas of Galaxies*. Washington, D.C.: Carnegie Institution, 1961. (A photographic atlas of galaxies of different types.)

*Sciama, D. W., *Modern Cosmology*. London: Cambridge University Press, 1971. (An excellent text in cosmology for the upper-division science student.)

Sciama, D. W., *The Unity of the Universe*. New York: Doubleday, 1959. (An account of the interplay of modern physics and cosmology for the intelligent layman.)

Shapley, H., *Galaxies*. Cambridge: Harvard University Press, 1972. (A semipopular description of the extragalactic universe by one of the great astronomers who played a role in its investigation.)

Shipman, H. L., *Black Holes, Quasars, and the Universe* (2nd ed.). Boston: Houghton Mifflin, 1980. (Probably the best popular or semipopular account of black holes, relativity, and active galactic nuclei available.)

Silk, J., *The Big Bang*. San Francisco: Freeman, 1980. (A new and authoritative book on cosmology for the intermediate undergraduate science student; highly recommended.)

Verschuur, G. L., *The Invisible Universe*. New York: Springer-Verlag, 1974. (An account of the role of radio astronomy in cosmology by a well-known radio astronomer; written for the general public.)

Weinberg, S., *The First Three Minutes*. New York: Basic Books, 1977. (One of the best popular accounts of modern cosmology, lucidly written by a Nobel Laureate-physicist who has been at the forefront in the study of the basic forces of nature.)

Life in the Universe

Berendzen, R., ed., *Life Beyond Earth and the Mind of Man*. Washington, D.C.: NASA, 1973. (A symposium of lectures by famous scientists and philosophers on the subject of life in the universe.)

Bracewell, R. N., *The Galactic Club*. Stanford, Calif.: Stanford Alumni Assn., 1974. (Speculations and scientific analyses of the kind of civilizations that could exist in the Galaxy.)

Condon, E. U., ed., *Scientific Study of Unidentified Flying Objects*. New York: Bantam Books, 1969. (The famous report of the study commissioned by the U.S. Air Force to investigate UFOs.)

Drake, F., *Intelligent Life in Space*. New York: Macmillan, 1962. (A perhaps optimistic account of the possibilities of life in the Galaxy, by the American scientist who pioneered the scientific study of the subject.)

Klass, P. J., *UFOs Explained*. New York: Vintage Books, 1976. (A revealing exposé of the UFO phenomenon by the foremost skeptical investigator of the subject.)

Ponnamperuma, C., *The Origins of Life*. New York: Dutton, 1972. (An account by one of the principal investigators of the way living organisms may have arisen.)

Ponnamperuma, C., and Cameron, A. G. W., *Interstellar Communication: Scientific Perspectives*. Boston: Houghton Mifflin, 1974. (The scientific possibility of communicating with extraterrestrial civilizations.)

Sagan, C., *The Cosmic Connection*. New York: Anchor Press-Doubleday, 1973. (A series of essays by America's best-known scientist, which establish a perspective of man's place in the universe.)

Sheaffer, R., *The UFO Verdict*. Buffalo, New York: Prometheus Book, 1981. (A skeptical survey of the evidence for extraterrestrial spaceships by a well-known UFO investigator.)

Shklovskii, I. S., and Sagan, C., *Intelligent Life in the Universe*. New York: Dell, 1966. (The classic work that investigates extraterrestrial intelligence in a thoughtful and authoritative manner.)

Star Atlases, Sky Guides, and Sky Lore

Allen, R. H., *Star Names*. New York: Dover, 1963. (An exhaustive reference on star names and their origins.)

Menzel, D. H., *A Field Guide to the Stars and Planets*. Boston: Houghton Mifflin, 1964. (A famous guide for the amateur.)

Minnaert, M., *The Nature of Light and Color in the Open Air*. New York: Dover, 1954. (The definitive book on the origin of optical phenomena in the sky, including rainbows, halos, shadow bands, and hundreds of others.)

Norton, W. W., *Sky Atlas*. Cambridge: Sky Publishing Company, 1971. (One of the most popular sky atlases for knowledgeable amateurs.)

Olcott, W. T., *Olcott's Field Book of the Skies*. New York: Putnam, 1954. (One of the standard amateur guide books.)

Rükl, A., *The Amateur Astronomer*. London: Octopus Books, Ltd., 1979. (English version of original Czech). (An excellent handbook for the beginner, with easy-to-use star charts and charts and tables of lunar and planetary information through the year 2000.)

Popular Journals on Astronomy

Astronomy, published monthly by AstroMedia Corp., 411 Mason St., P. O. Box 92788, Milwaukee, Wisconsin 53202.

The Griffith Observer, published monthly by the Griffith Observatory, 2800 East Observatory Road, Los Angeles, California 90027.

Mercury, published bimonthly by the Astronomical Society of the Pacific, 1290 24th Ave., San Francisco, California 94122.

Sky and Telescope, published monthly by Sky Publishing Corporation, 49 Bay State Rd., Cambridge, Massachusetts 02238.

Popular articles on astronomy also appear in
Scientific American, published monthly by Scientific American, Inc., 415 Madison Ave., New York, New York 10017.

Science 82 (or 83, 84, etc., with the year), published by the American Association for the Advancement of Sciences, 1515 Massachusetts Ave., Washington D.C. 20005.

Career Information

Information about a career in astronomy is available from the Education Officer, American Astronomical Society, Sharp Laboratory, University of Delaware, Newark, Delaware 19711.

The Executive Officer of the American Astronomical Society can be reached at

American Astronomical Society
1717 Massachusetts Avenue, N.W.
Suite 603
Washington, D.C. 20036
Telephone: (202) 232-3077

Slides and Photographs

Catalogues of slides and photographs of astronomical objects can be obtained from

Kitt Peak National Observatory
Tucson, Arizona 85726
(602) 327-5511

Lick Observatory
University of California, Santa Cruz
Santa Cruz, California 95140
(408) 274-5062

Palomar Observatory, California Institute of Technology
Bookstore
1201 East California Street
Pasadena, California 91125

Yerkes Observatory
Williams Bay, Wisconsin 53191
(414) 245-5555

Also, commercial outlets that offer NASA photographs and slides for sale.

Appendix 2 GLOSSARY

abberration (of starlight). Apparent displacement in the direction of a star due to the earth's orbital motion.

ablation. Fragmentation and vaporization of a meteorite upon entering the atmosphere.

absolute magnitude. Apparent magnitude a star would have at a distance of 10 pc.

absolute zero. A temperature of $-273°C$ (or 0 K), where all molecular motion stops.

absorption spectrum. Dark lines superimposed on a continuous spectrum.

accelerate. To change velocity; either to speed up, slow down, or change direction.

acceleration of gravity. Numerical value of the acceleration produced by the gravitational attraction on an object at the surface of a planet or star.

accretion. Gradual accumulation of mass, as by a planet forming by the building up of colliding particles in the solar nebula.

achromatic. Free of chromatic aberration.

active galactic nucleus. A violent event in the nucleus of a Galaxy; for example, a Seyfert Galaxy or a quasar.

active sun. The sun during times of unusual solar activity—spots, flares, and associated phenomena.

Age of Aquarius. Period (about 2000 years) during which the vernal equinox, moving because of precession, passes through the constellation of Aquarius.

airglow. Fluorescence in the upper atmosphere.

albedo. The fraction of incident sunlight that a planet or minor planet reflects.

almanac. A book or table listing astronomical events.

alpha particle. The nucleus of a helium atom, consisting of two protons and two neutrons.

altitude. Angular distance above or below the horizon, measured along a vertical circle, to a celestial object.

amplitude. The range in variability, as in the light from a variable star.

angstrom (Å). A unit of length equal to 10^{-8} cm.

angular diameter. Angle subtended by the diameter of an object.

angular momentum. A measure of the momentum associated with motion about an axis or fixed point.

annular eclipse. An eclipse of the sun in which the moon is too distant to appear to cover the sun completely, so that a ring of sunlight shows around the moon.

anomalistic month. The period of revolution of the moon about the earth with respect to its line of apsides, or to the perigee point.

anomalistic year. The period of revolution of the earth about the sun with respect to its line of apsides, or to the perihelion point.

Antarctic Circle. Parallel of latitude $66\frac{1}{2}°$ S; at this latitude the noon altitude of the sun is $0°$ on the date of the summer solstice.

antimatter. Matter consisting of antiparticles: *antiprotons* (protons with negative rather than positive charge), *positrons* (positively-charged electrons), and *antineutrons*.

apastron. The point of closest approach of two stars, as in a binary star orbit.

aperture. The diameter of an opening, or of the primary lens or mirror of a telescope.

aphelion. Point in its orbit where a planet is farthest from the sun.

apogee. Point in its orbit where an earth satellite is farthest from the earth.

Apollo asteroid (minor planet). An asteroid (minor planet) whose orbit brings it closer to the sun than to the earth.

Apollo program. The American program to land men on the moon (1961–1972).

apparent magnitude. A measure of the observed light flux received from a star or other object at the earth.

apparent relative orbit. The projection onto a plane perpendicular to the line of sight of the relative orbit of the fainter of the two components of a visual binary star about the brighter.

apparent solar day. The interval between two successive transits of the sun's center across the meridian.

apparent solar time. The hour angle of the sun's center *plus* 12 hours.

apse (or apsis; pl. apsides). The point in a body's orbit where it is nearest or farthest from the object it revolves about.

Arctic Circle. Parallel of latitude $66\frac{1}{2}°$ N; at this latitude the noon altitude of the sun is $0°$ on the date of the winter solstice.

argument of perifocus (or of perihelion or perigee). Angle at the focus of the orbit of a body (or at the center of the sun or earth), measured in the orbital plane and in the direction of motion of the body, from the ascending node to the perifocus (or perihelion or perigee) point.

artificial satellite. A manmade object put into a closed orbit about the earth.

ascending node. The point along the orbit of a body where it crosses from the south to the north of some reference plane, usually the plane of the celestial equator or of the ecliptic.

aspect. The configuration of the sun, moon, or planets with respect to one another.

association. A loose cluster of stars whose spectral types, motions, or positions in the sky indicate that they have probably had a common origin.

asteroid. A synonym for "minor planet."

asthenosphere. The mantle beneath the lithosphere of the earth.

astigmatism. A defect in an optical system whereby pairs of light rays in different planes do not focus at the same place.

astrology. The pseudoscience that treats with supposed influences of the configurations and locations in the sky on the sun, moon, and planets on human destiny; a primitive religion having its origin in ancient Babylonia.

astrometric binary. A binary star in which one component is not observed, but its presence is deduced from the orbital motion of the visible component.

astrometry. That branch of astronomy that deals with the determination of precise positions and motions of celestial bodies.

astronautics. The science of the laws and methods of space flight.

astronomical unit (AU). Originally meant to be the semimajor axis of the orbit of the earth; now defined as the semimajor axis of the orbit of a hypothetical body with the mass and period that Gauss assumed for the earth. The semimajor axis of the orbit of the earth is 1.000 000 230 AU.

astronomy. The branch of science that treats of the physics and morphology of that part of the universe that lies beyond the earth's atmosphere.

astrophysics. The part of astronomy that deals principally with the physics of stars, stellar systems, and interstellar material. Astrophysics also deals, however, with the structures and atmospheres of the sun and planets.

atmospheric refraction. The bending, or refraction, of light rays from celestial objects by the earth's atmosphere.

atom. The smallest particle of an element that retains the properties that characterize that element.

atomic clock. A time-keeping device regulated by the natural frequency of the emission or absorption of radiation of a particular kind of atom.

atomic mass unit. *Chemical:* one sixteenth of the mean mass of an oxygen atom. *Physical:* one twelfth of the mass of an atom of the most common isotope of carbon. The atomic mass unit is approximately the mass of a hydrogen atom, 1.67×10^{-24} g.

atomic number. The number of protons in each atom of a particular element.

atomic time. The time kept by a cesium (atomic) clock, based on the **atomic second**—the time required for 9,192,631,770 cycles of the radiation emitted or absorbed in a particular transition of an atom of cesium-133.

atomic transition. A change in the state of energy of an atom; the atom may gain or lose energy by collision with another particle or by the emission or absorption of a photon.

atomic weight. The mean mass of an atom of a particular element in atomic mass units.

aurora. Light radiated by atoms and ions in the ionosphere, mostly in the magnetic polar regions.

autumnal equinox. The intersection of the ecliptic and celestial equator where the sun crosses the equator from north to south.

azimuth. The angle along the celestial horizon, measured eastward from the north point, to the intersection of the horizon with the vertical circle passing through an object.

Baily's beads. Small "beads" of sunlight seen passing through valleys along the limb of the moon in the instant preceding and following totality in a solar eclipse.

ballistic missile. A missile or rocket that is given its entire thrust during a brief period at the beginning of its flight, and that subsequently "coasts" to its target along an orbit.

Balmer lines. Emission or absorption lines in the spectrum of hydrogen that arise from transitions between the second (or first excited) and higher energy states of the hydrogen atoms.

bands (in spectra). Emission or absorption lines, usually in the spectra of chemical compounds or radicals, so numerous and closely spaced that they coalesce into broad emission or absorption bands.

barred spiral galaxy. Spiral galaxy in which the spiral arms begin from the ends of a "bar" running through the nucleus rather than from the nucleus itself.

barycenter. The center of mass of two mutually revolving bodies.

baryons (and antibaryons). The heavy atomic nuclear particles, such as protons and neutrons.

base line. That side of a triangle used in triangulation or surveying whose length is known (or can be measured), and which is included between two angles that are known (or can be measured).

Be star. A spectral type B star with emission lines in its spectrum, which are presumed to arise from material ejected from or surrounding the star.

"big bang" theory. A theory of cosmology in which the expansion of the universe is presumed to have begun with a primeval explosion.

billion. In the United States and France, one thousand million (10^9); in Great Britain and Germany, one million million (10^{12}).

binary star. A double star; two stars revolving about each other.

binding energy. The energy required to completely separate the constituent parts of an atomic nucleus.

biosphere. That part of the earth (its surface, atmosphere, and oceans) where life can exist.

black body. A hypothetical perfect radiator, which absorbs and reemits all radiation incident upon it.

black dwarf. A presumed final state of evolution for a star, in which all of its energy sources are exhausted and it no longer emits radiation.

black hole. A hypothetical body whose velocity of escape is equal to or greater than the speed of light; thus no radiation can escape from it.

blink microscope (or **comparator**). A microscope in which the user's view is shifted rapidly back and forth between the corresponding portions of two different photographs of the same region of the sky.

Bode's law. A scheme by which a sequence of numbers can be obtained that give the approximate distances of the planets from the sun in astronomical units.

Bohr atom. A particular model of an atom, invented by Niels Bohr, in which the electrons are described as revolving about the nucleus in circular orbits.

bolide. A very bright fireball or meteor; sometimes defined as a fireball accompanied by sound.

bolometric correction. The difference between the visual (or photovisual) and bolometric magnitudes of a star.

bolometric magnitude. A measure of the flux of radiation from a star or other object received just outside the earth's atmosphere, as it would be detected by a device sensitive to *all* forms of electromagnetic energy.

bremsstrahlung. Radiation from free-free transitions in which electrons gain or lose energy while being accelerated in the field of an atomic nucleus or ion.

bubble chamber. A chamber in which bubbles from along the electrically charged path of a high-energy charged particle, rendering the track of that particle visible.

burnout. The instant when a rocket stops firing.

burster. A source of sudden bursts of X rays, believed to be a neutron star accreting mass from a companion star, and suddenly igniting that material in nuclear explosions.

calculus. A branch of mathematics that permits computations involving rates of change (*differential* calculus) or of the contribution of an infinite number of infinitesimal quantities (*integral* calculus).

carbon cycle. A series of nuclear reactions involving carbon as a catalyst, by which hydrogen is transformed to helium.

cardinal points. The four principal points of the compass: North, East, South, and West.

Cassegrain focus. An optical arrangement in a reflecting telescope in which light is reflected by a second mirror to a point behind the objective mirror.

CBR. See *cosmic background radiation*.

cD galaxy. A supergiant elliptical galaxy frequently found at the center of a cluster of galaxies.

celestial equator. A great circle on the celestial sphere 90° from the celestial poles; the circle of intersection of the celestial sphere with the plane of the earth's equator.

celestial mechanics. That branch of astronomy which deals with the motions and gravitational influences of the members of the solar system.

celestial navigation. The art of navigation at sea or in the air from sightings of the sun, moon, planets, and stars.

celestial poles. Points about which the celestial sphere appears to rotate, intersections of the celestial sphere with the earth's polar axis.

celestial sphere. Apparent sphere of the sky; a sphere of large radius centered on the observer. Directions of objects in the sky can be denoted by the position of those objects on the celestial sphere.

center of gravity. Center of mass.

center of mass. The mean position of the various mass elements of a body or system, weighted according to their distances from that center of mass; that point in an isolated system which moves with constant velocity, according to Newton's first law of motion.

centrifugal force (or **acceleration**). An imaginary force (or acceleration) that is often introduced to account for the illusion that a body moving on a curved path tends to accelerate radially from the center of curvature. The force present is the one that diverts the body's motion from a straight line and is directed *toward* the center of curvature. It is, however, legitimate to introduce a fictitious centrifugal force field in a rotating (and hence noninertial) coordinate system.

centripetal force (or **acceleration**). The force required to divert a body from a straight path into a curved path (or the acceleration experienced by the body); it is directed toward the center of curvature.

cepheid variable. A star that belongs to one of two classes (type I and type II) of yellow supergiant pulsating stars.

Ceres. Largest of the minor plants and first to be discovered.

cesium clock. An atomic clock that utilizes a transition in the atomic isotope cesium-133.

charged-coupled device (CCD). An array of electronic detectors of electromagnetic radiation, used at the focus of a telescope (or camera lens). A CCD acts like a photographic plate of very high sensitivity.

chondrite. A stony meteorite that contains small spherical particles called *chondrules*.

chromatic aberration. A defect of optical systems whereby light of different colors is focused at different places.

chromosphere. That part of the solar atmosphere that lies immediately above the photospheric layers.

chronograph. A device for recording and measuring the times of events.

chronometer. An accurate clock.

circle of position. A small circle on the surface of the earth, centered at the substellar point of some star, and of radius equal to the coaltitude of that star as seen by some observer; the observer must, therefore, be somewhere on this circle of position.

circular velocity. The critical speed with which a revolving body can have a circular orbit.

circumpolar regions. Portions of the celestial sphere near the celestial poles that are either always above or always below the horizon.

circumstellar dust. Solid particles surrounding a star, which either condensed from material ejected from the star or are part of the interstellar material from which the star is forming.

cloud chamber. A chamber in which droplets of liquid condense along the electrically charged path of a high-energy charged particle, rendering the track of that particle visible.

Clouds of Magellan. Two neighboring galaxies visible to the naked eye from southern latitudes.

cluster of galaxies. A system of galaxies containing from several to thousands of member galaxies.

cluster variable (RR Lyrae variable). A member of a certain large class of pulsating variable stars, all with periods less than one day. These stars are often present in globular star clusters.

coherent radiation. Electromagnetic radiation of a particular wavelength traveling in the same direction and in which the waves are all in phase with each other.

cold cloud (cold molecular cloud). Region of the interstellar medium where molecules, excited by collisions, radiate microwave and infrared radiation away, thereby cooling the region. Subsequent collapse of a cold cloud is believed to be a first step toward star formation.

color excess. The amount by which the color index of a star is increased when its light is reddened in passing through interstellar absorbing material.

color index. Difference between the magnitudes of a star or other object measured in light of two different spectral regions, for example, photographic *minus* photovisual magnitudes.

color-magnitude diagram. Plot of the magnitudes (apparent or absolute) of the stars in a cluster against their color indices.

coma. A defect in an optical system in which off-axis rays of light striking different parts of the objective do not focus in the same place.

coma (of comet). The diffuse gaseous component of the head of a comet.

comet. A small body of icy and dusty matter, which revolves about the sun. When a comet comes near the sun, some of its material vaporizes, forming a large *coma* of tenuous gas, and often a *tail*.

commensurability. State of a quantity (such as the period of a minor planet) that can be reduced to another quantity (such as the period of another minor planet) by multiplication by a ratio of two small whole numbers.

compact galaxy. A galaxy of small size and high surface brightness.

comparison spectrum. The spectrum of a vaporized element (such as iron) photographed beside the image of a stellar spectrum, and with the same camera, for purposes of comparison of wavelengths.

compound. A substance composed of two or more chemical elements.

compound nucleus. An excited nucleus, usually temporary, formed by nuclei of two or more simpler atoms.

conduction. The transfer of energy by the direct passing of energy or electrons from atom to atom.

configuration. Any one of several particular orientations in the sky of the moon or a planet with respect to the sun.

conic section. The curve of intersection between a circular cone and a plane; these curves can be ellipses, circles, parabolas, or hyperbolas.

conjunction. The configuration of a planet when it has the same celestial longitude as the sun, or the configuration when any two celestial bodies have the same celestial longitude or right ascension.

conservation of angular momentum. The law that angular momentum is conserved in the absence of any force not directed toward or away from the point or axis about which the angular momentum is referred—that is, in the absence of a torque.

constellation. A configuration of stars named for a particular object, person, or animal; or the area of the sky assigned to a particular configuration.

contacts (of eclipses). The instants when certain stages of an eclipse begin.

continental drift. A gradual drift of the continents over the surface of the earth due to *plate tectonics*.

continuous spectrum. A spectrum of light comprised of radiation of a continuous range of wavelengths or colors rather than only certain discrete wavelengths.

convection. The transfer of energy by moving currents of a fluid containing that energy.

Coordinated Universal Time. Greenwich Mean Time standardized and regulated by an international agency, the *Bureau Internationale de l'Huere*, on the basis of astronomical observations reported from around the world.

Copernicus satellite. An artificial satellite with instrumentation especially designed for ultraviolet observations in space.

core (of earth). The central part of the earth, believed to be a liquid of high density.

coriolis effect. The deflection (with respect to the ground) of projectiles moving across the surface of the rotating earth.

corona. Outer atmosphere of the sun.

corona of Galaxy. The extension of the nuclear bulge of the Galaxy on either side of the plane of the Milky Way; a region containing hot gas that emits X rays.

coronagraph. An instrument for photographing the chromosphere and corona of the sun outside of eclipse.

coronal hole. A region in the sun's outer atmosphere where visible coronal streamers are absent.

corpuscular radiation. Charged particles, mostly atomic nuclei and electrons, emitted into space by the sun and possibly other objects.

cosine. One of the trigonometric functions of an angle. In a right triangle, the ratio of the lengths of the shorter side adjacent to the angle and the hypotenuse.

cosmic background radiation (CBR). The microwave radiation coming from all directions that is believed to be the redshifted glow of the big bang.

cosmic rays. Atomic nuclei (mostly protons) that are observed to strike the earth's atmosphere with exceedingly high energies.

cosmogony. The study of the origin of the world or universe.

cosmological constant. A term that arises in the development of the field equations of general relativity, which represents a repulsive force in the universe. The cosmological constant is often assumed to be zero.

cosmological model. A specific model, or theory, of the organization and evolution of the universe.

cosmological principle. The assumption that, on the large scale, the universe at any given time is the same everywhere.

cosmology. The study of the organization and evolution of the universe.

coudé focus. An optical arrangement in a reflecting telescope whereby light is reflected by two or more secondary mirrors down the polar axis of the telescope to a focus at a place separate from the moving parts of the telescope.

covariance function. A measure of the probability of finding a celestial object within a given angular separation from another object of the same or another given type.

Crab nebula. The expanding mass of gas that is the remnant of the supernova of 1054.

crater (lunar). A more or less circular depression in the surface of the moon.

crater (meteoritic). A crater on the earth caused by the collision of a meteoroid with the earth, and a subsequent explosion.

crescent moon. One of the phases of the moon when its elongation is less than 90° from the sun and it appears less than half full.

crust (of earth). The outer layer of the earth.

cyclonic motion. A counterclockwise circular circulation of winds (in the northern hemisphere) that results from the coriolis effect.

Cyclops project. A proposed system of radio telescopes for the primary purpose of detecting radio signals from extraterrestrial civilizations.

dark nebula. A cloud of interstellar dust that obscures the light of more distant stars and appears as an opaque curtain.

daylight saving time. A time one hour more advanced than standard time, usually adopted in spring and summer to take advantage of long evening twilights.

deceleration parameter (q_o). A quantity that characterizes the future evolution of the various models of the universe based on general relativity.

declination. Angular distance north or south of the celestial equator to some object, measured along an hour circle passing through that object.

deferent. A stationary circle in the Ptolemaic system along which moves the center of another circle (epicycle), along which moves an object or another epicycle.

degenerate gas. A gas in which the allowable states for the electrons have been filled; it behaves according to different laws from those that apply to "perfect" gases.

density. The ratio of the mass of an object to its volume.

density-wave theory. A theory proposed by C. C. Lin and his associates for the spiral structure of the Galaxy.

descending node. The point along the orbit of a body where it crosses from the north to the south of some reference plane, usually the plane of the celestial equator or of the ecliptic.

deuterium. A "heavy" form of hydrogen, in which the nucleus of each atom consists of one proton and one neutron.

diamond ring. A flash of sunlight at the instants before and after totality in a solar eclipse while the corona is visible as a complete ring of light around the moon.

differential equation. An equation that involves derivatives or rates of change.

differential galactic rotation. The rotation of the galaxy, not as a solid wheel, but so that parts adjacent to each other do not always stay close together.

differential gravitational force. The difference between the respective gravitational forces exerted on two bodies near each other by a third, more distant body.

differentiation (geological). A separation or segregation of different kinds of material in different layers in the interior of a planet.

diffraction. The spreading out of light in passing the edge of an opaque body.

diffraction grating. A system of closely spaced equidistant slits or reflecting strips which, by diffraction and interference, produce a spectrum.

diffraction pattern. A pattern of bright and dark fringes produced by the interference of light rays, diffracted by different amounts, with each other.

diffuse nebula. A reflection or emission nebula produced by interstellar matter (not a planetary nebula).

digicon. A device for electronically detecting and recording light from faint objects observed with a telescope.

disk (of planet or other object). The apparent circular shape that a planet (or the sun, or moon, or a star) displays when seen in the sky or viewed telescopically.

disk of Galaxy. The central disk or "wheel" of our Galaxy, superimposed on the spiral structure.

dispersion. Separation, from white light, of different wavelengths being refracted by different amounts.

distance modulus. Difference between the apparent and absolute magnitudes of an object, which provides a measure of its distance through the inverse-square law of light.

diurnal. Daily.

diurnal circle. Apparent path of a star in the sky during a complete day due to the earth's rotation.

diurnal libration. The phenomenon whereby slightly different hemispheres of the moon can be observed during a day because of the motion of the observer caused by the earth's rotation.

diurnal motion. Motion during one day.

diurnal parallax. Apparent change in direction of an object caused by a displacement of the observer due to the earth's rotation.

Doppler shift. Apparent change in wavelength of the radiation from a source due to its relative motion in the line of sight.

draconic month. The period of revolution of the moon about the earth with respect to the nodes of the moon's orbit.

dwarf (star). A main-sequence star (as opposed to a giant or supergiant).

dynamical parallax. A distance (or parallax) for a binary star derived from the period of mutual revolution, the mass-luminosity relation, and the laws of mechanics.

dyne. The metric unit of force; the force required to accelerate a mass of 1 gram in the amount 1 centimeter per second per second.

east point. The point on the horizon 90° from the north point (measured clockwise as seen from the zenith).

eccentric. The off-center position of the earth in the presumed circular orbits of the sun, moon, and planets in the Ptolemaic system.

eccentricity (of ellipse). Ratio of the distance between the foci to the major axis.

eclipse. The cutting off of all or part of the light of one body by another passing in front of it.

eclipse path. The track along the earth's surface swept out by the tip of the shadow of the moon (or the extension of its shadow) during a total (or annular) solar eclipse.

eclipse season. A period during the year when an eclipse of the sun or moon is possible.

eclipsing binary star. A binary star in which the plane of revolution of the two stars is nearly edge on to our line of sight, so that the light of one star is periodically diminished by the other passing in front of it.

ecliptic. The apparent annual path of the sun on the celestial sphere.

ecliptic limit. The maximum angular distance from a node where the moon can be for an eclipse to take place.

Einstein telescope. The orbiting X-ray telescope (HEAO 2) launched in late 1978.

electromagnetic force. One of the four fundamental forces of nature; the force that acts between charges and binds atoms and molecules together.

electomagnetic radiation. Radiation consisting of waves propagated through the building up and breaking down of electric and magnetic fields; these include radio, infrared, light, ultraviolet, X rays, and gamma rays.

electromagnetic spectrum. The whole array or family of electromagnetic waves.

electron. A negatively charged subatomic particle that normally moves about the nucleus of an atom.

electron volt. The kinetic energy acquired by an electron that is accelerated through an electric potential of 1 volt; 1 electron volt is 1.60207×10^{-12} erg.

element. A substance that cannot be decomposed, by chemical means, into simpler substances.

elements (of orbit). Any of several quantities that describe the size, shape, and orientation of the orbit of a body.

ellipse. A conic section: the curve of intersection of a circular cone and a plane cutting completely through the cone.

elliptical galaxy. A galaxy whose apparent photometric contours are ellipses, and which contains no conspicuous interstellar material.

ellipticity. The ratio (in an ellipse) of the major axis *minus* the minor axis to the major axis.

elongation. The difference between the celestial longitudes of a planet and the sun.

emission line. A discrete bright spectral line.

emission nebula. A gaseous nebula that derives its visible light from the fluorescence of ultraviolet light from a star in or near the nebula.

emission spectrum. A spectrum consisting of emission lines.

encounter (gravitational). A near passing, on hyperbolic orbits, of two objects that influence each other gravitationally.

energy. The ability to do work.

energy equation. See *vis viva* equation.

energy level (in an atom or ion). A particular level, or amount, of energy possessed by an atom or ion above the energy it possesses in its least energetic state.

energy spectrum. A table or plot showing the relative numbers of particles (in cosmic rays or corpuscular radiation) of various energies.

ephemeris A table that gives the positions of a celestial body at various times, or other astronomical data.

ephemeris time. A kind of time that passes at a strictly uniform rate; used to compute the instants of various astronomical events.

epicycle. A circular orbit of a body in the Ptolemaic system, the center of which revolves about another circle (the deferent).

equant. A stationary point in the Ptolemaic system not at the center of a circular orbit about which a body (or the center of an epicycle) revolves with uniform angular velocity.

equation of state. An equation relating the pressure, temperature, and density of a substance (usually a gas).

equation of time. The difference between apparent and mean solar time.

equator. A great circle on the earth, 90° from its poles.

equatorial mount. A mounting for a telescope, one axis of which is parallel to the earth's axis, so that a motion of the telescope about the axis can compensate for the earth's rotation.

equinox. One of the intersections of the ecliptic and celestial equator.

equivalent width. A measure of the strength of a spectral line; the width of an absorption line of rectangular profile and zero intensity at its center.

erg. The metric unit of energy; the work done by a force of one dyne acting through a distance of one centimeter.

eruptive variable. A variable star whose changes in light are erratic or explosive.

establishment of port. The time interval, at a particular port, between the meridian passage of the moon and high tide.

ether. A hypothetical medium once supposed to exist in space and to transport electromagnetic radiation.

Euclidean. Pertaining to Euclidean geometry, or *flat space*.

event. A point in four-dimensional spacetime.

event horizon. The surface through which a collapsing star is hypothesized to pass when its velocity of es-

cape is equal to the speed of light, that is, when the star becomes a black hole.

evolutionary cosmology. A theory of cosmology that assumes that all parts of the universe have a common age and evolve together.

excitation. The process of imparting to an atom or an ion an amount of energy greater than that it has in its normal or least-energy state.

exclusion principle. See *Pauli exclusion principle*.

extinction. Attenuation of light from a celestial body produced by the earth's atmosphere, or by interstellar absorption.

extragalactic. Beyond the Galaxy.

eyepiece. A magnifying lens used to view the image produced by the objective of a telescope.

faculus (pl. faculae). Bright region near the limb of the sun.

Faraday rotation. The rotation of the plane of polarization of polarized radiation passing through ionized gas.

fermions. Fundamental particles, such as protons, electrons, and neutrons, all of spin $\frac{1}{2}$, that obey certain laws formulated by Enrico Fermi.

field equations. A set of equations in general relativity that describe the curvature of spacetime in the presence of matter.

filtergram. A photograph of the sun (or part of it) taken through a special narrow-bandpass filter.

fireball. A spectacular meteor.

First Point of Aries. The vernal equinox.

fission. The breakup of a heavy atomic nucleus into two or more lighter ones.

flare. A sudden and temporary outburst of light from an extended region of the solar surface.

flare star. A member of a class of stars that show occasional, sudden, unpredicted increases in light.

flash spectrum. The spectrum of the very limb of the sun obtained in the instant before or after totality in a solar eclipse.

flocculus (pl. flocculi). A bright region of the solar surface observed in the monochromatic light of some spectral line; flocculi are now usually called *plages*.

fluorescence. The absorption of light of one wavelength and reemission of it at another wavelength; especially the conversion of ultraviolet into visible light.

flux. The rate at which energy or matter crosses a unit area of a surface.

fluxions. Name given by Newton to the calculus.

focal length. The distance from a lens or mirror to the point where light converged by it comes to a focus.

focal ratio (speed). Ratio of the focal length of a lens or mirror to its aperture.

focus. Point where the rays of light converged by a mirror or lens meet.

focus of a conic section. Mathematical point associated with a conic section, whose distance to any point on the conic bears a constant ratio to the distance from that point to a straight line known as the *directrix*.

forbidden lines. Spectral lines that are not usually observed under laboratory conditions because they result from atomic transitions that are highly improbable.

force. That which can change the momentum of a body; numerically, the rate at which the body's momentum changes.

Foucault pendulum. A pendulum that seems to change direction of oscillation as the earth turns. The experiment was first conducted in 1851 by Jean Foucault to demonstrate the rotation of the earth.

Fraunhofer line. An absorption line in the spectrum of the sun or of a star.

Fraunhofer spectrum. The array of absorption lines in the spectrum of the sun or of a star.

free-free transition. An atomic transition in which the energy associated with an atom or ion and passing electron changes during the encounter, but without capture of the electron by the atom or ion.

frequency. Number of vibrations per unit time; number of waves that cross a given point per unit time (in radiation).

fringes (interference). Successive dark and light lines, caused by interference of light waves with each other before they strike a screen or detecting device and are observed.

full moon. That phase of the moon when it is at opposition (180° from the sun) and its full daylight hemisphere is visible from the earth.

fusion. The building up of heavier atomic nuclei from lighter ones.

galactic cluster. An "open" cluster of stars located in the spiral arms or disk of the Galaxy.

galactic equator. Intersection of the principal plane of the Milky Way with the celestial sphere.

galactic latitude. Angular distance north or south of the galactic equator to an object, measured along a great circle passing through that object and the galactic poles.

galactic longitude. Angular distance, measured eastward along the galactic equator from the galactic center, to the intersection of the galactic equator with a great circle passing through the galactic poles and an object.

galactic poles. The poles of the galactic equator; the intersections with the celestial sphere of a line through the observer that is perpendicular to the plane of the galactic equator.

galactic rotation. Rotation of the Galaxy.

galaxy. A large assemblage of stars; a typical galaxy contains millions to hundreds of thousands of millions of stars.

Galaxy. The galaxy to which the sun and our neighbor-

ing stars belong; the Milky Way is light from remote stars in the Galaxy.

Galilean satellite. Any of the four largest of Jupiter's satellites, discovered by Galileo.

gamma rays. Photons (of electromagnetic radiation) of energy higher than those of X rays; the most energetic form of electromagnetic radiation.

gauss. A unit of magnetic flux density.

gegenschein (counterglow). A very faint, diffuse glow of light opposite the sun in the sky, believed to be caused by sunlight reflected from interplanetary particles.

Geiger counter. A device for counting high-energy charged particles and hence for measuring the intensity of corpuscular radiation.

geodesic. The path of a body in spacetime.

geodesic equations. A set of equations in general relativity by which the paths of objects in spacetime can be calculated.

geomagnetic. Referring to the earth's magnetic field.

geomagnetic poles. The poles of a hypothetical bar magnet whose magnetic field most nearly matches that of the earth.

giant (star). A star of large luminosity and radius.

gibbous moon. One of the phases of the moon in which more than half, but not all, of the moon's daylight hemisphere is visible from the earth.

Gliese catalogue. A catalogue of nearby stars compiled by the astronomer, W. Gliese.

globular cluster. One of about 120 large star clusters that form a system of clusters centered on the center of the Galaxy.

globule. A small, dense, dark nebula; believed to be a possible site of star formation.

gluon. The hypothetical particle that mediates the strong nuclear force between quarks.

granulation. The "rice-grain" like structure of the solar photosphere.

gravitation. The tendency of matter to attract itself.

gravitational constant, G. The constant of proportionality in Newton's law of gravitation; in metric units G has the value 6.672×10^{-8} dyne $\cdot$ cm^2/gm^2.

gravitational energy. Energy that can be released by the gravitational collapse, or partial collapse, of a system.

gravitational lens. A configuration of celestial objects, one of which provides one or more images of the other by gravitationally deflecting its light.

gravitational redshift. The redshift caused by a gravitational field. The slowing of clocks in a gravitational field.

gravitational waves. Oscillations in spacetime, propagated by changes in the distribution of matter.

great circle. Circle on the surface of a sphere that is the curve of intersection of the sphere with a plane passing through its center.

greatest elongation (east or west). The largest separation in celestial longitude (to the east or west) that an inferior planet can have from the sun.

greenhouse effect. The blanketing of infrared radiation near the surface of a planet by, for example, carbon dioxide in its atmosphere.

Greenwich meridian. The meridian of longitude passing through the site of the old Royal Greenwich Observatory, near London; origin of longitude on the earth.

Gregorian calendar. A calendar (now in common use) introduced by Pope Gregory XIII in 1582.

H. See *Hubble constant*.

H I region. Region of neutral hydrogen in interstellar space.

H II region. Region of ionized hydrogen in interstellar space.

hadron. A subnuclear particle; one of hundreds now known to exist, of mass from somewhat less to considerably more than that of the proton.

half-life. The time required for half of the radioactive atoms in a sample to disintegrate.

halo (around sun or moon). A ring of light around the sun or moon caused by refraction by the ice crystals of cirrus clouds.

halo (of galaxy). The outermost extent of our Galaxy or another, containing a sparse distribution of stars and globular clusters in a more or less spherical distribution.

harmonic law. Kepler's third law of planetary motion: the cubes of the semimajor axes of the planetary orbits are in proportion to the squares of the sidereal periods of the planets' revolutions about the sun.

harvest moon. The full moon nearest the time of the autumnal equinox.

Hayashi line. Track of evolution on the Hertzsprung-Russell diagram of a completely convective star.

head (of comet). The main part of a comet, consisting of its nucleus and coma.

HEAO. High Energy Astronomy Observatory; one of a series of artificial satellites carrying X-ray or gamma-ray detectors. HEAO 2 is the Einstein telescope.

"heavy" elements. In astronomy, usually those elements of greater atomic number than helium.

Heisenberg uncertainty principle. A principle of quantum mechanics that places a limit on the precision with which the simultaneous position and momentum of a body or particle can be specified.

heliacal rising. The rising of a star or planet the first time that it can be observed in the morning before sunrise.

helio-. Prefix referring to the sun.

heliocentric. Centered on the sun.

helium flash. The nearly explosive ignition of helium in the triple-alpha process in the dense core of a red giant star.

Helmholtz-Kelvin contraction. The gradual gravitational contraction of a cloud or a star, with the release of gravitational potential energy.

Hertz. A unit of frequency: one cycle per second. Named for Heinrich Hertz, who first produced radio radiation.

Hertzsprung gap. A V-shaped gap in the upper part of the Hertzsprung-Russell diagram where few stable stars are found.

Hertzsprung-Russell (H-R) diagram. A plot of absolute magnitude against temperature (or spectral class or color index) for a group of stars.

high-velocity star (or object). A star (or object) with high space motion; generally an object that does not share the high orbital velocity of the sun about the galactic nucleus.

homogeneous star (or stellar model). A star (or theoretical model of a star) whose chemical composition is the same throughout its interior.

horizon (astronomical). A great circle on the celestial sphere 90° from the zenith.

horizon system. A system of celestial coordinates (altitude and azimuth) based on the astronomical horizon and the north point.

horizontal branch. A sequence of stars on the Hertzsprung-Russell diagram of a typical globular cluster of approximately constant absolute magnitude (near $M_v = 0$).

horizontal parallax. The angle by which an object appears displaced (after correction for atmospheric refraction) when viewed on the horizon from a place on the earth's equator, compared with its direction if it were viewed from the center of the earth.

horoscope. A chart showing the positions along the zodiac and in the sky of the sun, moon, and planets at some given instant and as seen from a particular place on earth—usually corresponding to the time and place of a person's birth.

hour angle. The angle measured westward along the celestial equator from the local meridian to the hour circle passing through an object.

hour circle. A great circle on the celestial sphere passing through the celestial poles.

house. A division or segment of the sky numbered according to its position with respect to the horizon and used by astrology in preparing a horoscope.

Hubble constant. Constant of proportionality between the velocities of remote galaxies and their distances. The Hubble constant is thought to lie in the range of 50 to 100 km/s per million parsecs.

Hubble law. The law of the redshifts.

hydrostatic equilibrium. A balance between the weights of various layers, as in a star or the earth's atmosphere, and the pressures that support them.

hyperbola. A conic section of eccentricity greater than 1.0; the curve of intersection between a circular cone and a plane that is at too small an angle with the axis of the cone to cut all of the way through it, and is not parallel to a line in the face of the cone.

hyperfine transition. A change in the energy state of an atom that involves a change in the spin of its nucleus.

hypothesis. A tentative theory or supposition, advanced to explain certain facts or phenomena, which is subject to further tests and verification.

image. The optical representation of an object produced by light rays from the object being refracted or reflected, as by a lens or mirror.

image tube. A device in which electrons, emitted from a photocathode surface exposed to light, are focused electronically.

inclination (of an orbit). The angle between the orbital plane of a revolving body and some fundamental plane—usually the plane of the celestial equator or of the ecliptic.

Index Catalogue, IC. The supplement to Dreyer's *New General Catalogue* of star clusters and nebulae.

index of refraction. A measure of the refracting power of a transparent substance; specifically, the ratio of the speed of light in a vacuum to its speed in the substance.

inertia. The property of matter that requires a force to act on it to change its state of motion; momentum is a measure of inertia.

inertial system. A system of coordinates that is not itself accelerated, but that is either at rest or is moving with constant velocity.

inferior conjunction. The configuration of an inferior planet when it has the same longitude as the sun, and is between the sun and earth.

inferior planet. A planet whose distance from the sun is less than the earth's.

infrared radiation. Electromagnetic radiation of wavelength longer than the longest (red) wavelengths that can be perceived by the eye, but shorter than radio wavelengths.

insolation. The rate at which all radiation from the sun is received per unit area on the ground.

intercalate. To insert, as a day in a calendar.

interference. A phenomenon of waves that mix together such that their crests and troughs can alternately reinforce and cancel each other.

interferometer (stellar). An optical device, making use of the principle of interference of light waves, with which small angles can be measured.

International Date Line. An arbitrary line on the surface of the earth near longitude 180° across which the date changes by one day.

interplanetary medium. The sparse distribution of gas and solid particles in the interplanetary space.

interstellar dust. Microscopic solid grains, believed to be mostly dielectric compounds of hydrogen and other common elements in interstellar space.

interstellar gas. Sparse gas in interstellar space.

interstellar lines. Absorption lines superimposed on stellar spectra, produced by the interstellar gas.

interstellar matter. Interstellar gas and dust.

ion. An atom that has become electrically charged by the addition or loss of one or more electrons.

ionization. The process by which an atom gains or loses electrons.

ionization potential. The energy required to remove an electron from an atom.

ionosphere. The upper region of the earth's atmosphere in which many of the atoms are ionized.

ion tail (of comet). The relatively straight tail of a comet produced by the interaction of solar wind with the ions in the comet.

irregular galaxy. A galaxy without rotational symmetry; neither a spiral nor elliptical galaxy.

irregular variable. A variable star whose light variations do not repeat with a regular period.

island universe. Historical synonym for galaxy.

isotope. Any of two or more forms of the same element, whose atoms all have the same number of protons but different numbers of neutrons.

isotropic. The same in all directions.

iteration. The "closing in" on the solution to a mathematical problem by repetitive calculations.

IUE (International Ultraviolet Explorer). An earth satellite dedicated to ultraviolet observations of celestial objects.

Jovian planet. Any of the planets Jupiter, Saturn, Uranus, and Neptune.

Julian calendar. A calendar introduced by Julius Caesar in 45 B.C.

Julian day. The number of the day in a running sequence beginning January 1, 4713 B.C.

Jupiter. The fifth planet from the sun in the solar system.

Kepler's laws. Three laws, discovered by J. Kepler, that describe the motions of the planets.

kiloparsec (kpc). 1000 parsecs, or about 3260 LY.

kinetic energy. Energy associated with motion; the kinetic energy of a body is one half the product of its mass and the square of its velocity.

kinetic theory (of gases). The science that treats the motions of the molecules that compose gases.

Kirkwood's gaps. Gaps in the spacing of the minor planets that arise from pertubations produced by the major planets.

Lagrangian points. Five points in the plane of revolution of two bodies, revolving mutually about each other in circular orbits, where a third body of negligible mass can remain in equilibrium with respect to the other two bodies.

Lagrangian surface. The surface in space between two objects revolving about each other on which a small particle is attracted equally to each object.

laser. An acronym for *light amplification by stimulated emission of radiation;* a device for amplifying a light signal at a particular wavelength into a coherent beam.

latitude. A north-south coordinate on the surface of the earth; the angular distance north or south of the equator measured along a meridian passing through a place.

launch window. A range of dates during which a space vehicle can be launched for a specific mission without exceeding the fuel capabilities of that system.

law. A statement of order or relation between phenomena that, under given conditions, is presumed to be invariable.

law of areas. Kepler's second law: the radius vector from the sun to any planet sweeps out equal areas in the planet's orbital plane in equal intervals of time.

law of the redshifts. The relation between the radial velocity and distance of a remote galaxy: the radial velocities are proportional to the distances of galaxies.

lead sulfide cell. A device used to measure infrared radiation.

leap year. A calendar year with 366 days, intercalated approximately every four years to make the average length of the calendar year as nearly equal as possible to the tropical year.

lepton. A subatomic particle of small mass, such as an electron or positron.

libration. Any of several phenomena by which an observer on earth, over a period of time, can see more than one hemisphere of the moon.

libration in latitude. Libration caused by the fact that the moon's axis of rotation is not perpendicular to its plane revolution.

libration in longitude. Libration caused by the regularity in the moon's rotation but irregularity in its orbital speed.

light. Electromagnetic radiation that is visible to the eye.

light curve. A graph that displays the time variation in light or magnitude of a variable or eclipsing binary star.

light year. The distance light travels in a vacuum in one year; 1 LY = 9.46×10^{17} cm, or about 6×10^{12} mi.

limb (of sun or moon). Apparent edge of the sun or moon as seen in the sky.

limb darkening. The phenomenon whereby the sun is less bright near its limb than near the center of its disk.

limiting magnitude. The faintest magnitude that can be observed with a given instrument or under given conditions.

line broadening. The phenomenon by which spectral lines are not precisely sharp but have finite widths.

line of apsides. The line connecting the apsides of an orbit (the perifocus and farthest from focus points); or the line along the major axis of the orbit.

line of nodes. The line connecting the nodes of an orbit.

line of position. A part of a circle of position, of so small an arc that it can be considered a straight line.

line profile. A plot of the intensity of light versus wavelength across a spectral line.

linear diameter. Actual diameter in units of length.

lithosphere. The upper layer of the earth, to a depth of 50 to 100 km, involved in plate tectonics.

Local Group. The cluster of galaxies to which our Galaxy belongs.

local oscillator. The old (classical) idea of an atom absorbing or emitting radiation by setting itself in oscillation or by reducing that oscillation. The local oscillator has been replaced by a different model in the modern quantum theory.

local standard of rest. A coordinate system that shares the average motion of the sun and its neighboring stars about the galactic center.

Local Supercluster. The supercluster of galaxies to which the Local Group belongs.

longitude. An east-west coordinate on the earth's surface; the angular distance, measured east or west along the equator from the Greenwich meridian, to the meridian passing through a place.

longitude of the ascending node. The angle measured eastward from a reference direction (usually the vernal equinox) in a fundamental plane (usually the plane of the celestial equator or of the ecliptic) to the ascending node of the orbit of a body.

low-velocity star (or object). A star (or object) that has low space velocity; generally an object that shares the sun's high orbital speed about the galactic center.

luminosity. The rate of radiation of electromagnetic energy into space by a star or other object.

luminosity class. A classification of a star according to its luminosity for a given spectral class.

luminosity function. The relative numbers of stars (or other objects) of various luminosities or absolute magnitudes.

luminous energy. Light.

lunar. Referring to the moon.

lunar eclipse. An eclipse of the moon.

Lyman lines. A series of absorption or emission lines in the spectrum of hydrogen that arise from transitions to and from the lowest energy states of the hydrogen atoms.

Magellanic Clouds. See Clouds of Magellan.

magnetic field. The region of space near a magnetized body within which magnetic forces can be detected.

magnetic pole. One of two points on a magnet (or the earth) at which the greatest density of lines of force emerge. A compass needle aligns itself along the local lines of force on the earth and points more or less toward the magnetic poles of the earth.

magnetometer. A device for measuring the strength of magnetic fields.

magnetosphere. The region around the earth or a planet occupied by its magnetic field.

magnifying power. The number of times larger (in angular diameter) an object appears through a telescope than with the naked eye.

magnitude. A measure of the amount of light flux received from a star or other luminous object.

main sequence. A sequence of stars on the Hertzsprung-Russell diagram, containing the majority of stars, that runs diagonally from the upper left to the lower right.

major axis (of ellipse). The maximum diameter of an ellipse.

major planet. A Jovian planet.

mantle (of earth). The greatest part of the earth's interior, lying between the crust and the core.

mare. Latin for "sea"; name applied to many of the "sealike" features on the moon or Mars.

Mariner space probes. A series of space probes launched in the 1960s and 1970s to explore the planets Mercury, Venus, and Mars.

Mars. Fourth planet from the sun in the solar system.

mascons. Mass concentrations beneath the surface of the moon; possibly submerged asteroids that impinged on the moon.

maser. An acronym for *microwave amplification of stimulated emission radiation;* a device for amplifying a microwave (radio) signal at a particular wavelength into a coherent beam.

mass. A measure of the total amount of material in a body; defined either by the inertial properties of the body or by its gravitational influence on other bodies.

mass defect. The amount by which the mass of an atomic nucleus is less than the sum of the masses of the individual nucleons that compose it.

mass function. A numerical relation between the masses of the components of a binary star and the inclination of their plane of mutual revolution to the plane of the sky; the mass function is determined from an analysis of the radial-velocity curve of a spectroscopic binary when the spectral lines of only one of the stars are visible and gives a lower limit to the mass of the star whose spectrum is *not* observed.

mass-luminosity relation. An empirical relation between the masses and luminosities of many (principally main-sequence) stars.

mass-radius relation (for white dwarfs). A theoretical relation between the masses and radii of white dwarf stars.

Maunder minimum. The interval from 1645 to 1715 when solar activity was very low or absent.

Maxwell's equations. A set of four equations that describe the fields around magnetic and electric charges, and how changes in those fields produce forces and electromagnetic radiation.

mean density of matter in the universe. The average density of the universe if all of its matter and energy could be smoothed out to absolute uniformity.

mean solar day. Interval between successive meridian passages of the mean sun; average length of the apparent solar day.

mean solar time. Local hour angle of the mean sun *plus* 12 hours.

mean sun. A fictitious body that moves eastward with uniform angular velocity along the celestial equator, completing one circuit of the sky with respect to the vernal equinox in a tropical year.

mechanics. That branch of physics which deals with the behavior of material bodies under the influence of, or in the absence of, forces.

Medical astrology. That branch of astrology that deals with supposed connections between planets and zodiacal signs and bodily organs and their diseases.

megaparsec (Mpc). One million (10^6) pc.

Mercury. Nearest planet to the sun in the solar system.

meridian (celestial). The great circle on the celestial sphere that passes through an observer's zenith and the north (or south) celestial pole.

meridian (terrestrial). The great circle on the surface of the earth that passes through a particular place and the north and south poles of the earth.

meson. A subatomic particle of mass intermediate between that of a proton and that of an electron.

mesosphere. The layer of the ionosphere immediately above the stratosphere.

Messier catalogue. A catalogue of nonstellar objects compiled by Charles Messier in 1787.

metastable level. An energy level in an atom from which there is a low probability of an atomic transition accompanied by the radiation of a photon.

meteor. The luminous phenomenon observed when a meteoroid enters the earth's atmosphere and burns up; popularly called a "shooting star."

meteor shower. Many meteors appearing to radiate from a common point in the sky caused by the collision of the earth with a swarm of meteoritic particles.

meteorite. A portion of a meteoroid that survives passage through the atmosphere and strikes the ground.

meteorite fall. The occurrence of a meteorite striking the ground.

meteoroid. A meteoritic particle in space before any encounter with the earth.

Michelson-Morley experiment. The classic (1887) experiment by A. A. Michelson and E. W. Morley, in which they tried to measure the speed of the earth in space by timing the speed of light in different directions. The failure of their experiment was later explained by special relativity.

micrometeorite. A meteoroid so small that, on entering the atmosphere of the earth, it is slowed quickly enough that it does not burn up or ablate but filters through the air to the ground.

micron. One millionth (10^{-6}) meter, or 10^{-3} mm.

microphotometer. A device for accurately recording variations in photographic density or transmission through the emulsion along a path in a photograph, especially a spectrogram.

microwave. Short-wave radio wavelengths.

Milky Way. The band of light encircling the sky, which is due to the many stars and diffuse nebulae lying near the plane of the Galaxy.

minor axis (of ellipse). The smallest or least diameter of an ellipse.

minor planet. One of several tens of thousands of small planets, ranging in size from a few hundred kilometers to less than one kilometer in diameter.

Mira Ceti-type variable star. Any of a large class of red-giant long-period or irregular pulsating variable stars, of which the star Mira is a prototype.

missile. A projectile, especially a rocket.

model atmosphere (or **photosphere**). The result of a theoretical calculation of the run of temperature, pressure, density, and so on, through the outer layers of the sun or a star.

molecule. A combination of two or more atoms bound together; the smallest particle of a chemical compound or substance that exhibits the chemical properties of that substance.

momentum. A measure of the inertia or state of motion of a body; the momentum of a body is the product of its mass and velocity. In the absence of a force, momentum is conserved.

monochromatic. Of one wavelength or color.

muon. A *mu meson,* a subatomic particle that behaves like an electron but which has about 200 times the electron's mass.

_n_-body problem. The problem of determining the positions and motions of more than two bodies in a system in which the bodies interact under the influence of their mutual gravitation.

N galaxy. A galaxy with a stellar-appearing nucleus with the remainder of the galaxy appearing as a surrounding faint haze. Most or all N galaxies are probably either Seyfert galaxies or quasars.

nadir. The point on the celestial sphere 180° from the zenith.

nanosecond. One thousand-millionth (10^{-9}) second.

nautical mile. The mean length of one minute of arc on the earth's surface along a meridian.

navigation. The art of finding one's position and course at sea or in the air.

neap tide. The lowest-ranging tide during the month, which occurs when the moon is near first or last quarter.

nebula. Cloud of interstellar gas or dust.

nebular hypothesis. The basic idea that the sun and planets formed from the same cloud of gas and dust in interstellar space.

Neptune. Eighth planet from the sun in the solar system.

neutrino. A fundamental particle that has little or no rest mass and no charge but that does have spin and energy.

neutron. A subatomic particle with no charge and with mass approximately equal to that of the proton.

neutron star. A star of extremely high density composed almost entirely of neutrons.

New General Catalogue (NGC). A catalogue of star clusters, nebulae, and galaxies compiled by J. L. E. Dreyer in 1888.

new moon. Phase of the moon when its longitude is the same as that of the sun.

Newtonian focus. An optical arrangement in a reflecting telescope, in which a flat mirror intercepts the light from the primary before it reaches the focus and reflects it to a focus at the side of the telescope tube.

Newton's laws. The laws of mechanics and gravitation formulated by Isaac Newton.

night sky light. The faint illumination of the night sky; the main source is usually fluorescence by atoms high in the atmosphere.

node. The intersection of the orbit of a body with a fundamental plane—usually the plane of the celestial equator or of the ecliptic.

nodical month. The period of revolution of the moon about the earth with respect to the line of nodes of the moon's orbit.

nodical (eclipse) year. Period of revolution of the earth about the sun with respect to the line of nodes of the moon's orbit.

nonthermal radiation. See synchrotron radiation.

north point. That intersection of the celestial meridian and astronomical horizon lying nearest the north celestial pole.

nova. A star that experiences a sudden outburst of radiant energy, temporarily increasing its luminosity by hundreds to thousands of times.

nuclear. Referring to the nucleus of the atom.

nuclear bulge. Central part of our Galaxy.

nuclear transformation. Transformation of one atomic nucleus into another.

nucleon. Any one of the subatomic particles that compose a nucleus.

nucleosynthesis. The building up of heavy elements from lighter ones by nuclear fusion.

nucleus (of atom). The heavy part of an atom, composed mostly of protons and neutrons, and about which the electrons revolve.

nucleus (of comet). The solid chunk of ice and dust in the head of a comet.

nucleus (of galaxy). Central concentration of matter at the center of a galaxy.

null geodesic. The path of a light ray in four-dimensional spacetime.

nutation. A "nodding" of the earth's polar axis; a small periodic motion of the earth's axis superimposed on precession. It comes about because of the regression of the moon's nodes with an 18-year period.

O-association. A stellar association in which the stars are predominantly of types O and B.

objective. The principal image-forming component of a telescope or other optical instrument.

objective prism. A prismatic lens that can be placed in front of a telescope objective to transform each star image into an image of its spectrum.

oblate spheroid. A solid formed by rotating an ellipse about its minor axis.

oblateness. A measure of the "flattening" of an oblate spheroid; numerically, the ratio of the difference between the major and minor diameters (or axes) to the major diameter (or axis).

obliquity of the ecliptic. Angle between the planes of the celestial equator and the ecliptic; about $23\frac{1}{2}°$.

obscuration (interstellar). Absorption of starlight by interstellar dust.

occultation. An eclipse of a star or planet by the moon or a planet.

ocular. Eyepiece.

opacity. Absorbing power; capacity to impede the passage of light.

open cluster. A comparatively loose or "open" cluster of stars, containing from a few dozen to a few thousand members, located in the spiral arms or disk of the Galaxy; galactic cluster.

opposition. Configuration of a planet when its elongation is 180°.

optical binary. Two stars at different distances nearly lined up in projection so that they appear close together, but which are not really dynamically associated.

optics. The branch of physics that deals with light and its properties.

orbit. The path of a body that is in revolution about another body or point.

outgassing. The process by which the gasses of a planetary atmosphere work their way out from the crust of the planet.

Pallas. Second minor planet to be discovered.

Pangaea. Name given to the hypothetical continent from which the present continents of the earth separated.

parabola. A conic section of eccentricity 1.0; the curve of intersection between a circular cone and a plane parallel to a straight line in the surface of the cone.

paraboloid. A parabola of revolution; a curved surface of parabolic cross section. Especially applied to the surface of the primary mirror in a standard reflecting telescope.

parallactic ellipse. A small ellipse that a comparatively nearby star appears to trace out in the sky, which results from the orbital motion of the earth about the sun.

parallax. An apparent displacement of an object due to a motion of the observer.

parallax (stellar). An apparent displacement of a nearby star that results from the motion of the earth around the sun; numerically, the angle subtended by 1 AU at the distance of a particular star.

parallelogram of forces. A geometrical construction that permits the determination of the resultant of two different forces.

parsec. The distance of an object that would have a stellar parallax of one second of arc; 1 parsec = 3.26 light-years.

partial eclipse. An eclipse of the sun or moon in which the eclipsed body does not appear completely obscured.

Pauli exclusion principle. Quantum-mechanical principle by which no two particles of the same kind can have the same position and momentum.

peculiar velocity. The velocity of a star with respect to the local standard of rest; that is, its space motion, corrected for the motion of the sun with respect to our neighboring stars.

penumbra. The portion of a shadow from which only part of the light source is occulted by an opaque body.

penumbral eclipse. A lunar eclipse in which the moon passes through the penumbra, but not the umbra, of the earth's shadow.

perfect cosmological principle. The assumption that, on the large scale, the universe appears the same from every place and at all times.

perfect gas. An "ideal" gas that obeys the perfect gas laws.

perfect gas laws. Certain laws that describe the behavior of an ideal gas; Charles' law, Boyle's law, and the equation of state for a perfect gas.

perfect radiator. Black body; a body that absorbs and subsequently reemits all radiation incident upon it.

periastron. The place in the orbit of a star in a binary star system where it is closest to its companion star.

perifocus. The place on an elliptical orbit that is closest to the focus occupied by the central force.

perigee. The place in the orbit of an earth satellite where it is closest to the center of the earth.

perihelion. The place in the orbit of an object revolving about the sun where it is closest to the center of the sun.

period. A time interval; for example, the time required for one complete revolution.

period-density relation. Proportionality between the period and the inverse square root of the mean density for a pulsating star.

period-luminosity relation. An empirical relation between the periods and luminosities of cepheid-variable stars.

periodic comet. A comet whose orbit has been determined to have an eccentricity of less than 1.0.

perturbation. The disturbing effect, when small, on the motion of a body as predicted by a simple theory, produced by a third body or other external agent.

phases of the moon. The progression of changes in the moon's appearance during the month that results from the moon's turning different portions of its illuminated hemisphere to our view.

photocathode. The photoemissive surface in a television camera or image tube or photocell from which electrons are dislodged when illuminated with an optical image.

photocell (photoelectric cell). An electron tube in which electrons are dislodged from the cathode when it is exposed to light and are accelerated to the anode, thus producing a current in the tube, whose strength serves as a measure of the light striking the cathode.

photoelectric effect. The emission of an electron by the absorption of a photon by a substance.

photographic magnitude. The magnitude of an object, as measured on the traditional, blue- and violet-sensitive photographic emulsions.

photometry. The measurement of light intensities.

photomultiplier. A photoelectric cell in which the electric current generated is amplified at several stages within the tube.

photon. A discrete unit of electromagnetic energy.

photon sphere. A surface surrounding a black hole, of radius about 1.4 times that of the event horizon, where a photon can have a closed circular orbit.

photosphere. The region of the solar (or a stellar) atmosphere from which radiation escapes into space.

photosynthesis. The formation of carbohydrates in the chlorophyll-containing tissues of plants exposed to sunlight. In the process, oxygen is released to the atmosphere.

photovisual magnitude. A magnitude corresponding to the spectral region to which the human eye is most sensitive, but measured by photographic methods with suitable green- and yellow-sensitive emulsions and filters.

pion. A particular kind of *meson* or subatomic particle of mass intermediate between that of a proton and electron.

Pioneer spacecraft. A series of spacecraft launched to Jupiter and more distant planets and to Venus in the 1970s.

pixel. An individual picture element in a detector; for example, a particular silicon diode in a CCD or a grain in a photographic emulsion.

plage. A bright region of the solar surface observed in the monochromatic light of some spectral line; flocculus.

Planck's constant. The constant of proportionality relating the energy of a photon to its frequency.

Planck's radiation law. A formula from which can be calculated the intensity of radiation at various wavelengths emitted by a black body.

planet. Any of nine solid bodies revolving about the sun.

planetarium. An optical device for projecting on a screen or domed ceiling the stars and planets and their apparent motions in the sky.

planetary nebula. A shell of gas ejected from, and enlarging about, a certain kind of extremely hot star.

planetoid. Synonym for minor planet.

plasma. A hot ionized gas.

plate tectonics. The motion of segments or plates of the outer layer of the earth over the underlying mantle.

Pluto. Ninth planet from the sun in the solar system.

polar axis. The axis of rotation of the earth; also, an axis in the mounting of a telescope that is parallel to the earth's axis.

polarization. A condition in which the planes of vibration (or the E vectors) of the various rays in a light beam are at least partially aligned.

polarized light. Light in which polarization is present.

Polaroid. Trade name for a transparent substance that produces polarization in light.

Population I and II. Two classes of stars (and systems of stars), classified according to their spectral characteristics, chemical compositions, radial velocities, ages, and locations in the Galaxy.

position angle. Direction in the sky of one celestial object from another; for example, the angle, measured to the east from the north, of the fainter component of a visual binary star in relation to the brighter component.

positron. An electron with a positive rather than negative charge; an antielectron.

postulate. An essential prerequisite to a hypothesis or theory.

potential energy. Stored energy that can be converted into other forms; especially gravitational energy.

Poynting-Robertson effect. An effect of the pressure of radiation from the sun on small particles that causes them to spiral slowly into the sun.

precession (of earth). A slow, conical motion of the earth's axis of rotation, caused principally by the gravitational torque of the moon and sun on the earth's equatorial bulge. *Lunisolar precession,* precession caused by the moon and sun only; *planetary precession,* a slow change in the orientation of the plane of the earth's orbit caused by planetary perturbations; *general precession,* the combination of these two effects on the motion of the earth's axis with respect to the stars.

precession of the equinoxes. Slow westward motion of the equinoxes along the ecliptic that results from precession.

primary cosmic rays. The cosmic-ray particles that arrive at the earth from beyond its atmosphere, as opposed to the secondary particles that are produced by collisions between primary cosmic rays and air molecules.

primary minimum (in the light curve of an eclipsing binary). The middle of the eclipse during which the most light is lost.

prime focus. The point in a telescope where the objective focuses the light.

prime meridian. The terrestrial meridian passing through the site of the old Royal Greenwich Observatory; longitude 0°.

primeval atom. A single mass whose explosion (in some cosmological theories) has been postulated to have resulted in all the matter now present in the universe.

primeval fireball. The extremely hot opaque gas that is presumed to have comprised the entire mass of the universe at the time of or immediately following the "big bang"; the exploding primeval atom.

Principia. Contraction of *Philosophiae Naturalis Principia Mathematica,* the great book by Newton in which he set forth his laws of motion and gravitation in 1687.

principle of equivalence. Principle that a gravitational force and a suitable acceleration are indistinguishable within a sufficiently local environment.

principle of relativity. Principle that all observers in uniform relative motion are equivalent; the laws of nature are the same for all, and no experiment can reveal an absolute motion or state of rest.

prism. A wedge-shaped piece of glass that is used to disperse white light into a spectrum.

prolate spheroid. The solid produced by the rotation of an ellipse about its major axis.

prominence. A phenomenon in the solar corona that commonly appears like a flame above the limb of the sun.

proper motion. The angular change in direction of a star per year as seen from the sun.

proton. A heavy subatomic particle that carries a positive charge; one of the two principal constituents of the atomic nucleus.

proton-proton chain. A chain of thermonuclear reactions by which nuclei of hydrogen are built up into nuclei of helium.

protoplanet (or -star or -galaxy). The original material from which a planet (or a star or galaxy) condensed.

pulsar. A variable radio source of small angular size that emits radio pulses in very regular periods that range from 0.03 to 5 seconds.

pulsating variable. A variable star that pulsates in size and luminosity.

q_o. See *deceleration parameter.*

quadrature. A configuration of a planet in which its elongation is 90°.

quantum mechanics. The branch of physics that deals with the structure of atoms and their interactions with each other and with radiation.

quark. A hypothetical subatomic particle. Quarks of from 1 to 6 different kinds, in various combinations, are presumed to make up all other particles in the atomic nucleus.

quarter moon. Either of the two phases of the moon when its longitude differs by 90° from that of the sun; the moon appears half full at these phases.

quasar. A stellar-appearing object of very high redshift, presumed to be extragalactic and highly luminous; an active galactic nucleus.

R Coronae Borealis variables. Eruptive variable stars that show sudden and irregular drops in brightness; the class is named for the prototype, R Coronae Borealis.

RR Lyrae variable. One of a class of giant pulsating stars with periods less than one day; a cluster variable.

RW Aurigae stars. Variable stars, generally associated with interstellar matter, that show rapid and irregular light variations.

radar. A technique for observing the reflection of radio waves from a distant object.

radial velocity. The component of relative velocity that lies in the line of sight.

radial velocity curve. A plot of the variation of radial velocity with time for a binary or variable star.

radiant (of meteor shower). The point in the sky from which the meteors belonging to a shower seem to radiate.

radiation. A mode of energy transport whereby energy is transmitted through a vacuum; also the transmitted energy itself, either electromagnetic or corpuscular.

radiation pressure. The transfer of momentum carried by electromagnetic radiation to a body that the radiation impinges upon.

radical. A bond of two or more atoms that does not, in itself, comprise a molecule, but that has characteristics of its own and enters into chemical reactions as if it were a single atom.

radio astronomy. The technique of making astronomical observations in radio wavelengths.

radio galaxy. A galaxy that emits greater amounts of radio radiation than average.

radio telescope. A telescope designed to make observations in radio wavelengths.

radioactive dating. The technique of determining the ages of rocks or other specimens by the amount of radioactive decay of certain radioactive elements contained therein.

radioactivity (radioactive decay). The process by which certain kinds of atomic nuclei naturally decompose with the spontaneous emission of subatomic particles and gamma rays.

range (of a rocket or space vehicle). Distance away.

range rate. Rate at which range is changing; radial velocity.

ray (lunar). Any of a system of bright elongated streaks, sometimes associated with a crater on the moon.

Rayleigh scattering. Scattering of light (photons) by molecules of a gas.

recurrent nova. A nova that has been known to erupt more than once.

red giant. A large, cool star of high luminosity; a star occupying the upper right portion of the Hertzsprung-Russell diagram.

reddening (interstellar). The reddening of starlight passing through interstellar dust, caused by the dust scattering blue light more effectively than red.

redshift. A shift to longer wavelengths of the light from remote galaxies; presumed to be produced by a Doppler shift.

reflecting telescope. A telescope in which the principal optical component (objective) is a concave mirror.

reflection. The return of light rays by an optical surface.

reflection nebula. A relatively dense dust cloud in interstellar space that is illuminated by starlight.

refracting telescope. A telescope in which the principal optical component (objective) is a lens or system of lenses.

refraction. The bending of light rays passing from one transparent medium (or a vacuum) to another.

regression of nodes. A consequence of certain perturbations on the orbit of a revolving body whereby the nodes of its orbit slide westward in the fundamental plane (usually the plane of the ecliptic or of the celestial equator).

relative orbit. The orbit of one of two mutually revolving bodies referred to the other body as origin.

relativistic particle (or **electron**). A particle (or electron) moving at nearly the speed of light.

relativity. A theory formulated by Einstein that describes the relations between measurements of physical phenomena by two different observers who are in relative motion at constant velocity (the *special theory of relativity*), or that describes how a gravitational field can be replaced by a curvature of space-time (the *general theory of relativity*).

resolution. The degree to which fine details in an image are separated or resolved.

resolving power. A measure of the ability of an optical system to resolve or separate fine details in the image it produces; in astronomy, the angle in the sky that can be resolved by a telescope.

rest mass. The mass of an object or particle as measured when it is at rest in the laboratory.

restricted three-body problem. The study of the motion of a body of negligible mass in the gravitational field of two other bodies revolving about each other in circular orbits.

Reticon. Trade name for a silicon chip containing an array of semiconducting diodes for the detection of faint images.

retrograde motion. An apparent westward motion of a planet on the celestial sphere or with respect to the stars.

revolution. The motion of one body around another.

right ascension. A coordinate for measuring the east-west positions of celestial bodies; the angle measured eastward along the celestial equator from the vernal equinox to the hour circle passing through a body.

rille (or **rill**). A crevasse or trenchlike depression in the moon's surface.

Roche's limit. The smallest distance from a planet or other body at which purely gravitational forces can hold together a satellite or secondary body of the same mean density as the primary; within this distance the tidal forces of the primary would break up the secondary.

rotation. Turning of a body about an axis running through it.

Russell-Vogt theorem. The theorem that the mass and chemical composition of a star determine its entire structure if it derives its energy entirely from thermonuclear reactions.

saros. A particular cycle of similar eclipses that recur at intervals of about 18 years.

satellite. A body that revolves about a larger one; for example, a moon of a planet.

Saturn. The sixth planet from the sun in the solar system.

scale (of telescope). The linear distance in the image corresponding to a particular angular distance in the sky; say, so many centimeters per degree.

Schmidt telescope. A type of reflecting telescope invented by B. Schmidt, in which certain aberrations produced by a spherical concave mirror are compensated for by a thin objective correcting lens.

Schwarzschild radius. See *event horizon*.

science. The attempt to find order in nature or to find laws that describe natural phenomena.

scientific method. A specific procedure in science: (1) the observation of phenomena or the results of experiments; (2) the formulation of hypotheses that describe these phenomena, and that are consistent with the body of knowledge available; (3) the testing of these hypotheses by noting whether or not they adequately predict and describe new phenomena or the results of new experiments.

scintillation counter. Device for recording primary or secondary cosmic-ray particles from flashes of light produced when these particles strike fluorescent materials.

Sculptor-type system. A dwarf elliptical galaxy, of which the system in Sculptor is a typical example.

secondary cosmic rays. Secondary particles produced by interactions between primary cosmic rays from space and the atomic nuclei in molecules of the earth's atmosphere.

secondary minimum (in an eclipsing binary light curve). The middle of the eclipse of the cooler star by the hotter, in which the light of the system diminishes less than during the eclipse of the hotter star by the cooler.

secular. Not periodic.

secular parallax. A mean parallax for a selection of stars, derived from the components of their proper motions that reflect the motion of the sun.

seeing. The unsteadiness of the earth's atmosphere, which blurs telescopic images.

seismic waves. Vibrations traveling through the earth's interior that result from earthquakes.

seismograph. An instrument used to record and measure seismic waves.

seismology. The study of earthquakes and the conditions that produce them and of the internal structure of the earth as deduced from analyses of seismic waves.

seleno-. Prefix referring to the moon.

semimajor axis. Half the major axis of a conic section.

semiregular variable. A variable star, usually a red giant or supergiant, whose period of pulsation is far from constant.

separation (in a visual binary). The angular separation of the two components of a visual binary star.

Seyfert galaxy. A galaxy belonging to the class of those with active galactic nuclei; one whose nucleus shows bright emission lines; one of a class of galaxies first described by C. Seyfert.

shadow cone. The umbra of the shadow of a spherical body (such as the earth) in sunlight.

shell star. A type of star, usually of spectral-type B to F, surrounded by a gaseous ring or shell.

shock wave. A surface of highly compressed gas moving through the medium with the speed of sound; it is produced by an object moving through the medium at a speed greater than that of sound.

shower (of cosmic rays). A large "rain" of secondary cosmic-ray particles produced by a very energetic primary particle impinging on the earth's atmosphere.

shower (meteor). Many meteors, all seeming to radiate from a common point in the sky, caused by the encounter by the earth of a swarm of meteoroids moving together through space.

sidereal astrology. Astrology in which the horoscope is based on the positions of the planets with respect to the fixed stars rather than with respect to signs that, as a consequence of precession, slide through the zodiac.

sidereal day. The interval between two successive meridian passages of the vernal equinox.

sidereal month. The period of the moon's revolution about the earth with respect to the stars.

sidereal period. The period of revolution of one body about another with respect to the stars.

sidereal time. The local hour angle of the vernal equinox.

sidereal year. Period of the earth's revolution about the sun with respect to the stars.

sign (of zodiac). Astrological term for any of twelve equal sections along the ecliptic, each of length 30°. Starting at the vernal equinox, and moving eastward, the signs are Aries, Taurus, Gemini, Cancer, Leo, Virgo, Libra, Scorpio, Sagittarius, Capricorn, Aquarius, and Pisces.

simultaneity. The occurrence of two events at the same time. In relativity, absolute simultaneity is seen not to have meaning, except for two simultaneous events occurring at the same place.

sine (of angle). One of the trigonometric functions; the sine of an angle (in a right triangle) is the ratio of the length of the side opposite the angle to that of the hypotenuse.

sine curve. A graph of the sine of an angle plotted against the angle.

Skylab. An orbiting scientific laboratory occupied by several successive teams of astronauts in the late 1960s and early 1970s.

small circle. Any circle on the surface of a sphere that is not a great circle.

solar activity. Phenomena of the solar atmosphere: sunspots, plages, and related phenomena.

solar antapex. Direction away from which the sun is moving with respect to the local standard of rest.

solar apex. The direction toward which the sun is moving with respect to the local standard of rest.

solar constant. Mean amount of solar radiation received per unit time, by a unit area, just outside the earth's atmosphere, and perpendicular to the direction of the sun; the numerical value is 1.37×10^6 ergs/cm$^2 \cdot$ s.

solar motion. Motion of the sun, or the velocity of the sun, with respect to the local standard of rest.

solar nebula. The cloud of gas and dust from which the solar system is presumed to have formed.

solar parallax. Angle subtended by the equatorial radius of the earth at a distance of 1 AU.

solar system. The system of the sun and the planets, their satellites, the minor planets, comets, meteoroids, and other objects revolving around the sun.

solar time. A time based on the sun; usually the hour angle of the sun *plus* 12 hours.

solar wind. A radial flow of corpuscular radiation leaving the sun.

solstice. Either of two points on the celestial sphere where the sun reaches its maximum distances north and south of the celestial equator.

south point. Intersection of the celestial meridian and astronomical horizon 180° from the north point.

space motion. The velocity of a star with respect to the sun.

space probe. An unmanned interplanetary rocket carrying scientific instruments to obtain data on other planets or on the interplanetary environment.

space technology. The applied science of the immediate space environment of the earth.

spacetime. A system of one time and three spatial coordinates, with respect to which the time and place of an *event* can be specified; also called *spacetime continuum*.

specific gravity. The ratio of the density of a body or substance to that of water.

spectral class (or type). A classification of a star according to the characteristics of its spectrum.

spectral sequence. The sequence of spectral classes of stars arranged in order of decreasing temperatures of stars of those classes.

spectrogram. A photograph of a spectrum.

spectrograph. An instrument for photographing a spectrum; usually attached to a telescope to photograph the spectrum of a star.

spectroheliogram. A photograph of the sun obtained with a spectroheliograph.

spectroheliograph. An instrument for photographing the sun, or part of the sun, in the monochromatic light of a particular spectral line.

spectrophotometry. The measurement of the intensity of light from a star or other source at different wavelengths.

spectroscope. An instrument for directly viewing the spectrum of a light source.

spectroscopic binary star. A binary star in which the components are not resolved optically, but whose binary nature is indicated by periodic variations in radial velocity, indicating orbital motion.

spectroscopic parallax. A parallax (or distance) of a star that is derived by comparing the apparent magnitude of the star with its absolute magnitude as deduced from its spectral characteristics.

spectroscopy. The study of spectra.

spectrum. The array of colors or wavelengths obtained when light from a source is dispersed, as in passing it through a prism or grating.

spectrum analysis. The study and analysis of spectra, especially stellar spectra.

spectrum binary. A binary star whose binary nature is revealed by spectral characteristics that can only result from the composite of the spectra of two different stars.

spectrum (α^2 Canum Venaticorum) variable. Any of a class of main-sequence spectral-type A stars that show anomalously strong lines of certain elements which vary periodically in intensity.

speed. The rate at which an object moves without regard to its direction of motion; the numerical or absolute value of velocity.

spherical aberration. A defect of optical systems whereby on-axis rays of light striking different parts of the objective do not focus at the same place.

spherical harmonics. A series of terms by which the shape of a body can be expressed mathematically to any desired degree of accuracy (by using enough terms in the series).

spicule. A narrow jet of rising material in the solar chromosphere.

spiral arms. Arms of interstellar material and young stars that wind out in a plane from the central nucleus of a spiral galaxy.

spiral galaxy. A flattened, rotating galaxy with pinwheel-like arms of interstellar material and young stars winding out from its nucleus.

sporadic meteor. A meteor that does not belong to a shower.

spring tide. The highest tidal range of the month, produced when the moon is near either the full or new phase.

Sputnik. Russian for "satellite," or "fellow traveler"; the name given to the first Soviet artificial satellite.

SS 433. An object in the Galaxy that ejects relativistic gas, believed to be from an accretion disk.

standard time. The local mean solar time of a standard meridian, adopted over a large region to avoid the inconvenience of continuous time changes around the earth.

star. A self-luminous sphere of gas.

star cluster. An assemblage of stars held together by their mutual gravitation.

statistical equilibrium. A condition in a system of particles in which, statistically, energy is divided equally among particles of all types.

statistical parallax. The mean parallax for a selection of stars, derived from the radial velocities of the stars and the components of their proper motions that cannot be affected by the solar motion.

steady state (theory of cosmology). A theory of cosmology embracing the perfect cosmological principle, and involving the continuous creation of matter.

Stefan's law or **Stefan-Boltzmann Law.** A formula from which the rate at which a black body radiates energy can be computed; the total rate of energy emission from a unit area of a black body is proportional to the fourth power of its absolute temperature.

stellar evolution. The changes that take place in the sizes, luminosities, structures, and so on, of stars as they age.

stellar model. The result of a theoretical calculation of the run of physical conditions in a stellar interior.

stellar parallax. The angle subtended by 1 AU at the distance of a star; usually measured in seconds of arc.

stimulated emission or **stimulated radiation.** Photons emitted by excited atoms undergoing downward transitions as a consequence of being stimulated by other photons of the same wavelength.

Stonehenge. An assemblage of upright stones in Salisbury Plain, England, believed to have been constructed by early people for astronomical observations connected with timekeeping and the calendar.

stony meteorite. A meteorite composed mostly of stony material.

stratosphere. The layer of the earth's atmosphere above the troposphere (where most weather takes place) and below the ionosphere.

Strömgren sphere. A region of ionized gas in interstellar space surrounding a hot star; an H II region.

strong nuclear force. The force that binds together the parts of the atomic nucleus.

subdwarf. A star of luminosity lower than that of main-sequence stars of the same spectral type.

subgiant. A star of luminosity intermediate between those of main-sequence stars and normal giants of the same spectral type.

subtend. To have or include a given angular size.

summer solstice. The point on the celestial sphere where the sun reaches its greatest distance north of the celestial equator.

sun. The star about which the earth and other planets revolve.

sundial. A device for keeping time by the shadow a marker (gnomon) casts in sunlight.

sunspot. A temporary cool region in the solar photosphere that appears dark by contrast against the surrounding hotter photosphere.

sunspot cycle. The semiregular 11-year period with which the frequency of sunspots fluctuates.

supercluster. A large region of space (50 to 100 million parsecs across) where matter is concentrated into galaxies, groups of galaxies and clusters of galaxies; a cluster of clusters of galaxies.

supergiant. A star of very high luminosity.

supergranulation. Large-scale convective patterns in the solar photosphere (up to 30,000 km in diameter).

superior conjunction. The configuration of a planet in which it and the sun have the same longitude, with the planet being more distant than the sun.

superior planet. A planet more distant from the sun than the earth.

supernova. A stellar outburst or explosion in which a star suddenly increases its luminosity by from hundreds of thousands to hundreds of millions of times.

surface gravity. The weight of a unit mass at the surface of a body.

surveying. The technique of measuring distances and relative positions of places over the surface of the earth (or elsewhere); generally accomplished by triangulation.

synchrotron radiation. The radiation emitted by charged particles being accelerated in magnetic fields and moving at speeds near that of light.

synodic month. The period of revolution of the moon with respect to the sun, or its cycle of phases.

synodic period. The interval between successive occurrences of the same configuration of a planet; for example, between successive oppositions or successive superior conjunctions.

syzygy. A configuration of the moon in which its elongation is 0° or 180° (new or full).

T Tauri stars. Variable stars associated with interstellar matter that show rapid and erratic changes in light.

tachyon. A hypothetical particle that always moves with a speed greater than that of light. (There is no evidence that tachyons exist.)

tail (of comet). Gases and solid particles ejected from the head of a comet and forced away from the sun by radiation pressure or corpuscular radiation.

tangent (of angle). One of the trigonometric functions; the tangent of an angle (in a right triangle) is the ratio of the length of the side opposite the angle to that of the shorter of the adjacent sides.

tangential (transverse) velocity. The component of a star's space velocity that lies in the plane of the sky.

target. The surface in a television camera or image tube where photoelectrons, dislodged from a photocathode, are electronically focused.

T-association. A stellar association containing T Tauri stars.

tau component (of proper motion). The component of a star's proper motion that lies perpendicular to a great circle passing through the star and the solar apex.

tectonics. See *plate tectonics*.

tektites. Rounded glassy bodies that some investigators suspect to be of meteoritic origin.

telescope. An optical instrument used to aid the eye in viewing or measuring, or to photograph distant objects.

telluric. Of terrestrial origin.

temperature (absolute). Temperature measured in centigrade (Celsius) degrees from absolute zero.

temperature (Celsius; formerly centigrade). Temperature measured on a scale where water freezes at 0° and boils at 100°.

temperature (color). The temperature of a star as estimated from the intensity of the stellar radiation at two or more colors or wavelengths.

temperature (effective). The temperature of a black body that would radiate the same total amount of energy that a particular body does.

temperature (excitation). The temperature of a star as estimated from the relative strengths of lines in its spectrum that originate from atoms in different stages of excitation.

temperature (Fahrenheit). Temperature measured on a scale where water freezes at 32° and boils at 212°.

temperature (ionization). The temperature of a star as estimated from the relative strengths of lines in its spectrum that originate from atoms in different stages of ionization.

temperature (Kelvin). Absolute temperature measured in centigrade degrees.

temperature (kinetic). A measure of the mean energy of the molecules in a substance.

temperature (radiation). The temperature of a black body that radiates the same amount of energy in a given spectral region as does a particular body.

tensor. A generalization of the concept of the vector, consisting of an array of numbers or quantities that transform according to specific rules.

terminator. The line of sunrise or sunset on the moon or a planet.

terrestrial planet. Any of the planets Mercury, Venus, Earth, Mars, and sometimes Pluto.

Tetrabiblos. A standard and widely used treatise on astrology by Ptolemy.

theory. A set of hypotheses and laws that have been well demonstrated as applying to a wide range of phenomena associated with a particular subject.

thermal energy. Energy associated with the motions of the molecules in a substance.

thermal equilibrium. A balance between the input and outflow of heat in a system.

thermal radiation. The radiation emitted by any body or gas that is not at absolute zero.

thermocouple. A device for measuring the intensity of infrared radiation.

thermodynamics. The branch of physics that deals with heat and heat transfer among bodies.

thermonuclear energy. Energy associated with thermonuclear reactions or that can be released through thermonuclear reactions.

thermonuclear reaction. A nuclear reaction or transformation that results from encounters between nuclear particles that are given high velocities (by heating them).

thermopile. A device consisting of a series or pile of thermocouples, which is used to measure the intensity of infrared radiation.

thermosphere. The region of the earth's atmosphere lying between the mesosphere and the exosphere.

tidal force. A differential gravitational force that tends to deform a body.

tide. Deformation of a body by the differential gravitational force exerted on it by another body; in the earth, the deformation of the ocean surface by the differential gravitational forces exerted by the moon and sun.

ton (American short). 2000 lb.

ton (English long). 2240 lb.

ton (metric). One million grams (2204.6 lb).

topography. The configuration or relief of the surface of the earth, moon, or a planet.

total eclipse. An eclipse of the sun in which the sun's photosphere is entirely hidden by the moon, or an eclipse of the moon in which it passes completely into the umbra of the earth's shadow.

train (of meteor). A temporarily luminous trail left in the wake of a meteor.

transit. An instrument for timing the exact instant a star or other object crosses the local meridian. Also, the passage of a celestial body across the meridian; or the passage of a small body (say, a planet) across the disk of a large one (say, the sun).

triangulation. The operation of measuring some of the elements of a triangle so that other ones can be calculated by the methods of trigonometry, thus determining distances to remote places without having to span them directly.

triaxial ellipsoid. A solid figure whose cross sections along three planes at right angles to each other, and all passing through its center, are ellipses of different sizes and eccentricities.

trigonometry. The branch of mathematics that deals with the analytical solutions of triangles.

trillion. In the United States, 1 thousand billions or million millions (10^{12}); in Great Britain, 1 million billions or million-million millions (10^{18}).

triple-alpha process. A series of two nuclear reactions by which three helium nuclei are built up into one carbon nucleus.

Trojan asteroid (or minor planet). One of a large number of asteroids that share Jupiter's orbit about the sun, but either preceding or following Jupiter by 60°.

Tropic of Cancer. Parallel of latitude $23\frac{1}{2}$° N.

Tropic of Capricorn. Parallel of latitude $23\frac{1}{2}$° S.

tropical astrology. The conventional practice of astrology, in which the horoscope is based on the signs that move through the zodiac with precession.

tropical year. Period of revolution of the earth about the sun with respect to the vernal equinox.

troposphere. Lowest level of the earth's atmosphere, where most weather takes place.

tsunami. A series of very fast waves of seismic origin traveling through the ocean; popularly called "tidal waves."

turbulence. Random motions of gas masses, as in the atmosphere of a star.

21-cm line. A spectral line of neutral hydrogen at the radio wavelength of 21 cm.

U, B, V **system.** A system of stellar magnitudes consisting of measures in the ultraviolet, blue, and green-yellow spectral regions.

Uhuru. An orbiting X-ray observatory, launched in 1970.

ultraviolet radiation. Electromagnetic radiation of wavelengths shorter than the shortest (violet) wavelengths to which the eye is sensitive; radiation of wavelengths in the approximate range 100 to 4000 angstroms.

umbra. The central, completely dark part of a shadow.

uncertainty principle. See *Heisenberg uncertainty principle*.

universal time. The local mean time of the prime meridian.

universe. The totality of all matter and radiation and the space occupied by same.

upsilon component (of proper motion). The component of a star's proper motion that lies along a great circle passing through the star and the solar apex.

Uranus. Seventh planet from the sun in the solar system.

UV Ceti stars. Main-sequence stars, mostly of spectral class M, which show occasional, unpredicted flare-ups in light; also called "flare stars."

Van Allen layer. Doughnut-shaped region surrounding the earth where many rapidly moving charged particles are trapped in its magnetic field.

variable star. A star that varies in luminosity.

variation of latitude. A slight semiperiodic change in the latitudes of places on the earth that results from a slight shifting of the body of the earth with respect to its axis of rotation.

vector. A quantity that has both magnitude and direction.

velocity. A vector that denotes both the speed and direction a body is moving.

velocity of escape. The speed with which an object must move in order to enter a parabolic orbit about another body (such as the earth), and hence move permanently away from the vicinity of that body.

Venus. The second planet from the sun in the solar system.

vernal equinox. The point on the celestial sphere where the sun crosses the celestial equator passing from south to north.

vertical circle. Any great circle passing through the zenith.

Very Large Array (VLA). An array of 27 radio telescopes near Socorro, New Mexico.

very long baseline interferometry (VLBI). A technique of radio astronomy whereby signals from telescopes thousands of kilometers apart are combined to obtain very high resolution with interferometry.

Viking space probes. Spacecraft that landed laboratories on Mars in 1976. The Viking landers were especially designed to search for life on Mars.

virial theorem. A relation between the potential and kinetic energies of a system of mutually gravitating bodies in statistical equilibrium.

visual binary star. A binary star in which the two components are telescopically resolved.

visual photometer. An instrument used with a telescope for visually measuring the light flux from a star.

vis viva **(energy) equation.** An equation that expresses the conservation of energy for two mutually revolving bodies; it relates their relative speed to their separation and the semimajor axis of their relative orbit.

volume. A measure of the total space occupied by a body.

von Jolly balance. A balance invented by von Jolly in 1881 to measure the mass of the earth.

Voyagers. A series of spacecraft that were launched by the United States in 1977 to explore Jupiter and more distant planets.

Vulcan. A hypothetical planet once believed to exist and have an orbit between that of Mercury and the sun; the existence of Vulcan is now generally discredited.

W Ursae Majoris star. Any of a class of eclipsing binaries whose components are nearly in contact and hence suffer tidal distortion and loss or transfer of matter.

W Virginis star (type II Cepheid). A variable star belonging to the relatively rare class of population II Cepheids.

walled plain. A large lunar crater.

wandering of the poles. A semiperiodic shift of the body of the earth relative to its axis of rotation; responsible for variation of latitude.

watt. A unit of power; 10 million ergs expended per second.

wavelength. The spacing of the crests or troughs in a wave train.

weak nuclear force. The nuclear force involved in radioactive decay. The weak force is characterized by the slow rate of certain nuclear reactions—such as the decay of the neutron, which occurs with a half-life of 11 min.

weight. A measure of the force due to gravitational attraction.

west point. The point on the horizon 270° around the horizon from the north point, measured in a clockwise direction as seen from the zenith.

whistles (of meteors). Brief interruptions in high-frequency radio broadcasting reception due to the ionization of the air by meteors.

white dwarf. A star that has exhausted most or all of its nuclear fuel and has collapsed to a very small size; believed to be near its final stage of evolution.

white hole. The hypothetical time reversal of a black hole, in which matter and radiation gush up. White holes are believed *not* to exist.

Widmanstätten figures. Crystalline structure that can be observed in cut and polished meteorites.

Wien's law. Formula that relates the temperature of a black body to the wavelength at which it emits the greatest intensity of radiation.

winter solstice. Point on the celestial sphere where the sun reaches its greatest distance south of the celestial equator.

wobble. Wandering of the poles; variation of latitude.

Wolf-Rayet star. One of a class of very hot stars that eject shells of gas at very high velocity.

X-ray bursters. See *bursters*.

X rays. Photons of wavelengths intermediate between those of ultraviolet radiation and gamma rays.

X-ray stars. Stars (other than the sun) that emit observable amounts of radiation at X-ray frequencies.

year. The period of revolution of the earth around the sun.

Zeeman effect. A splitting or broadening of spectral lines due to magnetic fields.

zenith. The point on the celestial sphere opposite to the direction of gravity; or the direction opposite to that indicated by a plumb bob.

zenith distance. Arc distance of a point on the celestial sphere from the zenith; 90° minus the altitude of the object.

zero-age main sequence. Main sequence for a system of stars that have completed their contraction from interstellar matter, are now deriving all their energy from nuclear reactions, but whose chemical composition has not yet been altered by nuclear reactions.

zodiac. A belt around the sky 18° wide centered on the ecliptic.

zodiacal light. A faint illumination along the zodiac, believed to be sunlight reflected and scattered by interplanetary dust.

zone of avoidance. A region near the Milky Way where obscuration by interstellar dust is so heavy that few or no exterior galaxies can be seen.

zone time. The time, kept in a zone 15° wide in longitude, that is the local mean time of the central meridian of that zone. Zone time is used at sea, but over land the boundaries are irregular to conform to political boundaries, and it is called *standard time*.

Appendix 3 SOME MATHEMATICAL NOTES

1. POWERS-OF-TEN NOTATION

It is often necessary to deal with very large or very small numbers. For example, the earth is 150,000,000 kilometers from the sun and the mass of the hydrogen atom is 0.000 000 000 000 000 000 000 001 67 g. Instead of writing and carrying so many zeros, the numbers are usually written as figures between 1 and 10 multiplied by the appropriate power of 10. For example, 150,000,000 is $1.5 \times 100,000,000$, or 1.5×10^8. Similarly, 0.000 000 000 000 000 000 000 001 67 is $1.67/1,000 000 000$ $000 000 000 000 000$ or $1.67/10^{24} = 1.67 \times 10^{-24}$. The rule in reading numbers written in this notation is that the exponent of 10 is the number of places the decimal point is to be moved to the right (if the exponent is positive) or to the left (if the exponent is negative).

Multiplication, division, and exponentiation of numbers are facilitated in powers-of-10 notation. Examples:

$$6,000,000 \times 400 = 6 \times 10^6 \times 4 \times 10^2$$
$$= (6 \times 4) \times (10^6 \times 10^2) = 24 \times 10^8$$
$$= 2.4 \times 10^9.$$
$$\frac{6 \times 10^{-26}}{9.3 \times 10^7} = \frac{6}{9.3} \times \frac{10^{-26}}{10^7} = \frac{6}{9.3} \times 10^{-26-7}$$
$$= 0.645 \times 10^{-33} = 6.45 \times 10^{-34}.$$
$$(4000)^3 = (4 \times 10^3)^3 = 4^3 \times (10^3)^3 = 64 \times 10^9$$
$$= 6.4 \times 10^{10}.$$
$$(64,000,000)^{1/2} = (64 \times 10^6)^{1/2} = 64^{1/2} \times (10^6)^{1/2}$$
$$= 8 \times 10^3.$$

2. ANGULAR MEASURE

The most common units of angular measure used in astronomy are the following:
1. arc measure:
   one circle contains 360 degrees = 360°;
   1° contains 60 minutes of arc = 60′;
   1′ contains 60 seconds of arc = 60″.
2. time measure:
   one circle contains 24 hours = 24^h;
   1^h contains 60 minutes of time = 60^m;
   1^m contains 60 seconds of time = 60^s.
3. radian measure:
   one circle contains 2π radians.

A radian is the angle at the center of a circle subtended by a length along the circumference of the circle equal to its radius. Since the circumference of a circle is 2π times its radius, there are 2π radians in a circle.

Relations between these different units of angular measure are given in the following table.

ARC MEASURE	TIME MEASURE	RADIANS	SECONDS OF ARC
57°.2958	3^h.820	1.0	206,264.806
15°	1^h	0.2618	54,000
1°	4^m	1.745×10^{-2}	3,600
15′	1^m	4.363×10^{-3}	900
1′	4^s	2.090×10^{-4}	60
15″	1^s	7.27×10^{-5}	15
1″	0^s.0667	4.85×10^{-6}	1

3. PROPERTIES OF CIRCLES AND SPHERES

The ratio of the circumference of a circle to its diameter is always the same, regardless of the size of the circle. This ratio is universally symbolized by the Greek letter pi (π). Because it cannot be expressed as a simple ratio of two integers, π is a type of number called an *irrational number*, and so its value can never be specified exactly. It can, however, be approximated to any desired degree of accuracy by methods of mathematical analysis. Even the Greeks had determined the value of π by geometrical means to considerable accuracy. For many purposes, a sufficient approximation to π is

$$\pi = 3\frac{1}{7} = \frac{22}{7}.$$

The value of π has been evaluated to hundreds of decimal places; however, it is seldom needed to greater accuracy than

$$\pi = 3.14159265.$$

In a circle of circumference C, diameter D, and radius R ($R = \frac{1}{2} D$), we have, from the definition of π,

$$\pi = \frac{C}{D}, \quad \text{or} \quad C = \pi D = 2\pi R.$$

The circumference of a sphere of radius R is the circumference of any circle on the surface of the sphere whose center coincides with the center of the sphere (a *great circle*), and hence which also has a radius R. Thus, also on a sphere

$$C = 2\pi R.$$

For example, the radius of the earth is about 6400 km. Its circumference is thus

$$C = 2\pi(6400) = 40{,}200 \text{ km.}$$

The *area* of a circle of radius R and diameter D is

$$A = \pi R^2 \quad \text{or} \quad A = \frac{\pi D^2}{4}.$$

The *surface area* of a sphere of radius R is

$$A = 4\pi R^2.$$

This is the total area over the outside surface and must not be confused with the volume. For example, the surface area of the earth is approximately:

$$A = 4\pi(6400)^2 = 5 \times 10^8 \text{ km}^2.$$

The *volume* of a sphere of radius R is

$$V = \frac{4}{3}\pi R^3.$$

Appendix 4 METRIC AND ENGLISH UNITS

In the English system of measure the fundamental units of length, mass, and time are the yard, pound*, and second, respectively. There are also, of course, larger and smaller units, which include the ton (2000 lb), the mile (1760 yd), the rod (16½ ft), the inch (¹/₃₆ yd), the ounce (¹/₁₆ lb), and so on. Such units are inconvenient for conversion and arithmetic computation.

In science, therefore, it is more usual to use the metric system, which has been adopted universally in nearly all countries. The fundamental units of the metric system are:

length: 1 meter (m)
mass: 1 kilogram (kg)
time: 1 second (sec)

A meter was originally intended to be 1 ten-millionth of the distance from the equator to the North Pole along the surface of the earth. It is about 1.1 yd. A kilogram is about 2.2 lb. The second is the same in metric and English units. The most commonly used quantities of length and mass of the metric system are the following:

length

1 km	= 1 kilometer	= 1000 meters	= 0.6214 mile	
1 m	= 1 meter	= 1.094 yards	= 39.37 inches	
1 cm	= 1 centimeter	= 0.01 meter	= 0.3937 inch	
1 mm	= 1 millimeter	= 0.001 meter	= 0.1 cm	= 0.03937 inch
1μ	= 1 micron	= 0.000 001 meter	= 0.0001 cm	= 3.937×10^{-5} inch

also: 1 mile = 1.6093 km
1 inch = 2.5400 cm

mass

1 metric ton	= 10^6 grams	= 1000 kg	= 2.2046×10^3 lb
1 kg	= 1000 grams	= 2.2046 lb	
1 g	= 1 gram	= 0.0022046 lb	= 0.0353 oz
1 mg	= 1 milligram	= 0.001 g	= 2.2046×10^{-6} lb

also: 1 lb = 453.6 g
1 oz = 28.3495 g

---

*A pound is also used as a unit of force—especially by engineers. Then the corresponding unit of mass is the *slug*, which is 14.6 kg. In this book we use the more familiar system whereby a pound is a mass unit (in this system, the corresponding force unit is the *poundal*).

Appendix 5 TEMPERATURE SCALES

Three temperature scales are in general use:
1. Fahrenheit (F); water freezes at 32°F and boils at 212°F.
2. Celsius or centigrade* (C); water freezes at 0°C and boils at 100°C.
3. Kelvin or absolute (K); water freezes at 273 K and boils at 373 K.
All molecular motion ceases at $-459°F = -273°C = 0$ K. Thus Kelvin temperature is measured from this lowest possible temperature, called *absolute zero*. It is the temperature scale most often used in astronomy. Kelvins are degrees that have the same value as centigrade or Celsius degrees, since the difference between the freezing and boiling points of water is 100 degrees in each.

On the Fahrenheit scale, water boils at 212 degrees and freezes at 32 degrees; the difference is 180 degrees. Thus to convert Celsius degrees or Kelvins to Fahrenheit it is necessary to multiply by $180/100 = 9/5$. To convert from Fahrenheit to Celsius degrees or Kelvins, it is necessary to multiply by $100/180 = 5/9$.

Example 1: What is 68°F in Celsius and in Kelvins?

$$68°F - 32°F = 36°F \text{ above freezing.}$$

$$\frac{5}{9} \times 36° = 20°;$$

thus,

$$68°F = 20°C = 293 \text{ K.}$$

Example 2: What is 37°C in Fahrenheit and in Kelvins?

$$37°C = 273° + 37° = 310 \text{ K;}$$

$$\frac{9}{5} \times 37° = 66.6 \text{ Fahrenheit degrees;}$$

thus,

$$37°C \text{ is } 66.6°F \text{ above freezing}$$

or

$$37°C = 32° + 66.6° = 98.6°F.$$

---

*Celsius is now the name used for centigrade temperature; it has a more modern standardization, but differs from the old centigrade scale by less than 0.1°.

Appendix 6 SOME USEFUL CONSTANTS

$$\pi = 3.1415926536$$
$$1 \text{ radian} = 57\overset{\circ}{.}2957795$$
$$= 3437\overset{'}{.}74677$$
$$= 206264\overset{''}{.}806$$

Number of square degrees on a sphere $= 41\ 252.96124$

PHYSICAL CONSTANTS

velocity of light	$c = 2.99792458 \times 10^{10}$ cm/s
constant of gravitation	$G = 6.672 \times 10^{-8}$ dyne $\cdot$ cm^2/g^2
Planck's constant	$h = 6.626 \times 10^{-27}$ erg $\cdot$ s
Boltzmann's constant	$k = 1.381 \times 10^{-16}$ erg/deg
mass of hydrogen atom	$m_H = 1.673 \times 10^{-24}$ g
mass of electron	$m_e = 9.1095 \times 10^{-28}$ g
charge on electron	$\varepsilon = 4.803 \times 10^{-10}$ electrostatic units
Stefan-Boltzmann constant	$\sigma = 5.670 \times 10^{-5}$ erg/cm^2 $\cdot$ deg^4 $\cdot$ s
constant in Wien's law	$\lambda_{max} T = 0.28979$ cm/deg
Rydberg's constant	$R = 1.09737 \times 10^{5}$ per cm
1 electron volt	$eV = 1.6022 \times 10^{-12}$ erg
1 angstrom	$\text{Å} = 10^{-8}$ cm
1 ton TNT	$= 4.2 \times 10^{16}$ erg

ASTRONOMICAL CONSTANTS

astronomical unit	$AU = 1.495978707 \times 10^{13}$ cm
parsec	$pc = 206265$ AU
	$= 3.262$ LY
	$= 3.086 \times 10^{18}$ cm
light year	$LY = 9.4605 \times 10^{17}$ cm
	$= 6.324 \times 10^{4}$ AU
tropical year	$= 365.242199$ ephemeris days
sidereal year	$= 365.256366$ ephemeris days
	$= 3.155815 \times 10^{7}$ s
mass of earth	$M_\oplus = 5.977 \times 10^{27}$ g
mass of sun	$M_\odot = 1.989 \times 10^{33}$ g
equatorial radius of earth	$R_\oplus = 6378$ km
radius of sun	$R_\odot = 6.960 \times 10^{10}$ cm
luminosity of sun	$L_\odot = 3.83 \times 10^{33}$ erg/s
solar constant	$S = 1.37 \times 10^{6}$ erg/cm^2 $\cdot$ s
obliquity of ecliptic (1900)	$\varepsilon = 23°27'8\overset{''}{.}26$
direction of galactic center (1950)	$\alpha = 17^h42^m4$
	$\delta = -28°55'$
direction of north galactic pole (1950)	$\alpha = 12^h49^m$
	$\delta = +27\overset{\circ}{.}4$

Appendix 7 ASTRONOMICAL COORDINATE SYSTEMS

Several astronomical coordinate systems are in common use. In each of these systems the position of an object in the sky, or on the celestial sphere, is denoted by two angles. These angles are referred to a *reference plane*, which contains the observer, and a *reference direction*, which is a direction from the observer to some arbitrary point lying in the reference plane. The intersection of the reference plane and the celestial sphere is a great circle, which defines the "equator" of the coordinate system. At two points, each 90° from this equator, are the "poles" of the coordinate system. Great circles passing through these poles intersect the equator of the system at right angles.

One of the two angular coordinates of each coordinate system is measured from the equator of the system to the object along the great circle passing through it and the poles. Angles on one side of the equator (or reference plane) are reckoned as positive; those on the opposite are negative. The other angular coordinate is measured along the equator from the reference direction to the intersection of the equator with the great circle passing through the object and the poles.

The system of terrestrial latitude and longitude provides an excellent analogue. Here the plane of the terrestrial equator is the fundamental plane, and the earth's equator is the equator of the system; the North and South terrestrial Poles are the poles of the system. One coordinate, the *latitude* of a place, is reckoned north (positive) or south (negative) of the equator along a meridian passing through the place. The other coordinate, *longitude*, is measured along the equator to the intersection of the equator and the meridian of the place from the intersection of the equator and the Greenwich meridian. The direction (from the center of the earth) to this latter intersection is the reference direction. Terrestrial longitude is either east or west (whichever is less), but the corresponding coordinate in celestial systems is generally reckoned in one direction from 0 to 360° (or, equivalently, from 0 to 24^h).

The following table lists the more important astronomical coordinate systems and defines how each of the angular coordinates is defined.

Astronomical Coordinate Systems

SYSTEM	REFERENCE PLANE	REFERENCE DIRECTION	"LATITUDE" COORDINATE	RANGE	"LONGITUDE" COORDINATE	RANGE
Horizon	Horizon plane	North point (formerly the south point was used by astronomers)	Altitude, h; toward the zenith (+) toward the nadir (−)	±90°	Azimuth, A; measured to the east along the horizon from the north point	0 to 360°
Equator	Plane of the celestial equator	Vernal equinox	Declination, δ; toward the north celestial pole (+) toward the south celestial pole (−)	±90°	Right ascension, α or R.A.; measured to the east along the celestial equator from the vernal equinox	0 to 24^h
Ecliptic	Plane of the earth's orbit (ecliptic)	Vernal equinox	Celestial latitude, β; toward the north ecliptic pole (+) toward the south ecliptic pole (−)	±90°	Celestial longitude, λ; measured to the east along the ecliptic from the vernal equinox	0 to 360°
Galactic	Mean plane of the Milky Way	Direction to the galactic center	Galactic latitude, b; toward the north galactic pole (+) toward the south galactic pole (−)	±90°	Galactic longitude, l; measured along the galactic equator to the east from the galactic center	0 to 360°

Appendix 8 SOME NUCLEAR REACTIONS OF IMPORTANCE IN ASTRONOMY

Given here are the series of thermonuclear reactions that are most important in stellar interiors. The subscript to the left of a nuclear symbol is the atomic number; the superscript to the left is the atomic mass number. The symbols for the positive electron (positron) and electron are e^+ and e^-, respectively, for the neutrino is ν, and for a photon (generally of gamma-ray energy) is γ.

1. The Proton-Proton Chains

(Important below 15×10^6 K)

There are three ways the proton-proton chain can be completed. The first (a_1, b_1, c_1) is the most important, but depending on the physical conditions in the stellar interior, some energy is released by one or both of the following alternatives: a_1, b_1, c_2, d_2, e_2, and a_1, b_1, c_2, d_3, e_3, f_3.

(a_1) $_1^1H + _1^1H \rightarrow _1^2H + e^+ + \nu$
(b_1) $_1^2H + _1^1H \rightarrow _2^3He + \gamma$
(c_1) $_2^3He + _2^3He \rightarrow _2^4He + 2_1^1H$

or (c_2) $_2^3He + _2^4He \rightarrow _4^7Be + \gamma$
(d_2) $_4^7Be + e^- \rightarrow _3^7Li + \nu$
(e_2) $_3^7Li + _1^1H \rightarrow 2_2^4He$

or (d_3) $_4^7Be + _1^1H \rightarrow _5^8B + \gamma$
(e_3) $_5^8B \rightarrow _4^8Be + e^+ + \nu$
(f_3) $_4^8Be \rightarrow 2_2^4He$

2. The Carbon-Nitrogen Cycle

(Important above 15×10^6 K)
(a) $_6^{12}C + _1^1H \rightarrow _7^{13}N + \gamma$
(b) $_7^{13}N \rightarrow _6^{13}C + e^+ + \nu$
(c) $_6^{13}C + _1^1H \rightarrow _7^{14}N + \gamma$
(d) $_7^{14}N + _1^1H \rightarrow _8^{15}O + \gamma$
(e) $_8^{15}O \rightarrow _7^{15}N + e^+ + \nu$
(f) $_7^{15}N + _1^1H \rightarrow _6^{12}C + _2^4He$

3. The Triple-Alpha Process

(Important above 10^8 K)
(a) $_2^4He + _2^4He \rightarrow _4^8Be + \gamma$
(b) $_2^4He + _4^8Be \rightarrow _6^{12}C + \gamma$

Appendix 9 ORBITAL DATA FOR THE PLANETS

PLANET	SYMBOL	SEMIMAJOR AXIS		SIDEREAL PERIOD		SYNODIC PERIOD (Days)	MEAN ORBITAL SPEED (km/s)	ORBITAL ECCENTRICITY	INCLINATION OF ORBIT TO ECLIPTIC
		AU	10^6 km	Tropical Years	Days				
Mercury	☿	0.3871	57.9	0.24085	87.97	115.88	47.9	0.206	7°004
Venus	♀	0.7233	108.2	0.061521	224.70	583.96	35.0	0.007	3.394
Earth	⊕	1.0000	149.6	1.000039	365.26	—	29.8	0.017	0.0
Mars	♂	1.5237	227.9	1.88089	686.98	779.87	24.1	0.093	1.850
(Ceres)	⊙	2.7671	414	4.603		466.6	17.9	0.077	10.6
Jupiter	♃	5.2028	778	11.86		399	13.1	0.048	1.308
Saturn	♄	9.588	1427	29.46		378	9.6	0.056	2.488
Uranus	♅ or ⛢	19.191	2871	84.07		370	6.8	0.046	0.774
Neptune	♆	30.061	4497	164.82		367	5.4	0.010	1.774
Pluto	♇	39.529	5913	248.6		367	4.7	0.248	17.15

Adapted from *The Astronomical Almanac* (U. S. Naval Observatory), 1981.

Appendix 10 PHYSICAL DATA FOR THE PLANETS

PLANET	DIAMETER		MASS (Earth = 1)	MEAN DENSITY (g/cm³)	ROTATION PERIOD (days)	INCLINATION OF EQUATOR TO ORBIT	OBLATE-NESS	SURFACE GRAVITY (Earth = 1)	ALBEDO	VISUAL MAGNITUDE AT MAXIMUM LIGHT*	VELOCITY OF ESCAPE (km/s)
	km	Earth = 1									
Mercury	4878	0.38	0.055	5.43	58.6	0°0	0	0.38	0.106	−1.9	4.3
Venus	12,104	0.95	0.82	5.24	−243.0	177.4	0	0.91	0.65	−4.4	10.4
Earth	12,756	1.00	1.00	5.52	0.9973	23.4	1/298.2	1.00	0.37	—	11.2
Mars	6794	0.53	0.107	3.9	1.026	25.2	1/164	0.38	0.15	−2.0	5.0
Jupiter	142,796	11.2	317.8	1.3	0.41	3.1	1/16	2.53	0.52	−2.7	60
Saturn	120,000	9.41	94.3	0.7	0.43	26.7	1/9.2	1.07	0.47	+0.7	36
Uranus	50,800	3.98	14.6	1.3	−0.65	97.9	1/30	0.92	0.50	+5.5	21
Neptune	50,450	3.81	17.2	1.5	0.77	29	1/40	1.18	0.5	+7.8	24
Pluto	3400	0.27	0.0023	0.5 (?)	6.387	90	?	0.03	0.5	+15.1	1

*At mean opposition for superior planets.
Adapted from *The Astronomical Almanac* (U. S. Naval Observatory), 1981.

Appendix 11 SATELLITES OF THE PLANETS

PLANET	SATELLITE	DISCOVERED BY	MEAN DISTANCE FROM PLANET (km)	SIDEREAL PERIOD (Days)	ORBITAL ECCEN-TRICITY	DIAMETER OF SATELLITE* (km)	MASS† (Planet = 1)	APPROXIMATE MAGNITUDE AT OPPOSITION
Earth	Moon	—	384,404	27.322	0.055	3476	0.0123	−12.5
Mars	Phobos	A. Hall (1877)	9,380	0.319	0.021	25	(2.7×10^{-8})	+12
	Diemos	A. Hall (1877)	23,500	1.262	0.003	13	4.8×10^{-9}	13
Jupiter‡	XIV	Voyager II (1979)	129,000	0.297	0	<40	(10^{-14})	18–19
	V Almalthea	Barnard (1892)	181,300	0.498	0.003	240	(2×10^{-9})	13
	I Io	Galileo (1610)	421,600	1.769	0.000	3640	4.7×10^{-5}	5
	II Europa	Galileo (1610)	670,900	3.551	0.000	3130	2.5×10^{-5}	6
	III Ganymede	Galileo (1610)	1,070,000	7.155	0.002	5270	7.8×10^{-5}	5
	IV Callisto	Galileo (1610)	1,880,000	16.689	0.008	4840	5.6×10^{-5}	6
	VI Himalia	Perrine (1904)	11,470,000	250.57	0.158	(170)	(8×10^{-10})	14
	VII Elara	Perrine (1905)	11,800,000	259.65	0.207	(40)	(4×10^{-11})	18
	X Lysithea	Nicholson (1938)	11,850,000	263.55	0.130	(10)	(1×10^{-12})	19
	XIII Leda	Kowal (1974)	11,110,000	239.2	0.147	(8)	(5×10^{-13})	20
	XII Aranke	Nicholson (1951)	21,200,000	631.1	0.169	(10)	(7×10^{-13})	18
	XI Carme	Nicholson (1938)	22,600,000	692.5	0.207	(15)	(2×10^{-12})	19
	VIII Pasiphae	Melotte (1908)	23,500,000	738.9	0.378	(25)	(8×10^{-12})	17
	IX Sinope	Nicholson (1914)	23,700,000	758	0.275	(15)	(2×10^{-12})	18
Saturn§	Mimas	W. Herschel (1789)	185,500	0.942	0.020	390	6.6×10^{-8}	13
	Enceladus	W. Herschel (1789)	237,900	1.370	0.004	500	1.3×10^{-7}	12
	Tethys	Cassini (1684)	294,700	1.888	0.000	1050	1.1×10^{-6}	10
	Dione	Cassini (1684)	377,400	2.737	0.002	1120	1.8×10^{-6}	10
	Rhea	Cassini (1672)	526,700	4.518	0.001	1530	4×10^{-6}	10
	Titan	Huygens (1655)	1,222,000	15.945	0.029	5120	2.3×10^{-4}	8
	Hyperion	Bond (1848)	1,481,000	21.277	0.104	310	(2×10^{-7})	14
	Iapetus	Cassini (1671)	3,560,000	79.331	0.028	1440	3.3×10^{-6}	11
	Phoebe	W. Pickering (1898)	12,930,000	550.45	0.163	40	(7×10^{-10})	16
Uranus	Miranda	Kuiper (1948)	123,000	1.414	0	(200)	1×10^{-6}	17
	Ariel	Lassell (1851)	191,700	2.520	0.003	(600)	1.5×10^{-5}	14
	Umbriel	Lassell (1851)	267,000	4.144	0.004	(400)	6×10^{-6}	15
	Titania	W. Herschel (1787)	438,000	8.706	0.002	(1000)	5×10^{-5}	14
	Oberon	W. Herschel (1787)	585,960	13.463	0.001	(900)	3×10^{-5}	14
Neptune	Triton	Lassell (1846)	353,400	5.877	0.000	6000	3×10^{-3}	13
	Nereid	Kuiper (1949)	5,560,000	359.881	0.749	(500)	(10^{-6})	19
Pluto	Charon	Christy (1978)	17,000	6.387	0	(1200)	(0.1)	17

*A diameter of a satellite given in parentheses is estimated from the amount of sunlight it reflects.
†A mass of a satellite given in parentheses is estimated from its size and an assumed density.
‡Does not include two satellites discovered by Voyager; see Chapter 18.
§Does not include at least seven satellites discovered by Voyager; see Table 18.2.
Orbital data from *The Astronomical Almanac* (U. S. Naval Observatory).
Other data compiled from various sources.

Appendix 12   TOTAL SOLAR ECLIPSES FROM 1952 THROUGH 2030

DATE	DURATION OF TOTALITY (*min*)	WHERE VISIBLE
1952 Feb. 25	3.0	Africa, Asia
1954 June 30	2.5	North-Central U.S. (Great Lakes), Canada, Scandinavia, U.S.S.R., Central Asia
1955 June 20	7.2	Southeast Asia
1958 Oct. 12	5.2	Pacific, Chile, Argentina
1959 Oct. 2	3.0	Northern and Central Africa
1961 Feb. 15	2.6	Southern Europe
1962 Feb. 5	4.1	Indonesia
1963 July 20	1.7	Japan, Alaska, Canada, Maine
1965 May 30	5.3	Pacific Ocean, Peru
1966 Nov. 12	1.9	South America
1970 March 7	3.3	Mexico, Florida, parts of U.S. Atlantic coastline
1972 July 10	2.7	Alaska, Northern Canada
1973 June 30	7.2	Atlantic Ocean, Africa
1974 June 20	5.3	Indian Ocean, Australia
1976 Oct. 23	4.9	Africa, Indian Ocean, Australia
1977 Oct. 12	2.8	Northern South America
1979 Feb. 26	2.7	Northwest U.S., Canada
1980 Feb. 16	4.3	Central Africa, India
1981 July 31	2.2	Siberia
1983 June 11	5.4	Indonesia
1984 Nov. 22	2.1	Indonesia, South America
1987 March 29	0.3	Central Africa
1988 March 18	4.0	Philippines, Indonesia
1990 July 22	2.6	Finland, Arctic Regions
1991 July 11	7.1	Hawaii, Central America, Brazil
1992 June 30	5.4	South Atlantic
1994 Nov. 3	4.6	South America
1995 Oct. 24	2.4	South Asia
1997 March 9	2.8	Siberia, Arctic
1998 Feb. 26	4.4	Central America
1999 Aug. 11	2.6	Central Europe, Central Asia
2001 June 21	4.9	Southern Africa
2002 Dec. 4	2.1	South Africa, Australia
2003 Nov. 23	2.0	Antarctica
2005 April 8	0.7	South Pacific Ocean
2006 March 29	4.1	Africa, Asia Minor, U.S.S.R.
2008 Aug. 1	2.4	Arctic Ocean, Siberia, China
2009 July 22	6.6	India, China, South Pacific
2010 July 11	5.3	South Pacific Ocean
2012 Nov. 13	4.0	Northern Australia, South Pacific
2013 Nov. 3	1.7	Atlantic Ocean, Central Africa
2015 March 20	4.1	North Atlantic, Arctic Ocean

Appendix 12 TOTAL SOLAR ECLIPSES FROM 1952 THROUGH 2030
(Continued)

DATE	DURATION OF TOTALITY (*min*)	WHERE VISIBLE
2016 March 9	4.5	Indonesia, Pacific Ocean
2017 Aug. 21	2.7	Pacific Ocean, U.S.A., Atlantic Ocean
2019 July 2	4.5	South Pacific, South America
2020 Dec. 14	2.2	South Pacific, South America, South Atlantic Ocean
2021 Dec. 4	1.9	Antarctica
2023 April 20	1.3	Indian Ocean, Indonesia
2024 April 8	4.5	South Pacific, Mexico, East U.S.A.
2026 Aug. 12	2.3	Arctic, Greenland, North Atlantic, Spain
2027 Aug. 2	6.4	North Africa, Arabia, Indian Ocean
2028 July 22	5.1	Indian Ocean, Australia, New Zealand
2030 Nov. 25	3.7	South Africa, Indian Ocean, Australia

Appendix 13 THE NEAREST STARS

STAR	RIGHT ASCENSION (1950) h m	DECLINATION (1950) ° ′	DISTANCE (pc)	PROPER MOTION ″	RADIAL VELOCITY (km/s)	SPECTRA OF COMPONENTS A	B	C	VISUAL MAGNITUDES OF COMPONENTS A	B	C	ABSOLUTE VISUAL MAGNITUDES OF COMPONENTS A	B	C
Sun						G2V			− 26.8			+ 4.8		
Proxima Centauri*	14 26.3	− 62 28	1.31	3.86	− 16	M5V			+ 11.05			+ 15.4		
α Centauri	14 36.2	− 60 38	1.35	3.68	− 22	G2V	K0V		− 0.01	+ 1.33		+ 4.4	+ 5.7	
Barnard's Star	17 55.4	+ 4 33	1.81	10.34	− 108	M5V			+ 9.54			+ 13.2		
Wolf 359	10 54.1	+ 7 19	2.35	4.70	+ 13	M8V			+ 13.53			+ 16.7		
Lalande 21185	11 00.6	+ 36 18	2.52	4.78	− 84	M2V			+ 7.50			+ 10.5		
Luyten 726-8	1 36.4	− 18 13	2.60	3.36	+ 30	M5.5V	M5.5V		+ 12.45	+ 12.95		+ 15.3	+ 15.8	
Sirius	6 42.9	− 16 39	2.65	1.33	− 8	A1V	wd		− 1.46	+ 8.68		+ 1.4	+ 11.6	
Ross 154	18 46.7	− 23 53	2.90	0.72	− 4	M4.5V			+ 10.6			+ 13.3		
Ross 248	23 39.4	+ 43 55	3.13	1.58	− 81	M6V			+ 12.29			+ 14.8		
ε Eridani	3 30.6	− 9 38	3.28	0.98	+ 16	K2V			+ 3.73			+ 6.1		
Ross 128	11 45.1	+ 1 6	3.31	1.37	− 13	M5V			+ 11.10			+ 13.5		
Luyten 789-6	22 35.7	− 15 36	3.31	3.26	− 60	M6V			+ 12.18			+ 14.6		
61 Cygni	21 4.7	+ 38 30	3.42	5.22	− 64	K5V	K7V		+ 5.22	+ 6.03		+ 7.6	+ 8.4	
ε Indi	21 59.6	− 57 00	3.44	4.69	− 40	K5V			+ 4.68			+ 7.0		
τ Ceti	1 41.7	− 16 12	3.46	1.92	− 16	G8V			+ 3.50			+ 5.7		
Procyon	7 36.7	+ 5 21	3.51	1.25	− 3	F5IV-V	wd		+ 0.37	+ 10.7		+ 2.6	+ 13.0	
BD + 59°1915	18 42.2	+ 59 33	3.52	2.28	+ 5	M4V	M5V		+ 8.90	+ 9.69		+ 11.2	+ 11.9	
BD + 43°44	0 15.5	+ 43 44	3.55	2.89	+ 17	M1V	M6V		+ 8.07	+ 11.04		+ 10.3	+ 13.3	
CD − 36°15693	23 2.6	− 36 9	3.58	6.90	+ 10	M2V			+ 7.36			+ 9.6		
G51-15	8 26.9	+ 26 57	3.66	1.26		MV			+ 14.8			+ 17.0		
Luyten 725-32	1 10.1	− 17 16	3.79	1.22		M5V			+ 11.5			+ 13.6		
BD + 5°1668	7 24.7	+ 5 23	3.79	3.73	+ 26	M5V			+ 9.82			+ 12.0		
CD − 39°14192	21 14.3	− 39 4	3.85	3.46	+ 21	M0V			+ 6.67			+ 8.8		
Kapteyn's Star	5 09.7	− 45 00	3.91	8.89	+ 245	M0V			+ 8.81			+ 10.8		
Kruger 60	22 26.3	+ 57 27	3.94	0.86	− 26	M3V	M4.5V		+ 9.85	+ 11.3		+ 11.9	+ 13.3	
Ross 614	6 26.8	− 2 46	4.12	0.99	+ 24	M7V	?		+ 11.07	+ 14.8		+ 13.1	+ 16.8	
BD − 12°4523	16 27.5	− 12 32	4.20	1.18	− 13	M5V			+ 10.12			+ 12.1		
Wolf 424	12 30.9	+ 9 18	4.27	1.75	− 5	M5.5V	M6V		+ 13.16	+ 13.4		+ 15.0	+ 15.2	
v. Maanen's Star	0 46.5	+ 5 9	4.31	2.95	+ 54	wd			+ 12.37			+ 14.3		
CD − 37°15492	0 2.5	− 37 36	4.44	6.08	+ 23	M3V			+ 8.63			+ 10.4		
Luyten 1159 − 16	1 57.4	+ 12 51	4.52	2.08		M8V			+ 12.27			+ 13.9		
BD + 50°1725	10 8.3	+ 49 52	4.61	1.45	− 26	K7V			+ 6.59			+ 8.3		
CD − 46°11540	17 24.9	− 46 51	4.63	1.13		M4V			+ 9.36			+ 11.0		
CD − 49°13515	21 30.2	− 49 13	4.67	0.81	+ 8	M3V			+ 8.67			+ 10.3		
CD − 44°11909	17 33.5	− 44 17	4.69	1.16		M5V			+ 11.2			+ 12.8		
BD + 68°946	17 36.7	+ 68 23	4.69	1.33	− 22	M3.5V			+ 9.15			+ 10.8		
G158 − 27	0 4.2	− 7 48	4.72	2.06		MV			+ 13.7			+ 15.3		
G208-44/45	19 53.3	+ 44 17	4.76	0.75		MV	MV		+ 13.4	+ 14.0		+ 15.0	+ 15.6	
Ross 780	22 50.6	− 14 31	4.78	1.16	+ 9	M5V			+ 10.7			+ 11.8		
40 Eridani	4 13.0	− 7 44	4.83	4.08	− 43	K0V	wd	M4.5V	+ 4.43	+ 9.53	+ 11.17	+ 6.0	+ 11.1	+ 12.7
Luyten 145-141	11 43.0	− 64 33	4.85	2.68		wd			+ 11.44			+ 13.0		
BD + 20°2465	10 16.9	+ 20 7	4.93	0.49	+ 11	M4.5V			+ 9.43			+ 11.0		
70 Ophiuchi	18 2.9	+ 2 31	4.93	1.13	− 7	K1V	K5V		+ 4.2	+ 6.0		+ 5.7	+ 7.5	
BD + 43°4305	22 44.7	+ 44 5	5.00	0.83	− 2	M4.5V			+ 10.2			+ 11.7		

*Proxima Centauri is sometimes considered an outlying member of the α Centauri system.
Adapted from data supplied by the U.S. Naval Observatory.

Appendix 14 THE TWENTY BRIGHTEST STARS

STAR	RIGHT ASCENSION (1950) h	m	DECLINATION (1950) °	′	DISTANCE* (pc)	PROPER MOTION ″	SPECTRA OF COMPONENTS A	B	C	VISUAL MAGNITUDES OF COMPONENTS A	B	C	ABSOLUTE VISUAL MAGNITUDES OF COMPONENTS A	B	C
Sirius	6	42.9	−16	39	2.7	1.33	A1V	wd		−1.46	+8.7		+1.4	+11.6	
Canopus	6	22.8	−52	40	30	0.02	F0Ib-II			−0.72			−3.1		
α Centauri	14	36.2	−60	38	1.3	3.68	G2V	K0V		−0.01	+1.3		+4.4	+5.7	
Arcturus	14	13.4	+19	27	11	2.28	K2IIIp			−0.06			−0.3		
Vega	18	35.2	+38	44	8.0	0.34	A0V			+0.04			+0.5		
Capella	5	13.0	+45	57	14	0.44	GIII	M1V	M5V	+0.05	+10.2	+13.7	−0.7	+9.5	+13
Rigel	5	12.1	−8	15	250	0.00	B8 Ia	B9		+0.14	+6.6		−6.8	−0.4	
Procyon	7	36.7	+5	21	3.5	1.25	F5IV-V	wd		+0.37	+10.7		+2.6	+13.0	
Betelgeuse	5	52.5	+7	24	150	0.03	M2Iab			+0.41v			−5.5		
Achernar	1	35.9	−57	29	20	0.10	B5V			+0.51			−1.0		
β Centauri	14	00.3	−60	08	90	0.04	B1III	?		+0.63	+4		−4.1	−0.8	
Altair	19	48.3	+8	44	5.1	0.66	A7IV-V			+0.77			+2.2		
α Crucis	12	23.8	−62	49	120	0.04	B1IV	B3		+1.39	+1.9		−4.0	−3.5	
Aldebaran	4	33.0	+16	25	16	0.20	K5III	M2V		+0.86	+13		−0.2	+12	
Spica	13	22.6	−10	54	80	0.05	B1V			+0.91v			−3.6		
Antares	16	26.3	−26	19	120	0.03	M1Ib	B4eV		+0.92v	+5.1		−4.5	−0.3	
Pollux	7	42.3	+28	09	12	0.62	K0III			+1.16			+0.8		
Fomalhaut	22	54.9	−29	53	7.0	0.37	A3V	K4V		+1.19	+6.5		+2.0	+7.3	
Deneb	20	39.7	+45	06	430	0.00	A2Ia			+1.26			−6.9		
β Crucis	12	44.8	−59	24	150	0.05	B0.5IV			+1.28v			−4.6		

*Distances of the more remote stars have been estimated from their spectral types and apparent magnitudes, and are only approximate.

Note: Several of the components listed are themselves spectroscopic binaries. A "v" after a magnitude denotes that the star is variable, in which case the magnitude at median light is given. A "p" after a spectral type indicates that the spectrum is peculiar. An "e" after a spectral type indicates that emission lines are present. When the luminosity classification is rather uncertain, a range is given.

Appendix 15 PULSATING VARIABLE STARS

TYPE OF VARIABLE	SPECTRA	PERIOD (*Days*)	MEDIAN ABSOLUTE MAGNITUDE	AMPLITUDE (*Magnitudes*)	DESCRIPTION	EXAMPLE	NUMBER KNOWN IN GALAXY*
Cepheids (type I)	F to G supergiants	3 to 50	−1.5 to −5	0.1 to 2	Regular pulsation; period-luminosity relation exists	δ Cep	706
Cepheids (type II)	F to G supergiants	5 to 30	0 to −3.5	0.1 to 2	Regular pulsation; period-luminosity relation exists	W Vir	(About 50; included with type I)
RV Tauri	G to K yellow and red bright giants	30 to 150	−2 to −3	Up to 3	Alternate large and small maxima	RV Tau	104
Long-period (Mira-type)	M red giants	80 to 600	+2 to −2	>2.5	Brighten more or less periodically	o Cet (Mira)	4566
Semiregular	M giants and supergiants	30 to 2000	0 to −3	1 to 2	Periodicity not dependable; often interrupted	α Ori	2221
Irregular	All types	Irregular	<0	Up to several magnitudes	No known periodicity; many may be semiregular, but too little data exist to classify them as such	π Gru	1687
RR Lyrae or cluster-type	A to F blue-giants	<1	0 to +1	<1 to 2	Very regular pulsations	RR Lyr	4433
β Cephei or β Canis Majoris	B blue giants	0.1 to 0.3	−2 to −4	0.1	Maximum light occurs at time of highest compression	β Cep	23
δ Scuti	F subgiants	<1	0 to +2	<0.25	Similar to, and possibly related to, RR Lyrae variables	δ Sct	17
Spectrum variables	A main sequence	1 to 25	0 to +1	0.1	Anomalously intense lines of Si, Sr, and Cr vary in intensity with same period as light; most have strong variable magnetic fields	α^2C Vn	28

*According to the 1968 edition of the Soviet *General Catalogue of Variable Stars*.

Appendix 16 ERUPTIVE VARIABLE STARS

TYPE OF VARIABLE	SPECTRA	DURATION OF INCREASED BRIGHTNESS	NORMAL ABSOLUTE MAGNI-TUDE	AMPLITUDE (*Magnitudes*)	DESCRIPTION	EXAMPLE	NUMBER KNOWN IN GALAXY*
Novae	O to A hot subdwarfs	Months to years	>0	7 to 16	Rapid rise to maximum; slow decline; ejection of gas shell	GK Per	166
Novalike variables or P Cygni stars	Hot B stars	Erratic	−3 to −6	Several magnitudes	Slow, erratic, and nova-like variations in light; may be unrelated to novae. Gas shell ejected	P Cyg	39
Supernovae	?	Months to years	?	15 or more	Sudden, violent flareup, followed by decline and ejection of gas shell	CM Tau (Crab nebula)	7
R Coronae Borealis	F to K supergiant	10 to several hundred days	−5	1 to 9	Sudden and irregular drops in brightness. Low in hydrogen abundance, but high in carbon abundance	R CrB	32
T Tauri or RW Aurigae	B to M main sequence and subgiants	Rapid and erratic	0 to +8	Up to a few magnitudes	Rapid and irregular light variations. Generally associated with interstellar material. Subtypes from G to M are called T Tauri variables	RW Aur, T Tau	1109
U Geminorum or SS Gygni or "dwarf novae"	A to F hot subdwarfs	Few days to few weeks	>0	2 to 6	Novalike outbursts at mean intervals which range from 20 to 600 days. Those with longer intervals between outbursts tend to have greater amplitudes. Many, if not all, are members of binary-star systems	SS Cyg, U Gem	215
Flare stars	M main sequence	Few minutes	>8	Up to 6	Sudden flareups in light; probably localized flares on surface of star	UV Cet	28
Z Camelopardalis variables	A to F hot subdwarfs	Few days	>0	2 to 5	Similar to U Geminorium, except that variations are sometimes interrupted by constant light for several cycles. Intervals between outbursts normally range from 10 to 40 days.	Z Cam	20

*According to the 1968 edition of the Soviet *General Catalogue of Variable Stars*.

Appendix 17 THE LOCAL GROUP OF GALAXIES

GALAXY	TYPE	RIGHT ASCENSION (1980) h	m	DECLINATION (1980) °	'	VISUAL MAGNITUDE (M_v)	DISTANCE (kpc)	(1000 LY)	DIAMETER (kpc)	(1000 LY)	ABSOLUTE MAGNITUDE (M_v)	RADIAL VELOCITY (km/s)	MASS (Solar Masses)
Our Galaxy	Sb	—	—	—		—	—	—	30	100	(−21)	—	2×10^{11}
Large Magellanic Cloud	Irr I	5	26	−69		0.9	48	160	10	30	−17.7	+276	2.5×10^{10}
Small Magellanic Cloud	Irr I	0	51	−73		2.5	56	180	8	25	−16.5	+168	
Ursa Minor system	E4 (dwarf)	15	8.6	+67	11		70	220	1	3	(−9)		
Sculptor system	E3 (dwarf)	0	58.9	−33	48	8.0	83	270	2.2	7	−11.8		(2 to 4 × 10⁶)
Draco system	E2 (dwarf)	17	19.9	+57	56		100	330	1.4	4.5	(−10)		
Carina system	E3 (dwarf)	6	41.1	−50	57		(170)	(550)	1.5	4.8	(−10)		
Fornax system	E3 (dwarf)	2	38.9	−34	36	8.4	250	800	4.5	15	−13.6	+39	(1.2 to 2 × 10⁷)
Leo II system	E0 (dwarf)	11	12.4	+22	16		230	750	1.6	5.2	−10.0		
Leo I system	E4 (dwarf)	10	7.4	+12	24	12.0	280	900	1.5	5	−10.4		(1.1 × 10⁶)
NGC 6822	Irr I	19	43.8	−14	50	8.9	460	1500	2.7	9	−14.8	−32	
NGC 147	E6	0	32.0	+48	23	9.73	570	1900	3	10	−14.5		
NGC 185	E2	0	37.8	+48	14	9.43	570	1900	2.3	8	−14.8	−305	
NGC 205	E5	0	39.2	+41	35	8.17	680	2200	5	16	−16.5	−239	
NGC 221 (M32)	E3	0	41.6	+40	46	8.16	680	2200	2.4	8	−16.5	−214	
IC 1613	Irr I	1	2.1	+1	51	9.61	680	2200	5	16	−14.7	−238	
Andromeda galaxy (NGC 224; M31)	Sb	0	41.6	+41	10	3.47	680	2200	40	130	−21.2	−266	3×10^{11}
And I	E0 (dwarf)	0	44.6	+37	54	(14)	(680)	(2200)	0.5	1.6	(−11)		
And II	E0 (dwarf)	1	15.2	+33	18	(14)	(680)	(2200)	0.7	2.3	(−11)		
And III	E3 (dwarf)	0	34.2	+36	24	(14)	(680)	(2200)	0.9	0.9	(−11)		
NGC 598 (M33)	Sc	1	32.7	+30	33	5.79	720	2300	17	60	−18.9	−189	8×10^9

Appendix 18 THE MESSIER CATALOGUE OF NEBULAE AND STAR CLUSTERS

M	NGC or (IC)	RIGHT ASCENSION (1980)		DECLI-NATION (1980)		APPARENT VISUAL MAGNITUDE	DESCRIPTION
		h	m	°	′		
1	1952	5	33.3	+22	01	8.4	"Crab" nebulae in Taurus; remains of SN 1054
2	7089	21	32.4	−0	54	6.4	Globular cluster in Aquarius
3	5272	13	41.2	+28	29	6.3	Globular cluster in Canes Venatici
4	6121	16	22.4	−26	28	6.5	Globular cluster in Scorpio
5	5904	15	17.5	+2	10	6.1	Globular cluster in Serpens
6	6405	17	38.8	−32	11	5.5	Open cluster in Scorpio
7	6475	17	52.7	−34	48	3.3	Open cluster in Scorpio
8	6523	18	02.4	−24	23	5.1	"Lagoon" nebula in Sagittarius
9	6333	17	18.1	−18	30	8.0	Globular cluster in Ophiuchus
10	6254	16	56.1	−4	05	6.7	Globular cluster in Ophiuchus
11	6705	18	50.0	−6	18	6.8	Open cluster in Scutum Sobieskii
12	6218	16	46.3	−1	55	6.6	Globular cluster in Ophiuchus
13	6205	16	41.0	+36	30	5.9	Globular cluster in Hercules
14	6402	17	36.6	−3	14	8.0	Globular cluster in Ophiuchus
15	7078	21	28.9	+12	05	6.4	Globular cluster in Pegasus
16	6611	18	17.8	−13	47	6.6	Open cluster with nebulosity in Serpens
17	6618	18	19.6	−16	11	7.5	"Swan" or "Omega" nebula in Sagittarius
18	6613	18	18.7	−17	08	7.2	Open cluster in Sagittarius
19	6273	17	01.4	−26	14	6.9	Globular cluster in Ophiuchus
20	6514	18	01.2	−23	02	8.5	"Trifid" nebula in Sagittarius
21	6531	18	03.4	−22	30	6.5	Open cluster in Sagittarius
22	6656	18	35.2	−23	56	5.6	Globular cluster in Sagittarius
23	6494	17	55.8	−19	00	5.9	Open cluster in Sagittarius
24	6603	18	17.3	−18	26	4.6	Open cluster in Sagittarius
25	(4725)	18	30.5	−19	16	6.2	Open cluster in Sagittarius
26	6694	18	44.1	−9	25	9.3	Open cluster in Scutum Sobieskii
27	6853	19	58.8	+22	40	8.2	"Dumbbell" planetary nebula in Vulpecula
28	6626	18	23.2	−24	52	7.6	Globular cluster in Sagittarius
29	6913	20	23.3	+38	27	8.0	Open cluster in Cygnus
30	7099	21	39.2	−23	16	7.7	Globular cluster in Capricornus
31	224	0	41.6	+41	10	3.5	Andromeda galaxy
32	221	0	41.6	+40	46	8.2	Elliptical galaxy; companion to M31
33	598	1	32.7	+30	33	5.8	Spiral galaxy in Triangulum
34	1039	2	40.7	+42	43	5.8	Open cluster in Perseus
35	2168	6	07.5	+24	21	5.6	Open cluster in Gemini
36	1960	5	35.0	+34	05	6.5	Open cluster in Auriga
37	2099	5	51.1	+32	33	6.2	Open cluster in Auriga
38	1912	5	27.3	+35	48	7.0	Open cluster in Auriga
39	7092	21	31.5	+48	21	5.3	Open cluster in Cygnus
40		12	21	+59			Close double star in Ursa Major

Table continues on next page.

Appendix 18 THE MESSIER CATALOGUE OF NEBULAE AND STAR CLUSTERS *(Continued)*

M	NGC or (IC)	RIGHT ASCENSION (1980)		DECLI-NATION (1980)		APPARENT VISUAL MAGNITUDE	DESCRIPTION
		h	m	°	′		
41	2287	6	46.2	−20	43	5.0	Loose open cluster in Canis Major
42	1976	5	34.4	−5	24	4	Orion nebula
43	1982	5	34.6	−5	18	9	Northeast portion of Orion nebula
44	2632	8	39	+20	04	3.9	Praesepe; open cluster in Cancer
45		3	46.3	+24	03	1.6	The Pleiades; open cluster in Taurus
46	2437	7	40.9	−14	46	6.6	Open cluster in Puppis
47	2422	7	35.7	−14	26	5	Loose group of stars in Puppis
48	2548	8	12.8	−5	44	6	"Cluster of very small stars";
49	4472	12	28.8	+8	06	8.5	Elliptical galaxy in Virgo
50	2323	7	02.0	−8	19	6.3	Loose open cluster in Monoceros
51	5194	13	29.1	+47	18	8.4	"Whirlpool" spiral galaxy in Canes Venatici
52	7654	23	23.3	+61	30	8.2	Loose open cluster in Cassiopeia
53	5024	13	12.0	+18	16	7.8	Globular cluster in Coma Berenices
54	6715	18	53.8	−30	30	7.8	Globular cluster in Sagittarius
55	6809	19	38.7	−30	59	6.2	Globular cluster in Sagittarius
56	6779	19	15.8	+30	08	8.7	Globular cluster in Lyra
57	6720	18	52.8	+33	00	9.0	"Ring" nebula; planetary nebula in Lyra
58	4579	12	36.7	+11	55	9.9	Spiral galaxy in Virgo
59	4621	12	41.0	+11	46	10.0	Spiral galaxy in Virgo
60	4649	12	42.6	+11	40	9.0	Elliptical galaxy in Virgo
61	4303	12	20.8	+4	35	9.6	Spiral galaxy in Virgo
62	6266	16	59.9	−30	05	6.6	Globular cluster in Scorpio
63	5055	13	14.8	+42	07	8.9	Spiral galaxy in Canes Venatici
64	4826	12	55.7	+21	39	8.5	Spiral galaxy in Coma Berenices
65	3623	11	17.9	+13	12	9.4	Spiral galaxy in Leo
66	3627	11	19.2	+13	06	9.0	Spiral galaxy in Leo; companion to M65
67	2682	8	50.0	+11	53	6.1	Open cluster in Cancer
68	4590	12	38.4	−26	39	8.2	Globular cluster in Hydra
69	6637	18	30.1	−32	23	8.0	Globular cluster in Sagittarius
70	6681	18	42.0	−32	18	8.1	Globular cluster in Sagittarius
71	6838	19	52.8	+18	44	7.6	Globular cluster in Sagitta
72	6981	20	52.3	−12	38	9.3	Globular cluster in Aquarius
73	6994	20	57.8	−12	43	9.1	Open cluster in Aquarius
74	628	1	35.6	+15	41	9.3	Spiral galaxy in Pisces
75	6864	20	04.9	−21	59	8.6	Globular cluster in Sagittarius
76	650	1	41.0	+51	28	11.4	Planetary nebula in Perseus
77	1068	2	41.6	−0	04	8.9	Spiral galaxy in Cetus
78	2068	5	45.7	0	03	8.3	Small emission nebula in Orion
79	1904	5	23.3	−24	32	7.5	Globular cluster in Lepus
80	6093	16	15.8	−22	56	7.5	Globular cluster in Scorpio

M	NGC or (IC)	RIGHT ASCENSION (1980)		DECLI- NATION (1980)		APPARENT VISUAL MAGNITUDE	DESCRIPTION
		h	m	°	′		
81	3031	9	54.2	+69	09	7.0	Spiral galaxy in Ursa Major
82	3034	9	54.4	+69	47	8.4	Irregular galaxy in Ursa Major
83	5236	13	35.4	−29	31	7.6	Spiral galaxy in Hydra
84	4374	12	24.1	+13	00	9.4	Elliptical galaxy in Virgo
85	4382	12	24.3	+18	18	9.3	Elliptical galaxy in Coma Berenices
86	4406	12	25.1	+13	03	9.2	Elliptical galaxy in Virgo
87	4486	12	29.7	+12	30	8.7	Elliptical galaxy in Virgo
88	4501	12	30.9	+14	32	9.5	Spiral galaxy in Coma Berenices
89	4552	12	34.6	+12	40	10.3	Elliptical galaxy in Virgo
90	4569	12	35.8	+13	16	9.6	Spiral galaxy in Virgo
91	omitted						
92	6341	17	16.5	+43	10	6.4	Globular cluster in Hercules
93	2447	7	43.7	−23	49	6.5	Open cluster in Puppis
94	4736	12	50.0	+41	14	8.3	Spiral galaxy in Canes Venatici
95	3351	10	42.9	+11	49	9.8	Barred spiral galaxy in Leo
96	3368	10	45.7	+11	56	9.3	Spiral galaxy in Leo
97	3587	11	13.7	+55	07	11.1	"Owl" nebula; planetary nebula in Ursa Major
98	4192	12	12.7	+15	01	10.2	Spiral galaxy in Coma Berenices
99	4254	12	17.8	+14	32	9.9	Spiral galaxy in Coma Berenices
100	4321	12	21.9	+15	56	9.4	Spiral galaxy in Coma Berenices
101	5457	14	02.5	+54	27	7.9	Spiral galaxy in Ursa Major
102	5866(?)	15	05.9	+55	50	10.5	Spiral galaxy (identification as M102 in doubt)
103	581	1	31.9	+60	35	6.9	Open cluster in Cassiopeia
104★	4594	12	39.0	−11	31	8.3	Spiral galaxy in Virgo
105★	3379	10	46.8	+12	51	9.7	Elliptical galaxy in Leo
106★	4258	12	18.0	+47	25	8.4	Spiral galaxy in Canes Venatici
107★	6171	16	31.4	−13	01	9.2	Globular cluster in Ophiuchus
108★	3556	11	10.5	+55	47	10.5	Spiral galaxy in Ursa Major
109★	3992	11	56.6	+53	29	10.0	Spiral galaxy in Ursa Major
110★	205	0	39.2	+41	35	9.4	Elliptical galaxy (companion to M31)

★Not in Messier's original (1781) list; added later by others.

Appendix 19 THE CHEMICAL ELEMENTS

ELEMENT	SYMBOL	ATOMIC NUMBER	ATOMIC WEIGHT* (Chemical Scale)	NUMBER OF ATOMS PER 10^{12} HYDROGEN ATOMS†
Hydrogen	H	1	1.0080	1×10^{12}
Helium	He	2	4.003	8×10^{10}
Lithium	Li	3	6.940	<10
Beryllium	Be	4	9.013	1.4×10^{1}
Boron	B	5	10.82	$<13 \times 10^{2}$
Carbon	C	6	12.011	4.2×10^{8}
Nitrogen	N	7	14.008	8.7×10^{7}
Oxygen	O	8	16.0000	6.9×10^{8}
Fluorine	F	9	19.00	3.6×10^{4}
Neon	Ne	10	20.183	1.3×10^{8}
Sodium	Na	11	22.991	1.9×10^{6}
Magnesium	Mg	12	24.32	3.2×10^{7}
Aluminum	Al	13	26.98	3.3×10^{6}
Silicon	Si	14	28.09	4.5×10^{7}
Phosphorus	P	15	30.975	3.2×10^{5}
Sulfur	S	16	32.066	1.6×10^{7}
Chlorine	Cl	17	35.457	3.2×10^{5}
Argon	Ar(A)	18	39.944	1.0×10^{6}
Potassium	K	19	39.100	1.4×10^{5}
Calcium	Ca	20	40.08	2.2×10^{6}
Scandium	Sc	21	44.96	1.1×10^{3}
Titanium	Ti	22	47.90	1.1×10^{5}
Vanadium	V	23	50.95	1.0×10^{4}
Chromium	Cr	24	52.01	5.1×10^{5}
Manganese	Mn	25	54.94	2.6×10^{5}
Iron	Fe	26	55.85	3.2×10^{7}
Cobalt	Co	27	58.94	3.2×10^{4}
Nickel	Ni	28	58.71	1.9×10^{6}
Copper	Cu	29	63.54	1.1×10^{4}
Zinc	Zn	30	65.38	2.8×10^{4}
Gallium	Ga	31	69.72	6.3×10^{2}
Germanium	Ge	32	72.60	3.2×10^{3}
Arsenic	As	33	74.91	2.5×10^{2}
Selenium	Se	34	78.96	2.5×10^{3}
Bromine	Br	35	79.916	5.0×10^{2}
Krypton	Kr	36	83.80	2.0×10^{3}
Rubidium	Rb	37	85.48	4.0×10^{2}
Strontium	Sr	38	87.63	7.9×10^{2}
Yttrium	Y	39	88.92	1.3×10^{2}
Zirconium	Zr	40	91.22	5.6×10^{2}
Niobium (Columbium)	Nb(Cb)	41	92.91	7.9×10^{1}
Molybdenum	Mo	42	95.95	1.4×10^{2}
Technetium	Tc(Ma)	43	(99)	—
Ruthenium	Ru	44	101.1	68
Rhodium	Rh	45	102.91	32
Palladium	Pd	46	106.4	32
Silver	Ag	47	107.880	4
Cadmium	Cd	48	112.41	71
Indium	In	49	114.82	45
Tin	Sn	50	118.70	100
Antimony	Sb	51	121.76	10

*Where mean atomic weights have not been well determined, the atomic mass numbers of the most stable isotopes are given in parentheses.
†Provided by L. H. Aller.

ELEMENT	SYMBOL	ATOMIC NUMBER	ATOMIC WEIGHT* *(Chemical Scale)*	NUMBER OF ATOMS PER 10^{12} HYDROGEN ATOMS†
Tellurium	Te	52	127.61	2.5×10^2
Iodine	I(J)	53	126.91	40
Xenon	Xe(X)	54	131.30	2.1×10^2
Cesium	Cs	55	132.91	<80
Barium	Ba	56	137.36	1.2×10^2
Lanthanum	La	57	138.92	13
Cerium	Ce	58	140.13	35
Praseodymium	Pr	59	140.92	4
Neodymium	Nd	60	144.27	18
Promethium	Pm	61	(147)	—
Samarium	Sm(Sa)	62	150.35	5
Europium	Eu	63	152.0	5
Gadolinium	Gd	64	157.26	13
Terbium	Tb	65	158.93	2
Dysprosium	Dy(Ds)	66	162.51	11
Holmium	Ho	67	164.94	3
Erbium	Er	68	167.27	7
Thulium	Tm(Tu)	69	168.94	2
Ytterbium	Yb	70	173.04	8
Lutecium	Lu(Cp)	71	174.99	6
Hafnium	Hf	72	178.50	6
Tantalum	Ta	73	180.95	1
Tungsten	W	74	183.86	50
Rhenium	Re	75	186.22	<2
Osmium	Os	76	190.2	5
Iridium	Ir	77	192.2	28
Platinum	Pt	78	195.09	56
Gold	Au	79	197.0	6
Mercury	Hg	80	200.61	$<10^2$
Thallium	Tl	81	204.39	8
Lead	Pb	82	207.21	85
Bismuth	Bi	83	209.00	<80
Polonium	Po	84	(209)	—
Astatine	At	85	(210)	—
Radon	Rn	86	(222)	—
Francium	Fr(Fa)	87	(223)	—
Radium	Ra	88	226.05	—
Actinium	Ac	89	(227)	—
Thorium	Th	90	232.12	2
Protoactinium	Pa	91	(231)	—
Uranium	U(Ur)	92	238.07	<4
Neptunium	Np	93	(237)	—
Plutonium	Pu	94	(244)	—
Americium	Am	95	(243)	—
Curium	Cm	96	(248)	—
Berkelium	Bk	97	(247)	—
Californium	Cf	98	(251)	—
Einsteinium	E	99	(254)	—
Fermium	Fm	100	(253)	—
Mendeleevium	Mv	101	(256)	—
Nobelium	No	102	(253)	—

*Where mean atomic weights have not been well determined, the atomic mass numbers of the most stable isotopes are given in parentheses.
†Provided by L. H. Aller.

Appendix 20 THE CONSTELLATIONS

CONSTELLATION (*Latin name*)	GENITIVE CASE ENDING	ENGLISH NAME OR DESCRIPTION	ABBRE-VIA-TION	APPROXIMATE POSITION	
				α	δ
				h	°
Andromeda	Andromedae	Princess of Ethiopia	And	1	+40
Antlia	Antliae	Air pump	Ant	10	−35
Apus	Apodis	Bird of Paradise	Aps	16	−75
Aquarius	Aquarii	Water bearer	Aqr	23	−15
Aquila	Aquilae	Eagle	Aql	20	+5
Ara	Arae	Altar	Ara	17	−55
Aries	Arietis	Ram	Ari	3	+20
Auriga	Aurigae	Charioteer	Aur	6	+40
Boötes	Boötis	Herdsman	Boo	15	+30
Caelum	Caeli	Graving tool	Cae	5	−40
Camelopardus	Camelopardis	Giraffe	Cam	6	+70
Cancer	Cancri	Crab	Cnc	9	+20
Canes Venatici	Canum Venaticorum	Hunting dogs	CVn	13	+40
Canis Major	Canis Majoris	Big dog	CMa	7	−20
Canis Minor	Canis Minoris	Little dog	CMi	8	+5
Capricornus	Capricorni	Sea goat	Cap	21	−20
Carina*	Carinae	Keel of Argonauts' ship	Car	9	−60
Cassiopeia	Cassiopeiae	Queen of Ethiopia	Cas	1	+60
Centaurus	Centauri	Centaur	Cen	13	−50
Cephus	Cephei	King of Ethiopia	Cep	22	+70
Cetus	Ceti	Sea monster (whale)	Cet	2	−10
Chamaeleon	Chamaeleontis	Chameleon	Cha	11	−80
Circinus	Circini	Compasses	Cir	15	−60
Columba	Columbae	Dove	Col	6	−35
Coma Berenices	Comae Berenices	Berenice's hair	Com	13	+20
Corona Australis	Coronae Australis	Southern crown	CrA	19	−40
Corona Borealis	Coronae Borealis	Northern crown	CrB	16	+30
Corvus	Corvi	Crow	Crv	12	−20
Crater	Crateris	Cup	Crt	11	−15
Crux	Crucis	Cross (southern)	Cru	12	−60
Cygnus	Cygni	Swan	Cyg	21	+40
Delphinus	Delphini	Porpoise	Del	21	+10
Dorado	Doradus	Swordfish	Dor	5	−65
Draco	Draconis	Dragon	Dra	17	+65
Equuleus	Equulei	Little horse	Equ	21	+10
Eridanus	Eridani	River	Eri	3	−20
Fornax	Fornacis	Furnace	For	3	−30
Gemini	Geminorum	Twins	Gem	7	+20
Grus	Gruis	Crane	Gru	22	−45
Hercules	Herculis	Hercules, son of Zeus	Her	17	+30
Horologium	Horologii	Clock	Hor	3	−60
Hydra	Hydrae	Sea serpent	Hya	10	−20
Hydrus	Hydri	Water snake	Hyi	2	−75
Indus	Indi	Indian	Ind	21	−55
Lacerta	Lacertae	Lizard	Lac	22	+45

Appendix 20 THE CONSTELLATIONS
(Continued)

CONSTELLATION (Latin name)	GENITIVE CASE ENDING	ENGLISH NAME OR DESCRIPTION	ABBRE-VIA-TION	APPROXIMATE POSITION	
				α (h)	δ (°)
Leo	Leonis	Lion	Leo	11	+15
Leo Minor	Leonis Minoris	Little lion	LMi	10	+35
Lepus	Leporis	Hare	Lep	6	−20
Libra	Librae	Balance	Lib	15	−15
Lupus	Lupi	Wolf	Lup	15	−45
Lynx	Lyncis	Lynx	Lyn	8	+45
Lyra	Lyrae	Lyre or harp	Lyr	19	+40
Mensa	Mensae	Table Mountain	Men	5	−80
Microscopium	Microscopii	Microscope	Mic	21	−35
Monoceros	Monocerotis	Unicorn	Mon	7	−5
Musca	Muscae	Fly	Mus	12	−70
Norma	Normae	Carpenter's level	Nor	16	−50
Octans	Octantis	Octant	Oct	22	−85
Ophiuchus	Ophiuchi	Holder of serpent	Oph	17	0
Orion	Orionis	Orion, the hunter	Ori	5	+5
Pavo	Pavonis	Peacock	Pav	20	−65
Pegasus	Pegasi	Pegasus, the winged horse	Peg	22	+20
Perseus	Persei	Perseus, hero who saved Andromeda	Per	3	+45
Phoenix	Phoenicis	Phoenix	Phe	1	−50
Pictor	Pictoris	Easel	Pic	6	−55
Pisces	Piscium	Fishes	Psc	1	+15
Piscis Austrinus	Piscis Austrini	Southern fish	PsA	22	−30
Puppis*	Puppis	Stern of the Argonauts' ship	Pup	8	−40
Pyxis* (= Malus)	Pyxidus	Compass on the Argonauts' ship	Pyx	9	−30
Reticulum	Reticuli	Net	Ret	4	−60
Sagitta	Sagittae	Arrow	Sge	20	+10
Sagittarius	Sagittarii	Archer	Sgr	19	−25
Scorpius	Scorpii	Scorpion	Sco	17	−40
Sculptor	Sculptoris	Sculptor's tools	Scl	0	−30
Scutum	Scuti	Shield	Sct	19	−10
Serpens	Serpentis	Serpent	Ser	17	0
Sextans	Sextantis	Sextant	Sex	10	0
Taurus	Tauri	Bull	Tau	4	+15
Telescopium	Telescopii	Telescope	Tel	19	−50
Triangulum	Trianguli	Triangle	Tri	2	+30
Triangulum Australe	Trianguli Australis	Southern triangle	TrA	16	−65
Tucana	Tucanae	Toucan	Tuc	0	−65
Ursa Major	Ursae Majoris	Big bear	UMa	11	+50
Ursa Minor	Ursae Minoris	Little bear	VMi	15	+70
Vela*	Velorum	Sail of the Argonauts' ship	Vel	9	−50
Virgo	Virginis	Virgin	Vir	13	0
Volans	Volantis	Flying fish	Vol	8	−70
Vulpecula	Vulpeculae	Fox	Vul	20	+25

*The four constellations Carina, Puppis, Pyxis, and Vela originally formed the single constellation, Argo Navis.

Appendix 21 STAR MAPS

The star maps, one for each month, are printed on six removable sheets at the back of this book. To learn the stars and constellations, the sheet containing the map for the current month should be taken outdoors and compared directly with the sky. The maps were designed for a latitude of about 35° N but are useful anywhere in the continental United States. Each map shows the appearance of the sky at about 9:00 P.M. (Standard Time) near the middle of the month for which it is intended; near the beginning and end of the month, it shows the sky as it appears about 10:00 P.M. and 8:00 P.M., respectively. To use a map, hold the sheet vertically, and turn it so that the direction you are facing is shown at the bottom. The middle of the map corresponds to your zenith (the point overhead), so the stars and constellations in the lower part of the chart (with the printing upright) will match, approximately, the sky in front of you.

These star maps were originally prepared by C. H. Cleminshaw for the *Griffith Observer* (published by the Griffith Observatory, P.O. Box 27787, Los Angeles, California, 90027), and are reproduced here by the very kind permission of the Griffith Observatory, Edwin C. Krupp, Director.

INDEX

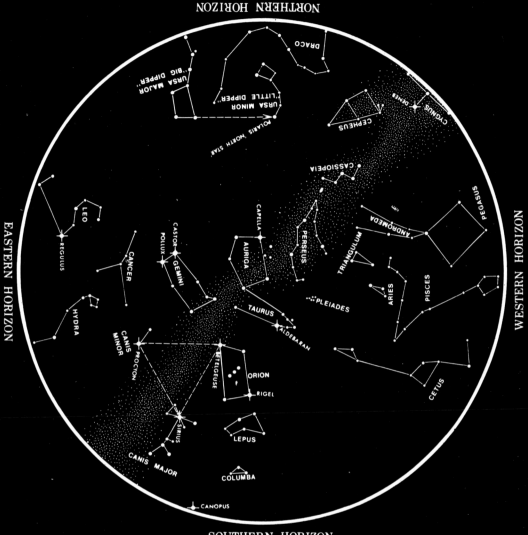

NORTHERN HORIZON

EASTERN HORIZON

WESTERN HORIZON

SOUTHERN HORIZON

THE NIGHT SKY IN JANUARY

Latitude of chart is 34°N, but it is practical throughout the continental United States.

To use: Hold chart vertically and turn it so the direction you are facing shows at the bottom.

Chart time (Local Standard):
10 p.m. First of month
9 p.m. Middle of month
8 p.m. Last of month

Star Chart from *GRIFFITH OBSERVER*, Griffith Observatory, Los Angeles

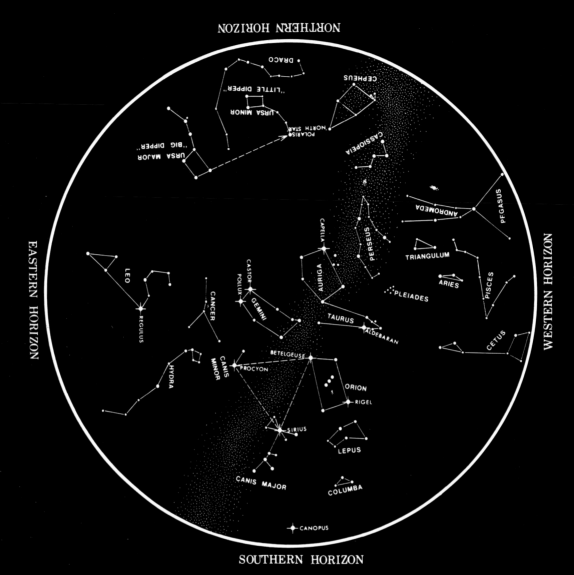

THE NIGHT SKY IN FEBRUARY

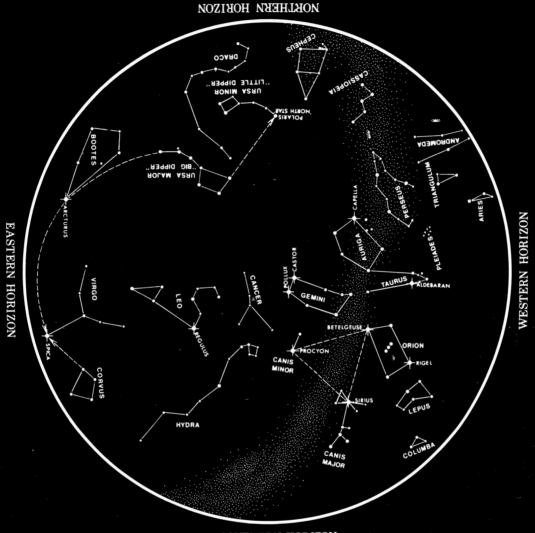

THE NIGHT SKY IN MARCH

Latitude of chart is 34°N, but it is practical throughout the continental United States.

To use: Hold chart vertically and turn it so the direction you are facing shows at the bottom.

Chart time (Local Standard):

10 p.m. First of month
9 p.m. Middle of month
8 p.m. Last of month

Star Chart from *GRIFFITH OBSERVER*, Griffith Observatory, Los Angeles

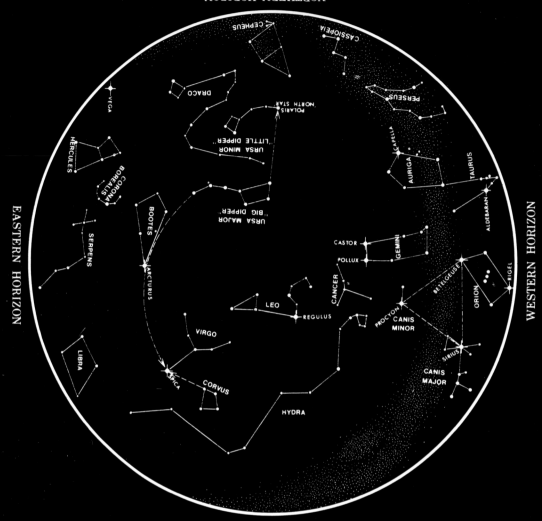

THE NIGHT SKY IN APRIL

Latitude of chart is 34°N, but it is
practical throughout the continental
United States.

To use: Hold chart vertically and turn
it so the direction you are facing
shows at the bottom.

Chart time (Local Standard):

10 p.m. First of month

9 p.m. Middle of month

8 p.m. Last of month

Star Chart from *GRIFFITH OBSERVER*, Griffith Observatory, Los Angeles

NORTHERN HORIZON

EASTERN HORIZON

WESTERN HORIZON

SOUTHERN HORIZON

THE NIGHT SKY IN MAY

Latitude of chart is 34°N, but it is practical throughout the continental United States.

To use: Hold chart vertically and turn it so the direction you are facing shows at the bottom.

Chart time (Local Standard):

10 p.m. First of month
9 p.m. Middle of month
8 p.m. Last of month

Star Chart from *GRIFFITH OBSERVER*, Griffith Observatory, Los Angeles

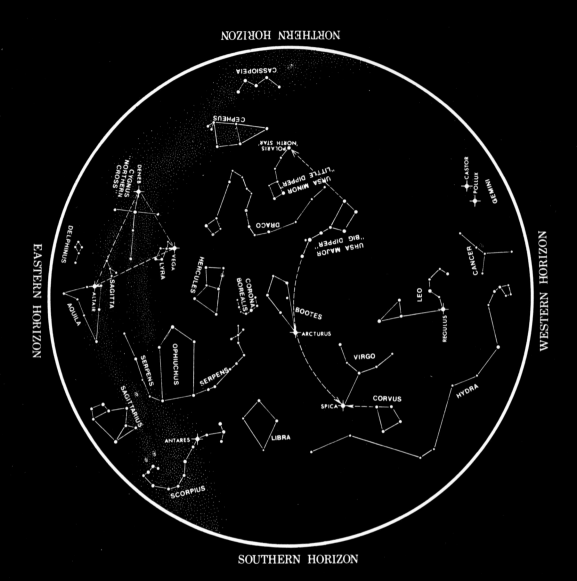

THE NIGHT SKY IN JUNE

Latitude of chart is 34°N, but it is practical throughout the continental United States.

To use: Hold chart vertically and turn it so the direction you are facing shows at the bottom.

Chart time (Local Standard):

10 p.m. First of month
9 p.m. Middle of month
8 p.m. Last of month

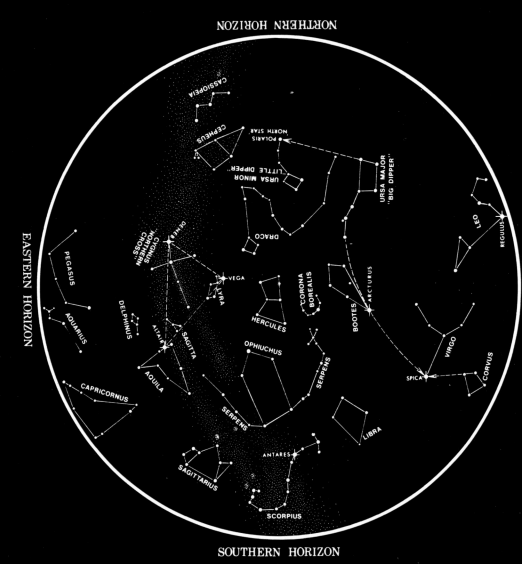

EASTERN HORIZON

SOUTHERN HORIZON

THE NIGHT SKY IN JULY

Latitude of chart is 34°N, but it is practical throughout the continental United States.

To use: Hold chart vertically and turn it so the direction you are facing shows at the bottom.

Chart time (Local Standard):

10 p.m. First of month
9 p.m. Middle of month
8 p.m. Last of month

Star Chart from *GRIFFITH OBSERVER*, Griffith Observatory, Los Angeles

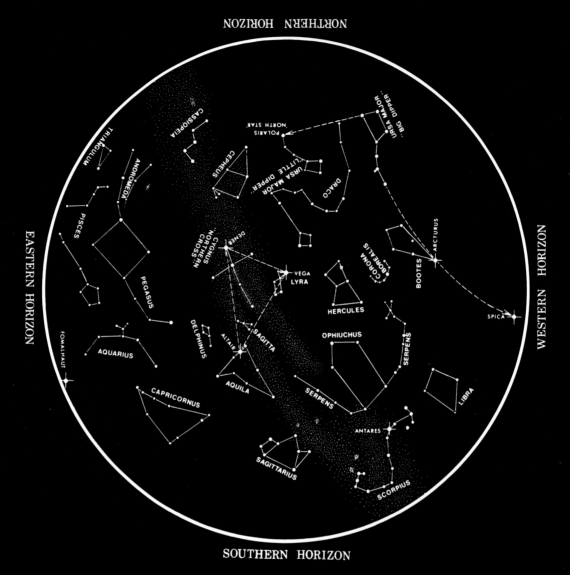

THE NIGHT SKY IN AUGUST

Latitude of chart is 34°N, but it is practical throughout the continental United States.

To use: Hold chart vertically and turn it so the direction you are facing shows at the bottom.

Chart time (Local Standard):

10 p.m. First of month
9 p.m. Middle of month
8 p.m. Last of month

Star Chart from *GRIFFITH OBSERVER*, Griffith Observatory, Los Angeles

NORTHERN HORIZON

EASTERN HORIZON

WESTERN HORIZON

SOUTHERN HORIZON

THE NIGHT SKY IN SEPTEMBER

Latitude of chart is 34°N, but it is practical throughout the continental United States.

To use: Hold chart vertically and turn it so the direction you are facing shows at the bottom.

Chart time (Local Standard):

10 p.m. First of month

9 p.m. Middle of month

8 p.m. Last of month

Star Chart from *GRIFFITH OBSERVER*, Griffith Observatory, Los Angeles

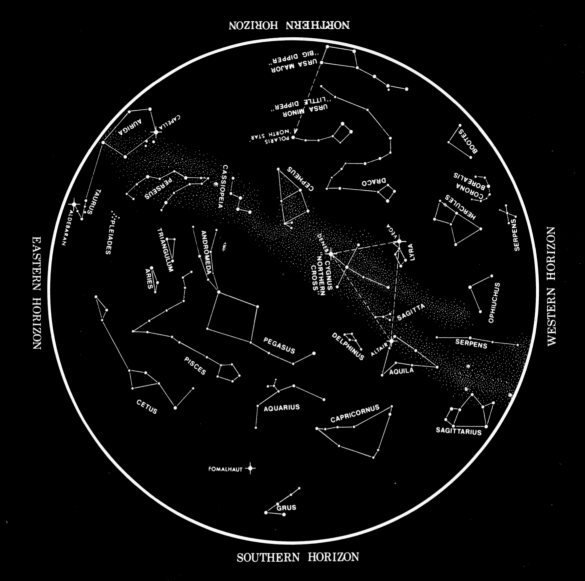

NORTHERN HORIZON

EASTERN HORIZON

WESTERN HORIZON

SOUTHERN HORIZON

THE NIGHT SKY IN OCTOBER

Latitude of chart is 34°N, but it is
practical throughout the continental
United States.

To use: Hold chart vertically and turn
it so the direction you are facing
shows at the bottom.

Chart time (Local Standard):

10 p.m. First of month

9 p.m. Middle of month

8 p.m. Last of month

Star Chart from *GRIFFITH OBSERVER*, Griffith Observatory, Los Angeles

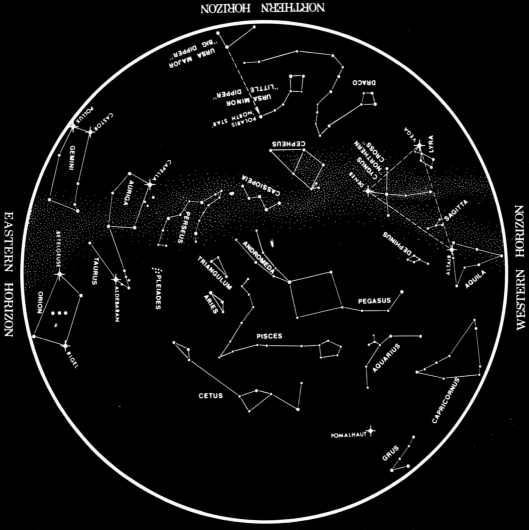

NORTHERN HORIZON

EASTERN HORIZON

WESTERN HORIZON

SOUTHERN HORIZON

THE NIGHT SKY IN NOVEMBER

Latitude of chart is 34°N, but it is practical throughout the continental United States.

To use: Hold chart vertically and turn it so the direction you are facing shows at the bottom.

Chart time (Local Standard):

10 p.m. First of month

9 p.m. Middle of month

8 p.m. Last of month

Star Chart from *GRIFFITH OBSERVER*, Griffith Observatory, Los Angeles

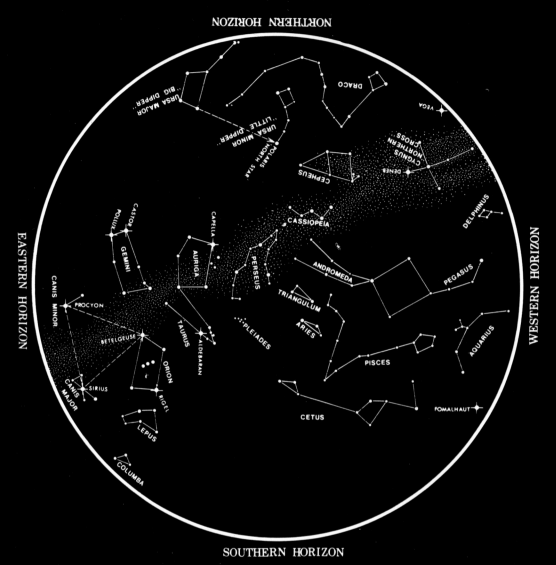

NORTHERN HORIZON

EASTERN HORIZON

WESTERN HORIZON

SOUTHERN HORIZON

THE NIGHT SKY IN DECEMBER

Latitude of chart is 34°N, but it is practical throughout the continental United States.

To use: Hold chart vertically and turn it so the direction you are facing shows at the bottom.

Chart time (Local Standard):

10 p.m. First of month
9 p.m. Middle of month
8 p.m. Last of month

Star Chart from *GRIFFITH OBSERVER*, Griffith Observatory, Los Angeles